TRACE METALS IN SEA WATER

NATO CONFERENCE SERIES

- I Ecology
- II Systems Science
- III Human Factors
- IV Marine Sciences
- V Air–Sea Interactions
- VI Materials Science

IV MARINE SCIENCES

Volume 1 Marine Natural Products Chemistry
edited by D. J. Faulkner and W. H. Fenical

Volume 2 Marine Organisms: Genetics, Ecology, and Evolution
edited by Bruno Battaglia and John A. Beardmore

Volume 3 Spatial Pattern in Plankton Communities
edited by John H. Steele

Volume 4 Fjord Oceanography
edited by Howard J. Freeland, David M. Farmer, and Colin D. Levings

Volume 5 Bottom-interacting Ocean Acoucstics
edited by William A. Kuperman and Finn B. Jensen

Volume 6 Marine Slides and Other Mass Movements
edited by Svend Saxov and J. K. Nieuwenhuis

Volume 7 The Role of Solar Ultraviolet Radiation in Marine Ecosystems
edited by John Calkins

Volume 8 Structure and Development of the Greenland–Scotland Ridge
edited by Martin H. P. Bott, Svend Saxov, Manik Talwani, and Jörn Thiede

Volume 9 Trace Metals in Sea Water
edited by C. S. Wong, Edward Boyle, Kenneth W. Bruland, J. D. Burton, and Edward D. Goldberg

TRACE METALS IN SEA WATER

Edited by

C. S. Wong
Institute of Ocean Sciences
Sidney, British Columbia, Canada

Edward Boyle
Massachusetts Institute of Technology
Cambridge, Massachusetts

Kenneth W. Bruland
University of California, Santa Cruz
Santa Cruz, California

J. D. Burton
University of Southampton
Southampton, United Kingdom

and

Edward D. Goldberg
University of California, San Diego
La Jolla, California

SPRINGER SCIENCE+BUSINESS MEDIA, LLC

Library of Congress Cataloging in Publication Data

NATO Advanced Research Institute on Trace Metals in Sea Water (1981: Ettore Majorana International Center for Scientific Culture)
Trace metals in sea water.

(NATO conference series. IV, Marine sciences; v. 9)
Proceedings of a NATO Advanced Research Institute on Trace Metals in Sea Water, held 3/30 – 4/3/81, at the Ettore Majorana Center for Scientific Culture in Erice, Sicily, Italy.
Published in cooperation with NATO Scientific Affairs Division.
Bibliography: p.
Includes index.
1. Trace elements—Analysis—Congresses. 2. Sea water—Analysis—Congresses. I. Wong, C. S. II. North Atlantic Treaty Organization. Scientific Affairs Division. III. Title. IV. Series.
GC117.T7N37 1981 511.46′01 82-18097
ISBN 978-1-4757-6866-4 ISBN 978-1-4757-6864-0 (eBook)
DOI 10.1007/978-1-4757-6864-0

Proceedings of a NATO Advanced Research Institute on Trace Metals in Sea Water, held March 30 – April 3, 1981, at the Ettore Majorana Center for Scientific Culture, in Erice, Sicily, Italy

Originally published by Plenum Press, New York in 1983
Softcover reprint of the hardcover 1st edition 1983

PREFACE

In recent years, rapid scientific advances have been shattering classical concepts of oceanic trace metals concentrations. Most of the data gathered before the mid-1970s have had to be discarded. Possible associations of organic and inorganic ligands with the metals were throwing views of metal speciation into great uncertainty. Biological effects of metals need to be re-examined after recent revelations of unsuspected metal contaminations in methodology. The investigations appear chaotic, yet exciting. It implies that a new order is going to replace the past. Now, an opportunity opens its door to a brave new world for the young generation of scientists to put metal chemistries in the oceans into perspective. This N.A.T.O. International Conference on "Trace Metals in Sea Water" hoped to catalyze this exciting process of unifying various aspects of trace metals in sea water in future years.

The Conference, in the form of an Advanced Research Institute supported by the Scientific Affairs Division of N.A.T.O. supplemented by further assistance of the U.S. Office of Naval Research, was held at the "Ettore Majorana" Center for Scientific Culture in the medieval town of Erice on the island of Sicily, Italy from March 30 to April 3, 1981. It was the first organized gathering of international scientists in this specialized field. Seventy scientists with various expertise in different aspects of the subject were present: including those from N.A.T.O. countries (Canada, France, F.R. Germany, Greece, Iceland, Italy, U.K. and U.S.A.) as well as from non-N.A.T.O. participants (China, Japan, Monaco, Sweden and Thailand).

The organizing committee, also served as the editorial board of the Conference Proceedings, consisted of: J.D. Burton, University of Southampton, U.K.; E.A. Boyle, Massachusetts Institute of Technology, Cambridge, U.S.A.; K.W. Bruland, University of California at Santa Cruz, U.S.A.; E.D. Goldberg, Scripps Institution of Oceanography, La Jolla, U.S.A. and C.S. Wong, Institute of Ocean Sciences, Sidney, B.C. Canada (Chairman).

Looking not too far back, we are shocked by our ignorance of metal levels and pathways in the marine environment during the

1950s. Then, we found ourselves handicapped by lack of adequate chemical tools. In the 1960s and 1970s we had great difficulties in establishing ocean metal baselines, particularly in unpolluted waters. The concept of a "clean laboratory" approach, pioneered by F. Haber in the 1920s in Germany and by C.C. Patterson of California Institute of Technology, has pointed out the prerequisite to a meaningful data base. Thus armed with analytical refinements and a common-sense approach towards avoidance of contamination in sampling and analysis, marine chemists have generated in the late 1970s and the beginning of the 1980s a new wealth of data and concepts. We needed a forum to summarize and exchange our views. The marine chemists have reached a watershed - it is time to pause and to look beyond before moving to a new fertile valley. Thus, the Erice Conference was born.

The Conference hoped to achieve three objectives. First, it would create a document, in the form of these Proceedings, which disseminates the state-of-the-art knowledge in accurate measurements of trace metals in sea water of interactions with the biosphere of their chemical speciation, of geochemical processes and of organic/particulate interactions. Second, it attempted to create awareness of different on-going and planned international programs on research and preparative work, such as international intercalibration activities. Third, it would document the widely scattered but crucial information on sampling, analysis and processing, in particular, the exceptional precautions required for accurate measurements of natural levels of trace metals in sea water, and on the meaningful interpretation of trace metals data, e.g. based on distributional processes and oceanographic consistency.

The editorial board shared a concern over the appearance of unjustifiable trace metal data in marine chemical and environmental journals because of either the lack of adequate refereeing or the lack of familiarity of the editors with the recent revolution in trace metal analysis. Papers in this volume have been subjected to a review process. Of course, in relatively exploratory studies, a balance has to be struck between the significance of the work and the controversy surrounding the validity of measurements and interpretation where the methodology is not universally endorsed. Thus, the board accepted without unanimity some of the papers on the interaction between organics and trace metals with this qualification. The proceedings capture the publishable material only. The ideas and discussion, freely flowing in the Erice Conference though not in print, have broken the barriers between the established and the young international scientists - this is the most important intellectual contribution to future fruition.

C.S. Wong
Editor-in-Chief
March 16, 1982, Sidney, B.C.

ACKNOWLEDGEMENT

I thank the Scientific Affairs Division of N.A.T.O. for financial support of this Advanced Research Institute and the production of the Conference Volume, the U.S. Office of Naval Research for travel support of some U.S. participants, Dr. A. Gabrielle, Director of the "Ettore Majorana" Centre for Scientific Culture for local arrangements of the Conference, and Mrs. J. Poulin of the Institute of Ocean Sciences at Sidney, B.C. for her secretarial work and assistance with the task of editing.

C.S. Wong

SESSION CHAIRMEN AND RAPPORTEURS

Program Session	Chairmen	Rapporteur
1. Metalloids and the hydride generation	E.D. Goldberg (U.S.A.)	H. Elderfield (U.K.)
2. Arctic Chemistry	D. Dyrssen (Sweden)	E.A. Boyle (U.S.A.)
3. Intercalibration Exercise	C.S. Wong (Canada)	E.A. Crecelius
4. Estuarine Processes Involving Metals	J.M. Bewers (Canada)	R.W. Macdonald (Canada)
5. Air/Sea Exchange and Coastal Processes Involving Metals	F. Millero (U.S.A.)	R.W. Macdonald (Canada)
6. Oceanic Distribution and Analysis	K. Kremling (F.R. Germany)	G. Klinkhammer (U.S.A.)
7. Chemical Speciation	J.D. Burton (U.K.)	C.I. Measures (U.S.A.)
8. Metals and the Biosphere	F. Morel (U.S.A.)	K.W. Bruland (U.S.A.)
9. The Future	E.D. Goldberg (U.S.A.)	All Rapporteurs reporting

CONTENTS

SESSION VII: CHEMICAL SPECIATION
Chairman: *J.D. Burton (U.K.)*
Rapporteur: *C.I. Measures (U.S.A.)*

THE DETERMINATION OF THE CHEMICAL SPECIES OF SOME OF THE "HYDRIDE ELEMENTS" (ARSENIC, ANTIMONY, TIN AND GERMANIUM) IN SEAWATER: METHODOLOGY AND RESULTS

Meinrat O. Andreae

Department of Oceanography
Florida State University
Tallahassee, FL 32306
U.S.A.

ABSTRACT

A number of elements in the fourth, fifth and sixth group of the periodic system form hydrides upon reduction with sodium borohydride, which are stable enough to be of use for chemical analysis (Ge, Sn, Pb, As, Sb, Se, Te). Of these elements, we have investigated in detail arsenic, antimony, germanium and tin. The inorganic and organometallic hydrides are separated by a type of temperature-programmed gas-chromatography. In most cases it is optimal to combine the functions of the cold trap and the chromatographic column in one device. The hydrides are quantified by a variety of detection systems, which take into account the specific analytical chemical properties of the elements under investigation. For arsenic, excellent detection limits ($\cong$40 pg) can be obtained with a quartz tube cuvette burner which is positioned in the beam of an atomic absorption spectrophotometer. For some of the methylarsines, similar sensitivity is available by an electron capture detector. The quartz-burner/AAS system has a detection limit of 90 pg for tin; for this element much lower limits ($\cong$10 pg) are possible with a flame photometric detection system, which uses the extremely intense emission of the SnH molecule at 609.5 nm. The formation of GeO at the temperatures of the quartz tube furnace makes this device quite insensitive for the determination of germanium. Excellent detection limits ($\cong$140 pg) can be reached for this element by the combination of the hydride generation system with a modified graphite furnace/AAS.

The application of these techniques has led to the discovery of a number of organometallic species of arsenic, tin and antimony in the marine environment. Germanium has not been observed to form organometallic compounds in nature. Some aspects of the geochemical cycles of these elements which have been elucidated by the use of these methods will be discussed.

INTRODUCTION

Some metals and metalloids in the fourth, fifth and sixth main groups of the periodic system form volatile hydrides which are reasonably stable against hydrolysis and decomposition at room temperature; among these elements are germanium, tin, lead, arsenic, antimony, selenium and tellurium. This fact was used in one of the earliest trace analytical procedures, the arsenic test of Marsh[1], which is based on the reduction of arsenate or arsenite to arsine (AsH_3) by zinc and sulfuric acid, the thermal decomposition of the arsine in a hydrogen flame, and the visual detection of the elemental arsenic formed when a glass plate is held into this flame. The detection limit of this method was about 2 ppm As_2O_3[2], the very frontier of the analytical art at the time.

More recent developments have expanded the use of Marsh's test from arsenic to the determination of all of the "hydride elements": germanium, tin, lead, arsenic, antimony, bismuth, selenium and tellurium. While the analytical procedures currently used still adhere to the principle of the Marsh test, some important modifications have been made: in 1972, Braman et al.[3] suggested the use of sodium borohydride ($NaBH_4$) as a reducing agent to replace metallic zinc, which is awkward to handle and often contains large blanks of the elements of interest. Sodium borohydride is now used almost exclusively in the various modifications of the hydride method. While the early procedures usually relied on collecting the metal hydrides together with the evolved hydrogen in some kind of a gas reservoir (including ox bladders and toy balloons), many of the recent methods make use of the condensation of the hydrides in a cold trap at liquid nitrogen temperature as originally described by Holak[4]. Braman and Foreback pioneered the use of a packed cold trap to serve both as a substrate to collect the hydrides at liquid nitrogen temperature, and to separate arsine and the methylarsines chromatographically by controlled heating of the trap. In the same paper, they described the differentiation between As(III) and As(V) by a prereduction step and by control of the pH at which the reduction takes place. The porcelain shard held into the flame on Marsh's apparatus has been replaced by a variety of highly sensitive detectors, many of which are element-selective. Most of the detectors commonly used for gas chromatography have been applied to the detection of the hydrides, among them the thermal conductivity, flame ionization and the electron capture detector[6]. A molecular emission detector has been used for tin[7]. Atomic emission spectrometric detectors based on DC discharges[3,5] and microwave induced plasmas[8] were applied to the speciation of arsenic in environmental samples. The currently most popular detection system is atomic absorption spectrometry in one of its numerous variants. The hydrides were at first introduced into a normal AA flame, but it was soon recognized that better detection limits could be achieved with enclosed atom reservoirs and with very small flames or with flameless systems. A number of heated quartz furnace devices without internal flames are now on the commercial market. The lowest detection limits were achieved by cold-trapping of the hydrides and subsequent introduction into either a quartz cuvette furnace[6] or into a commercial graphite furnace[9]. This paper will discuss the methodology of the determination of arsenic, antimony, germanium and tin on these systems, and its application to the investigation of the marine and estuarine chemistry of these elements.

METHODOLOGY

The determination of the hydride element species consists of five steps: 1) the reduction of the element species to the hydrides; 2) the removal of interferent volatiles from the gas stream; 3) the cold-trapping of the hydrides; 4) the separation of the substituted and unsubstituted hydrides from each other and from interfering compounds; and 5) the quantitative detection of the hydrides. A typical instrumental configuration to accomplish these steps is shown in Figure 1 for the borohydride reduction/flame photometric detection system for tin speciation analysis.

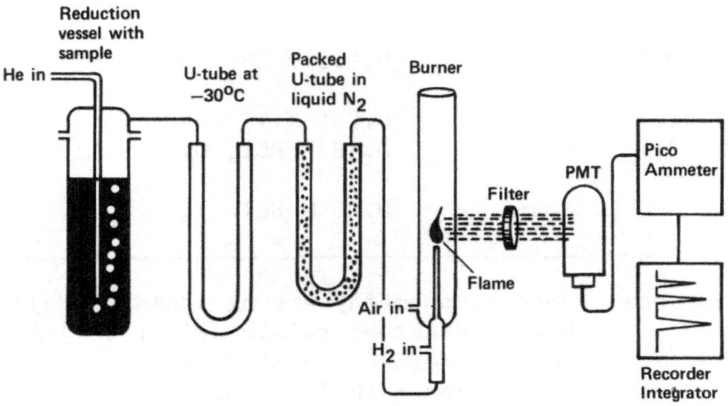

Fig. 1 Schematic design of the $NaBH_4$-reduction/flame photometric detection system for the determination of tin species in natural waters.

Reduction of the element species to the hydrides

Most of the hydride elements occur in a number of different species. The optimum reduction conditions vary from element to element, and between different species of the same element, e.g. Sb(III), Sb(V), methylstibonic acid [$(CH_3)SbO(OH)_2$] and dimethylstibinic acid [$(CH_3)_2SbO(OH)$]. The conditions under which the element species are being reduced have been optimized in our laboratory to give maximal reduction yield, species selectivity, minimal reagent blank, speed and convenience of manipulation. The reaction conditions currently used in our laboratory are summarized in Table 1.

With the exception of Sb(V), which requires the presence of iodide for its reduction, all species can be reduced in an acid medium at pH of 1-2. However, the reduction of some species,

Table 1 Reaction conditions for the reduction of various "hydride element" species to the corresponding hydrides.

Species[1]	pK_a	pH	Composition of reaction medium	$NaBH_4$[2]
As(III)	9.2	6-7	0.05 M TRIS-HCl	1
As(V)	2.3)			
MMAA	3.6)	∿ 1	0.12 M HCl	3
DMAA	6.2)			
Sn(II)	9.5)			
Sn(IV)	∿ 10)	2-8[3]	0.01 M HNO_3	1
Me_xSn	11.7[4])			
Ge(IV)	9.3	∿ 6	0.095 M TRIS-HCl	3
Sb(III)	11.0	∿ 6	0.095 M TRIS-HCl	2
Sb(V)	2.7	∿ 1	0.18 M HCl, 0.15 M KI	3
MMSA	-)			
DMSA	-)	1.5-2	0.06 M HCl	2

[1]Abbreviations: MMAA = monomethylarsonic acid [$(CH_3)AsO(OH)_2$]
DMAA = dimethylarsinic acid [$(CH_3)_2AsO(OH)$]
Me_xSn = $MeSn^{3+}$, Me_2Sn^{2+}, Me_3Sn^+
MMSA = monomethylstibonic acid [$(CH_3)SbO(OH)_2$]
DMSA = dimethylstibinic acid [$(CH_3)_2SbO(OH)$]

[2]ml of 4% $NaBH_4$ solution per 100 ml sample.

[3]Increases during the reaction.

[4]Data available only for $Me_2Sn(OH)_2$.

including Sb(III) and As(III) and all tin species, will also proceed at higher pH, where As(V) and Sb(V) are not converted to their hydrides. This effect permits the selective determination of the different oxidation states of these elements[6,10]. In the case of tin, reduction can be achieved at the pH of the TRIS-HCl buffer ($\sim$6-7), but due to the tin contamination in commercially available TRIS-HCl we prefer to perform the reaction in a medium containing a small amount of nitric acid. This addition results in a solution pH of about 8 after the injection of the $NaBH_4$; without it the pH would rise above 10 and the reduction to the stannanes would be inhibited. Nitric acid is used, as it is available with a tin blank below the limit of detection (most HCl contains detectable tin blanks). We normally add the acid to the sample immediately after it has been taken; it then serves both to stabilize the solution and to control the pH of the analytical reaction[11]. We are not able at this time to differentiate between Sn(II) and Sn(IV); both species are reduced with the same yield under our operating conditions. We assume that the quantity listed as "Sn(IV)" by Braman and Tompkins[7], who work with a TRIS-HCl buffer, also represents the sum of Sn(II) and Sn(IV).

In contrast to the findings of Foreback[12] and Tompkins[13], we were not able to reduce antimony(V) quantitatively at pH 1.5-2.0 without the addition of potassium iodide. A concentration of at least 0.15 M KI in the final solution at a pH less than 1.0 was necessary to achieve complete reduction[10]. This is in agreement with the work of Fleming and Ide[14] who suggested an addition of ca. 0.2 M KI per liter to ensure the reduction of Sb(V). Under the conditions used by Foreback[12] we find only partial reduction of both Sb(III) and Sb(V) (about 30% for both species!). Tompkins[13] appears to rely on a reference to Foreback[12] for the speciation of antimony; he does not state in his thesis if he had investigated the reduction of Sb(V).

Germanium can be reduced through a wide range of pH[9]. The optimum pH is in the near-neutral region, as the efficiency of Ge-reduction decreases at lower pH, probably due to the competitive acid-catalyzed hydrolysis of the borohydride ion. At a pH above 8, the yield also decreases, both due to the decrease of the reducing power of borohydride with increasing pH and to the lack of hydrogen evolution at high pH (the hydrogen gas evolving in the solution helps to strip out the hydrides more efficiently than the relatively large helium bubbles passing through the solution).

A comparison of the reaction pH and the pK_a of the species to be reduced shows that the reduction is usually performed just a few pH units below the pK_a of the species of interest. We hypothesize that only neutral or cationic species, but not anionic ones, are subject to reduction by the negatively charged borohydride ion. The difference between the dissociation constants of the antimony and arsenic acids of oxidation states 3+ and 5+ thus permits their species-selective determination, while the lack of a significant pH-difference in the formation of anionic species of Sn(II) and Sn(IV) prevents the selective determination of the oxidation states of tin by the borohydride method.

Removal of interferent volatiles from the gas stream

Depending on the detector used, some volatile compounds which are formed or released during the hydride generation step may interfere with the detection of the hydrides of interest. Most prominent among them are water, CO_2, and, in the case of anoxic water samples, H_2S. The atomic absorption detector is insensitive toward these compounds; thus no precautions need to be taken when this detector is used. It has been found convenient in some applications, however, to remove most of the water before it enters the cold-trap/column which serves to condense and separate the hydrides. This can be accomplished by passing the gas stream through a larger cold trap cooled by a dry-ice/alcohol mixture or by an immersion-cooling system[6]. This method was also used with water-sensitive detectors, e.g. the electron-capture detector for methylarsines[6], or with plasma discharge detectors (e.g., Crecelius[15]). CO_2 produces an interfering peak on the plasma discharge detectors and the flame photometric detector for tin. If the separation by the column used is adequate, no additional precautions are necessary to remove CO_2 interference. Otherwise, a small tube filled with granulated NaOH can be included in the gas stream to absorb CO_2. Samples of marine anoxic water often contain large amounts of H_2S. This causes a significant interference in a number of different detectors. It can be removed by passing the gas through a tube filled with lead acetate[15].

Cold-trapping of the hydrides

Only when the very contamination-sensitive electron-capture detector is used is it necessary to provide separate gas streams, one for the reaction and stripping part of the system, the other for the carrier gas stream of the column and detector. Otherwise, the same gas stream can be used to strip the hydrides from solution and to carry them into the detector, which greatly simplifies the apparatus. This is of considerable significance, as each additional surface and joint in the apparatus increases the possibility of irreversible adsorption of the sensitive hydrides and thus is a potential contributor to analytical error. The cold trap then serves both to collect the hydrides from the reaction gas stream and to chromatographically separate them as it is heated up. Initially, column packings of glass beads[5] or glass wool[6] were used. These packings produce poor separation of the methylated species from each other and badly tailing peaks, however. We are therefore using a standard gas chromatographic packing (15% OV-3 on Chromosorb W/AWDMCS, 60-80 Mesh) in our U-tubes for the separation of the inorganic and alkyl-species of arsenic, antimony and tin. This packing is quite insensitive to water and produces sharp and well separated peaks, as demonstrated in Figure 2 for stibine, methylstibine and dimethylstibine in a standard and a seawater sample. The retention times can be regulated by winding a heating wire around the outside of the U-tube and controlling the current supplied to this heating coil.

Passivation of the internal surfaces of the apparatus

The most persistent difficulty of ultratrace analysis by the borohydride method has been the loss of significant fractions of the hydrides to the internal walls of the apparatus by irreversible adsorption. Some hydrides, e.g. dimethylarsine, are especi-

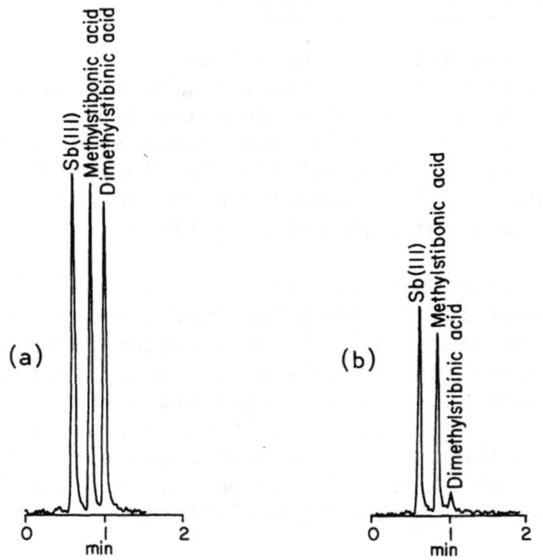

Fig. 2 a) Chromatogram of stibine, methylstibine and dimethylstibine as separated by the OV-3 trap/column with the quartz furnace/AA detector. The stibines were prepared by the $NaBH_4$ reduction of 2 ng Sb each as Sb(III), methlystibonic and dimethylstibinic acid. b) Analysis of a sample of seawater (100 ml) from the open Gulf of Mexico (Station SN4-3-1, 12 March, 1981, Lat. 27° 15.16'N, Long. 96° 29.88'W).

ally susceptible to this process. The consequences are significant random and systematic errors. Two steps must be taken to minimize this problem: 1) strive for the most simple apparatus design possible, using as few joints as necessary, and only relatively inert materials (Pyrex and Teflon); and 2) treat the internal surfaces of the apparatus with passivating reagents. The first goal is accomplished by using the simple, straight-through apparatus shown in Figure 1 and by making all connections glass-to-glass, with 6 mm o.d. Pyrex tubing held together by short 1/4-inch o.d. Teflon tubing sleeves. We treat the inside of all components with silanizing reagents. The trap/column is conditioned in a GC-oven at 150°C, and then at the same temperature injected with several aliquots of Silyl-8 (Pierce Chemical Co., Rockford, IL), and conditioned for a few more hours. All other parts of the system are filled at room temperature with a solution of dimethyldichlorosilane in toluene, washed with toluene and methanol and dried at 110°C. This treatment eliminates the irreversible adsorption of the hydrides and leads to well-shaped peaks with little tailing.

Quantitative detection of the hydrides

Four different detectors have been used in our investigations: the electron-capture detector (for the methylarsines), the quartz-cuvette atomic absorption detector (for arsenic and antimony species), the graphite-furnace atomic absorption detector (for germanium and tin species) and the flame photometric detector (for tin species). Their performance in the borohydride analysis system is evaluated and compared in Table 2.

The electron-capture detector was originally found to be a sensitive detector for the methylarsines[6], and was used extensively for the determination of the methylarsenicals in natural waters[16,17]. After improvements of the atomic absorption detectors had been made (especially concerning adsorptive losses and peak shapes of the methylarsines) it was found that this detector could be used to replace the electron-capture detector, which because of its lack of specificity and its sensitivity to contamination and changes in operating conditions was very inconvenient to work with.

The flame photometric detector for the determination of tin was first described by Braman and Tompkins[7]. This detector, which measures the emission of the diatomic species SnH in a molecular emission band with a bandhead at 610 nm, is highly sensitive to tin when the emission from a "reversed flame" (a jet of air burning in an environment of hydrogen) is being monitored. It is subject to interference by CO_2 and some other gases, but the chromatographic separation obtained on the OV-3 trap described

Table 2 Detection limits in nanograms of the element for the species of As, Sn, Ge and Sb with different detection systems.

Species	QCAA[1]	GFAA[1]	FPD[1]	ECD[1]
As(III), As(V)	0.03	0.09	-	-
MMAA	0.03	0.09	-	0.4
DMAA	0.05	0.15	-	0.2
Sn(II), Sn(IV)	0.05	0.05	0.02	-
Me_xSn	0.06	0.06	0.015	-
Ge(IV)	3	0.14	-	-
Sb(V)	0.05	0.15	-	-
Sb(III))				
MMSA)	0.04	0.12	-	-
DMSA)				

[1]QCAA = quartz cuvette/atomic absorption; GFAA = graphite furnace/atomic absorption; FPD = flame photometric detector; ECD = electron capture detector.

above is adequate to resolve these interferences if most of the CO_2 is removed by sparging the solution for a few minutes after the addition of the acid and before the trap is immersed in liquid nitrogen[11].

The most versatile system is the combination of hydride generation with atomic absorption spectroscopy. Here, the objective is to introduce the hydrides into an atom reservoir aligned in the beam of the instrument and to dissociate them to produce a population of the atoms of interest. This can be either achieved in a fuel-rich hydrogen/air flame in a quartz tube (cuvette) as described by Andreae[6] or in a standard graphite furnace by electrothermal atomization[9]. The quartz cuvette has a higher sensitivity than the graphite furnace for arsenic and antimony; it is therefore preferred for the determination of these two elements. When organotin compounds are analyzed using the quartz cuvette system, spurious peaks are sometimes seen eluting with the methyltins. The origin of these peaks is not clear. This interference can be avoided by using the graphite furnace system. Here, the hydrides are introduced with the carrier gas stream, to which some argon has been added, into the internal purge inlet of the graphite furnace. They have to pass through the graphite tube and leave through the internal purge outlet. The heating cycle of the furnace is timed so it reaches the required atomization temperature shortly before the arrival of the unsubstituted hydride and is held at temperature until the last alkyl-substituted hydride has eluted. With this system, probably due to the higher operating temperature, no spurious tin peaks are present.

The graphite furnace system was originally developed when we found that the quartz cuvette gave only very poor sensitivity for germanium. We attribute this to the formation of GeO, a very stable diatomic species, at the relatively low temperatures of the quartz cuvette. At the higher temperatures available with the graphite furnace (we use 2600°C for the determination of Ge) we were able to obtain a sensitivity for germanium comparable to that of the other hydride elements.

Outlook

The full potential of the hydride methods has not yet been realized. Some workers are investigating its application to other elements (e.g. Cutter[18] for Se), and we plan to eventually extend our work to cover the remaining elements bismuth, tellurium and lead. The latter element is of great potential interest, but apparently the reduction to plumbane is impeded by kinetic factors, and most authors have obtained very low yields. We are working on an approach to overcome this problem.

Other detection systems have not yet been adequately explored. Some work on using microwave-induced plasmas (Denham, pers. comm.) for the determination of germanium shows great promise. The use of an atomic-emission based detection system will eventually open the way to the simultaneous multi-element detection of the hydride elements by a polychromator system[19].

RESULTS

The methods outlined above have been applied to the investigation of the marine and estuarine chemistry of the elements arsenic, tin, germanium and antimony. Some of the results of this work will be discussed in the following paragraphs.

Arsenic

I originally investigated the marine chemistry of arsenic in the search for that elusive process which leads to the large excess of some elements (including arsenic and the rest of the hydride elements) in atmospheric particulates over the amounts expected from crustal element ratios. Wood[20] had suggested that methylated species which would be volatile enough to escape into the atmosphere could be formed by biomethylation. I decided to investigate the biogeochemical cycle Wood had proposed for arsenic (similar to the one he described for mercury) by analyzing the chemical speciation of arsenic in seawater, sedimentary pore waters, and marine rain. I found that methylated species of arsenic were indeed present, but in the form of the highly water soluble methylarsenic acids (methylarsonic and dimethylarsinic acid). These species are not volatile enough to make a significant contribution to the atmospheric cycle of arsenic. In further contrast to the model described by Wood, the methylarsenicals were not formed by reductive bacterial biomethylation in the anoxic environment, but by the biosynthesis of large molecules containing organoarsenic structures by marine phytoplankton. This was borne out by the absence of any methylarsenicals in anoxic pore waters, and by the presence of the methylarsenic acids in all samples from the marine euphotic zone. The methylarsenicals, together with As(III), display a very characteristic profile: they occur at almost constant concentrations in the euphotic zone, reflecting primary productivity, and decline sharply in concentration near the level of 1% light penetration to values near or below their detection limits. Arsenic(V), on the other hand, shows a nutrient-like profile with depletion in the upper waters and regeneration at depth. In contrast to the "real" nutrients, however, this depletion amounts to only 10-30% of the deep water value[16,17].

To document the role of marine planktonic algae in the biogeochemical cycle of arsenic, we investigated its behavior in pure, axenic cultures of marine phytoplankton species[21]. We found substantial changes in the arsenic speciation of the growth medium as a consequence of algal activity in all of nine plankton species investigated. These changes included the removal of As(V), and the release of As(III) and of the methylarsenic acids. The relative amounts of the biologically released As species varied considerably between algal species. We found a significant number of organoarsenic species to occur both in the water-soluble and the lipid-soluble fractions of cell extracts of the algal species investigated. The most important one of these compounds was identified by Cooney et al.[22] as a trimethylarsonium-lactic acid phospholipid.

The simple mono- and dimethyl arsenic acids were found to be only minor constituents of the cell extracts. They are, on the other hand, the only organoarsenicals found by the borohydride method in natural waters. We therefore investigated if the larger organoarsenicals could be a significant fraction of the total arsenic in natural waters by subjecting samples of seawater containing natural levels of inorganic arsenic and the methylarsenicals in measured concentrations to rigorous base hydrolysis (Andreae and Bailey, unpublished data). This treatment should cleave all arsenic-carbon bonds, and the presence of previously unidentified organoarsenicals would show up as an increase of the total arsenic measured by the borohydride procedure[23]. In no case was a measurable increase of total arsenic observed after hydrolysis. This precludes the presence of organoarsenicals (at least of the type found in marine algae and crustaceans) in natural waters at levels higher than about two percent of total arsenic. These observations suggest that the organoarsenicals which are the major products of algal biosynthesis are either very short-lived outside the cell and are rapidly hydrolyzed to the simple methylarsenic acids after their release from the cells, or that the large organoarsenic molecules are enzymatically broken down to the methylarsenic acids within the cell and then rapidly excreted.

Tin

Braman and Tompkins[7] were the first to describe the presence of the methylated tin species [monomethyl-, dimethyl- and trimethyltin, probably present as the hydroxo-species, e.g. $(CH_3)_2Sn(OH)_2$ etc.[24]] in natural waters. The formation of methyltin compounds from inorganic tin by a strain of mercury-adapted bacteria (*Pseudomonas* sp.) from Chesapeake Bay was demonstrated by Huey et al.[25]. We have investigated a large number of riverine waters from the Gulf Coast of the United States (Byrd and Andreae, unpublished data) and have observed the presence of methyltins at the ng l^{-1} level in almost all samples, with dimethyltin usually being the most abundant species. Inorganic tin is present at levels of a few ng l^{-1} in the dissolved state in most of the rivers studied, with the exception of the Mississippi River, which showed a concentration of 123 ng Sn l^{-1}.

In several estuarine profiles in the Ochlockonee River (Florida), we have been able to document consumption of inorganic tin in the estuarine environment (Figure 3). No formation of organotin species was evident within the estuary, so that scavenging by particles is the most likely sink for inorganic tin. The saline end member of this profile shows a concentration of ca. 4.4 ng Sn l^{-1}. This is in good agreement with our determinations of the tin content of surface waters in the open Gulf of Mexico, which showed values of about 4.6 ng Sn l^{-1}. We have analyzed a vertical profile of tin species concentration to 250 m depth in the western Gulf of Mexico, and did not observe a consistent change in the concentration of the inorganic or the organic species of tin throughout this depth range. In this profile, the abundance of the methyltins decreased in the sequence trimethyl-, dimethyl-, monomethyltin. We sought tetramethyltin in seawater by sparging water samples without the addition of borohydride; this species was not present above the limit of detection of about 0.2 ng l^{-1}.

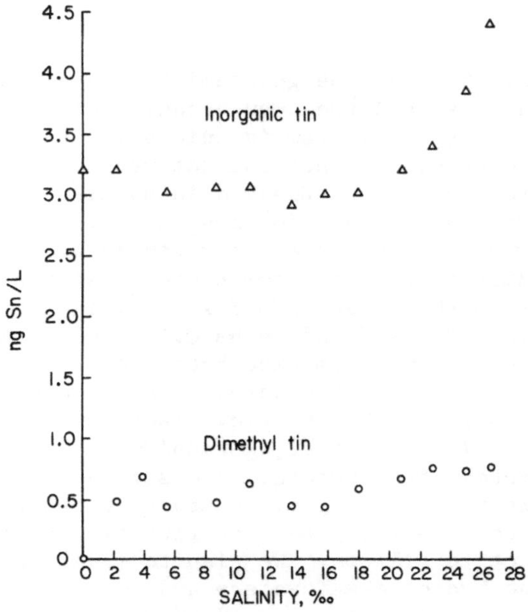

Fig. 3 Estuarine profile of inorganic and dimethyl tin from the estuary of the Ochlockonee River, Florida (15 Dec. 1980).

In our work on the geochemistry of tin we have been continuously plagued by the problem of sample integrity. Both contamination of the samples and analyte loss (probably to the walls of the sample container) can be very serious. Many commonly used sampling devices are made from polyvinyl chloride. This material contains large amounts of organotin stabilizers and contaminates samples with amounts of inorganic and methyl- and butyl-tin which may be several orders of magnitude in excess of natural values. On the other hand, most commonly used container materials, including polyethylene, adsorb significant amounts of inorganic tin in a relatively short time. We have observed losses of ca. 70% of the inorganic tin from river samples over some two hours of sample storage. For most samples, this problem can be overcome by acidification, but we must emphasize that analyte loss is a serious problem, and consequently checks should regularly be made with the particular kind of container and the type of water studied in an investigation[11]. The use of the flame photometric detection system for tin is of considerable advantage in this respect, as it can be easily carried in a van or be brought on board a research vessel (in contrast to an atomic absorption spectrophotometer), so that analyses can be made practically immediately after sample recovery.

Germanium

We have investigated the geochemistry of dissolved germanium in rivers, estuaries, and the open ocean. Among the elements discussed in this paper, germanium is unique insofar as it occurs only in a single species in natural waters, $Ge(OH)_4^0$. No organo-germanium species have been observed in nature. The geochemical behavior of germanium closely follows that of silicon[26]. Oceanic profiles of germanium are (except for the larger scatter in the case of Ge) indistinguishable from those of silica (Figure 4). The ratio Ge:Si in the ocean is 0.7×10^{-6}; in the estuary of the Ochlockonee River it was found to be 0.6×10^{-6}. A phytoplankton bloom in this estuary was removing both elements at the ratio at which they were present in the water. In the exogenic cycle, germanium behaves like a "super heavy isotope of silicon". In hydrothermal solutions, however, germanium has a more chalcophilic behavior and becomes incorporated into sulfide minerals. Thus the geothermal effluents, which provide about one-third of the global flux of Si to the oceans may be significantly depleted in germanium with respect to silicon. We plan to investigate the Ge/Si ratios in marine hydrothermal waters and possible use of this ratio in siliceous oozes as a record of variations in the relative

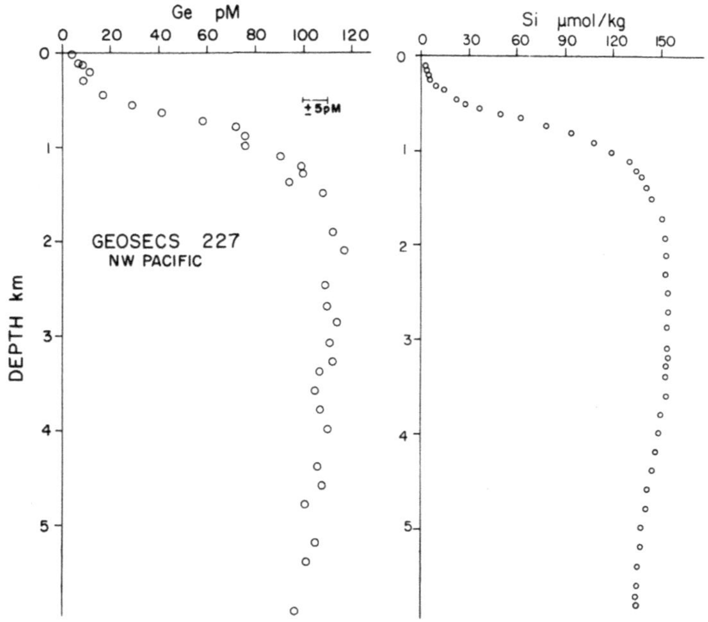

Fig. 4 Vertical profile of germanium and silicon at GEOSECS Station 227 in the North Pacific (25° 00' N, 170° 05' E).

rates of the input of silica into the ocean from continental weathering and from hydrothermal activity.

Antimony

The first and last systematic investigation into the marine chemistry of antimony was published by Schutz and Turekian[27]. They reported values in the range of 0.12 to 1.1 μg l^{-1}, with an average of 0.33 μg l^{-1} in seawater from a large number of stations in the World Ocean. In contrast, Tompkins[13] measured concentrations of some tens of ng l^{-1} of antimony in coastal water of the Gulf of Mexico. When using the same reduction conditions as described by Tompkins, we initially obtained similar values for inorganic Sb. However, when we investigated the reaction conditions for Sb(III) and Sb(V) in detail, it became clear that it is necessary to add potassium iodide to the reaction medium in order to obtain complete reduction of Sb(III) to the hydride. This resulted in much higher concentration values for inorganic antimony and a partial resolution of the discrepancy between the findings of Schutz and Turekian and those of Tompkins.

In further contrast to the work of Tompkins, we identified two additional peaks eluting after inorganic antimony. We were able to obtain authentic samples of methylstibonic acid and dimethylstibinic acid from Dr. H.A. Meinema in Utrecht (The Netherlands) who had prepared these compounds as part of his Ph.D. dissertation[28]. When subjected to borohydride reduction, these compounds were converted to the corresponding methyl- and dimethyl-stibines, which co-chromatograph on two different chromatographic packings with the peaks observed from samples of natural waters. Together with the selectivity of the atomic absorption detector, this is convincing evidence for the presence of these compounds in nature[10]. A profile along the salt gradient in the Ochlockonee River estuary shows the presence of all four antimony species (Figure 5). Sb(III) shows a considerable amount of scatter, probably partly due to some oxidation to Sb(V) during sample storage. Its concentration increases from ca. 3 ng l^{-1} in the river to near 11 ng l^{-1} in the sea, with no identifiable deviation from linearity. The methylated forms, on the other hand, show evidence of consumption in the estuary, decreasing from 12.5 and 1.5 ng Sb l^{-1} in the form of methylstibonic and dimethylstibinic acid, respectively, to values below 1 ng l^{-1}. (It is interesting to note that in the case of arsenic, the dimethyl form is about ten times as abundant as the monomethyl form, while the reverse is true for antimony.) The dominant form of antimony is Sb(V), increasing from 12 to 132 ng l^{-1}, in agreement with the low end of the range of antimony concentrations measured by Schutz and Turekian. There is a slight upward bulge in the profile of Sb(V), suggesting production of this species in the estuary. This could

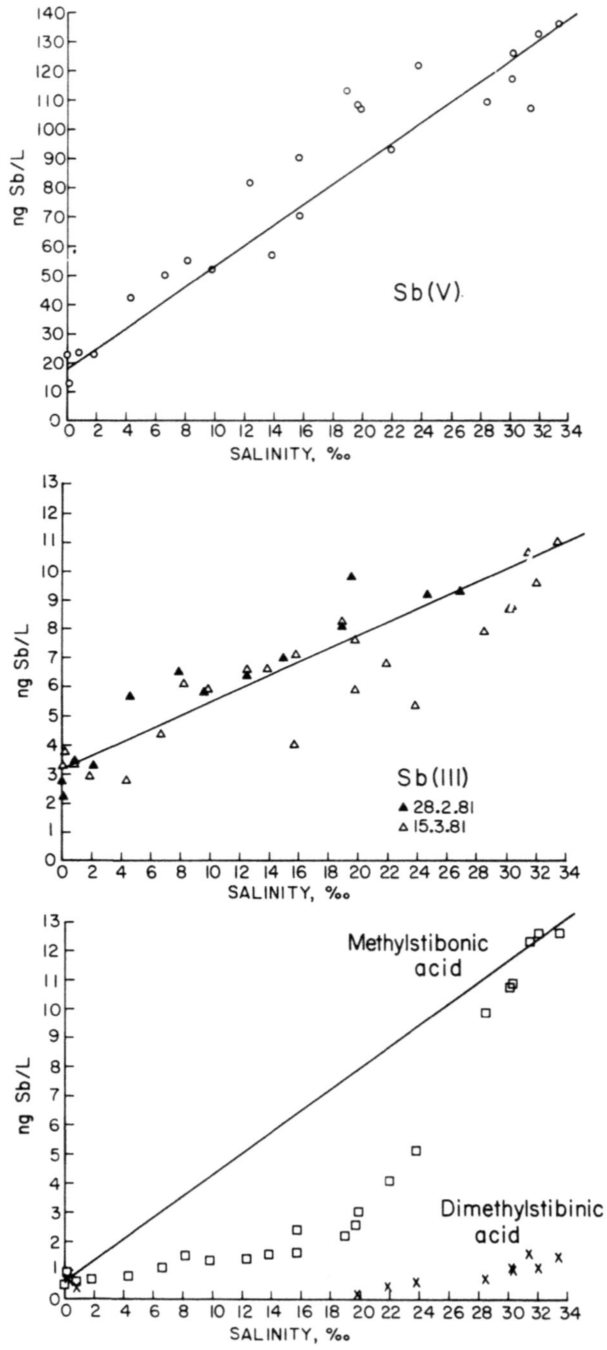

ng Sb/L
Sb(V)
SALINITY, ‰
ng Sb/L
Sb(III)
▲ 28.2.81
△ 15.3.81
SALINITY, ‰
ng Sb/L
Methylstibonic acid
Dimethylstibinic acid
SALINITY, ‰

be the result of the demethylation of the methylantimony acids suggested by their distribtuion. In water samples from a vertical profile in the Pacific Ocean (GEOSECS Station 227) values between 92 and 141 ng l^{-1} Sb(V) were found. No clear trends in the vertical distribution were identifiable. We suspect that some changes may have occurred during sample storage.

ACKNOWLEDGEMENTS

I am grateful for the time, effort and enthusiasm my students and co-workers have invested in the work that forms the basis of this review and progress report. Special thanks go to J.-F. Asmodé, K. Bailey, J.T. Byrd, P. Foster, P.N. Froelich and L. Van 't dack. I thank H.A. Meinema for the specimens of the methylantimony acids. This work was supported in part by the office of Naval Research (USN N000 14-75-C-0152), by the National Science Foundation (OCE 79-20183 and OCE 81-25223) and the Petroleum Research Fund, administered by the American Chemical Society (PRF-12144-GZ).

REFERENCES

1. Marsh, J., 1837: Beschreibung eines neuen Verfahrens, um kleine Quantitaten Arsenik von den Substanzen abzuscheiden womit er gemischt ist. Ann. Chem., 23, 207.
2. Mohr, F., 1837: Zusätze zu Marsh's Arsenikentdeckung. Ann. Chem., 23, 247.
3. Braman, R.S., L.L. Justen and C.C. Foreback, 1972: Direct volatilization-spectral emission type detection system for nanogram amounts of arsenic and antimony. Anal. Chem., 44, 2195.
4. Holak, W., 1969: Gas-sampling technique for arsenic determination by atomic absorption spectrophotometry. Anal. Chem., 41, 1712.

Fig. 5 Estuarine profile of the antimony species [Sb(V), Sb(III), methylstibonic and dimethylstibinic acid]. All samples were collected in the Ochlockonee River estuary on 15 March, 1981, with the exception of some samples only analyzed for Sb(III) (full black triangles) which were taken on 28 February, 1981.

5. Braman, R.S. and C.C. Foreback, 1973: Methylated forms of arsenic in the environment. Science, 182, 1247.
6. Andreae, M.O., 1977: Determination of arsenic species in natural waters. Anal. Chem., 49, 820.
7. Braman, R.S. and M.A. Tompkins, 1979: Separation and determination of nanogram amounts of inorganic tin and methyltin compounds in the environment. Anal. Chem., 51, 12.
8. Talmi, Y. and D.T. Bostick, 1975: Determination of alkylarsenic acids in pesticide and environmental samples by gas chromatography with a microwave emission spectrometric detection system. Anal. Chem., 47, 2145.
9. Andreae, M.O. and P.N. Froelich, Jr., 1981: Determination of germanium in natural waters by graphite furnace atomic absorption spectrometry with hydride generation. Anal. Chem., 53, 287.
10. Andreae, M.O., J.-F. Asmodé, P. Foster and L. Van 't dack, 1981: Determination of antimony(III), antimony(V), and methylantimony species in natural waters by atomic absorption spectrometry with hydride generation. Anal. Chem., 53, 1766.
11. Andreae, M.O. and J.T. Byrd, 1981: Determination of tin and methyltin species in natural waters. Submitted to Anal. Chem.
12. Foreback, C.C., 1973: Some studies on the detection and determination of mercury, arsenic and antimony in gas discharges. Thesis, Univ. of South Florida, Tampa.
13. Tompkins, M.A., 1977: Environmental-analytical studies of antimony, germanium and tin. Thesis, Univ. of South Florida, Tampa.
14. Fleming, H.H. and R.G. Ide, 1976: Determination of volatile hydride-forming metals in steel by atomic absorption spectrometry. Anal. Chim. Acta., 83, 67.
15. Crecelius, E.A., 1978: Modification of the arsenic speciation technique using hydride generation. Anal. Chem., 50, 826.
16. Andreae, M.O., 1978: Distribution and speciation of arsenic in natural waters and some marine algae. Deep-Sea Res., 25, 391.
17. Andreae, M.O., 1979: Arsenic speciation in seawater and interstitial waters: the influence of biological-chemical interactions on the chemistry of a trace element. Limnol. Oceanogr., 24, 440.
18. Cutter, G.A., 1978: Species determination of selenium in natural waters. Anal. Chim. Acta., 98, 59.
19. Robbins, W.B. and J.A. Caruso, 1979: Development of hydride generation methods for atomic spectroscopic analysis. Anal. Chem., 51, 890A.
20. Wood, J.M., 1974: Biological cycles for toxic elements in the environment. Science, 183, 1049.

21. Andreae, M.O. and D. Klumpp, 1979: Biosynthesis and release of organoarsenic compounds by marine algae. Environ. Sci. Technol., 13, 738.
22. Cooney, R.V., R.O. Mumma and A.A. Benson, 1978: Arsonium-phospholipid in algae. Proc. Natl. Acad. Sci. USA 75, 4262.
23. Edmonds, J.S. and K.A. Francesconi, 1977: Methylated arsenic from marine fauna. Nature, 265, 436.
24. Baes, C.F. and R.E. Mesmer, 1976: "The Hydrolysis of Cations" Wiley, New York.
25. Huey, C., F.E. Brinckman, S. Grim and W.P. Iverson, 1974: The role of tin in bacterial methylation of mercury. Proc. Internatl. Conf. on Transport of Persistent Chemicals in Aquatic Ecosystems, Natl. Research Council of Canada, Ottawa.
26. Froelich, P.N. and M.O. Andreae, 1981: The marine geochemistry of germanium: Eka-Silicon. Science, 213, 205.
27. Schutz, D.F. and K.K. Turekian, 1965: The investigation of the geographical and vertical distribution of several trace elements in sea water using neutron activation analysis. Geochim. Cosmochim. Acta., 29, 259.
28. Meinema, H.A. and J.G. Noltes, 1972: Investigations on organoantimony compounds.VI. Preparation and properties of thermally stable dialkylantimony(V) compounds of the types $R_2Sb(OR')_3$, $R_2Sb(OAc)_3$ and $R_2Sb(O)OH$. J. Organomet. Chem., 36, 313.

CONVERSION TABLE

1 ng Sb l^{-1} = 8.21 pmol l^{-1}
1 g As l^{-1} = 13.35 nmol l^{-1}
1 ng Sn l^{-1} = 8.43 pmol l^{-1}

ANTIMONY CONTENT AND SPECIATION IN THE WATER COLUMN AND INTERSTITIAL WATERS OF SAANICH INLET

K.K. Bertine and Dong Soo Lee*

Department of Geological Sciences
San Diego State University
San Diego, California, U.S.A.

*Scripps Institution of Oceanography
La Jolla, California 92093 U.S.A.

ABSTRACT

A hydride generation method was developed for the determination of Sb(III), Sb(III)-S and Sb(V) species in natural waters, and has been applied to the waters and interstitial waters of Saanich Inlet. In the oxic waters, dissolved antimony exists primarily as Sb(V) (1.2-1.4 nM) with a few percent Sb(III) (.01-.07 nM). Sb(III) production occurs in the photic zone and probably accounts for its concentration at levels greater than those predicted from equilibrium considerations. Even in the low H_2S bottom waters, there is no significant reduction of Sb(V) and Sb(III). However, up to 0.15 nM Sb(III)-S species is found there. In the upper 10 cm of the reducing interstitial waters there is a very large release of Sb from the particulate matter. At least 50% of the Sb exists as Sb(III) + Sb(III)-S. The remainder most probably exists as a thioantimonate. This dissolved Sb diffuses to the overlying waters where it is rapidly oxidized to Sb(V). In the deeper portions of the sediment, Sb is resedimented perhaps as a sulfide or associated with iron sulfide precipitation. Dimethylstibonic acid occurs at low concentrations (.01-.03 nM) in the water column. In the methane zone of the pore waters, there is a significant production of methylstibonic acid reaching levels of up to 4.9 nM.

INTRODUCTION

Saanich Inlet is a reducing fjord located on the southeast coast of Vancouver Island, Canada (Fig. 1). It is about 25 km long, 0.4 to 7.6 km wide with a maximum depth of 235 m and a sill depth of about 70 m. It has a seasonal anoxic cycle described by Anderson and Devol[1]. In late summer or early fall dense, oxygenated water originating from upwelling along the Washington-Oregon coast reaches the sill and displaces the deep water in the inlet. During late winter to summer a strong vertical pycnocline isolates deep water of this productive basin resulting in oxygen depletion and sulfide production in the deeper water column. Hence, this fjord is particularly suited for investigations of redox processes. Previously, Emerson, Cranston and Liss[2] studied ten elements involved in redox reactions during June, 1977 when the sulfide layer was well established in the water column. Herein, an additional redox couple (Sb(III)/Sb(V)) is investigated. Water column samples were taken twice, one when the oxygen-hydrogen sulfide boundary was at the sediment-water interface, and once when the boundary was in the water column. Interstitial water samples were collected only on the first cruise.

METHODS

Samples were collected from two stations (SI-3 and SI-5, Figure 1) on January 8 and 9, 1981. Water depth at SI-3 was 225 m

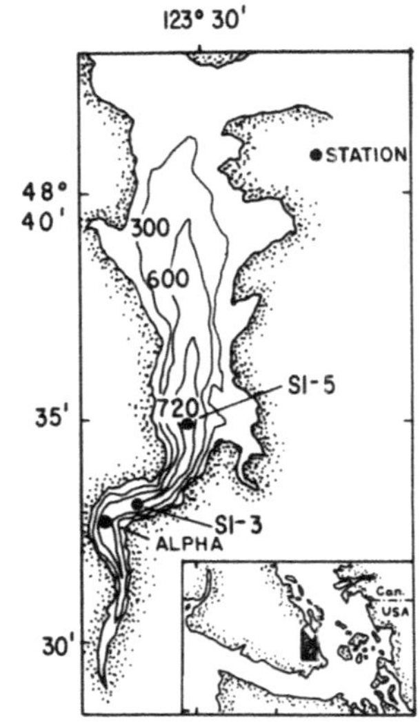

Fig. 1 Location map and sampling locations in Saanich Inlet, British Columbia, Canada.

and at SI-5 235 m. Water column samples were collected using modified 2 l PVC Niskin bottles (and one 5 l Go-Flo sampler for comparison) which were hung on a stainless steel hydrowire. Samplers were soaked in dilute nitric acid, then distilled water prior to their use. Interstitial water samples were collected in conjunction with Ross Barnes of Walla Walla College using his 205 cm harpoon sampler[3] which has a 45 cm surface probe with closely-spaced inlets, designed to slide so that it does not bury itself readily in soft sediments. The 205 cm main probe has sample ports overlapping the surface probe so that the samples from the two probes can be compared. The stainless steel filter elements were soaked in diluted nitric acid, then distilled water. Prior to use, deionized water was flushed through the sampling apparatus.

Blank determinations for Sb contamination were made on the surface probe by immersing the probe in Patricia Bay water and manually passing 25 ml of water through the device. Antimony blanks, resulting from passing through the interstitial water sampler, were then determined by the difference in Sb concentrations between the waters passing through the probe and the Patricia Bay water.

Oxygen, hydrogen sulfide, silicate, phosphate, nitrate, nitrite, temperature and salinity were measured on the water column samples. Silicate, phosphate, nitrate, nitrite and sulfate were determined on the interstitial waters. Waters for nutrient analyses were quick frozen and stored for about a month prior to analysis. Samples for phosphate and silicate analysis were acidified and shaken immediately after defrosting to prevent precipitation of iron phosphates. It was found upon melting that a white, turbid precipitate had formed. Since it interfered with the auto-analyzer analysis, the samples were centrifuged prior to analysis. This resulted in a loss of phosphate and silicate when compared to previous analyses[4]. Oxygen was run immediately after collection by a modified Winkler method and hydrogen sulfide was analyzed spectrophotometrically.

Water column samples were collected for us by Steve Emerson (University of Washington) on May 11-17, 1981 from Station Alpha (48° 32.3'N, 123°32.7'W) fairly close to station SI-3 (Fig. 1). At this time the O_2/H_2S boundary was at 170 m water depth (Emerson, personal communication).

Antimony Analyses

New methods were developed for the analyses of Sb(V), Sb(III) and "Sb-S" species by hydride generation techniques.

Total Antimony. Total antimony content was analyzed by a hydride generation technique utilizing a quartz burner with a hydrogen flame. The water sample was made 2N in HCl with a final volume of 100 ml. Two ml of 20% (w/v) KI were added and the sample degassed using He as a carrier gas for 100 seconds. A silanized glass wool trap on which to collect the SbH_3 was then placed in liquid nitrogen and 2 ml of 5% NaBH were slowly injected over a time period of 100 seconds. The sample was stripped for 300 seconds, then the trap was removed from the liquid nitrogen and the hydride was carried to an electrically heated quartz burner with a hydrogen flame. The Sb concentration was measured using atomic absorption spectrophotometer. Detection limits of about 0.01 ng are obtainable.

Both the HCl and $NaBH_4$ contributed to the blank. The Sb(V) in the 12 N HCl was removed by uptake on Dowex 1-X8 anion exchange resin. $NaBH_4$ was purified, after dissolution, by addition of 0.5 ml NaOH (50%) to 200 ml of 5% $NaBH_4$ and subsequent filtration through a HCl precleaned 0.45 μm millipore membrane.

Hydrogen sulfide gas interfered in the determination of antimony since, after the addition of hydrogen sulfide, a peak comes a few seconds after the Sb peak. It was found that either degassing the sample for 300 seconds rather than 100 seconds or placement of lead acetate in the line as suggested by Andreae (this volume) eliminated the problem without interfering with the Sb determination.

Sb(III). 2 ml of citrate buffer (purified by an Fe-APDC precipitation) were added to maintain a pH of 5-6. The sample was degassed for 100 seconds prior to the injection of 2 ml of 5% $NaBH_4$. The sample was stripped for 300 seconds. This procedure gives a complete extraction of Sb(III) and no extraction of Sb(V) even in the presence of 100-fold higher concentrations of Sb(V).

At a pH of 5-6, hydrogen sulfide evolution by degassing proceeds at a much slower rate. Background correction using a hydrogen lamp or lead acetate placed in line was able to remove any interference from the amount remaining after 400 seconds. However, it was found that the extraction yield of Sb(III) standards prepared in Na_2S (at H_2S levels comparable to those found in Saanich Inlet waters) was not only incomplete but the yield was also inversely proportional to the amount of sulfide added to the standards. It has been hypothesized that in sulfide-rich waters an antimony sulfide complex may exist. Cotton and Wilkenson[5] propose that in excess sulfide SbS_3^{-3} most probably exists. Thermodynamic data exist for SbS_2^- (Ref 6). By experimentation, it was discovered that a complete yield of Sb(III) in Na_2S could be attained by making a 1-2 ml sample 2N in HCl, degassing for 5 minutes, bringing the volume to 100 ml, adding sufficient Tris buffer to bring the pH to 6, then proceeding with the hydride generation method as above. No Sb(V), even in 1000-fold excess, was detected by the above method.

Methyl antimony species. In the chromatograms resulting from the hydride generation technique, one to two peaks were frequently observed following the inorganic Sb peak. Andreae (this volume) reports the occurrence of methylstibonic and dimethylstibonic acids in estuarine and sea waters using a similar hydride generation technique. By using his relationships between inorganic Sb and methylstibonic and dimethylstibonic acid yields, we are able to estimate the amount of these species in our waters.

RESULTS AND DISCUSSION

Water column. The Sb (total inorganic) and Sb(III) contents in Saanich Inlet waters are similar to values found by Andreae (this volume). The total Sb contents are slightly less than those (1.64-2.46 nmol l^{-1}) found by Schutz and Turekian[7], Brewer, Spencer and Robertson[8], and Gohda[9]. They are higher than Bramen and Tompkins'[10] values which were also based on a hydride generation method. However, their method only analyzed a small part of the Sb(V) at a pH of 1-2 with no addition of KI. Thus, the most prevalent species in oxygenated waters was only partially analyzed. For Sb(III), Gohda[9] found values to average between 0.41 and 0.82 nmol l^{-1}. Compared to our values in Saanich Inlet, Scripps Pier sea water (0.07 nmol l^{-1}) and Andreae's values, their values are clearly suspect.

At the time of the January, 1981 sampling, Saanich Inlet had been flushed and oxygen was found in all samples throughout the water column (Table 1) and as observed by a submersible, squat lobsters (*Munida quadrispina*) were resting on the bottom. The 0 m depth samples in profiles SI-3 and SI-5 were diluted by fresh water (rainfall and/or river runoff) as evidenced by their low salinities of 11-12°/oo compared to 31°/oo in the bottom waters. The total Sb contents of these samples (Table 1 and Fig. 2) were 0.59 and 0.74 nmol l^{-1}. Based on their salinity contents and the average Sb contents in the water column, this would correspond to Sb contents of 0.30 and 0.53 nmol l^{-1} in the fresh water components with the higher value closer to the estuary head. Andreae (this volume) found approximately 0.12 nmol l^{-1} total inorganic Sb in a Florida river. Rainfall in the San Diego area ranged from 0.74 nmol l^{-1} to less than 0.02 nmol l^{-1} with progressively lower values as the rains continued (Suleyman, personal communication). Thus, if the above values pertain to the study area, the low Sb surface values could represent dilution by either or both rainfall and river water.

Table 1. Water Column Chemistry

Depth	T°C	Salinity ‰	O_2 µM	NO_3 µM	NO_2 µM	PO_4 µM	SiO_3 µM	Sb(III) nM	Sb(V)[a] nM	Total Inorganic Sb nM	Methylstibonic Acid nM[b]
SI-3											
0	6.82	11.22	356	16.5	0.23	0.91	16.3	0.03	0.72	0.72,0.76	0.02
10	8.62	28.30	236	27.7	0.17	2.46	55.0	0.04	1.05	1.08	0.03
25	8.69	28.94	211	28.2	0.12	2.53	54.0	0.07	1.28	1.40,1.30	0.03
50	8.99	29.50	191	29.0	0.12	2.58	51.9	0.10	0.97	1.02,1.12	0.03
75	9.20	30.23	160	29.1	0.15	2.75	56.8	0.03	1.31	1.31,1.35	0.03
100	9.29	30.84	25	21.6	0.10	4.40	85.9	0.05	1.15	1.22,1.18	0.03
125	9.10	31.11	4	9.9	0.09	5.45	61.0	0.03	1.12	1.15	0.03
150	9.08	31.14	5	11.8	0.08	5.46	76.5	0.04	1.05	1.21,1.05	0.03,0.03
175	9.15	31.17	9	17.3	0.13	5.17	72.1	0.03	1.02	1.05	0.02
200	9.22	31.22	5	20.1	0.09	5.10	81.3	0.04	0.85	0.90,0.89	0.03
210	9.22	31.22	6	20.3	0.09	5.19	70.1	0.06	1.49	1.59,1.54	0.03
215	9.25	31.24	4	17.0	0.12	5.58	89.1	0.09	1.47	1.56	0.02
220	--	31.24	4	16.3	0.09	5.68	83.7	0.07	1.62	1.69	0.03
Bottom water				11.6	0.16	6.8	117	0.35	4.42	4.76	--
SI-5											
0	6.56	12.05	328	--	0.24	1.03	--	<0.004	0.59	0.59	<0.004
10	8.60	28.41	220	27.6	0.16	2.49	55.6	0.03	0.99	1.02	0.02
25	8.65	29.02	207	30.1	0.15	2.51	55.6	0.04	0.90	0.97,0.91	0.02
50	8.98	29.52	192	28.4	0.10	2.61	55.6	0.07	1.02	1.08	0.02
75	9.06	30.22	174	28.7	0.09	2.68	55.1	0.05	0.84	0.89	0.02
100	9.30	30.91	15	23.6	0.10	4.39	84.8	0.05	1.09	1.13,1.15	0.02
125	9.16	31.10	2	12.8	0.09	5.35	75.3	0.02	1.04	1.05	0.01
150	9.18	31.16	14	18.0	0.09	4.84	79.4	0.02	1.13	1.15	0.02
175	9.19	31.20	19	21.5	0.09	4.73	82.4	0.03	0.92	0.95	0.02
200	9.20	31.23	5	22.1	0.08	4.97	85.2	0.03	0.87	0.89,0.90	0.02
210	9.21	31.23	3	21.0	0.13	5.13	100.2	0.03	1.27	1.30	0.02
215	9.21	31.23	4	19.7	0.10	5.25	87.4	0.05	1.17	1.22	0.02
220	9.20	31.24	3	17.8	0.15	5.52	105.9	0.05	1.50	1.50,1.61	0.03
225	--	31.25	1	14.9	0.13	5.90	101.2	0.07	1.81	1.87	0.02
Bottom water				2.0(?)	0.18	7.8	121	4.13,3.89	2.21,2.40	6.28,6.34	--

(a) By difference
(b) Concentrations based on Sb(III) Standard

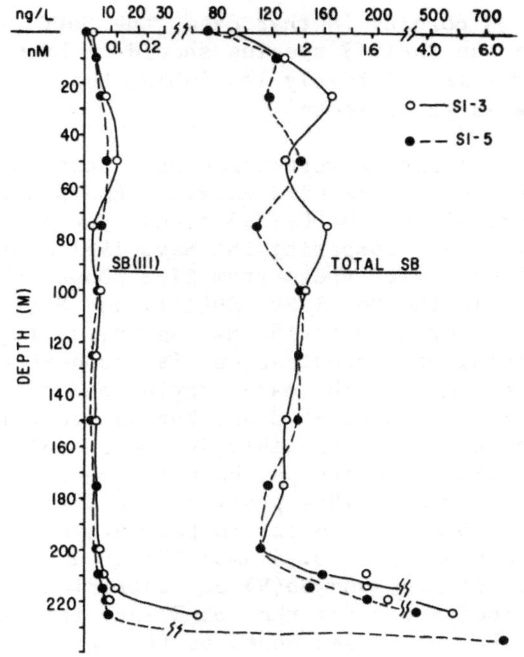

Fig. 2 Sb(III) and total inorganic Sb concentrations $\frac{ng\ l^{-1}}{nM}$ in the water column at SI-3 and SI-5.

From 10 m to 200 m water depth the Sb contents vary from 0.89 to 1.35 nmol l^{-1}. The variation in adjacent samples may be partly due to the analytical error (although duplicates of the same sample were generally within 0.08 nmol l^{-1} of each other) but also may be due to the horizontal flushing processes in Saanich Inlet, especially of the deep waters. For instance, at 175 to 200 m the lower Sb contents correspond to a water which had recently flowed in over the sill as evidenced by its higher oxygen and nitrate contents and lower phosphate and nitrite contents.

Below 200 m the Sb concentrations increase toward the bottom reaching 6.31 nmol l^{-1} 70 cm above the bottom. Phosphate, silicate and nitrite increase in this interval whereas nitrate and oxygen decrease.

With the exception of the two samples taken just above the sediment surface, the percentage of Sb(III) in the water column ranges between 0 and 10% of the total Sb (average 4%) with the Sb(III) concentration ranging between 0 and 0.10 nmol l^{-1} (Fig. 2). Thus, the predominant form of inorganic Sb in oxygenated waters is Sb(V) in agreement with the theoretical thermodynamic data (Fig. 3). According to that data, the Sb(V) species should be $Sb(OH)_6^-$ and the Sb(III) species should be $HSbO_2$. In chloride-rich waters SbOCl is relatively insoluble, but its role in Sb geochemistry is as yet unknown.

The May, 1981 samples were taken at a station fairly close to SI-3 when the bottom waters were anoxic. Emerson (personal communication) indicates that the O_2-H_2S boundary was between 170-175 m water depth (Fig. 4). Comparing the May, 1981 antimony profiles (Fig. 5 and Table 2) with those from SI-3 shows that very good agreement exists in the total Sb, Sb(III) and Sb(V) concentrations between the depths of 25 m to 150 m. Again, a slight minimum in the Sb(V) and total Sb concentrations is found at the base of the euphotic zone (40-50 m). The major deviations between the two profiles occur in the upper 25 m and the waters below 160 m. The surface difference is most probably due to the dilution with low antimony, low salinity waters as shown in the January waters. The bottom waters in the May, 1981 profile have been influenced by the presence of up to 5 μM H_2S in the bottom waters. This has resulted in the presence of up to 0.15 nmol l^{-1} of an Sb(III)-S complex. Since the total antimony and Sb(V) concentrations also increase in this interval, the source for the "Sb-S" and the extra Sb(V) must either be from the bottom sediments or from release from the particulate matter falling down to the anoxic zone. However, throughout the water column, there is no significant difference in Sb concentrations in the filtered and unfiltered samples (Table 2).

The most significant conclusion from these H_2S-rich waters, however, is that there is no major reduction of Sb(V) to either

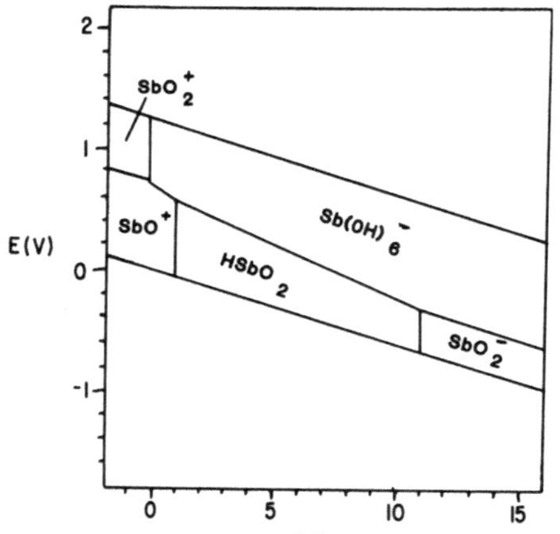

Fig. 3 E(v) - pH diagram for H_2O - Sb. From Pourbaix[14].

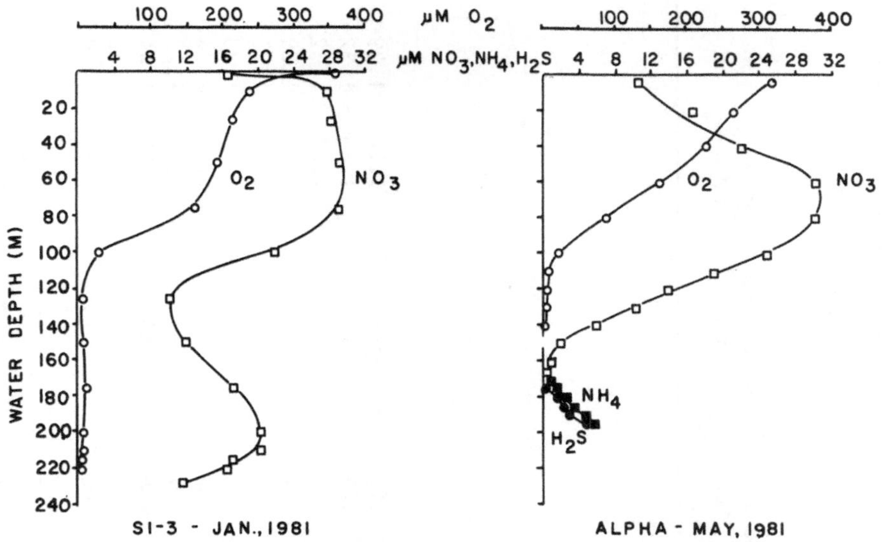

Fig. 4 Comparison of the NO_3^-, O_2, H_2S and NH_4^+ concentrations in the water column between January, 1981 and May, 1981.

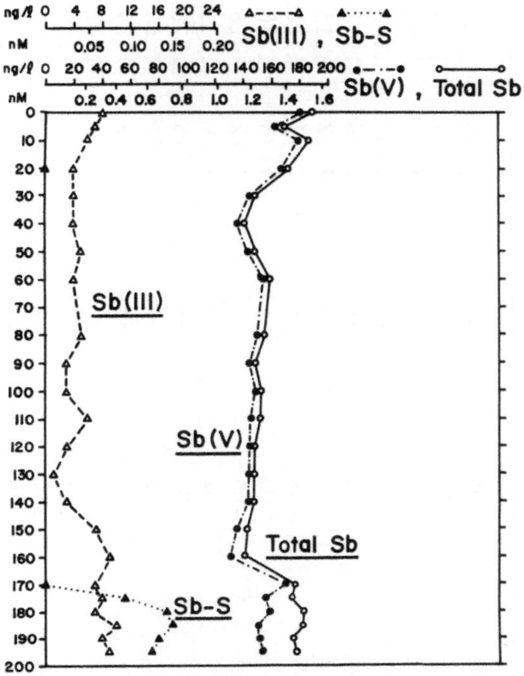

Fig. 5 Sb(III), Sb(III)-S, Sb(V) and total inorganic Sb concentrations, May, 1981.

Table 2 Water Column Chemistry, May, 1981.

	Unfiltered				Filtered			
Depth (m)	Sb(III) nM	Sb(III)-S nM	Sb(V) nM	Total (a)Inorganic Sb nM	Sb(III) nM	Sb(III)-S nM	Sb(V) nM	Total (a)Inorganic Sb nM
0	.07	--	1.48	1.54				
5	.06	--	1.33	1.39				
10	.05	--	1.47	1.52				
20	.03	ND	1.37	1.40				
30	.03	--	1.18	1.22				
40	.03	--	1.12	1.15				
50	.04	--	1.17	1.22				
60	.03	--	1.26	1.30				
80	.04	--	1.22	1.26				
90	.02	--	1.29	1.31	.02	--	1.26	1.29
100	.02	--	1.25	1.26	.02	--	1.30	1.32
110	.05	--	1.39	1.44	.05	--	1.27	1.32
120	.03	--	1.26	1.29	.02	--	1.26	1.29
130	.02	--	1.29	1.31	.01	--	1.26	1.27
140	.11	--	1.17?	1.29	.02	--	1.26	1.28
150	.14	--	1.13?	1.27	.06	--	1.18	1.24
160	.06	--	1.22	1.27	.07	--	1.16	1.23
170	.07	ND	1.45	1.52	.06	ND	1.39	1.44
175	--	--	--	--	.07	.09	1.27	1.43
180	.04	∿ .14	∿ 1.30	1.48	.06	.14	1.30	1.49
185	.03	∿ .17	∿ 1.29	1.49	.08	.15	1.23	1.46
190	.07	.14	1.30	1.50	.07	∿ .13	∿ 1.24	1.44
195	.07	.14	1.29	1.50	.07	.12	1.26	1.45

Sb(III) or the Sb(III)-S complex in disagreement with the theoretical equilibrium predictions as shown in Figure 3.

In the atomic absorption spectra, one peak was frequently observed after the inorganic Sb peak which most probably represents methylstibonic acid (Andreae, this volume). Due to the long storage of the samples and lack of appropriate standards, quantification is difficult. However, based on Andreae's finding that the sensitivity of Sb(III) and methylstibonic acid are similar, values of <0.004 to 0.03 nmol l^{-1} are computed. No variation with depth in the water column was observed.

Interstitial waters. Total antimony contents were determined at SI-3 and SI-5 (Table 3). Blank values ranged between 0.16 and 1.81 nmol l^{-1} with the higher values corresponding to the older filter elements. The depths as represented in this table are approximate since it is not certain the extent to which the sampler sunk into the soft mud. The uncertainty is about 5 cm (Ross Barnes, personal communication). As seen in Figures 6 and 7, there is a large increase in the total Sb concentration reaching a maximum of 140 nmol l^{-1} in SI-3 at a depth of 5 cm and 102 nmol l^{-1} in SI-5 at a depth of 8 cm. In SI-3, values decrease below this peak reaching values of 8.2-12.3 nmol l^{-1} at a depth of 145 cm. Total Sb in SI-5 exhibits the same pattern except that lower background values of about 5.3 nmol l^{-1} are reached at 45 cm rather than at 145 cm. The difference in background values, maximum values and the depths at which they are attained may be due to the relative amounts of organic matter in the deposits at the two locations. Presley et al.[11] reported that the concentrations of organic matter in Saanich Inlet sediments increased toward the estuary head. If the release of Sb is related to organic matter decomposition, then higher organic matter might be expected to produce higher amounts of Sb as found at SI-3.

Sb(III) contents in the interstitial waters range from 0.41 to 53.4 nmol l^{-1}. These values may be minimal since it was observed in some samples (note H_2S and 9 S in SI-3) that with successive meltings and analysis, the Sb(III) contents decreased as the H S smell disappeared. All SI-5 waters had been melted and analyzed for total Sb prior to Sb(III) analysis, whereas the waters from SI-3 were analyzed first for Sb(III). However, these samples had been stored for four months in a frozen state prior to analysis and the effects of such storage are uncertain. Some elemental sulfur, indicating some oxidation, was observed in all samples prior to analysis. The values of Sb(III) are generally lower in SI-5 as compared to SI-3. This situation is particularly evident in the deeper waters where methane was found and could be due either to oxidation or to the relative amounts of organic matter at the sites. Assuming that Sb(III) had not become oxidized prior to analysis, the Sb(III) percentage ranges from 3% to

Table 3. Interstitial Water Chemistry

	Depth (cm)	NO_3 µM	NO_2 µM	PO_4 µM	SO_4 µM	Sb(III) nM	Sb(V)(f) nM	Sb(Total)(a) Inorganic nM	Methylstibonic Acid nM
SI3									
10 S	2		.30	246	11.4	52.5,39.1(d)	50.6	103	ND
9 S	5		.81	233	8.4	53.4,18.5(d)	86.5	140	ND
8 S	8		1.27	284	7.3	40.9	78.4	119	ND
7 S	12		1.02	320	6.8	51.1	46.5	97.6	ND
6 S	16.5		.41	375	2.4	36.6	44.6	81.2	ND
5 S	21.5		.40	287	<1.0	37.8	36.9	74.7	ND
4 S	26.5		.32	435		30.0	31.1	61.1	ND
3 S	31.5		.59	459		40.7	20.5	61.2	ND
2 S	36.5		.50	470		25.1(d)	27.9(d)	53.0	ND
1 S	43		.53	657		37.0	10.5	47.5	ND
12 M	5		.21	264	13.3	18.9(?)	49.7	68.6	ND
11 M	15		.38	187	16.7	28.8	48.9	77.7	ND
10 M	25		.29	176	12.6	26.7	32.0	58.7	1.2
9 M	45			479	<1.0	13.6	16.4	30.0	1.8
8 M	65			440	<1.0	3.7(b)	30.8(b)	34.5(b)	4.9
6 M	105			426		11.5	40.7	52.2	3.5
4 M	145			482		3.7	10.0	13.7	2.8
3 M	165			476		4.7	17.0	21.7	4.0
1 M	205			507		1.6	10.3	11.9	2.9
SI5									
10 S	2	36	1.61	238	4.9	37.4(d)	38.3	75.7	ND
9 S	5	3.2	1.41	299	<1.0	37.0(d)	48.1	85.1	ND
8 S	8	2.8	.94	333	<1.0	26.5(d)	75.1	102	ND
7 S	12	9.6	.88	303		26.7(d)	29.2	55.9	ND
6 S	16.5	6.4	.29	308		9.5(b)(d)	43.6(b)	53.1(b)	ND
5 S	21.5	7.2	.35	374		7.8(d)	11.0	18.8	0.78
4 S	26.5	4.8	.33	398		3.4(d)	5.9	9.3	0.56
3 S	31.5	7.2	.44	390		1.9(b)(d)	7.3(b)	9.2(b)	0.48
2 S	36.5	10.0	.39	472		0.58(c)(d)	4.5(c)	5.1(c)	0.58
1 S	43	6.8	.30	486		1.0(d)	2.7	3.7	0.33
12 M	5	1.6	.36	299	<1.0	17.3(d)	54.1	71.4	ND
11 M	15	4.8	.97	285	<1.0	23.7	62.5	86.2	ND
10 M	25	12.8	.32	367		4.6(d)	53.8	58.4	0.76
9 M	45	8.4	.32	367		2.9(d)	19.7	22.6	0.82
3 M	65			322		0.82(d)	9.3	10.1	0.41
6 M	105			478		0.66(d)	21.1	21.8,21.8	0.33
4 M	145			565		0.41(d)	4.9	5.3	0.62
3 M	165			539		0.41(d)	5.8	6.2	0.82
1 M	205			641		1.0(d)	11.0	12.0	0.99

(a) Main probe corrected by using average blank of surface probe
(b) Backflushed slightly
(c) Contaminated with black precipitate
(d) Samples previously analyzed for Sb (total)
(e) Concentrations based on Sb(III) Standard
(f) By difference
ND = Not Detected

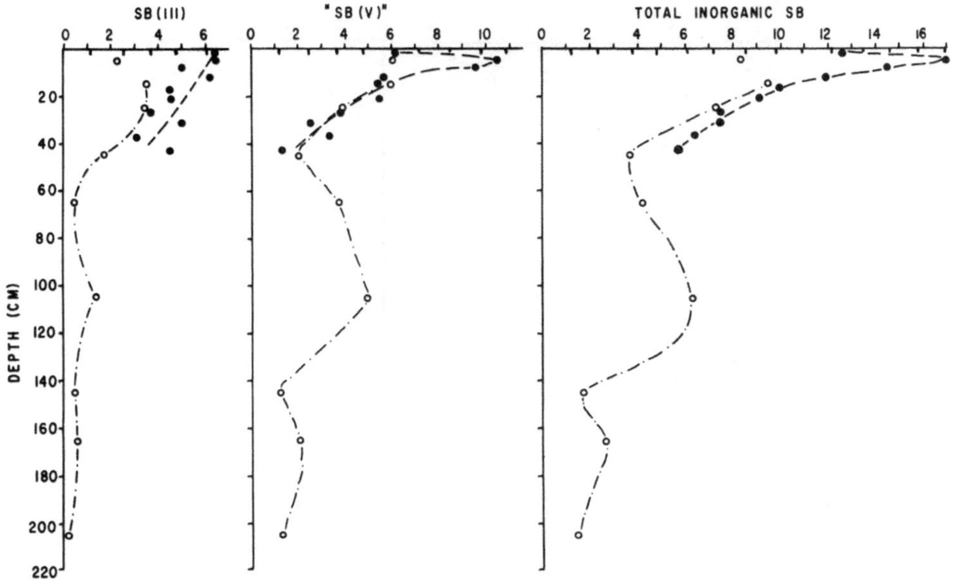

Fig. 6 Sb(III), "Sb(V)" and total inorganic Sb in SI-3 interstitial waters. "Sb(V)" from difference between total Sb and Sb(III). ● - surface probe, o main probe. (μg l^{-1})/nM. 10 nM units.

78% of total Sb. The waters to depths of 45 cm in SI-3 and 25 cm in SI-5 average around 45-50% Sb(III), with a few waters from SI-5 having values around 25% Sb(III), probably a result of oxidation. Such correspondence suggests that, at least for some of the upper waters with high H_2S contents, little or no oxidation of Sb has probably occurred and some of the values may be real.

The question then remains as to what are the other species accounting for half of the inorganic Sb in these interstitial waters. Sb_2S_5 solids had been shown by Mössbauer spectroscopy to contain only Sb(III)[12]. Thus, we doubted that the difference was due to Sb(V). One possibility was an Sb(III)-S complex such as SbS_2^- or SbS_3^{-3} since we had found that the addition of Na_2S to Sb(III) standard prevented the hydride formation under the normal Sb(III) analytical procedure. Thus, the presence of such a complex was sought in the interstitial waters as described in the Methods section. Values so obtained were in general less than or equal to Sb(III) values. Considerably lower values were found in the deeper portions of the cores and are probably due to oxidation of Sb(III) after melting and exposure to oxygen since no H_2S smell was evident in any sample.

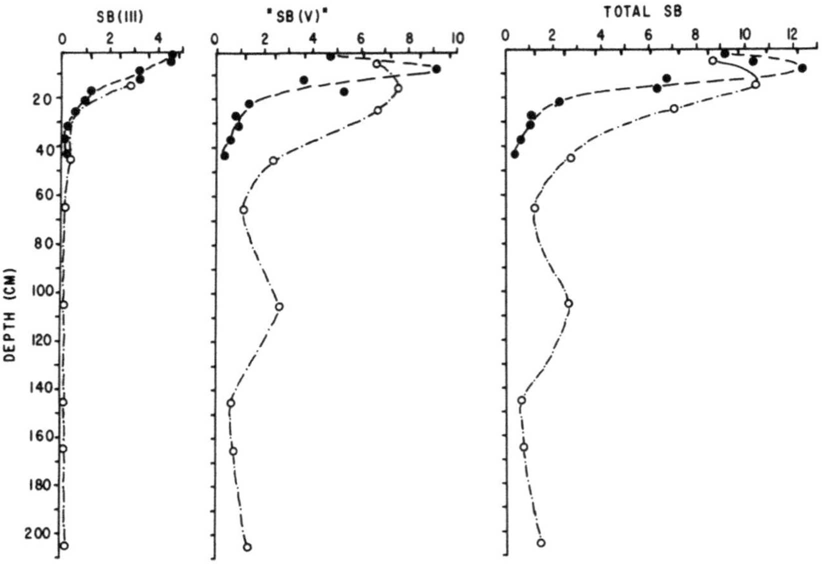

Fig. 7 Sb(III), "Sb(V)" and total inorganic Sb in SI-5 interstitial waters. "Sb(V)" from difference between total Sb and Sb(III). ● - surface probe, o main probe. ($\mu g\ l^{-1}$)/nM. 10 nM units.

To further elucidate the behavior of Sb(V) in the presence of sulfide waters, the following experiment was performed. Scripps Pier sea water was spiked with 131 nmol l^{-1} of Sb(V). To aliquots of this solution, 1 μM to 10 mM Na_2S were added, pH adjusted to 7, and the solutions allowed to stand for five days at room temperature. The concentrations of Sb(III), Sb(III)-S complex and total Sb were then determined. In the 1 mM solution, 9% existed as the Sb(III)-S complex after 5 days with the remainder as Sb(V). In the 10 mM solution, 38% had been converted to Sb(III), 50% to Sb(III)-S complex and only 12% remained as Sb(V). At lower sulfide concentrations no Sb(III) or Sb(III)-S was detected. Thus, on the time scale of this experiment, and under the experimental conditions, Sb(V) is slowly reduced to Sb(III)-S, then Sb(III) depending upon the sulfide concentration of the water. The results of this experiment suggest that, at the up to 4 μM sulfide values found in the overlying waters during the May, 1981 cruise, the Sb(III)-S complex found in the bottom waters most probably originated from diffusion out of the pore waters rather than being produced in the water column.

Sb(V) has been postulated to form thioantimonates (presumably SbS_4^{-3}) in alkali solutions with excess sulfide. In the analytical method used here, we do not differentiate between antimonate and thioantimonate species. Therefore, it is probable that any Sb(V) species present in the experiment and in Saanich Inlet sulfide-rich pore waters may be a thioantimonate. This species may also explain the higher concentrations of Sb(V) in the sulfide-rich bottom waters of the May, 1981 cruise.

The Sb profile with depth found here suggests the existence of several processes. First, there must be a large release of antimony from the sedimented materials to produce the 100-fold increase in the pore waters compared to the overlying water column. Presley et al.[11] found that enrichment factors for most metals in the interstitial water of Saanich Inlet relative to sea water were limited to a 2 to 5-fold increase with the exceptions of Fe and Mn. This generation of dissolved Sb occurs where rapid degradation of the organic matter is taking place. In SI-5, phosphate and silicate contents are increasing, nitrate is found only in the top sample, nitrite is measurable to 12 cm depth, sulfate has disappeared by 5 cm depth and hydrogen sulfide was evident in all samples. SI-3 shows similar profiles with depth. Background values of nitrate and nitrite in the lower portions of the core are due to the turbidity of the samples giving spurious values during analysis.

From the gradient in the upper pore water samples and the overlying water column in the January, 1981 profiles, it is evident that Sb is diffusing out of the sediment into the overlying water where it is oxidized to Sb(V). Lyalikova[13] showed the existence of antimony oxidizing bacteria (Stibiobacter) capable of oxidizing Sb_2S_3 to Sb(V) minerals in the oxidizing zone overlying antimony ore deposits. Emerson, Cranston and Liss[2] found that the particulate layer at the redox boundary was primarily (80%) bacteria and the remainder manganese oxides. During the May, 1981 cruise, bacteria were collected from the water column at various depths and analyzed for their ability to oxidize Mn and Co. Brad Tebo (personal communication) found no oxidation of Mn at 140 m water depth and below, no Co oxidation at 130 m, low Mn oxidation by bacteria at 120 m water depth. Peak Mn oxidation occurred at 105 m water depth corresponding to the zone of the O_2 increase. Since the Sb(III)/Sb(V) ratio is essentially unchanged over this interval, apparently the bacteria responsible for the Mn and Co oxidation do not affect the Sb speciation. Rather, Sb oxidation takes place near the sediment-water interface, regardless of whether the bottom waters are oxygenated or H_2S bearing. If Sb oxidation takes place in H_2S-rich waters, it most probably represents the conversion of $Sb(III)S_2^-$ to $Sb(V)S_4^{-3}$. Conversely, perhaps oxidation occurs during flushing when the bottom waters are slightly oxygenated, and the higher Sb(V) in the H_2S waters of the May, 1981 cruise are an artifact of a prior flushing event.

Below 8 cm the Sb decrease in the pore waters is probably due to two processes: (1) diffusion from the zone of maximal Sb production and (2) sorption onto the precipitating iron sulfides or perhaps formation of Sb_2S_3. Murray, Grundmanis and Smethie[4] found a maximal production of dissolved iron just above the dissolved sulfide maximum. Fe then decreased, presumably by inorganic removal of FeS.

Approximate methylstibonic acid concentrations in the interstitial waters were determined using the Sb(III) standard. Values ranged from less than detection limits using a 1 ml sample to values higher than Sb(III) in the lower portions of the core. Higher values were associated with waters containing significant amounts of methane, and their formation may be related to methane production. Levels were generally higher in the higher organic matter core (SI-3) than in SI-5, similar to results for Sb(III) and total inorganic Sb. The interesting observation is that below 145 cm depth, the methylstibonic acid seems to become the second most important Sb species.

CONCLUSIONS

In oxic waters the dissolved antimony exists primarily as Sb(V) with a few percent Sb(III) which is thermodynamically out of equilibrium. Some Sb(III) production occurs in the photic zone in Saanich Inlet. Even in the H_2S waters of the May, 1981 cruise, there is no significant reduction of Sb(V) to Sb(III). In these waters there is the formation of a Sb(III)-S species (presumably SbS_2^-).

In the upper 10 cms of the reducing interstitial waters there is a large release of Sb, minimum 50% Sb(III) possibly as SbS_2^-, and the remainder unknown, probably thioantimonate (SbS_4^{-3}) to the pore waters from the sediments at depths corresponding to intense degradation of the organic matter and hydrogen sulfide production. The Sb diffuses upward and downward out of this zone. In the overlying waters Sb(III) is rapidly oxidized to Sb(V), perhaps bacterially mediated. Below this zone, Sb(III), Sb(V) and Sb (total inorganic) decrease markedly suggesting that Sb is sedimented out perhaps as a sulfide or associated with iron sulfide precipitation. Methylstibonic acid becomes more important in the lower portions of the core associated with methane production.

ACKNOWLEDGEMENTS

We would especially like to thank C.S. Wong, Institute of Ocean Sciences, Sidney, British Columbia for use of facilities during K.K. Bertine's sabbatical leave and for the collection and partial analysis of the samples. This work was further sponsored by the Office of Naval Research, Contract USN NOOO 14-75-C-0152. F. Whitney, J. Thompson, K. Johnson, F. McLaughlin, W. Richardson and J. Gieskes aided in the collection of samples and analysis of the nutrients, oxygen, T, salinity, and sulfate. E. Goldberg, J. Thompson and C.S. Wong aided in interpretation of the results and development of the method.

We are particularly grateful to Ross Barnes, Walla Walla College, Washington for the use of his interstitial water sampler and his invaluable aid in obtaining these samples and interpreting the results.

REFERENCES

1. Anderson, J.J. and A.H. Devol, 1973: Deep water renewal in Saanich Inlet, an intermittently anoxic basin. Estuar. Coastal Mar. Sci., 1, 1-10.
2. Emerson, S., R.E. Cranston and P.S. Liss, 1979: Redox species in a reducing fjord: equilibrium and kinetic considerations. Deep Sea Res., 26A, 859-878.
3. Barnes, R.O., 1973: An in situ interstitial water sampler for use in unconsolidated sediments. Deep Sea Res., 20, 1125-1128.
4. Murray, I.W., V. Grundmanis and W.M. Smethie Jr., 1978: Interstitial water chemistry in the sediments of Saanich Inlet. Geochim. Cosmochim. Acta., 42, 1011-1026.
5. Cotton, F.A. and G. Wilkenson, 1972: Inorganic Chemistry. John Wiley & Sons, New York.
6. Garrels, R.M. and C.L. Christ, 1965: Solution, Minerals and Equilibrium. Harper & Row, New York, 450.
7. Schutz, D.F. and K.K. Turekian, 1965: The investigation of the geographical and vertical distribution of several trace elements in sea water using neutron activation analysis. Geochim. Cosmochim. Acta., 29, 259-313.
8. Brewer, P.G. and D.W. Spencer, 1972: Trace element profiles from the GEOSECS-11 test station in the Sargasso Sea. Earth Planet. Sci. Lett., 16, 111-116.
9. Gohda, S., 1974: The content and the oxidation state of As and Sb in coastal water of Japan. J. Ocean. Soc. Japan, 30, 163-167.
10. Bramen, R.S. and M.A. Tompkins, 1978: Atomic emission spectrometric determination of antimony, germanium and methylgermanium compounds in the environment. Anal. Chem., 50, 1088-1093.
11. Presley, B.J., Y. Kolodny, A. Nissenbaum and I.R. Kaplan, 1972: Early diagenesis in a reducing fjord, Saanich Inlet, British Columbia - II. Trace element distribu-

tion in interstitial water and sediment. Geochim. Cosmochim. Acta., 36, 1073-1090.

12. Birchall, T. and E. Della Valle, 1970: The non-existence of Sb_2S_5. Chem. Comm., 675.
13. Lyalikova, M.N., 1978: Antimony-oxidizing bacteria and their geochemical activity. In: "Environmental Biogeochemistry and Geomicrobiology", vol. 3, W.E. Krumbein, ed. Ann Arbor Science, Ann Arbor, Mich., 929-936.
14. Pourbaix, M., 1966: Atlas of Electrochemical Equilibrium in Aqueous Solutions. Pergamon Press, London, 644.

ULTRATRACE SPECIATION AND BIOGENESIS OF METHYLTIN TRANSPORT SPECIES IN ESTUARINE WATERS

F.E. Brinckman, J.A. Jackson, W.R. Blair, G.J. Olson and W.P. Iverson

Chemical and Biodegradation Processes Group
National Bureau of Standards
Washington, D.C. 20234 U.S.A.

ABSTRACT

Environmental tin, widely dispersed at low concentrations in waters, sediments, and biota, is shown to be a bioactive element susceptible to methylation and even hydridization by marine bacteria. The redox cycle of tin in natural waters is poorly understood and recent advances in tin-specific molecular characterization fail to speciate Sn(II) and Sn(IV) reliably. Nonetheless, such rapid developments in speciation methodology now permit growing numbers of studies of organotin distributions in aquatic systems, raising the question of the "natural" biogeochemical flux of methylstannanes in relation to increased anthropogenic organotin influx from industry and shipping. New methods for direct speciation of aquated or involatile organotins by liquid chromatography are compared with advances in purge-and-trap sampling of volatile or hydrophobic organotins speciated by gas chromatography. The work in our laboratory indicates that effective models for estuarine formation and transport may ultimately be developed, but that basic roadblocks to progress stem from inadequate descriptive aqueous organometallic chemistry and knowledge of critical kinetic parameters for the lifetimes of key organotin species in sea water, occurring at sub-nanomolar concentrations.

INTRODUCTION

Tin is very widely dispersed in the aquatic environment, appearing as trace amounts in sediments, particulates, and biota (mg kg^{-1}), and at much lower levels in waters (ng l^{-1})[1]. The concentration of tin decreases notably from estuarine and near shore sites to oceanic environments [2], in a manner seen for other metals or metalloids subject to significant anthropogenic inputs and rapid geochemical turnovers and transport[3].

The chemistry of tin in natural waters is generally presumed to involve stannic or tin(IV) species, but its redox chemistry in sea water remains largely unclear, as does its biogeochemistry. This last aspect has recently assumed great importance in understanding the mechanisms by which aquatic tin is mobilized (by atmospheric or aquatic transport) and in questioning both bioavailability and bioaccumulation of tin[4,5]. Several reports on the biomethylation of tin in both inorganic and organometallic forms have now appeared [2,6,7,8,9], and ubiquitous distributions of methyltin(IV) species at ultratrace concentrations (ng l^{-1}) in natural waters have been observed[10,11,12]. This evidence suggests that such molelcules may take part in a global tin cycle. Unfortunately, the matter is obscured by the sharp increase over the past twenty-five years of anthropogenic introduction of both methyl- and other organotins into the environment by diverse commercial materials[13].

Interactions of man-made organotin compounds intrude into the "natural" biogeochemical cycle of tin[14] in special ways. Use of organotins mainly occurs as plastics stabilizers and biocides; these, in turn, find principal applications in ways that involve environmental exposure, especially in fresh and estuarine discharges. We have undertaken studies to speciate tin and to assess its dynamics in the marine environment, but in view of the foregoing situation we are examining the tin concentration gradient from a well-known estuarine source, the Chesapeake Bay, to oceanic sites. In this context, we will discuss progress in current research to solve fundamental problems impeding progress toward resolving the role of tin in the sea. The objectives of this research have been to: (1) establish a firm data base for the aqueous chemistry of organotin(IV), tin(IV), and tin(II) species, including dynamics of their interactions with other trace aquated metal ions; (2) develop and apply reliable ultratrace molecular speciation techniques to the measurement of aquatic tin in its various forms, including key, volative transport species such as tetramethyltin; and (3) assess the role of marine microbiota in generating or retarding formation of tin transport species, by biosynthesis, biodegradation, and uptake.

BIOAVAILABILITY OF TIN IN RELATION TO OTHER BIOACTIVE ELEMENTS

Several estimates of global transport cycles suggest[15,16] that atmophilic properties of certain elements may, in part, depend upon biological mediation. For a given element, such volatilization processes basically must rely upon release from terrestrial or aquatic sediment reservoirs of hydrophobic gases such as elemental mercury, Hg^{o}, or coordinatively saturated methylelements, Me_nE^{o}, where n is the principal oxidation state[5]. In essence, the underlying chemical processes governing retentivity or transport to/from sediments or particulates to/from water columns to/from atmosphere are,

$$Me_xE^{(n-x)+}{}_{(s)}\downarrow \rightleftharpoons Me_xE^{(n-x)+}{}_{(aq)} \rightleftharpoons Me_xEL_y^{(n-x-y)}{}_{(gas)}\uparrow, \quad (1)$$

where L = some anionic ligand imparting volatility, such as H^-, Me^-, Cl^-, or OH^-. Below, we shall review the features of such derivatization chemistry as a means to produce water-air partition coefficients favorable for development of analytical methods employing chromatographic separations coupled with element-selective detection[17]. A perspective on mobilization of tin is gained by comparing its non-specific (total) availability in common environmental reservoirs in relation with other bioactive elements known to undergo transformations of the kind implied by equation (1).

Potential Availability of Tin in Environmental Media*

Table 1 summarizes data from various sources which suggest that considerable amplification of many such active elements can occur. Sewage sludges from 16 American cities show substantial industrial and microbial accretion of many heavy elements from common sources (cf. cow manure) into forms that may offer enhanced availability to subsequent biota during translocation via stream loadings to estuaries and then to the sea. It is tempting to associate the very high concentrations of biogenic cobalamins (vitamin B_{12} derivatives) observed during sewage sludge treatment (1.9 to 27.1 μg g^{-1})[24] with potential involvement of a powerful methylator, methylcobalamin or CH_3-B_{12}[14], in metal release. Though not yet identified in sewage sludges, presence of CH_3-B_{12}, especially in anaerobic sediments, is reasonable. Consequently, the effectiveness with which this exocellular metabolite methylates a broad range of metal ions in solution[14,25] suggests an

*The concentration terms used in this paper, e.g., μg g^{-1} or ng l^{-1}, do not represent SI units, but are those employed by the various sources cited. As concentrations of speciated marine organotins and other metal-containing solutes are increasingly reported, preferred units will be mol l^{-1}. Thus, in the present work, detection limits for organotins were approximately 0.15 to 0.3 n<u>M</u> by GC-P/T-FPD and 0.4 μ<u>M</u> by HPLC-GFAA. Conversions to SI are given in Table 5.

Table 1. Compositional Data[a] for Tin and Some Other Bioactive Elements in Various Environmental Sources

Source	As	Hg	Pb	Se	Sn
Cow manure[b]	4.0	0.2	16.2	2.4	3.8
Sewage sludge[b]	3.0-30.0	3.4-18.0	136.0-7627.0	1.7-8.7	111.0-492.0
Urban dust[c]	115.0	--	0.655[d]	24.0[e]	--
U.S. water supplies[f]	0-100.0	0.01-30.0	1.0-400.0	0-10.0	0.3-30.0
Chesapeake Bay					
Baltimore Harbor sed.	1.1[g]	<0.01-0.31[h]	<1 -1.38[d,h]	--	1.8[g]-240.0[i]
Baltimore Harbor waters	0.29[g]	<0.071[g]	1.9[g]	--	0.19[g]-<2.0
Channel sed.[j]	0.92-10.0	<0.065-0.37	0.54-75.0	--	0.19[g]-1.7
Channel waters	<0.079-0.37	<0.038-0.16	0.10-1.9	--	<0.13-0.47
Organic colloids[l]	3.0-4.0	--	--	0.65-2.0	54.0-57.0

[a] Individual, mean, or range values in $\mu g\ g^{-1}$, dry weight (solids), or $\mu g\ l^{-1}$ (waters).

[b] Ref. 18

[c] NBS Standard Reference Material 1648

[d] Given as % Pb

[e] Not certified

[f] Ref. 19

[g] Ref. 20, outer harbor station

[h] Ref. 21

[i] Ref. 2

[j] Ref. 22, S→N Bay axis, 22 stations

[k] Unfiltered

[l] Ref. 23

important new line of investigation for metal release from sewage outfalls. Moreover, Thayer[26] reports that CH_3-B_{12} is also capable of dissolving many refractory metal oxides, including highly insoluble SnO_2, to possibly form methylmetal species in aqueous solution.

Similar considerations, doubtless for small chemical concentration gradients, probably are valid for some of the other tin sources listed in Table 1. The very high tin concentrations (also associated with heavy loadings of organic matter) in the Baltimore Harbor sites focused the direction of our present work. Moderately high concentrations of total tin in sub-micron materials from the Chesapeake Bay channel represent favorable surfaces for attachment and transformation of metals by marine bacteria[27]. In Table 1, no mention of another very important microenvironment for amplification of metals is given. Heavy elements are well known to greatly concentrate in oily surface microlayers[28], but data for total or speciated tin in the Chesapeake Bay are not yet available.

Anthropogenic Introduction of Organotins into Environmental Media

In the foregoing we have discussed the typical total tin content of environmental compartments involved in transmission of this element (and related bioactive elements) from land to sea. The rich chemistry of tin has made this element one of great importance to man's industries for millenia, and in recent decades its remarkable range of properties as organotin compounds has generated totally new markets and distribution patterns in the environment[13].

The contribution of all anthropogenic tin compounds represents the bulk of the tin discharge system to oceans[16]. Over four percent of this contribution involves organotins of widely varying toxic properties, particularly toward aquatic biota. Table 2 indicates the spectrum of organotin formulations involved. The R Sn- class is the most toxic generally, but the R_2Sn= species still represent lipophilic moieties that can probably undergo transformations <u>in vivo</u> to create toxic responses within biota, as we shall see later.

It is well known that, in addition to the degree of substitution on tin(IV), the precise molecular structure of the substituent group R covalently bound to the organotin influences its lethal action in different categories of organisms[29]. Nonetheless, most investigators now agree that biodegradation of even toxic R_3Sn-containing materials can occur in soils and sediments, presumably caused by bacteria that successively cleave tin-carbon bonds to ultimately produce inorganic tin:

Table 2 Production of Organotins and Their Uses [a,b]

Uses	Chemical Form [c]	Relative %
PVC Plastics (heat stabilization)	R_2SnL_2	66.4
Catalysts (polyurethanes, silicone elastomers)	R_2SnL_2	20.7
Biocides (pesticides, antifoulants)	R_3SnL and $-[L]-SnR_3$	8.3
Miscellaneous (research, fiber stabilization, export)	R_3SnL	3.3
Poultry (anthelmintics, coccidiostats)	R_2SnL_2	1.2

[a] Ref. 13
[b] Represents 4.2% of 1976 total world production, 2.26×10^{11}g
[c] R = butyl, cyclo-hexyl, phenyl, octyl, methyl, and propyl; L = F, Cl, laurate, maleate, and thioglycolate.

$$R_3Sn- \rightarrow R_2Sn= \rightarrow RSn \equiv \rightarrow Sn(IV)\downarrow . \qquad (2)$$

In practical fact, these steps are assumed, there having been no clearcut demonstration of the entire sequence in the environment. Moreover, the new evidence for microbial methylation of inorganic tin(IV), tin(II), and common intermediate organotins suggests - just as was shown earlier for the sedimentary cycle for methylmercury [26,30,31] - that a counter mechanism for sustaining methyltin(IV) species in environmental media is available. The reported [10,11], widespread incidence of highly diluted solvated methyltin(IV) ions, $Me_nSn^{(4-n)+}(aq)$, in United States coastal waters may therefore simply reflect a steady state concentration. The idea that this situation is reached only through biological mediation is not valid, however. Brinckman [5] has pointed out that among the methyltin species identified in marine waters, $Me_3Sn^+(aq)$ is itself a potent methylator of many aquatic metal ions at bimolecular rates which might significantly shift the apparent "steady-state" concentration. For example, with aqueous mercuric ion, the reaction

$$Me_3Sn^+_{(aq)} + Hg^{2+}_{(aq)} \rightarrow Me_2Sn^{2+}_{(aq)} + MeHg^+_{(aq)} , \qquad (3)$$

proceeds to completion at a rate first-order for each metal and is dependent on the [Cl^-] and ionic strength of the solution [32]. Based on available concentrations of free $Hg^{2+}(aq)$ and $Me_3Sn^+(aq)$, along with typical pCl, pH, and ionic strengths for coastal waters, it was estimated[5] that the half-life for leakage of Me_3Sn^+ from that compartment was about 6 x 10^5 yr. This residence is in surprising agreement with that estimated for oceanic total tin, 1 x 10^5 yr[33], but must be fortuitous since many other abiotic transmethylation pathways for forming and decomposing methyltins are known. We shall consider several of these possibilities in the following section, both in the context of biomethylation and abiotic transmethylation.

CHEMICAL AND BIOLOGICAL INTERACTIONS OF METHYLTINS WITH ELEMENTS

Within a microenvironment, concurrent events involving both chemical and biological reduction and methylation of metals may transpire individually or in synergic fashion. Figure 1 illustrates a combination of in vitro and environmental experience for three heavy elements of great technological importance that undergo microbial methylation in marine sediments[26]. No evidence is available for common enzymes among the bacterial populations that methylate all three elements, but it is reported that pure strains of genus Pseudomonas, for example, can both reduce and methylate As(V) to As(III) and form Me_nAsH_{3-n} gases[34], reduce Hg(II) to Hg^o[35], and both methylate and reduce Sn(IV) to form Me_nSnH_{4-n} gases[5,12]. Together, these metabolites or their precursor metal species may also interact in situ. The case for a mercury-tin "crossover" model has been described[6,36] where combined stresses of Sn(IV) and Hg(II) (a much more realistic representation of polluted sediments!) were metabolized by a metal-tolerant

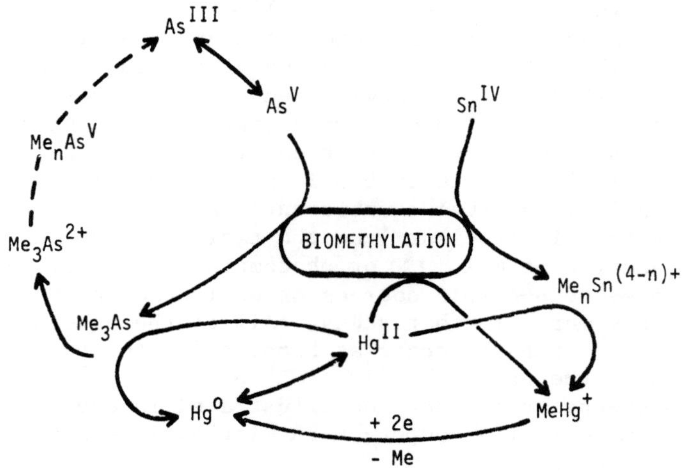

Fig. 1 Associated abiotic chemistries of three elements known to undergo biomethylation in marine sediments are compared.

strain of Pseudomonas isolated from the Chesapeake Bay[35]. The metabolic reduction of mercury to Hg^{o} gas and the biomethylation of tin(IV), seen to occur for the respective metals as individual stresses, were seen to occur with the combined stress. A new species was formed only during metabolism of the dual metal stress. This species was found to be $MeHg^{+}$, presumably formed by the reaction given in equation (3), where it was concluded that both concurrent biomethylation of tin(IV) and its mediation (abiotically) in $MeHg^{+}$ production were kinetically competitive[6].

Other dual metal crossovers are possible in principle. We examined the possibility for $Me_3As^{2+}(aq)$, an isoelectronic analog of $Me_3Sn^{+}(aq)$, for its ability to methylate metal electrophiles, in particular Hg^{2+} [37]. As depicted in Figure 1, only a redox reaction in the reverse sense occurred reducing arsenic volatility,

$$Me_3As + Hg^{2+} \rightarrow Me_3As^{2+} + Hg^{o} \uparrow, \tag{4}$$

which was highly dependent upon pH and ligands bound to mercury[38]. Such chemistry is important to considerations of microbial production of Me_3As[34] in sediments where high concentrations of both As and Hg are commonly found (cf. Table 1). Similar experiments with arsenic and tin species in saline waters at these pH and pCl conditions gave no conclusive reactions.

The abiotic organometallic reactions noted above which readily occur in aqueous solutions represent only a few of many similar redox and transmethylation processes recently described for other metals in this expanding field[5, 26]. To aid in understanding our recent results in speciation and dynamics of methyltins in estuarine waters, brief mention of a group of abiotic reactions described for both environmental and laboratory conditions is desirable. In Table 3, we summarize some representative reactions which either favor formation of new methylelement species potentially capable of biotic uptake, or elimination of "active" methyl groups from aquatic environments and potential biotic activity. It is clear from the table that ubiquitous coastal production of MeX [15,44] or photomethylation via metabolic acetates represent possible sources of environmental methylelements, and thus merit further study. All of the aqueous reactions cited in Table 3 probably occur at bimolecular (or pseudo first-order) rates of the order, $k \sim 10^{-3}\ M^{-1}\ S^{-1}$ or less[5], but like the related transmethylation reaction illustrated in equation (3) probably can kinetically compete with microbial methylation and demethylation reactions [26].

Table 3 Some Non-biological Reactions[a] Providing Formation or Removal of "Active" Methyl Groups from Aquated Metal Ions.

Formation Reactions	Reference
$MeX^{b} + Sn^{2+} \rightarrow Me_nSn^{(4-n)+} + nX-$	
(X = F, Cl, $MeCO_2$, SO_4) (n = 1,2)	7
$MeI + Pb^{2+} \rightarrow [MePbI^{2+}] \rightarrow \rightarrow Me_4Pb$	39
$MeX^{b} + Me_3E \rightarrow Me_4E^{+} + X^{-}$	
(X = Br, I; E = As, Sb)	40
$MeCOO^{-c} + Hg^{2+} + h\nu^{d} \rightarrow MeHg^{+} + CO_2$	41,42

Removal Reactions	
$Me_3E^{+} + M'Cl_4^{2-} \rightarrow M'^{o} + MeCl + Me\text{-}Me + Me_2E^{2+}$	
(E = Sn, Pb; M' = Pd, Pt)	42
$Me_3E^{+} + AuCl_4^{-} \rightarrow Au^{o} + MeCl + Me_2E^{2+}$ (E = Sn, Pb)	42
$Me_4Pb + Cu^{2+} \rightarrow CuCl + MeCl + Me_3Pb^{+}$	43
$MeHg^{+} + PdCl_4^{2-} \rightarrow Hg^{2+} + MeCl + Pd^{o}$	
$42Me_3E^{+} + TlCl_n^{(3-n)+} \rightarrow Tl^{+} + MeCl + Me_2E^{2+}$ (E = Sn, Pb)	
$Me_2Hg + h\nu^{d} \rightarrow Hg^{o} + Me\text{-}Me$	41,42

[a] In saline solutions
[b] A common metabolite (X = Cl, Br, I) from bacteria and algae in coastal sea water, see Ref. 44
[c] Common bacterial metabolite
[d] Sunlight or ultraviolet irradiation

SPECIATION OF ORGANOTINS IN SEA WATER*

Requirements for Nondestructive Methods

Beyond the usual difficulties encountered with determinations of metals at ultratrace (ng l^{-1}) concentrations in biological

*Complete descriptions of equipment and methods employed at NBS are given in references 12, 51, or 49,50 for gas and liquid speciation, respectively.

fluids and aquatic media[45], creation of a realistic framework of data for establishing the role of tin in the sea requires molecular speciation as well. We have examined the interdependence of biological and chemical processes for transformation and mobilization of tin and other metals and have noted that rate-determining events depend upon very specific features of molecular architecture and oxidation state of the associated metal ion. Thus, the number and kinds of organic ligands bonded to tin(IV) are not only diagnostic of toxicity, but they also provide a molecularly distinctive analyte for elucidation of trace chemistry, if suitable nondestructive chemical or physical means for isolation, concentration, and unambiguous detection can be applied.

Two basically different approaches are possible, and have been adopted in our present work. We recognize from equation (1) that organotin species of interest to us, $R_nSn^{(4-n)+}$, may exist in equilibrium between aqueous solution and complexes on solid substrates, such as inorganic particulates or biological detritus. Apart from those species already in homogeneous or "free" solution, a choice is possible for treatment with selective ligands, L, that can either cause desorption of solid-bound organotins, or, alternatively, form new organotin derivatives possessing necessary volatility (hydrophobicity) for recovery from solution as gases. In both of the methods to be described, we directly speciate aquated organotins - those in "free" solution - without pretreatment. For those organotin molecules of sufficient volatility, e.g., Me_4Sn, a gas chromatographic separation step is employed; for involatile solvated organotins, a direct liquid chromatographic separation process is utilized, and, when appropriate, hydride derivatization can be exploited with the gas chromatograph. Both methods give high precision retention times (RSD < 3 percent) with authentic organotins, a measurement prerequisite for reliable speciation.

Gas Phase Speciation of Aquatic Organotins

Braman[10] and Goldberg[11] with their students successfully speciated aquatic organotins (R = Me and Bu), respectively, with flame emission (SnH) and atomic absorption detectors, following gas sparging of large (100 ml) water samples treated with a hydriding agent,

$$R_nSn^{(4-n)+}{}_{(aq)} + \text{excess } BH_4^- \rightarrow R_nSnH_{4-n}\uparrow . \tag{5}$$

The resulting stannanes (n = 0 to 3) were cryogenically trapped, then distilled into the detector. Though extremely sensitive (< 1 ng l^{-1}) the methods failed to isolate Me_4Sn or other volatile organotins not requiring hydride volatilization. We have modified commerical equipment to provide a system which avoids the above limitation, though it is not as sensitive.

Our system employs a "Tenax-GC" polymeric sorbent[46] in an automatic purge and trap (P/T) sampler coupled to a conventional glass column gas chromatograph equipped with a flame photometric detector (FPD). A schematic of the P/T-GC-FPD assembly with typical operating conditions is depicted in Figure 2. Flame conditions in the FPD were tuned to permit maximum response to SnH emission in a H-rich plasma, as detected through narrow band-pass interference filters (610 $\pm$ 5 nm)[47]. Two modes of analysis were used: (1) volatile stannanes were trapped directly from sparged 10 to 50 ml water samples with no pretreatment, and (2) volatilized tin species were trapped from the same or replicate water samples following rapid injection of aqueous excess $NaBH_4$ solution directly into the P/T sparging vessel immediately prior to beginning the P/T cycle[12].

Direct Speciation of Organotins in Solution

The liquid-solid solvation process illustrated in equation (1) also governs several desirable molecular separation processes offered by high-performance liquid chromatography, HPLC[48]. For either ion exchange resolution of aqueous cations, $R_nSn^{(4-n)+}_{aq}$,[49], or their separation as ion pairs, $[R_nSn^{(4-n)+}X^-_{4-n}]^0$, on reverse bonded-phase columns[50], the method is restricted to "free" tin analytes. Unlike the vigorous hydride derivatization used in the GC-FPD method, common HPLC solvent combinations or their ionic

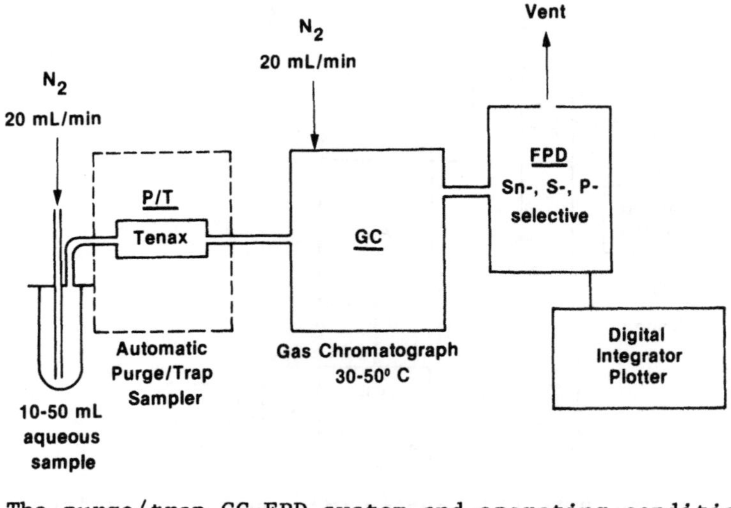

Fig. 2 The purge/trap GC-FPD system and operating conditions[12].

addends usually will not provide sufficient coordination strength to labilize organotin ions strongly bound to solids in environmental samples. Moreover, the HPLC separations require that injected samples be free of particulates that may clog the column or pumping system.

On the other hand, HPLC, if coupled with a sensitive element-specific detection system such as atomic absorption spectrometry (AA), offers a valuable tool for organotin speciation in complex fluids (especially with high organic loadings) not readily amenable to gas phase derivatization methods. Figure 3 shows a schematic of our basic HPLC setup coupled to a graphite furnace AA (GFAA) in a manner giving automatic periodic (typically 45 s intervals) sampling of the resolved eluents for tin-specific determination [50]. Injected sample volumes may vary from 10 to 500 μl. Consequently, system sensitivity is broad and samples can be very representative. For example, direct examination of soil or sewage leachates, process waters, solutions exposed to organotin (antifoulant) controlled-release agents, or microbial growth media, only requires separation of particulates (or cells) prior to speciation. This is simply done by decantation, syringe fil-

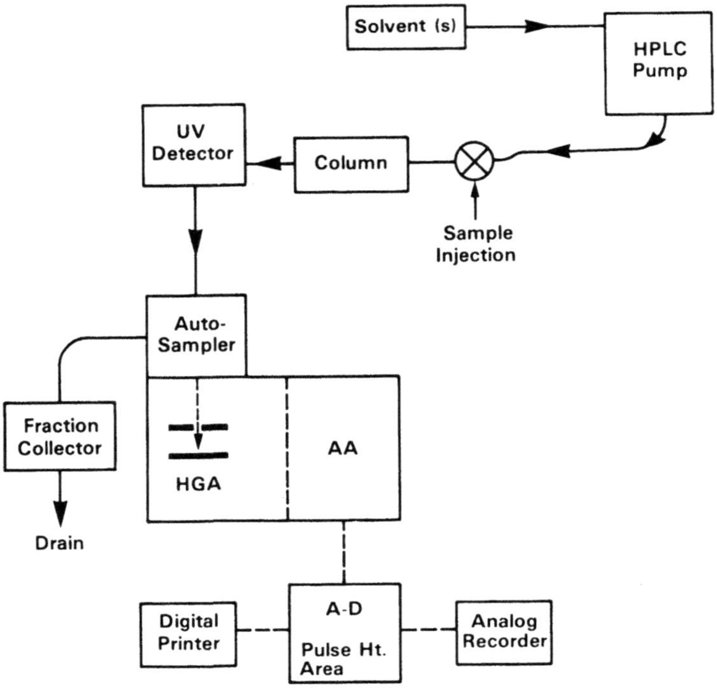

Fig. 3 The HPLC-GFAA system with automated peripherals [50].

tration, or ultracentrifugation before direct injection of the solution into the HPLC-GFAA system. Thereby, needed information on the nature and true concentrations of "free", involatile transport or leachate organotin species becomes available without recourse to chemical treatment.

Comparison of Analytical Results from GC-FPD and HPLC-GFAA

Before turning to our current organotin speciation studies in the Chesapeake Bay estuary, it is useful to compare the nature of experimental data, their form, and relative reliability, which, in turn, depend on the physico-chemical differences between the two methods. Figures 4 and 5 compare calibration curves for organotins from both anthropogenic and environmental sources. System detection limits, δ (95 percent confidence level)[51], for biocidal triorganotins representative of commercial pesticides in widespread use[13] are shown in Figure 4. Mixtures of R_3Sn^+ compounds (R = n-butyl, phenyl, cyclo-hexyl) were separated by ion exchange HPLC-GFAA[49]. The small spread in calibration slopes in Figure 4 signifies similar efficiencies for their separation and column recovery, as well as GFAA sensitivites. The δ values (ng) shown

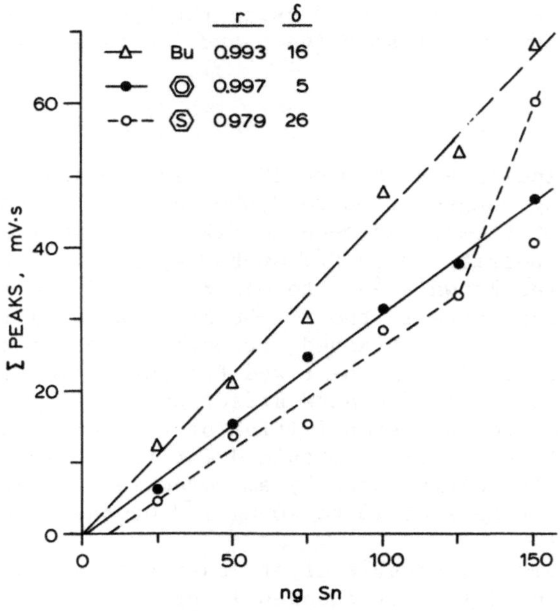

Fig. 4 Calibration curves for R_3Sn^+ (R = butyl, phenyl, c-hexyl) separated by HPLC-GFAA with strong cation exchange (SCX) columns using $MeOH/H_2O/NH_4OAc$ eluents[40] are shown with respective correlation coefficients (r) and system detection limits (δ) (95% confidence level[51]).

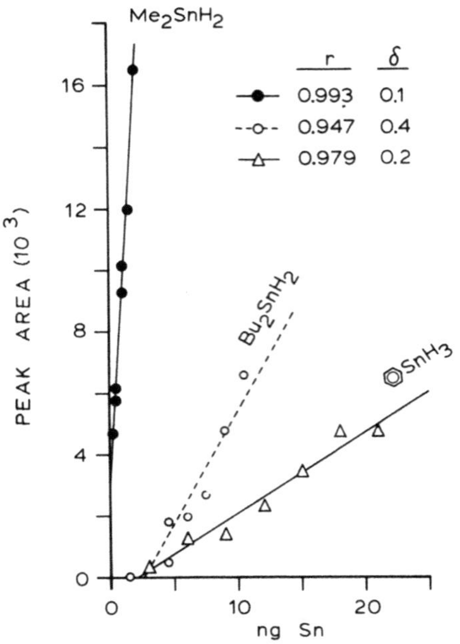

Fig. 5 Calibration curves for three aqueous organotins indicated separated by P/T-GC-FPD using the hydride generation mode[12] are shown with respective r and δ estimates.

in the figure indicate that with 100 μl sample injections, nominal solution working ranges of 50 to 2,600 μg l^{-1} may readily be speciated and quantitated. As seen in Figure 5, considerably more sensitivity is possible with P/T-GC-FPD speciation of related organotin species known[5,10,11] to occur in environmental media. Much greater divergence in the P/T-GC-FPD system calibration slopes (ratios > 25) is obtained, probably a result of different hydridization rates during the fixed P/T purge time (10 min), different partition coefficients affecting sparging rates of each species[12], or different retentivities on the Tenax-GC sorbent[46]. On the basis of the δ values obtained for the GC method (Fig. 5), with 10 ml sample volumes usually employed in our laboratory, nominal working ranges of 10 to 40 ng l^{-1} organotin are feasible.

Both systems are capable of at least a ten-fold increase in sensitivity with only minor changes in procedure and equipment. For HPLC-GFAA, this can be achieved by both increasing injected sample size and optimizing flow rates with a GFAA thermal program designed to give maximum atomization efficiency for a specific organotin analyte[50,51]. For P/T-GC-GFD, improvements are realized by adjusting purge flow rate and time while altering $NaBH_4$ additions to optimize evolution of a given organotin analyte[12]. Also, both increasing sample volumes[10,11] and operating the Tenax-GC trap at subambient temperatures[2] will yield lower working ranges.

Ultimately, such concerns for widely different system responses to various organotin analytes are very crucial to reliable environmental speciation and quantitation by coupled chromatographic-element-selective detection methods because: (a) each known species analyzed must have its (same or different) calibration curve established, and (b) unknown species must either be unequivocally identified and calibrated or their calibration curves inferred from similar molecular types or guessed. Under ideal conditions, or course, a perfect chromatograph-detector system will always yield identical calibration curves for any separated molecule containing the diagnostic element of interest[17,51]. It should be recognized that in some circumstances where robust chemical derivatiziation is called for, as with formation of volatile tin hydrides from sea water for FPD measurements (equation (5)), possibilities exist for unexpected chemical transformations of unknown substrates into forms taken as characteristic of known analytes. In concluding this section, we will present an example for the tin(II)-tin(IV) redox system which we now feel casts uncertainty on the reported speciation and quantitation of tin as Sn(IV) in marine waters[10,11].

The Question of Sn(IV) and Sn(II) Speciation in Aquatic Media

Most workers regard "free" and particulate-bound tin in the sea to be in its +4 oxidation state. Based upon available free-energy data[52] and usual oceanic values of pH = 8.1 and pE = 12 (E^o = 0.7394 V)[53], the thermodynamic distributions of supposed principal tin species are given by the following equilibria:

$$\log_{10} \frac{[SnO_3^{2-}]}{[HSnO_2^{-}]} = 37, \text{ and } \log_{10} \frac{[SnO_3^{2-}]}{[Sn^{4+}]} = 45. \qquad (6)$$

In anaerobic environments (reducing conditions), Sn(II) and other tin(IV) species, including chloro-hydroxy derivatives[5,32], may become important. Additionally, we recognize that the steady-state condition and concentrations of Sn(II) and Sn(IV) in aquatic systems may be kinetically, not thermodynamically, controlled by biological activity. This was found to be the case for observed non-thermodynamic equilibrium concentrations favoring arsenite over arsenate produced by bacterial reduction[54,55] coupled with relatively long lifetimes of As(III) in aerobic waters[56]. We are, therefore, left with the possibility that in some marine waters, especially highly polluted estuarine sites, that Sn(II) species may occur. The question arises: does our current use of hydride generation insure that the SnH_4 generated from aquatic samples is a faithful representation of the presence only of Sn(IV) species or can Sn(II) interfere with and obscure the quantitation?

Concern for substantial interference by Sn(II) in the reduction of Sn^{4+}_{aq} via equation (5) is justified. Schaeffer and Emilius[57] showed that yields of SnH_4 as high as 84 percent (based on Sn) were obtained from BH_4^- treatment of stannous chloride in aqueous 0.6 N HCl. Yields were dependent on pH, and smaller yields of distannane were noted. Jolly and Drake[58] subsequently refined this procedure to provide a preferred general method for aqueous generation of Ge, As, Sb, and Sn hydrides. For tin(II) the reaction

$$4HSnO_2^- + 3BH_4^- + 7H^+ + H_2O \rightarrow 3H_3BO_3 + 4SnH_4\uparrow, \quad (7)$$

involves oxidation of the metal to Sn(IV) with concurrent redox of H^- and H^+ (possibly as $H\cdot$). The basic problem for our present methods employing borohydride treatments of water samples for Sn(IV) "speciation" is clear. In a recent series of experiments we have attempted to determine the extent of the problem. Our results are summarized in Table 4.

Calibration curves were generated by the P/T-GC-PFD method for borohydride reductions of Sn(IV), Sn(II), and Me_2Sn^{2+} species to SnH_4, SnH_4, or Me_3SnH_2, respectively, in three types of media, as tabulated. All three analytes showed substantial increases in their calibration slopes in going from distilled water to 0.2 M NaCl solution, the latter approximating the salinity and ionic strength common to estuarine waters. Presumably, these effects could arise from formation of chlorohydroxy tin species favoring more rapid hydridization (cf. equations (5) and (7))[5,32], as well as the more propitious partition coefficients for dynamic gas stripping of the volatile tin hydrides from saline solutions[12,46]. In the typical laboratory distilled water calibration solutions only 16 percent of Sn(II) was recovered as SnH_4, compared with Sn(IV), though this sensitivity ratio can probably be altered somewhat with pH changes[10,57]. However, in spiking anaerobic pre-purged Chesapeake Bay water with these three tin species, a striking reversal occurred in overall relative sensitivities, i.e., calibration slopes. We found that not only was Me_2SnH_2 generation repressed by 50 percent, but that, very significantly, SnH_4 formation from Sn(IV) was reduced by 15-fold, as compared with the NaCl medium. Production of SnH_4 from Sn(II) was decreased by 50 percent in going from NaCl to Bay waters. Clearly, dissolved or particulate substances in the Bay water, not involving Cl^- primarily, were responsible for severe interferences in the P/T-GC-FPD analyses of the three tin species, the most inhibitory

Table 4. Hydridization of Sn(IV, Sn(II), and $Me_2Sn(IV)$ in Aqueous Media

Species[a]	Medium	Detection Limit, ng l^{-1}[b]	Regression Slope ± S.E.	Number of Runs, N	R	Relative Sensitivity[c]
Sn(IV)	H_2O[d]	19	210.4 ± 16.4	7	0.985	1.00
Sn(II)	"	36	32.6 ± 4.9	7	0.949	0.16
Me_2Sn^{2+}	"	18	5322.0 ± 382.0	7	0.987	25.3
Sn(IV)	0.2 M NaCl[d]	29	318.5 ± 40.2	10	0.942	1.51
Sn(II)	"	23	117.9 ± 9.6	11	0.972	0.56
Me_2Sn^{2+}	"	21	10,997.0 ± 1016.0	10	0.968	52.3
Sn(IV)	Bay water[e]	30	21.6 ± 2.8	10	0.937	0.10
Sn(II)	"	10	51.4 ± 2.2	9	0.993	0.24
Me_2Sn^{2+}	"	37	5086.0 ± 589.0	4	0.987	24.2

[a]Added as solutions of $SnCl_4 \cdot 5H_2O$, $SnCl_2$, or Me_2SnCl_2, respectively, to 10 ml P/T samples
[b]Estimated at 95 percent confidence level by the method in Ref. 51; [c]Ratio of regression slopes;
[d]In distilled, deionized water (18 MΩ·cm); [e]In Jones Falls, Baltimore Harbor water, pH ∼ 8, S°/oo ∼ 11, pre-purged with zero N_2 (1 hr at 50 ml min^{-1}) before spiking with tin species.

effect being seen with the strongest Lewis acid, Sn(IV). Less diminution (about one-half) occurred with the weaker acids, Me_2Sn^{2+} and Sn(II), which either form less stable chlorocomplexes (vide supra) or are less likely to form strong bonds to ligands and substrates present in estuarine waters. The overall effect by estuarine water on the hydridization process is thus one of reducing yields of the three tin species tested. We expect that not only the dissolved and particulate organics and Cl^- influence formation of Sn-H bonds, but that other aquated metal ions play an important role, too. Several workers have reported that, for example, As(III), As(V), Cu(II), Co(II), Ni(II), Hg(II), Pb(II), and Ag(I) interfere by unknown means at low concentrations[10,59]. Several of these metals are commonly found in the polluted Baltimore Harbor waters (see Table 1), and their presence may explain, in part, our results. In solutions of 0.01M HNO_3, we found that yields for Me_2Sn^{2+} reduction with $NaBH_4$ were consistent with the values obtained in distilled water and pre-purged Bay water. However, stannane yields from Sn(II) and Sn(IV) in 0.01M HNO_3 diminished considerably as a result of either insufficient trapping from rapid evolution of stannane caused by HNO_3 or of a repressed reaction caused by HNO_3. In addressing the former consideration, cryogenic trapping would improve recovery of SnH_4 formed by hydride reduction.

In summary, the hydride generation method cannot adequately differentiate between aquated Sn(IV) and Sn(II) which may coexist in certain, especially anaerobic, environments found in marine waters. Previous reports[10,11] of inorganic tin, speciated as "tin(IV)", should probably be regarded as "total reducible inorganic tin" until more discriminatory techniques become available.

OCCURRENCE AND FATE OF METHYLTINS IN ESTUARINE WATERS

Biomethylation of Tin in Sediments

Recently, three groups have independently reported on the methylation of both inorganic tin and organotin substrates by the mixed populations of microbial flora present in sediments collected from a Canadian fresh water lake[7], and from estuarine sites in San Francisco Bay[8] and Chesapeake Bay[2,9,60]. Biogenesis of Me_4Sn was seen only to occur with additions of Me_3Sn^+ to incubated sediments[7,8], but the redistribution reactions of intermediate methyltins to form Me_4Sn by non-biological pathways must be noted as competitive events in such experiments[4,8]. The concentrations of tin compounds added to incubated sediments were consistent with values found in polluted sediments (Table 1). The influence of other bioactive pollutant metals also commonly found in such sediments was not investigated. We already noted that a pure culture of a metal-tolerant *Pseudomonas* species isolated from the Chesapeake Bay [35] methylates Sn(IV) in the presence of mercury, and subsequent abiotic demethylation of $Me_nSn^{(4-n)+}$ (n = 3, 4) by Hg^{2+} occurs to give MeHg $^{+}$ [6]. This reaction (equation (3)) and other possible demethylation routes (cf. Table 3) may importantly influence methyltin distribution patterns emanating from sediments. It, therefore, becomes very important in future incubated sediment experiments to establish the relationship between biogenic $Me_nSn^{(4-n)+}$ distributions and the presence of Hg and other bioactive heavy elements.

Estuarine Degradation, Methylation, and Uptake of Man-made Organotins

Sediment incubation studies implied that commercial organotin biocides, particularly those used prevalently as marine antifoulants, could undergo transformations which replace existing R-Sn bonds with methyl groups [2,7]

$$(Bu_3Sn)_2O \rightarrow Me_3Sn^+_{aq}, \text{ and } (phenyl)_3SnOAc \rightarrow Me_3Sn^+_{aq}. \qquad (8)$$

No evidence of intermediate products, $R_nSnMe_{3-n}{}^+$, though plausible, was given. In one sense, this is a desirable environmental biodegradation of spent man-made biocides according to equation (2) if such processes occur in estuaries. On the other hand, the extent (rate) of conversion of R_3Sn^+ to Me_3Sn^+ is unknown. Hence methyltins formed by release of R_3Sn species from widely dispersed vessel antifouling paints cannot be discounted as an active pathway for uptake of physiologically active tin in local food chains. Preliminary evidence points to a bioaccumulation trend for total tin in bottom feeder mussels (Mytilus edulis) from vessel-related activities in Southern California harbors [61]. Thus, we see that bioconcentration of either man-made forms of organotins or biologically transformed organotins may occur to some degree by transfer from one trophic level to the next higher, as has been demonstrated for Hg(II) in a three compartment Chesapeake Bay food web comprising bacteria → ciliated protozoa → copepods [62]. Such studies involve determination of the species and the flux of organotins resulting from anthropogenic releases of organo-tin biocides into estuarine waters, as well as identification of assessment of the microbial flora responsible for uptake of bioactive tins into the primary trophic level of estuarine food webs.

We recently examined the form and extent of organotins leached into water from typical commercial marine antifouling paints [49] and a prototype organotin polymer formulation under development by the United States Navy as a long-term controlled release antifoulant. Such materials are regarded as sources of man-made organotin discharge into navigable waters, especially from ships at dock in harbors. In comparison, we also examined a typical blasting grit from a Chesapeake Bay shipyard where it was used to remove old or "spent" antifouling paint. Here, our aim was to speciate the chief leachable forms of weathered or degraded organotins discharged into sluice waters in the shipyard operations or directly into harbor sediments from ship's hulls. Figure 6 depicts direct speciation of such leachable organotins by the HPLC-GFAA method. In Figure 6a we see that the organotin polymer (OMP) controlled-release formulation slowly releases both biocidal thiorganotin moieties as free R_3Sn^+ species in water. This chromatogram was taken from leachates in solution during the late, zeroth-order release rate exhibited by this OMP. At this point,

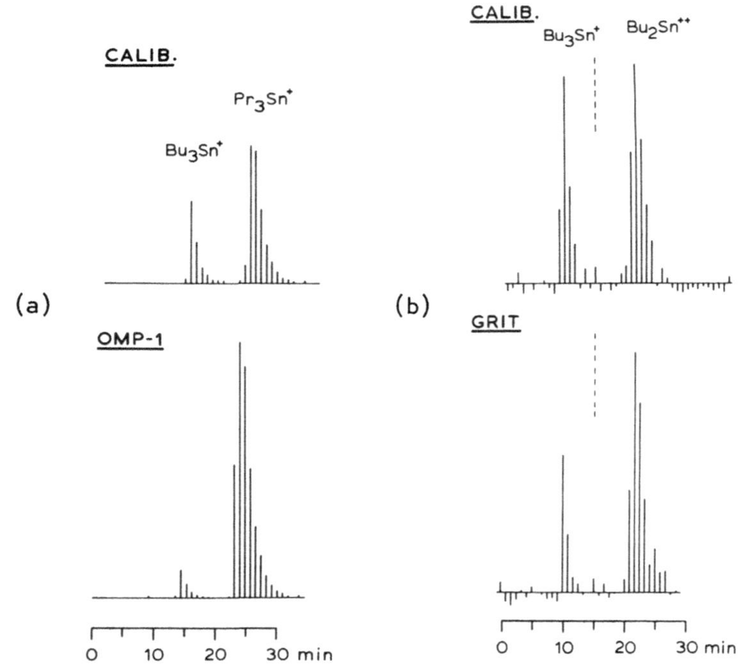

Fig. 6 Typical chromatograms obtained by tin-specific HPLC-GFAA employing a strong cation exchange column with isocratic $MeOH-H_2O-NH_4OAc$ mobile phase at 1 ml min^{-1} show separations of di- and triorganotins following direct injections (100 to 200 µl) of aqueous leachates obtained from: (a) a candidate Navy antifouling co-polymer of tri-n-butyltin and tri-n-propyltin methacrylates (1:1)[49]; and (b) water shaken with shipyard grits used for removal of weathered antifouling paints[49]. Areas under chromatographic "peaks" were accurately estimated by summing all individual GFAA peak heights comprising each HPLC "peak" shown[50]. The vertical-dashed "break" in (b) signifies a flow rate change to 1.5 ml min^{-1}.

the major release toxicant is Pr_3Sn, implying that mixed formulations of controlled-release agents may discharge their ingredients at differing rates and thereby yield variable toxic effects on target microfouling organisms and the marine environment. Below, we will suggest that steady release of these triorganotin cations may find rapid uptake on both viable and dead cells in estuarine waters.

Figure 6b offers direct evidence for degradation of Bu_3Sn-containing antifouling paints following field service on ships hulls. We cannot ascertain yet if the degradation of original Bu_3Sn^+ to Bu_2Sn^{2+} occurred by purely chemical weathering forces (light, oxidants, or marine chemicals), or by biological agents such as bacteria. The obvious important consequence is that such widespread use of these paints can discharge leachable forms of organotins into estuarine harbor locales. The tri- and dibutyltin leachates speciated here were present at 3 and 6 mg l^{-1}, respectively. No evidence for other organotin ions, such as methylbutyltin species was found[49].

Tin-resistant bacteria were recently estimated to represent 17 percent of the total bacterial population at nine diverse sites in the Chesapeake Bay[2,9,60]. We have similarly isolated a Bu_3Sn-resistant pure bacterial strains, including <u>Pseudomonas</u> 244 [6,35], from four stations in Baltimore Harbor and the Bay[63]. All were gram negative rods which grew on Nelson's agar[35] containing 50 mg l^{-1} tin as Bu_3SnCl or 100 mg l^{-1} tin as Sn(IV). Tin uptake was measured for nine of the isolates as a function of time: at time zero Bu_3Sn^+ was added (10 mg l^{-1} final concentration) to flasks containing starved cells in buffer (pH = 7.4) solutions with or without glucose (10 m$\underline{M}$). Aliquots of cell suspensions were removed after 1.5 hr, filtered, rinsed, and digested in hot HNO_3 for total tin analysis by GFAA. For 9 isolates, the total tin uptake as a dry weight concentration factor (mg per g cells/mg per ml medium) was 621 $\pm$ 179 (no glucose) and 602 $\pm$ 274 (glucose added). This suggests that Bu_3Sn^+ accumulation was not an energy-requiring process since glucose did not stimulate starved cells to greater tin uptake. For several isolates, aliquots of cells exposed to Bu_3Sn^+ solutions were collected at more frequent intervals. Generally, we found the rate of Bu_3Sn^+ uptake to be rapid, apparently reaching steady state after 2 hr. Cell-bound Bu_3Sn^+ was substantially removed by 1 m$\underline{M}$ EDTA or methanol washes, and this procedure afforded a method for direct solution speciation by HPLC-GFAA of both the organotins taken up by cells and the residual organotins remaining in growth media. Typical results from the speciation work are illustrated in Figure 7, where it was seen that virtually no biotransformation of original Bu_3Sn^+ into a logical degraded product, Bu_2Sn^{2+} occurred. We also noted that killed whole cells accumulated almost twice as much Bu_3Sn^+ as live cells, suggesting that exclusion or volatilization of tin might

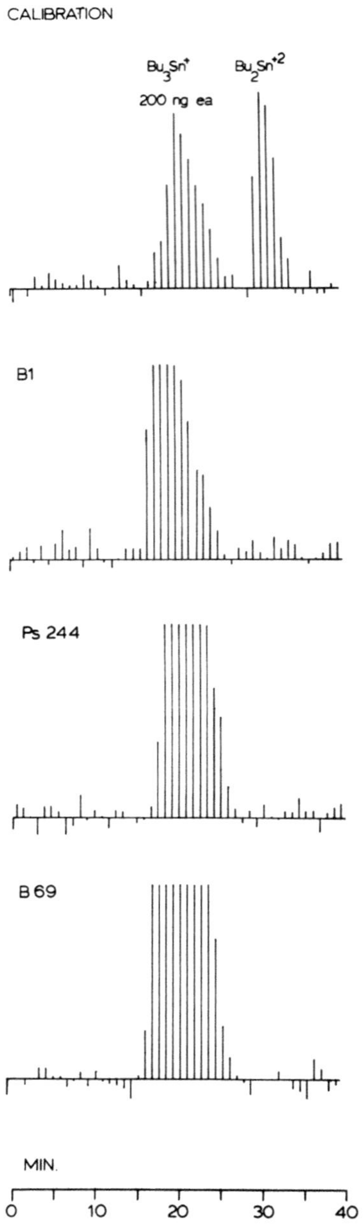
CALIBRATION
Bu_3Sn^+
Bu_2Sn^{+2}
200 ng ea
B1
Ps 244
B 69
MIN.
0
10
20
30
40

be a mechanism of resistance and that cellular detritus in estuarine waters may be an important "sink" for organotins[63].

TETRAMETHYLTIN AND METHYLTIN HYDRIDES IN CHESAPEAKE BAY

During 1980 we surveyed unfiltered surface and bottom waters at sites in Baltimore Harbor and the Bay channel, selected for their respective decrease in anthropogenic tin influx (Table 1)[2,20,21]. Employing the P/T-GC-FPD system in both modes, e.g., with or without $NaBH_4$ additions to replicate samples, we established[12] for the first time the presence of tetramethyltin in natural waters. We additionally demonstrated the unexpected presence of methyltin hydrides, $Me_n SnH_{4-n}$ (n = 2, 3), also heretofore unknown in the environment. Main features and interpretation of the experimental results are illustrated by the chromatograms in Figure 8.

Use of two modes with P/T analyses provides an effective tool for speciating both solvated or involatile $Me_n Sn^{(4-n)+}{}_{aq}$ and volatile coordinatively saturated $Me_n SnH_{4-n}$ according to equation (1). Figure 8a depicts an example of such clearcut separation, and the future useful fact that $Me_4 Sn$ is recovered virtually quantitatively after BH_4^- treatment. With Chesapeake Bay waters, several important observations were therefore possible. In both surface and bottom unfiltered waters, significant differences in the involative/volatile methyltin distributions were noted, though highly variable. In Figure 8b, for example, in surface water sample from a heavily polluted urban runoff in Baltimore Harbor, comparison of replicate samples by the two P/T modes showed that both Me_4Sn and Me_2SnH_2 were present as volatile, whereas only "reducible inorganic Sn" and some $BuSn^{3+}aq$ were present as involа-

---

Fig. 7 HPLC-GFAA chromatograms showing speciation of cell-bound organotin on three examples of Chesapeake Bay tin-resistant bacteria are compared with a calibration run for 200 ng each of Bu_3Sn and Bu_2Sn species. The isolates were grown in Nelson's broth containing 10 mg l^{-1} Bu_3SnCl, harvested (10 ml) by ultra-centrifugation, and washed 3 times in distilled water. A methanol extract (100 μl) was injected directly into the HPLC-GFAA system. No significant (95% confidence level) formation of dibutyltin degradation products resulting from cellular extracts (shown) or residual growth solu-free of cells was observed. The highest GFAA peaks in the chromatograms represent off-scale values collected with the digitizer. Based on overall system sensitivity, biotransformation to Bu_2Sn of 1% Bu_3Sn in growth solutions or 3% of cell-bound Bu Sn would have been detected.[63]

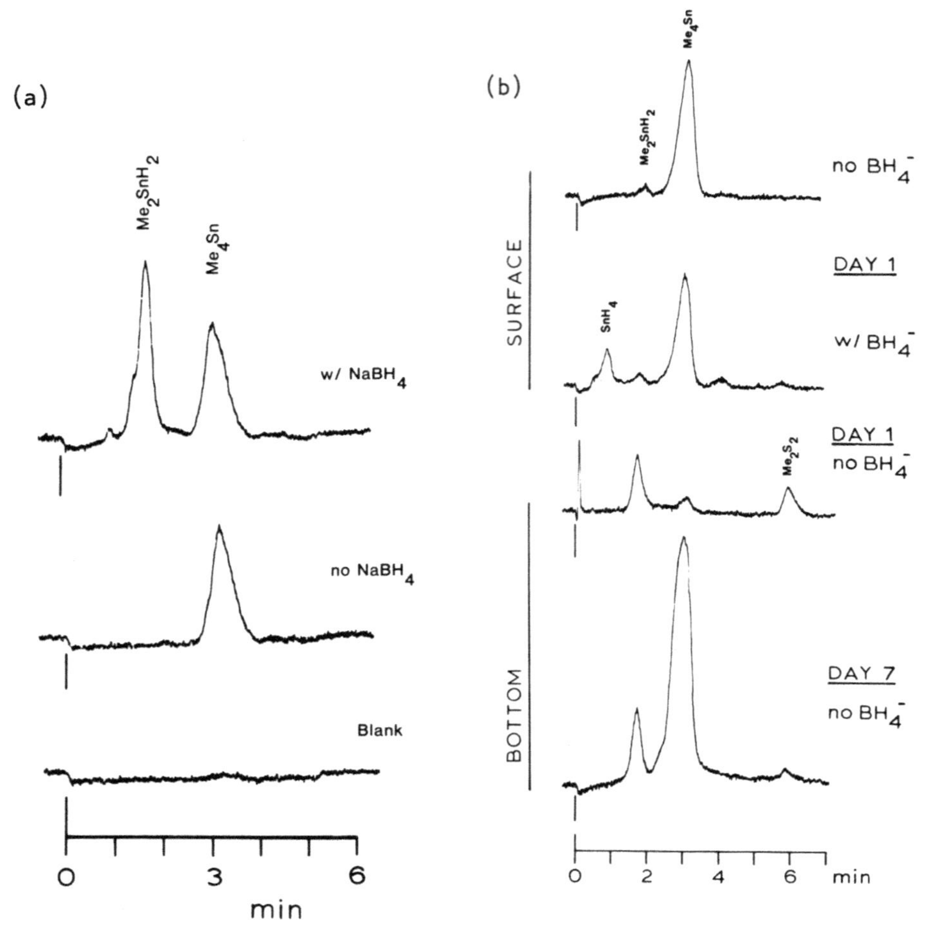

Fig. 8 Application of P/T-GC-FPD system for speciation of organotins: (a) replicate calibration solutions containing Me_2Sn^{2+} plus Me_4Sn run by Mode II (addition of $NaBH_4$) (top) and by Mode I (no addition of $NaBH_4$) (middle) are compared with blank run by Mode II; (b) representative samples from surface and bottom waters at Jones Falls sewage outfall (Baltimore Harbor), fresh (day 1) samples were stored at 2 to 4°C and run (Mode I) within 24 h after collection, aged (day 7) sample was maintained at 2 to 4°C then run by Mode I.

tiles. Also in Figure 8b, comparison of surface and bottom waters at this station shown characteristic increases in concentrations of Me_2SnH_2 and Me_2S_2 [12], and decreased Me_4Sn in bottom water.

We have noted that reproducible results were obtained only with water samples maintained at 2 to 4°C and analyzed by P/T methods within 24 h following collection. As figure 8b illustrates, samples kept refrigerated for seven days undergo substantial alteration with significant increases of Me_4Sn induced by either chemical or biological methylation, or both. These results presage lines of investigation to establish optimum sampling protocols for ultratrace aquatic organotin (organometal) speciation, as well as determining the mechanisms by which continued methylation of tin occur *in vitro*.

In Chesapeake Bay, we generally found increasing methyltin concentrations with increased anthropogenic tin influx (Table 1), in accord with Hallas' contention[2,60]. Concentration values obtained for individual organotin species often varied more than ten-fold within the water column at a given site and from visit to visit. We found much higher (10 X) organotin concentrations in Baltimore Harbor than in the Bay channel. The highest concentrations observed over nine months (spring to fall) occurred at Harbor locations with greatest organic loadings (Jones Falls), viz. Me_2SnH_2 (0.20 μg l^{-1}), Me_3SnH (0.40), and Me_4Sn (0.48); a maximum value of 0.93 μg l^{-1} for Me_4Sn was obtained at Colgate Creek which is primarily an urban-industrial run-off site. We conducted a sustained three-week study on Baltimore Harbor in order to compare methyltin levels with those available for other American estuaries. The data are summarized in Table 5.

We did not detect $MeSnH_3$, either as an involatile or volatile species, in Baltimore Harbor. Presumably, from its reported presence in San Diego[11] and Florida[10] estuaries at ranges below 0.008 μg l^{-1}, it may have been present in the Bay waters below our P/T system detection limit (∿ 0.005 μg l^{-1}). Though butyltins were not detected in San Diego Bay, the same workers[11] detected very high concentrations of $BuSn^{3+}aq$ (1.22 μg l^{-1}) and $Bu_2Sn^{2+}aq$ (1.60 μg l^{-1}) in surface waters of Lake Michigan which they ascribe to atmospheric inputs or possible contamination from PVC water samplers (cf. Table 2). We found BuSn species regularly in our three-week study, both as volatile $BuSnH_3$ and solvated $BuSn_{aq}{}^{3+}$. We saw a trend toward higher concentrations in surface water (Table 5), though not as pronounced as in Lake Michigan. Our P/T methods employed in the present studies, did not reliably speciate Bu_2Sn^{2+} or higher homologs, but it seems more likely to us that such organotins will be anthropogenic and probably introduced from antifouling paints on vessels. More detailed studies on the distribution patterns of the entire sequence $Bu_nSn^{(4-n)+}$ (equation (2)) based on water movements and shipping patterns are now required to resolve this point.

Finally, we can estimate the ratio of organotins to total tin in the suspended particulates found in Baltimore Harbor surface

Table 5. Methyltins in Estuarine Waters[a]

Site	SnH_4	$MeSnH_3$	Me_2SnH_2	Me_3SnH	Me_4Sn	Other	Ref.
Baltimore Harbor[b]							
Surface	nd[c]	nd[c]	<0.005-0.02[c]	nd[c]	<0.01-0.30[c]	0.05-0.10[c,e]	12
	nd-4[d]	nd[d]	<0.005-0.02[d]	<0.005-0.01[d]	<0.01-0.30[d]	0.05-0.30[d,e]	
Bottom	nd[c]	nd[c]	<0.005-0.02	nd[c]	0.05-0.30[c]	nd-0.10[c,e]	
	0.80[d]	nd[d]	<0.005-0.10[d]	nd-0.02[d]	nd-0.30[d]	nd-0.10[d,e]	
San Diego Bay							
Surface[d,f]	0.009-0.038	0.002-0.008	0.015-0.045	--	--	0[e,g]	11
Florida Estuaries							
Surface[d,h]	0.009	0.002	0.002	<0.001	--	--	10

[a]Concentrations in μg l^{-1} for analytes shown. Multiply by 8.4 for n<u>M</u> concentrations; [b]Range for three collections during three weeks at two sites; [c]Mode I, no addition of $NaBH_4$; [e]$BuSnH_3$; [f]Range for five sites collected over one day; [g]Bu_2SnH_2; [h]Average of eleven sites collected over various periods of time.

water. Nichols[20] observed that unfiltered surface water from this locale contained 0.19 $\mu g\ l^{-1}$ (with 8.7 $\mu g\ g^{-1}$ Sn in filtered particulates) (Table 1). Our estimates of total organotins, including methyl- and butyltins for Baltimore Harbor are represented by the low values of the ranges tabulated in Table 5. Thus we derive <0.07 $\mu g\ l^{-1}$ as an upper sum which, in turn, implies that [(<0.07) x 100]/(0.19) = <37 percent of Baltimore Harbor surface water suspensions occur as organotins. This remarkable extent of bioactive tin materials seems large, and may reflect the deficiency of adequate field data at this time. In any event, both the means and needs to further explore this question are at hand[13]. In neither Nichols'[20] nor our present studies has the surface micro-layer been examined. This oleaginous film also supports amplified microbiological activity and amplifies concentrations of heavy metals[5,28]. "Free" or volatile organotin species would be expected to partition into this lipophilic layer, there to be accumulated, transformed, and possibly released into the atmosphere.

CONCERNING AQUATIC METHYLATION AND HYDRIDIC METAL (TIN) CYCLES

From the foregoing discussion of our recent results and those from other laboratories, we see that tin joins a growing family of bioactive elements for which evidence of biological and chemical methylation supporting global transport cycles widens[5,64]. In large part, substantial advances in ultratrace molecular speciation methodology and accessible commerical equipment[17,45] have spurred many creative efforts to comprehend both the biological and chemical factors that govern the likelihood, forms, and lifetimes of such organotins in the aquatic environment. We can now reasonably ask what will be the shape of future advances, and what are the roadblocks we face?

Though exciting new biogenic organometals, such as the methyltin hydrides in Chesapeake Bay, will continue to be unexpectedly discovered, more rigorous and predictive considerations must guide future progress. Recognition[5,14,15,26,64] that many aquatic organometallic reactions occur by second-order, or at least pseudo first-order, kinetic pathways limits any selection of useful model reactions. Moreover, we cannot presume that the wealth of chemical literature for organometallic reactions, typically obtained at high concentrations in non-aqueous solvents, can be successfully translated to environmental action levels extant at $\ll 10^{-10}$ $\underline{M}$ in aqueous saline media (cf. Table 5). With current speciation techniques of the sort described in this paper, the limits of detection of reaction products of model reactions is perhaps 4×10^{-11} $\underline{M}$, i.e., 5 ng l^{-1} as $MeSn^{3+}_{aq}$. For example, in aqueous oxidative additions of MeI to Sn^{2+}_{aq} or Pb^{2+}_{aq}, which are bimolecular S_N2 reactions (Table 3) we estimate $k_2 < 10^{-3}$ $\underline{M}^{-1}$ s^{-1} [5,7,26,39]. Thus, for this tin methylation, starting with both reactants at 10^{-6} M, kinetic measurements could begin after 11 h during which time the reaction will have proceeded only 0.004 percent; starting with 10^{-8} M reactants, product $MeSn^{3+}$ could be detected after 0.4 percent reaction, but this would now require 12.7 years! Clearly, the need for careful selection of reaction conditions is paramount given the constraints of available detection limits for tin species or those of any other metal or metalloid.

We must also consider the thermodynamic bases upon which we assess relative populations or stability domains for aquatic ultratrace metal-containing molecules. Figure 9 presents a highly simplified two-electron redox cycle for an element E in the chemical or biological synthesis (or degradation) or methyl- or hydrido- derivatives in an aquatic system. Here, either methyl carbanion (Me^-) or carbonium ion (Me^+) and hydride (H^-) or proton (H^+) can act as two-electron carriers in both enzymatic [65] or abiotic[15,26] reactions. The relative stability of a given element toward redox is indicated by the usual thermodynamic expression for $E_{ox} = E_{red}$ shown as a vertical potential gradient. Approximate equilibrium concentrations for the respective species, viz. As(V) - As(III), Sn(IV) - Sn(II), or Hg(II) - Hg(0), can thereby be estimated, given appropriate environmental parameters of pH, E_h, pCl, etc. However, such calculations fall far short of approximating the true nature and concentrations of redox species in natural environments[52,55]. The relative concentrations of organometallic and inorganic species are greatly determined by non-equilibrium, catalytically induced kinetic processes which involve both chemical and biological factors[54].

For a methylation (electrons added, ↑) or a demethylation (electrons removed, ↓), as well as an analogous hydridization, the crucial knowledge about the stability or lifetime of such species formed is prerequisite to assessing the extent of redox. Beyond this, knowledge about lifetimes of such methylated/hydridized metal species provides guidance for designing relevant approaches to their speciation in active environmental compartments. For

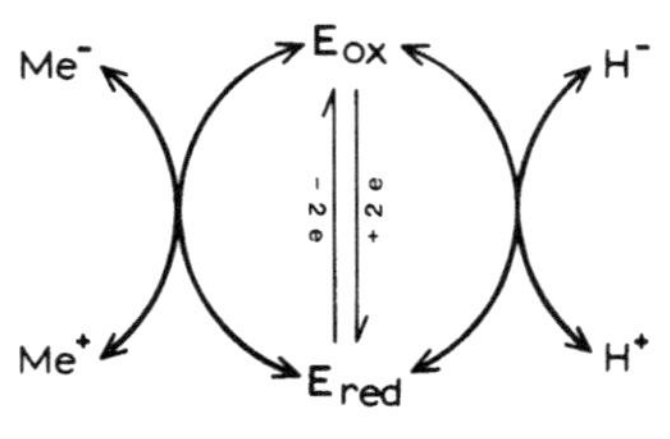

Fig. 9 Basic features of aquatic synthesis of methyl- and hydrido- derivatives of elements (E), and their possible stabilities are summarized in a two-electron redox cycle.

example, we expect methylation of mercury(II) to result in formation of relatively long-lived[5] $MeHg^+_{aq}$ species which we can (and do) detect in marine waters, whereas any possible HgH^+, HgH_2^o, or $MeHgH^o$ species are not expected[66] to survive sufficiently long for analysis (speciation), but eject H_2 or MeH to form Hg^o[36]. For many other metals and metalloids, such mobile metabolites can exist in natural waters for periods long enough for environmental speciation. Our discovery of methyltin hydrides, Me_nSnH_{4-n} (n = 2, 3) in polluted Chesapeake Bay waters[12] is one good example, as was the detection of bacterial formation of methylarsines, Me_nAsH_{3-n}, in soils[34]. At the concentrations involved, and in the presence of a multitude of potential reactants, we conclude at this point that such hydrides may possess aquatic lifetimes far longer than suggested by the meager literature available. To a great extent, future progress in this new and important area of trace metals in the sea will depend upon careful measurements of such species' lifetimes in saline solutions and characterization of the pathways for their disappearance.

ACKNOWLEDGEMENTS

Contributions from the National Bureau of Standards are not subject to copyright. Major portions of this work were supported by the Office of Naval Research, to whom we are grateful for sustained interest and encouragement. We thank the Chesapeake Bay Institute and Johns Hopkins University for providing ship time on the R/V Ridgely Warfield. Our colleagues, R.R. Colwell, J.J. Cooney, J.M. Bellama, L.E. Hallas, Y.K. Chau, K.L. Jewett, M. Nichols, and W.M. Coleman were generous with ideas and comments, and graciously shared unpublished data with us; we thank them for their invaluable help. G.J. Olson was supported as a National Research Council-National Bureau of Standards Postdoctoral Research Associate, 1979-1981. Identification of certain commercial equipment or materials in this paper does not imply any endorsement of quality or recommendation by the National Bureau of Standards. Rather, such specifications are made only to adequately detail experimental procedures.

REFERENCES

1. Smith, J.D. and J.D. Barton, 1972: The occurrence and distribution of tin with particular reference to marine environments. Geochim. Cosmochim. Acta., 36, 621-629.
2. Hallas, L.E., 1981: Ph.D. Dissertation, University of Maryland.
3. Helz, G.R., 1976: Trace element inventory for the northern Chesapeake Bay with emphasis on the influence of man. Geochim. Cosmochim. Acta., 40, 573-580.

4. Craig, P.J., 1980: The feasibility of environmental methylation for tin - An assessment. Environ. Technol. Lett., 1, 225-234.

5. Brinckman, F.E., 1981: Environmental organotin chemistry today: Experiences in the field and laboratory. J. Organometal. Chem. Library, 12, 343-376.

6. Huey, C., F.E. Brinckman, S. Grim and W.P. Iverson, 1974: The role of tin in bacterial methylation of mercury. In: "Proc. Internat. Conf. on Transport of Persistent Chemicals in Aquatic Ecosystems", A.S.W. deFreitas, D.J. Kushner and S.U. Quadri, eds. National Research Council, Ottawa, 1173-1178.

7. Chau, Y.K., 1980: Biological methylation of tin compounds in the aquatic environment. 3rd Internat. Conf. Organometal. Coordinat. Chem. Germanium, Tin, Lead, Univ. Dortmund, West Germany, July, 1980. 30 pp.

8. Guard, H.E., A.B. Cobet and W.M. Coleman, 1981: Methylation of trimethyltin compounds by estuarine sediments. Science, 213, 770-771.

9. Hallas, L.E., J.C. Means and J.J. Cooney, 1982: Microbial transformations of tin by estuarine microorganisms. Science, 213, 1505-1506.

10. Braman, R.S. and M.A. Tompkins, 1979: Separation and determination of nanogram amounts of inorganic tin and methyltin compounds in the environment. Anal. Chem., 51, 12-19.

11. Hodge, V.F., S.L. Seidel and E.D. Goldberg, 1979: Determination of tin(IV) and organotin compounds in natural waters, coastal sediments, and macro algae by atomic absorption spectrometry. Anal. Chem., 51, 1256-1259.

12. Jackson, J.A., W.R. Blair, F.E. Brinckman and W.P. Iverson, 1982: Gas chromatographic speciation of methylstannanes in the Chesapeake Bay using purge and trap sampling with a tin-selective detector. Environ. Sci. Technol., 16, 110-119.

13. Zuckerman, J.J., R.P. Reisdorf, H.V. Ellis and R.R. Wilkinson, 1978: Organotins in biology and the environment. In: "Organometals and Organometalloids: Occurence and fate in the environment", F.E. Brinckman and J.M. Ballama, eds. American Chemical Society, Washington, D.C., 388-422.

14. Ridley, W.P., L.J. Dizikes and J.M. Wood, 1977: Biomethylation of toxic elements in the environment. Science, 197, 329-332.

15. Wood, J.M., H.J. Segall, W.P. Ridley, A. Cheh, W. Chudyk and J.S. Thayer, 1977: Metabolic cycles for toxic elements in the environment. In: "Proc. Internat. Conf. on Heavy Metals in the Environment", T.C. Hutchinson, ed. University of Toronto, Toronto, 49-68.

16. Lantzy, R.J. and F.T. Mackenzie, 1979: Atmospheric trace metals: Global cycles and assessment of man's impact. Geochim. Cosmochim. Acta., 43, 511-525.
17. Von Loon, J.C., 1979: Metal speciation by chromatography/atomic spectrometry. Anal. Chem., 51, 1139A-1150A.
18. Furr, A.K., A.W. Lawrence, S.S.C. Tong, M.C. Grandolfo, R.A. Hofstader, C.A. Bache, W.H. Gutenmann and D.J. Lisk, 1976: Multielement and chlorinated hydrocarbon analysis of municipal sewage sludges of American cities. Environ. Sci. Technol., 10, 683-687.
19. Brezonik, P.L., 1974: Analysis and speciation of trace metals in water supplies. In: "Aqueous-environmental Chemistry of Metals", A.J. Rubin, ed. Ann Arbor Science, Ann Arbor, 167-191.
20. Nichols, M.: Virginia Inst. Marine Sci., personal communication.
21. Villa, O. and P.G. Johnson, 1974: Distribution of metals in Baltimore Harbor sediments. Annapolis Field Office Tech. Report 59, U.S. Environmental Protection Agency, Washington, D.C.
22. Harris, R., M. Nichols and G. Thompson, 1979: heavy metal inventory of suspended sediment and fluid mud in Chesapeake Bay. Special Sci. Report 99, Virginia Inst. Marine Sci., Gloucester Point, Virginia.
23. Sigleo, A.C., G.R. Helz and W.H. Zoller, 1980: Organic-rich colloidal material in estuaries and its alteration by chlorination. Environ. Sci. Technol., 14, 673-679.
24. Beck, R.A. and J.J. Brink, 1978: Production of cobalamins during activated sewage sludge treatment. Environ. Sci. Technol., 12, 435-438.
25. Dodd, D. and M.D. Johnson, 1973: The organic compounds of cobalt(III). Organometal. Chem. Rev., 52, 1-232.
26. Thayer, J.S. and F.E. Brinckman, 1982: The biological methylation of metals and metalloids. In: "Advances in Organometallic Chemistry", F.G.A. Stone and R. West, eds. Vol. 20, Academic Press, Inc., New York, 313-356.
27. Daniels, S.L., 1980: Mechanisms involved in sorption of microorganisms to solid surfaces. In: "Adsorption of Microorganisms to Surfaces", G. Bitton and K.C. Marshall, eds. Wiley-Interscience, New York, 7-58.
28. Piotrowicz, S.R., B.J. Ray, G.L. Hoffman and R.A. Duce, 1972: Trace metal enrichment in the sea-surface microlayer. J. Geophys. Res., 77, 5243-5254.
29. Smith, P.J., 1978: Toxicological data on organotin compounds. Internat. Tin Res. Inst., London, 16 pp.
30. Jernelöv, A. and A. Martin, 1975: Ecological implications of metal metabolism by microorganisms. Ann. Rev. Microbiol., 29, 61-77.
31. Spangler, W.J., J.L. Spigarelli, J.M. Rose, R.S. Flippen and H.H. Miller, 1973: Degradation of methylmercury by

bacteria isolated from environmental samples. Appl. Microbiol., 25, 488-493.

32. Jewett, K.L., F.E. Brinckman and J.M. Bellama, 1978: Influence of environmental parameters on transmethylation between aquated metal ions. In: "Organometals and Organometalloids: Occurrence and Fate in the Environment", F.E. Brinckman and J.M. Bellama, eds. American Chem. Soc., Washington, D.C., 158-187.
33. Goldberg, E.D., 1965: In: "Chemical Oceanography", J.P. Riley and G. Skirrow, eds. Academic Press, London; E.D. Goldberg, 1963: In: "The Sea", M.N. Hills, ed. Wiley-Interscience, New York.
34. Cheng, C.N. and D.D. Focht, 1979: Production of arsine and methylarsines in soil and in culture. Appl. Environ. Microbiol., 38, 494-498.
35. Nelson, J.D. and R.R. Colwell, 1975: The ecology of mercury-resistant bacteria in the Chesapeake Bay. Microbiol. Ecol., 1, 191-218.
36. Brinckman, F.E. and W.P. Iverson, 1975: Chemical and bacterial cycling of heavy metals in the estuarine system. In: "Marine Chemistry in the Coastal Environment", T.M. Church, ed. American Chem. Soc., Washington, D.C., 319-342.
37. Brinckman, F.E., G.E. Parris, W.R. Blair, K.L. Jewett, W.P. Iverson and J.M. Bellama, 1977: Questions concerning environmental mobility of arsenic: Needs for a chemical data base and means for speciation of trace organoarsenicals. Environ. Health Perspectives, 19, 11-24.
38. Parris, G.E. and F.E. Brinckman, unpublished results.
39. Ahmad, I., Y.K. Chau, P.T.S. Wong, A.J. Carty and L. Taylor, 1980: Chemical alkylation of lead(II) salts to tetraalkyllead(IV) in aqueous solution. Nature, 287, 715-717.
40. Parris, G.E. and F.E. Brinckman, 1975: Reactions which relate to the environmental mobility of arsenic and antimony. I. Quaternization of trimethylarsine and trimethylstibine. J. Org. Chem., 40, 3801-3803.
41. Akagi, H. and E. Takabatake, 1973: Photochemical formation of methylmercuric compounds from mercuric acetate. Chemosphere, 2, 131-133.
42. Jewett, K.L., 1978: Ph.D. Dissertation, University of Maryland.
43. Clinton, N.A. and J.K. Kochi, 1973: Alkylation as a route to reduction of copper(II) by tetraalkyllead. J. Organometal Chem., 56, 243-254.
44. Lovelock, J.E., 1975: Natural halocarbons in the air and in the sea. Nature, 256, 193-194.
45. Dulka, J.J. and T.H. Risby, 1976: Ultratrace metals in some environmental and biological systems. Anal. Chem., 48, 640A-653A.

46. Michael, L.C., M.D. Erickson, S.P. Parks and E.D. Pellizzari, 1980: Volatile environmental pollutants in biological matrices with a headspace purge technique. Anal. Chem., 52, 1836-1841.
47. Aue, W.A. and C.G. Flinn, 1977: A photometric tin detector for gas chromatography. J. Chromatogr., 142, 145-154.
48. Snyder, L.R. and J.J. Kirkland, 1979: "Introduction to Modern Liquid Chromatography", 2nd Ed., Wiley & Sons, New York.
49. Jewett, K.L. and F.E. Brinckman, 1981: Speciation of trace di- and triorganotins in water by ion exchange HPLC-GFAA. J. Chrmoatogr. Sci., 19, 583-593.
50. Brinckman, F.E., W.R. Blair, K.L. Jewett and W.P. Iverson, 1977: Application of a liquid chromatograph coupled with a flameless atomic absorption detector for speciation of trace organometallic compounds. J. Chromatogr. Sci., 15, 493-503.
51. Parris, G.E., W.R. Blair and F.E. Brinckman, 1977: Chemical and physical considerations in the use of atomic absorption detectors coupled with a gas chromatograph for determination of trace organometallic gases. Anal. Chem., 49, 378-386.
52. Pourbaix, M., 1966: "Atlas of Electrochemical Equilibria". Pergamon Press, New York.
53. Sillén, L.G., 1961: The physical chemistry of sea water. In: "Oceanography", M. Sears, ed. Amer. Assoc, Adv. Sci., Washington, D.C.
54. Johnson, D.L., 1972: Bacterial reduction of arsenate in sea water. Nature, 240, 44-45.
55. Andreae, M.O., 1979: Arsenic speciation in sea water and interstitial waters: The influence of biological-chemical interactions on the chemistry of a trace element. Limnol. Oceanogr., 24, 440-452.
56. Tallman, D.E. and A.U. Shaikh, 1980: Redox stability of inorganic arsenic(III) and arsenic(V) in aqueous solution. Anal. Chem., 52, 196-199.
57. Schaeffer, G.W. and M. Emilius, 1954: The preparation of stannane. J. Amer. Chem. Soc., 76, 1203-1204.
58. Jolly, W.L. and J.E. Drake, 1963: Hydrides of germanium, tin, arsenic, and antimony. Inorg. Synth., 7, 34-44.
59. Evans, W.H., F.J. Jackson and D. Deller, 1979: Evaluation of a method for determination of total antimony, arsenic, and tin in foodstuffs using measurement by atomic-absorption spectrophotometry with atomization in a silica tube using the hydride generation technique. Analyst, 104, 16-34.
60. Hallas, L.E. and J.J. Cooney, 1981: Tin and tin-resistant microorganisms in Chesapeake Bay. Appl. Environ. Microbiol., 41, 466-471.
61. Young, D.R., G.V. Alexander and D. McDermott-Ehrlich, 1979: Vessel related contamination of Southern California

harbors by copper and other trace metals. Mar. Pollut. Bull., 10, 50-56.

62. Berk, S.G. and R.R. Colwell, 1981: Transfer of mercury through a marine microbial food web. J. Exp. Mar. Biol. Ecol., 52, 157-172.

63. Blair, W.R., C.J. Olson, F.E. Brinckman and W.P. Iverson, 1982: Uptake and fate of tributyltin cation in estuarine bacteria. Microbial. Ecol., submitted.

64. Craig, P.J., 1980: Metal cycles and biological methylation. In: "The Handbook of Environmental Chemistry", Vol. 1, Part A., O. Hutzinger, ed. Springer-Verlag, New York, 169-227.

65. Williams, R.J.P., 1980: On first looking into nature's chemistry. Part I. The role of small molecules and ions: The transport of the elements. Chem. Soc. Rev., 9, 281-364.

66. Jensen, F.R. and B. Rickborn, 1968: "Electrophilic Substitution of Organomercurials", McGraw-Hill Book Co., New York.

THE RELATIONSHIP OF THE DISTRIBUTION OF DISSOLVED SELENIUM IV AND VI IN THREE OCEANS TO PHYSICAL AND BIOLOGICAL PROCESSES

C.I. Measures, B.C. Grant, B.J. Mangum and J.M. Edmond

Department of Earth and Planetary Sciences
Massachusetts Institute of Technology
Cambridge, MA 02139 U.S.A.

ABSTRACT

Over 500 determinations each of selenium IV and selenium VI have been made on ocean water samples collected from the Atlantic, Pacific and Indian oceans. All three oceans show similar distributions for both oxidation states which resembles those of the nutrients phosphate and silicate. Pacific and Indian ocean samples show selenium IV values of 50 pmol/l in surface waters rising to 790 pmol/l in deep waters and selenium VI values of 500 pmol/l in surface waters rising to 1400 pmol/l in deep waters. Atlantic ocean concentrations are 30-40% lower.

Departures from normal profiles are readily understood in terms of hydrographic features. One particular anomaly, low Se VI values, is believed to trace advection of water masses from regions of intense oxygen minima.

INTRODUCTION

The measurement of the distribution of a trace element in ocean water is useful in identifying the sources and sinks of the element. In some cases the observed distributions may serve a much wider purpose in constraining the processes responsible for the control of the dissolved constituents of the oceans. The applicability of a particular element to this last role is largely determined by the precision and ease with which determinations may be performed and of course the extent to which the element is involved in these processes.

Selenium is an element which satisfies many of these criteria. It exists in ocean waters in two different oxidation states (IV and VI) and is intimately involved in the biogeochemical cycle. Through the development of a rapid and precise technique[1] it has been possible to perform a total of more than 500 determinations each of SeIV and SeVI in the Atlantic, Pacific and Indian Oceans (Fig. 1, Table 1). The method involves the formation of 4 nitro-2,1,3-benzoselenadiazole in aqueous solution and its extraction into toluene, the determination being performed by electron capture detection - gas chromatography. The method has a detection limit of approximately 10 pmol/l using a 100 ml sample and a precision of approximately 3-5% at 400 pmol/l. SeIV is measured directly and SeVI is calculated from the difference between total selenium and selenium IV.

DISTRIBUTIONS

The general behaviour of selenium in the oceans is typified by the profile observed at GEOSECS 1 REOCCUPATION[2] (Fig. 2). SeIV concentrations in surface water are low or undetectable rising sharply at the thermocline and with fairly uniform concentration in deep water (approx. 790 pmol/l).

SeVI values show a very similar pattern except that the concentrations observed throughout the water column are much higher i.e. 500 pmol/l in surface water and 1400 pmol/l in deep water. This pattern of distribution is observed in the Pacific, Atlantic and Indian Oceans with only minor local variations. Despite this consistent pattern there are differences in absolute concentrations, with the Pacific and Indian Oceans showing similar levels while the Atlantic SeIV and VI values are some 30-40% lower.

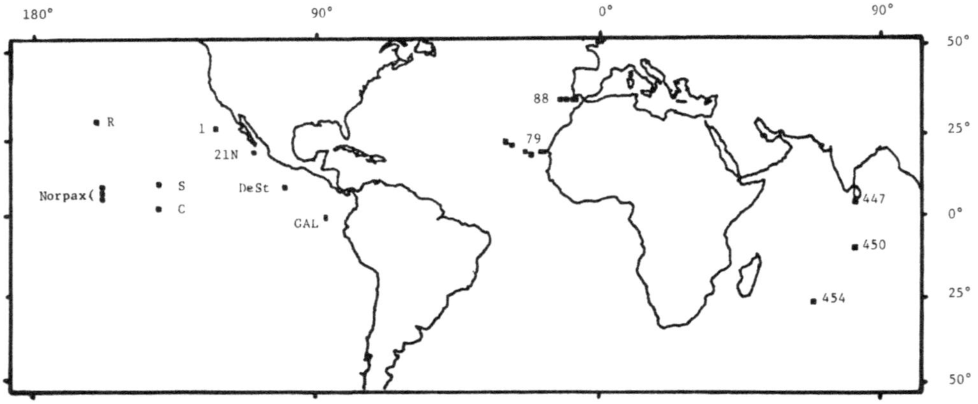

Fig. 1 Location of selenium profiles in the Pacific, Atlantic and Indian Oceans.

Table 1 Location of sampling stations

Map designation	Description		Location	
	Pacific Ocean			
1	GEOSECS 1 Re-occupation		28°29'N	122°13'W
R	Manganese Nodule	R	30°N	160°W
S	Program station	S	11°02'N	140°05'W
C		C	1°03'N	138°55'W
DeSt	USNS DeSteiguer		12°54'N	100°4'W
21 N	East Pacific Rise		20°50'N	109°06'W
GAL	Galapagos spreading center			
NORPAX	FGGE Cruise 15		5°59'N	158°0'W
			8°0'N	158°01'W
			10°0'N	158°01'W
	Atlantic Ocean			
79	R.R.S. Discovery Cruise 79		24°17'N	30°27'W
			23°6'N	27°52'W
			21°30'N	24°29'W
			20°19'N	21°49'W
			20°43'N	18°53'W
			20°47'N	18°14'W
88	R.R.S. Discovery Cruise 88		36°1'N	7°54'W
			36°0'N	8°55'W
			36°1'N	10°49'W
			36°0'N	13°9'W
	Indian Ocean			
447			4°59'N	79°57'E
450	GEOSECS Stations		10°0'S	79°59'E
454			26°59'	67°5'E

The existence of two oxidation states raises the question of whether the observed partitioning is a reflection of the ambient oxidation potential. The data show however that the SeIV/SeVI ratio in surface water (0.1) yields a pE of 6.7, considerably lower than the predictions of either Breck[3] or Sillen[4]. In deep waters where the SeIV/VI ratio is higher (.55), the pE would have to be lower still. Clearly without lowering the presumed pE of seawater considerably it would be impossible to bring the observed oxidation state distribution of selenium into the regime of thermodynamic control.

The observed distribution has been interpreted[5] as that shown by an element intimately involved in the biogeochemical cycle with reductive biological removal from surface water into particulate material followed by regeneration at depth in both the shallow and deep cycles representative of the soft (tissue) and hard (skele-

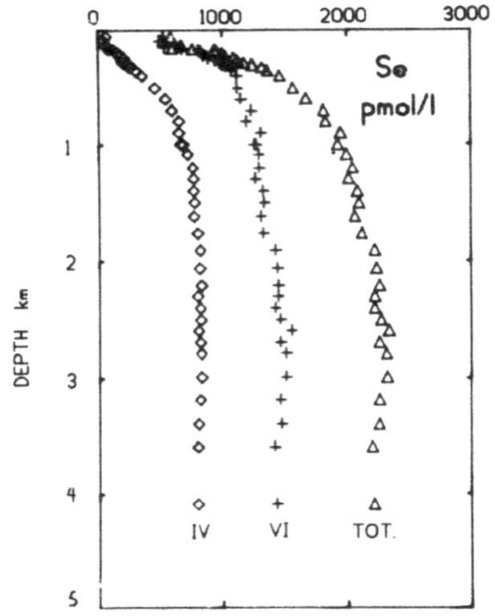

Fig. 2 Selenium at GEOSECS 1 re-occupation.

tal) components of biological material, these two processes being traced by the nutrients phosphate and silicate respectively. This mechanism would provide the continual addition of selenium IV to the deep water where it persists as a result of kinetic stability. This distribution pattern is somewhat different from that observed for other multiple oxidation state elements, e.g. I[6], Cr[7] and As[8], where the reduced forms are only encountered in surface layers or at the oxygen minima.

NUTRIENT COVARIANCE

Multiple linear regression analysis of all data for SeIV and VI (in pmol/l) against silicate and phosphate (in micro moles/l) yields the following equations:

$$SeIV = 63 + 4.21\ Si + 50\ PO_4 \qquad r^2 = .94 \quad n = 466$$
$$SeVI = 448 + 3.32\ Si + 181\ PO_4 \qquad r^2 = .83 \quad n = 409$$

Numerically the coefficients imply that the main source of SeIV regeneration is in processes traced by silicate (skeletal) and for SeVI the sources are about equal for processes traced by phosphate (tissue) and silicate.

Tempting as it is to use these coefficients to further elucidate the transport mechanism of these species certain drawbacks must be born in mind.

Firstly, the distribution of the species observed is the net result of uptake and regeneration during which process a considerable amount of interconversion of oxidation states may occur.

Secondly, the distributions of silicate and phosphate are not independent and are, relative to each other, somewhat similar in most of the world's oceans.

Thirdly, this type of approach largely ignores the problem of advection which by the introduction of water masses of different origin with possible differences in the ratio of preformed selenium to nutrient ratios may well bias the calculation over certain depth intervals. This last point can be demonstrated by the fact that selenium IV and VI data from stations near the mouth of the Mediterranean can be satisfied by equations involving only silicate and phosphate respectively. With these warnings in mind it is interesting to note that the SeIV data plot well against silicate (Fig. 3) and the SeVI against phosphate (Fig. 4) the scatter for both being generally within the $\pm$ 2σ precision of the selenium measurements.

ADVECTION

As mentioned earlier departures from the normal profile seen at some stations are readily explained in terms of advection of particular water masses. A good example of this is found in the

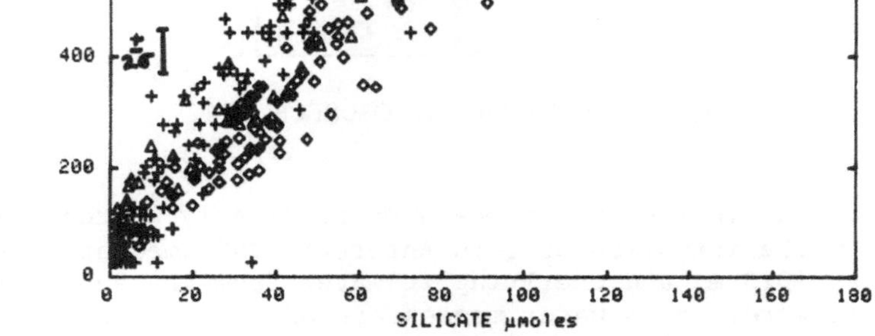

Fig. 3. SeIV vs silicate all oceans, ◇ Pacific, △ Indian, + Atlantic

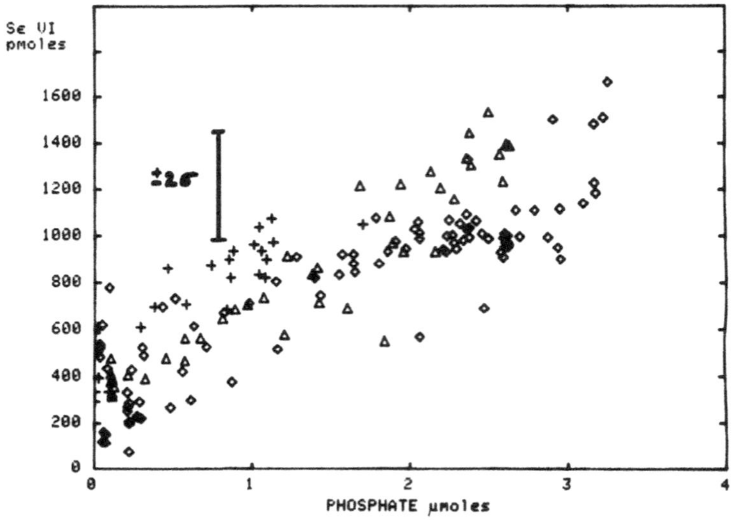

Fig. 4 SeVI vs phosphate all oceans, ◇ Pacific, △ Indian, + Atlantic

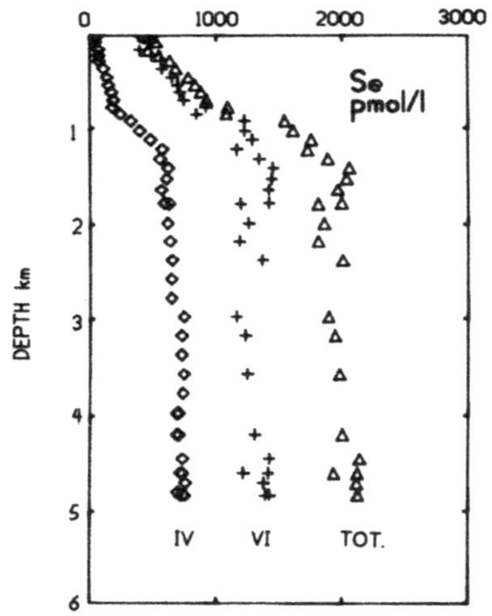

Fig. 5 Selenium at GEOSECS 454.

GEOSECS stations in the Indian Ocean where closely spaced samples clearly show the intrusion of both Antarctic Intermediate Water (centered at 1023 m) and overlying it water from the subtropical anticyclonic gyre. Both water masses are of recent surface origin and show characteristic impoverishments in both nutrients and selenium species to a much greater depth than is normally observed (Figs. 5, 6). Even such small features as Wyrtki[9] layer J water

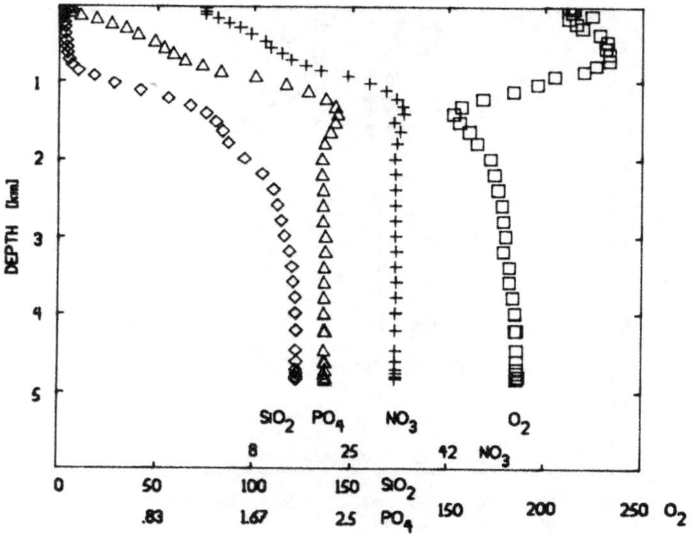

Fig. 6 Nutrients at GEOSECS 454. All units are μmoles/l.

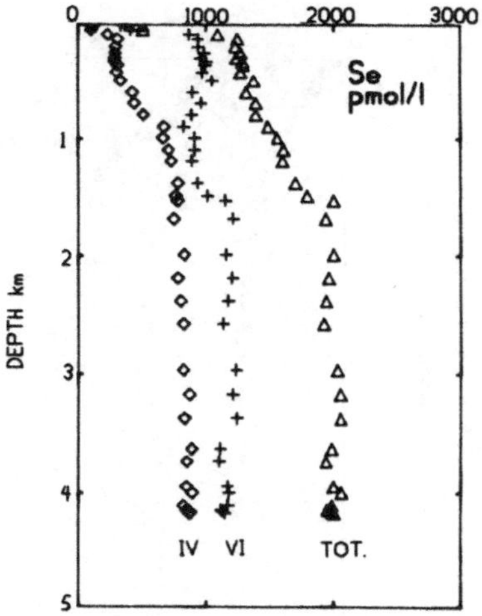

Fig. 7 Selenium at GEOSECS 447.

(318-405 m) are visible in station 447 (Fig. 7) where the selenium data closely follow those of the nutrients (Fig. 8).

DESELENIFICATION

Station 447 (Fig. 8) also shows one of the most intriguing features observed in a selenium profile: a decrease of SeVI with

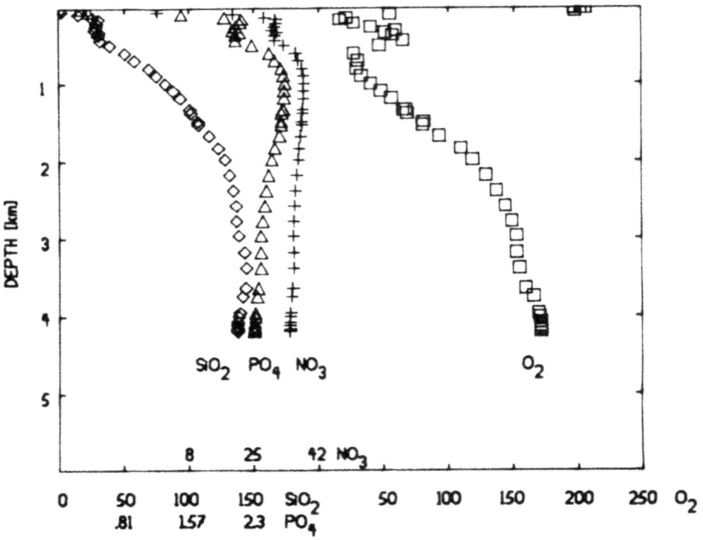

Fig. 8 Nutrients at GEOSECS 447. All units are μmoles/l.

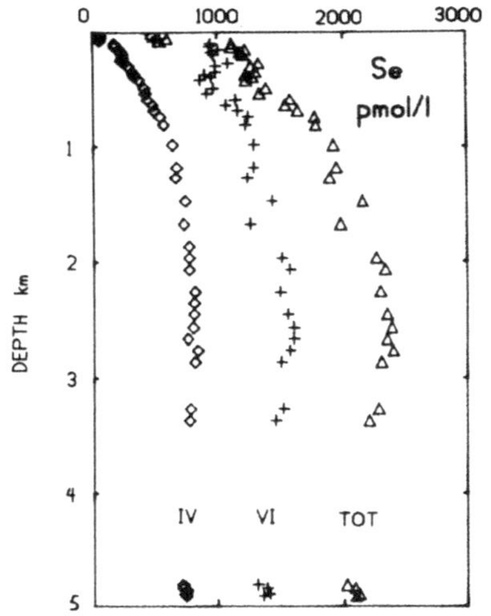

Fig. 9 Selenium at MANOP S.

depth over a small depth interval (600-1400 m). This phenomenon has also been observed in a Pacific profile at Manop station S (Fig. 9). In both profiles the decreases observed are analytically significant and are identified by 7 or 8 sequential samples. We have named this feature deselenification by analogy to dentrification which it resembles. It is then a relatively small depth interval over which normally increasing or static selenium VI levels are seen to actually decrease without a concommitant increase or decrease in selenium IV levels. In the case of the Indian ocean station the depth range of the phenomena coincides well with the advection of Red Sea water and its associated oxygen minima. In the Pacific station the feature is again associated with the oxygen minimum which appears as part of northern oxygen minimum core "spreading" westwards from the eastern tropical Pacific [10].

In both cases therefore the feature appears to be associated with the oxygen minimum. However other stations in both the Indian and Pacific oceans with oxygen minima of comparable intensity do not show this feature. A connecting component in both cases is that the water masses in which the phenomenon is observed have in their history been subjected to an extreme oxygen minimum i.e. 0.1 μm/l for Red Sea water in the northern Arabian Sea and 1 μm/l in the Eastern Tropical Pacific.

One of the problems with interpreting a feature such as this (which is presumed to be advective) is that the feature itself may be an artifact of advective influences above and below the area of interest i.e. a sandwich of normal water between two unusual slices of advected water. This problem may be overcome by plotting selenium VI values normalised to phosphate. In terms of a one dimensional model "normal" SeVI/phosphate ratios would be expected to reduce from high levels in surface waters (high preformed SeVI/phosphate ratio) to some constant value in deep water with any deselenification features appearing as a minimum in the SeVI/phosphate ratios. A plot of this parameter vs depth for the Indian Ocean Stations (Fig. 10) shows this "normal" trend with only station 447 exhibiting the anomaly. A similar plot for the Pacific Ocean (Fig. 11) shows that many more stations in the Pacific exhibit the deselenification feature and that the SeVI/phosphate minima in all stations occur at about the $\sigma_\Theta = 27.00$ surface corresponding to the oxygen minimum tongue Manop R which does not show the feature at this σ_Θ value is the only station out of this area. The Manop R station however does exhibit a small minima at $\sigma_\Theta = 27.3$, corresponding to another low oxygen tongue[10].

POSSIBLE CAUSES OF THE LOW SELENIUM VI FEATURE

Given that the low SeVI or the more widespread SeVI/phosphate anomaly is not an artifact of the normalising procedure which can be shown by the fact that both silicate and phosphate values in the affected areas are not unusual there are then only 3 possible explanations. 1) Selenium VI is being removed in situ; 2) The selenium VI/phosphate ratio of regeneration is unusually low either in situ or in the advected water mass; 3) The advected water mass had an unusually low preformed selenium VI/phosphate ratio.

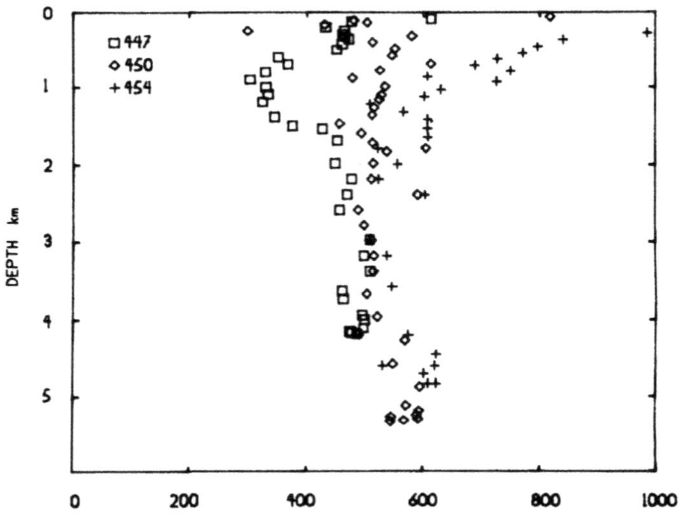

Fig. 10 SeVI/PO_4 ratio as a function of depth Indian Ocean. Ratios are $X10^6$.

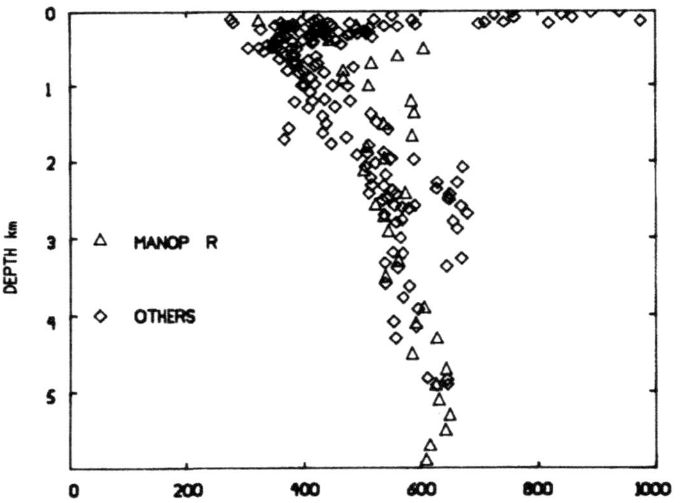

Fig. 11 SeVI/PO_4 ratio as a function of depth Pacific Ocean. Ratios are $X10^6$.

The first possibility is unlikely as a plot of the minima in the selenium VI/phosphate ratio vs the observed oxygen shows no coherent trend. The second and third possibilities cannot be distinguished between as information about the preformed SeVI/ phosphate levels in the source waters is not available. It is interesting to speculate however that deselenification by analogy with denitrification is a signature of extremely low oxygen levels, making selenium VI values a useful tracer for advected oxygen minima.

CONCLUSIONS

Selenium IV and selenium VI distributions are essentially similar in all three oceans and resemble those of nutrients. Absolute concentration differences exist with Atlantic values some 30-40% lower than those of the Pacific and Indian oceans. Departures from normal profiles are readily understood in terms of hydrographic features. Selenium VI values and selenium VI/phosphate ratios may be useful in identifying water masses advected from areas of intense oxygen minima.

REFERENCES

1. Measures, C.I. and J.D. Burton, 1980: Gas chromatographic method for the determination of selenite and total selenium in sea water. Anal. Chim. Acta., 120, 177-186.
2. Measures, C.I., R.E. McDuff and J.M. Edmond, 1980: Selenium redox chemistry at GEOSECS I re-occupation. Earth Planet. Sci. Lett., 49, 102-108.
3. Breck, W.G., 1974: Redox levels in the sea. In: "The sea" 5, E.D. Goldberg, ed. Wiley, New York.
4. Sillen, L.G., 1961: The physical chemistry of seawater. In: "Oceanography", M. Sears, ed. AAAS, WAshington, D.C.
5. Measures, C.I. and J.D. Burton, 1980: The vertical distribution and oxidation states of dissolved selenium in the northeast Atlantic Ocean and their relationship to biological processes. Earth Planet. Sci. Lett., 46, 385-396.
6. Elderfield, H. and V.W. Truesdale, 1980: On the biophilic nature of iodine in seawater. Earth Planet. Sci. Lett., 50, 105-114.
7. Cranston, R.E. and J.W. Murray, 1978: The determination of chromium species in natural waters. Anal. Chim. Acta., 99, 275-282.
8. Andreae, M.O., 1978: Distribution and speciation of arsenic in natural waters and some marine algae. Deep-Sea Res., 25, 391-402.
9. Wyrtki, K., 1971: "Oceanographic Atlas of the International Indian Ocean Expedition". National Science Foundation/ I.D.O.E., Washington, D.C.
10. Reid, J.L., Jr., 1965: "Intermediate Waters of the Pacific Ocean". The Johns Hopkins Press, Baltimore.

TRACE METALS IN THE ARCTIC OCEAN

Lars-Göran Danielsson and Stig Westerlund

Department of Analytical and Marine Chemistry
Chalmers University of Technology and
University of Gothenburg, S-412, 96
Gothenburg, Sweden

ABSTRACT

Results from trace metal determinations on samples from a station in the Eastern Arctic Ocean are presented. Levels observed in subsurface water are Cd, 0.14-0.20 nM, Cu, 1.9-2.3 nM, Ni, 3.5-4.1 nM and for Zn, 0.86-3.9 nM. For these elements the surface depletion-bottom enrichment is weakly developed compared to that normally found in ocean waters. For iron, low levels were found in directly extracted samples (1.2-9.9 nM) while much higher levels (6.8-<63 nM) were found after storage under acid conditions.

INTRODUCTION

A growing interest for scientific studies in the Arctic has been evident during recent years. This is partly due to an anticipation of an increase in exploitation of the natural resources in this area. At present the region surrounding the North Pole is, however, probably the most pristine on the northern hemisphere. Thus measurements of pollutant levels in this area are expected to give baseline levels.

Earlier investigations in the area have been carried out mainly from land areas such as Greenland or Svalbard or from drifting ice islands. Ice-free areas have been investigated from ships, but the vast pack-ice areas between the solid ice of the north and open water to the south have not been accessible by these methods. Studies in this area are important in understanding the connection between ice, water and the biological activity.

Few determinations of trace metal concentrations have been undertaken in the Arctic Ocean[1]. Investigations concerning trace metals in coastal Arctic waters have on the other hand been performed[2,3].

During the YMER-80 expedition a powerful Swedish ice breaker (HMS YMER, 7800 ton, 25000 hp) as a research platform made it possible to work effectively in the pack ice for the first time. As HMS YMER has no laboratories, working places for the scientists were arranged in containers placed onboard. The first leg of the YMER-80 expedition covered the area north of Spitsbergen - Franz Josef land and in Barents Sea. Trace metal concentrations from a station at 82° 31.6'N, 43° 57.9'E in the Arctic Basin, as indicated in Fig. 1, will be discussed. Oceanographic data[4] from this station are shown in Fig. 2. A characteristic feature is the layer of less saline water at the surface due to fresh water inflow through rivers and from melting ice. Furthermore, on the temperature profile the inflow of warm Atlantic water at a depth of 200-600 m can be clearly seen.

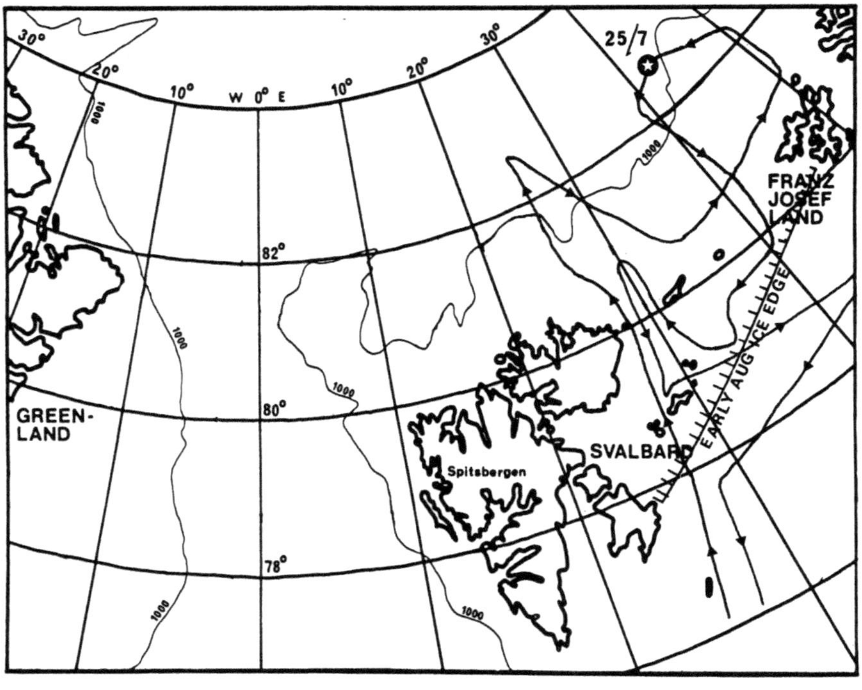

Fig. 1. Cruise track for the first leg of YMER-80 expedition. Position of Arctic basin station indicated.

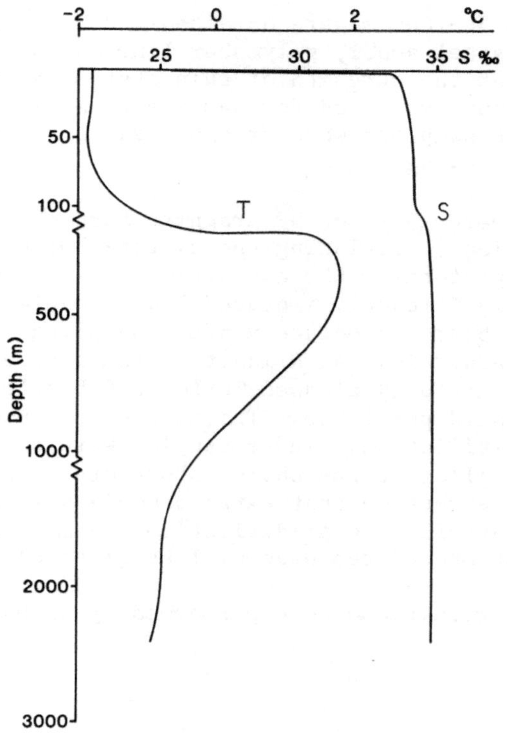

Fig. 2. Salinity and temperature profiles for Arctic basin station.

SAMPLING AND ANALYSIS

All sampling, except surface water, was done with GO-FLO samplers (General Oceanics, Miami, Florida), modified by coating with teflon and replacing the delrine valves with teflon counterparts. Initially the samplers failed to close due to low Arctic temperatures, necessitating the replacement of rubber tubing by stainless steel springs.

To avoid contamination by the HMS YMER herself, or intensive mixing by her prop-wash, all samples from the top 50 m were taken at least 200 m to windward from the ship.

Sampling commenced with the collection of surface waters from the edge of the ice-floe with the aid of arm-long gloves. One liter Teflon FEP bottles were lowered into the water, opened, filled and recapped below the surface. Pools on the ice-floes were sampled in the same way. The other surface samples were then taken by hand with the GO-FLO sampler hung on a polyester line weighted with lead encased in plexiglass. The sampler was tripped with a teflon-coated messenger.

Samples from 50-1000 m were obtained from a ship mounted winch with a stainless steel spool, polyester line and a nylon block. Due to doubts about the strength of this system, a teflon coated Rosette sampler mount was used for samples from deeper than 1000 m. After sampling, the samplers were transported to the clean room container for processing.

The samples were analysed by graphite furnace - AAS after a two stage extraction[5]. Replacing the citrate buffer, used earlier, with acetate and performing the extraction in a clean room with teflon utensils significantly improved blank levels. Extractions were performed on board immediately after sampling and the extracts brought home for analysis. An aliquot of the sample was also transferred into carefully cleaned Teflon FEP bottles and acidified with 1 ml nitric acid per litre. The nitric acid had been purified by sub-boiling distillation. These samples were extracted about 2 months after sampling at the shore laboratory. The same method was used with the exception that extra ammonia was added to the buffer to compensate for the acidification. A more detailed description of the procedures used will be given elsewhere.

Nutrient determinations were performed by standard methods[6].

RESULTS

Cadmium

Results of cadmium determinations are given together with nutrient data in Table 1. The relatively high nutrient concentrations in surface waters indicate that nutrients brought there by upwelling under the pack-ice had not yet been used up by the primary producers[7].

Cadmium concentrations increase with depth albeit to a lesser extent than for other ocean areas[8,9,10,11,12]. The covariation between nutrients and cadmium found by several workers [9,10,11,12] holds also for the Arctic Ocean as can be seen in Figs. 3a and b. The slope, $\frac{\Delta Cd}{\Delta P}$, of 1.4×10^{-4} found here for phosphate concentrations between 0.35-1 μM is low when compared to other ocean areas. Values between 1.9×10^{-4} and 3.5×10^{-4} have been reported for the Atlantic, Pacific and Indian Oceans [9,11,12,13]. Our low $\frac{\Delta Cd}{\Delta P}$ ratio may result from the rapid growth rate of plankton during the short productive season since such conditions have been shown to contribute to lower metal uptakes by plankton[14].

Table 1. Concentrations of cadmium and nutrients at 82°31.6'N, 43°57.9'E on July 25, 1981

Depth	Cd* nM	PO_4 µM	NO_3 µM	$Si(OH)_4$ µM
0	0.133	0.37	4.55	2.54
3	0.125	0.40	5.00	2.70
20	0.151	0.55	7.00	3.96
50	0.156	0.66	8.55	4.24
100	0.178	0.70	9.50	4.72
700	0.191	0.76	10.0	5.10
600	0.191	0.80	11.0	6.40
1000	0.222	0.89	12.2	8.96
2000	0.200	0.97	12.2	11.0
2400	0.205	0.97	12.2	11.3

* mean of 2 determinations

Copper

The depth profile for copper, given in Fig. 4, shows the relatively good agreement between the two extractions. The large difference between surface duplicates may be partly caused by the fact that different samples were used. However, this explanation cannot account for the difference seen at 100 m where the high value is probably the result of contamination during extraction.

Earlier investigations have found marked increases in copper concentration with depth and a correlation with silicate concentration[12,15]. Here, however, the copper concentration increased only slightly with depth.

Using the $\frac{\Delta Cu}{\Delta Si}$ ratio of 2.4×10^{-5} found for the Indian Ocean[12] and the silicate concentrations from Table 1, a plausible deep water copper enrichment can be calculated. The difference in copper concentration between surface and deep water estimated in this manner amounts to 0.21 nM, a difference too small to be confidently measured with available methods.

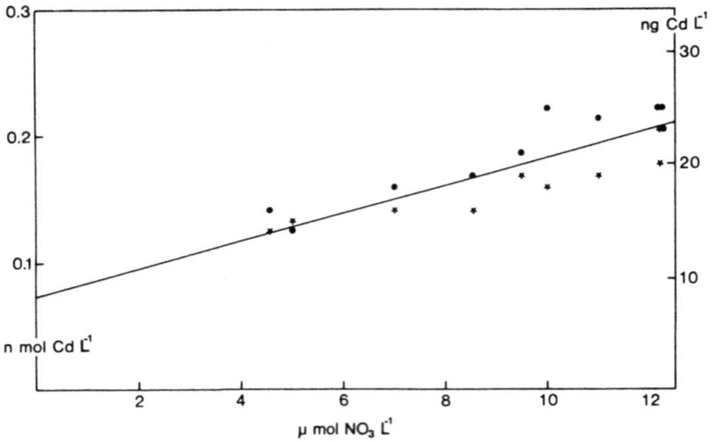

a

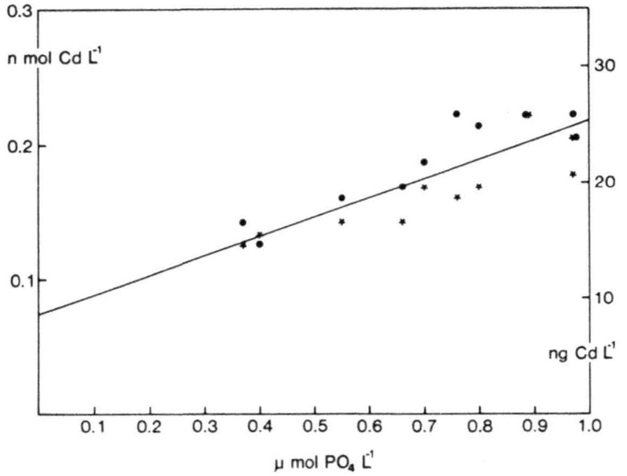

b

Fig. 3. Cadmium - nutrient correlations for Arctic basic sample. ★ samples extracted after 2 month storage. ● samples extracted on board. (a) Cadmium versus nitrate Cd(nM) = 0.075 + 0.011 NO_3(μM). (b) Cadmium versus phosphate Cd(nM) = 0.075 + 0.14 PO_4(μM). Regression lines have been calculated from the means of the cadmium determinations as reported in Table 1.

Iron

For iron, completely different results were obtained from the two extractions. This is due to the existence of a large part of the total iron in sea water in the form of suspended particles probably of colloidal size. The particulate iron is not chelated

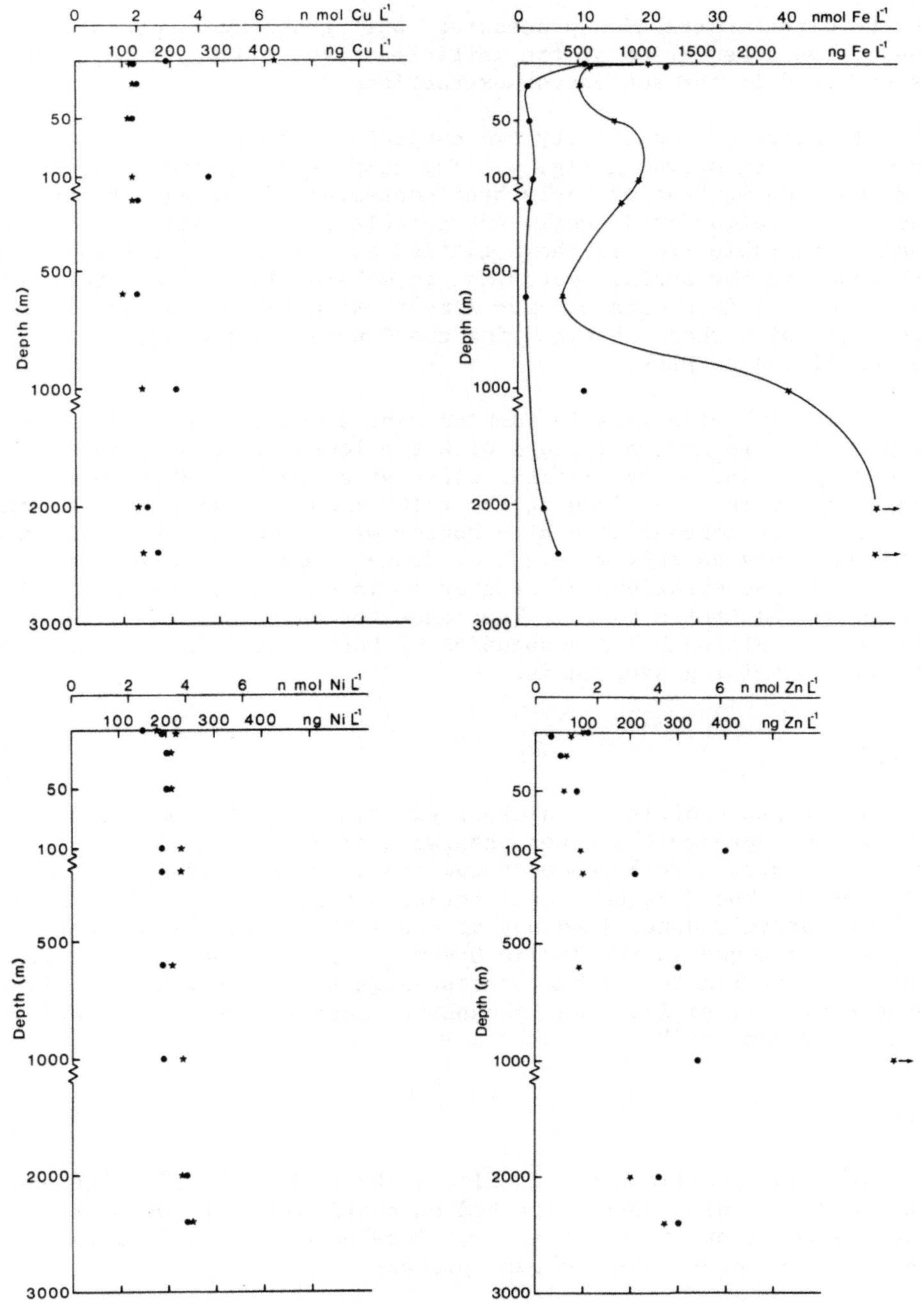

Fig. 4. Trace element profiles from the Arctic basin. ★ Samples extracted after two months storage. ● Samples extracted directly on board.

in the direct extraction procedure. During storage under acid conditions dissolution of the particles takes place and the iron is chelated in the subsequent extraction.

To increase readability two conjectural depth profiles for iron have been drawn in Fig. 4. Two samples extracted on board, 3 m and 1000 m, have probably been contaminated during extraction. The concentration of directly extractable iron is consistantly lower than extractable iron in the acidified samples and slight increases are found in the surface water and in waters close to the bottom. The concentrations obtained for direct extractable iron are in good agreement with those obtained for the Central Arctic Ocean[2], in non-acidified samples.

Iron concentrations in samples stored under acid conditions display much larger variations with the lowest concentrations occurring in inflowing Atlantic water at a depth of 200-600 m. A weak maximum in total iron at about 100 m may result from a fluvial source. Iron concentrations in bottom waters could not be determined with certainty as they were out of range. However, it is clear that high concentrations of iron exist in suspension in the deep waters of the Arctic basin. The cause for this is not yet fully understood, although a resuspension of bottom sediments seems to be the most promising hypothesis.

Nickel

The depth profile for nickel, given in Fig. 4, shows consistently good agreement between the two extractions and, as in the case of copper, a weak tendency towards increasing concentration with depth. The 0.39 nM nickel enrichment found here is too small to be accurately determined but agrees with the nickel-silicate correlation found in the Indian Ocean[12]. It is, however, smaller than the approximately 1 nM increase suggested by the covariation between nickel, silicate and phosphate found in the Pacific and Atlantic Oceans[11,16].

Zinc

The zinc profile given in Fig. 4 shows that the 100, 200 and 600 m samples which were extracted on board, and both samples at 1000 m were severely contaminated. A release of zinc from these samplers was later shown to take place.

The other data are consistent between extractions, and although it is evident that zinc increases toward the bottom, a good correlation with nutrients was not found.

DISCUSSION

According to Palosuo[7] 10.5 μM NO_3, 0.75 μM PO_4 and 5.3 μM $Si(OH)_4$ are probable winter concentrations of nutrients in the surface waters of the Arctic basin. The nutrient concentrations given in Table 1 show that perhaps half of the nutrient pool has been used up. At this station, production rate was high although the biomass was considerably lower than later in the summer[17].

Some of the trace metals studied are thought to be at least partly involved in the biological production of the sea, and this is reflected in a surface water depletion and a deep water enrichment similar to that found for nutrients. By correlating trace metal concentrations with nutrient concentrations mean metal/nutrient ratios can be determined here as they have been for other ocean areas. Using these and the nutrient concentrations found during the second leg of the expedition, trace metal concentrations at the end of the production period can be estimated. These calculations show that the cadmium concentration of the surface water should decrease further to about 0.075 nM (cf. Fig. 3). Concentrations of copper, nickel and zinc should, however, not be significantly affected and will therefore show less variation with depth in the Arctic Ocean than in other ocean areas. The reason is that this general region is one of the birthplaces for North Atlantic Deep Water, NADW. The deep water of this area has sunk recently and not yet accumulated nutrients and trace metals from sinking debris.

Comparison of deep water concentrations of trace metals in the Arctic Ocean with corresponding values from the Atlantic show moderate increases. Moore[15] found 2.7 nM Cu for depths greater than 1500 m in the eastern Atlantic. Bender and Gagner[8] found 3.2 nM Cu in NADW. These values can be compared with the values for the Arctic Ocean presented in this paper, 2.5 nM Cu. Corresponding values for cadmium are 0.20 nM in the Arctic basin and 0.22 nM in NADW sampled in the Sargasso Sea[8]. Nickel concentration in Atlantic deep water has been determined to be 7.0 nM by Sclater et al.[16], and 3.4 nM by Bender and Gagner[8] compared with our value of 3.9 nM. No reliable zinc values for Atlantic deep waters have been found. The concentration found by Bruland et al.[18] in Pacific deep water, 9.3 nM, is higher than our value but that is to be expected since their value represents a much older bottom water.

CONCLUSIONS

Compared to other ocean deep waters the water of the Arctic basin displays low concentrations of cadmium, copper, nickel and zinc. These concentrations are representative for newly formed NADW.

Direct extractable iron is fairly constant throughout the water column. After acid leaching, much higher concentrations are found and a strong increase in the deepest layers is evident.

ACKNOWLEDGEMENTS

We thank the commander of YMER, Captain Anders Billström and his crew for skilful navigation and good support during sampling operations. Thanks also to the scientific leaders of the cruise, Professor Valter Schytt and Hans Dahlin.

This investigation was supported by the Natural Science Research Council of Sweden and by the Royal Society of Arts and Sciences of Göteborg.

REFERENCES

1. R.M. Moore, 1981: Trace metal distributions and behaviour in the Central Arctic Ocean. This volume.
2. D.C. Burrel, 1975: Environmental assessment of the north-eastern Gulf of Alaska: Chemical Oceanography (Trace metals) First year final reprint to the NOAA. Contract No. 03-5-022-36.
3. I.A. Campbell and D.H. Loring, 1980: Baseline levels of heavy metals in the waters and sediments of Baffin Bay. Mar. Poll. Bull. 11, 257-261.
4. B. Rudels and P.I. Sehlstedt, 1981: Personal communication.
5. L-G. Danielsson, B. Magnusson and S. Westerlund, 1978: An improved metal extraction procedure for the determination of trace metals in sea water by atomic absorption spectrometry with electrothermal atomization. Anal. Chim. Acta 98, 47-57.
6. J. Valderaama, 1981: Personnal communication.
7. E. Palosuo, 1981: The biologically important areas in the Arctic Ocean. Paper presented at POAC 81, Quebec, Canada.
8. M.L. Bender and C. Gagner, 1976: Dissolved copper, nickel and cadmium in the Sargasso Sea. J. of Mar. Res. 34,3; 327-339.
9. E.A. Boyle, F. Sclater and J.M. Edmond, 1976: On the marine geochemistry of cadmium. Nature, 263, 42-44.
10. K.W. Bruland, R.P. Franks, G.A. Knauer and J.H. Martin, 1979: Sampling and analytical methods for the determination of copper, cadmium, zinc and nickel at the nanogram per liter level in sea water. Anal. Chim. Acta, 105, 233-245.
11. K.W. Bruland, 1980: Oceanographic distributions of cadmium, zinc, nickel and copper in the North Pacific. Earth and Planetary Science Letters, 47, 176-198.

12. L-G. Danielsson, 1980: Cadmium, cobalt, copper, iron, lead, nickel, and zinc in Indian Ocean water. Mar. Chem. 8, 199-215.
13. J.H. Martin, K.W. Bruland and W.W. Broenhow, 1976: Cadmium transport in the California current. In: H.L. Windom R.A. Duce (eds.), "Marine Pollutant Transfer", Lexington Books, pp. 159-185.
14. G.A. Knauer, J.H. Martin, 1973: Seasonal variations of cadmium, copper, manganese, lead and zinc in water and phytoplankton in Montherey Bay, California. Limnol. and Oceanog. 18, 597-604.
15. R.M. Moore, 1978: The distribution of dissolved copper in the eastern Atlantic Ocean. Earth and Planetary Science Letters, 41, 461-468.
16. F.R. Sclater, E. Boyle and J.M. Edmond, 1976: On the marine geochemistry of nickel. Earth and Planetary Science Letters, 31, 119-128.
17. L. Edler, 1981: Personal communication.
18. K.W. Bruland, G.A. Knauer and J.H. Martin, 1978: Zinc in north-east Pacific water. Nature, 271, 741-743.

COPPER IN SUB-ARCTIC WATERS OF THE PACIFIC NORTHWEST

David T. Heggie

Graduate School of Oceanography
University of Rhode Island
Narragansett, Rhode Island 02882 U.S.A.

ABSTRACT

Copper concentration data from the north-central continental shelf-slope of the Bering Sea and nearshore regions of the western Gulf of Alaska are summarized. Copper concentrations which increase toward the sediments were found in a frontally confined continental shelf water mass of the Bering Sea, and also in deep waters of restricted circulation of a fjord on the north-western Gulf of Alaska coastline. These data are indicative of a benthic flux of copper to overlying waters. Mass balance estimates suggest that the benthic remineralization and flux to the overlying waters in the shallow shelf of the Bering Sea contributes significantly to the copper input to the deep Bering basin. However, in the fjord basin of the Gulf of Alaska, remineralized copper from inshore surface sediments was estimated at less than 20% of copper added to surface waters. Also, most copper input could be accounted for trapped in accumulating fjord sediments.

INTRODUCTION

The movement of elements through various reservoirs, where removal and remobilization reactions compete in estuarine and continental shelf areas, serves as a control on the flux from the continents to the ocean. The mobility of some metals such as manganese from estuarine sediments is well known [1,2]. The overall sense is that estuarine and coastal sediments by acting as efficient traps of most metals, exert a significant control on metal mobilities through interactions of those metals with particles[3]. This note summarizes copper data measured in the Bering Sea[4] and nearshore regions of the Gulf of Alaska[5] during 1975-77. Copper mass balances have been estimated and the two sites compared and contrasted with respect to the important processes controlling copper concentrations, distributions and the mass balances. Data demonstrate remobilizations from surface sediments in both regions but it is only in the Bering Sea where the shallow shelf sediment flux of copper is apparently important in the mass balance.

Samples were collected from twenty four stations (Fig. 1) in the north-central Bering Sea during July of 1977. Five samples to depths of 75 m were collected from each station with a Rosette sampler. Samples were filtered on board, via a non-contaminating Masterflex tube pump, through pre-cleaned in-line 0.4 μm Nuclepore filters. Samples were acidified to pH 2.5 immediately after filtering and analyzed at sea by the differential pulsed mode of anodic stripping voltammetry (DPASV) using in-situ plating techniques with triple-distilled mercury[6]. Gulf of Alaska (Fig. 1) samples were collected from two stations on seven occasions over a twelve month period 1975-76. Niskin bottles were hung on a CSTD hydrographic cable and 10-15 samples collected at depths up to 290 m. The depth-sampling frequency was increased toward the bottom to define gradients there. Samples were immediately filtered with enclosed polycarbonate filter towers through pre-cleaned 0.4 μm Nuclepore filters into pre-cleaned linear polyethylene bottles and acidified to a pH ∿ 2.5 with ultra-clean nitric acid. Copper determinations were made in the shore lab by DPASV with a PAR model 174 analyzer using rotating glassy carbon electrodes[5]. The procedural blank was generally less than 0.3 nmol kg^{-1}. The precision of the analyses at copper concentrations of 2-30 nmol kg^{-1} was generally ± 10%.

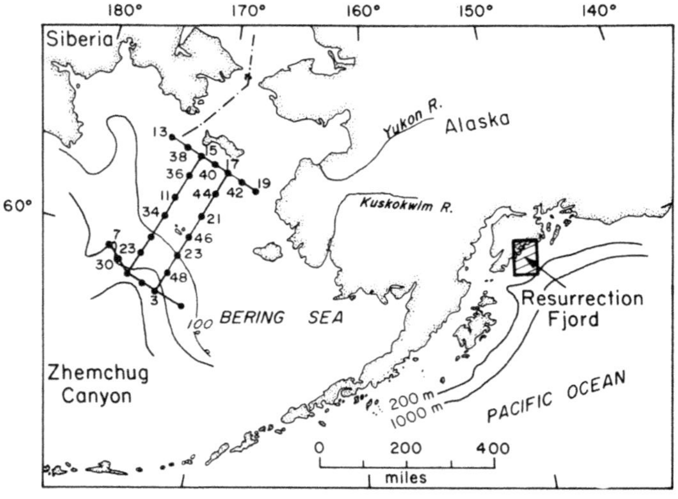

Fig. 1 The Pacific Northwest showing study area locations.

The Bering Sea Region

Copper concentrations in samples collected from the polygon study area have been presented previously[4]. That note made the following observations. 1) Copper concentrations over the Zhemchug Canyon where water depths are in excess of 1000 m were 2-4 nmol kg^{-1}. 2) Copper concentrations on the shallow, < 100 m, continental shelf were more variable 2-25 nmol kg^{-1} but were significantly higher than those measured offshore. 3) Copper-depth profiles on the shallow shelf were dominated by concentration increases toward the sediments.

Fig. 2a and b plot selected copper profiles from the slope and shelf waters. Fig. 2c plots mean and range of copper concentrations at all stations, and Fig. 3 plots "across-shelf" spatial copper concentration variations. This note examines those data further to evaluate a copper mass balance for the continental shelf-slope region. The hydrographic data from the north-central Bering Sea indicate two water types separated by a hydrographic front at 100 m [4]. Water properties in the study area are similar to those measured further south in the Bristol Bay region. Bristol Bay has been the focus of several years work from which has developed an understanding of the processes controlling the continental shelf property distributions and circulation[7,8,9,10,11,12]. The principal points from the above works which are pertinent here are summarized below.

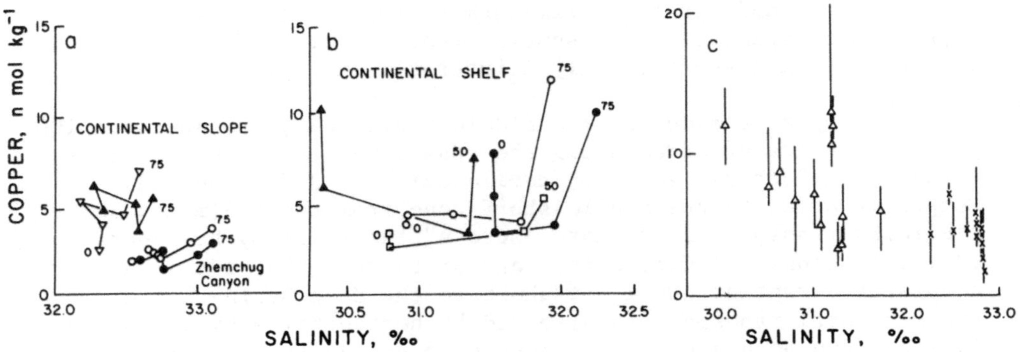

Fig. 2 a) Salinity-copper continental slope water mass S>32.5 ‰ stations 23,9; Zhemchug Canyon station 3,5. b) Salinity-copper continental shelf S<32.5 ‰ depth < 100 m, stations 11, 15, 34, 17. c) Salinity-copper data summary, copper mean and range for all stations; Δ continental shelf; X continental slope.

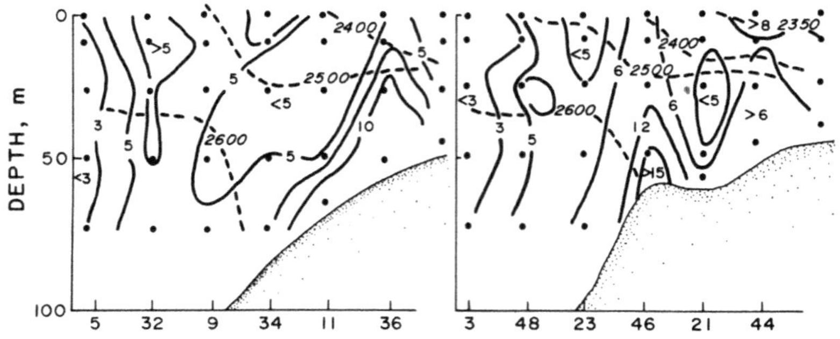

Fig. 3 Bering Sea "across shelf" copper profiles. Dashed lines are sigma-t isopleths.

1) The broad expanse ∿ 550 km of the continental shelf is divided into distinct hydrographic regimes separated by semi-permanent frontal systems. 2) A broad front ∿ 50 km wide exists over the 100 m isobath and separates the middle shelf zone, a two-layered stratified system, from Alaska Stream/Bering sea water offshore. 3) The front at 100 m controls the material transport between the middle shelf zone and Alaska Stream/Bering sea water. 4) Water which enters the Bering Sea passes between the Aleutian Islands and upwells on the outer shelf to drift parallel to the shelf break in a north-westerly direction. Water exits from the Bering Sea north through the Bering Strait. 5) There is no on-shelf advection. The lateral onshore flux of salt and offshore flux of freshwater are effected by horizontal eddy diffusion. 6) The cross-frontal exchange of water properties is effected by horizontal eddy diffusion across intrusions between water masses.

From the hydrographic framework a simple model evaluates a copper mass balance. The essential features of the model are listed below and schematically illustrated in Fig. 4.

1) Copper is added to the middle shelf zone, shoreward of 100 m, with freshwater added along the coastline. Riverine copper is transported "across" shelf by horizontal eddy diffusion. 2) Copper is added to the middle shelf zone from the sediments and similarly transported offshore "across" the shelf by horizontal eddy diffusion. 3) Copper is transported from the middle shelf, across the front at 100 m, offshore to the deep Bering basin. Cross-frontal transport is effected by horizontal eddy diffusion. 4) There are no additions nor losses of copper from the region in the "along" shelf direction.

For a steady state condition the transport of soluble copper across the front at 100 m is maintained by riverine and sediment-seawater inputs:

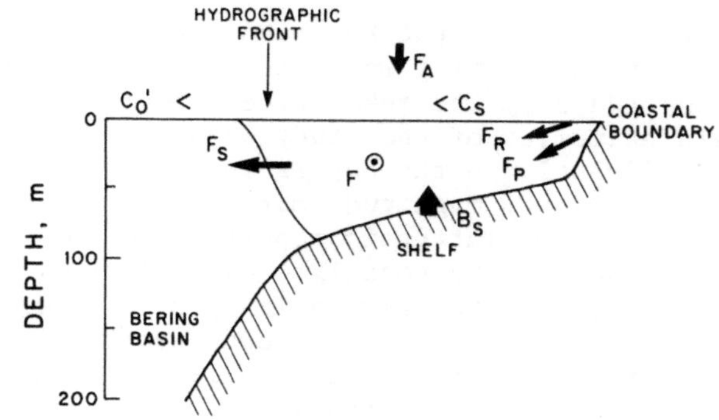

Fig. 4 Schematic of shelf-slope copper mass balance: River, F_r; River particulate, F_p; Benthic, B_s; Atmospheric, F_a; Cross-frontal, F_s; Along shelf, F; Middle Shelf copper, C_s; Oceanic copper, C_o.

$$F_s = F_r + B_s$$

where F_s is cross-frontal copper output from the middle-shelf to the deep Bering basin; F_r is riverine input to the shelf, and B_s is benthic input to the middle-shelf. Each of the above are evaluated below. The horizontal transport of copper from the shelf (F_s) to the deep Bering basin, in the absence of cross-frontal advection is:

$$F_s = \rho K_s \cdot {}^{ds}/_{dx} \cdot {}^{dc}/_{ds}$$

where K_s is cross-shelf horizontal eddy diffusion coefficient; ds/dx is cross-shelf salinity gradient; dc/ds is cross-shelf copper-salinity (Fig. 2c); and ρ is sea water density. Salt and freshwater mass balances have been used from several years data to compute cross-shelf[10] and cross-frontal[12] horizontal eddy diffusion coefficients. Median values of between 2 and 5 x 10^6 cm^2 s^{-1} were computed for the upper and lower layers of the water column[12] Because it is intended here only to describe the large features of the copper data, i.e. the concentration difference between the continental shelf and slope waters, a single value for the horizontal exchange coefficient has been chosen at 3 $\pm$ 1 x 10^6 cm^2 s^{-1}. For typical cross-shelf salinity gradients of 33°/oo - 30°/oo, or 3 g kg^{-1} and a copper-salinity ratio of 3 $\pm$ 1 nmol g^{-1} (Fig. 2c), the horizontal flux of copper to the deep Bering is estimated at 1.6 $\pm$ 0.7 x 10^4 nmol cm^{-2} y^{-1}. The total copper input to the Bering basin is computed assuming the flux from the study area to be representative of the perimeter of the Bering sea shelf ∿ 1000 km. Copper added across the 100 m contour therefore is 1.6 x 10^4 nmol cm^{-2} y^{-1} x 10^{12} cm^2, or 1.6 $\pm$ 0.7 x 10^{16} nmol y^{-1}.

Riverine copper (F_r) input has been estimated by two methods. Freshwater input to the Bering was computed[13] at 1.3×10^4 m^3 s^{-1}; 80% of which was added by the Yukon river about 100 km to the east of the northern extremity of the study area. For lesser rivers to contribute significantly to the copper input their concentrations would need be an order of magnitude more than that of the Yukon. The concentration of the latter is reported[14,29] at 31 nmol kg^{-1}. Most freshwater added at the coastline exits through the Bering Strait to the north, and the mean salinity of the broad middle shelf $\sim 31°/oo$ requires that marine source water, $S \sim 33°/oo$, be diluted only 5% with freshwater. One estimate of riverine input to the Bering basin is computed from:

$$F_R = \rho R \,(1 - S_s/S_o)\, Cr$$

where R is the annual freshwater flow; S_s the mean shelf salinity; S_o the oceanic salinity; C_r river copper concentration; and ρ the freshwater density. River copper contribution by this estimation therefore is 7.3×10^{14} nmol y^{-1}.

The cross-shelf flux of freshwater is by eddy diffusion[10,12] and the cross-shelf transport of copper associated with the freshwater transport computed from:

$$F_R^{1} = \rho K_s \frac{d[(1 - S_s/S_o)Cr]}{dx}$$

where K_s is the cross-shelf eddy diffusion coefficient, the bracketed term represents the cross-shelf copper gradient associated with freshwater content and X is the cross-shelf distance. For values of $K_s = 3 \pm 1 \times 10^6$ cm^2 s^{-1}, riverine copper (C_r) 31 nmol kg^{-1}, and a cross-shelf salinity difference of 3 g kg^{-1}, $S = 33°/oo$, $S_s = 30°/oo$ this estimate of cross-shelf copper input associated with freshwater is $4.6 \pm 1.5 \times 10^3$ nmol cm^{-2} y^{-1}, or $4.6 \pm 1.5 \times 10^{15}$ nmol y^{-1}. The above provide estimates of the river copper contribution at 5-30% of the estimate of the copper output from the shelf to the deep slope waters across the front at 100 m. From the shelf-slope mass balance equation the benthic input is computed to contribute not less than 1.2×10^{16} nmol y^{-1}. The continental shelf break bisects the Bering Sea into nearly equal areas of deep basin and shelf regimes estimated at 5×10^{15} cm^2. The shelf benthic copper flux is therefore estimated at not less than 2.4 nmol cm^{-2} y^{-1}, more appropriately 2-3 nmol cm^{-2} y^{-1}.

The deep Bering basin bounded to the north, east and west by the North American and Eurasian continents and contiguous shelf areas there, and to the south by the Aleutian Island chain is now considered. Integrated inputs from all sources must be balanced by burial removal in deep Bering Sea basin sediments for the maintenance of a steady state Bering basin seawater concentration. The principal sources of copper to the Bering Sea basin, it is assumed here, are the perimeter 1000 km of the broad continental shelf reaches and atmospheric deposition. The mass balance equation is:

$$A = F_s + F_a$$

where A is the accumulation rate of copper in deep reducing muds of the Bering basin > 3600 m; F_s is total dissolved input from the continental shelf, including riverine contribution, and F_a is atmospheric deposition. Copper in deep reducing layers of Bering Sea siliceous muds 53°32.4'N, 177°15.8'W has been measured[16] at about 0.2 ± 0.1 μmol g^{-1}. A mass accumulation rate, from ^{226}Ra and ^{14}C methods has been determined[17] for Bering Sea siliceous muds 57°07.0'N, 176°56.4'W, of 2×10^{-2} g cm^{-2} y^{-1}. A copper accumulation rate in deep Bering Sea basin sediments was estimated[4] from these data at 4.5 ± 2.1 nmol $cm^{-2} y^{-1}$. This latter constrains copper input to the Bering Sea from all sources. One estimate of atmospheric deposition of copper in Arctic latitudes has been made[31] at about 0.5 nmol cm^{-2} y^{-1}, and the contribution from the shelf* estimated at 3.2 ± 1.4 nmol $cm^{-2} y^{-1}$.

The Pacific Ocean as a source of copper to the Bering Sea is difficult to assess. Flow from the Pacific into the Bering is compensated for by outflow through the Bering Strait[11] estimated to be $1 - 2 \times 10^6$ m^3 s^{-1}. If the Pacific was the sole source of copper to the Bering Sea, the copper accumulating in deep sediments of the Bering must be accounted for by an enrichment of copper in Pacific waters over Bering Sea waters, i.e.

$$(C_p - C_B)\, T\rho = A \,.\, B$$

where C_p is the Pacific copper; C_B is Bering copper; T is mass transport, A is copper accumulation rate in deep Bering sea sediments; B is deep Bering basin surface area; and ρ is sea water density. The computed necessary enrichment of the principal source of Pacific water to the Bering Sea is approximately 0.7 - 1.5 nmol kg^{-1}. There are insufficient data to evaluate this further.

The comparisons here suggest that copper remobilized from shallow shelf sediments may be the significant contributor to the total copper input to the Bering Sea. The source of remobilized copper must be in particulate matter added at the coastline. The following, from a limited data set, estimates the particulate

* Total copper input off of the shelf $1.6 \pm 0.7 \times 10^{16}$ nmol y^{-1}; area of Bering deep basin estimated 5×10^{15} cm^2.

copper input to the north-eastern Bering Sea shelf to test if this latter can support the estimated shelf benthic flux of copper. Particulate matter copper has been measured in the Yukon River at approximately 5 mmol kg^{-1},[14] and in the Yukon River estuary[18] (S = 0-15°/oo) at 0.9 mmol kg^{-1}, and 1.0 mmol kg^{-1} (S = 15-25°/oo) and, in the northeastern Bering Sea in the vicinity of the eastern extremity of the study area at between 0.6 and 1.0 mmol kg^{-1}. Mass sedimentation rates in the northeastern Bering Sea, Norton Sound region, have been estimated[19] between 0.05 and 0.17 g cm^{-2} y^{-1} (median = 0.1 g cm^{-2} y^{-1}). These data provide a crude but upper limit to the particulate copper input to the sediments of the northeastern Bering Sea shelf of the order of 5 x 10^{18} nmol y^{-1} which is significantly greater than the estimated benthic input of copper, 1.2 x 10^{16} nmol y^{-1}. As the potentially exchangeable and available copper in Yukon River suspended matter has been reported[20] at approximately 13% of total copper, the particulate riverine copper input can readily support the sediment-seawater exchange of dissolved copper. Copper concentrations from various sources in different Bering Sea reservoirs are listed in Table 1, and components of the mass balance in Table 2.

Table 1 Copper in Bering Sea reservoirs.

Reservoir	Concentration	Reference
Seawaters		
GEOSECS 219 53°06'N, 177°17'W		
0 - 500 m	2.6 - 3.1 nmol kg^{-1}	30
500 m - 3700 m	2.8 - 6.0 nmol kg^{-1}	30
Zhemchug Canyon S>32.5°/oo	2 - 4 nmol kg^{-1}	4
water depth > 1000 m		
Cont. Slope water S>32.5°/oo	2 - 11 nmol kg^{-1}	4
water depth < 200 m		
Cont. Shelf water S<32.5°/oo	2 - 25 nmol kg^{-1}	4
water depth < 100 m		
Rivers		
Yukon	31 nmol kg^{-1}	14,29
Suspended Particulate		
Yukon River	5.0 μmol g^{-1}	14
Yukon estuary S, 0-15°/oo	0.9 μmol g^{-1}	18
Yukon estuary S, 15-25°/oo	1.0 μmol g^{-1}	18
North-east Bering shelf	0.6 - 1.0 μmol g^{-1}	18
Sediments		
Deep basin (53°32.4'N, 177°15.8'W)		
water depth 3930 m		
Deep reducing layer > 487 cm		
i) acid soluble	78.7 ± 66.1 nmol g^{-1}	16
ii) reducible	137 ± 85 nmol g^{-1}	

Table 2 Bering Sea copper mass balance.

Component	Integrated Value nmol y^{-1}	Flux nmol cm^{-2} y^{-1}
Input of copper from shelf to deep Bering basin (F_s)	$1.6 \pm 0.7 \times 10^{16}$	3.2 ± 1.4
Riverine copper input (F_r, F_r')	$0.7 - 4.6 \times 10^{15}$	0.1 - 0.9
Benthic input from shelf sediments (B_s)	$1 - 1.5 \times 10^{16}$	2 - 3
Atmospheric input (F_a)	2.5×10^{15}	0.5
Copper accumulation in deep basin Bering Sea sediments > 3600 m	$- 2.2 \times 10^{16}$	$- 4.5 \pm 2.1$
Yukon River particulate matter input (F_p)	5×10^{18}	1000

Area of Bering Sea deep basin 5×10^{15} cm^2
Area of Bering Sea shelf 5×10^{15} cm^2

Resurrection Fjord-Gulf of Alaska

The coastal regions of the western Gulf of Alaska are hydrographically distinct from those of the Bering Sea. Whereas the continental shelf of the latter is comprised of semi-permanent frontal systems, where cross-shelf transport of freshwater, salt, nutrients and copper is principally by horizontal eddy diffusion, the waters of the nearshore continental shelf of the Gulf of Alaska are advectively renewed year round. A coastal convergence and downwelling condition during winter maintains deep and bottom waters > 150 m well-flushed. During summer months these same waters are renewed by advection of more dense water from the continental slope onto the shelf[21,22]. This latter occurs as a result of the relaxation of the winter downwelling condition[22]. Also, during summer months freshwater added at the coastline significantly dilutes surface waters and remains trapped in a narrow lens, ∿ 10 km in width, adjacent to the coastline in a cyclonic motion around the Gulf[23]. The above hydrographic conditions exert controls on the seasonal and spatial copper concentrations in semi-enclosed fjord basins and the near shore shelf of the western Gulf of Alaska. Data from an inshore fjord and continental shelf stations have been described[5]. This section excerpts that work, computes estimates of copper input to surface waters and copper accumulating in fjord sediments.

Resurrection fjord is long ∿ 30 km and narrow ∿ 4 km. An inner basin 290 m is separated from the outer fjord and continental shelf by a sill at 183 m. The Resurrection river adds freshwater to the head of the fjord during oceanographic summer months, May - October. Representative seasonal copper and salinity data from the outer fjord-shelf station RES 4, and the inner fjord station RES-2.5 are shown in Fig. 5 and summarized below.

1) Copper concentrations are highest in surface waters S < 30 ‰ during summer-fall months of high freshwater runoff; 10-20 nmol kg^{-1}, and lowest 2-5 nmol kg^{-1} during oceanographic winter months March-April. 2) Copper concentrations in shelf and fjord bottom waters > 150 m were ∿ 4 nmol kg^{-1} during summer months as a result of bottom water renewals[21] by inflows of dense (sigma-t > 26.0) water from the outer continental shelf. 3) Seasonally, deep and bottom water copper concentrations on the shelf (RES-4) are ∿ 4 nmol kg^{-1}. The coastal convergence maintains the neashore shelf well-flushed[22]. 4) At the inner fjord station (RES-2.5) in contrast, isolated from the winter coastal downwelling condition by a sill barrier at 183 m, a more restricted circulation results[21] Copper concentrations increased in bottom waters from ∿ 4 nmol kg^{-1} subsequent to bottom water renewals in October to nearly 12 nmol kg^{-1} toward the end of the bottom water isolation period some six months later. 5) Surface sediment pore water concentrations were measured[5] on five occasions over a twelve month period and varied between 78 and 152 nmol (mean = 122 nmol kg^{-1}) about an order of magnitude higher than bottom water concentrations. These latter and overlying water column data increasing toward the sediments were indicative of a flux of copper from the mildly reducing surface sediments. The latter was estimated[5] from depth integration of bottom water copper concentration variations at 32 $\pm$ 12 nmol cm^{-2} y^{-1}.

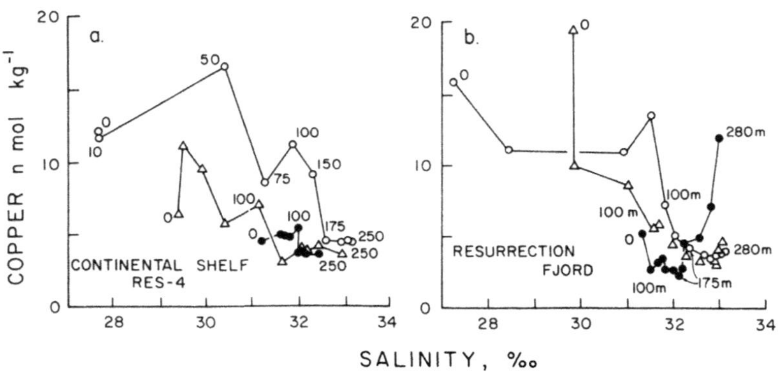

Fig. 5 a) Salinity-copper continental shelf Station RES-4; O Oct.; Δ Dec.; ● April. b) Salinity-copper Resurrection fjord. Station RES-2.5; O Oct.; Δ Dec.; ● April.

The following simple model, by comparing the measured copper concentration with the computed copper contribution from fjord source waters, computes an "excess" of copper in fjord surface < 150 m waters. The "excess" is that which cannot be accounted for by fjord source water contributions, and estimates total dissolved copper added annually to surface waters including copper contributed by freshwater runoffs. Within this estuarine system fjord source water was defined at the depth of minimum annual density variation $\sim$150 m[24]. Above 150 m surface waters are diluted during summer by riverine runoff, and below 150 m fjord bottom waters are advectively renewed also during summer months. Similarly, seasonal copper variations are minimal around mid-depths 100-150 m. Copper-salinity for samples collected from depths < 150 m are summarized in Fig. 6. Although the possibility of random contamination cannot be ruled out - two samples collected in July from the outer fjord had copper concentrations in excess of 40 nmol kg^{-1} - there are systematic features in these data. During the winter months of March-April when fjord waters are not stratified copper-depth concentration variations are minimal, and surface water copper concentrations approximate the year round source water concentrations Table 3. Winter surface water data and source water data are tightly grouped to the left hand corners of plots on Fig. 6. The fraction of source water in any segment of surface water < 150 is:

$$fo = S_n/S_o$$

where S_n is measured salinity; and S_o is source water salinity. Copper contributed by source water to any segment of surface water is fo Co, where Co is the mean annual copper concentration of fjord source water. At any given observation period the depth integrated excess of copper in surface waters therefore is:

$$Ct = \rho \sum_{150}^{o} (Cn - foCo)\,\Delta Z$$

where: Ct is "excess" copper nmol cm^{-2}; Cn is measured copper content of given depth interval; Co is source water copper, and ρ is seawater density.

Integration of "excess" copper computed that for two stations studied between 220 and 411 nmol cm^{-2} of copper was added annually to fjord surface waters. Copper added to estuarine waters may have one of two probable fates; 1) to be buried in fjord sediment, or 2) to escape rapid burial via recycling processes and be transported to the deep ocean where it ultimately will be trapped and buried in the sediments there. Because of insufficient data on flow in the Resurrection fjord system no estimate of copper transported from the fjord can be made. However, copper in the combined acid-reducible, and exchangeable fractions[5] and organic fractions[25] of deep 1 m Resurrection Bay sediments is computed at

Table 3 Copper in Gulf of Alaska-Resurrection Fjord.

Reservoir	Concentration	Reference
Seawaters		
Gulf of Alaska 400-1500 m (N = 5)	2.2 nmol kg^{-1}	5
Resurrection Fjord source water 100-150 m (N = 30)	5.2 ± 2.3 nmol kg^{-1}	5
Resurrection Fjord winter surfacewater < 100 m March-April (N = 15)	5.9 ± 3.3 nmol kg^{-1}	5
Resurrection Fjord summer surfacewater < 100 m June-October (N = 32)	12.8 ± 10.1 nmol kg^{-1}	5
River		
Resurrection River summer (N = 9)	1.9 - 16.2 nmol kg^{-1} mean 7.3 ± 5.3 nmol kg^{-1}	5
Resurrection River winter (N = 4)	0.8 - 1.4 nmol kg^{-1} mean 1.1 ± 0.3 nmol kg^{-1}	5
Pore waters		
Resurrection Fjord pore waters 0-5 cm	78 - 152 nmol kg^{-1} mean 122 ± 23 nmol kg^{-1}	5
Sediments		
*Resurrection Fjord duplicate analyses samples < 90 cm	335 - 369 nmol g^{-1} mean 334 ± 1 nmol g^{-1}	5
**Resurrection Fjord duplicate analyses 4 samples < 50 cm	90 - 110 nmol g^{-1}	25
*Gulf of Alaska Continental shelf16 samples duplicate analyses surface < 5 cm	178 ± 14 nmol g^{-1}	5

*Combined hydroxylamine-hydrochloride and acetic acid extracts[32].
**hydrogen peroxide extract[25].

0.44 μmol g^{-1} (Table 3) and these data combined with sedimentation rates in Alaskan fjords[25] of ∿ 1 cm y^{-1} have been used to estimate the copper accumulation rate in fjord sediments:

$$A = C.S. (1-\phi)D$$

where: A is copper accumulation rate; C is copper concentration in sediments; S is sedimentation rate; ϕ is porosity; and D is sediment density. For values of porosity 0.7 and sediment density 2.65 g cm^{-3} the copper accumulation rate in Resurrection Bay sedi-

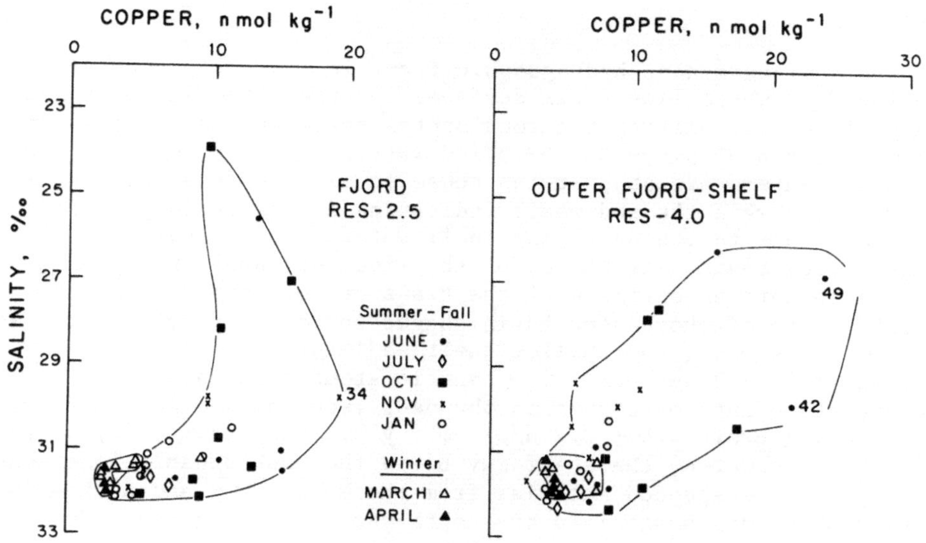

Fig. 6 Seasonal surface water < 150 m copper-salinity, RES-4 and RES-2.5.

ments therefore is approximately 350 nmol cm^{-2} y^{-1}, a value comparable to the estimated inputs to surface waters of 220-411 nmol cm^{-2} y^{-1}. The comparison suggests that most added copper is removed to the fjord sediments. Copper concentrations in the fjord sediments are about double those of the adjacent continental shelf, Table 3. The cyclonic flow of freshwater at the coastline[23] may provide the physical barrier to trap copper added with freshwater in the nearshore region. The chemical removal mechanism is unknown, but effective and efficient removal mechanisms by iron-organic colloids in estuaries has been demonstrated[15]. Copper in reservoirs of the Resurrection fjord system are summarized in Table 3, and the mass balance components in Table 4.

Table 4 Resurrection Fjord copper mass balance.

Component	Flux (nmol cm^{-2} y^{-1})
Annual input to surface water < 150 m	220 - 411
Benthic input inner fjord (RES-2.5)	32 $\pm$ 12
Accumulation rate in fjord sediments	$\sim$ 350

SUMMARY

A semi-permanent hydrographic front in the north-central Bering Sea, where iso-pycnal surfaces intersect the sediment-sea-water interface distinguish continental shelf waters $S < 32.5^{o}/oo$ from continental slope $S > 32.5^{o}/oo$ waters. Copper concentrations in shelf waters are about twice those in slope waters and copper increases toward the sediments indicating a flux of copper from the sediments to the overlying shelf waters and offshore to the deep Bering basin. Estimates of the riverine input of copper cannot support an estimate of the transport of copper from the shelf to the offshore deep basin suggesting a contribution from an additional source, the shallow shelf sediments. This latter is estimated at 2-3 nmol $cm^{-2}\ y^{-1}$, and is about comparable to the accumulation rate of copper in the deep reducing sediments of the deep Bering basin 4.5 ± 2.2 nmol $cm^{-2}\ y^{-1}$. The sedimentary source of mobile copper on the shelf may be in the exchangeable fractions of copper in suspended material from the Yukon [14,18] and Kuskokwim rivers which discharge into the north-eastern Bering Sea. Total particulate matter copper input was estimated at more than sufficient to support the estimated benthic flux. Migration of dissolved chemical species, whose source is in continental sediments, from shelf areas to oceanic basins has been documented in the case of ^{228}Ra[26,27]. Further cross-shelf metal profiles, ^{228}Ra distributions, pore water copper profiles and overlying water ^{222}Rn profiles would aid in testing the preliminary mass balance calculations presented here.

Most copper input to the western Gulf of Alaska, assuming the work in Resurrection bay is typical of this coastline, appears associated with fluvial input during summer months. Although there is evidence for a remobilization and vigorous recycling of copper from surface sediments of inshore fjord basins during winter months[5] the benthic flux is estimated at less than 20% of the copper input to surface waters during summer months (220-411 nmol $cm^{-2}y^{-1}$). Further, surface water input is comparable to an estimate of the copper accumulation rate in fjord sediments of ~ 350 nmol $cm^{-2}y^{-1}$. Apparently little of the integrated copper input to surface fjord water escapes the coastal zone. The copper accumulation rate in Pacific oceanic sediments has been estimated[28] at ~ 3 nmol $cm^{-2}y^{-1}$.

The isolated waters of deep fjord basins of the Alaska coastline are similar to the frontally confined shallow shelf waters of the Bering Sea in that recycling of copper from surface sediments was evident from both areas. The shallow shelf sediments of the Bering Sea contribute significantly to the dissolved copper input to the deep Bering basin but in contrast most added, including remobilized copper, apparently remains trapped in deep fjord basins and nearshore regions of the western Gulf of Alaska.

REFERENCES

1. Graham. W.F., M.E. Bender and G.P. Klinkhammer, 1976: Manganese in Narragansett Bay. Limnol. Oceanog., 21, 665-673.
2. Eaton, A.: The impact of anoxia on Mn fluxes in the Chesapeake Bay. Geochim. et Cosmochim. Acta., 43, 429-432.
3. Turekian, K.K., 1977: The fate of metals in the oceans. Geochim. et Cosmochim. Acta., 41, 1139-1144.
4. Heggie., D.T.: Copper in surface waters of the Bering Sea. (Accepted for publication Geochim. et Cosmochim. Acta.)
5. Heggie, D.T., 1977: Copper in the sea: a physical-chemical study of reservoirs fluxes and pathways in an Alaskan fjord. Ph.D. dissertation, U. of Alaska. 217 pp.
6. Florence, T.M., 1970: Anodic stripping voltammetry with a glassy carbon electrode mercury plated in-situ. J. Electroanal. Chem., 27, 273-281.
7. Kinder, T.H., L.K. Coachman and J.A. Galt, 1975: The Bering Sea current system. J. Phys. Oceanog., 5, 231-244.
8. Kinder, T.H. and L.K. Coachman, 1978: The front overlaying the continental slope in the eastern Bering Sea. J. Geophys. Res., 83, 4551-4559.
9. Schumacher, J.D., T.H. Kinder, D.J. Pashinski and R.L. Charnell, 1979: A structural front over the continental shelf of the eastern Bering Sea. J. Phys. Oceanog., 9, 79-87.
10. Coachman, L.K. and R.L. Charnell, 1979: On lateral water mass interaction - a case study, Bristol Bay, Alaska. J. Phys. Oceanog., 9, 278-297.
11. Coachman, L.K., K. Aagaard and R.B. Tripp, 1975: Bering Strait: The Regional Physical Oceanography. 172 pp. U. of Washington Press, Seattle.
12. Coachman, L.K. and J.J. Walsh, 1981: A diffusion model of cross-shelf exchange of nutrients in the south-eastern Bering Sea. Deep Sea Res., 28A, No. 8, 819-846.
13. Roden, G.I., 1967: On river discharge into the northeastern Pacific and Bering Sea. J. Geophys. Res., 72, 5613-5629.
14. Gibbs, R.J., 1977: Transport phases of transition metals in the Amazon and Yukon Rivers. Geol. Soc. Amer. Bull., 88, 829-843.
15. Sholkovitz, E.R., 1978: The flocculation of dissolved Fe, Mn, Al, Cu, Ni, Co and Cd during estuarine mixing. Earth Planet. Sci. Letters, 41, 77-86.
16. Tsunogai, S., I. Yonemarie and M. Kusakake, 1979: Post depositional migration of Cu, Zn, Ni, Co and Pb and Ba in deep sea sediments. Geochem. J., 13, 239-252.
17. Tsunogai, S. and Masatoshi Yomada, 1979: ^{226}Ra in Bering Sea sediment and its application as a geochronometer. Geochem. J., 13, 231-238.

18. Feely, R.A., G.J. Massoth and A.J. Paulson, 1981: The distribution and elemental composition of suspended particulate matter in Norton Sound and the Northeastern Bering Sea shelf: Implications for Mn and Zn recycling in coastal waters. In: "The Eastern Bering Sea Shelf, Oceanography and Resources". D.W. Hood and A.J. Calder, eds. Vol. 1, U.S. Dept. of Commerce, Washington, D.C., pp. 321-338.
19. Nelson, C.H. and J.S. Creager, 1977: Displacement of Yukon-derived sediment from Bering Sea to Chukchi Sea during Halocene Time. Geology, 5, 141-146.
20. Gibbs, R.J., 1973: Mechanisms of trace metal transport in rivers. Science, 180, 71-73.
21. Heggie, D.T. and D.C. Burrell, 1981: Deep-water renewals and oxygen consumption in an Alaskan fjord. Est. Coastal Shelf. Sci., 13, 83-99.
22. Royer, T.C., 1975: Seasonal variations of waters in the northern Gulf of Alaska. Deep Sea Res., 22, 403-416.
23. Royer, T.C., 1979: On the effect of precipitation and runoff on coastal circulation in the Gulf of Alaska. J. Phys. Oceanogr., 9, 555-563.
24. Muench, R.D. and D.T. Heggie, 1978: Deep Water Exchange in Alaskan Sub-arctic Fjords. In: "Estuarine Transport Processes", B. Kjerfve, ed. Belle Baruch Library Marine Sci. No. 7, U. South Carolina. pp. 239-268.
25. Burrel, D.C.: personal communication.
26. Feely, H.W., G.W. Kipphut, R.M. Frier and C. Kent, 1980: ^{228}Ra and ^{228}Th in coastal waters. Est. Coastal Mar. Sci., 11, 179-205.
27. Li, Y.H., H.W. Feely and P. Santshi, 1979: ^{228}Th - ^{228}Ra radioactive disequilibrium on the New York Bight and its implications for coastal pollution. Earth Planet. Sci. Letters, 42, 13-26.
28. Bostrom, K., T. Kraesser and S. Gantner, 1973: Provenance and accumulation rates of opaline silica, Al, Ti, Fe, Mn, Cu, Ni and Co in Pacific pelagic sediments. Chem. Geol., 11, 123-148.
29. Boyle, E.A., 1979: Copper in natural waters. In: "Copper in the Environment, Part I", J.O. Nriagru, ed. John Wiley & Sons, pp. 77-88.
30. Boyle, E.A., 1977: The distribution of dissolved copper in the Pacific. Earth Planet. Sci. Letters, 37, 38-54.
31. Rahn, K.A.: personal communication.
32. Chester, R. and M.J. Hughes, 1967: A chemical technique for the separation of ferro-manganese minerals, carbonate minerals and absorbed trace metal elements from pelagic sediments. Chem. Geol., 2, 249-262.

LOW LEVEL DETERMINATION OF TRACE METALS IN ARCTIC SEA WATER AND SNOW BY DIFFERENTIAL PULSE ANODIC STRIPPING VOLTAMMETRY

Léon Mart, Hans-Wolfgang Nürnberg and David Dyrssen*

Institute of Applied Physical Chemistry
Chemistry Department, Nuclear Research Center
Juelich, Federal Republic of Germany

*Department of Analytical and Marine Chemistry
Chalmers University of Technology and University of Gothenburg, Göteborg, Sweden

ABSTRACT

The work described took place on board the Swedish icebreaker "Ymer" during the second leg of the arctic expedition "Ymer 80", from August 9th to September 25th, 1980. Sea water from different depths at various locations between Greenland and Franz Josef Land was sampled. Surface water and samples from 10-60 m depth were taken from ice floes. Deep-sea samples were collected by an automatic device working after the principle of the CIT sampler. Snow samples, mainly fresh fallen snow, were collected into widemouth bottles of 0.5 l volume.

Analysis was performed directly on board in a clean room container. An advanced electrochemical determination procedure, differential pulse anodic stripping voltammetry at a rotating glassy carbon mercury film electrode was used for Cd, Pb and Cu analysis. Surface samples of the Arctic Sea with a mean of 0.078 nmol kg^{-1} Cd, 0.072 nmol kg^{-1} Pb, 1.49 nmol kg^{-1} Cu and 1.70 nmol kg^{-1} Ni fit into the usual pattern of unpolluted oceanographic areas. Cadmium correlates well with nutrients and shows a 2-3 fold increase with depth, thus behaving in general analogously as in the Atlantic.

Fresh fallen snow yielded trace metal results by far lower then data reported from comparable areas like Greenland. Only data of condensed and agglomerated snow from the last winter season 79/80 agreed well with Pb results from other authors.

All the reported nickel and cobalt results have been analyzed at home, using a new, very sensitive voltammetric method based on interfacial preconcentration as adsorbed dimethylglyoxime chelates.

INTRODUCTION

Scientists have noted anthropogenic pollution via the atmosphere from one pole to the other. One of the nuclear bomb tests at Bikini-Eniwetok in 1954 can be read by its radioactive fall-out in contaminated snow layers of Central Antarctica. Analyses by Murozumi et al.[1] demonstrate that the lead content of annual layers of ice from the interior of Antarctica and Greenland is increasing sharply after 1940, coinciding fairly well with the introduction of high octane leaded fuel on the American market. Another proof for the necessity to consider the flux of trace metals into the atmosphere on a global scale is given by the Arctic Research Laboratory at Barrow, Alaska. High trace metal pollution values are found during winter time, obviously originating from highly industrialized areas in Western Europe.

Further information is needed to assess the long range transport of pollutant trace metals. The Arctic Ocean, although less influenced by human activities than most other regions of the world, is an extremely sensitive environment with little ecological regeneration capacity, due to its cold climate. Studies on trace metals in snow and ice layers in Arctic regions[2,3,4,5] have been carried out in Greenland. As proposed by Boutron[5] such results could be possibly influenced by landborne dust particles of local origin, e.g. from ice-free coastal areas of Greenland. This would suggest that historical records of trace metals in snow and ice layers on Greenland are not as representative of the

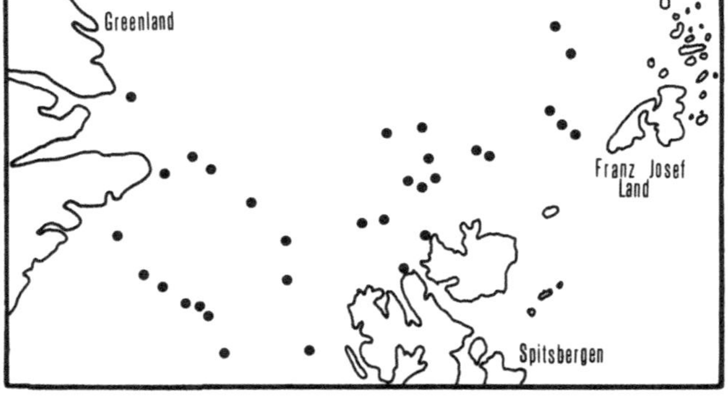

Fig. 1 Sampling stations in Arctic Oceans

Arctic as thought until now. The "Ymer 80" expedition offered for the first time the possibility to collect fresh fallen snow from ice floes in regions that have not yet been touched by man and where landborne influences are minimized. Abundant snowfall over the first weeks of the beginning winter season could be registrated.

The Arctic Ocean receives Pacific water with low salinity and high silica through the Bering Strait and North Atlantic water (Gulf Stream water) by the West Spitsbergen current. The outflows into the North Atlantic occur through the Canadian Archipelago into the Baffin Bay and by the East Greenland current. The main ice export takes place by the East Greenland current. There is a slow drift of ice over the North Pole towards Greenland, but some of this ice is caught in the Beaufort gyre north of Canada. The surface water in the Arctic Ocean is a mixture of river water, Bering Strait water and Gulf Stream water, which is cooled by the ice to -1.8 to -1.9°. Below the surface water, which extends down to 100 to 200 m, one finds North Atlantic water with higher temperatures and salinities. The deep water in the Eurasian Basin is probably formed wintertime in the Greenland Sea. Ice formation also creates heavy water by brine formation. Little is known about the vertical exchange, vital for growing of plankton and about the distribution of trace metals within the water column.

During "Ymer 80" an automatic sampler, working after the principles of the CIT sampler designed by Schaule and Patterson[6], was used for collecting water samples. As could be proved in previous expeditions[7], such samples for trace metal analysis are by far more reliable and consistent than deep sea samples collected by conventional, commercially available samplers.

EXPERIMENTAL

Cleaning Procedures

Sampling bottles and plastic bags, both made of high pressure polyethylene, were rinsed after a procedure described in detail elsewhere[8] and whose sequence is given here as a summary. First clean with detergent in a laboratory washing machine, rinse with deionized water, soak in hot (about 60°) acid bath, beginning with 20% hydrochloric acid, reagent grade, followed by two further acid baths of lower concentration, the last being of Merck, Suprapur quality or equivalent. The bottles are then filled with dilute hydrochloric acid, Merck, Suprapur, this operation being carried out under a clean bench. They are soaked once more in dilute acid and heated up. Empty bottles under clean bench, rinse and fill them up with very pure water (pH 2). Bottles are wrapped into two polyethylene bags. For transport purposes, lots of ten bottles are enclosed hermetically into a larger bag.

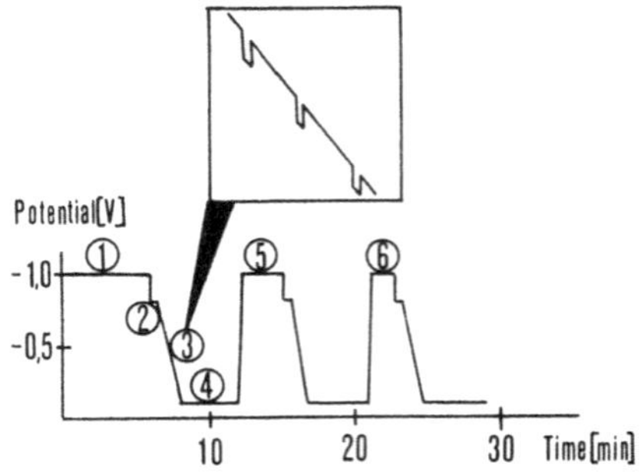

Fig. 2 Voltage sequence and details of differential pulse scan. 1. Cathodic deposition period. 2. Rest time. 3. Linear voltage ramp with pulses (viz. detail) of 50 mV height; pulse duration: 29 ms; pulse repetition time: 240 ms. 4. Standard addition followed by voltage patterns with systematically reduced deposition times (5,6).

Analytical Methods

Electrochemical determination of Cd, Pb and Cu (Bi could not be detected in any sample) was performed by DPASV, e.g. differential pulse anodic stripping voltammetry. After a previous plating of the trace metals into a mercury film on a rotating electrode with highly polished glassy carbon as substrate, they are stripped in the differential pulse mode. In-situ plating, as proposed first by Florence[7], was used.

Analysis on Board

The polarograph used on board "Ymer" was a PAR 174 A, Princeton Applied Research, Princeton, N.J., and a y-t recorder BD8 Kipp & Zonen, Delft, Netherlands. The polarograph had been adapted to analysis of trace metals at extremely low levels, as has been described elsewhere[10]. Two rotating electrodes of our own construction[11] could be connected to the poloargraph. One electrode was used for preparative purposes like polishing and outgassing of a sample, whilst the other was connected to the polarograph for a running analysis. The electrodes were placed under a clean bench, class 100 (Fig. 3).

For contamination check of the cell and electrodes, a "blank" with ultrapure water was run with a short plating time of 5 min. If the Pb blank was below 1 or 2 ng kg^{-1}, the cell was filled with the sample to be analyzed.

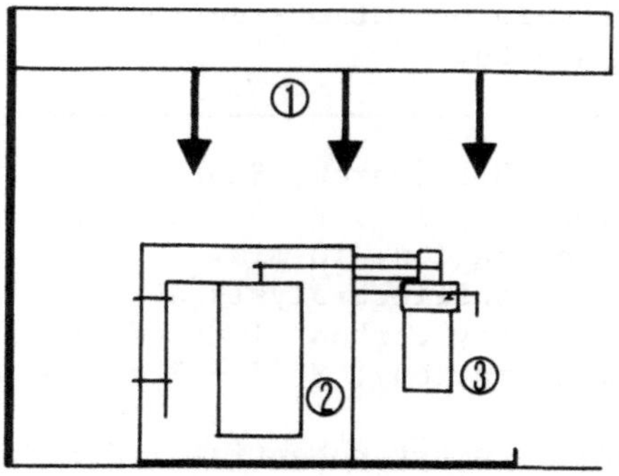

Fig. 3 Principle of rotating electrode designed for clean bench working. 1. Clean air current. 2. Motor and electric connections completely separated from clean air area. 3. Voltammetric cell. 4. Operator's place.

The film formed during the blank test was left on the glassy carbon surface. Under addition of further mercury nitrate, 20 to 50 1 of a 5000 mg l^{-1} solution are added to 50 ml, analysis was performed. For samples where copper was expected to be close to the determination limit of 10 ng kg^{-1}, only the film plated during the blank test run was used, thus avoiding an increasing slope by the growth of the mercury film. Details of the whole analytical procedure have been described elsewhere[11] and are summarized in Table 1.

The advantage of the voltammetric method, besides its accuracy and extremely high sensitivity, is that it could be run while the ship was moving, even under ice breaking conditions. The only problem arose from the mercury contact with the rotating axle of the electrode (detail 6 in Fig. 4) which literally jumped out of its housing at hard ice contacts.

Analytical Work in Home Laboratory

A new very sensitive voltammetric method for the determination of nickel and cobalt, developed only recently[12] has been applied.

The pH of the sample is adjusted to 9.2-9.3 by adding a NH_3 /NH_4Cl-buffer. Optimal ammonia buffer concentration is 0.1 M for nickel concentrations below 10 µg kg^{-1}. 20 µl of a 0.1 M dimethylglyoxime solution in ethanol is added to a 50 ml sample. The analyte is deaerated for 10 min. At the working electrode, a hanging mercury drop electrode, a potential of -0.7 V is adjusted and the

Table 1 Sequence of main operations during voltammetric analysis for Pb, Cd and Cu.

1) Sample acidified to pH 2 (Merck, Suprapur HCl); spike with $Hg(NO_3)_2$: 20 µl (5000 mg l^{-1}) to 50 ml sample.
2) Purge with N_2 for deaeration, 10 min.
3) Potential-controlled preelectrolysis at - 1 V for a definite time, e.g. 6 min; glassy carbon electrode rotates; Hg^{++} + 2 e^- ► Hg^o (film formation); Me^{++} + 2 e^- ► Me^o; Hg^o + Me^o ► Me-Amalgam.
4) Stop rotation, adjust start potential: -0.8 V; wait 30 sec.
5) Start potential scan from -0.8 V to -0.1 V; differential pulse mode. Me-Amalgam ► Me^{++} + Hg^o + 2 e^-; voltammogramme is recorded.
6) Clean Hg-film at -0.1 V; electrode rotates; add known amount of Me^{++}.
7) Repeat preelectrolysis: step 3 to step 6. Reduce plating time to one half, e.g. 3 min.
8) Repeat preelectrolysis: step 3 to step 6. Reduce plating time to one third, e.g. 2 min.
9) Evaluation.

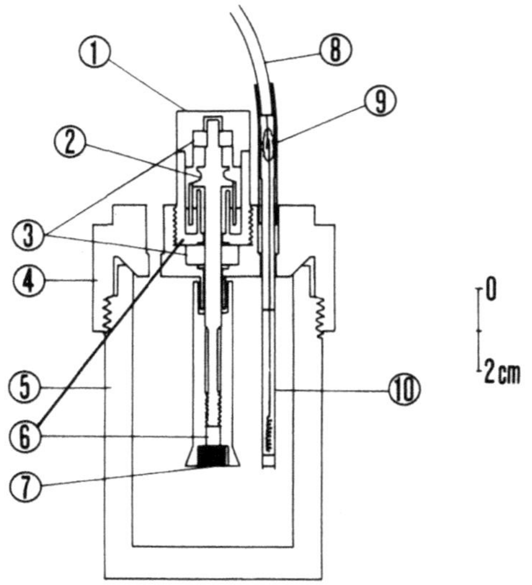

Fig. 4 Details of voltammetric cell, corresponding to 3. in Fig. 3. 1. Axle cover removable for change of driving belt. 2. Stainless steel axle with driving wheel. 3. Ball bearings. 4. Cell cover, machined teflon. 5. Voltammetric cell, machined teflon. 6. Mercury for electric contact. 8. Shielded electric connection to counter electrode. 9. Soldered connection enclosed with glue for avoidance of pollution of electrolyte in counter electrode tubing. 10. Heat-shrinking teflon tubing with inserted Vycor frit.

nickel-dimethylglyoxime complex is adsorbed at the mercury surface. To speed up mass transfer the solution is stirred with a magnetic bar. Depending on the concentration of nickel (and cobalt) 5 to 10 min. of adsorption time are needed. After a rest period of 30 s the voltammogram is recorded by scanning the potential into negative direction. Concentrations have to be evaluated by standard addition. Details of this new method are presented elsewhere[12].

A preparative step, that could not be carried out on board the ship, is UV irradiation of samples in order to decompose strong organic complexes and to leach the trace metals trapped in them. As experience has shown from previous expeditions, only river, estuarine water and waters with high biological activity need this preparative step. Deep sea waters and molten snow could be measured directly. Details about this procedure have been described elsewhere [11]. The principle of the UV-irradiation, avoiding pollution by the ambient atmosphere, is depicted in Fig. 5.

The cell is filled under the clean bench, covered and separated from the outer atmosphere by a water bath. It is irradiated and opened again under clean bench.

All the surface water samples collected during "Ymer 80" had to be irradiated by this method.

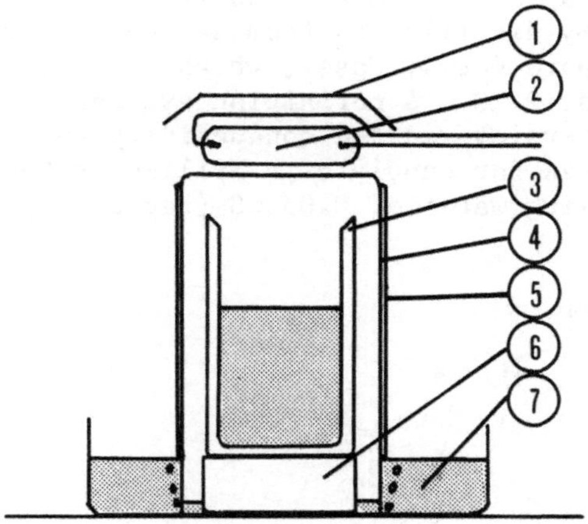

Fig. 5 UV irradiation device. 1. Reflector. 2. UV-lamp. 3. Voltammetric cell filled with sample. 4. Quartz beaker. 5. Aluminium foil, the area of the foil determines the temperature under the beaker. 6. Distance block from teflon. 7. Glass dish with water for separating the cell from the outer atmosphere.

Contamination Control

As already mentioned before contamination by the atmosphere of the ship is one of the most serious problems in both sampling and analyzing. All the preparative steps prior to field missions and to analysis were run in a clean air container laboratory. The analysis itself was carried out under a clean bench. Dust penetration was lowered by a positive pressure (4 mm water column) of filtrated air.

Direct analysis of most samples and above all, control of each preparative and sampling step, permitted to detect at once potential contamination sources. Thus, for example, collection of snow while the ship was moving in comparison to collection of the same fresh fallen snow from an ice floe demonstrated that it is extremely difficult to find a convenient place on board a ship for reliable sampling of rain and snow. Therefore, snow was only collected during station time away from the ship on ice floes.

After sampling, a control of the inner precleaned bag and the outside of the sampling bottle was made by shaking the whole with 25 ml of pure acidified water (pH 1). After dilution to 50 ml, several analyses of this rinsing water were run. In 90% of these analyses, the amount of leached lead, the most severe contaminant, was below 1.5 ng.

On previous expeditions, a commercially available double quartz still was used for clean water supply, the system being somewhat problematic under rough sea conditions. Thus, a more comfortable and rapid procedure of water deionization was used. Tap water, made by distillation from sea water, fed a Milli-RO unit, from Millipore Corp., Mass., which supplied water with a conductivity about 2 μS. A normal ion exchange cartridge fed by this Milli-RO water lowered the conductivity to about 0.2 μS. This water was used for supply of a Milli-Q system, Millipore Corp., Mass., giving water of 0.05 μS (begin of journey) to 0.09

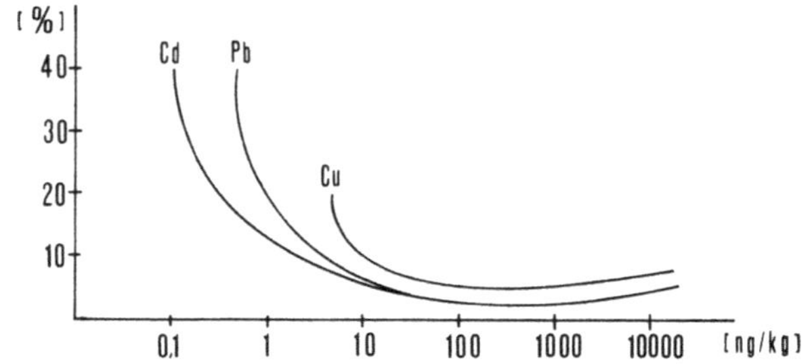

Fig. 6 Precision and actual determination limits.

Table 2 Snow sampled on board "Ymer" and on an ice floe outside the area polluted by the ship. Position: 81° 43'N: 03° 31'W.

	Cd		Pb		Cu	
	ng kg^{-1}	nmol kg^{-1}	ng kg^{-1}	nmol kg^{-1}	ng kg^{-1}	nmol kg^{-1}
on board "Ymer"	3	0.026	380	1.83	400	6.34
on ice floe	0.3	0.003	12	0.06	15	0.24

Table 3 Trace metal contents in deionized water.

	Cd		Pb		Cu	
	ng kg^{-1}	nmol kg^{-1}	ng kg^{-1}	nmol kg^{-1}	ng kg^{-1}	nmol kg^{-1}
begin of journey	0.1	0.9	0.8	3.9	10	158
end of journey	0.3	2.7	2.0	9.7	10	158

µS. The trace metal content of this water is obviously depending on the quality of the charge of the ion exchange cartridges. Results of analyses are listed in Table 3.

The sampling bottles and the electrolysis cells were always given a final rinsing with the sample itself. Thus the trace metal content of the deionized water did not contribute a measurable analytical blank.

SAMPLING PROCEDURES

For facilitating ice work, sampling of snow and taking air samples during station, the icebreaker ran its bow upwind onto a large ice floe. During a station, oceanographic work could be done by the hydrographic winches through an area kept more or less ice free by the ship's rear propellers. For ice work and sampling purposes, scientists could leave the ship by means of an elevator situated in the front of the ship. The ship was left upwind into a distance of about 200 m. Larger distances were prohibited because of the presence of polar bears.

Water samples were taken in open leads of various sizes, taking into account the wind direction. Surface waters were collected with a polyethylene bottle, 0.5 l, within a holder at the end of a telescopic bar that could be extended to a maximum length of about 3 m (Fig. 7). Filling the bottles directly by hand out from the edge of a floe was generally too risky. The holder at the end of the bar was loaded with a clean polyethylene bottle that had been enclosed into two polyethylene bags. Although three bags might be safer for reducing gradually the contamination of the outer parts of the bottle, two of them were merely enough to cope with them in the prevailing strong wind. The bottle was immersed quickly through the surface layer and, most important, moved steadily through a water layer of 0.5 to 1 m depth. After rinsing twice, the bottle was filled definitely. Each of the three immersions had to be effectuated at a new place, thus avoiding to collect contamination left by the bar itself. During manipulations, the bottle was always kept into the wind to avoid contamination originating from the clothing of the operator. Contamination by unavoidable close contact was reduced by wearing clean polyethylene gloves and clean room arm sheets. Intermediate waters down to 60 m were obtained with a commercially available Go-Flo sampler from General Oceanics, Miami. The external rubber spring had been replaced by a steel spring. The sampler was cleaned, prepared and loaded within the clean room container and transported in a plastic bag. At the sampling site, two plexiglas-encasted weights at the end of the hydro-line were fixed at the lower third of the sampler, so that the opening of the sampler was the lowest point during launching. The hydro-line was operated by hand and the Go-Flo was triggered during continuous lowering, thus, approaching the sampling principle of Patterson and Schaule[6,13], i.e. sampling dynamically at the end of the hydrowire. The closing of the sampler could be felt distinctly and the actual sampling depth could be read from the hydro-line. Generally the sampler was emptied in the clean room area.

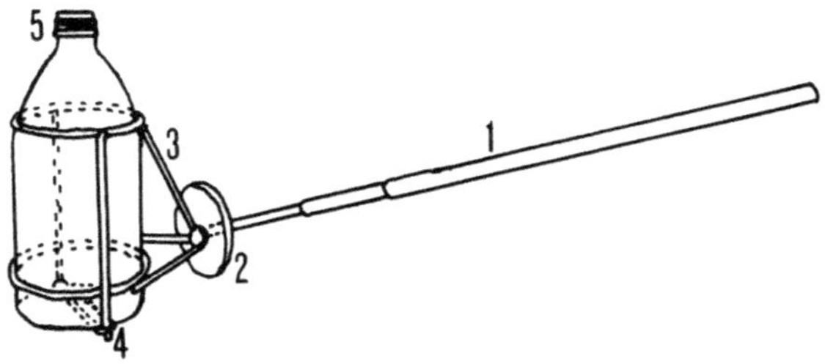

Fig. 7 Sampling gear. 1. Telescope bar, fiberglass, maximum length 3.5 m. 2. Polyethylene disk. 3. Holder coated with nylon. 4. Fastener made of silicone tubing. 5. Sampling bottle.

Deep sea water samples were collected by a sampling device that worked after the principles described by Schaule and Patterson[6,13]. Technical details have been published elsewhere[14]. At a given depth, the sampler is triggered automatically. A protective cup moves sidewards, whilst clean water is pressed through the entry port. A piston moves up, sucking water into a polyethylene bag. After filling, the protective cup moves back, closing the entry port. This sequence of movements is triggered in three different depths. Sampling of water polluted by the steel frame is avoided by moving the sampler continuously down through uncontaminated water while filling. As proposed by Schaule[13], the lowering speed is reduced from about 100 m per minute to 30 m per minute during sampling, thus yielding a sample that represents an integration of a 15 m water column. Handling on a generally

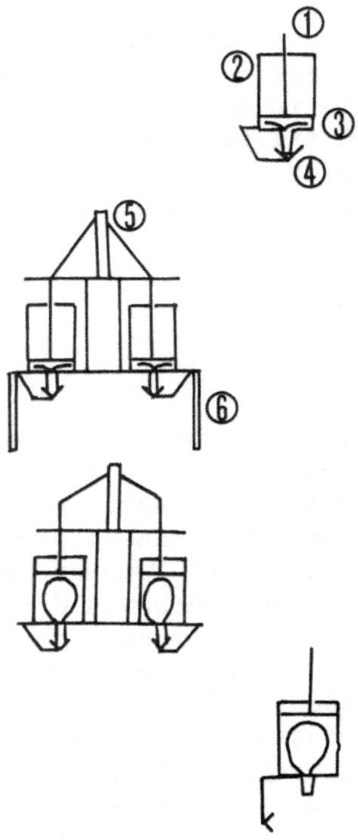

Clean laboratory: preparation of sampling units.
1. piston 2. cylinder
3. sampling bag, collapsed
4. protective cup, closed.

Ship deck: insertion of 3 units into automatic main frame.
5. main frame 6. retractable legs.

Deep sea: sampling at 3 different depths during moving down into clean waters. Protective cup opens, bags are filled, cup closes.

Clean laboratory: aliquotation of samples.

Fig. 8 Deep sea sampler baptized "Moonlander" during "Ymer 80" expedition. Main operations and function. Protective cup and filling into plastic bags, as well as sampling dynamically at the end of the hydrowire after Schaule. The driving force for all automatic movements is the weight of the sampler itself (about 180 kg in sea water). For execution of one complete sequence of movements, the sampler creeps down 30 cm at the main wire.

heavily polluted ship deck has been reduced to the insertion of three hermetically closed sampling units into the main frame. The sampling units, easily transportable, could be prepared and also emptied in the clean area. Sampling bottles are filled by connecting them directly to the entry port. The Teflon checkvalve, retaining water into the filled bag, is lifted by a thin Teflon tube.

Snow was sampled into a wide-mouth polyethylene bottle, 0.5 l, yielding about 60 to 100 ml of water after melting. Using this size of bottles is not merely a trial to reduce sampling volume, as suspected sometimes by insiders of analytical methods needing large sample volumes. Obviously a 0.5 l bottle is easy to clean inside and outside and, by far more important, easy to handle and to keep clean during sampling under usually problematic conditions in field mission. The operator wearing long clean room arm protectors and clean plastic gloves, opened the bottle rinsing the thread with the clean acidified (pH 1) water in it. Kneeling down into the snow, the bottle was held at extended arms length upwind and snow was collected by moving the bottle in quarter circle movements through the upper layer of fresh fallen snow. Each filling movement was followed by a proceeding upwind, so that only virgin snow came into contact with the bottle and was sampled.

PRELIMINARY RESULTS AND DISCUSSION

Snow

Throughout the Arctic, snowfall is light, because the low temperatures limit the moisture-bearing capacity of the air. Average snow depths at the time of maximum cover in March or April range from 20 to 50 cm over the frozen ocean. The annual precipitation of 10 to 15 cm is mostly associated with late summer and fall frontal cyclones. Taking into account the southward drift of the ice floes and their medium life span of 2 to 4 years, no definite stratifications over more than 2 years can be expected. Thus, efforts were concentrated on collection of fresh fallen snow, as conditions for this purpose were best. As far as possible only definite layers of snow, avoiding accumulations of drifted snow of unknown fate, were collected. Mostly temperatures were close to 0°, so that the snowfall was moist, this simplifying the collection considerably.

The first snowfall was registrated on August 19 and the last station with sampling of snow was on September 18. Within this month, a medium precipitation of 10 cm snow was noticed, with, of course, drift snow up to 80 cm. The sampling areas can be seen in Fig. 9. Short-range meteorological data could be used for rough determination of direction and possible origin of air masses.

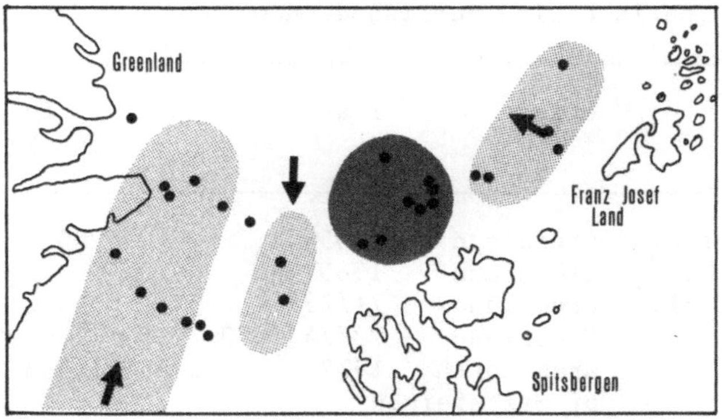

Fig. 9 Snow sampling stations. Dark shaded area with collection of aggromerated snow from last winter season. Light shaded areas indicate zones of fresh fallen snow. Arrows give approximate wind direction.

In the area north of Spitsbergen only old snow, obviously from last winter season, could be collected. After melting, the snow agglomerated to grains of 2-3 mm diameter. Trace metal results from these layers fit into the range of results published by other authors.

In sharp contrast to these data all until now analyzed fresh fallen snow samples, most of them having been collected during snowfall or within the following day, have trace metal data lower by about a factor 10. Three main sampling areas can be distinguished. One area east of North Greenland was supplied by air from the Atlantic. A more restricted area, that has been crossed during the journey back to Spitsbergen, was covered by snow from air masses crossing the North Pole. At the end of the expedition, in the vicinity of Franz Josef Land, snow fell from air originating in Northern Russia.

Exceptional local situations can be excluded considering the sampling area from Greenland to Franz Josef Land, the different meteorological situations and the sampling time over four weeks from mid-August to mid-September. This season is representative for the formation of about half the annual snow layer.

From these extremely low trace level data and medium values from previous year it can be concluded that during winter and spring snow must have higher contents of trace metals in the range of several hundred ng kg^{-1}.

Table 4 Comparison with published snow data

Author	Origin	Year	Cd	Pb (ng kg^{-1})	Cu
M Murozumi et al.[1]	Greenland	1965	-	200	-
H. Weiss et al.[2]	Greenland	1965	3	-	-
M. Herron et al.[3]	Greenland	71/73	3	145-221	-
C. Boutron[5]	Greenland	73/74	0.7-85	128-899	34-102
this work	area north of Spitsbergen	1979	3-8	116-378	80-120

Table 5 Range of trace metals in fresh fallen snow (number of determinations n = 10).

	Cd	Pb	Cu	Ni	Co
ng kg^{-1}	0.2-3.8	8.2-43	10-25	17-70	5
pmol kg^{-1}	1.7-33.9	39.6-207	158-397	293-1207	85

Further details are needed for completing the picture. Correlation with data from sampling of dust particles, carried out all along "Ymer 80", could give information about dry deposition, which, of course, can be expected to be very low.

Sea Water

A previous summary of surface water results, without distinction of Atlantic or Arctic waters, is given in Table 6. All values have been taken into account without using the $\pm$ 3 s.d. rule. Possibly 2 or 3 samples have been contaminated, not by the sampling procedure, but by water movements from the ship's propellers during maneuvering its bow onto an ice floe. The surface water values fit well into the pattern given by clean ocean areas[7]. They are lower than open Atlantic water levels, but distinctly higher then some ocean waters from the Pacific, e.g. Aitutaki Passage[7,15] with lead levels of 5 ng kg^{-1}.

Randomly taken samples from surface have been reanalysed in the home laboratory after prior UV-irradiation. It turned out that frequently the samples not subjected to this pretreatment, but only acidified to pH 2, yielded at least 50% lower trace metal

Table 6 Arctic surface water (0.5 m depth); n = 17

	Cd	Pb	Cu	Ni
$\bar{x}$	8.7	15	95	100
Range ng kg^{-1}	5.1-18	4.4-26	50-183	79-131
$\bar{x}$	0.078	0.072	1.496	1.703
Range nmol kg^{-1}	0.045-0.16	0.021-0.125	0.787-2.882	1.34-2.23

concentrations. This is obviously caused by the binding of a certain amount of the trace metals in inert organic species formed by exudates and dissolved products from the frequently high algae content in surface waters of the Arctic Ocean. In addition the performance of the voltammetric responses is affected adversively, due to inhibition effects. UV-irradiation combined with heating of the acidified sample close to boiling temperature for about 2 hours led to complete decomposition of the interfering organic matter.

Although contamination by the outer parts of the Go-Flo sampler and by the hydro-line had been avoided by dynamic sampling, the results of intermediate waters of 20 to 60 m seem to be partially random. Some samples have lead values up to 100 ng kg^{-1} and copper values up to 480 ng kg^{-1}, results that can be only explained by contamination, probably originating from seals of the sampler. Contamination of samples by the Go-Flo sampler has already been described by Schaule et al.[6].

Deep sea water samples, collected with the "Moonlander" could be analyzed at once after acidification and warming up to room temperature. Repetition of several analyses at home with prior UV-irradiation gave results not differing within the margin of relative standard deviation (viz. Fig. 6) so that the values represent total trace metal values and prior UV-irradiation is not needed.

One deep sea profile for lead, cadmium and phosphate can be seen in Fig. 10. Results from two deep sea stations in close vicinity were combined to a profile from surface to 1500 m depth. Of course, the profiles from surface to 300 m must be considered with caution, as there could be possibly higher values, at least for lead within this water column. Lead decreases with depth, as is known[6] from other oceans. Cadmium correlates well with phosphate. Its increase with depth corresponds only to about one fifth of the levels expected from deep sea samples from the Pacific[15,16] but agrees roughly with data from the Atlantic. A certain deviation from the correlation equation: Cd (nmol kg^{-1}) =

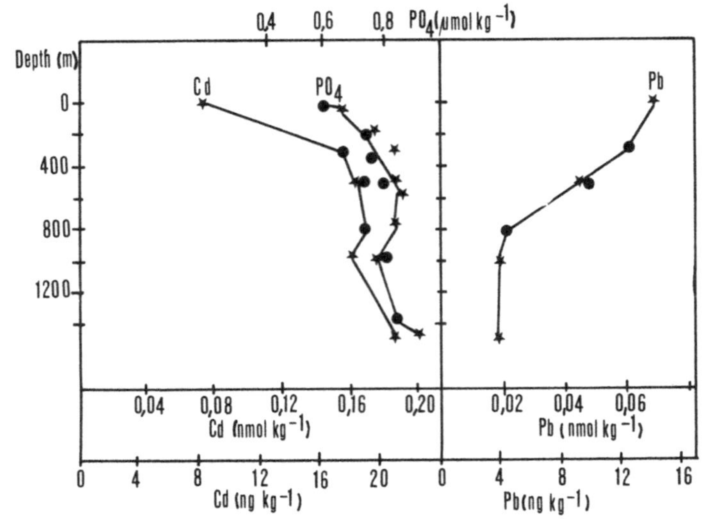

Fig. 10 Two deep sea stations in Arctic Sea. Positions: 81°43'N, 08°51'W★; 81°41'N, 09°05'W ●.

-0.032 + 0.31 PO_4 (nmol kg $^{-1}$) after Bruland[16] can be observed. Cd is about 20% lower than predicted by this equation.

Independent measurements carried out by Danielsson and Westerlund[17] on the first leg of the "Ymer 80" expedition gave the equation: Cd (nmol kg^{-1}) = 0.075 + 0.14 PO_4 (μmol kg^{-1}), which agrees well with our data.

CONCLUSION

The electroanalytical method of voltammetry has proved to be valuable in ultratrace metal analysis of natural samples. Prominent advantages are the possibility to utilize rotating electrodes while the ship is moving, the compactness of electroanalytical instrumentation and thus a continuous control of contamination during all stages of operation from sampling to analysis itself.

ACKNOWLEDGEMENTS

We wish specially to acknowledge the good cooperation with Lars-Göran Danielsson and Leif Andersson, Department of Analytical and Marine Chemistry, University of Göteborg. We also thank Hans Christen Hansson, Department of Nuclear Physics, University of Lund, for his kind help during sampling. We are indebted to the scientific leaders of the "Ymer 80" cruise, Valter Schytt and Hans

Dahlin. We greatly appreciated the skilful navigation of the commander of "Ymer", Captain Anders Billström, and the cooperation of his crew during sampling operations.

REFERENCES

1. Murozumi, M., T.J. Chow and C. Patterson, 1969: Chemical concentration of pollutant lead aerosols, terrestrial dusts and sea salts in Greenland and Antarctic snow strata. Geochim. Cosmochim. Acta., 33, 1247-1294.
2. Weiss, H., K. Bertine, M. Koide and E.D. Goldberg, 1975: The chemical composition of a Greenland glacier. Geochim. Cosmochim. Acta., 39, 1-10.
3. Herron, M.M., C.C. Langway, H.V. Weiss and J.H. Cragin, 1977: Atmospheric trace metals and sulfate in the Greenland Ice Sheet. Geochim. Cosmochim. Acta., 41, 915-920.
4. Boutron, C., 1979: Past and present day tropospheric fallout fluxes of Pb, Cd, Cu, Zn and Ag in Antarctica and Greenland. Geophys. Res. Let., 6, 159-162.
5. Boutron, C., 1979: Trace element content of Greenland snow along an east-west transect. Geochim. Cosmochim. Acta., 43, 1253-1258.
6. Schaule, B. and C.C. Patterson, 1980: The occurence of lead in the northeast Pacific and the effects of anthropogenic inputs. In: "Lead in the Marine Environment", M. Branica and Z. Konrad, eds. Pergamon Press, Oxford.
7. Mart, L., H. Rützel, P. Klahre, L. Sipos, J. Golimowski, U. Platzek, P. Valenta and H.W. Nürnberg: Comparative studies on the distribution of trace metals in the oceans and in coastal waters. Sci. Tot. Env., in press.
8. Mart, L., 1979: Prevention of contamination and other accuracy risks in voltammetric trace metal analysis of natural waters. Fresenius Z. Anal. Chem., 296, 350-357.
9. Florence, T.M., 1970: Anodic stripping voltammetry with a glassy carbon electrode mercury-plated in situ. J. Electroanal. Chem., 27, 273-281.
10. Valenta, P., L. Mart and H. Rützel, 1977: New potentialities in polarographic ultra trace analysis. J. Electroanal. Chem., 82, 327-343.
11. Mart, L., H.W. Nürnberg and P. Valenta, 1980: Prevention of contamination and other accuracy risks in voltammetric trace metal analysis of natural waters. Fresenius Z. Anal. Chem., 300, 350-362.
12. Pihlar, B., P. Valenta and H.W. Nürnberg, 1981: New high-performance analytical procedure for the voltammetric determination of nickel in routine analysis of waters, biological materials and food. Fresenius Z. Anal. Chem., 307, 337-346.

13. Schaule, B., 1979: Zur Ozeanischen Geochemie des Bleis - Nachweis einer anthropogenen Störung anhand kontaminationskontrollierter Bleikonzentrationsmessung im offenen Nord-Pazifik. Doctor Thesis, Univ. Heidelberg.
14. Haas, H. and L. Mart, 1980: Konstruktion und Funktionsweise eines automatischen Wasserprobennehmers für den Tiefsee-Einsatz. Jül-Report, 1698, Kernforschungsanlage Jülich.
15. Mart, L., T. Thijssen, L. Sipos, G.P. Glasby, B. Pihlar, H.W. Nürnberg and G. Friedrich: The determination of Co, Ni, Cu, Cd and Pb in surface and bottom sea water samples from manganese nodule-rich areas of the equatorial and southern Pacific using a new sampling device. In preparation.
16. Bruland, K.W., G.A. Knauer and J.H. Martin, 1979: Cadmium in northeast Pacific waters. Limnol. Oceanogr., 23, 618-625.
17. Danielsson, L.-G. and S. Westerlund: Trace metals in the Arctic Ocean. This volume.

THE RELATIONSHIP BETWEEN DISTRIBUTIONS OF DISSOLVED CADMIUM, IRON AND ALUMINIUM AND HYDROGRAPHY IN THE CENTRAL ARCTIC OCEAN

Robert M. Moore

Department of Oceanography
Dalhousie University
Halifax, Nova Scotia
Canada B3H 4J1

INTRODUCTION

Over many years there has been considerable interest in the role of iron as a micronutrient element[1] with attention focussing on its solubility[2,3], speciation, both inorganic[3] and organic[4], its availability to plankton[5,6,7] and modification of its speciation by organisms[6,7,8,9], and the possible ecological effects of its concentration and chemical form.

Studies of the biological roles, solubility and speciation of iron in seawater have not, however, been matched by studies giving a coherent picture of the distribution of this element in the world ocean. The reasons for this are probably related to the ubiquitous terrestrial distribution of the element and its very low solubility in seawater which combine to give contamination problems equalled by few other elements. Added to this, the distribution of the metal between dissolved and particulate forms, with the latter strongly favoured, results in some difficulty in comparing data sets which make different divisions of the total iron pool. These divisions result from the procedures used for storage, filtration and analysis.

This paper describes the vertical distribution of iron in the central Arctic Ocean and compares this with profiles of Cd and Al, relating each to the hydrography.

Sampling, Storage and Analysis

Water samples were collected from a drifting ice station between 88°N, $139^{\circ}50'$W and $89^{\circ}9'$N, $97^{\circ}7'$W in the central Arctic Ocean (Fig. 1). The samples for trace metal analysis were taken with 5 l Go-flo bottles attached to a new stainless steel hydrographic wire. Samples were not filtered since this procedure is liable to contaminate them, also, for most of the elements analyzed the concentration of particulate matter in open ocean waters is too low to make a significant contribution to the total. This may not be the case for iron which probably has a colloidal form with uncertain behaviour on filtration. In view of this, measurement of the total iron in these waters should provide a useful starting point for understanding its geochemistry in the Arctic Ocean. Immediately after collection, the samples in acid washed polyethylene bottles were frozen for storage.

The analytical method is essentially the solvent extraction procedure described by Danielsson et al.[11] but with some modifications designed to reduce handling of the sample. A closely similar method was used for analyzing samples for Cd, Zn and Cu[9,12] The sample was thawed and to a 200 ml aliquot was added 1 ml of a solution containing sodium diethyldithiocarbamate (1% w/v) and ammonium pyrrolidine dithiocarbamate (1% w/v) in 1% ammonia solu-

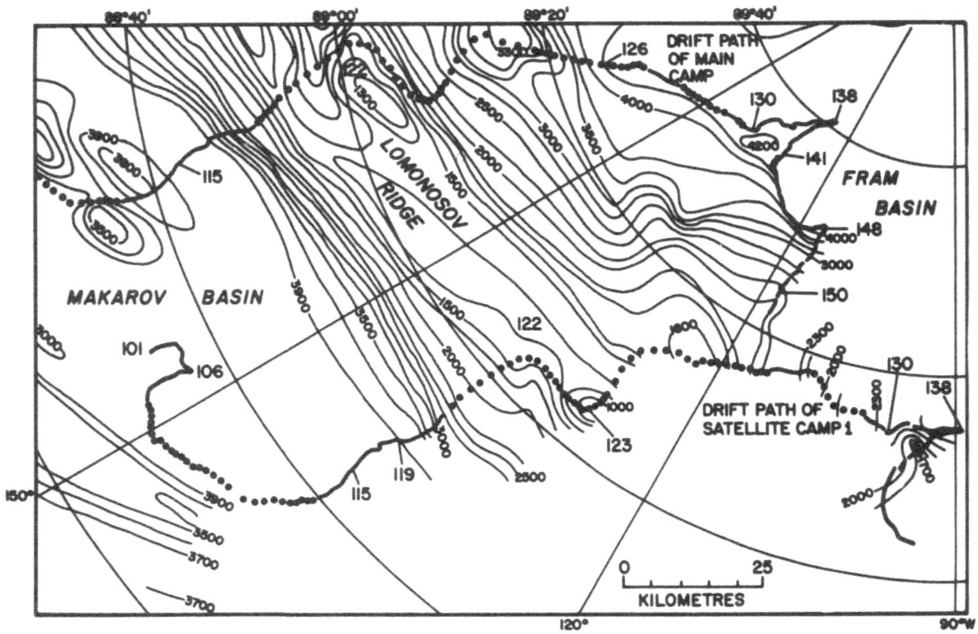

Fig. 1 Drift path of LOREX satellite camp 1. Numbers along path indicate Julian days.

tion, and 0.650 ml of 1 M hydrochloric acid to bring the pH to 4. The solution was extracted three times with 5 ml volumes of 1,1,2 trichloro-1,2,2 trifluorethane and the organic phase evaporated dryness in a silica vial and treated with 0.1 ml of Ultrex hydrogen peroxide (30%) to initiate the decomposition of organic matter present. After an hour or more, 0.5 ml of 0.1 M hydrochloric acid was added and the solution irradiated with a 1000 W Hanovia medium pressure mercury vapour discharge tube at a distance of 4 cm for 18 minutes. The iron in the concentrate was then compared with standards in 0.1 M hydrochloric acid using a Perkin-Elmer Model 403 Spectrophotometer fitted with a Perkin-Elmer graphite furnace (HGA 2200).

The blank, determined by repeating this procedure on extracted sea water, was typically 5 ng Fe, (equivalent to 25 ng l^{-1}).

The coefficient of variation of analyses was 21% for seven subsamples containing 1.6 nmol Fe l^{-1}, and 30% for eight subsamples at 0.6 nmol Fe l^{-1}. The detection limit was estimated to be 0.2 nmol Fe l^{-1}.

The efficiency of the extraction procedure was tested using sea water spiked with iron-59, which indicated a recovery of 97%, and with stable iron of 86%.

Aluminium analyses were made by a slightly modified version[12] of the procedure described by Hydes and Liss[13].

Results and discussion

Salinity, temperature and silicate profiles for the central Arctic Ocean are given in Figures 2-5. In the following discussion the distributions of iron, cadmium and aluminium, illustrated in Figures 6-8, are related to the hydrography and to addition and removal processes.

The central Arctic Ocean is characterized by a low salinity (ca. 30 $^o/oo$) well-mixed surface layer extending to about 50 m, below which the salinity increases very sharply (Fig. 2) reaching 34 $^o/oo$ by 130 m. Stable oxygen isotope ratios were linearly correlated with salinity, the relationship being described by the equation $\delta^{18}O = -23.2 + 0.656\ S\ ^o/oo$: this indicates that the fresh water component is primarily meteoric rather than sea ice meltwater (F. Tan, personal communication). The fresh water input comprises contributions from river flow and direct precipitation in a ratio of 3:1[14]. The trace metal composition of the surface water should therefore reflect that of Arctic river waters with modification by an atmospheric flux and by biological and inorganic removal processes.

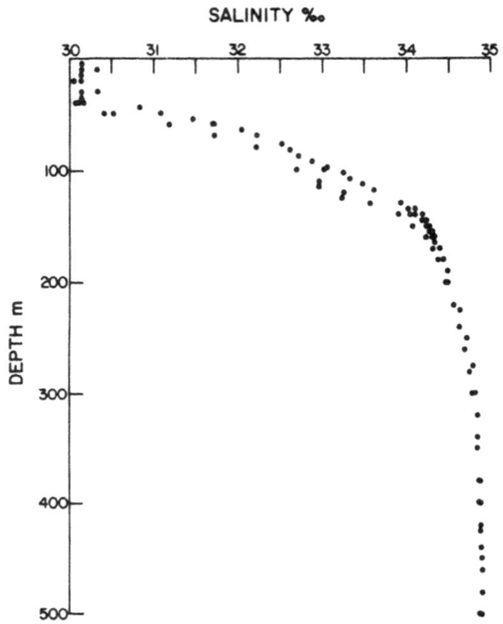

Fig. 2 Salinity profile to 500 m. Filled circles indicate profile from the Makarov Basin; open circles, profile from Fram Basin.

It is found that the surface layer has, relative to most surface ocean waters, high levels of silicate (7 μmol l^{-1}), phosphate (1 μmol l^{-1}) which are matched by relatively high cadmium (0.3 nmol l^{-1}). Unless the Arctic rivers carry unusually high concentrations of cadmium, the surface values must be accounted for by the admixture of the water that lies just beneath the mixed layer. This would also account for the surface silicon concentrations without needing a significant contribution from rivers. The high concentrations of both nutrients and those trace metals that are subject to biological control are allowed to persist in the surface waters on account of the very low primary productivity. This has been estimated by Ryther[15] to be only 1 g C m^{-2} yr^{-1}, much lower than even the oligotrophic waters of subtropical ocean regions (ca. 25 g C m^{-2} yr^{-1})[16], the reason is that the light flux is severely attenuated by the covering of ice.

Neither iron nor aluminium (Fig. 7, 8) show any enhancement of concentration in the surface waters that might be expected from the river inputs. In the case of aluminium it has been proposed[12] that the surface levels must be accounted for by an inorganic removal process possibly occurring in the nearshore or shallow

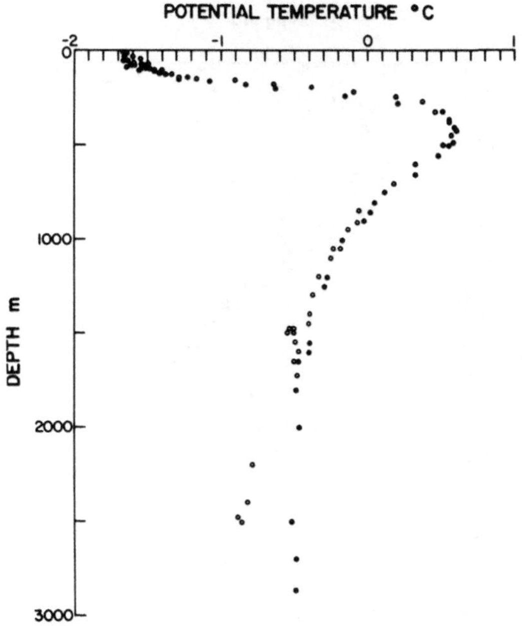

Fig. 3 Potential temperature profiles to 3000 m. Symbols as on Fig. 1.

continental shelf region. Similarly, in the absence of efficient biological stripping of trace elements from the surface waters, it may be argued that the iron levels are also controlled by inorganic removal processes. There is good evidence for inorganic removal of iron in the estuarine zone[17] but it is not yet clear whether similar processes continue to operate in the coastal zone.

Neither the iron nor aluminium distribution shows any sign of an atmospheric flux even though these elements are major components of atmospheric particles[18,19]. Some consideration should be given to the mechanism of atmospheric supply of elements to the Arctic Ocean. During much of the year there can be no addition of atmospherically derived trace elements to the water column on account of the complete ice cover. During the summer months the upper surface of the sea ice which has accumulated the atmospheric deposition of the foregoing winter is melted and the fresh water added to the surface layer with which it mixes. Additionally Rahn[14] suggests that a major part of the atmospheric deposition occurs in the summer, since most of the Arctic Ocean precipitation is rain. The atmospherically derived metals are therefore added in a pulse when temperatures rise above freezing and the addition continues until temperatures fall again below zero. Data available so far[20,21]

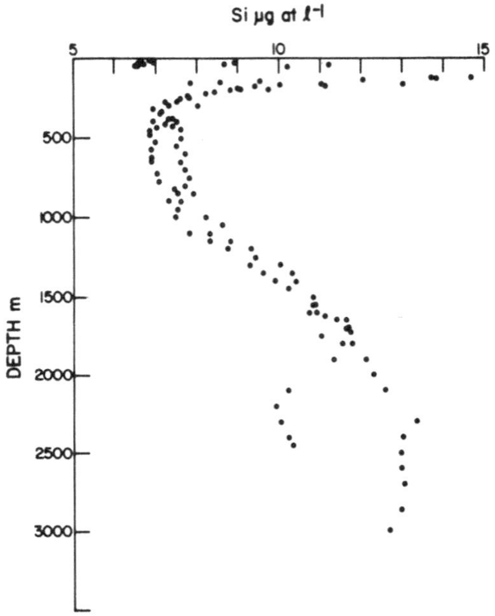

Fig. 4 Silicate profiles to 3000 m. Maximum values at ca. 100 m are not shown. Symbols as in Fig. 1.

would suggest that for both of the elements only a small fraction (1-4%) of the atmospheric flux is soluble.

A pronounced hydrographic feature below the surface mixed layer is a well-defined nutrient maximum centred at about 100 m (Fig. 4). This has been identified[22,10] as water originating in the Bering Sea entering the Arctic Ocean through the Bering Strait. Associated with the coincident silicate and phosphate maxima are peaks in the cadmium and zinc profiles[12]. This is in agreement with the association between these metals and nutrients which is well documented in the case of the Pacific Ocean[23,24].

Iron and aluminium profiles show no feature in the region of this advected water mass. This is in accordance with the suggestion that both elements have distributions strongly regulated by inorganic processes. It is worth mentioning at this point that radionuclide measurements[25] made on samples collected at the time of this work suggest that complete removal of Am-241 occurs while this water mass is in the shallow waters of the Bering Strait and adjoining shelf areas.

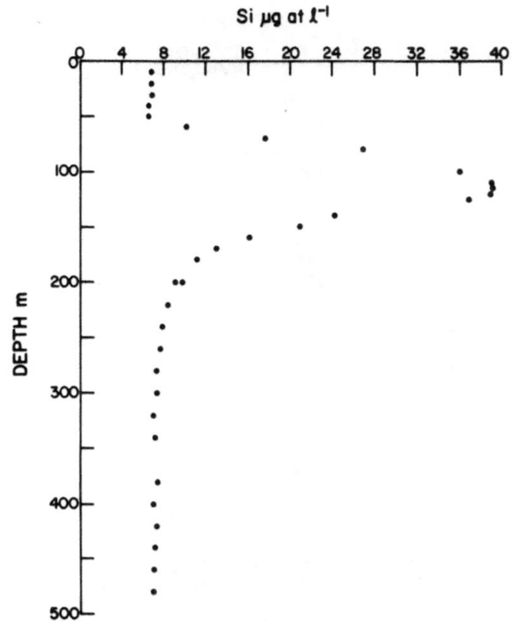

Fig. 5. Silicate profile to 500 m from Makarov Basin.

Below the layer of water from the Bering Sea lies a layer of Atlantic water characterized by a temperature maximum (Fig. 3), below this is Arctic Bottom Water arbitrarily defined by a potential temperature of below 0°. Cadmium levels drop sharply below the nutrient maximum and then show little variation with depth. This is consistent with the nutrient distribution in Atlantic and Arctic bottom water layers where there is only a gradual increase in nutrients with depth indicative of a low flux of biogenic material and moderate residence time of the deep water.

The increase in aluminium with depth, Fig. 8, is similar to that reported by Hydes[26] for depths below 1000 m in the North Atlantic. While such a profile can frequently be related to transport of the element concerned in biogenic material along with the nutrients, this simple explanation is inappropriate here since the shallow nutrient maximum is not matched by a maximum of dissolved aluminium.

The deep water concentrations of iron are not significantly different from the surface and intermediate water values. The samples analyzed for iron in this work were not filtered and so the measured concentrations of around 1×10^{-9} molar include an

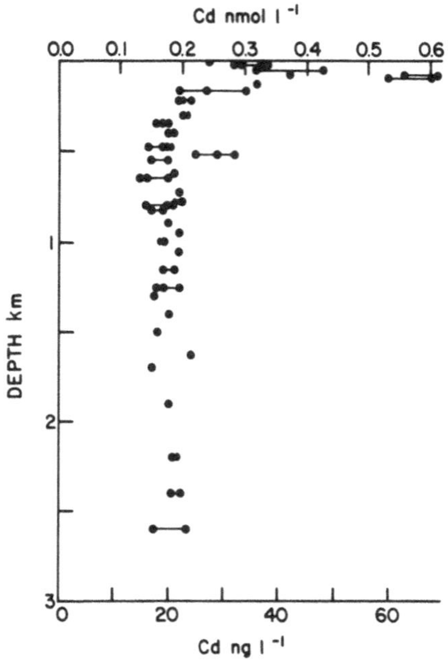

Fig. 6 Cadmium profile comprising samples from both sides of Lomonosov Ridge. From Moore (1981) with permission of Pergamon Press.

unknown contribution of reactive particulate iron. There appear to be no data on the distribution of particulate matter in the waters of the Arctic Ocean but in view of the low productivity of the waters and turbulence in the shelf regions reduced as a result of ice cover it may be supposed that the suspended load is relatively low, and that the particulate iron is correspondingly low.

CONCLUSIONS

A preliminary look at the distribution of reactive iron in the water column of the central Arctic Ocean indicates a marine geochemistry differing markedly from that of a number of trace metals that have been more thoroughly studied. Its distribution does not reflect the influence of river or atmospheric inputs nor the influence of advected water masses having quite different trace element compositions. It appears that inorganic processes are dominant in the removal of river-derived iron from the surface waters of the Arctic Ocean; this also applies to dissolved aluminium. The distribution of iron in the water column is consistent

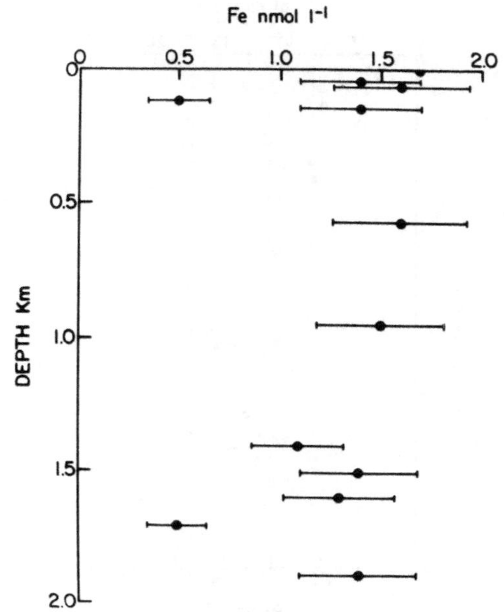

Fig. 7 Reactive iron concentrations, samples taken from both sides of Lomonosov Ridge. Error bars indicate $\pm$ 1 standard deviation.

with control by equilibrium with solid phases. Perhaps the simplest among the possible explanations for the dissolved aluminium profile is also equilibrium with a solid phase, though in this case a pressure dependence of the solubility of the controlling solid phase would be necessary to account for the increasing concentration with depth.

The observations of cadmium distribution are in accord with the interpretation of earlier studies of the metal in the other oceans; this work does, however, illustrate that the correlation between cadmium and phosphate can be maintained in shallow waters where inorganic removal processes are dominant.

It is not possible to ascertain how general are the conclusions that may be drawn from this specific study; the Arctic Ocean provides an environment differing greatly from the more temperate ice-free oceans, it is strongly stratified, has a large freshwater input, a periodic input of atmospherically derived materials, low productivity, unique light cycle and hence productivity cycle, and an unusually extensive shelf area. The study of trace element geochemistry in such an environment provides data that constrain the hypotheses constructed to explain observations made in more usual oceanic conditions.

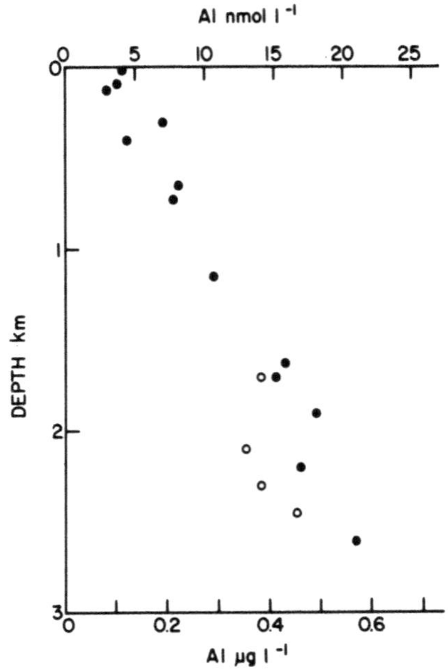

Fig. 8 Reactive aluminium concentrations. Open circles represent samples from below depth of ridge crest (ca. 1500 m) in the Fram Basin. From Moore (1981) with permission of Pergamon Press.

ACKNOWLEDGEMENTS

The valuable assistance of M.G. Lowings in the field work is gratefully acknowledged, also the support provided by the Natural Sciences and Engineering Research Council, Canada, the Earth Physics Branch of Energy, Mines and Resources, Canada, and the Polar Continental Shelf Project. The author wishes to thank H. Gote Ostlund of the tritium laboratory, University of Miami for his support of the project. This paper is LOREX contribution #8.

REFERENCES

1. Menzel, D.W. and J.H. Ryther, 1961: Nutrients limiting the production of phytoplankton in the Sargasso Sea, with special reference to iron. Deep-Sea Res., 7, 276-281.
2. Kester, D.R., T.P. O'Connor and R.H. Byrne, 1975: Solution chemistry, solubility, and adsorption equilibria of iron, cobalt and copper in marine systems. Thalassia Jugoslavica, 11, 121-134.

3. Byrne, R.H. and D.R. Kester, 1976: Solubility of hydrous ferric oxide and iron speciation in seawater. Mar. Chem., 4, 255-276.
4. Sugimura, Y., Y. Suzuki and Y. Miyake, 1978: The dissolved organic iron in seawater. Deep-Sea Res., 25, 309-314.
5. Levandowsky, M. and S.H. Hutner, 1975: Utilization of Fe^{3+} by the inshore colorless marine dinoflagellate Crypthe-conium cohnii. Annals New York Academy of Sciences, 245, 16-25.
6. Davies, A.G., 1970: Iron, chelation and the growth of marine phytoplankton 1. Growth kinetics and chlorophyll production in cultures of the euryhaline flagellate Dunalliela tertiolecta under iron-limiting conditions. J. Mar. Biol. Ass. U.K., 50, 65-86.
7. Lewin, J. and C.H. Chen, 1971: Available iron: a limiting factor for marine phytoplankton. Limnol. Oceanogr., 16, 670-675.
8. Sorokin, Y.I. and Y.A. Bogdanov, 1971: Transformation of iron during bacterial decomposition of planktonic organic matter. Hydrobiol. J., 7, 89-90.
9. Gonye, E.R. and E.J. Carpenter, 1974: Production of iron-binding compounds by marine microorganisms. Limnol. Oceanogr., 19, 840-842.
10. Moore, R.M. and M.G. Lowings, in preparation: Hydrographic features and nutrients at the LOREX station in the central Arctic Ocean.
11. Danielsson, L.-G., B. Magnusson and S. Westerlund, 1978: An improved metal extraction procedure for the determination of trace metals in sea water by atomic absorption with electrothermal atomization. Anal. Chim. Acta., 98, 47-57.
12. Moore, R.M., 1981: Oceanographic distributions of zinc, cadmium, copper and aluminium in waters of the central Arctic. Geochim. Cosmochim. Acta., 45, 2475-2482.
13. Hydes, D.J. and P.S. Liss, 1976: Fluorimetric method for the determination of low concentrations of dissolved aluminium in natural waters. Analyst, 101, 922-931.
14. Rahn, K.A.: Atmospheric, riverine and oceanic transport of trace elements to the Arctic Ocean. Paper presented at the Conference on the Arctic Ocean, London, 11-12 March, 1980 sponsored by the Arctic Committee of Monaco and the Royal Geographic Society.
15. Ryther, J.H.: Geographic variations in productivity. In, "The Sea", vol. 2. M.N. Hill (ed.) Interscience.
16. Koblentz-Mishke, O.J., V.V. Volkovinsky and J.G. Kabanova, 1970: Plankton primary production of the world ocean. In, "Scientific exploration of the South Pacific", W.S. Wooster (ed.), Washington, U.S. Nat. Acad. of Sciences.

17. Holliday, L.M. and P.S. Liss, 1976: The behaviour of dissolved iron, manganese and zinc in the Beaulieu Estuary. Estuarine Coastal Mar. Sci., 4, 349-353.
18. Chester, R. and J.H. Stoner, 1974: The distribution of Mn, Fe, Cu, Ni, Co, Ga, Cr, V, Ba, Sr, Sn, Zn, and Pb, in some soil-sized particulates from the lower troposphere over the world ocean. Mar. Chem., 2, 157-188.
19. Duce, R.A., B.J. Ray, G.L. Hoffman and P.R. Walsh, 1976: Trace metal concentration as a function of particle size in marine aerosols from Bermuda. Geophys. Res. Lett., 3, 339-342.
20. Hodge, V., S.R. Johnson and E.D. Goldberg, 1978: Influence of atmospherically transported aerosols on surface ocean water composition. Geochem. J., 12, 7-20.
21. Crecelius, E.A., 1980: The solubility of coal fly ash and marine aerosols in seawater. Mar. Chem., 8, 245-250.
22. Kinney, P., M.E. Arhelger and D.C. Burrell, 1970: Chemical characteristics of water masses in the Amerasian Basin of the Arctic Ocean. Geophys. Res., 75, 4097-4104.
23. Bruland, K.W., G.A. Knauer and J.H. Martin, 1978: Cadmium in northeast Pacific waters. Limnol. Oceanogr., 23, 618-625.
24. Bruland, K.W., G.A. Knauer and J.H. Martin, 1978: Zinc in northeast Pacific waters. Nature, 271, 741-743.
25. Livingston, H.D., S.L. Kupferman, V.T. Bowen and R.M. Moore: Vertical profile of artificial radionuclide concentrations in the Arctic Ocean. Submitted to Geochim. Cosmochim. Acta.
26. Hydes, D.J., 1979: Aluminium in seawater: control by inorganic processes. Science, 205, 1260-1262.

INTERCOMPARISON OF SEAWATER SAMPLING DEVICES FOR TRACE METALS

J.M. Bewers[1] and H.L. Windom[2]

[1] Bedford Institute of Oceanography
P.O. Box 1006, Dartmouth
Nova Scotia, Canada B2Y 4A2

[2] Skidaway Institute of Oceanography
P.O. Box 13687
Savannah, Georgia 31404 U.S.A.

EXTENDED ABSTRACT

Recently several round-robin intercalibrations for trace metals in seawater[1-5] have demonstrated a marked improvement in both analytical precisions and numerical agreement of results among different laboratories. However, it has often been claimed that spurious results for the determination of metals in seawater can arise unless certain sampling devices and particular methods of sampler deployment are applied to the collection of seawater samples. It is, therefore, desirable that the biases arising through the use of different, commonly-used, sampling techniques be assessed in order to decide upon the most appropriate technique(s) for both oceanic baseline and nearshore pollution studies.

Two international organizations, the International Council for the Exploration of the Sea (ICES) and the Intergovernmental Oceanographic Commission (IOC) have recently sponsored activities aimed at improving the determination of trace constituents in seawater through intercalibrations. Since 1975, ICES has conducted a series of trace metal intercalibrations, to assess the comparability of data from several tens of laboratories. These exercises have included the analyses of both standard solutions and real seawater samples[1-6]. The considerable improvement in the precisions and relative agreement between laboratories has been reflected in the results of these intercalibrations. By 1979 it had been concluded that sufficient laboratories were capable of conducting high-precision analyses of seawater for several metals to allow an examination of the differences between commonly-used sampling techniques for seawater sample collection.

In early 1980, the IOC, with the support of the World Meteorological Organization (WMO) and the United Nations Environment Program (UNEP), organized a Workshop on the Intercalibration of Sampling Procedures at the Bermuda Biological Station during which the most commonly-used sampling bottles and hydrowires were to be intercompared. This exercise forms part of the IOC/WMO/UNEP Pilot Project on Monitoring Background Levels of Selected Pollutants in Open-Ocean Waters. Windom[7] had already conducted a survey of the seawater sampling and analytical techniques used by marine laboratories and the conclusions of this survey were largely used for the selection of sampling devices to be intercompared. The bottles selected for comparison in Bermuda were modified and unmodified GO-FLO(R) samplers, modified Niskin(R) bottles and unmodified Hydro-Bios(R) bottles. GO-FLO samplers are the most widely used sampling device for trace metals in seawater. The other two devices continue to be used by several marine laboratories. Windom's[7] 1979 survey established that the most common method of sampler deployment was on hydrowires, as opposed to the use of rosette systems. The hydrowires selected for intercomparison were Kevlar(R), stainless steel and plastic coated steel. Kevlar and plastic-coated steel were selected because they are widely used in continental shelf and nearshore environments and are believed to be relatively 'clean'. Stainless steel was the most commonly-used hydrowire until the late seventies and, therefore, deserved inclusion to determine its suitability for trace metal sample collection. This wire is still in use by a few marine laboratories for both inshore and offshore work.

The method of intercomparison of the various devices was to deploy pairs of sampler types on different hydrowires to collect water samples from a homogeneous body of deep water at Ocean Station S ('Panulirus Station') near Bermuda. Originally it was intended to collect samples from within the North Atlantic Deep Water at about 1750 metres depth (Fig. 1). The water at this depth has characteristics of 3.97 $\pm$0.05^{o}C temperature and 35.01 $\pm$0.02 ‰ salinity for the month of January[8]. However, the restricted length of Kevlar hydrowire available necessitated the collection of samples in the lower thermocline at depths between 1150 and 1250 metres. Although the vertical and temporal variability in major properties, particularly temperature, is more pronounced at these shallower depths it is still sufficiently small to give reasonable assurance of homogeneity in the distribution of trace metals over the depth and time scales employed for sample collection. Nevertheless the assumption was tested for several metals (Cd, Co, Fe, Mn, Ni, and Zn) by the collection of samples, at the temporal and spatial extremities of the two samp-

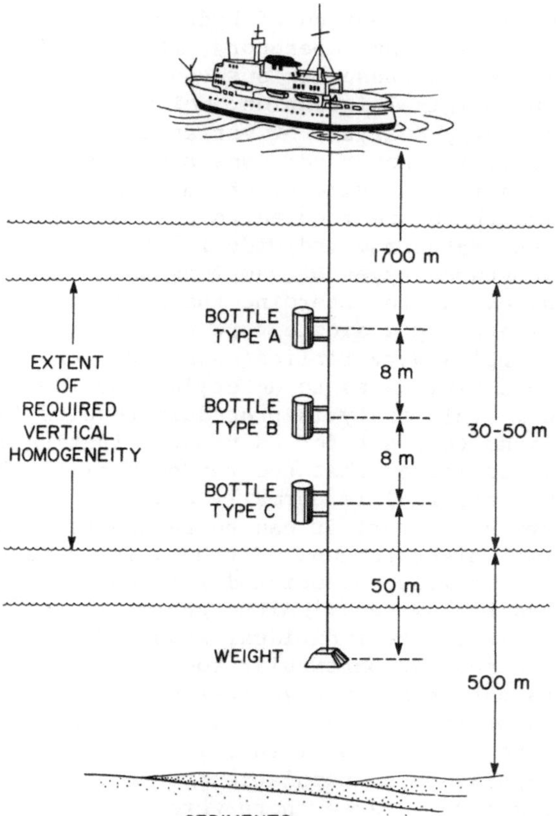

Fig. 1 Sampling Strategy

ling sequences, for later analysis by three participants having relatively high precisions as determined in a preceding ICES experiment[1].

In addition to heterogeneity in the water body used for the collection of samples, the identifiable sources of variation within the experiment are as follows: (1) Sampling bottles; (2) Hydrowires; (3) Analytical differences (extraction and instrumental methods); (4) Operators and within-laboratory variations, e.g. contamination; and (5) Analytical precision. It would have been possible to eliminate sources (3) and (4) by utilizing one laboratory having relatively good methodological precision to determine the variations associated with the other sources, but this would have necessitated the placing of a great deal of confidence in a single laboratory. Alternatively, if a complete or systematic sampling design had been used, involving the collection of samples from each combination of hydrowire and sampling bottle type for each participating laboratory, it would have been possible to investigate all these various sources of error. Indeed this had been the original intention, which would have enabled the conduct of a three-way analysis of variance on the entire data set but, unfortunately, weather conditions and shiptime limitations forced a reduction in the symmetry of the experimental design. Data analysis was therefore reduced to a separate one-way analysis of variance on the data from individual laboratories to examine the differences between types of sampling bottle on a single (common) hydrowire, and to determine the influences of the three types of hydrowire using a single type of sampling bottle (modified GO-FLO). Samples were replicated so that there were, in all cases, two or more replicates to determine the lowest level and analytical error. This analysis procedure results in thirteen sets of conclusions (one set for each participating laboratory) and requires an assumption that the random variable being estimated (sampling variation) is normally distributed. Since this is a form of measurement error, it can be reasonably expected to adhere to a normal distribution. Further, it is assumed that there is a common variance associated with each treatment (sampling bottles and hydrowires). Although we did not test this assumption explicitly, the individual standard deviations associated with the means for particular combinations of devices reflect its reasonableness in most instances. Analysis of variance was used in preference to t-tests or non-parametric tests, for estimating the equality of means, in order to maintain consistency in the comparisons of populations - three of which were usually being compared because there were three types of hydrowire and four types of sampling bottle.

Replicate (4) unfiltered water samples were collected for each participant for the comparison of pairs of sampling bottles on different hydrowires. Modified GO-FLO bottles were employed on each of the three hydrowires and this permitted a comparison of the three types of hydrowire. Only in the cases of iron and manganese were there indications of inhomogeneity at levels that might invalidate the intercomparisons. This is assumed to be due to inhomogeneity in the distribution of suspended particulate material that will influence metals that have major fractions in the particulate phase. The samples used for independent comparisons of bottles and hydrowires by each participant were collected by a common sampling team and then returned to the participants' own laboratories for analysis. All shipboard sub-sampling was carried out in a clean laboratory supplied by the US National Oceanographic and Atmospheric Administration that also supplied the research vessel. Sub-samples provided to the participants were labelled with coded identifiers. The experiment was therefore conducted 'blind' with the participants being unaware of the identities of individual sub-samples until after the return of these analyses to the coordinators.

The participants were drawn from the following 9 countries; Canada, Federal Republic of Germany, German Democratic Republic, Iceland, Japan, Korea, Malaysia, the Netherlands and the U.S.A. (Table 1). Results were provided by the participants for the following metals; cadmium, copper, iron, manganese, mercury, molybdenum, nickel, vanadium and zinc. The results from individual participants were then analysed to determine the numerical and statistical differences between the sampling bottles and hydrowires using the BREAKDOWN Subprogram of the Statistical Package for the Social Sciences[9] (as examples, results for nickel and copper in Tables 2 - 5). The differences between sampling devices for particular metals, as determined by each participant, are given when these are significant at the 95% and 90% levels of confidence (see Tables 3 and 5). Only the results of participants that had acceptable analytical performance, as measured by preci-

Table 1 List of Participants

Canada	Bedford Institute of Oceanography Dartmouth, Nova Scotia	P.A. Yeats J.M. Bewers
Federal Republic of Germany	Deutsches Hydrographischés Institute, Hamburg	D. Schmidt
German Democratic Republic	Institut fur Meereskunde Warnemunde-Rostock	L. Brügmann
Iceland	Marine Research Institute, Reykjavik	J. Olafsson
Japan	National Institute for Environmental Studies, Yatabe	C. McLeod
	Sagami Chemical Research Centre, Kanawaga	M. Ambe
Korea	Korea Ocean Research and Development Institute, Seoul	K. Lee
Malaysia	University Sains Malaysia, Penang	M. Sivalingam
Netherlands	Netherlands Institute for Sea Research, Texel	J.C. Duinker
U.S.A.	Skidaway Institute of Oceanography, Savannah, Georgia	R. Smith F. Storti H.L. Windom
	Texas A&M University, College Station, Texas	B.J. Presley
	University of Connecticut, Groton, Connecticut	D. Waslenchuk
	University of Delaware, Newark, Delaware	T. Church J. Tramontano
	Naval Research Laboratory, Washington, D.C.	R. Pellenbarg

sion and agreement with contemporary consensus values for deep North Atlantic waters (see Table 6), were used in drawing conclusions. A full report of the experiment is provided elsewhere[10].

Table 2 Numerical Comparisons for Nickel*. Units µg/l.

Wire		PCS	PCS	SS mod	SS exw	KEV	KEV	PCS	SS	KEV
Bottle		HB	MGF	GF	GF	NIS	MGF	MGF	MGF	MGF
Laboratory										
1	m	0.224	0.205	0.209	0.243	0.233	0.207	0.201	0.226	0.207
	sd	0.015	0.020	0.023	0.013	0.023	0.025	0.029	0.025	0.025
2	m	0.298	0.278	0.235	0.235	0.218	0.150	0.223	0.235	0.231
	sd	0.031	0.123	0.077	0.048	0.019	0.018	0.128	0.059	0.119
4	m	0.18	0.47	0.47	0.35			0.47	0.41	0.23
	sd	0.07	-	-	0.17			-	0.12	-
5	m	0.478	0.221	0.240	0.235	0.273	0.237	0.193	0.238	0.237
	sd	0.066	0.026	0.014	0.012	0.021	0.016	0.048	0.012	0.016
6	m	0.340	0.159	0.220	0.238	0.237	0.232	0.159	0.230	0.230
	sd	0.052	0.035	0.030	0.021	0.023	0.026	0.035	0.027	0.024
7	m	1.93	1.63	1.60	1.75			1.63	1.68	1.72
	sd	0.24	0.10	0.37	0.13			0.10	0.27	0.25
8A	m	0.185	0.100	0.123	0.113	0.160	0.105	0.100	0.119	0.123
	sd	0.041	-	0.005	0.012	0.054	0.010	-	0.009	0.041
10	m	0.737	0.357	0.634	0.353	0.385	0.461	0.413	0.493	0.462
	sd	0.078	0.077	0.309	0.131	0.064	0.162	0.159	0.262	0.162
11	m	0.511	0.367	0.349	0.365	0.421	0.393	0.367	0.357	0.404
	sd	0.034	0.008	0.034	0.009	0.009	0.027	0.008	0.025	0.037
12	m	0.230	0.165							
	sd	0.024	0.036							
13	m	0.230	0.200	0.238	0.265	0.204	0.236	0.200	0.250	0.236
	sd	0.036	0.020	0.045	0.007	0.056	0.007	0.020	0.035	0.007

* Number of figures given in table results from common computer analyses and not all such figures will be necessarily significant.

Key:

PCS	Plastic-coated steel hydrowire
SS	Stainless steel (type 302 unlubricated) hydrowire
KEV	Kevlar^R hydrowire
HB	Hydro-Bios sampler
MGF	Modified GO-FLO sampler
mod GF	Modified GO-FLO sampler
exw GF	Unmodified GO-FLO sampler
NIS	Modified Niskin sampler
m	mean
sd	standard deviation

Table 3 Statistical Comparisons for Nickel

Base Comparison	PCS MGF/MGF	PCS HB/MGF	SS MGF/GF	KEV MGF/MGF	KEV NIS/MGF	MGF WIRES
Lab No.						
1	NS	Sig HB>MGF	Sig GF>MGF	NS	Sig NIS>MGF	Sig SS>KEV>PCS
2	NS	NS	NS	Sig	Sig NIS>MGF	NS
4		Sig MGF>HB	NS			NS
5		Sig HB>MGF	NS		90 NIS>MGF	Sig SS>KEV>PCS
6		Sig HB>MGF	Sig GF>MGF	NS	NS	Sig KEV>SS>PCS
8A		Sig HB>MGF				
10	NS	Sig HB>MGF	NS	Sig	NS	NS
11		Sig HB>MGF	NS	NS	NS	Sig KEV>PCS>SS
12	90	Sig HB>MGF				
13		NS	NS	NS	NS	90 SS>KEV>PCS

Key:
- PCS Plastic-coated steel hydrowire
- SS Stainless steel (type 302 unlubricated) hydrowire
- KEV Kevlar[R] hydrowire
- HB Hydro-Bios sampler
- MGF Modified GO-FLO sampler
- GF Unmodified GO-FLO sampler
- NIS Modified Niskin sampler
- Sig Difference is significant ($P<0.05$)
- 90 Difference is significant ($P<0.1$)
- NS Not significant ($P>0.1$)

The experiment reveals that the differences between results obtained through the use of various combinations of hydrowires and samplers are not large and in no case can they account for the recent decline in the oceanic concentrations of trace metals reported in the literature. Nevertheless, for several metals, most notably copper, nickel and zinc, significant differences are evident between both bottles and hydrowires. For deep ocean studies the best combination of those tested is undoubtedly modified GO-FLO samplers and plastic-coated steel hydrowire. Except

Table 4 Numerical Comparisons for Copper*. Units μg/l.

Wire		PCS	PCS	SS	SS	KEV	KEV	PCS	SS	KEV
				mod	exw					
Bottle		HB	MGF	GF	GF	NIS	MGF	MGF	MGF	MGF
Laboratory										
1	m	0.094	0.092	0.095	0.103	0.111	0.131	0.093	0.099	0.120
	sd	0.007	0.009	0.012	0.012	0.011	0.012	0.011	0.012	0.021
2	m	1.000	0.765	0.553	0.620	0.455	0.272	0.650	0.586	0.403
	sd	0.857	0.289	0.261	0.100	0.487	0.185	0.291	0.186	0.298
3	m	0.437	0.180	0.211	0.205	0.447	0.550	0.233	0.208	0.455
	sd	0.347	0.084	0.067	0.034	0.540	0.317	0.081	0.051	0.314
4	m	0.533	0.435	1.25	1.065			0.435	1.158	1.065
	sd	0.163	0.177	0.35	0.177			0.177	0.252	0.177
5	m	0.188	0.063	0.064	0.142	0.101	0.072	0.101	0.103	0.072
	sd	0.108	0.003	0.004	0.010	0.049	0.012	0.059	0.043	0.012
6	m	0.615†	0.070	0.074	0.083	0.121	0.120	0.070	0.079	0.120
	sd	0.560	0.005	0.003	0.004	0.039	0.022	0.005	0.006	0.026
7	m	0.35	0.27	0.71	0.28			0.27	0.50	0.32
	sd	0.29	0.12	0.59	0.23			0.12	0.47	0.37
8B	m	0.155	0.045	0.163	0.278	0.133	0.160	0.045	0.220	0.140
	sd	0.076	0.006	0.044	0.059	0.030	0.037	0.006	0.078	0.038
9	m	0.84	0.32			0.35	0.55	0.32		0.44
	sd	0.79	0.03			0.02	0.21	0.03		0.17
10	m	0.123	0.135	0.158	0.119	0.096	0.100	0.130	0.138	0.101
	sd	0.015	0.003	0.033	0.032	0.015	0.019	0.024	0.037	0.019
11	m	0.195	0.137	0.102	0.106	0.109	0.132	0.137	0.104	0.149
	sd	0.089	0.027	0.005	0.001	0.013	0.019	0.027	0.004	0.073
12	m	0.059†	0.172							
	sd	0.325	0.040							
13	m	0.168	0.101	0.105	0.292	0.133	0.121	0.101	0.186	0.121
	sd	0.063	0.028	0.013	0.006	0.009	0.020	0.028	0.100	0.200

* Number of figures given in table results from common computer analyses and not all such figures will be necessarily significant.
† Suspected contamination.
All other symbols are the same as those used in Table 2.

in the cases of mercury and manganese, Hydro-Bios samplers appear to yield higher metal values than modified GO-FLO samplers. In contrast, Niskin bottles, modified by the replacement of the internal spring by silicone tubing, are capable of collecting samples of comparable quality to those collected by modified GO-FLO sampler for all metals except zinc. Modification to factory-

Table 5 Statistical Comparisons for Copper

Base Comparison	PCS MGF/MGF	PCS HB/MGF	SS exwGF/ modGF	KEV MGF/MGF	KEV NIS/MGF	MGF WIRES
Lab No.						
1		NS	NS	Sig	NS	Sig KEV>SS>PCS
2	NS	NS	NS	NS	NS	NS
3	Sig	90 HB>GF	NS	NS	NS	Sig KEV>PCS>SS
4		NS	NS			Sig SS>KEV>PCS
5		90 HB>GF	Sig EXW>MOD		NS	NS
6		Sig HB>GF	Sig EXW>MOD	NS	NS	Sig KEV>SS>PCS
7		NS	NS			NS
8B		Sig HB>GF	Sig EXW>MOD	NS	NS	Sig SS>KEV>PCS
9		NS		NS	NS	NS (KEV & PCS only)
10		NS	NS	Sig	NS	Sig SS>PCS>KEV
11		NS	NS	NS	NS	NS
12	Sig	Sig HB>GF				
13		NS	Sig EXW>MOD	NS	NS	NS

Symbols are the same as those used in Table 3.

supplied Teflon$^{(R)}$ coated GO-FLO bottles (i.e., replacement of 'O' rings with silicone equivalents* and the substitution of all-Teflon drain cocks for those originally supplied), do appear to result in a significant reduction in the levels of most metals in seawater samples collected with them. Kevlar and stainless steel hydrowires generally yield measurably greater concentrations of most metals than does plastic-coated steel. These differences, however, are small enough to suggest that these hydrowires are still suitable for trace metal studies of all but the most metal-depleted waters if proper precautions are taken[11-14].

* GO-FLO bottles equipped with silicone 'O' rings are also available from the manufacturer (General Oceanics Inc.)

Table 6 Results of Sampling Bottle and Hydrowire Intercomparisons

Metal	Mean ± S.D. µg/l	(N_{labs})	Best Combined Sampling/Analytical Precisions µg/l	Comparisons: Hydrowires	Comparisons: Samplers
Cd	0.035±0.016	(12)	0.001	PCS<(KEV ≈ SS)	(MGF ≈ NIS)<HB<GF
Cu	0.13 ±0.04	(6)	0.010	PCS<(KEV ≈ SS)	(MGF ≈ NIS)<HB<GF
	0.51 ±0.28	(6)			
Ni	0.21 ±0.05	(7)	0.02	PCS<(KEV ≈ SS)	(MGF ≈ NIS ≈ GF)<HB
	0.42 ±0.11	(3)			
Zn	0.35 ±0.18	(5)	0.05	PCS<(KEV ≈ SS)	MGF<(NIS ≈ HB ≈ GF)
Fe	0.41 ±0.29	(3)	0.05	PCS<KEV<SS	(MGF ≈ NIS)<GF<HB
Mn	0.064±0.038	(2)	0.010	(PCS ≈ KEV ≈ SS)	(MGF ≈ NIS ≈ GF ≈ HB)
	0.012±0.006	(1)	0.003		
Hg	0.007±0.002	(2)	0.002	INSUFFICIENT COMPARISONS	
	0.001	(1)			

The objectives of the IOC experiment were to assess the suitability of sampling devices for deep ocean baseline studies. Nevertheless the conclusions can be extended to cover continental shelf and nearshore waters that may be of concern in relation to pollution. If reasonable precautions are taken in preparation, deployment and recovery, modified Niskin and GO-FLO bottles may be used to collect viable samples in nearshore and estuarine waters as well as in deep ocean areas. In metal-depleted surface waters, in both continental shelf and deep marine environments, the combination of modified GO-FLO bottles on plastic-coated steel hydrowire appears to be the safest choice. However, it should be stressed that the suitability of any combination of bottles and hydrowire at very low metal concentrations (i.e. those prevailing for biologically-active metals in the upper mixed layer) still needs to be determined.

One other major conclusion of the Bermuda experiment is that the use of differing sampling devices and hydrowires only accounts for a small portion of the differences between trace metal results from different laboratories. It appears, from this experiment, and from a previous ICES intercalibration[1], that the major contributions to such differences are analytical artifacts. Thus we should stress that, although the sampling tools available to marine goechemists appear adequate for the measurement of metal distributions in the ocean, the execution of cooperative monitoring programs for metals should be preceded by a mandatory intercomparison of sample storage and analytical procedures.

ACKNOWLEDGEMENTS

We thank the participants, listed in Table 1, without whom this experiment could not have been conducted. Equal thanks are due to the Intergovernmental Oceanographic Commission (IOC), the United Nations Environment Programme (UNEP) and the World Meteorological Organization (WMO) for sponsoring this work and to the U.S. National Oceanographic and Atmospheric Administration, particularly Dr. Don Atwood, for the provision of the Research Vessel '<u>Kelez</u>' and shipboard clean laboratory facilities. We gratefully acknowledge the assistance of Dr. Graham Topping and John L. Barron in the organization of the experiment and the analysis of the data. This project was an endeavour of the IOC Group of Experts on Methods, Standards and Intercalibration (GEMSI) under the chairmanship of Dr. Neil R. Andersen. This is a contribution of the Bermuda Biological Station for Research.

REFERENCES

1. Bewers, J.M., J. Dalziel, P.A. Yeats and J.L. Barron, 1981: An intercalibration for trace metals in seawater. Marine Chemistry, 10, 173-193.
2. Olafsson, J., 1978: Report on the ICES international intercalibration of mercury in sea water. Marine Chemistry, 6, 87-95.
3. Olafsson, J., 1980: "A preliminary report on ICES intercalibration of mercury in seawater for the Joint Monitoring Group of the Oslo and Paris Commissions", submitted to the Marine Chemistry Working Group of ICES, Feb. 1980.
4. Thibaud, Y., 1980: "Exercise d'intercalibration CIEM, 1979, cadmium en eau de mer", Report submitted to the Marine Chemistry Working Group of ICES, Feb. 1980.
5. Jones, P.G.W., 1977: "A preliminary report on the ICES intercalibration of sea water samples for the analyses of trace metals". ICES CM1977/E:16.
6. Jones, P.W.G., 1976: "An ICES intercalibration exercise for trace metal standard solutions". ICES CM1979/E:15.
7. Windom, H.L., 1979: "Report on the results of the ICES questionnaire on sampling and analysis of seawater for trace elements", Submitted to the First Meeting of the Marine Chemistry Working Group, Lisbon, May 1979.
8. Pocklington, R., 1972: Variability of the ocean off Bermuda. Bedford Institute of Oceanography Report - Series BI-R-72-3.
9. Nie, H.N., C.H. Hull, J.G. Jenkins, K. Steinbrenner and D.H. Bent, 1975: Statistical package for the social sciences. Second Edition, McGraw-Hill Inc.
10. IOC, 1981: The Scientific report of the intercalibration exercise of the IOC/WMO/UNEP Pilot Project on Monitoring Background Levels of Selected Pollutants in Open-ocean Waters; IOC Technical Series No. 23 (in press).
11. Boyle, E.A., F.R. Sclater and J.M. Edmond, 1977: The distribution of dissolved copper in the Pacific. Earth Planet. Sci. Lett., 37, 38-54.
12. Bruland, K.W., G.A. Knauer and J.H. Martin, 1978a: Cadmium in northeast Pacific waters. Limnol. Oceanogr., 23, 618-625.
13. Bruland, K.W., G.A. Knauer and J.H. Martin, 1978b: Zinc in northeast Pacific water. Nature, 271, 741-743.
14. Sclater, F.R., E. Boyle and J.M. Edmond, 1976: On the marine geochemistry of nickel. Earth Planet. Sci. Lett., 31, 119-128.

THE ANALYSIS OF TRACE METALS IN BIOLOGICAL REFERENCE MATERIALS: A DISCUSSION OF THE RESULTS OF THE INTERCOMPARISON STUDIES CONDUCTED BY THE INTERNATIONAL COUNCIL FOR THE EXPLORATION OF THE SEA

G. Topping

DAFS
Marine Laboratory
Victoria Road
Aberdeen, Scotland

ABSTRACT

The reference materials used in these exercises were prepared from fish and shellfish tissue and distributed in the main to laboratories within the ICES framework which were participating in the ICES fish and shellfish pollution monitoring programme.

The results of these exercises showed that there was a progressive improvement in the comparability of analytical results of Cu, Zn and Hg at concentrations normally found in fish and shellfish tissue i.e. 30-280 μmol kg^{-1} (dry wt), 350-1000 μmol Zn kg^{-1} (dry wt) and 0.25-4.0 μmol Hg kg^{-1} (dry wt). However, the results of the Cd and Pb analysts were in general not as encouraging. The range of values reported by participants for these metals covered more than an order of magnitude at tissue concentrations in the ranges 0.009-0.9 μmol Cd kg^{-1} (dry wt) and 0.024-4.8 μmol Pb kg^{-1}. The most recent exercise for Cd and Pb showed that comparable data was obtained for tissue concentrations of 10 μmol kg^{-1} (dry wt) and 12 μmol kg^{-1} (dry wt) respectively i.e. concentrations normally found in shellfish tissue.

INTRODUCTION

The International Council for the Exploration of the Sea (ICES) is an intergovernmental organisation, consisting of 18 member states, whose main function is to encourage and coordinate research on the sea and its living resources. Its area of coverage is the North Atlantic and adjacent seas (e.g. North Sea,

Barents Sea and Baltic) and member countries include those with coastlines bordering these water bodies. In the late sixties, in response to general concern over the potential impact of man-made wastes discharged to the sea, ICES established a scientific working group with responsibility for studying this problem. One of the first tasks of this group was to collect data on the concentrations of contaminants in commercial fish and shellfish in order to assess the risk to the consumer. In 1971 the group began this work by conducting a baseline study in the North Sea; this study was extended to the North Atlantic in 1975. Analytical intercomparison exercises formed an important component of these studies [1,2,3] and the subsequent routine monitoring programmes [4,5,6] which concentrated on areas and species, identified in the baseline studies as those which exhibitied enhanced levels of contaminants.

During the last 10 years, ICES has organised and conducted five intercomparison exercises for selected metals in biological tissue. A full account of these exercises (preparation and distribution of reference samples, names and addresses of participants and brief description of analytical procedures) together with a discussion of the results of analysis is presented elsewhere [7,8,9]. In these reports the results submitted by each participant are clearly identified and the extent of agreement between analysts has been assessed by an analysis of variance and a multiple range test [10].

This paper summarises the findings of these intercomparison exercises and discusses them in relation to monitoring programmes for metals in fish and shellfish.

REFERENCE MATERIALS

Ideally the reference materials used in this type of study should be identical, or very similar, to samples analysed by participants in monitoring programmes i.e. wet tissue of fish or shellfish. It is generally accepted however that the preparation and distribution of a homogeneous wet tissue is extremely difficult. It was agreed therefore that the reference materials used in these exercises should be a dried homogeneous material prepared from fish or shellfish. Since no single reference material can adequately test the range of concentrations of elements encountered in the allied monitoring programmes, a number of different preparations were made during the course of the programme.

Descriptions of the reference materials used in these exercises are given in Table 1. Further details of the preparation of these materials and the manner of their distribution can be found elsewhere [7,8,9].

Table 1 Reference materials examined in ICES intercomparison exercise

Exercise (Year)	Raw Material	Brief description of preparation of Reference Material	Elements under Study
1 (1971)	Commercial fish meal	Air dried; repeatedly ground in a hammer mill to a fine flour	Cu, Zn, Hg, Cd and Pb
2 (1973)			
(a)	Fish fillet (cod, unskinned)	Steamed cooked; air dried for 24 hours, minced and repeatedly ground in a hammer mill to a fine flour	Cu, Zn, Hg, Cd and Pb
(b)	Stock standard metal solution (1000 ppm in IN acid)	Dilution of stock solution using IN acid. Cu, Zn, Cd and Pb stored in plastic phials, glass phials used for Hg solutions	Cu, Zn, Hg, Cd and Pb
3 (1975)	Fish fillet (cod, skinned)	As for 2a	Cu, Zn, Hg, Cd and Pb
4 (1978)			
(a)	Fish fillet (cod, skinned)	Wet tissue cut into small pieces (3 cm x 3 cm); blast frozen, freeze dried and repeatedly ground in a hammer mill to a fine flour	Cu, Zn, Hg, Cd, Pb and As
(b)	Fish fillet (cod, skinned)	Chopped wet tissue washed with dilute acid to reduce Hg content. Freeze dried and ground into flour as above	Hg only
5 (1980)			
(a)	White meat of edible crab	As for 4a	Cd only
(b)	Commercial fish meat	As for 4a	Pb only
(c)	Hepatopancreas of lobster	Prepared in the form of acetone power	Pb only

PARTICIPANTS

Although the intercomparison exercises were conducted primarily for the benefit of laboratories within ICES member countries, other laboratories and countries (e.g. Australia) not directly associated with these monitoring programmes also took part. It is worth noting that there was a steady increase in the numbers of participants with each successive exercise, particularly since all analytical data, both 'good' and 'bad', are openly documented and evaluated in the detailed reports of these exercises. Lack of space here precludes the publication of names and addresses of all participants. This information can be obtained from the detailed reports on the exercises referred to above.

ANALYTICAL METHODS

It was agreed from the outset that no attempts would be made to impose a standard analytical procedure for the measurement of metals in tissue. It was understood, however, that analysts who took part in the monitoring programmes and who produced results which were not comparable with the majority of results would need to re-examine their analytical methods with a view to replacing or modifying their techniques.

Most participants used atomic absorption spectrometry (AAS) procedures for the measurement of metals in marine tissue. In the first two exercises of the intercomparison programme analysts employed flame techniques for the analysis of copper, zinc, cadmium and lead and cold vapour technique for the analysis of mercury. The introduction of heated graphite furnace attachments in 1974 resulted in an increasing number of analysts employing flameless techniques for the analysis of cadmium and lead and a few used a chelation/extraction procedure to isolate these elements from the matrix. In the fourth exercise analysts were asked to include arsenic in their suite of elements; most analysts employed a hydride generation technqiue coupled with either a flame procedure (H_2/N_2 or H_2/Ar) or a flameless procedure utilising a heated quartz tube.

RESULTS

The results of the analyses of fish flesh and other tissue reference materials used in the five exercises are summarised in Table 2. The information presented in this table consists of (a) ranges of mean values submitted by participants, grand mean values and interlaboratory standard deviations (SD) and coefficients of variation (CV). The results of the analyses of acidified reference metal solutions, distributed in the second exercise, are presented in Table 3.

Table 2. Summary of the results of the analysis of reference materials distributed in the ICES metals intercomparison exercise (1971-1980)

Elements	Exercise	No. of Participants	Range of Mean Values submitted (μmol kg^{-1})	Grand Mean (μmol kg^{-1})	S.D.*	C.V.$^{+}$	Outliers (or Qualifi-cations)
Copper	1	6	173 - 314	267	47	18	None
	2a	7	130 - 159	144	11	8	None
	3	20	42 - 89	60	11	18	None
	4a	36	<6 - 63	28	11	39	<6 as 6
Zinc	1	6	597 - 1220	1010	214	21	None
	2a	7	352 - 490	413	46	11	None
	3	21	428 - 811	581	92	16	None
	4a	36	199 - 566	352	61	18	None
Mercury	1	7	0.45 - 5.5	0.80	0.3	37	One high value (5.5) omitted
	2a	8	3.0 - 4.2	3.5	0.35	10	None
	3	16	3.7 - 6.3	4.3	0.3	7	One high value (6.3) omitted
	4a	33	0.25 - 1.9	1.05	0.35	33	None
	4b	34	<0.05 - 1.3	0.3	0.15	50	All < values (two) and two high values (1.05 and 1.25) omitted

Table 2 (continued)

Elements	Exercise	No. of Participants	Range of Mean Values submitted (μmol kg^{-1})	Grand Mean (μmol kg^{-1})	S.D.*	C.V.+	Outliers (or Qualifi-
Cadmium	1	5	9.8 - 22.3	15.1	6.2	41	None
	2a	7	<0.18 - 9.8	6.6	2.8	43	One < value omitted
	3	21	<0.27 - 16.0	0.7	0.5	75	All < values (four) omitted
	4a	35	0.05 - 8.8	0.29	0.24	87	All < values (five) and four high values (2.5, 2.8, 3.5 and 8.8) omitted
	5a	52	4.7 - 9.9	7.1	4.8	17	None
Lead	1	6	4.8 - 43.2	26.4	16.3	62	None
	2a	5	6.2 - 14.4	8.6	2.4	28	One < value omitted
	3	21	0.77 - 19.2	5.3	4.3	82	One high value (19.2) omitted

	4a	32	0.96 - 36.0	1.0	0.7	71	All < values (seven) and two high values (13.4, 36.0) omitted
	5b	52	1.06 - 37.4	13.0	6.1	47	Two high values (29.3 and 37.4) omitted
	5c	32	0.53 - 15.4	3.6	2.5	71	One high value (15.4) omitted
Arsenic	4a	16	6.7 - 275	196	56	28	Three low values (6.7, 8.4 and 21.3) omitted

* SD = interlaboratory standard deviation
\+ CV = interlaboratory coefficient of variation

Table 3 Second ICES trace metal intercomparison exercise. Analysis of acidified metal solutions.

Labor-atory*	Mean value (μmol kg^{-1}) and SD^{+}				
	Copper	Zinc	Mercury	Cadmium	Lead
1	6.1±0.16	9.3±0.61	0.90±0.02	0.53±0.005	1.2±0.02
8	6.3±0.48	8.1±0.61	0.70±0.03	1.1 ±0.06	
9	7.9±0.32	8.3±0.61	0.50±0.04	0.62±0.09	0.3±0.10
10			0.60±0.01		
14	7.2±0.32	8.4±0.02	0.55±0.02	1.8 ±0.04	1.6±0.05
15	7.5	7.7	0.60	1.3	2.4
True value	6.3	7.7	0.50	0.9	1.4

+ intra laboratory standard deviation
* see Topping and Holden (1978) for laboratory identification

DISCUSSION

On the basis of the results obtained in the first exercise it was concluded that most of the participants produced comparable data for copper, zinc, cadmium and mercury but that they experienced difficulty in the analysis of lead. The second exercise showed that there was significant improvement in the comparability of the analytical results for copper, zinc and mercury analysis but little or no improvement was made in cadmium and lead analyses. The analysis of the acidified metal solutions revealed however that the majority of analysts were employing inaccurate working standards. An examination of the mercury data for both fish flour and acidified solution shows that there is a positive correlation between these two sets of results (Fig. 1). If the results of the fish flour analyses are adjusted to take into account the apparent differences in the strengths of the individual laboratory working standards, the reported range of values for mercury is reduced from 2.3-4.2 μmol kg^{-1} to 2.3-2.9 μmol kg^{-1} and the grand mean value is reduced from 3.5 μmol kg^{-1} to 2.9 μmol kg^{-1}. Similar calculations performed on the copper and zinc data reduce the ranges of values (mean value) as follows: copper, from 135-159 μmol kg^{-1} (148 μmol/kg) to 118-151 μmol kg^{-1} (135μmol/kg^{-1}) and zinc, from 352-490 μmol kg^{-1} (398 μmol kg^{-1}) to 337-428 μgmol kg^{-1} (382 μmol kg^{-1}).

The statistical examination of data reported in the third exercise showed that there were significant differences between results of copper, zinc and mercury (Table 4). The results submitted by laboratories which has participated in all three

Table 4. ICES third intercalibration exercise - Results of multiple range tests* - copper, zinc and mercury data

Copper (μmol kg^{-1})

Lab. No.	6	21	2	1	10	4	17	3	7	18	16	15	9	19	11	20	14	13	5
Mean value	42.2	47.3	47.9	48.5	52.6	53.1	53.1	53.7	55.4	56.2	57.6	58.4	60.0	60.3	60.6	65.0	66.3	68.6	89.2

Zinc (mol kg^{-1})

Lab. No.	21	17	7	15	19a	2	3	4	11	10	1	9	16	20	14	13
Mean value	474	480	516	532	545	549	560	565	575	588	589	600	617	640	655	806

Mercury (mol kg^{-1})

Lab. No.	13	17	1	9	14	15	3	4	8	16	18	10	6
Mean value	3.7	4.0	4.0	4.1	4.1	4.2	4.4	4.5	4.5	4.5	4.5	4.7	4.7

* see O'Neill and Wetherill (1971)
(Key to Data: any two or more mean values underscored by the same line are not significantly different)

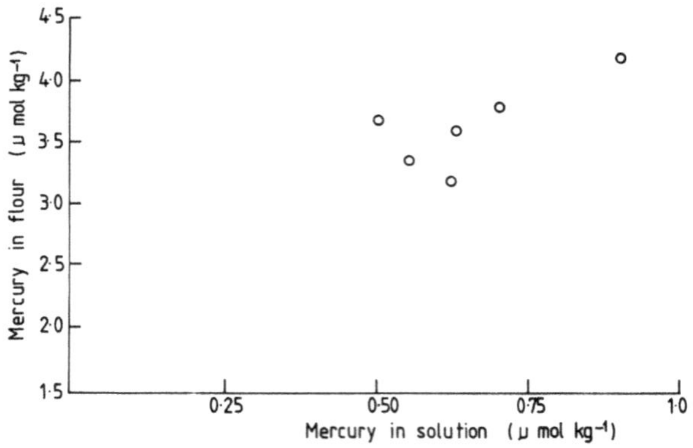

Fig. 1 Mercury concentrations in fish flour and in acidified reference solutions reported by participants in the second ICES exercise.

exercises revealed that there had been a progressive improvement in the comparability of their results (Table 5) e.g. the inter-laboratory coefficient of variation for mercury analysis being reduced from 38% to 5% between the first and third exercises. The results for the analysis of cadmium and lead however were not so encouraging and it was evident that all participants, including the regular ones, were still experiencing some difficulties in producing comparable data for relatively low tissue concentrations of these metals; i.e. those found in most fish muscle and some shellfish tissues. Following an examination of the methods employed by participants for these metals it was concluded that most of the procedures did not have a detection limit low enough to produce accurate data at these tissue concentrations. In anticipation of the fourth exercise, analysts were asked to consider changing or modifying their procedures to improve their accuracy at relatively low concentrations of cadmium and lead.

Despite the large number of participants in the fourth exercise the results for the analysis of copper, zinc and mercury demonstrated that most analysts were continuing to produce reasonably comparable and accurate data for these metals at levels typical of those found in fish muscle and shellfish tissue. The results for mercury were particularly good in view of the relatively low concentrations in samples A and B.

The analysis of arsenic appears to have posed problems for some of the analysts in view of the wide range of values reported in the fourth exercise i.e. 6.27-275 $\mu mol\ kg^{-1}$. An independent check of arsenic in the sample by analysts at the Atomic Energy Research Establishment, Hårwell, England, who used neutron activa-

Table 5 Summary table of copper, zinc and mercury results for regular participants* in the first, second and third exercises.

Exercise	Results Mean values (μmol kg^{-1}) and CV+ Copper	Zinc	Mercury
1	264 (21%)	979 (23%)	0.8 (39%)
2a	144 (8%)	413 (12%)	3.5 (12%)
3	58 (10%)	581 (8%)	4.2 (5%)
4a	25 (17%)	337 (17%)	0.9 (20%)
4b	-	-	0.15 (21%)

+ CV = interlaboratory coefficient of variance
* See Table 11 Topping and Holden (1978) for laboratory identification

tion analysis (NAA), produced a mean value of 200 µmol kg^{-1} was a CV of 6%. With the exception of one analyst, who used X-ray fluorescence (mean arsenic concentration of 216 µmol kg^{-1}), all analysts employed a similar, but individually modified, procedure for the analysis of arsenic, i.e. following destruction of the organic matter by wet digestion or dry ashing the arsenic was liberated from the resultant matrix as $As\ H_3$ and then measured by either flame and flameless AAS or colorimetry. (A brief description of the methods used in this exercise is given in Table 6). If it is assumed that the results produced by X-ray fluorescence and neutron activation analysis represent the 'true concentration' of arsenic in the reference material then the low results produced by some participants are incorrect. It follows that the methods used by these analysts may suffer from some form of matrix interference. From an analysis of the arsenic methodology it appears that the root of the analytical problem may lie with the choice of technique for the destruction of organic matter. This is suggested by the fact that all methods incorporating a dry ashing step produced high values (> 133 µmol kg^{-1}) whereas some methods employing a wet digestion step produced very low values, in the range 6.7-119 µmol kg^{-1}. Three of the five wet digestion procedures which produced high values appear to have overcome the effects of matrix interference by either the addition of nickel salts to the digest before measurement by the flameless AAS or by utilising a much stronger reducing agent at the arsine generation state. It appears that some component(s) of the matrix, which is destroyed or eliminated during dry ashing but not during wet

Table 6 Analysis of Arsenic - Methods employed by analysts in fourth exercise.

Arsenic Mean Value $\mu mol\ kg^{-1}$	Summary of method
6.7	WD/Reduction/FAA-heated quartz tube
8.4	WD/Reduction/Colorimetric using molybdate
21.0	WD/Reduction/FAA-heated quartz tube
70.1	WD/Reduction*/extraction/FAA-furnace
118	Wd/FAA (no nickel salts present)
162	DA/Reduction/colorimetric using molybdate
172	DA/Reduction-AsH_3 collected in cold trap/FAA-heated graphite furnace
181	WD/Reduction/FAA-heated quartz tube
186	DA/Reduction/colorimetric using Ag diethyldithio-carbamate
206	WD/FAA in the presence of nickel salts
209	WD/Reduction**/AA-argon/hydrogen flame
216	X-ray fluorescence
226	WD/Reduction/AA-nitrogen/hydrogen flame
269	DA/Reduction/AA-argon/hydrogen flame
275	WD/FAA in the presence of nickel salts
200	Neutron Activation Analysis

Key: WD - Wet Digestion; DA - Dry Ashing in the presence of MgO and Mg $(NO_3)_2$; Reduction - reduce As to AsH_3 using $NaBH_4$; Reduction* - reduction of As^{5+} to As^{3+} by ascorbic acid; Reduction** - very strong concentration of $NaBH_4$ used to reduce As to AsH_3; FAA - flameless atomic absorption; AA - Atomic absorption.

digestion, depresses the release of arsenic as AsH_3 and also suppresses the arsenic signal in flame and flameless AAS, unless nickel salts are added to the digest prior to measurement. In the light of these findings some of the analysts who participated in this exercise have agreed to change or modify their methods to bring them into line with the rest of the ICES analysts.

Although there was some evidence of improvement by particular laboratories in relation to the analysis of cadmium and lead the standard of data reported by most analysts was poor in terms of comparability and accuracy. The problem appears to lie in the choice of analytical procedure i.e. most analysts continued to use techniques which are not good enough to meet the required criteria of precision and accuracy at concentrations of 0.0X μmol kg^{-1} and 0.X μmol kg^{-1} for cadmium and lead respectively. The reluctance or inability of most analysts to improve or change analytical procedures in the intervening years is considered to have serious implications in relation to the monitoring of these elements in fish muscle. Not only are most of these analysts unable to produce comparable data for these metals in fish muscle samples, which are collected in a coordinated or cooperative monitoring exercise, but few analysts appear to be capable of measuring changes in the concentration of these elements in these samples.

The results of the fifth exercise show that the majority of participants can produce comparable (i.e. inter-laboratory CV of 10%), and accurate data for cadmium at a tissue concentration of ca 10 μmol kg^{-1} which is typically encountered in shellfish monitoring programmes. Unfortunately the results of the analysis of lead in samples B and C, with tissue concentrations of ca 12 μmol kg^{-1} and ca 2.5 μmol kg^{-1}, indicate that some participants still experience difficulties in producing comparable and accurate data at concentrations which cover the range of values encountered in shellfish. An examination of the results from 10 laboratories which participated in the third, fourth and fifth exercises and which are also participating in the ICES coordinated monitoring programme has been made to assess the analytical performance over a period of five years (Table 7). This study shows that the regular participants have achieved only slightly better level of agreement than the group as a whole for exercises 3, 4a and 5b but a much better level of agreement than the group as a whole for exercise 5c. Despite this superior performance the regular participants do not appear to have a sufficiently high standard of performance to enable the ICES working group realistically to compare lead data collected in shellfish monitoring programmes. It should be noted that a test of homogeneity on sample B, carried out in the author's laboratory, did not find evidence of gross inhomogeneity in this sample i.e. mean values and CVs for aliquots from 10 separate packets of this material was found to be 12 μmol kg^{-1} and 4% respectively (comparable values for cadmium in sample A are 7.0 μmol kg^{-1} and 4%). It appears therefore that differences between laboratories are attributable to the use of inaccurate analytical techniques.

The accuracy of an analytical technique is normally estimated by the analysis of a standard reference material e.g. National Bureau of Standards (NBS) Orchard leaves. These materials are certified on the basis of analytical results from two or more analysts employing two or more different analytical techniques. The results have to agree within certain limits for the sample to be classed as a standard reference material. The lack of appropriate standard marine reference materials[11,12] meant that the author in

Table 7 Analysis of lead in exercises 3, 4a, 5b and 5c: Comparison of results produced by regular participants with those of the entire group.

Exercise	Participants (Number)	Range of values (μmol kg^{-1})	Mean (μmol kg^{-1})	CV*	Outliers
3	All (21)	0.77-14.4	5.28	77	one high value
	Regulars (10)	0.77-10.1	3.79	75	one high value
4a	All (33)	0.086-3.41	1.01	72	all<values (7) and 2 high values
	Regulars^{+} (10)	0.086-2.16	0.86	70	one high value
5b	All (52)	1.06-22.3	12.1	38	two high values
	Regulars (10)	3.55-15.7	10.9	30	none
5c	All (39)	0.53-15.4	3.94	81	none
	Regulars (10)	1.44-14.4	3.36	47	none

* interlaboratory coefficient of variation
\+ Regular participants consist of Laboratory Nos. 3, 4, 15, 19, 24, 30, 34, 36, 45 and 46 (see Topping (1982)).

the past has been unable to state categorically which laboratories have produced the most accurate data for lead or cadmium in the ICES exercises. Fortunately, one of the participants in the fifth exercise produced four sets of analyses for lead in sample B, and three sets of analysis for sample C using different analytical procedures i.e. an "NBS type standardization". The methods used were (a) atomic absorption, (b) inductively coupled plasma emission spectroscopy and (c) isotope dilution solid-source mass spectrometry. Using these procedures the analyst produced the following over-all mean values and CVs for lead, 12.8 μmol kg^{-1} and 17% (sample B) and 1.6 μmol kg^{-1} and 18% (sample C). If one assumes that these values represent the 'true' concentrations of lead in these samples then 31 participants out of 49 in exercise 5b, and 9 participants out of 32 in exercise 5c produced mean values which fell within the limits for the true concentrations in the respective samples.

An alternative approach to the identification of laboratories with problems of lead analysis has been suggested by Dr. J. Uthe (personal communication). He states that a Youden plot of the individual laboratory's lead data for samples B and C (see Fig. 2) can be informative since it separates laboratories which produce consistently higher or lower values than the median values. He stresses that the plot as used here cannot differentiate between poor laboratories or poor methods but one must question a laboratory which produces either a high B value/low C value or a low B value/high C value. It is evident from Figure 2 that some of these laboratories must examine their methods and change or modify them if they wish to produce accurate data for lead at these concentrations in the future.

A breakdown of the analytical results for lead in sample B and C in relation to the analytical procedures reported by analysts in the fifth exercise is presented in Table 8. The data in this table shows that the analysts incorporating a chelation/extraction step (e.g APDC/MIBK or Dithizone/$CHCl_3$) in their atomic absorption procedure produced more comparable results than those employing the more commonly adopted wet digestion/atomic absorption procedure. It is worth noting that two analysts in the former group who were the only ones in this group to receive sample C produced a mean value for lead in sample C which was similar to that obtained by the analyst employing isotope dilution solid source mass spectrometry.

A comparison was made of the results obtained in the third, fourth and fifth ICES exercises with those from recent intercomparison exercises conducted by the International Atomic Energy Agency (IAEA), Monaco[13,14]. The IAEA exercises consisted of three

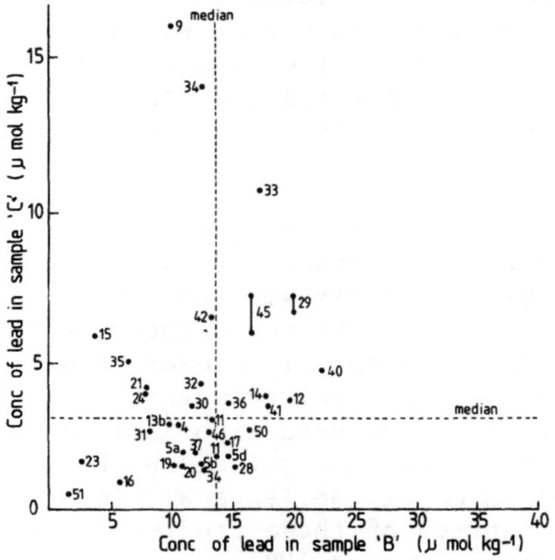

Fig. 2 Youden plot of the concentrations of lead in samples B and C (fifth exercise). (See Topping, (1982) for laboratory identification).

Table 8 Results of the analysis of lead in Samples B and C (fifth exercise) in relation to analytical technique.

Technique	Sample B No. of analysts	Sample B Mean Value (μmol kg^{-1})	Sample B CV*	Sample C No. of analysts	Sample C Mean Value (μmol kg^{-1})	Sample C CV*
Wet digestion /AAS+	30	13.0	58	24	4.42	86
Dry ashing/ AAS	9	13.6	24	3	3.36	76
Wet digestion /chelation /extraction /AAS	7	12.3	14	2	1.44	2
IDSSMS++	1	13.9	5**	1	1.68	18**

* intralaboratory coefficient of variation based on mean values submitted by analysts
** intralaboratory coefficient of variation based on six replicate analyses
\+ Atomic Absorption Spectrometry
\++ Isotope dilution solid-source mass spectrometry

reference materials prepared from fish flesh, a sea weed and zooplankton (copepod) and were conducted during 1978. This comparison (Table 9) reveals that participants in both intercomparison programmes achieved a similar level of performance for analysis of the five commonly examined trace metals. Since the IAEA exercise included participants from a wider geographic area than that of the ICES programme it illustrated that the problems associated with the analysis of cadmium and lead are experienced on a global basis.

In a recent excellent, review paper[15], the author who has considerable experience and expertise in the field of analytical chemistry deals comprehensively with the question of analytical accuracy and comparability. He emphasises the importance of regular participation in intercomparison exercises in relation to harmonization of methods and believes that these exercises can be of considerable value to analysts with limited experience. The results obtained and experience gained in the ICES intercomparison exercises have convinced all analysts within the ICES framework of the need for continuation of these studies, particularly in relation to lead analysis. During 1981 a small sub-group of the ICES analysts, working individually in the respective laboratories and

Table 9 Comparison of results of intercomparison exercises using marine reference materials

Element	Marine Reference Material	No. of Participants	Range of values µmol kg^{-1}	Mean µmol kg^{-1}	SD	CV	Reference
Copper	Fish flesh	64	7.9–276	70.7	37.7	53	IAEA (1980)
	Fish flesh	20	42.4–89.5	59.7	11.0	18	Topping and Holden (1978)
	Fish flesh	36	6.3–62.8	28.3	11.0	39	Holden and Topping (1981)
	Sea plant	67	51.8–675	198	50.2	26	IAEA (1978)
	Copepod	56	70.7–256	121	23.6	19	IAEA (1978)
Zinc	Fish flesh	74	61.2–3121	505	122	26	IAEA (1980)
	Fish flesh	21	428–811	581	92	16	Topping and Holden (1978)
	Fish flesh	36	199–566	352	61	18	Holden and Topping (1981)
	Sea plant	75	23–3443	979	260	27	IAEA (1978)
	Copepod	66	64.3–3779	2433	367	15	IAEA (1978)
Mercury	Fish flesh	52	0.9–17.5	2.5	0.7	29	IAEA (1980)
	Fish flesh	16	3.4–6.3	4.3	0.3	7	Topping and Holden (1978)
	Fish flesh	33	0.27–1.85	1.05	0.4	33	Holden and Topping (1981)
	Fish flesh	34	0.05–1.25	0.3	0.2	50	Holden and Topping (1981)
	Sea plant	75	0.25–28	1.7	0.7	43	IAEA (1978)
	Copepod	43	0.20–18	1.4	0.7	48	IAEA (1978)
Cadmium	Fish flesh	41	0.18–9.8	1.5	2.3	150	IAEA (1980)
	Fish flesh	21	0.27–16.0	0.71	0.53	75	Topping and Holden (1978)
	Fish flesh	35	0.05–8.8	0.29	0.24	87	Holden and Topping (1981)
	Crab meat	52	4.7–9.9	7.1	1.2	17	Topping (1982)
	Sea plant	46	1.0–214	6.2	6.2	96	IAEA (1978)
	Copepod	43	3.0–26.7	6.7	1.7	26	IAEA (1978)
Lead	Fish flesh	33	0.48–24.5	3.8	2.9	71	IAEA (1980)
	Fish flesh	21	0.77–19.2	5.3	4.3	82	Topping and Holden (1978)
	Fish flesh	32	0.10–36.0	1.0	0.7	71	Holden and Topping (1981)
	Fish meal	52	1.06–37.4	13.0	6.1	47	Topping (1982)
	Lobster liver	32	0.53–15.4	3.6	2.5	71	Topping (1982)

communicating by correspondence, will systematically evaluate one method of lead analysis i.e. wet digestion of tissue followed by measurement by AAS. They will examine, evaluate and make recommendations on the following technical points (a) digestion of tissue, (b) elimination or avoidance of matrix interference, (c) factors affecting blank values, (d) reduction of background contamination and (e) calibration of method. In due course the recommended analytical procedure(s) for lead in tissue will be tested by ICES analysts in the next round of its intercomparison programme.

SUMMARY

During 1971-1980 ICES conducted five intercomparison exercises for metals using specially prepared fish flour reference materials. The over-all results of this study compare favourably with the results obtained by IAEA in a similar exercise conducted during 1977-1979.

Throughout this study the participants, particularly those who took part in all of the exercises, showed a progressive improvement in the analysis of copper, zinc and mercury. On the basis of these results it is concluded that the analytical data for these metals produced by these participants in a fish and shellfish monitoring programme are comparable.

The identification of significant differences in the working standards concentrations and the subsequent adoption of a common procedure for the preparation of these solutions are considered to be important factors in the achievement of this improved comparability for the above metals.

The study revealed that the participants were unable to produce comparable, and in most cases accurate, data for lead and cadmium at low tissue concentrations i.e. in the range 0.024-4.8 μmol kg^{-1} and 0.009-0.89 μmol kg^{-1} respectively. However at relatively high tissue concentration (2.5-12.0 μmol kg^{-1} and 10 μmol kg^{-1} respectively) the majority of analysts experienced little difficulty in producing accurate data for cadmium but that the analysis of lead presented some problems for a minority of the participants. On the basis of these results it is considered that the participants in ICES fish and shellfish monitoring progammes can produce comparable data for cadmium in shellfish tissue but not for cadmium in fish muscle or lead in both fish muscle and shellfish tissue.

Further intercomparison exercises are required periodically to assess the comparability and accuracy of analytical data for copper, zinc, mercury and cadmium in future ICES monitoring programmes and to examine any improvements made in lead analysis following the adoption of more accurate analytical procedures.

REFERENCES

1. Anon., 1974: Report of Working Group for the International Study of the Pollution of the North Sea and its Effects on Living Resources and their Exploitation. ICES Coop. Res. Rep. No. 39.
2. Anon., 1977: A baseline study of the level of contaminating substances in living resources of the North Atlantic. ICES Coop. Res. Rep. No. 69.
3. Anon., 1980: Extensions to the baseline study of contaminant levels in living resources of the North Atlantic. ICES Coop. Res. Rep. No. 95.
4. Anon., 1977a: The ICES Coordinated Monitoring Programme in the North Sea, 1974. ICES Coop. Res. Rep. No. 58.
5. Anon., 1977b: The ICES Coordinated Monitoring Programmes, 1975 and 1976. ICES Coop. Res. Rep. No. 72.
6. Anon., 1980: The ICES Coordinated Monitoring Programme, 1977. ICES Coop. Res. Rep. No. 98.
7. Topping, G. and A.V. Holden, 1978: Report on intercalibration analysis in ICES North Sea and North Atlantic baseline studies. ICES Coop. Res. Rep. No. 80.
8. Topping, G., 1982: Report on 6th ICES trace metal intercomparison exercise for cadmium and lead in biological tissue. ICES Coop. Res. Rep. No. 111.
9. Holden, A.V. and G. Topping, 1981: Review on further intercalibration analysis in ICES Pollution Monitoring and Baseline Studies. ICES Coop. Res. Rep. No. 108.
10. O'Neill, R. and G.B. Wetherill, 1971: The present state of multiple comparison methods. J. Roy. Stat. Soc. Series B., 33, 218-250.
11. Anon., 1972: Baseline studies of pollutants in the marine environment and research recommendations. The IDOE Baseline Conference, May 24-26, 1972. New York, 1972.
12. Beckert, W.F., 1978: Mercury, lead, arsenic and cadmium in biological tissue: the need for adequate standard reference materials. US Dept. of Commerce. National Technical Information Service PB-288-198.
13. IAEA, 1978: International Atomic Energy Agency, Monaco: Intercalibration of analytical methods on marine environmental samples. Progress Report No. 19. (November, 1978).
14. IAEA, 1980: International Atomic Energy Agency, Monaco: Intercalibration of analytical methods on marine environmental samples. Progress Report No. 20. (April, 1978).
15. Hamilton, E.I., 1976: Review of the Chemical Elements and Environmental Chemistry - Strategies and Tactics. Sci. Total Environ., 5, (1976), 1-62.

AN INTERCOMPARISON OF SAMPLING DEVICES AND ANALYTICAL TECHNIQUES USING SEA WATER FROM A CEPEX ENCLOSURE

C.S. Wong[1], K. Kremling[2], J.P. Riley[3], W.K. Johnson[1], V. Stukas[1], P.G. Berrang[1]*, P. Erickson[1]*, D. Thomas[1]*, H. Petersen[2], and B. Imber[3]

[1] Ocean Chemistry Division, Institute of Ocean Sciences, P.O. Box 6000 Sidney, B.C., V8L 4B2, Canada

*[1] Ocean Chemistry Division contract to SEAKEM Oceanography Ltd., Sidney, B.C. Canada

[2] Marine Chemistry Department, Institut für Meereskunde and der Universität, Kiel, Dunsternbrooker Weg 20, 2300 Kiel, F.R. Germany

[3] Department of Oceanography, University of Liverpool, P.O. Box 147, Liverpool, L69 3BX, U.K.

ABSTRACT

An intercomparison of sampling devices was conducted using sea water at 9 m. in a plastic enclosure of 65 m^3 in Saanich Inlet, B.C., Canada. The sampling methods were (i) peristaltic pumping with teflon tubing, (ii) Niskin PVC sampler, (iii) Go-Flow sampler, (iv) Close-open-close sampler, and (v) teflon-piston sampler. Sampling was conducted for 4 days: Day 1 (2 August, 1978) for mercury, Day 2 for lead, cadmium, copper, cobalt and nickel by Chelex extraction and differential pulse polarography (D.P.P.) as well as manganese by Chelex and flameless atomic absorptiometry (F.A.A.), Day 3 for lead by isotope dilution and Day 4 for cadmium, copper, iron, lead, nickel and zinc by freon extraction and F.A.A. Samples were processed in clean rooms in the shore laboratory within 30 minutes of sampling. Results indicated the feasibility of intercalibrating using the enclosure approach, the availability of chemical techniques of sufficient precision in the cases of copper, nickel, lead and cobalt for sampler intercomparison and storage tests, a problem in subsampling from the captured sea water in a sampler, and the difficulty of commonly used samplers to sample sea water in an uncontaminated way at the desired depth.

INTRODUCTION

Accurate measurements of trace metals in sea water have been the concern of marine chemists and geochemists in recent years in their quest for establishing the baseline levels of trace metals in sea water and in particular, in the unpolluted open ocean. The "apparent baseline level" of these metals, e.g. mercury, lead, zinc and copper, has been decreasing since earlier measurements in the 1950s-70s. This is not due to decreasing pollution, but to increasing experience of marine chemists in tackling the immense analytical and sampling problems. The disagreement between reported values may be due to one or more factors: lack of sensitive detection techniques, contamination introduced during various stages of sample processing and analysis, failure to obtain representative samples from the ocean, and the changes occurring after sampling and storage.

Intercalibration is an important tool to sharpen our approach. Successful international intercomparisons include in particular the work on lead in sea water by Patterson[1], mercury and cadmium by Sugawara[2], the Baltic trace metals by Kremling[3], the I.C.E.S. sea water intercomparison reported by Jones[4] and more recently the I.O.C. intercalibration reported by Bewers et al.[5,6]. Participation of a large number of laboratories usually created logistic problems in performance, in transporting and setting up gear in an unfamiliar working environment and in time-tabling of participants from various stages of gearing up. The N.A.T.O. funded approach described here is to develop a sea water reference material for trace metals using deep water from ocean weather station P (50°N 145°W) so that such a reference sea water may be stored and distributed to interested laboratories. It depends on work on stability of storage[7], ability to sample uncontaminated sea water[8], guarantee of cleanliness during processing[8,9], and detection methods of sufficient accuracy[10,11,12].

METHODOLOGY

Sea water used in the intercalibration exercise was sampled from a CEPEX (Controlled Ecosystem Pollution Experiment) enclosure[13] of sea water, approximately 65 m^3 in volume inside a polyethylene bag, 2.5 m diameter x 16 m length, suspended from a lucite flotation module. The bag was launched by divers at a 20 m depth with the folded bag being stretched and capturing the sea water column while the divers were swimming to the surface within 2 minutes.

Five sampling devices were used: a peristaltic pumping system, a Niskin sampler, a Go-Flow sampler, a close-open-close sampler and a teflon-piston sampler. The peristaltic pumping system consists of a peristaltic pump (Little Giant Model, teflon tubing of 0.64 cm diameter I.D., precleaned by immersion in 0.05% HNO_3, and a gas-operated generator to power the pump. A vinyl coated hydro-wire with a stainless steel weight covered with polyethylene was used to lower the four water bottles to the desired depth. A 5 l Niskin sampler (General Oceanics) of PVC was used with a teflon coated stainless steel spring closure mechanism and a Delrin drain spigot and tripping rod. The sampler was closed by means of a teflon block containing three stainless steel plugs sealed with screwin teflon caps. The metallic exposed parts were the spring, wing nuts and release pins in stainless steel. The Go-Flow sampler (General Oceanics) was all PVC construction for the cylindrical body, ball valves, with Delrin stopcocks and push rod, stainless steel wing nuts from the wire clamp, Latex external spring and Viton O-ring seals. The sampler was lowered through the surface microlayer with the ball valves closed; it was opened by hydrostatic pressure at about 8 m depth. The sampler at the desired depth was then closed by a triggering mechanism using the teflon messenger. The closed-open-closed sampler (Hydro-Bios) was made of transparent and shock-resistant polycarbonate, with PVC ball valves with teflon seals, a tripping mechanism of stainless steel, amber Latex tubing, stainless steel wire fasteners and plastic endcaps and spigots. It was operated by lowering the sampler down in a closed position to protect the interior from contamination through the surface microlayer. It was then opened by pulling on a line attached to a stainless steel eye on the tipping mechanism, and after equilibration, closed by sending down a teflon messenger to trigger a closing mechanism so that the bottle would rotate 90° to close the ball valve. The 2.5 l teflon-piston sampler (SEAKEM Oceanography Ltd.) is all plastic construction with a main body made of virgin TFE teflon, with the triggering mechanism made of either nylon or delrin and the spring mechanism of a surgical tubing enclosed inside silicon tubing of low metal content. The inside of the cylinder was protected from surface contamination by means of a diposable plastic bag. It was lowered through the surface waters in a protective closed position and triggered by a messenger so that the teflon plunger could pull in sea water via a one-way valve, through the tension of the surgical tubing.

CLEANING PROCEDURES

The teflon tubing used in the pumping system, the Niskin sampler and the Go-Flow sampler were cleaned by immersion in 0.05% HNO_3 for the tubing and by soaking the inside of the samplers in 0.05% HNO_3 overnight, rinsing with distilled water and repeating the dilute acid/distilled water cycle. The close-open-close sampler was cleaned at Kiel by 0.1N HNO_3 overnight, then rinsed with distilled water till the blank was acceptable. The teflon-piston sampler was cleaned by sucking in 0.05% HNO_3 and standing overnight. This deviated from the previous cleaning procedures involving a complete dissassembling, and washing of the TFE teflon parts in Sparkleen soap solution, then rinsing with methanol and distilled water, heating for three days in HNO_3 at 60°C (this step omitted for nylon parts), rinsing with Milli-Q water, soaking again for three days at 60°C with 3% (v/v) solution of HNO_3 (Aristar grade) in Milli-Q water, with a final soaking in a fresh dilute HNO_3 solution for one day. In the case of the poly bag liner used in the teflon-piston sampler, HCl was used instead of HNO_3. This departure in cleaning might have caused problems in this sampler. After cleaning, the samplers or tubing were stored in wooden boxes painted with aqua pon epoxy 2 part paint (Pittsburg Paints, using low metal TiO white base). The Kiel sampler was enclosed in a polyethylene bag at Kiel prior to sampling.

The storage bottles were cleaned as follows. The Pyrex bottles (2 l) were used for mercury samples only. They were cleaned by filling with a solution of 0.1% $KMnO_4$, 0.1% $K_2S_2O_8$ and 2% HNO_3 heating to 80°C for two hours, and after cooling and rinsing, stored filled with 2% HNO_3 containing 0.01% $K_2Cr_2O_7$ until ready for use. Conventional polyethylene (CPE) bottles of 1 or 2 l sizes, were used for the other metal samples. They were cleaned by Patterson's method[9] which involved rinsing with $CHCl_3$, soaking for 3 days in hot HCl, filling with dilute 0.05% HCl and heating for one day, filling again with dilute HCl and heating for five days, and finally stored at room temperature with 0.05% HCl. All bottles were stored inside two or three plastic bags to prevent contamination.

SAMPLING PROCEDURES

Sampling was carried out from a 2.7 m x 4 m raft secured to the lucite plexiglass structure. A wooden tripod on the Lucite structure was used to support a plastic block, carrying a vinyl-coated stainless steel wire from which a stainless steel weight coated with thin polyethylene was suspended. A sampling depth of 9 m was chosen because chemical parameters such as O_2, pH, salinity and temperature showed the least rate of change there. Sixty casts were performed over four days, with the collection of 3 sets of samples for each sampling procedure for a given day with one aliquot for salinity and two aliquots for trace metals.

For the pumping system, sea water was pumped up from 9 m and collected in the appropriate bottles on the raft and returned to the shore clean laboratory for preservation and/or analysis. For the other four sampling devices, the sampler was lowered to 9 m, allowed to equilibrate for 10 minutes, closed by a triggering mechanism activated by the teflon messenger, raised to the surface, transferred into the container, transported back by boat and trucked back to the shore clean laboratory, where the subsamples were drawn. The time between messenger activation and subsampling was about 30 minutes. For handling of the samples, messengers, teflon tubing, vinyl-coated hydrowires and sampling devices, all personnel wore polyethylene gloves to avoid contamination.

CLEAN ROOM FACILITIES

The clean laboratory for trace metals at Ocean Chemistry Division consists of 6 rooms with about 78 m^2 area. It was divided into three areas: entrance laboratory, instrument laboratory and the ultra-clean sample preparation laboratory, all under positive pressure with filtered air. A Mark Hot system supplies up to 290 $m^3\ min^{-1}$ of filtered air through a system of a prefilter, an activated charcoal filter and a series of absolute (or high efficiency) filters, with 99.97% removal of 0.3 μm sized particles at a static pressure of 10.8 cm Hg and with an air exchange of six times per hour. Personnel using the clean rooms are required to wear hair caps, polyethylene gloves, laboratory coats and designated shoes. These items are worn <u>only</u> in the clean rooms. A small changing room with sliding doors between the entrance laboratory and instrument laboratory was used for switching shoes and coats in preparation for entry into a clean area (instrument laboratory) and superclean area (the sample preparation laboratory). The background levels in the clean air and types of water used are shown in Table 1.

ANALYTICAL PROCEDURES

Sea water samples for mercury determinations were collected on the first day, subsampled into 1 liter bottles in the clean room and preservatives (20 ml conc. HNO_3 and 2 ml. 5% $K_2Cr_2O_7$) were added. The samples were stored for approximately 1 month so that mercury combined with the organic matter in sea water would be released as a labile form[2]. At the end of the storage period, 10 ml of 5% $KMnO_4$ and 10 ml of 5% $K_2S_2O_8$ were added to the sample inside the original 1 liter bottle and the contents then heated at 80°C for 2 hours. Aliquots of 100 ml were transferred to 250 ml Erlenmeyer flasks and reduced by 5 ml of 20% $SnCl_2$. The flask was immediately connected to the bubbler apparatus and Hg was purged with N_2 gas through the 30 cm path cell of the cold vapor mercury analyzer at 254 μm which was calibrated by standard additions on selected samples. For each

Table 1(a). Trace metal concentrations in purified water at the Institute of Ocean Sciences (ng/kg).

	Analysis	Lead		Cadmium				Nickel		Zinc		Iron	
		1978	1980	1978	1980	1978	1980	1978	1980	1978	1980	1978	1980
Glass Distilled Ag-11	Direct-AA		10		9		20		20				100
Reverse Osmosis	Direct-AA	100		<40		1000				2400			
Ion Exchange	Ex-AA	8		0.4				<10		300		400	
Milli Q	Direct-AA					<400							
Double Milli Q	Ex-AA		31		6		15		20				200
(in series)	Ex-MS		0.8		1.0		4				33		170
Quartz Sub-	Ex-AA	1	<1	0.2	0.2	0.6	<2		10	300	<10	600	<10
boiling Distill-ation	Ex-MS	0.11	0.05		0.08		0.8		4		1.6		4

Direct-AA: Analysed directly by flameless atomic absorption
Ex-AA : Extracted and concentrated 50 times before atomic absorption analysis
Ex-AA : Extracted and analysed by isotopic dilution mass spectrometry

Note: AA and MS samples are not duplicates and were analysed at different times of the year

Table 1(b). Levels of contamination from laboratory air: metal enhancement in nanograms per day of exposure[1].

	Cadmium		Copper		Iron		Lead		Nickel		Zinc	
	1979	1980	1979	1980	1979	1980	1979	1980	1979	1980	1979	1980
General Labs												
Office 2414			<.1		6.0		0.68		0.2		0.9	
Mercury Lab	<.01		0.4		3.5		0.30		0.2		1.1	
Lab 2418	<.01	0.01	0.2	0.2	7.1		0.50	0.15	0.3	0.7	2.7	2.6
Marine Lab		0.02		<.1				0.89		0.3		8.8
Clean Labs												
Entrance Lab	<.01	0.01	27.	<.1	9.3		1.5	0.11	<.2	0.3	3.6	
Change Room	0.11	0.02	0.8	0.4	5.6		0.43	0.11	<.2	1.0	1.6	2.3
Prep. Lab	<.01	0.01	<.1	<.1	2.2		0.18	0.08	<.2	0.3	1.2	0.4
Prep. Lab[2]							0.06					
Extraction Lab[2]							0.18					
Extraction Lab	<.01		0.1		2.0		0.27		<.2		0.9	
Plexiglass Hood		<.01		<.1				0.13		<.2		0.7
Hot Plate Area		0.02		0.1				0.08		0.2		1.7
M.S. Prep. Lab	0.01		<.1		2.7		0.03		<.2		0.6	
M.S. Prep. Lab[2]							0.05					
Still Area		0.01		<.1				0.08		0.5		
Clean Hood	0.04	<.01	<.1	0.1	0.1		0.02	0.05	<.2	<.2	0.4	1.4
Clean Hood[2]		0.0001				0.069	0.003	0.002		0.007		0.010

[1] Teflon beakers with slightly acidified water were exposed to air for six to ten days

[2] These values are from IDMS analysis, all others from flameless AA

sample, 5 to 10 aliquots were analyzed, as detailed in Bothner and Robertson[14].

Lead in sea water was analyzed by a stable isotope dilution technique[1,15,16]. After subsampling in the clean room, 0.1% NBS HCl and a spike of 100 µl of Pb^{206} was added to the sample inside a laminar flow clean hood. After digesting the sample at 55°C for 24 hours, the sample was placed inside 3 CPE plastic bags and stored frozen until analysis. Upon thawing, the sample was neutralized to pH8.0±0.2 with purified conc. ammonia solution then extracted with 35 ml 0.001% dithizone in $CHCl_3$ (purified by sub-boiling distillation). The lead was back-extracted with 2.5 ml 1% HCl (NBS sub-boiling distilled). The HCl solution was evaporated down to a small volume and loaded on a rhenium filament with a silica gel substrate. The entire operation was performed inside laminar flow clean hoods inside the clean room. All reagents were analyzed for acceptability of the blank prior to use. The total procedure blank, i.e. for reagents and handling, in extracting about 2 l of sea water was about 5.0 ng in 1977 and 0.75 ng in 1978. The latter value was close to the assessed reagent contributions, thus indicating negligible handling blank. A Nuclide 12-90-SU mass spectrometer was used for the isotope determinations with a precision of 0.5%, based on a mass discrimination of 0.28%/a.m.u., and a precision in isotope-ratio determinations (1σ) of 0.08% to 0.12% on the spike and the 1978 blank.

Measurement of Cu, Pb, Cd, Ni, and Co was performed by differential pulsed polarography (D.P.P.) following concentration by Chelex 100 ion exchange resin. After arrival in the clean room, samples were decanted into CPE bottles, precleaned by the Patterson technique[1] and immediately passed through 5 x 1 cm colums of Chelex 100 ion exchange resin in the calcium form. After washing the columns with quartz-distilled water, the metals were eluted with 50 ml of 2N HNO_3 into 60 ml precleaned polyethylene bottles. The concentrates were transferred to P.T.F.E. bombs and refluxed for 12 hours on a hot plate at 68°C to reduce the organic concentration. The tops of the bombs were then removed and the samples evaporated to dryness at 80°C. The residues were then taken up in 20 ml 0.1N HCl and again evaporated to dryness. The final residues were taken up in 10 ml 0.1N HCl. For Cu, Pb, Cd, Ni and Co, the polarographic procedure of Abdullah and co-workers[17,18] was applied using a dropping mercury electrode, with the modification of using 2 ml of concentrate from the sample, 0.5 ml of ammonia solution and 0.2 ml of dimethylglyoxine (DMG). Each concentrate was first spiked with three increments of Cu, Pb and Cd standard solutions. Ammonia solution was then added, followed by three increments of Co standard. Manganese was determined by flameless atomic absorption spectrophotometry (FAAS) using a Perkin-Elmer 503 instrument.

Table 2. Precision of the Danielsson Procedure For Cu, Ni, Cd, Fe, and Pb as applied in Ocean Chemistry Clean Room

Metal	Sea water Concentration nmol kg^{-1}	Blank nmol	Relative standard deviation at test level (average of 10 analyses)	% Recovery
Cd	1.16	0.009	6%	83-98
Cu	13.0	0.08	2%	95
Ni	14.8	0.12	2%	95
Zn	32.3	0.69	2%	95
Fe	7.8	0.54	3%	90
Pb	0.10	0.02	30%	85-100

The metals Cu, Ni, Cd, Zn, Pb and Fe were also determined by the freon TF extraction technique of Danielsson et al.[19] with modifications, using F.A.A. 500 ml acidified sea water was transferred to a teflon separatory funnel. 3 ml of purified buffer of 0.5 M diammonium hydrogen citrate, and 3 ml of a solution of 1% w/v each of ammonium pyrrolidinedithiocarbamate (APDC) and of diethylammonium diethyldithiocarbamate (DDDC) was added, followed immediately by 20 ml of Freon TF (1,1, 2-trichloro-1,2,2,-trifluoroethane), and the separatory funnel was shaken for 150 seconds vigorously. The lower organic phase, after allowing to settle, was transferred to a stoppered test tube. Ten ml of Freon TF was added and the funnel was shaken for 30 seconds, and the settled organic phase again combined into the first extract in the tube. A volume of 0.4 ml of conc. HNO_3 (NBS), a quantity different from the original Danielsson procedure, was added and shaken for 20 seconds. The aqueous extract was used for determination of Cu, Ni, Cd, Zn, Pb and Fe by FAAS, using a Perkin Elmer-503 atomic absorption spectrometer and an HGA-2100 graphite furnace. The samples were standardized by spiking a previously extracted sample with a mixed standard solution and re-extracting. A mean value of four such spiking operations was used to calculate the concentration of the metals in the original samples. The above operations were carried out in clean rooms, possibly an improved laboratory environment over that in Danielsson et al.[19]. The precision of the procedures under clean room conditions is shown in Table 2.

RESULTS AND DISCUSSIONS

The results [16] are plotted as determined concentrations vs time, for mercury in various collecting systems (Fig. 1) for lead analyzed by isotope dilution (Fig. 2), for Pb, Cu, Ni, Cd, Fe, Zn analyzed by freon extraction and F.A.A. (Fig. 3) and for Pb, Cu, Ni, Cd, Co, Mn determined by Chelex-100 extraction and D.P.P. analysis (Fig. 4).

The results showed that the enclosure approach was feasible for the purpose of sampler intercalibration. It may even have advantages over open-ocean intercalibration since the criteria of success are homogeneity of samples and speedy access to a contaminant-free clean laboratory. The exercise also confirmed the extra precautions required in attaining the true levels of metals, and in particular lead, in sea water. The lead level in the enclosure is an order of magnitude above the ambient level outside the bag. The isotope ratio of the lead in the enclosure sea water was found to be very different from the ratio in the local sea water, i.e. about 1.199 for $^{206}Pb/^{207}Pb$, and was very similar to that of gasoline lead. The elevated level appeared to come from the exhaust of motor boats frequenting the area during the two weeks with the bag open to the atmosphere. Only 0.05 nmol of lead would be required to elevate the lead concentration in 65,000 liters of sea water from 0.10 nmol kg^{-1} to 0.72 nmol kg^{-1}, and lead is present in regular gasoline at 0.5 nmol l^{-1}. For sampler intercalibration, such contamination though undesirable should not affect the practicality of intercomparison since the same body of homogeneous water is available. However small amounts of contamination that would not be detectable at these elevated levels might cause problems at natural levels.

An enclosed body of water can be manipulated to minimize interfering factors, thus representing an advantage over open-ocean conditions. To minimize the necessity to filter off detritus, organic detritus was removed in the following manner. After launching the system, the enclosure was allowed to stand in the open for about 4 weeks prior to the study. It was not covered for the first two weeks to permit the sunlight to promote photosynthesis. The plankton grew and died off. Detritus sank to the bottom of the system and the material that settled out was removed by pumping through a ball valve. In the second two weeks prior to sampling, the system was covered with a sheet of black plastic to prevent introduction of droppings from seagulls and exhausts from motor boats. The lead values were high, as discussed. The Hg values are between 0.06 and 0.12 nmol kg^{-1} (excluding the samples from the teflon sampler). The Hg values are high if compared with recent values of 0.02 - 0.04 nmol kg^{-1} from Saanich Inlet[16], although within the range of data in other parts of the world ocean. The Zn, Cd and Cu levels in the enclosure water are in reasonable agreement with values for

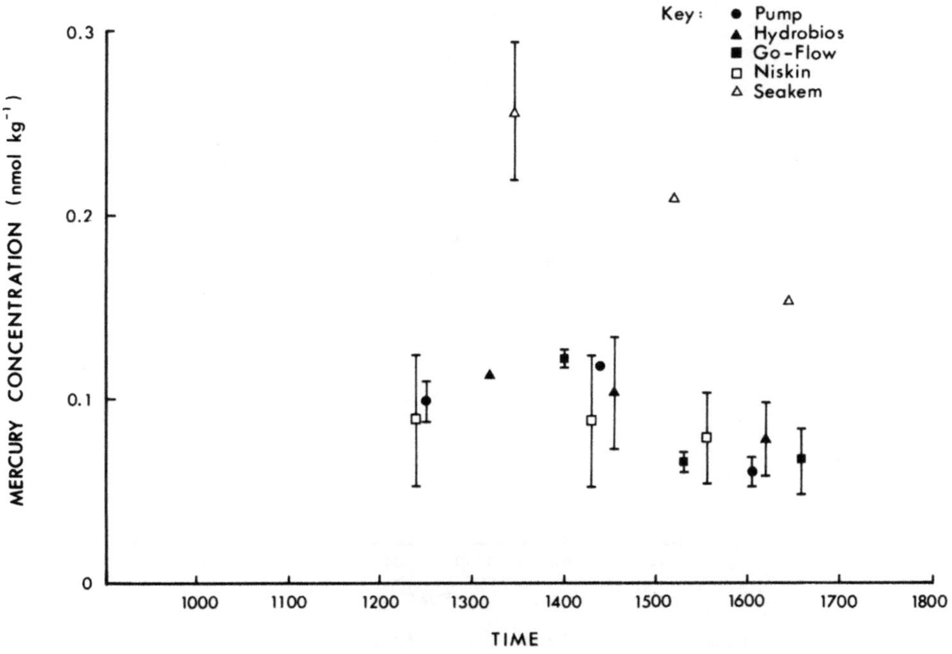

Fig. 1. Comparison of mercury concentrations found in a CEPEX enclosure using five different sampling methods.

Saanich Inlet 100 m deep water of 8.1-10.9 nmol kg^{-1} for Zn, 0.53 nmol kg^{-1} for Cd and 0.39-0.57 nmol kg^{-1} for Cu obtained with the teflon sampler previously[19]. Thus, the enclosure approach may yield reasonable representation of natural coastal waters for sampler intercomparison purposes, though further refinements are necessary, such as covering the system all the time, venting the exhaust from pumps and the power generating system near the site away from the bag using a submerged pipe, and use of manually powered boat. Atmospheric input of metal contaminants can also be circumvented by a new generation of enclosures CHEMCELL containing 500,000 l of seawater. This system is submerged 2 m below the surface with a sampling tube, normally closed, extending to the surface, while the bottom of the enclosure is attached to the sea bed.

The salinity data for samples from the enclosure revealed some shortcomings of existing samplers. Collection of sea water by peristaltic pumping at 9 m gave a rather uniform salinity of $29.114 \pm 0.0039 \times 10^{-3}$ over the 4 days of sampling. The Niskin sampler in general gave good agreement with the salinity from pumped samples but showed an occasional higher value during the last two days. The Go-Flow showed good agreement in the first

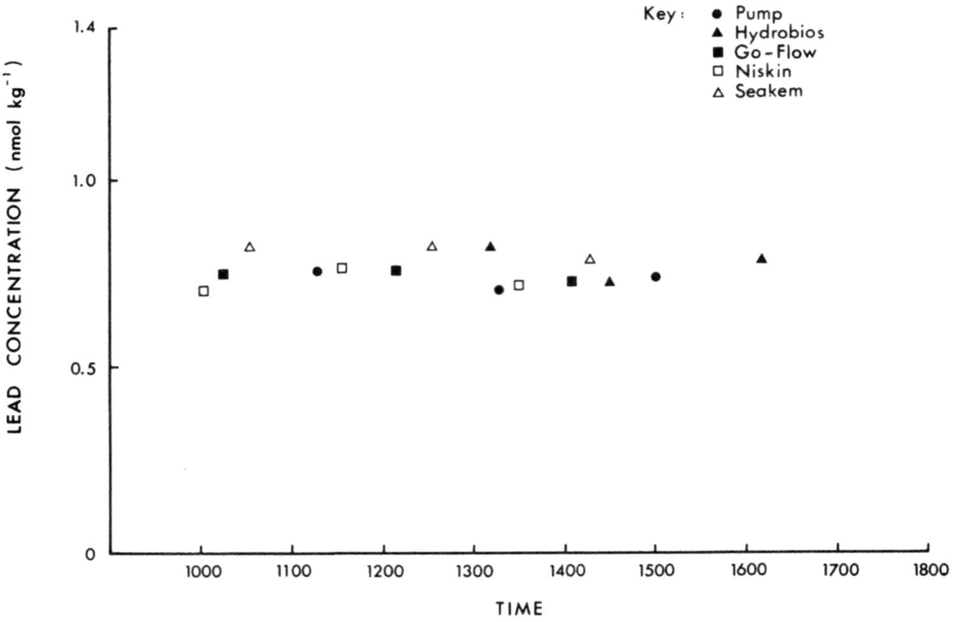

Fig. 2. Comparison of lead concentrations found in a CEPEX enclosure using five different sampling methods and IDMS analysis.

two days but again marginally higher salinity in the last two days. The close-open-close sampler, which operated on the same principle as the Go-Flow, agreed well with pumped samples but with a spread of 0.01×10^{-3}. The overall salinity average of each of these samples was slightly above that of the pump. This data suggests that these commercial samplers may be contaminated by adjacent waters a few meters above the desired depth. Another possibility is that salt may have been concentrated in the 'O' ring grooves and crevices of the samplers during the period between successive samples (approximately 1.5 hr). Such sampling inaccuracy is often masked by flushing action when the sampling bottle is being lowered into deeper waters. In the open ocean, turbulence and wave action tend to minimize these problems. The teflon-piston sampler showed erratic behaviour with abnormally low salinity on the first and second day and an abnormally high value on the fourth day. These anomalies are most likely due to the fact that this sampler does not flush and is therefore affected by water which may have remained in the sampler from previous use (i.e. cleaning, rinsing or sample water).

The average mercury contents obtained by pumping, Niskin sampler, Go-Flow sampler and the close-open-close device are

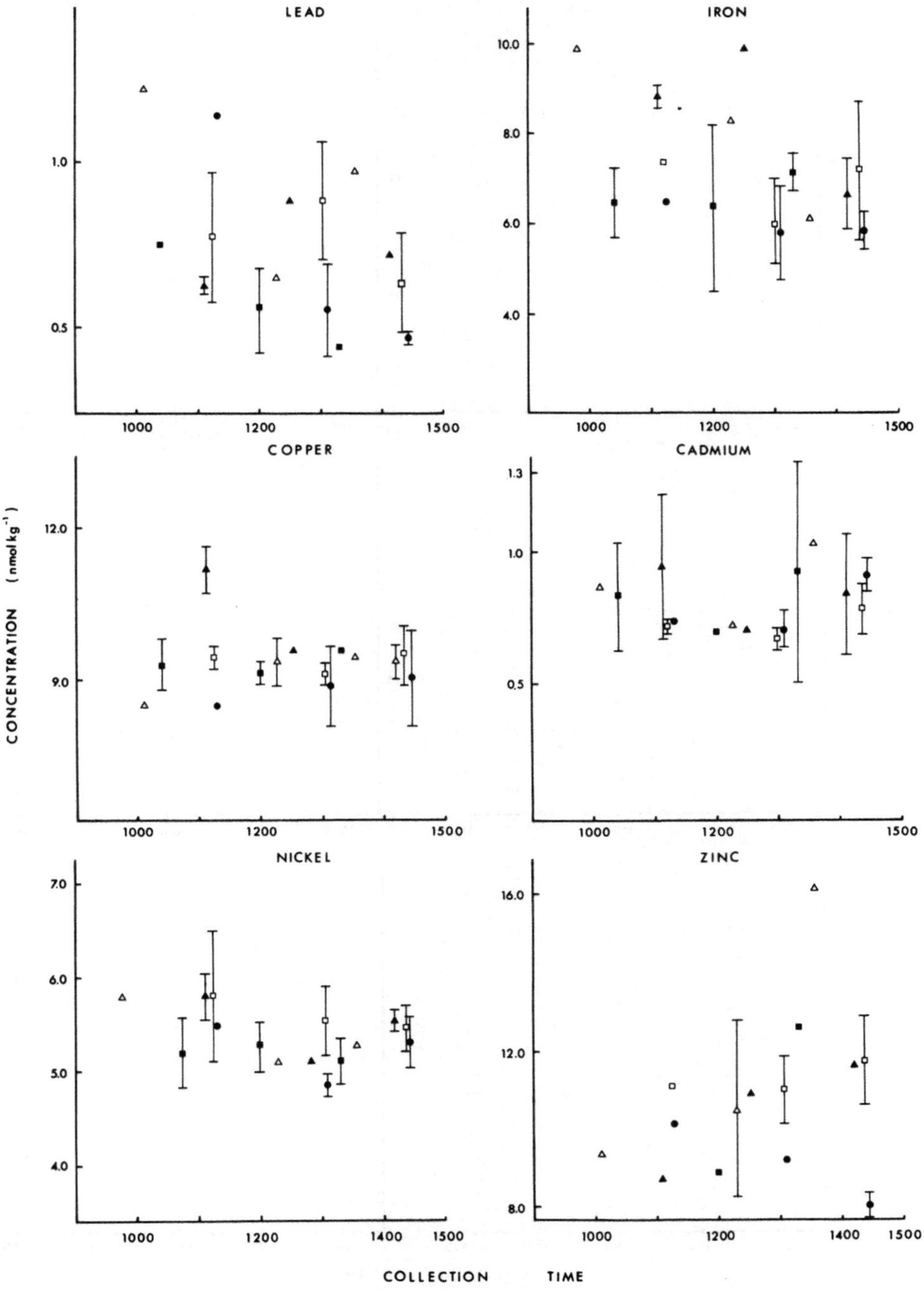

Fig. 3. Comparison of metal concentrations found in a CEPEX enclosure using five different sampling methods. Analyzed by the freon extraction method and F.A.A. analysis.

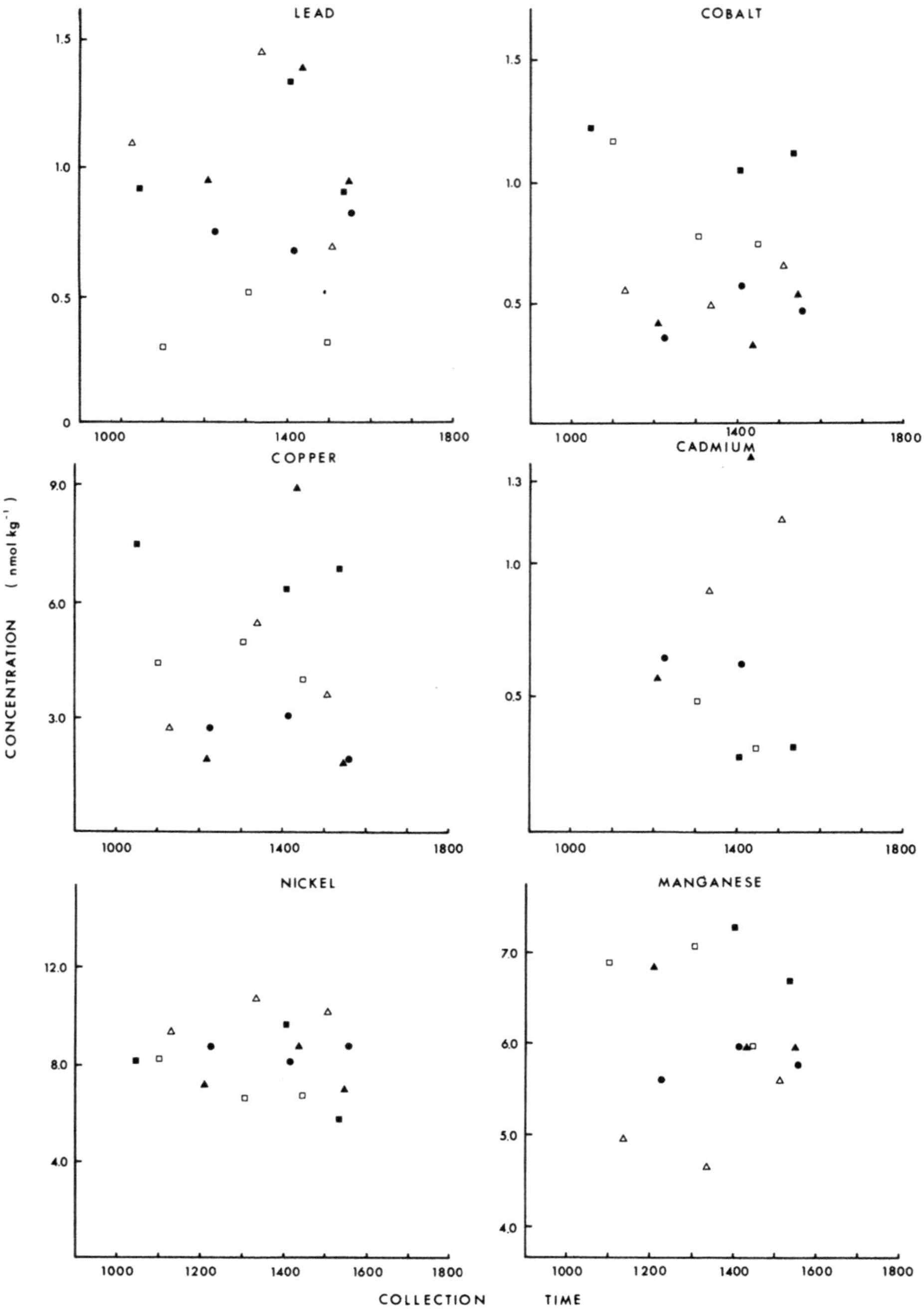

Fig. 4. Comparison of metal concentrations found in a CEPEX enclosure using five different sampling methods. Analyzed by the chelex extraction method and D.P.P. analysis.

0.09±.03, 0.08±.01, 0.08±.03 and 0.10±.02 nmols kg^{-1} respectively. The mercury values obtained by the teflon-piston sampler were high at 0.21±.05 nmol kg^{-1} due to malfunction with incomplete filling and previous contamination as indicated by the very low salinity in this set. The values inside the bag were higher than those outside, measured about one month after the intercomparison to be 0.02, 0.03 and 0.04 nmol kg^{-1}. There was a subsampling problem. The first and second draw of the sampling bottle usually showed a very wide spread in values, as much as 0.07 nmol kg^{-1} e.g. between 0.05 and 0.12 nmol kg^{-1}. This difference was real since the technique of cold vapor atomic absorption should be capable of detecting difference in subsamples from the same digested sample in a pyrex bottle. The peristaltic pumping method appears to yield the best agreement between subsamples: a difference of 0.02, 0.00 and 0.01 nmol kg^{-1} between subsamples from the three casts. The average Hg values for each sampler appeared to converge towards lower values on repeated casts within the same day. Further work is required to clarify contamination in mercury sampling.

The lead concentrations in the enclosure sea water were measured on three separate occasions: Day-2 samples by Chelex extraction and D.P.P., Day-3 samples by isotope dilution and mass spectrometry and Day-4 samples by freon extraction and F.A.A. As in the case of mercury, the lead values in the CEPEX enclosure are higher than those outside in the Saanich Inlet by a factor of 5, as determined by isotope dilution technique. Even at this high value of 0.73±0.02 nmol kg^{-1} (for peristaltic pump collection and mass spectrometry), isotope dilution and mass spectrometry clearly has the better sensitivity and accuracy to test the sampler difference. This technique showed the lead values to be 0.73±.02, 0.72±.03, 0.75±.02, 0.78±.05 and 0.81±.03 nmol kg^{-1} for sampling by peristaltic pump, Niskin sampler, Go-Flow sampler, close-open-close sampler and teflon-piston sampler, respectively. For the other two techniques, the teflon-piston sampler showed considerable variability and statistically much higher values. The results were not used in the comparison. The freon extraction and F.A.A. approach showed the same range of values as the isotope dilution approach, i.e. 0.71±.36, 0.76±.13,0.73±.13 nmol kg^{-1} for the pumping, Niskin sampler and close-open-close sampler respectively, with the exception of the Go-Flow sampler with a low value of 0.58±.15 nmol kg^{-1}. However, the range of values was wide, e.g. for the peristaltic pumping, 1.12 nmol kg^{-1} for the first cast dropping to 0.46 nmol kg^{-1} for the third cast. Chelex extraction and D.P.P. showed an even larger spread from 0.38±.12 nmol kg^{-1} for the three casts with the Niskin sampler to 1.09±.26 nmol kg^{-1} for the close-open-close sampler.

Copper concentrations in the CEPEX enclosure sea water showed rather good agreement for all the samplers as measured by the freon extraction followed by F.A.A. with values mostly within the range of 8.7 -9.4 nmol kg^{-1}. Pumping produced the lowest average of 8.8±0.3 nmol kg^{-1} while both the Niskin sampler and the Go-Flow sampler were about equal in performance for subsamples within the same cast 0 to ± 6% and for sampler averages of 9.4±0.2 nmol kg^{-1} for the Niskin and 9.3±0.2 nmol kg^{-1} for the Go-Flow. In general, the Chelex extraction plus D.P.P. approach produced values 2-4 times lower than the freon extraction plus F.A.A. technique. This might be the result of the difference in chemical species as determined by the two methods. The D.P.P. method measures only the "dissolved electrochemically labile" metal fraction, while the solvent extraction procedure detects the "total dissolved" metal.

Nickel concentrations are in good agreement for the freon method, with F.A.A., as in the case of copper. Subsamples within the same cast agreed usually to within 10% or less. The Go-Flow sampler gave an average of 5.2±0.1 nmol kg^{-1} for the 3 casts, matched by the peristaltic pumping with 5.2±0.3 nmol kg^{-1}. The Niskin close-open-close and the teflon-piston samplers gave slightly higher values of 5.6±0.2, 5.5±0.3 and 5.5 10.3 nmol kg^{-1} respectively as averages of 3 casts. Nickel values obtained by using chelex with D.P.P. in contrast to the case of copper, were 50% to 100% higher than the values found using freon with F.A.A. Zinc concentrations were obtained by freon extraction and F.A.A. only. Peristaltic pumping gave both superior subsample agreements of 0 and 6% differences and the lowest sampler average of 9.1±1.0 nmol kg^{-1}. The Niskin sampler showed a higher sampler average of 11.3±0.4 nmol kg^{-1} and larger difference between subsamples from the same cast. Much higher values of greater than 15.3 nmol kg^{-1} were shown in subsamples for the Go-Flow, close-open-close and teflon-piston samplers, thus suggesting contamination problems. By discarding the higher value, the close-open-close sampler would have an average of 10.4±1.5 nmol kg^{-1}. The values obtained by pumping, Niskin sampler and close-open-close sampler (discarding contaminated values) were in good agreement with the previous Saanich Inlet sea water value of 9.8 nmol kg^{-1} obtained in testing the teflon-piston sampler.

Cadmium concentrations obtained by freon extraction and F.A.A. were within the range of 0.71 to 0.80 nmol kg^{-1} for pumping, Niskin sampler, Go-Flow and close-open-close sampler. The teflon-piston sampler gave a value of 0.95 nmol kg^{-1} due mainly to a high value in the 4th cast. Cadmium values from Chelex extraction and D.P.P. were 79%, 54%, 36%, 115% and 106% of that by freon extraction and F.A.A., for the above samplers in the same order. It is quite possible that as with Cu, the Chelex and D.P.P. technique "sees" only the dissolved electrochemically labile species.

Cobalt was measured by the chelex extraction technique with D.P.P. Pumping and the close-open-close sampler gave low values of 0.46±0.10 and 0.42±0.10 nmol kg^{-1} followed by the teflon-piston sampler at 0.56±0.08, the Niskin sampler at 0.88±0.24 and the Go-Flow at 1.12±0.08 nmol kg^{-1}.

Iron as measured by the freon extraction and F.A.A. technique gave fairly good agreement for peristaltic pumping, Niskin sampler and Go-Flow sampler at 6.0±0.4, 6.8±0.7 and 6.6±0.4 nmol kg^{-1}, respectively. The close-open-close sampler and the teflon-piston sampler tended to give higher values of 8.4±1.6 and 7.7±1.6 nmol kg^{-1} respectively.

For manganese, as measured by the Chelex extraction plus D.P.P. technique, excellent agreement was obtained between values by peristaltic pumping, 5.8±0.2 nmol kg^{-1}, and the close-open-close sampler, 5.6±0.5 nmol kg^{-1}. The Niskin and Go-Flow samplers gave high values of 6.6±0.6 and 7.0±0.4 nmol kg^{-1}, respectively, and the teflon-piston sampler a low value of 5.0±0.5 nmol kg^{-1}.

CONCLUSIONS

(1) It is feasible to capture a large volume of sea water in the range of 65,000 liters by the CEPEX approach for the purpose of sampler intercomparison. It is possible by artificial stimulation of a plankton bloom and detritus removal to produce a reasonably homogeneous body of sea water for the study. Proximity of the in situ enclosure for the experiment and shore clean laboratory facilities eliminate errors introduced by shipboard contamination under less than ideal conditions on cruises.

(2) The following analytical techniques seem to be adequate for the concentrations under consideration: copper and nickel by freon extraction and F.A.A., cobalt by chelex extraction and D.P.P., mercury by cold vapor absorptiometry, and lead by isotope dilution plus clean room manipulation and mass spectrometry. These techniques may be used to detect changes in the above elements for storage tests: Cu at 8 nmol kg^{-1}, Ni at 5 nmol kg^{-1}, Co at 0.5 nmol kg^{-1}, Hg at 0.1 nmol kg^{-1} and Pb at 0.7 nmol kg^{-1}.

(3) Salinity of sea water captured by various sampling devices in the CEPEX enclosure indicates problems not revealed in the usual oceanographic sampling situation. Relative to peristaltic pumping, all samplers exhibited some salinity anomality. Inadequate flushing to rinse the sampler of any concentrated brine or entrapped seawater is thought to be the problem.

(4) Logistics and cleaning procedures are important factors in successful sampler intercomparisons. It is not desirable or

possible to endorse or to condemn the performance of a certain type of sampler or analytical technique based on results of one set of tests, especially if procedures are changed.

(5) The problems of subsampling from the same sea water sample has to be studied in greater detail.

(6) A long-term but sustained effort on sampler intercomparisons by a relatively small group would be advantageous in identifying problems.

ACKNOWLEDGEMENT

The authors thank the N.A.T.O. Scientific Affairs Division for support of the study under N.A.T.O. Grant No. SGR.15(77)TK/1234 and the U.S. National Science Foundation and C.E.P.E.X. (Controlled Ecosystem Pollution Experiment) Steering Committee for permission to use enclosure facilities for intercomparison work.

REFERENCES

1. Participants in the IDOE interlaboratory analyses workshop, 1975 (1976). Comparison determinations of lead by investigators analyzing individual samples of sea water in both their home laboratory and in an isotope dilution standardization laboratory. Marine Chemistry, 4, 389-392.
2. Sugawara, K. (1978). Interlaboratory comparison of the determination of mercury and cadmium in sea and fresh waters. Deep-Sea Res., 25, 323-332.
3. Kremling, K. (1977). Report of Group 2 (Trace Metals). In: Baltic Intercalibration Workshop (7-19, March, 1977).
4. Jones, P.G.W. (1976). A preliminary report on the ICES intercalibration of sea water samples for the analysis of trace metals. ICES CM 1977/E:16.
5. Bewers, J.M., G. Topping and H. Windom. (1978). Status and plans regarding ICES intercalibrations for trace metals in sea water. ICES CM 1978/E:27.
6. Bewers, J.M., P.A. Yeats, J. Dalziel and J.L. Baron (1979). Report of the fourth round intercalibration for trace metals in sea water. ICES CM 1979/E:37.
7. Wong, C.S., W.K. Johnson, and V.J. Stukas (Unpublished manuscript): Storage study of sea water for trace metals.
8. Schaule, B. and C.C. Patterson (1978). The occurrence of lead in the Northeast Pacific, and the effects of anthropogenic inputs. In: M. Branic (Editor),

Proceedings of an international expert discussion on lead: "Occurrence, fate and pollution in the marine environment"; Rudjer Boskovic Institute, Centre of Marine Research, Rovinj. Oct. 1977, Press, Oxford.

9. Patterson, C.C. and D.M. Settle (1976). The reduction of orders of magnitude errors in lead analysis. In: P.D. La Fleur (Editor). "Accuracy in Trace Analysis: Sampling, Sample Handing, Analysis." NBS Special Publication 422, 321-363.
10. Boyle. E., J.M. Edmond (1975). Copper in surface waters south of New Zealand. Nature 253, 107-109.
11. Boyle, E., F. Sclater and J.M. Edmond (1976). On the marine geochemistry of cadmium. Nature 263, 42-44.
12. Bruland, K.W., G.A. Knauer and J.H. Martin (1978). Zinc in northeast Pacific water. Nature 271, 741-743.
13. Menzel, D.M. and J. Case (1977). Concept and design: controlled ecosystem pollution experiment. Bull. Mar. Sci., 27, 1-7.
14. Bothner, M.H. and D.E. Robertson (1975). Mercury contamination of sea water samples stored in poloyethylene containers. Anal. Chem. 47, 592-595.
15. Stukas, V.J. and C.S. Wong. Stable lead isotopes as a tracer in coastal waters. Science, 211, 1424-1427,
16. Wong, C.S., K. Kremling, J.P. Riley, W.K. Johnson, V. Stukas, P.G. Berrang, P. Erickson, D. Thoms, H. Petersen and B. Imber (1979). Accurate Measurement of trace metals in sea water: an intercomparison of sampling devices and analytical techniques using CEPEX enclosure of sea water. Unpublished manuscript report, NATO study funded by NATO Scientific Affairs Division.
17. Abdullah M.I., O.A. el-Rayis and J.P. Riley (1976). Re-assessment of chelating ion-exchange resins for trace metal analysis of sea water. Anal. Chimica Acta, 84, 363-368.
18. Abdullah, M.I. and L.G. Royale (1972). The determination of copper, lead, cadmium, nickel, zinc and cobalt in natural waters by pulse polarography. Anal.Chim. Acta. 58,
19. Danielsson, L.G., B. Magnusson and S. Westerlund (1978). An improved metal extraction procedure for the determination of trace metals in sea water by atomic absorption spectrometry with electrothermal atomization. Anal. Chimica Acta. 98, 47-57.

ROLE OF FRESH WATER/SEA WATER MIXING ON TRACE METAL ADSORPTION PHENOMENA

Alain C. M. Bourg

Institute of Inorg., Anal. & Phys. Chemistry
University of Bern
Freiestrasse 3
CH-3000 Bern 9, Switzerland

ABSTRACT

In estuaries many phenomena affect the distribution of trace metals between dissolved and particulate forms. Laboratory experiments which are performed under carefully controlled conditions allow the study of any of these processes one by one or under conditions where one prevails.

The estuarine water/suspended solid surface behavior of Cd, Zn and Cu in the presence and in the absence of organic matter is described for SiO_2, γ-Al_2O_3, 2 clay minerals (kaolinite and montmorillonite) and 3 sediments (Rhone River, Garonne River and Gironde estuary). The nature and concentration of suspended solid surfaces and the water pH are more important than "pure" salinity effects (increase in ionic strength, increases in concentration of competitors for adsorption and of complexing agents).

The results are used to improve the estuarine chemical speciation model of Mantoura, Dickson and Riley[1] by adding the solid phase as a surface ligand.

INTRODUCTION

When trace metals transported by rivers enter estuaries they are exposed to an environment with increasingly different physical and chemical characteristics. Many phenomena may affect the distribution of these metals between dissolved and particulate forms. (1) Flocculation of colloids into bigger particles may take along trace metals in the transformation from "dissolved" (size smaller than 0.45 μm) to particulate state. (2) Precipitation (for example, of hydroxides and carbonates) may occur. (3) Finally, metals can move across the solid particle-solution interface by adsorption or desorption reactions. This last phenomenon will be investigated in this paper.

Duinker [2] reviewed the contradictory reports from the literature suggesting, depending on the authors and the aquatic systems, that trace metals are either adsorbed or desorbed under estuarine conditions. Field observations show no general trend (with the exception of Fe which is always removed from the dissolved phase). Laboratory experiments which allow the study of single phenomenon under somewhat well controlled conditions have been criticized because they may not represent the dynamic conditions present in estuaries. In addition, they can be interpreted with contradictory conclusions. Krauskopf's work[3] indicates that $Fe_2O_3.nH_2O$, apatite, clay, plankton and peat moss all adsorb trace metals (Zn, Cu, Pb, Ni, Co, Hg(II)) from sea water. On the other hand, the laboratory mixing experiments of Kharkar et al.[4] show mobilization of Co, Se and Ag from selected adsorbents like clays and oxides. In a step further, Van Den Weijden et al.[5] desorbed successfully Cr, Mn, Co, Ni, Cu, Zn and Cd (but not Pb and Fe) by mixing Rhine River suspended matter with artificial sea waters. For Co, Cu, and Cd, their "nitrate" sea water (salinity of sea water made up of only $NaNO_3$) was not as effective as full artificial sea water. The mixing experiments of Salomons[6] (also with Rhine River suspended matter) recently emphasized the effect of the suspended matter upon the adsorption behavior of trace metals under estuarine conditions. An increase in the concentration of surface sites (as encountered in a turbidity maximum) may result in an increase of the fraction of particulate metal even if the salinity is increasing.

Are metals adsorbed or desorbed upon mixing with sea water? Field observations and laboratory experiments simulating natural conditions show that both conditions can occur and I am going to demonstrate here that this apparent contradiction can be explained by surface chemistry concepts. The importance of suspended matter and even more of pH will be shown.

EXPERIMENTAL

The solid surfaces studied are described in Table 1. The pH_{ZPC} represents the pH of 1 g. of solid material in equilibrium with 1 liter of 0.001 mol l^{-1} $NaNO_3$. In order to remove all (or at least most of) the trace metals at the surface of the river and estuarine sediments, these were washed with EDTA (1 mol l^{-1}) and HNO_3 (1 mol l^{-1}) prior to the adsorption batch experiments.

Table 1 Description of the samples

Solid surface	Characteristics	pH_{ZPC}
amorphous SiO_2	"Aerosil 380", Degussa	6.44
γ-Al_2O_3	Merck	8.82
kaolinite	KGa-1, well crystallized kaolin, Georgia, USA, Clays Minerals Soc.	5.49
montmorillonite	SWy-1, Na-Montmorillonite, Wyoming, USA, Clays Minerals Soc.	9.77
Gironde sediment	PK 67, Gironde estuary, salinity 9.5°/oo, suspended matter (crème de vase)	4.84
Garonne sediment	freshly deposited sediment, Garonne River, Caudrot, France	5.45
Rhône sediment	freshly deposited sediment, Rhône River, Tain, France	9.42

Solutions of 2.5×10^{-5} mol l^{-1} of nitrate salts of Cu, Cd and Zn (present simultaneously) were prepared in various media (0.001 mol l^{-1} and 0.1 mol l^{-1} $NaNO_3$, 0.1 mol l^{-1} NaCl and 0.1 mol l^{-1} NaCl + 0.01 mol l^{-1} $CaCl_2$). Where applicable enough soil fulvic acid was added (freeze-dried) to give a concentration of 400 mg l^{-1} (M.W. ca. 4500).

Aliquots of 10 ml of these solutions were mixed with 50 mg of solid and 40 ml of solutions of various pH values (prepared with HNO_3, HCl and NaOH as required) in the same Na medium. The final concentrations were 5×10^{-6} mol l^{-1}, 1 g l^{-1} and 80 mg l^{-1} for the metals, the solids and the fulvic acid, respectively. All solutions were carefully made carbonate-free by degassing with purified N_2. The mixtures were agitated and equilibrated at 25°C. Equilibrium was assumed to be achieved after 36 hours because the pH was then constant with time. The pH was measured and the suspensions were subsequently centrifuged (10,000 rpm for 15 min).

The fulvic acid whose solutions were brought to pH 7 with a phosphate buffer and the metals left in solution were measured by u.v. spectroscopy (254 nm) and atomic absorption, respectively.

RESULTS AND DISCUSSION

The adsorption behavior of trace metals on solid oxides can be described by the formation of surface complexes[7-9]. The phenomenon is understood in terms of competition for surface sites with protons and other cations. More recently this model was extended to the adsorption of complexes of metals[10]. This concept leads to the determination of surface complex formation constants, corresponding to interface reactions such as (SOH : surface, M : metal):

$$SOH + M^{z+} \underset{}{\overset{\beta^{surf}}{\rightleftharpoons}} SOM^{(z-1)+} + H^{+} \quad (1)$$

As far as modeling is concerned, the surface sites S can then be regarded as more or less classical ligands[10-12]. Such models have been used to calculate the speciation of trace metals in aquatic systems[10-12] and also to explain the residence time of trace metals in the oceans[13]. I have already presented a crude application of one of these surface complex models to the mobilization of trace metals in the Gironde estuary (France)[14] and I am now reporting the preliminary results of an investigation which should allow the realisation of better models (especially by comparing the properties of model surfaces such as solid oxides and clay minerals to those of real sediments).

At this point the reader must be warned that even if in some of the following figures the increase in adsorbed metal does not occur within the pH range of natural waters, this does not preclude the environmental significance of the observed phenomena. The position of the adsorption curves is dependent upon the total concentration of the surface involved. Only most typical results of the 3 metals and 7 surfaces investigated are presented.

The adsorption of the trace metals on the 7 surfaces, as expected from equation (1), depends very much upon the pH. (See Figure 1 for the example of Cd). Montmorillonite is an exception since up to 80% of the surface sites originate from charge deficiencies caused by lattice isomorphous substitutions and thus are not pH-dependent. Figure 2 shows for 2 model surfaces and 2 sediments that the adsorption behaviors of the 3 metals investigated are very similar. An adsorption edge situated at a lower pH means a metal more competitive with H^+ and thus a larger surface complex formation constant. The general trend in adsorption power is Cu > Zn > Cd.

Effect of ionic strength (I)

As in solution chemistry, activity coefficients can affect the value of the surface constants. However one can see that in reaction (1) the effect should not be large for $I < 0.1$. This is observed on amorphous silica for the adsorption of Cu but not for Cd (Figure 3). The difference can readily be explained by the much lower surface constants of the second metal. The adsorption

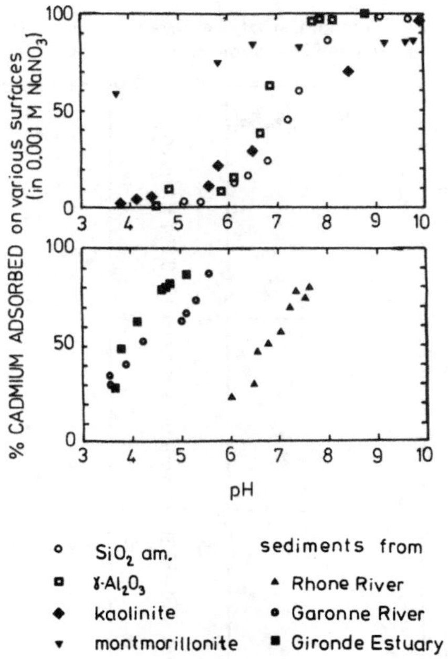

Fig. 1 Adsorption of Cd on various surfaces

of Cu is not affected by a greater number of Na^+ and NO_3^- support electrolyte ions. But Cd is not as competitive with Na^+ with respect to adsorption. In the case of Cd for a given pH value an increasing amount of Na will cause desorption.

Effect of complexing agent (inorganic)

The adsorption curve of Cu on SiO_2 is not displaced by Na^+, but when NO_3^- is replaced by Cl^- the formation of chloro complexes competes with adsorption (Figure 4). The same effect is observed for the system Cd/γAl_2O_3 (Figure 5).

Effect of competing adsorbing species

The addition of $CaCl_2$ to the medium provides an additional effect, the competition (from Ca^{2+}) for surface sites, displacing the adsorption curve to the right (Figure 5). Of the 3 metals investigated, the effect presented in this figure is greatest for Cd because it is the metal with the smallest surface complex formation constant and the largest chloro complex formation constant.

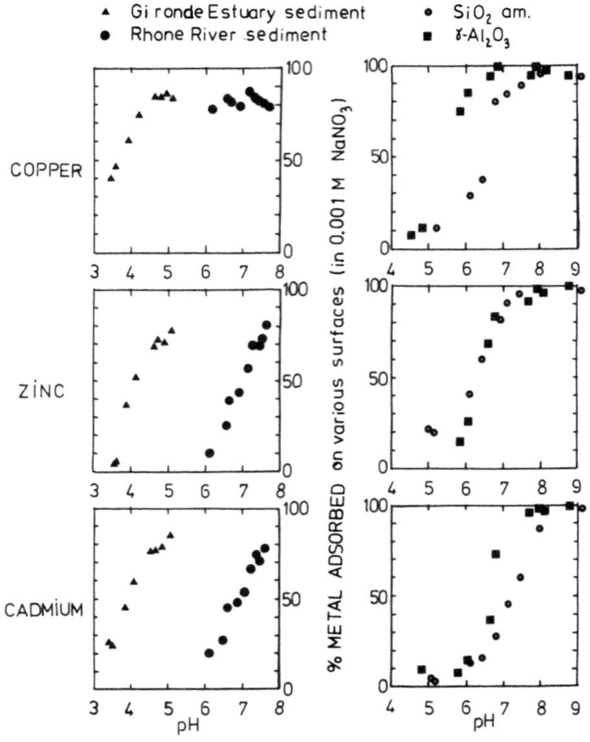

Fig. 2 A comparison of the adsorption of Cu, Zn and Cd

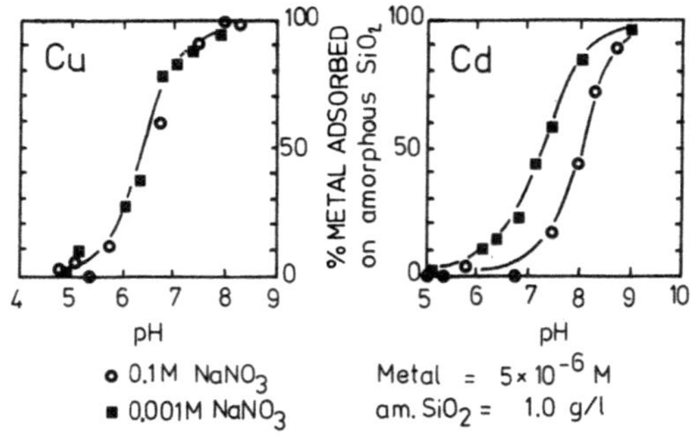

Fig. 3 Adsorption of Cu and Cd on amorphous silica

This behavior is not limited to Al and Si oxides. Estuarine and river sediments behave in the same fashion (Figure 6). The Cl^- complexation and the Ca^{2+} competition are felt more strongly by Zn than by Cu because, for Cu, surface constants are greater and the chloro complex constants are smaller. It is interesting to note that the Garonne River and the Gironde (estuary of the Garonne

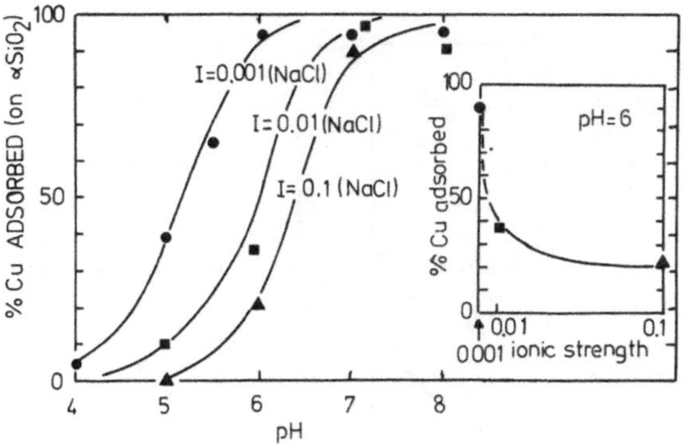

Fig. 4 Adsorption of Cu on α-SiO_2 ($Cu_{total} = 10^{-6}$ mol l^{-1}). (adapted from J. Vuceta, Ph.D. Thesis, Cal. Inst. Techn., Pasadena, Calif. USA, 1976).

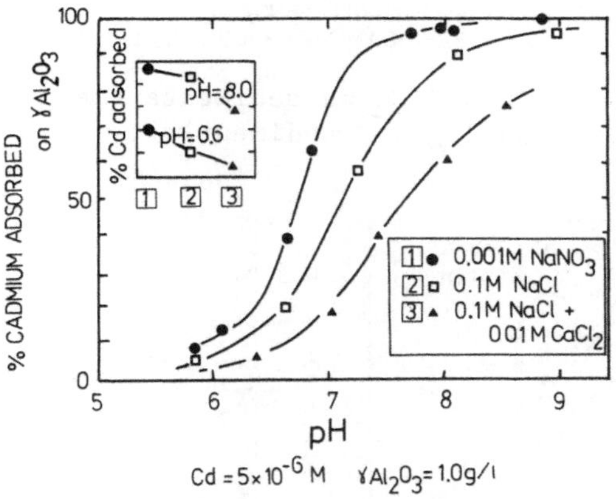

Fig. 5 Adsorption of Cd on γ-Al_2O_3

River) sediments exhibit similar surface behavior. The surrounding waters of greatly different salinities do not seem to affect the intrinsic surface properties.

Effect of organic matter

Salicylic acid is an organic ligand which exhibits adsorption properties similar to those of natural organic matter. As seen in Figure 7, the pH is important in regulating surface/solution distribution. When the surface is positively charged ($pH < 8.8$, for γ-Al_2O_3) increasing salinity promotes desorption. The value of pH_{ZPC} (Table 1) represents the pH for which the overall surface charge is equal to zero. For $pH < pH_{ZPC}$ the surface is positive and for $pH > pH_{ZPC}$ it is negative. As seen in Table 1, model and natural surfaces have pH_{ZPC} values scattered slightly above, within and slightly below the pH range of natural waters. Mixtures of these

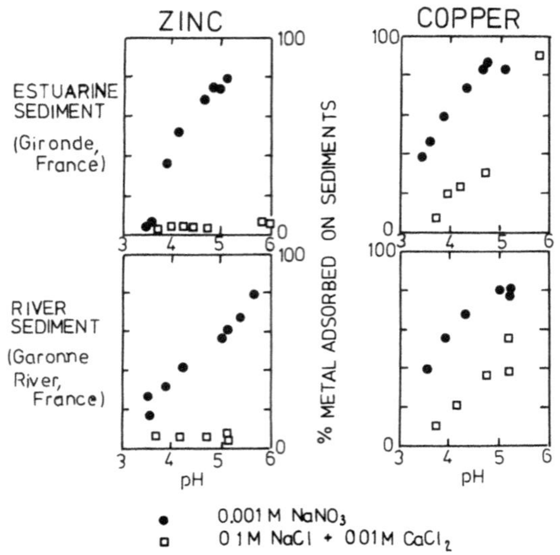

Fig. 6 Adsorption of Zn and Cu on sediments (Metal total = 5 x 10^{-6} mol l^{-1} and 1 g l^{-1} sediment).

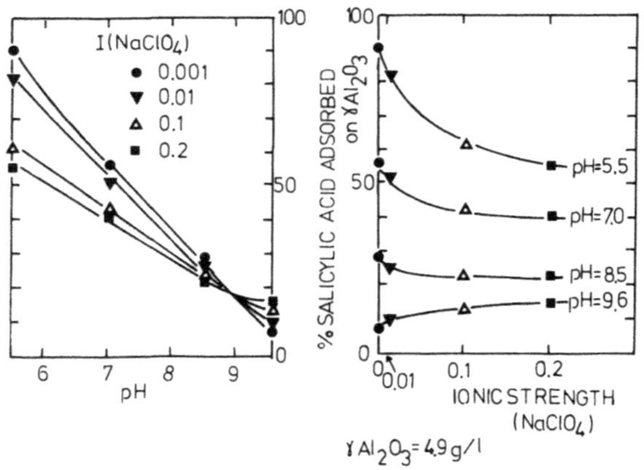

Fig. 7 Adsorption of salicylic acid on γ-Al_2O_3 (from R. Kummert, Doctoral diss., ETH, Zürich, Switzerland, 1979)

surfaces will of course complicate the understanding of the overall fluxes at the solid-solution interface. In principle, in presence of increasing salinity, the organic matter (especially small molecules) will adsorb when pH > pH_{ZPC} and desorb when pH < pH_{ZPC}.

The natural sediments studied here do contain natural particulate organic matter but the Figures 1, 2 and 6 show that their adsorption behavior is not different from other surfaces except in some cases by displacement of the pH curve, or in other words, by different (greater) surface stability constants.

The presence of dissolved organic matter complicates matters further especially in the case of Cu. Since the amount of Cu adsorbed (on γ-Al_2O_3) follows that of adsorbed fulvic acid (FA) (see Figure 8) one would expect the metal uptake to increase with salinity for pH values greater than the pH_{ZPC}. These FA-metal systems are not yet well understood and further investigations are in progress. In the following model speciation calculations involving only the modification of adsorption behavior of metals by dissolved (or adsorbed via the metal as ternary surface complex of the form surface-metal-organic matter) organic matter will be taken into account.

Estuarine modelling

After having demonstrated the existence and the complexity of the adsorption behavior of trace metals under carefully controlled

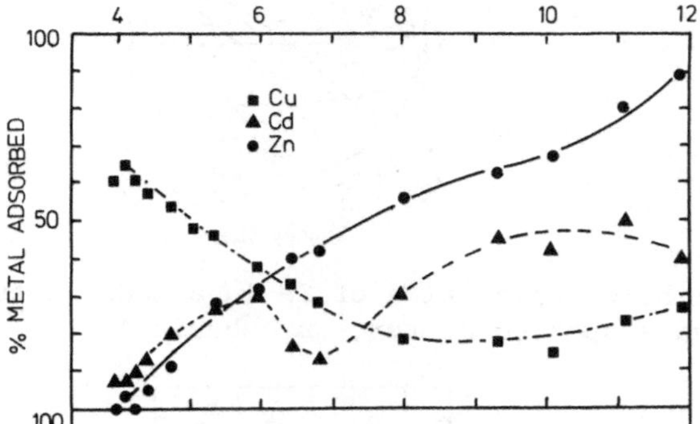

Fig. 8 Adsorption of Cd, Cu, Zn and Fulvic acid (FA) on γ-Al_2O_3 ($Cd_{tot} = Cu_{tot} = Zn_{tot} = 5 \times 10^{-6}$ mol l^{-1}, FA_{tot} = 80 mg l^{-1} and 1 g l^{-1} alumina in 0.001 mol l^{-1} $NaNO_3$).

simulated estuarine conditions, I now add these processes to the mathematical model of the speciation of the dissolved phase of Mantoura et al.[1]. There is much room for improvement of this type of model. This is one step in that direction. The surface constants of the natural sediments studied were not calculated because the total concentrations of surface sites were not yet determined. It was therefore assumed that natural suspended matter behaves like silica. Figures 9 to 12 present a detailed description of the speciation of trace metals for a constant pH value of 8.0 and for a constant concentration of suspended surface sites (having the surface properties of silica) of 30 mg l^{-1} or 1.8×10^{-5} mol l^{-1}. Figure 13 shows that at a constant pH value the diagrams of the adsorption in the presence of increasing salinity of, for example, Zn resembles that of the suspended matter concentration.

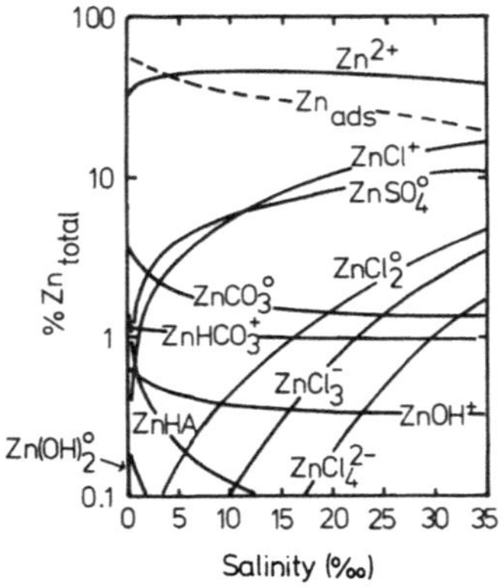

Fig. 9 Equilibrium speciation of Zn in a model estuary (constant pH and suspended matter, HA : humics)

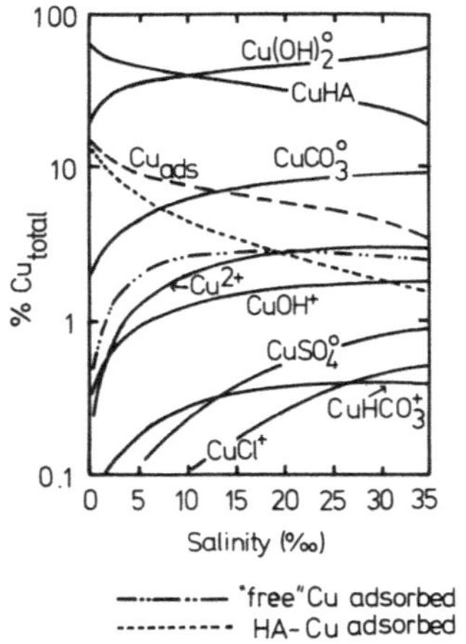

Fig. 10 Equilibrium speciation of Cu in a model estuary (constant pH and suspended matter, HA : humics)

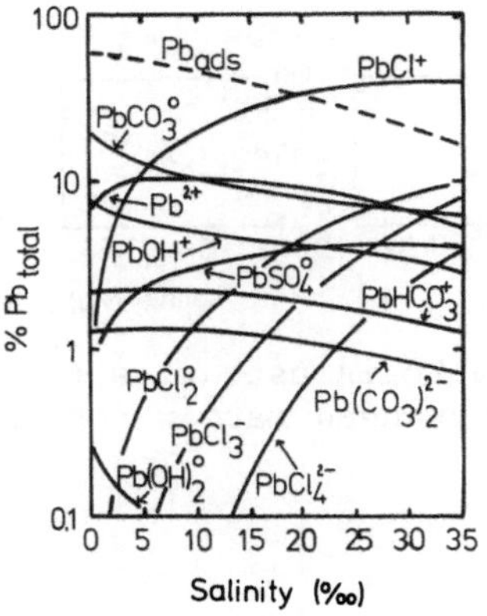

Fig. 11 Equilibrium speciation of Pb in a model estuary (constant pH and suspended matter. Pb-HA species are not considered, not because they do not exist but rather because of the lack of values of the constants).

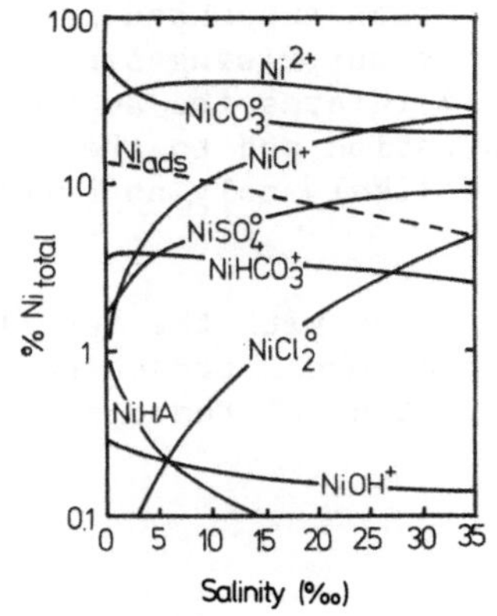

Fig. 12 Equilibrium speciation of Ni in a model estuary (constant pH and suspended matter, HA : humics)

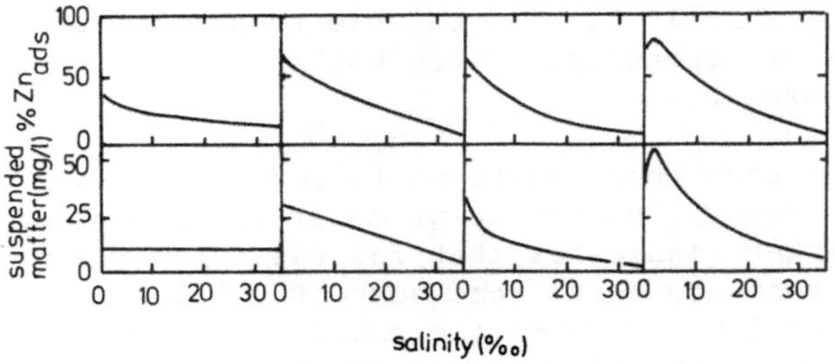

Fig. 13 Adsorption of Zn in estuaries (model calculations under several cases of variations of the concentration of suspended matter; pH has a constant value of 8.0)

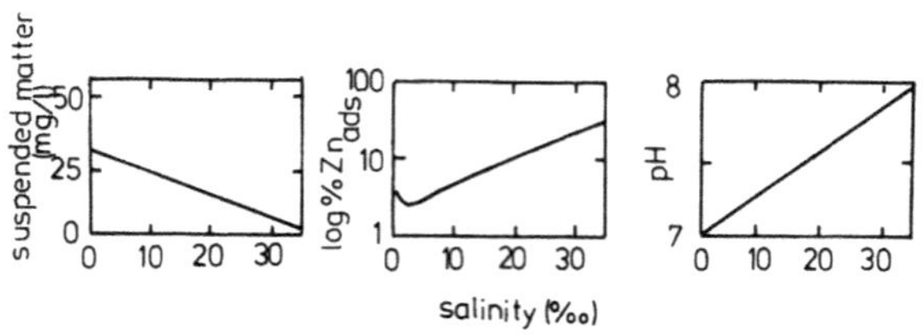

Fig. 14 Adsorption of Zn in estuaries (model calculations under decreasing suspended matter concentration and increasing pH).

In Figure 14, for pH and suspended matter conditions typical of an estuary with low alkalinity river water, the logarithm of the fraction of Zn adsorbed is compared to the water pH (also a logarithmic expression). It can readily be seen that pH is a parameter more important than suspended matter concentration. These observations are not surprising since the adsorption model described by equation (1) relates the amount of metal adsorbed to the surface site concentration and to the pH. The latter is more important but it is most likely only an artefact due to its logarithmic nature.

These observations agree with the results and the interpretation of Salomons[6]. But the model presented here should allow a calculation of the combination of the effects of salinity, pH and suspended matter.

CONCLUSION

Adsorption and desorption of trace metals in estuaries are complex phenomena which depend on many variables. Only a multicomponent analysis, and not schemes of the type "dissolved concentration versus salinity", will be able to provide some information on the important mechanisms. (See Wollast et al.[15] for an example of such an approach).

Depending upon the conditions (water and surface parameters), adsorption or desorption may occur under increasing salinity. The review of Duinker[2] indicates that for estuaries taken on a global scale, there seems to exist some sort of a balance between adsorption and desorption. However the preceding observations show that a slight change in pH and in suspended matter concentration (by physical resuspension or by change in flocculation dynamics) may drastically disrupt this state of balance.

The importance of the water pH can be shown by the following consideration (in the absence of large turbidity maxima). Estuaries of low alkalinity rivers will present a large increase in pH (at least one pH unit) and thus adsorption conditions (removal from the dissolved phase) are likely to occur. Estuaries of high alkalinity rivers do not present such a pH gradient and desorption (remobilization) may occur.

In conclusion any estuarine survey intended for the study of trace metals geochemical cycles should include suspended matter content and important water parameters (especially pH).

ACKNOWLEDGEMENTS

This work was partially supported by the Swiss National Foundation for Scientific Research (Grant 2.464-0.79) and by the Institut de Géologie du Bassin d'Aquitaine (IGBA) of the University of Bordeaux, France. Gironde samples were collected from the R/V Ebalia (IGBA). I thank P.W. Schindler (Bern), C. Latouche and O. Donard (Bordeaux) for their support and help and A.K. Bourg for typing the manuscript presented in Erice (Sicily).

REFERENCES

1. Mantoura, R.F.C., A. Dickson and J.P. Riley, 1978: The complexation of metals with humic materials in natural waters. Estuar. Coastal. Mar. Sci., 6, 387-408.
2. Duinker, J.C., 1980: Suspended matter in estuaries: Adsorption and desorption processes. In: "Chemistry and biogeochemistry of estuaries", E. Olausson and I. Cato, eds., John Wiley & Sons, New York.
3. Krauskopf, K.B., 1956: Factors controlling the concentrations of thirteen rare metals in sea-water. Geochim. Cosmochim. Acta., 9, 1-32 B.
4. Kharkar, D.P., K.K. Turekian and K.K. Bertine, 1968: Stream supply of dissolved Ag, Mo, Sb, Se, Cr, Co, Rb and Cs to the oceans. Geochim. Cosmochim. Acta., 32, 285-298.
5. Van Den Weijden, C.H., M.J.M.L. Arnoldus and C.J. Meurs, 1977: Desorption of metals from suspended material in the Rhine estuary. Neth. J. Sea Res., 11, 130-145.
6. Salomons, W., 1980: Adsorption processes and hydrodynamic conditions in estuaries. Environ. Technol. Letters, 1, 356-365.
7. Schindler, P.W., B. Fuerst, P.U. Wolf and R. Dick, 1976: Ligand Properties of surface silanol groups. J. Colloid Interface Sci., 55, 469-475.
8. Hohl, H. and W. Stumm, 1976: Interaction of Pb^{2+} with hydrous γ-Al_2O_3. J. Colloid Interface Sci., 55, 281-288.

9. Davis, J.A. and J.O. Leckie, 1978: Surface ionization and complexation at the oxide/water interface: surface properties of amorphous iron oxyhydroxide and adsorption of metal ions. J. Colloid Interface Sci., 67, 90-107.
10. Bourg, A.C.M., 1979: Effect of ligands at the solid-solution interface upon the speciation of heavy metals in aquatic systems. Proc. Internat. Conf. Heavy Metals in the Environment (London), 446-449.
11. Bourg, A.C.M., 1979: Spéciation chimiques des métaux traces dans les systèmes aquatiques: importance de l'interface solide-solution. J. Fr. Hydrologie, 10, 159-164.
12. Bourg, A.C.M., 1981: Reactions at the water-solid particulate matter interface. IUPAC Pub., in preparation.
13. Balistrieri, L., P.G. Brewer and J.W. Murray, 1981: Scavenging residence times and surface chemistry. Deep Sea Res., 28A, 101-121.
14. Bourg, A.C.M., H. Etcheber and J.M. Jouanneau, 1979: Mobilisation des métaux traces associés aux matières en suspension dans le complexe fluvio-estuarien de la Gironde. Rev. Biol. Ecol. Mediter., VI, 161-166.
15. Wollast, R., G. Billen and J.C. Duinker, 1979: Behaviour of Mn in the Rhine and Scheldt estuaries. Estuar. Coastal Mar. Sci., 9, 161-169.

EFFECTS OF PARTICLE SIZE AND DENSITY ON THE TRANSPORT OF METALS TO THE OCEANS

Jan C. Duinker

Netherlands Institute for Sea Research
POB 59, den Burg, Texel
The Netherlands

ABSTRACT

Details of the composition of suspended matter in seawater are interpreted by a model describing suspended matter as a mixture of a permanently suspended fraction, consisting of small/low density particles and a fraction consisting of larger/denser particles, derived from the bottom. These fractions have different settling velocities and different composition.

This model assists in explaining data on particulate trace metals in the Rhine estuary and the adjacent coastal area and the role of these environments as barriers for the transport of metals from land to oceans in dissolved and particulate forms.

INTRODUCTION

Metals in natural waters are transported in solution and in particulate forms. In estuaries, modifications in the distribution of riverborne components over these phases can occur in response to large gradients in important parameters as salinity, pH, concentrations of dissolved oxygen and suspended matter.

Various mechanisms have been reported to be involved, such as physical mixing of river and sea water[1,2,3], sedimentation and resuspension (often resulting in reverse sedimentation)[4,5], variations in grain size distribution as result of particle sorting[6], formation of new particles or agglomeration of existing particles into larger units[7,8,9], removal of dissolved components into particulate forms[10,11,12], desorption from particles into solution[13,14,15] and contributions from interstitially dissolved components to the overlying water[16,17]. The net effect is an often complex distribution pattern.

Analytical distinction of the various chemical forms of each component will assist in understanding the fate of riverborne components in the marine environment. In this paper we shall analyze some problems that we have identified with respect to the existence of different size/density particulate matter fractions in natural waters.

COMPOSITION OF SUSPENDED MATTER

Chemical Leaching Techniques

Ideally, for a detailed analysis of metals in particulate matter, it would be desirable to separate the sample into the various mineral and organic constituents in order to allow analysis of each individual component in each available particle size fraction. This is difficult if not impossible to achieve. A major problem is the common occurrence in natural waters of clay minerals, quartz, iron and manganese oxides, organic matter and other components in the form of associations. The properties of such aggregates are usually quite different from the individual components. For example, clay minerals interact with oxide coatings and humic material[18,19], and humics interact with metal ions and oxides[20]; also organic coatings will affect the properties of particles such as size and settling velocities[21].

Chemical leaching procedures may extract elements from specific binding sites in bottom sediments[22,23]. These methods may have limited quantitative value for suspended matter because modifications in the natural state of aggregation are expected to occur during sampling and separation processes. However, they can supply at least qualitatively useful data. A simple leaching technique such as treatment of suspended matter with a weak acid solution followed by total digestion has proven to be useful for distinguishing metal fractions that are readily available (adsorbed) and fractions that may become available after chemical changes likely to occur in the environment (organically bound and in oxide coatings) from forms that are relatively unavailable for release (in crystal structures).

The percentage leachable fraction of a particular metal with respect to its total metal content may vary with its total contents in particulate matter and this can be used to distinguish sediment sources. The leaching data can also be used to detect enrichment of metals in particulate matter resulting from their contributions from originally dissolved components during estuarine mixing (see below).

Variations of Seston Composition with Total Seston Concentration in Natural Waters

The 'size' of metal fractions in natural waters that we are interested in, covers the whole range from truly dissolved species up to sand-sized fractions, including the important colloidal material. The analytical distinction of metal fractions in solution and in particulate forms is not trivial. Samples requiring phase separation are usually filtered. This introduces an ambiguity because of the change of effective pore size below the nominal pore size during filtration. So, the average size of the particles collected on the filter will depend on the total concentration as well as on the size distribution of the particles in suspension.

Small particles may not be included in particulate matter collected by methods for phase separation, that are based on other mechanisms e.g. centrifugation and settling techniques (discriminating phases on the basis of size and density properties). Differences in efficiency of these methods might be useful to understand details of the composition of suspensions in natural waters.

Identical subsamples from coastal waters with a range of low and high suspended matter concentrations were subjected to filtration, continuous centrifugation and settling in containers, under conditions resulting, theoretically, in the collection of particles with effective diameter $> 0.45\ \mu m$ (density in the different fractions assumed to be constant). Particulate matter obtained was analyzed for several major and trace elements. The contents (weight/weight) of each element in material collected by any of the techniques had a characteristic dependence on suspended matter concentration[24,25]. Three different types of dependence can be distinguished (Fig. 1a). The data showed differences between the techniques. These can be understood on the basis of the different mechanisms involved in phase separation. For instance, the content of copper was considerably higher in filtered than in settled or centrifuged material (effect of small particles) and in filtered material, copper content was higher at low than at high suspended matter concentrations ($> 5\text{-}10\ mg\ l^{-1}$). These data support a model that considers suspended matter to be composed of small/low density particles that are permanently in suspension and larger/denser particles, belonging to suspension or bottom sediment, depending on hydrodynamical conditions. The relative contribution of the permanently suspended fraction to total suspended matter is larger at low than at high suspended matter concentrations and the larger/denser particles dominate at high

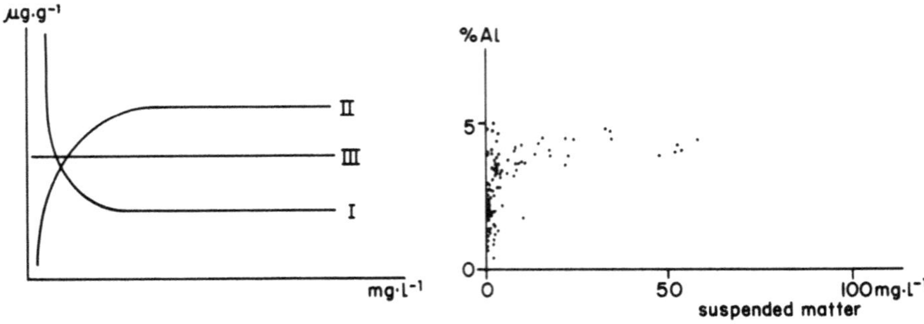

Fig. 1 Left: schematic relation between the element contents in seston ($\mu g\ g^{-1}$) and seston concentration (mg l^{-1}) for group I, II and III elements (see text). Right: actual data for Al in seston off the Dutch coast.

concentrations (typically in the order of several mg l^{-1}). The fractions have different settling velocities and also different compositions. The permanently suspended fraction has higher contents of Cu, Cd, Zn, Pb and organic C and N (group I elements, Fig. 1a). The larger/denser fraction has higher contents of Fe, Mn, Al, K and Ti (group II elements). No preference was found for Ca, Si and Cr (group III). The behaviour of Mg shows regional variations. The model, based on measurements in a series of samples, including a range of low and high suspended matter concentrations, is supported by recent measurements in individual size/density fractions isolated with a modified continuous centrifugation technique [26].

The relation between element contents and total suspended matter concentrations can be represented in simple plots with characteristics for the groups I, II and III (Fig. 1a). An example of to what extent this schematic representation reflects the actual data is given in Fig. 1b for Al.

The variation of element contents with total suspended matter concentrations as result of varying contributions of larger/denser particles and thus information on particle sources may remain undetected when particulate suspended concentrations are studied on a volume basis only, rather than on a relative weight basis as well. These considerations might assist in the interpretation of data in estuaries [27], and coastal [28] and open ocean [29] regions.

Element ratios in total suspended matter have been used to study transport processes, elemental partition and sources of seston in the sea [30,31]. It will be demonstrated that element: aluminum ratios (X/Al) can be used conveniently to study the relative importance of aluminum-silicates with respect to other minerals and organic matter as sites for metals, in particular in relation to the continuously suspended fraction and the larger/denser fractions that we have described above.

Plots of X/Al ratios vs. seston concentration for seston from the North Sea appear to be similar for all elements considered here in groups I-III (Fig. 2). The hyperbolic-type relation with a strong increase at low seston concentrations is expected for group I and III elements because of increasing (or constant) value of the nominator and decreasing value of the denominator in the X/Al ratios. Obviously, for the group II elements Fe, Mn and K, the rate of change of their content in seston present at low concentrations differs from that of Al.

Aluminum is closely related to total suspended matter; it may be considered as almost entirely associated with Al-silicates, mainly clay minerals. The content of an element X in seston is altered when varying amounts of components with low content of X (e.g. quartz) are mixed with fine-grained Al-minerals with higher contents of X. However, X/Al ratios are not altered.

The X/Al ratios above 5-10 mg l^{-1} for the major elements: Fe/Al 0.5, Mn/Al 0.02, K/Al 0.3 and Mg/Al 0.2, are very similar to the values reported for fine-grained sediments; 0.5, 0.01, 0.29 and 0.17[32,33]. This suggests that suspended matter at high seston concentrations contains a high fraction of bottom derived material. This is supported by observations on the chemical and mineral composition of seston resuspended and redeposited in the estuary of the river Varde Å by tidal currents[34].

ELEMENTAL PARTITION

The high X/Al ratios in the continuously suspended fraction suggest that the element contents are determined primarily by

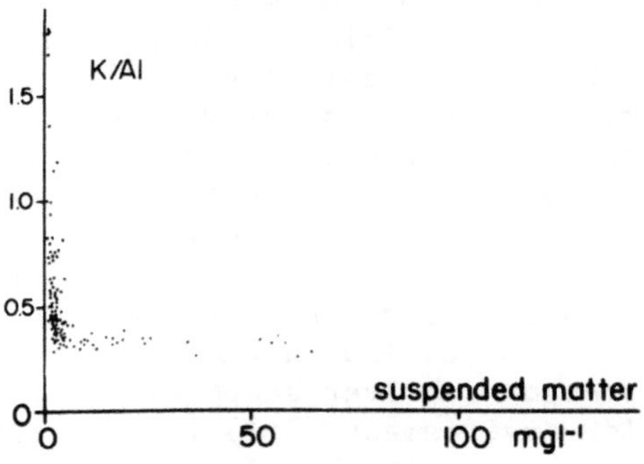

Fig. 2 Relation between K/Al ratios in seston and seston concentration as in Fig. 1.

components such as Fe and Mn oxides and organic matter rather than by Al-minerals.

The strong positive correlation between organic carbon and group I elements, and the negative correlation with group II elements might be used to suggest the association of group I elements with organic carbon. Alternatively these elements may be associated with Fe and Mn oxide coatings, occurring in association with organic matter. The importance of this association has been demonstrated by several authors [35-37].

Since analyses of pure phytoplankton samples from the Southern Bight have suggested that living phytoplankton is not likely to sorb trace metals [2], detrital organic matter is potentially a more significant sorption site. The sorption of trace metals to dead organic matter has been reported for deeper water of the NW Atlantic [27]. A strong correlation between organic matter and trace metals was observed in bottom sediments in the North Sea [38].

However, we have not been able to distinguish between the relative importance of Fe and Mn hydrous oxides and detrital organic matter as sites for trace elements.

METALS AND ESTUARINE MIXING

Mixing curves of dissolved trace metal concentrations, measured over the whole estuarine salinity range against a suitable index of mixing have been used to distinguish conservative from nonconservative behaviour of dissolved components. The main problems in the interpretation of these curves relate to variation in end-member concentrations, the input of material into the estuary from other sources and the correct choice of end-members [39]. Fig. 3a presents characteristic data obtained in the Rhine estuary, with the considerations mentioned above taken into account. The concentrations of Fe, Cd, Cu and Zn fall below the ideal dilution line, indicating the removal of these elements from solution into particulate forms at low salinities. The concentrations of Mn at low salinities are above the ideal dilution line; the behaviour of Mn will be given special attention below. The transition of metals from solution into particulates does not result in a positive deviation from linearity in the corresponding plots of particulate suspended trace metals vs. salinity (Fig. 3b). In fact, a negative deviation was observed. Its detailed interpretation is complicated because various sources contribute to the concentration and composition of particles in the estuary, such as i) particles originally present in river suspension, ii) particles derived from the marine environment, iii) flocs formed from smaller particles originally present in the river, including colloidal material, iv) particles derived from the bottom by wind, currents

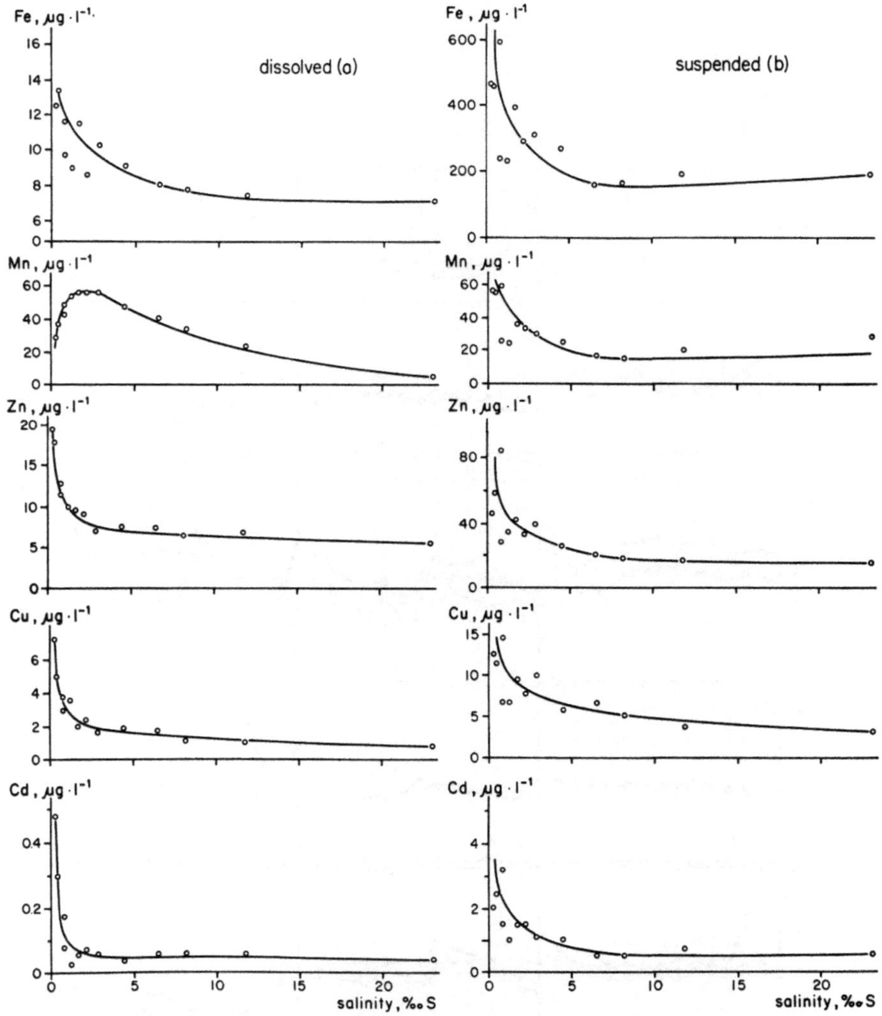

Fig. 3 Dissolved (a) and particulate leachable (b) metal-salinity plots along a longitudinal section of the Rhine estuary, 22-29 September 1975.

and organisms and v) components taken up from solution in the estuary (desorption may also take place).

The relative importance of these contributions may be studied by observations in longitudinal sections and/or at carefully selected fixed positions in the estuary at regular time intervals during tidal cycles. The usefulness of the latter approach may be illustrated by an example from the Rhine estuary. Fig. 4 repre-

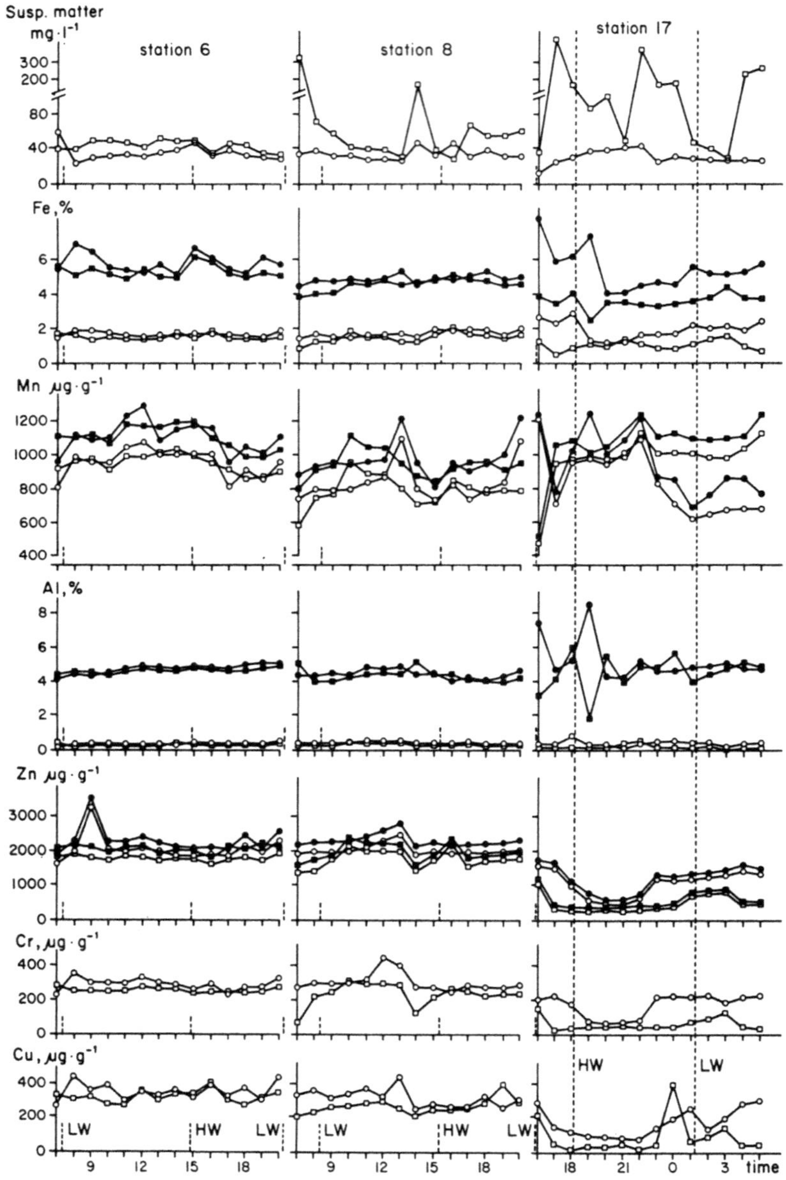
Susp. matter
mg·l⁻¹
station 6
station 8
station 17
Fe, %
Mn µg·g⁻¹
Al, %
Zn µg·g⁻¹
Cr, µg·g⁻¹
Cu, µg·g⁻¹
LW
HW
LW
LW
HW
LW
HW
LW
9
12
15
18
9
12
15
18
18
21
0
3
time

sents data on the content or particulate suspended matter and seston concentrations in samples of the surface and near-bottom layer (1 m above the bottom) taken at 1-hour intervals during complete tidal cycles at three different stations at fixed positions. Station 6 is a fresh water station upstream of the maximum landward transport of particles from the lower estuary. At station 8, salinity increases - slightly - above the river value at high tide and station 17, situated near the mouth of the river, is characterized by permanent stratification, salinity in the surface layer ranging from the characteristic fresh water value during low water to $20^{o}/oo$ at high water. A westerly gale was blowing during sampling at this station as well as during the 3 preceding days.

Fig. 5 gives the relation between leachable contents of Fe, Mn, Zn, Cr and Cu and salinity for the data in the mixing zone. It happens that the relation, including both surface and near-bottom layer data, is practically linear. This must be due to the storm conditions, causing the resuspension of particles from the near-bottom layer into the surface layer. A few near-bottom data for Zn, Cr and Cu are at high salinities above the mixing line. These are associated with the low seston concentrations at slack water. This is consistent with the model given earlier for group I elements and the particle sorting effect has been observed before . The contribution of near-bottom sediment particles to the concentrations in the surface layer was dominated by the coarser fraction.

The values obtained by extrapolating the relationships to the fresh water salinity, differ from the metal contents measured in seston of the freshwater samples at stations 6 and 8 (given as their range at $0.4^{o}/oo$ salinity). The extrapolated values are lower for Mn, Zn, Cr and Cu. For Zn, Cr and Cu this results from the contribution of larger, denser particles from the lower estuary (Fig. 4). The higher value for iron results from the relatively large contribution of dissolved iron that is transformed into particulate form upon the first contact of fresh water with

Fig. 4 Concentrations of seston (mg l^{-1}) and particulate Fe, Mn, Al, Zn, Cr and Cu for samples obtained at fixed positions in the Rhine estuary 11-19 December 1974. For salinity regimes, see text. Leachable surface (o) and near-bottom layer (□) concentration, and total surface (●) and near-bottom (■) concentrations on a weight per sediment dry weight basis (μg g^{-1}, %). Times of high and low water indicated. (Accurate total Cr and Cu concentrations not available: the fractions of the total concentrations that are leached by the method used vary between 60 and 80% for Cu and between 50 and 70% for Cr.)

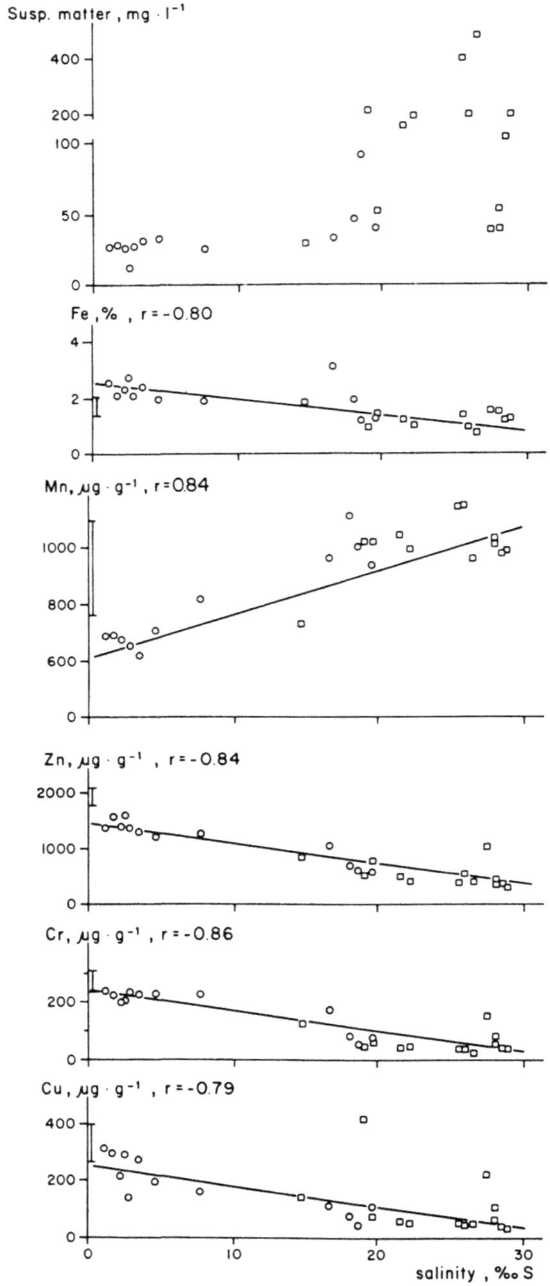

Fig. 5 Relation between salinity and particulate leachable Fe, Mn, Zn, Cr and Cu contents ($\mu g\ g^{-1}$, %) in surface (o) and near-bottom (□) samples of station 17 in the mixing zone of the Rhine estuary (Fig. 4). The data of the freshwater stations 6 and 8 (Fig. 4) are included as a range of values at $0.4^{o}/oo$ S.

salt water, increasing the leachable content of Fe in seston in the mixing zone to values above the river values. The slope for Mn in Fig. 5 is positive in contrast to the situation for the other elements. This is another indication for the unique behaviour of manganese in the Rhine estuary, characterized by i) a positive deviation from the ideal dilution line for the dissolved form at low salinities, ii) a negative deviation at high salinities, iii) high contents of Mn in particulate matter in the lower estuary and iv) higher contents of Mn in particulate matter in the near-bottom layer of the mixing zone than in the surface layer[2,14,15]. Several of these features have been observed in other estuaries as well[40,41,42].

An equilibrium model of the behaviour of Mn during estuarine mixing was established on the basis of known thermodynamic data. The chemical control parameters taken into consideration are ionic strength, pH, oxygen concentration and dissolved carbonate content[14]. This model was shown to account for at least the general trends of dissolved manganese variations observed in the estuaries of Rhine and Scheldt. The close approach of the thermodynamic equilibrium is a consequence of the high turbidity and the intense microbiological activity in these heavily polluted estuaries. The overall behaviour of Mn is determined by cyclic processes: removal from solution at salinities above 10-15 ^{o}/oo S (Fig. 3) accounts for the high suspended Mn contents in the lower estuary and along the Belgian and Dutch coasts. Part of the newly formed particulate Mn is transported back into the upper estuary. Once in contact with low pH or low Eh estuarine water, or after deposition in bottom sediment and subsequent reduction, Mn may return into the dissolved compartment within the water column[15,17]. An attempt was made to establish a mass balance for manganese in the Scheldt estuary which suggests that a large fraction of the riverine load accumulates in bottom sediments[14].

METALS AND COASTAL PROCESSES

The composition of the material leaving the estuary is different from the original composition in the river. Components in solution and associated with small, low density particles can escape while larger, denser particles are trapped in the estuary. Thus, the relative amounts of a metal in solution and suspension in the lower estuary and in the coastal area may be different from that in the river and the upper estuary. Fig. 6 illustrates this distribution for Zn and Cd in the Rhine and the adjacent coastal area. The concentrations of Zn in solution exceed those in suspension both in river and sea water. The dissolved concentrations of Cd, being less than those in suspension in the river, exceed the suspended concentrations in sea water.

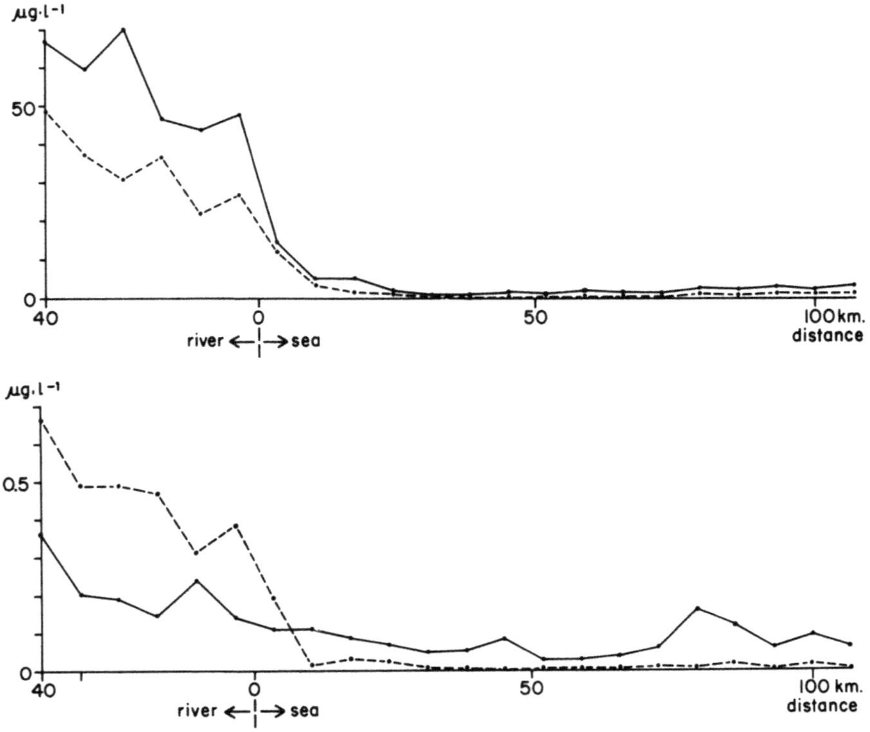

Fig. 6 Dissolved (——) and suspended (- - -) concentrations ($\mu g\ l^{-1}$) in a transect from the freshwater region of the Rhine to the British coast. Upper graph Zn, lower graph Cd.

This observation relates to the observation that particles accumulate in the coastal area including the Wadden Sea, similar to estuarine circulation processes. Fig. 7 represents the characteristic distributions of salinity and seston concentrations in the Southern Bight, the area of minimum seston concentration shifted toward the Dutch coast with respect to the axis of maximum salinity thus indicating a barrier for particles to escape from the coastal area. The strong interaction of particles between bottom sediment and suspension in shallow areas results in extremely variable seston concentrations (both in space and time). This is an additional reason why the effects of material derived from the estuary on the composition of coastal suspended matter may be obscure. According to what has been presented earlier, the highly variable metal contents of these samples, obtained over large areas in the coastal area can be represented by plots of metal content vs seston concentration. These can also serve to

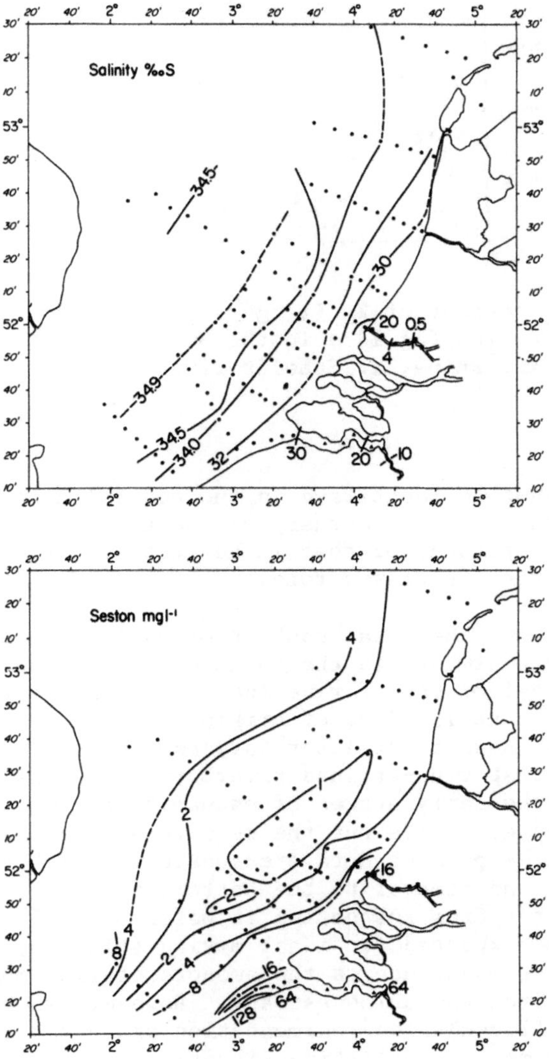

Fig. 7 Distribution of salinity (°/oo S) and seston concentration (mg l^{-1}) in the surface layer of the Southern Bight, October 1975.

distinguish whether unmodified river seston has any measurable effect on the composition of the coastal suspension. Fig. 8 is a plot of particulate zinc content in seston of the Dutch coastal area vs the suspended load. The corresponding data for riverine material is also indicated. It is obvious that the presence of

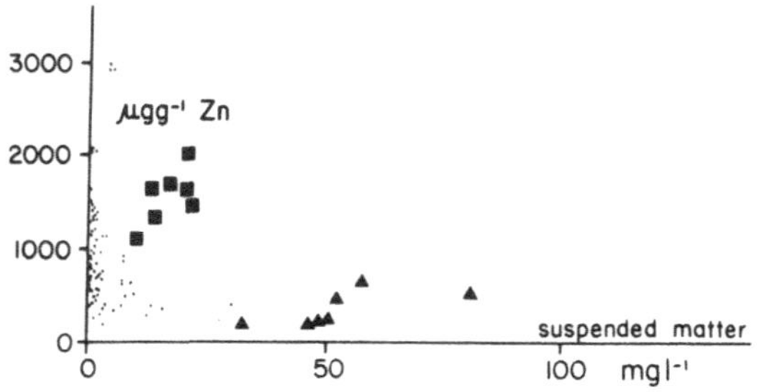

Fig. 8 Element content of Zn ($\mu g\ g^{-1}$) in relation to seston concentration ($mg\ l^{-1}$) in the Southern Bight, October 1975. ■ : Rhine, ▲: Scheldt river values.

this matter cannot be identified in the coastal data, suggesting that the material from the estuary may be too much diluted by the material already present or that it has been modified in the estuary. Both factors play a role.

In conclusion, the metal content of particulates in suspension and bottom sediments in the Rhine estuary is determined primarily by physical mixing of riverine and marine particles[1,2], with additional contributions of originally dissolved and colloidal forms[11]. Results of laboratory experiments in the past have suggested that fine-grained riverborne material would be flocculated in the early stages of estuarine mixing. It is likely to occur in many estuaries but the results are partly conflicting[5,7,43-45]. The present data are consistent with the occurrence of flocculation and removal to the bottom, resulting in a loss of particulate matter from suspension in water of 5-15 $^{o}/oo$ S of about 50% of the value expected for ideal mixing (Fig. 3b). This is of the same order of magnitude as the amounts of fine material dredged yearly from the Rhine estuary. However, we have found the net effects of flocculation-sedimentation processes to be highly variable with time[2]. Conflicting situations of this kind may be related to the fact that the conditions for effective flocculation are not always met, such as sufficient seston concentration, a large proportion of particles with appropriate surface properties and suitable flow conditions that allow formation and settling out of suspension. Still, the Rhine estuary acts as a barrier for the transport of particulate trace metals out of the estuary into the marine environment. This includes large fractions of originally dissolved riverborne components (Fe, Cu, Zn and Cd). Further modifications in the distribution of metals over solution and particulates, expected to continue in the coastal area, have only been identified for manganese, taking into account the effects of variations in size/density properties of particulates.

REFERENCES

1. Müller, G. and U. Förstner, 1975: Heavy metals in sediments of the Rhine and Elbe estuaries: mobilization of mixing effect Environ. Geol., 1, 33-39.
2. Duinker, J.C. and R.F. Nolting, 1976: Distribution model for particulate trace metals in the Rhine estuary, Southern Bight and Dutch Wadden Sea. Neth. J. Sea Res., 10 71-102.
3. Moore, R.M., J.D. Burton, P.J. LeB. Williams and M.J. Young, 1979: The behaviour of dissolved organic material, iron and manganese in estuarine mixing. Geochim. Cosmochim. Acta., 43, 919-926.
4. Postma, H., 1967: Sediment transport and sedimentation in the estuarine environment. In: "G.H. Lauff, Estuaries". Am. Ass. Adv. Sci. Publ., 83, 158-179.
5. Meade, R.H., 1972: Transport and deposition of sediments in estuaries. Mem. Geol. Soc. Am., 133, 91-120.
6. Schubel, J.R., 1971: Tidal variation of the size distribution of suspended sediment at a station in the Chesapeake Bay turbidity maximum. Neth. J. Sea Res., 5, 252-266.
7. Edzwald, J.K., J.B. Upchurch and C.R. O'Meila, 1974: Coagulation in estuaries. Environ. Sci. Technol., 8, 58-63.
8. Sholkovitz, E.R., 1976: Flocculation of dissolved and inorganic matter during mixing of river water and sea water. Geochim. Cosmochim. Acta., 40, 831-845.
9. Eisma, D., J. Kalf and M. Veenhuis, 1980: The formation of small particles and aggregates in the Rhine estuary. Neth. J. Sea Res., 14, 172-191.
10. Coonley, L.S., L.S. Baker and H.D. Holland, 1971: Iron in the Mullica river and in Great Bay, New Jersey. Chem. Geol., 7, 51-63.
11. Duinker, J.C. and R.F. Nolting, 1978: Mixing, removal and mobilization of trace metals in the Rhine estuary. Neth. J. Sea Res., 12, 205-223.
12. Boyle, E.A., J.M. Edmond and E.R. Sholkovitz, 1977: The mechanism of iron removal in estuaries. Geochim. Cosmochim. Acta., 41, 1313-1324.
13. Evans, D.W. and N.H. Cutshall, 1973: Effects of ocean water on the soluble-suspended distribution of Columbia River radionuclides. In: "Radioactive contamination of the marine environment". Int. Atomic Energy Agency, Vienna, 125-140.
14. Wollast, R., G. Billen and J.C. Duinker, 1979: Manganese in the Rhine and Scheldt estuaries. I. Physico-chemical behaviour. Estuar. Coast. Mar. Sci., 9, 161-169.
15. Duinker, J.C., R. Wollast and G. Billen, 1979: Manganese in the Rhine and Scheldt estuaries. II. Geochemical cycling. Estuar. Coast Mar. Sci., 9, 727-738.

16. Troup, N.B. and O.P. Bricker, 1975: Processes affecting the transport of materials from continents to oceans. In: "T.M. Church. Marine chemistry in the coastal environment". ACS Symp. Ser. 18. Am. Chem. Soc., Washington, D.C., 133-151.
17. Elderfield, H. and A. Hepworth, 1975: Diagenesis, metals and pollution in estuaries. Mar. Pollut. Bull., (NS) 6, 85-87.
18. Aston, S.R. and R. Chester, 1973: The influence of suspended particles on the precipitation of iron in natural waters. Estuar. Coast. Mar. Sci., 1, 225-231.
19. Johnson, R.G., 1974: Particulate matter at the sediment-water interface in coastal environments. J. Mar. Res., 32, 313-330.
20. Schnitzer, M. and S.U. Kahn, 1972: Humic substances in the environment. Dekker.
21. Kranck, K., 1974: The role of flocculation in the transport of particulate pollutants in the marine environment. In: "Proc. Int. Conf. of Persistent Chemicals in Aquatic Ecosystems". Ottawa Nat. Res. Council, Section 1, 41-46.
22. Patchineelam, S.R. and U. Förstner, 1977: Bindungsformen van Schwermetallen in marinen sedimenten. Senckenberg. Marit., 9, 75-104.
23. Tessier, A., P.G.C. Campbell and M. Bisson, 1979: Sequential extraction procedure for the speciation of particulate trace metals. Analyt. Chem., 51, 844-851.
24. Duinker, J.C., G.T.M. van Eck and R.F. Nolting, 1974: On the behaviour of copper, zinc, iron and manganese in the Dutch Wadden Sea; evidence for mobilization processes. Neth. J. Sea Res., 8, 214-239.
25. Duinker, J.C., R.F. Nolting and H.A. van der Sloot, 1979: The determination of suspended metals in coastal waters by different sampling and processing techniques (filtration, centrifugation). Neth. J. Sea Res., 13, 282-297.
26. van der Sloot, H.A. and J.C. Duinker, in press: Isolation of different suspended matter fractions and their trace metal contents. Environ. Technol. Lett., 2, 511-520.
27. Sholkovitz, E.R., 1979: Chemical and physical processes controlling the chemical composition of suspended material in the river Tay estuary. Estuar. Coast. Mar. Sci., 8, 523-545.
28. Sundby,B. and D.H. Loring, 1978: Geochemistry of suspended particulate matter in the Saguenay Fjord. Can. J. Earth Sci., 15, 1002-1011.
29. Wallace, G.T., G.L. Hoffman and R.A. Duce, 1977: The influence of organic matter and atmospheric deposition on the particulate trace metal concentration of Northwest Atlantic surface seawater. Mar. Chem., 5, 143-170.

30. Spencer, D.W. and P.L. Sachs, 1970: Some aspects of the distribution, chemistry, and mineralogy of suspended matter in the Gulf of Main. Mar. Geol., 9, 117-136.
31. Price, N.B. and S.E. Calvert, 1973: A study of the geochemistry of suspended particulate matter in coastal waters. Mar. Chem., 1, 169-189.
32. Goldschmidt, V.M., 1954: Geochemistry. E. Muir, ed., Oxford Univ. Press, Oxford.
33. Krauskopf, K.B., 1965: Factors controlling the concentrations of thirteen rare metals in sea water. Geochim. Cosmochim. Acta., 9, 1-32B.
34. Duinker, J.C., M.T.J. Hillebrand, R.F. Nolting, S. Wellershaus and N. Kingo Jacobsen, 1980: The river Varde A: processes affecting the behaviour of metals and organochlorines during estuarine mixing. Neth. J. Sea. Res., 14, 237-267.
35. Pillai, T.N.V., M.V.M. Desai, E. Mathew, S. Ganapathy and A.K. Ganguly, 1971: Organic materials in the marine environment and the associated metallic elements. Current Sci., 40, 75-81.
36. Neihof, R.H. and G.I. Loeb, 1972: The surface charge of particulate matter in seawater. Limnol. Oceanogr., 17, 7-16.
37. Senesi, N., S.M. Griffith, M. Schnitzer and M.G. Townsend, 1977: Binding of Fe^{3+} by humic materials. Geochim. Cosmochim. Acta., 41, 969-976.
38. Wollast, R., 1979: Heavy metals in the suspended matter and in the recent sediments of the Southern Bight. International meeting on holocene marine sedimentation in the North Sea basin. IAS abstracts, Texel, Sept. 17-23, 68.
39. Boyle, E.A., R. Collier, A.T. Dengler, J.M. Edmond, A.C. Ng and R.F. Stallard, 1974: On the chemical mass-balance in estuaries. Geochim. Cosmochim. Acta., 38, 1719-1728.
40. Evans, D.W., N.H. Cutshall, F.A. Cross and D.A. Wolfe, 1977: Manganese cycling in the Newport estuary, North Carolina. Estuar. Coast. Mar. Sci., 5, 71-80.
41. Graham, W.F., M.L. Bender and G.P. Klinkhammer, 1976: Manganese in Narraganset Bay. Limnol. Oceanogr., 21, 665-673.
42. Holliday, L.M. and P.S. Liss, 1976: The behaviour of dissolved iron, manganese and zinc in the Beaulieu estuary, S. England. Estuar. Coast. Mar. Sci., 4, 349-353.
43. Postma, H, 1961: Transport and accumulation of suspended matter in the Dutch Wadden Sea. Neth. J. Sea Res., 1, 148-190.
44. Nichols, M.M. and G. Poor, 1967: Sediment transport in a coastal plain estuary. Proc. Am. Soc. Civil Eng., 93WW4, 83-95.

45. Sheldon, R.W., 1968: Sedimentation in the estuary of the river Crouch, Essex, England. Limnol. Oceanogr., 13, 72-83.

THE EFFECT OF SEWAGE EFFLUENTS ON THE FLOCCULATION OF MAJOR AND TRACE ELEMENTS IN A STRATIFIED ESTUARY

Richard A. Feely, Gary J. Massoth and Marilyn F. Lamb

Pacific Marine Environmental Laboratory
Environmental Research Laboratories, NOAA
7600 Sand Point Way, N.E.
Seattle, Washington 98115
U.S.A.

ABSTRACT

In order to study the effects of sewage effluents on the flocculation of trace elements in a stratified estuary, water samples were collected from the Duwamish River Estuary at stations located above and below the Renton Sewage Disposal Site. The water samples were filtered through 0.2-μm pore size Nuclepore filters, the filtrate mixed with various amounts of filtered sea water from Elliott Bay, and the products refiltered onto 0.2-μm filters and analyzed for major and minor elements by x-ray secondary emission spectrometry. The results show that the downstream samples produced two to three times more flocculated material and three to five times more P, Ca, Fe, Ni, Cu and Zn than the upstream samples. In contrast, little differences were found for Al and Ti. Similar experiments utilizing the effluent from the Renton Sewage Treatment Plant also showed the same trends, suggesting that the sewage effluent enhanced the flocculation of these elements in the estuary. Major element analysis of the flocculated material indicates that it is mainly composed of organic material.

INTRODUCTION

The influence of trace elements from waste-water effluents on the ecology of coastal environments is a growing concern of environmental policy makers. In populated areas concentrations of toxic trace elements in sewage effluents have been sufficiently high to increase their levels in coastal sediments and cause abnormally high concentrations of these constituents in benthic organisms[1,2,3,4]. Rational management of sewage treatment and disposal procedures is made difficult by a lack of understanding of the mechanisms that control the fate of trace elements in estuarine and coastal waters.

A major fraction of trace elements in waste-water effluents reacts with estuarine particulate matter. Upon entering the estuary, the particulate matter undergoes a number of physico-chemical interactions, resulting in remobilization of some associated trace elements and flocculation of others[5,6]. Further remobilization of trace elements can occur when the particles reach the seafloor and become part of the sediment column[7]. In coastal environments where circulation is restricted by land forms and/or incomplete flushing, these processes can have a significant effect on the distributions of trace elements. Knowledge of the processes controlling the uptake and release of trace elements in association with suspended material is important, therefore, for predicting the fate and effects of these constituents in coastal marine ecosystems.

Although the processes involved in the uptake and release of trace elements associated with particles found in coastal waters have been studied to some extent, the results are confusing and in some cases contradictory. Krauskopf[8] conducted a theoretical and experimental evaluation of several processes controlling the concentrations of a number of trace elements in seawater and concluded that hydrous manganese and ferric oxides, organic detritus, and certain clay minerals are all effective in removing metals from seawater. Using estuarine mixing simulation experiments (product-mode experiments), Sholkovitz[6] observed the flocculation of several trace elements when filtered river water and seawater were mixed. However, laboratory investigations by other researchers have indicated that suspended solids, which are typical of riverine materials, release trace elements after exposure to seawater[9,10]. Similar results for trace elements associated with particles from waste-water effluents were demonstrated by the laboratory studies of Rohatgi and Chen[5]. In this paper, we report on the results of both laboratory and field investigations of flocculation processes involving trace elements in the Duwamish River estuary in Washington State. We will show that trace elements from sewage effluents are flocculated during estuarine mixing and that this process involves organic matter.

The Study Region

The physical characteristics of the Duwamish River estuary have been described by several authors[11,12,13,14]. The combined

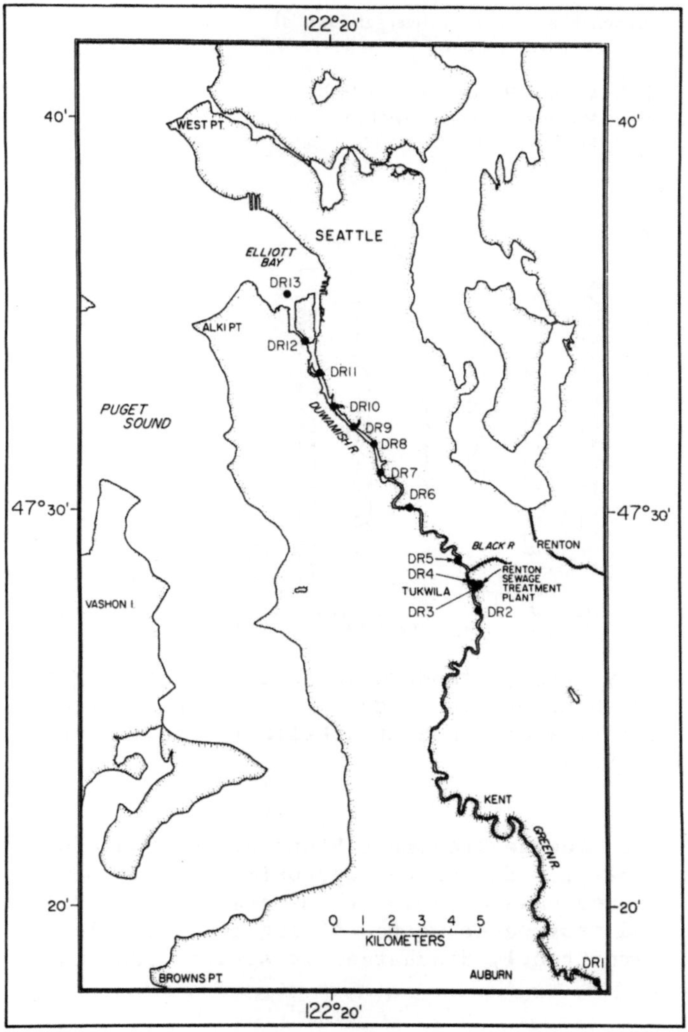

Fig. 1 Locations of sampling stations in the Duwamish River and Elliott Bay; samples were collected on August 11, 1979, during an ebb tide.

Green-Duwamish River system extends from the western slopes of the Cascade Mountains to Elliott Bay in Puget Sound. The Green River flows westward through forests, pastureland and farmland until it is joined by the Black River to form the Duwamish River. This river continues to meander to the northwest through the heavily industrialized regions of Renton, Tukwila and Seattle, Washington (Fig. 1). Figure 2 gives the discharge data for the Duwamish River at Tukwila.

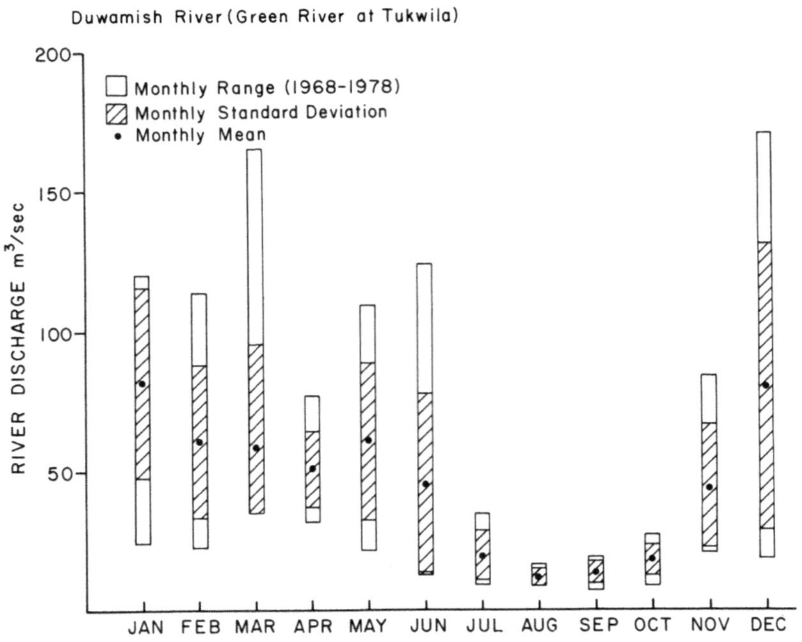

Fig. 2 Monthly means, standard deviations, and ranges for the Duwamish River discharge. (Data compiled from U.S.G.S. stream flow obtained at Tukwila for period of record: 1968-1978).

The Renton Sewage Treatment Plant of the Municipality of Metropolitan Seattle discharges approximately 136,000 m^3 day^{-1} of secondary treated sewage, 20.5 kilometers from the river mouth. Industrial and stormwater wastes of significantly lesser amounts are also intermittently discharged at several locations along the lower river.

The lower 10 kilometers of the Duwamish River, dredged and straightened by the U.S. Army Corps of Engineers, forms a two-layered estuary (type 2B of the Hansen-Rattray convention). The upper layer consists of mixed salt water and freshwater, and the lower layer is mostly unmixed salt water. The seaward end has a maximum tidal range of approximately 4 meters.

EXPERIMENTAL PROCEDURES

Sample Handling Methods

The Duwamish River estuary was chosen for this study because the waste-water discharge from the Renton Sewage Treatment Plant represents approximately 2-25 percent of the total river discharge and because it has been shown by Santos and Stoner [12] to affect the overall water quality of the lower estuary. Water samples were collected on August 11, 1979, from 13 stations in the Duwamish River Estuary (Fig. 1). For the suspended-matter studies water samples were collected in General Oceanics 5-L PVC Niskin bottles. To avoid loss of rapidly settling particles [15,16], aliquots from each Niskin bottle were rapidly withdrawn (within ten minutes of collection) into 1-L acid-cleaned polyethylene bottles. At the laboratory the samples were shaken and vacuum-filtered through 0.2-µm pore size Nuclepore polycarbonate filters (47 mm in diameter for suspended-matter concentration determination and 25 mm in diameter for elemental analyses other than C and N) and precombusted 0.2-µm pore size Selas silver filters (25 mm in diameter for C and N analyses). All samples were rinsed with three 10-ml aliquots of deionized, membrane-filtered water (adjusted to pH 8.0), placed in individual polycarbonate petri dishes with lids slightly ajar for a 24-hour desiccation period over sodium hydroxide, and then sealed and stored for subsequent analysis.

For the product-mode experiments, water samples were collected from stations 1, 2, 4, 5 and 13 in acid-cleaned 5-gal polyethylene carboys and filtered through 0.2-µm pore size Nuclepore filters to remove the particulate matter. The filtrates were then stored in acid-cleaned polyethylene bottles at $1^{o}C$ until the product-mode experiments were initiated, approximately three days later. Figure 3 shows the flow diagram for the product-mode experiments. 100 ml of filtered water from stations 1, 2, 4 and 5 were mixed with varying amounts of filtered seawater from station 13 to provide samples with salinities in the range $2.5^{o}/oo$ - $27.5^{o}/oo$. The mixtures were gently shaken at $20^{o}C$ for 48 hours, after which the newly flocculated material was filtered through 0.2-µm Nuclepore filters and preserved for analysis of major and trace elements.

Analytical Methods

Total concentrations of suspended matter were determined gravimetrically. The weighing precision ($2\sigma = \pm 0.011$ mg) and volume reading error ($\pm$ 10 ml) yield a combined coefficient of variation in suspended matter concentrations of approximately one percent. This variability is probably overshadowed, however, by that associated with the sampling precision, reported by Feely et al. [17] to be in the range 5 - 25%.

The major (Al, P, Si, K, Ca, Ti and Fe) and trace (Cr, Mn, Ni, Cu and Zn) inorganic elements in the suspended-matter and product-mode samples were determined by x-ray secondary-emission (fluorescence) spectrometry using a Kevex Model 0810-5100 x-ray

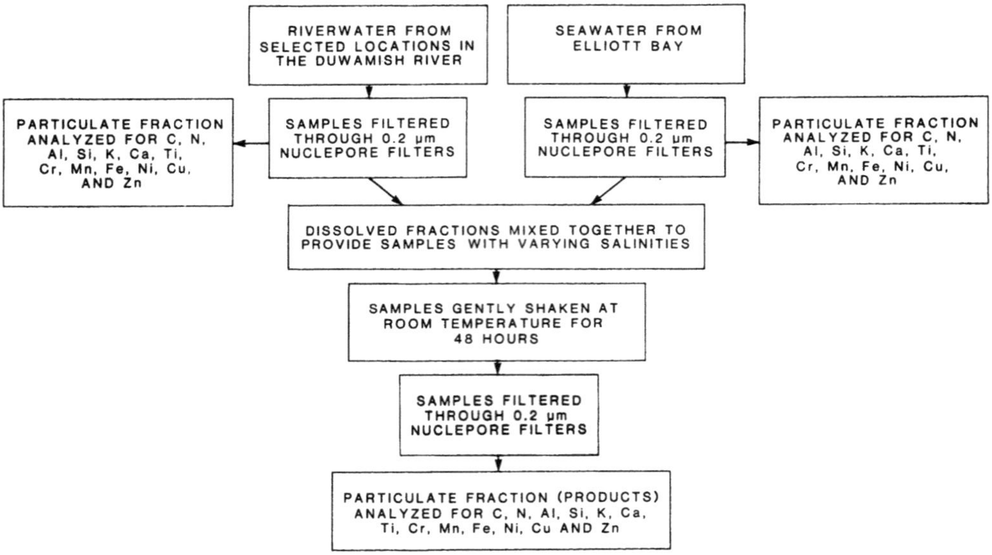

Fig. 3 Flow chart for collection and analysis of suspended and flocculated materials using samples from the Duwamish River and Elliott Bay.

energy spectrometer and the thin-film technique[18]. A silver x-ray tube (operated at 50 kV, 40 mA) was used to excite a sequence of secondary targets (Fe target for Al through Cr; Se target for the remaining elements), which efficiently fluoresced the range of elements in the sample. Standards were prepared from suspensions of finely ground U.S.G.S. Standard Rocks (W-1, BCR-1, AGV-1 and GSP-1; 90 percent by volume were less than 15 μm in diameter as determined by scanning electron microscopy) collected on Nuclepore filters identical to those used for sample acquisition. At a filter loading of 325 $\mu g\ cm^{-2}$ the determination limits were less than 0.25% by weight and 13 ppm for the major and trace elements, respectively. The relative standard deviations resulting from ten replicate analyses of a sample with similar weight distribution were less than 3% for major elements and less than 8% for trace elements.[19]

Analysis of total particulate carbon and nitrogen in the suspended matter was performed with a Hewlett Packard 185B CHN analyzer. In this procedure, particulate carbon and nitrogen compounds were combusted to CO_2 and N_2 (micro Pregl-Dumus method) chromatographed on Poropak Q, and detected sequentially with a thermal conductivity detector following a modification of the procedure outlined by Sharp[20]. NBS acetanilide was used for standardization. Analyses of replicate field samples yield relative standard deviations ranging from 2% to 10% for carbon and 7% to 14% for nitrogen.[19]

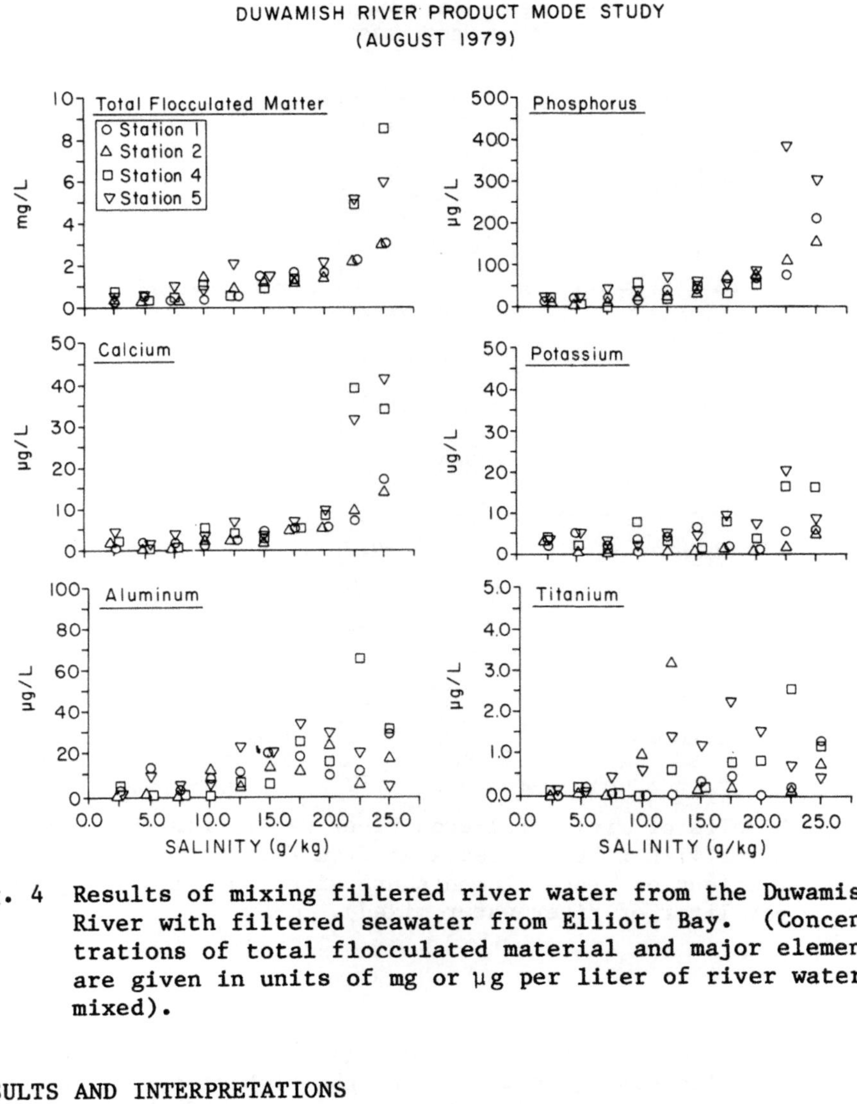

Fig. 4 Results of mixing filtered river water from the Duwamish River with filtered seawater from Elliott Bay. (Concentrations of total flocculated material and major elements are given in units of mg or µg per liter of river water mixed).

RESULTS AND INTERPRETATIONS

Produce Mode Experiments

The results of the product mode experiments are graphically represented in Figures 4 and 5. Figure 4 shows the results for total flocculated material and major elements. The concentrations have been blank corrected for the zero salinity in the amount of flocculated material per liter of river water§. The data for

§The concentration of flocculated materials is expressed in units of mg per liter of river water. This may be converted to mg per kilogram by multiplying by the factor 1.0013.

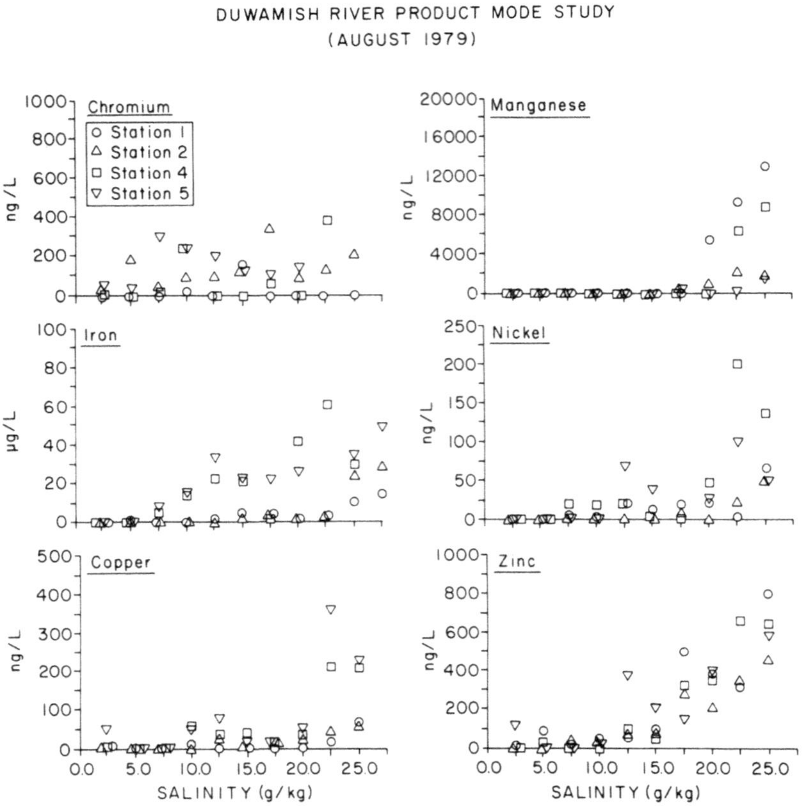

Fig. 5 Results of mixing filtered river water from the Duwamish River with filtered seawater from Elliott Bay. (Concentrations of trace elements are given in units of μg or ng per liter of river water mixed).

total flocculated material indicate a general increase in concentration with increasing salinity. At the highest salinities, samples from stations 1 and 2, located upstream of the Renton Sewage Treatment Plant, produced 2-3 mg of newly flocculated material per liter of river water, whereas, the high-salinity samples from stations 4 and 5, located downstream of the sewage treatment facility, produced 2-3 times more flocculated material. These data suggest that the effluent from the Renton Sewage Treatment Plant significantly increases the total amount of flocculated material produced in the experiments. Similarly, the data for P, K and Ca show two- to fourfold enrichments in the downstream samples relative to the upstream samples. In contrast, Al and Ti do not show a consistent pattern of enrichment.

Figure 5 shows the data for the trace elements. Two- to fivefold enrichments in the downstream samples were found for Cr, Fe, Ni, Cu and Zn, whereas Mn showed no significant enrichment in the downstream samples relative to those collected upstream of the sewage treatment plant. In fact, the data for Mn were anomalous in that no flocculation was observed below a salinity of $20.0^{\circ}/oo$. The reasons for this effect are unknown, but may be in part a result of the effects of increased salinity on the kinetics of Mn precipitation. These results indicate that, with the exception of Mn, flocculation of the trace elements studied was significantly enhanced by the sewage effluents. This is due, in part, to the increased concentrations of trace elements and organic matter in the sewage effluents[21]. The flocculated trace elements amount to approximately 18-62% of the total dissolved trace-element content of the lower river (i.e., Mn--40%, Fe--62%, Ni--18%, Cu--53% and Zn--34%) and contribute to the decrease in the concentration of dissolved trace elements in the estuary.[21,22]

In order to verify the results of the product-mode studies, we conducted a similar experiment using filtered sewage effluents in place of river water. The effluent was collected as a 24-hour composite sample on the same day the river samples were collected. The data demonstrate that for most of the elements in the effluent, flocculation occurs over the entire range of salinities (Fig. 6). Phosphorus shows the greatest amount of flocculation.§ This element is followed by Si, Ca and Fe. These elements are primary constituents of organic matter and, therefore, a flocculation mechanism involving colloidal organic matter is probable. In an attempt to further explore this point, we have calculated the concentrations of the elements in the newly flocculated material from stations 2 and 5 for salinity ranges of $2.5\text{-}10.0^{\circ}/oo$, $12.5\text{-}20.0^{\circ}/oo$ and $22.5\text{-}27.5^{\circ}/oo$. The data (Table 1) are representative of what was found for all the experiments. The results indicate that organic matter is the major phase in the flocculated matter. C, N, Si, P, Fe and Ca, which are the major elemental constituents of organic matter, are also the most abundant elements in the flocculated matter. Al and Ti, which are usually associated with inorganic materials, comprise only about 1-3% of the flocculated material.

With respect to the major elements one other trend is worth noting. With increasing salinity, P concentrations increase while the concentrations of Al and Ti are nearly constant or decrease slightly. Moreover, this trend is more dramatic for station 5 than for station 2. These data suggest that as the salinity of the estuarine water increases flocculation of organic materials predominates over flocculation of inorganic materials, especially

---

§ Carbon and nitrogen were not measured in these samples.

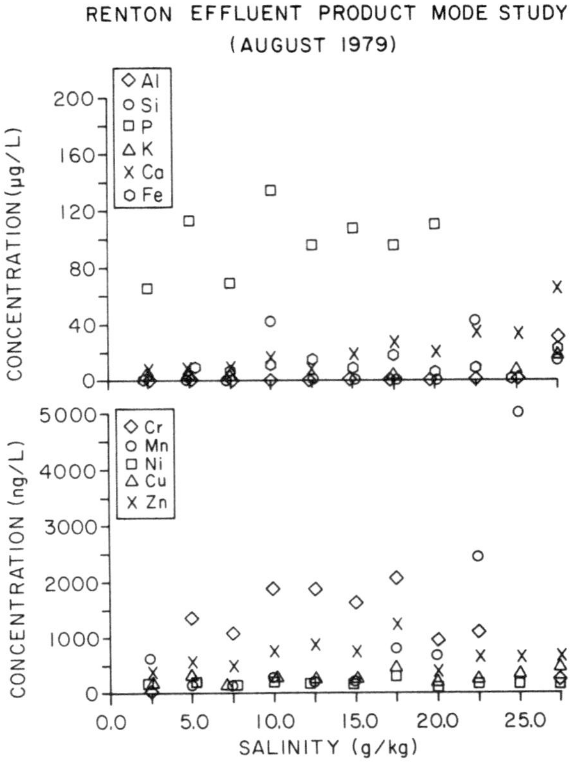

Fig. 6 Results of mixing filtered effluent from the Renton Sewage Treatment Plant with filtered seawater from Elliott Bay. (Concentrations are given in µg or ng per liter of effluent mixed. A composite sample was collected on August 11, 1979).

when sewage effluent is present. These trends are consistent with the general conclusions of Sholkovitz[23,6], who stated that flocculation processes in Scottish estuaries are dominated by interactions involving organic matter.

The trace-element data show large variability between samples but in general the station 5 samples contain more Cr, Cu and possibly Zn than the station 2 samples. These data are interpreted as evidence for concentration of these elements in the flocculated material, with the concentration effect being enhanced by the sewage effluents.

Table 1. Chemical Composition of Flocculated Material From Stations 2 and 5 at Salinity Ranges of 2.5-10.0°/oo, 12.5-20.0°/oo and 22.5-27.5°/oo.

ELEMENT	2.5-10.0°/oo Sta. 2	2.5-10.0°/oo Sta. 5	12.5-20.0°/oo Sta. 2	12.5-20.0°/oo Sta. 5	22.5-27.5°/oo Sta. 2	22.5-27.5°/oo Sta. 5
N† (Wt. %)	8.8 ±2.0	8.9 ±4.1	8.8 ±1.5	3.4 ±0.8	7.7 ± .4	7.2 ±1.2
C† (Wt. %)	ND	ND	62.0	77.0	67.0 ±15.9	74.5 ±21.1
P (Wt. %)	1.77 ±0.70	2.85 ±1.50	3.90 ±0.27	3.47 ±0.27	4.78 ±0.07	6.18 ±1.76
Al (Wt. %)	1.19 ±1.02	2.25 ±0.57	1.42 ±0.25	2.09 ±0.45	0.64 ±0.33	0.49 ±0.17
Si (Wt. %)	6.66 ±6.28	7.47 ±4.53	7.98 ±7.09	9.50 ±1.49	2.64 ±1.54	3.15 ±3.19
K (Wt. %)	0.23 ±0.27	0.39 ±0.24	0.08 ±0.03	0.33 ±0.11	0.11 ±0.07	0.26 ±0.14
Ca (Wt. %)	2.45 ±0.96	2.11 ±0.58	1.41 ±0.17	1.25 ±0.20	0.93 ±0.16	0.88 ±0.10
Ti (Wt. %)	0.15 ±0.10	0.11 ±0.05	0.13 ±0.16	0.13 ±0.03	0.04 ±0.02	0.03 ±0.01
Cr (ppm)	104 ±74	332 ± 61	152 ± 82	439 ±562	80 ± 5	244 ±199
Mn (ppm)	ND	ND	507 ±595	ND	987 ±197	538 ±109
Fe (Wt. %)	2.08 ±1.23	2.80 ±0.93	1.35 ±0.11	2.56 ±0.37	1.06 ±0.29	2.13 ±1.67
Ni (ppm)	ND	57 ± 25	ND	60 ± 50	13 ± 4	15 ± 6
Cu (ppm)	ND	38 ± 30	25 + 9	91 ±118	31 ± 12	53 ± 16
Zn (ppm)	136 ±128	179 ±127	195 ± 62	182 ± 47	162 ± 38	278 ±234

ND = Not detected above blank values.

† = Weight percentages of N and C were determined using samples collected with Nuclepore filters. The uncertainty of the blanks in the filters were ±2.8 μg N and ±100μgC.

Table 2. Comparison of the elemental composition of suspended matter from the Duwamish River with the composition of suspended material from the Duwamish River Estuary (samples collected August 11, 1979).

Sample Description	No. of Samples	C† Wt.% ±1σ	Mg Wt.% ±1σ	Al Wt.% ±1σ	Si Wt.% ±1σ	P Wt.% ±1σ	K Wt.% ±1σ	Ca Wt.% ±1σ	Ti Wt.% ±1σ	Cr Wt.% ±1σ	Mn Wt.% ±1σ	Fe Wt.% ±1σ	Ni ppm ±1σ	Cu ppm ±1σ	Zn ppm ±1σ
Duwamish River	3	12.9 ±1.7	0.92 ±0.35	3.15 ±0.88	14.73 ±4.52	1.96 ±0.67	0.43 ±0.14	1.76 ±0.28	0.26 ±0.08	103 ±22	1602 ±230	8.39 ±1.93	29 -	41 -	138 ±5
Duwamish River Estuary (0-10 $^{o}/oo$)	3	17.5 ±10.4	1.53 ±0.41	5.97 ±0.77	19.03 ±3.94	2.91 ±2.08	0.61 ±0.11	1.84 ±0.16	0.38 ±0.07	138 ±53	1620 ±535	11.13 ±4.13	43 ±10	76 ±12	277 ±90
Duwamish River Estuary (10-20 $^{o}/oo$)	11	11.5 ±3.2	1.77 ±0.51	5.93 ±0.77	20.21 ±3.53	3.61 ±0.75	0.63 ±0.07	1.91 ±0.20	0.37 ±0.04	210 ±57	1419 ±345	11.15 ±1.20	50 ±10	86 ±18	369 ±59
Duwamish River Estuary (20-30 $^{o}/oo$)	3	16.3 ±0.7	1.36 ±1.55	5.33 ±2.35	23.48 ±3.98	0.90 ±0.76	0.51 ±0.25	2.47 ±0.68	0.29 ±0.10	155 ±60	3786 ±511	3.15 ±2.29	70 -	- -	385 ±17
Duwamish Estuary Sediment††				8.01	24.00		1.30	2.00	0.50	94	590	5.30	31	124	227

† Weight percentages of C were determined using two different filter types (Selas silver filters and Nuclepore filters) and, therefore, are subject to a greater number of errors than the data obtained for the other elements, which were obtained from a single filter type.

†† Sample collected from the west channel of the Duwamish Waterway (after Riley et al., 1980).

Elemental Composition of the Particulate Matter

In order to determine whether or not the flocculation reactions observed in the laboratory studies with filtered samples actually play a role in causing changes in the elemental composition of the suspended matter in the Duwamish River estuary, the particulate samples collected on August 11, 1979, were analyzed for their major and trace-element contents. The resulting data are represented graphically in Figures 7 and 8 and summarized in Table 2. The riverine data are averaged from samples located below the sewage outfall (stations 4, 5 and 6). Figure 7 shows the relationship between total suspended matter and salinity for the entire estuary. The figure shows evidence for flocculation and sedimentation throughout most of the estuary and, in particular, at the very beginning of the salinity increase. In this region (0-10°/oo salinity), striking increases in the concentrations of particulate C, P, Al, Si, Fe, Cu and Zn were observed (Table 2). The largest increases were observed for C and Si, respectively, indicating a flocculant containing organic matter. These results agree well with the chemical data for the flocculated material (Table 1) in that the major constituents of flocculated matter were also C, Si, Al, P and Fe. They are also consistent with the geochemical data of Murray and Gill[24] for flocculation of dissolved Fe in the Duwamish River. As soon as the salinity of the estuary begins to increase, an Fe-rich organic phase starts to flocculate onto the riverine suspended matter. The organic flocculant is enriched in trace metals to the extent that the total trace-element content of the particulate matter is enriched by a factor of about 1.5 to 2.8 (Fig. 8). Mn in the particulate matter shows no enrichment until a salinity of about 20°/oo had been reached, similar to the results of the laboratory experiments (Fig. 5).

A significant difference between the laboratory experiments and the analysis of the field samples is the timing of flocculation in relation to salinity. Field samples indicated flocculation immediately after the salinity begins to increase, whereas the laboratory experiments show flocculation begins at higher

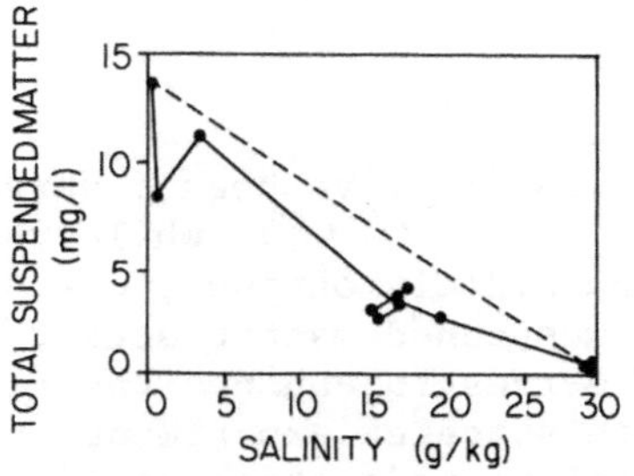

Fig. 7 Plot of the relationship between near-surface total suspended matter and salinity in the Duwamish River estuary. The dashed line shows the theoretical dilution curve.

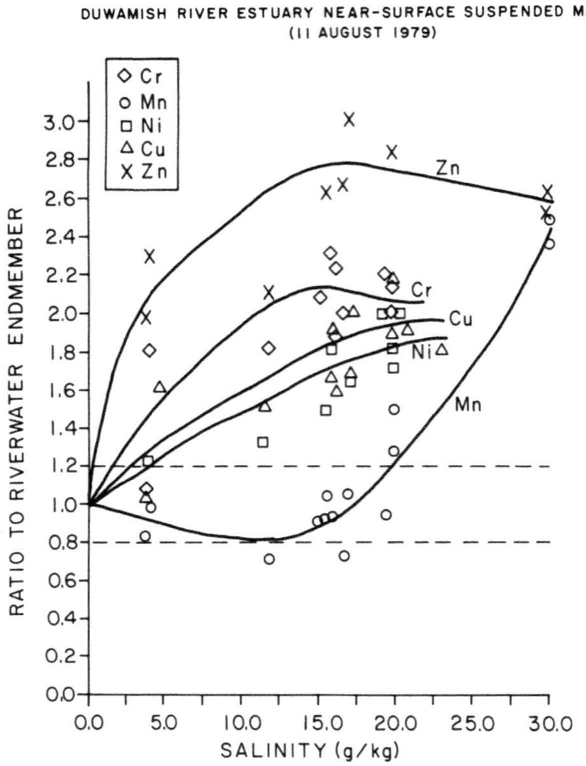

Fig. 8 Ratio of trace-element content of suspended matter from the Duwamish River estuary to the trace-element content of suspended matter from the Duwamish River immediately above the zone of interaction with seawater as a function of salinity. The dashed lines show the upper limits of the sampling and analytical errors.

salinities. Similar results have been observed for field samples in other estuaries [25,26,23,6,27,28]. While the reasons for this difference are not immediately obvious, one possible explanation is that the riverine suspended matter acts to catalyze the flocculation processes and serves to enhance the reactions under natural conditions. More sophisticated experiments than reported here are required before a complete understanding of this difference can be gained.

DISCUSSION

The flocculation of trace metals in the Duwamish River estuary is a major factor affecting the goechemistry and, very probably, the ecology of the surrounding local marine environment. As stated earlier, the product-mode experiments indicate that anywhere from 18% to 62% of the dissolved trace-metal burden of the Duwamish River are transformed from a dissolved state to an organic-rich flocculant during estuarine mixing. For some elements (i.e., Fe and Mn) this estimate may indeed be a minimum since the catalytic effect of the riverine particulate matter is removed in the laboratory experiments. For example, the Fe data of Murray and Gill[23] indicate up to 90% removal by flocculation in the Duwamish River. A large fraction of the flocculated material probably settles to the seafloor of the lower estuary and Elliott Bay and significantly increases the trace-metal burden of these sediments as indicated by the enrichments of the concentrations of Cr, Cu and Zn in the suspended matter (Table 2). Malins et al.[29] and Riley et al.[30] reported enrichments of trace metals in the sediments from the Duwamish River and Elliott Bay ranging from about 100 to 900% compared with previously reported concentrations of near-shore sediments from Puget Sound and other coastal areas. These enrichments were significantly higher than corresponding enrichments for the central basin of Puget Sound,[31] indicating rapid removal of the flocculated material by sedimentation processes. Thus, the net effect of the flocculation reactions is to decrease the concentrations of the dissolved trace metals in the estuary and to concentrate the metals in the suspended matter and ultimately in the local estuarine and near-shore sediments.

The possible ecological consequences of these findings should be viewed in the light of a number of recent studies. From laboratory studies with benthic bivalves, Louma and Jenne[32] reported that the organisms accumulated trace metals in direct proportion to the concentrations of organically bound metals in the sediments. Similar results were indicated by other investigators who studied metal concentrations in organisms feeding on polluted sediments and sewage wastewater[32,4,3,33]. Malins et al.[29] compared trace metal concentrations in bottom sediment from several Puget Sound embayments including the Duwamish River and Elliott Bay with a number of fish and shellfish population indices for benthic organisms sampled from the same locations and found significant correlations between the abundances of several trace elements in the sediments and taxon richness values for the benthic communities, indicating a possible causal relationship. If the organic-rich flocculants from the Duwamish River are the major source of the enriched trace metals in the sediments, then it is probable that the flocculants also provide trace metals to the organisms as well. The marine organisms sampled by these authors were also found to have the largest percentages of incidences of lesions and tumors. While the exact cause-effect relationships for these diseases have not been established, they are cause for concern and point out the need for a better understanding of how toxic metals become concentrated in coastal marine waters, sediments and organisms.

ACKNOWLEDGEMENTS

The authors wish to express their appreciation to Richard Finger, Process Control Supervisor of the Renton Sewage Treatment Plant, for providing samples and pertinent information relative to its operation. The critical reviews and comments of Herbert Curl, Jr., and Susan Hamilton are also gratefully appreciated. The discussions and data provided by Pierre Appriou and Anthony Paulson are also acknowledged.

This study was supported by the National Oceanic and Atmospheric Administration through the Office of Marine Pollution Assessment.

REFERENCES

1. Galloway, J.N., 1972: Man's alteration of the natural geochemical cycle of selected trace metals, Ph.D. dissertation, University of California, San Diego.
2. Bruland, K.W., K. Bertine, M. Koide and E.D. Goldberg, 1974: History of metal pollution in southern California Coastal zone. Environ. Sci. Tech., 8, 425-432.
3. Eganhouse, R.P. and D.R. Young, 1976: Mercury in tissues of mussels off southern California. Mar. Pol. Bul., 7(8), 145-147.
4. Eganhouse, R.P., D.R. Young and J.N. Johnson, 1978: Geochemistry of mercury in Palos Verdes sediments. Environ. Sci. and Tech., 12(10), 1151-1157.
5. Rohatgi, N. and R.Y. Chen, 1975: Transport of trace metals by suspended particulates on mixing with seawater. Journal of Water Pollution Control Federation, 47, 2298-2316.
6. Sholkovitz, E.R., 1978: The flocculation of dissolved Fe, Mn, Al, Cu, Ni, Co and Cd during estuarine mixing. Earth Planet Sci. Lett., 41, 77-86.
7. Elderfield, H., H. and A. Hepworth, 1975: Diagenesis, metals and pollution in estuaries. Mar. Pollut. Bull., 6, 85-87.
8. Krauskopf, K.B., 1956: Factors controlling the concentrations of thirteen rare metals in seawater. Geochim. Cosmochim. Acta., 9, 1-32.

9. Kharkar, D.P., K.K. Turekian and K.K. Bertine, 1968: Stream supply of dissolved silver, molybdenum, antimony, selenium, chromium, cobalt, rabidium and cesium to the oceans. Geochim. Cosmochim. Acta., 32, 285-298.

10. Murray, C.N. and L. Murray, 1973: Adsorption-desorption equilibria of some radionuclides in sediment-fresh water and sediment-seawater systems. In: "Radioactive Contamination of the Marine Environment", pp. 105-124, IAEA, Vienna.

11. Dawson, W.A. and L.J. Tilley, 1972: Measurement of salt wedge excursion distance in the Duwamish River Estuary, Seattle, Washington, by means of the dissolved-oxygen gradient. U.S. Geological Survey Water Supply Paper, 1873-D, 27 pp.

12. Santos, J.F. and J.D. Stoner, 1972: Physical, chemical and biological aspects of the Duwamish River estuary, King County, Washington, 1963-1967. U.S. Geological Survey Water Supply Paper, 1873-C, 74 pp.

13. Gardner, G.B. and J.D. Smith, 1978: Turbulent mixing in a salt wedge estuary. In: "Hydrodynamics of Estuaries and Fjords", (ed. J.C.J. Nicoul), pp. 79-106, Elsevier Publishing Co., Amsterdam, Netherlands.

14. Hamilton, S.E. and J. Cline, 1981: Hydrocarbons associated with suspended matter in the Green River, Washington. NOAA Tech. Mem. ERL PMEL-30, 116 pp.

15. Gardner, W.D., 1977: Incomplete extraction of rapidly settling particles from water samples. Limnol. Oceanogr., 22(4), 764-768.

16. Calvert, S.E. and M.J. McCartney, 1979: The effect of incomplete recovery of large particles from water samples on the chemical composition of oceanic particulate matter. Limnol. Oceanogr., 24(3), 532-636.

17. Feely, R.A., E.T. Baker, J.D. Schumacher, G.J. Massoth and W.D. Landing, 1979: Processes affecting the distribution and transport of suspended matter in the northeast Gulf of Alaska. Deep-Sea Res., 26(4A), 445-464.

18. Baker, E.T. and D.Z. Piper, 1976: Suspended particulate matter: Collection by pressure filtration and elemental analysis by thin-film x-ray fluorescence. Deep-Sea Res., 23, 1181-1186.

19. Feely, R.A., G.J. Massoth and A.J. Paulson, 1981: Distribution and elemental composition of suspended matter in Norton Sound and the northeastern Bering Sea Shelf: Implication for Mn and Zn cycling in coastal waters. In: "The Eastern Bering Sea Shelf: Oceanography and Resources", (eds. D.W. Hood and J.A. Calder), pp 321-338. U.S. Department of Commerce, Office of Marine Pollution Assessment. Washington, D. C.

20. Sharp, J.H., 1974: Improved analysis for particulate organic carbon and nitrogen from seawater. Limnol. Oceanogr., 6(19), 984-989.

21. Paulson, A.J., (n.d.). Unpublished data.
22. Appriou, P.Y., 1980: Analyse des technologies de dosage des métaux lourds en trace. Final Scientific Report 78/1886, Université de Bretagne Occidentale, Brest, France, 56 pp.
23. Sholkovitz, E.R., 1976: Flocculation of dissolved organic and inorganic matter during the mixing of river water and seawater. Geochim. Cosmochim. Acta., 40, 831-845.
24. Murray, J.W. and G. Gill, 1978: The geochemistry of iron in Puget Sound. Geochim. Cosmochim. Acta., 43, 9-19.
25. Aston, S.R. and R. Chester, 1973: The influence of suspended particles on the precipiatation of iron in natural waters. Est. Coast. Mar. Sci., 1, 225-231.
26. Holliday, L.M. and P.S. Liss, 1976: The behavior of dissolved iron, manganese and zinc in the Beaulieu estuary, S. England. Est. Coast. Mar. Sci., 4, 349-353.
27. Eaton, A., 1979: Observations on the geochemistry of soluble copper, iron, nickel and zinc in the San Francisco Bay estuary. Environ. Sci. Technol., 13, 425-432.
28. Sholkovitz, E.R. and N.B. Price, 1980: The major-element chemistry of suspended matter in the Amazon estuary. Geochim. Cosmochim. Acta., 44, 163-171.
29. Malins, D.C., B.B. McCain, D.W. Brown, A.K. Sparks and H.O. Hodgins, 1980: Chemical contaminants and biological abnormalities in central and southern Puget Sound. NOAA Tech. Memo., OMPA-2, Boulder, CO.
30. Riley, R.G., E.A. Crecelius, D.C. Mann, K.G. Abel, B.L. Thomas and R.M. Bean, 1980: Quantitation of pollutants in suspended matter and water from Puget Sound. NOAA Tech. Memo., ERL MESA-49, 99 pp.
31. Schell, W.R., A. Nevissi, D. Piper, G. Christian, J.W. Murray, D.S. Spyradukis, S. Olsen, D. Hantaner, E. Knudson and D. Zafiropoulos, 1977: Heavy metals near the West Point outfall and in the central basin of Puget Sound. Report to the Municipality of Metropolitan Seattle, 174 pp.
32. Louma, S.N. and E.A. Jenne, 1977: The availability of sediment-bound cobalt, silver and zinc to a deposit-feeding clam. In: "Biological Implications of Metals in the Environment", (eds. H. Drucker and R.E. Wildung), pp. 213-231, U.S. NTIS Conf. 750729, Springfield, VA.
33. Louma, S.N. and G.W. Bryan, 1978: Factors controlling the availability of sediment-bound lead to the estuarine bivalve *Scorbicalaria plana*. J. Mar. Biol. Ass. U.K., 58, 793-802.
34. Furr, A.K., T.F. Parkinson, J. Ryther, C.A. Bache, W.H. Butenmann, I.S. Pakrala and D.J. Lisk, 1981: Concentrations of elements in a marine food chain cultured in sewage wastewater. Bull. Environ. Contam. Toxicol., 26, 54-59.

IMPOVERISHMENT AND DECREASE OF METALLIC ELEMENTS ASSOCIATED WITH SUSPENDED MATTER IN THE GIRONDE ESTUARY

J.M. Jouanneau, H. Etcheber and C. Latouche

Institut de Géologie du Bassin d'Aquitaine -
Université de Bordeaux I - 351, Cours de la Libération
33405 Talence Cedex, France

ABSTRACT

It has been shown that in temperate climate estuaries, there exists an important upstream-downstream decrease of various metallic trace-elements linked with suspended matter (S.M.). Researchers have attributed this either to mixing processes between metal rich particles and impoverished particles, or to a solubilization process caused by a number of physico-chemical factors, salinity increase being the most popular.

During the past five years a number of our studies in the Garonne-Gironde estuarine system demonstrate the decrease and lead us to a new outlook on this phenomenon.

A study of some trace-elements (Zn, Ni, Cu, Pb) was carried out monthly. It revealed a drastic decrease of trace metal content of S.M. in the upper part of the estuary. This zone, while under the influence of the dynamic tide, is situated above the saline intrusion. The decrease here can account for up to 70% of the entire decrease in the estuary system and becomes more significant during high river flow when the saline intrusion and turbidity maximum are forced to the central estuary.

Geochemical fractionation has shown that Zn, for example, is mainly linked with the organic phase. Moreover, the upstream-downstream distribution of particulate organic carbon (P.O.C.) in the estuary corresponds to particulate metals. Therefore there is a Zn-POC seasonal cycle at the entrance to the estuary system and a considerable POC decrease independent of physico-chemical (S $^{o}/oo$, T^{o}) or hydrodynamical (turbidity maximum) parameters.

A detailed study of the POC-particulate metal decrease in the upper part of the estuary suggests that this is caused by a dilution with impoverished particles of fluvial origin. These come either from periods of low flow: directly from turbidity maximum (which at that time is located upstream) or from periods of high flow: resuspension of estuary material previously deposited in the upstream zone of the Bordeaux.

Without transport of suspended matter from the ocean, particles impoverished in metals and POC in the central part of the estuary imply that there exists an impoverishment process in this part of the estuary. Therefore, the decrease which occurs in the upper estuary and impoverishment do not take place in the same zone nor are they of the same nature. Thus, it is of the utmost importance to distinguish clearly between the two phenomena.

RESUME

Dans les estuaires des zones tempérées, une décroissance amont-aval importante de divers oligo éléments métalliques liés à la phase particulaire en suspension a pu être mise en évidence. Elle a été attribuée, soit à des processus de mélange entre particules riches en métaux et particules appauvries, soit à une solubilisation provoquée par un certain nombre de facteurs physico-chimiques parmi lesquels l'accroissement de la salinité est le plus souvent invoqué.

Dans le système estuarien Garrone-Gironde un certain nombre d'études menées depuis près de 5 ans viennent préciser ce schéma général dans un système estuarien de type macro-tidal non ou peu perturbé par les facteurs anthropiques. Elles apportent un éclairage nouveau sur les mecanismes naturels regissant le comportement des métaux dans les interfaces eau douce/eau salée.

L'étude de quelques trace-éléments (Zn, Ni, Cu, Pb) liés aux M.E.S. conduite mensuellement depuis la limite amont de la marée dynamique jusqu'à l'embouchure, révèle l'existence d'une zone de décroissance très rapide dans la partie "fluvio-estuarienne". Cette zone qui est soumise à la marée dynamique se situe par contre en amont de l'intrusion saline. La décroissance, peut y atteindre 70% de la décroissance totale observée sur l'ensemble du système; elle est plus nette en période de crues, alors que l'intrusion saline et le bouchon vaseux estuarien sont alors repoussés dans l' estuaire moyen.

L'étude par fractionnement séquentiel des différentes liaisons métal/support particulaire a permis de prouver, qu'en ce qui concerne le zinc, par exemple, c'est la phase organique qui est la plus riche des phases extraites. D'autre part la répartition amont-aval du C.O.P. dans l'estuaire se calque bien sur celle des métaux. C'est ainsi que, dans la zone "fluvio-estuarienne" on observe d'une part un cycle saisonnier Zn-COP à l'entrée du système estuarien, d'autre part une décroissance rapide de ce C.O.P. et ce, indépendement des paramètres physico-chimique (S $^{o}/oo$, T^{o}) et hydrodynamiques (position du bouchon vaseux).

Une étude détaillée de la décroissance des métaux et du carbone organique sous forme particulaire dans la partie supérieure de l'estuaire montre que cette décroissance peut être expliquée par un mélange de 2 types de particules: des particules fraiches apportées par le fleuve, riches en métaux et carbone; des particules plus anciennes appauvries en ces éléments. Ces dernieres proviennent - en période d'étiage, du bouchon estuarien alors rémonte vers l'amont - en période de crue, de la remise en suspension des matériaux deposés au fond lors de l'étiage précédent.

En l'absence de remontée de M.E.S. d'origine marine dans l'estuaire central, la presence de particules appauvries en métaux et C.O.P. implique un processus d'appauvrissement se produisant dans cette partie de l'éstuaire elle-même. Il en résulte que la décroissance - qui elle, apparaît dans le haut estuaire - et l'appauvrissement ne se produisent pas dans la même zone et sont de nature différente. Il est donc particulièrement important de bien distinguer les deux phénomènes.

INTRODUCTION

River systems constitute one of the major sources of material to the ocean. Many studies[1,2] have shown that in certain transitional fresh/salt water environments, the dissolved and suspended matter contents vary greatly in space and time. The causes of metal variation within estuaries have, however, been subject to controversy. They are attributed either to geochemical or mixing processes.

Pollution in estuaries of industrialized countries may, in fact, cause disturbance in the material supply of upstream/downstream elements, such as metals, Thus, the reconstruction of natural cycles of metallic elements is difficult to determine.

A five-year research program has been undertaken within the Gironde Estuary to collect information on the cycling of several trace metals. This estuary is the largest in France and drains the Aquitaine Basin, one of the least industrialized regions in France and perhaps Western Europe. As such it can be considered as an excellent example of large estuary undergoing few or no changes.

This report concerns trace metals associated with suspended matter (S.M.). Results involve only those metals displaying a significant upstream/downstream decrease (Zn, Pb, Cu). Particular attention is paid here to the upper part of the estuary extending between La Réole and Ambés (fig. 1). As the discussion on the behaviour of metals within this zone cannot be treated on its own, a minimum of information is obtained for the estuarine environment as a whole.

METHODS

Suspended sediment samples have been collected through centrifugation or filtering of large quantities of water (100-200 l). The response of S.M. and its representativeness after this treatment has been discussed earlier[3].

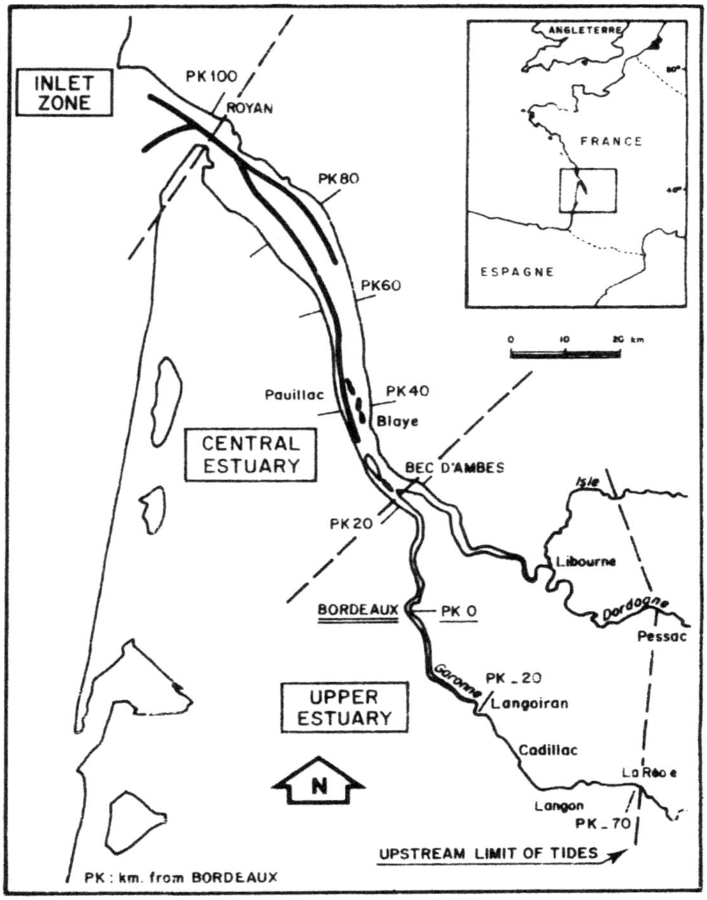

Fig. 1. Geomorphological framework of the Gironde Estuary.

Concentrations of suspended sediments in water were measured by 0.45 micron filtration for small water quantities (1 liter) and by weighing after drying the filters.

Particulate organic carbon was analyzed by back titration after having treated the sediment with a sulfo-chromic mixture[4,5].

Bulk dried and crushed samples were analyzed by X-ray fluorescence after LAPAQUELLERIE method[6], and thus avoided inter-element effects of adsorption and enhancement.

MORPHOLOGY AND HYDRODYNAMIC CHARACTERISTICS OF THE GIRONDE ESTUARY

The Gironde Estuary has been the object of many studies; indeed, its hydrological and sedimentological characteristics have been discussed at great length[7]. However, a review will be given only of essential data for the interpretation of the results presented here. Below the junction between the Garonne and Dordogne rivers at the Bec d'Ambés (fig. 1) this estuary covers at high tide an area of 625 km^2. Total average river discharge of both rivers is about 766 $m^3\ s^{-1}$. The inlet tidal prism (marine water volume introduced into the estuary by current flow) is estimated at $2 \times 10^9 m^3$ (spring tide) and at $1.1 \times 10^9 m^3$ (neap tide). It decreases exponentially towards the upper estuary. The region of the 2 rivers which is subject to the dynamic tide but not the saline one is considered as a fluvio-estuarine zone (Garonne upstream limit: La Réole; Dordogne upstream limit: Pessac). In this zone the limit of saline intrusion (salinity range > 0.5 $^o/oo$) reaches Bordeaux at low water level.

As in most estuaries there exists in the Gironde Estuary a residual circulation system, which is a major cause of the turbidity maximum (water masses loaded with suspended sediment: 0.1 to 10 g l^{-1}) and fluid mud (very turbid water lenses: 100 to 300 g l^{-1}). The total suspended sediment[9] reaches approximately 4.2 - 5.3 10^6t. Both turbidity maximum (1.1 - 2.0 10^6t) and fluid mud (2.2 - 2 10^6t) are subject to fluctuations, cyclical displacements (associated with tides) and especially seasonal effects (associated with river discharge). During flooding, turbidity maximum and fluid mud may go further than 80 km downstream of Bordeaux, whereas at low water level they go as far as Cadillac (approximately 50 km at the upper part of Bordeaux). Further upstream, suspended matter becomes less important (about 10 mg l^{-1}) with a maximum of 100 mg l^{-1} during flooding.

Sedimentological studies[7,10] have shown that for suspended sediments, the downstream estuary supply by marine mineral material is minimal. Evidence of marine sand bed load transport has been reported, but only close to the estuary inlet[11,12].

BEHAVIOUR OF TRACE METALS

We have shown earlier[4,13,14] that as in most other estuaries, trace metals in the Gironde Estuary are transported mainly in particulate forms. Furthermore, it has been noticed that an important particulate metal decrease occurs between fluvial and estuarine suspended matter.

The average metal contents of particulates established at a series of stations over a 5 year period are shown in figure 3 while standard deviations are listed in table 1. Figure 3 shows the variation of metal content of particulates within the estuary.

Table 1. Mean metal contents in suspended sediments (1975-1979). World rivers as reference after Martin and Meybeck[15].

Metals ($\mu g\ g^{-1}$)	Gironde Estuary: Entry (La Réole)*	Gironde Estuary: Central Estuary (PK*30 96 included)	References: World Rivers
Measurements	51	250	
Zn	862 ± 150	230 ± 20	350
Pb	144 ± 60	48 ± 5	150
Cu	83 ± 40	22 ± 3	100
Ni	50 ± 20	34 ± 2	90

* See figure 1.

From table 1 and figures 2 and 3 it is obvious that the highest metal contents are found in fluvial inputs, especially for Zn. These very high Zn contents could indicate pollution or be due to the existence of a Zn geological formation in the Aquitaine Basin[16].

Conversely, metal contents in the central estuary are normal and comparable to values reported in the estuaries considered to be slightly or not polluted.

Irrespective of the origin of these metals, natural or man-induced, the salient point is the marked content decrease between the Gironde Estuary entry and its central part. Within this distance, metal contents diminish: Zn by 73%, Pb by 66%, Cu by 74% and Ni by 32%. It appears, therefore, that the estuary acts as a regulator for fluvial inputs to the ocean.

Nevertheless, the mean content does not provide a truthful picture of this decrease due to variations in space and time. As

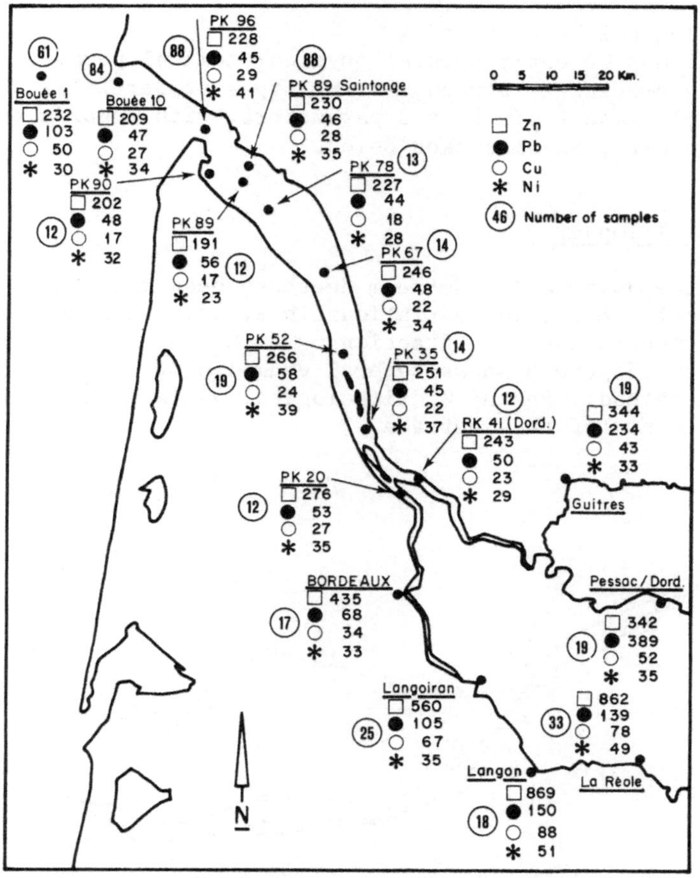

Fig. 2. Location of sampling stations and mean particulate metallic contents (μg g^{-1}) - 1975-1979.

indicated earlier[17] the Zn decrease appears more important when metal contents at the entry of the estuary are higher. Indeed, at the entry, contents vary in proportion from 1 to 10 (maximum during flooding periods) whereas in the centre they vary from 1 to 2. Similar relationships (fig. 4) are to be noticed for Pb and to a lesser extent for Cu. Most of this decrease takes place between La Réole and Ambés i.e. between the limits of the dynamic tide and salinity tide[4].

POSSIBLE FACTORS INDUCING THE DECREASE

Observations that the upstream/downstream variations in metal concentrations associated with estuarine S.M. and sediments could be the product of 3 main factors:

1. lithological heterogeneity
2. variations of environmental physico-chemical parameters
3. mixing processes between several types of particles.

The possible impact of these 3 parameters, with regards to the Gironde Estuary, is discussed below.

Lithological Impact

Had the upstream/downstream decrease been the result of lithological composition variations in S.M. (granulometry) then normalization to the fine fraction or Al_2O_3 values would remove the effect. Figure 5 shows several examples of results obtained from this method. Pb and Zn lithological variations apparently do not account for the decrease.

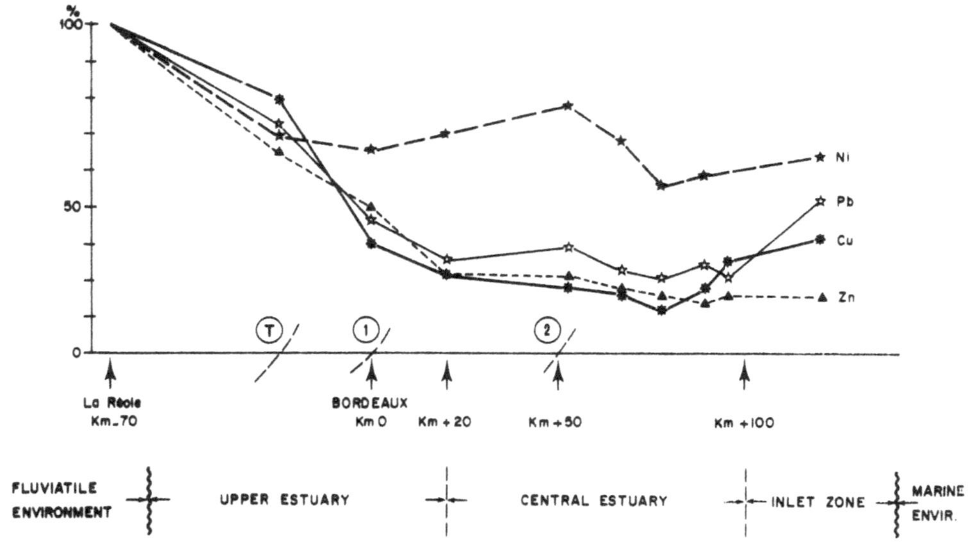

Fig. 3. Upstream/downstream variations of metallic content in suspended matter within the Gironde Estuary. (Expressed in % of metallic content at the entry of the estuary (up-stream inputs): La Réole, based on mean values from 1975-1979).
1. Highest upstream limit of saline intrusion during low water level (0.5‰).
2. Highest upstream limit of saline intrusion during flooding (0.5‰).
T. Highest upstream limit of turbidity maximum during flooding.

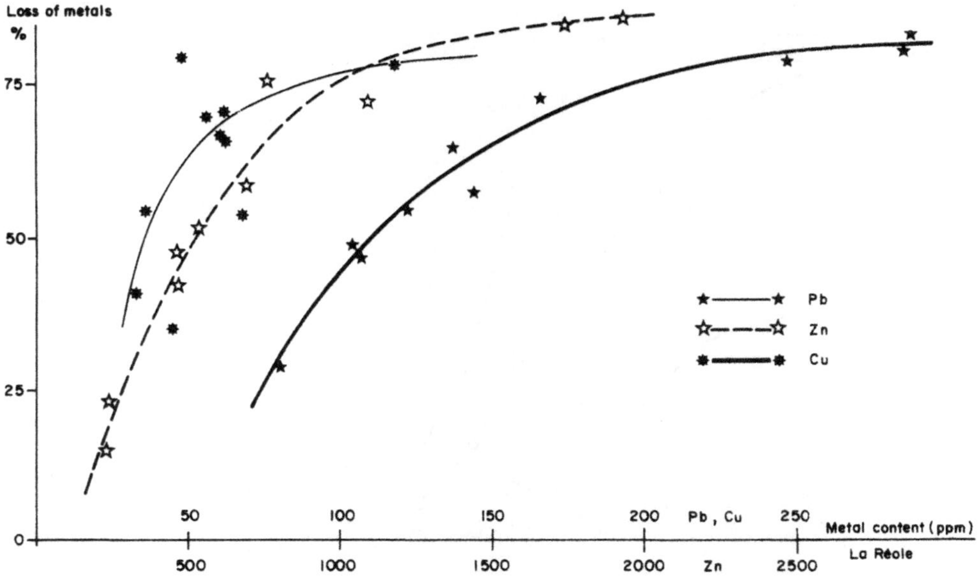

Fig. 4. Particulate metal loss in the central estuary with respect to initial content ratio at La Réole; loss % =

$$\frac{\text{La Réole content} - \text{mean content in the central estuary} \times 100}{\text{La Réole content}}$$

(each point represents a series of samples between 1975 and 1979).

Earlier work[18,19,20] has shown that a large proportion (>50%) of the metals described here are linked to the organic fraction of S.M. It is, therefore, instructive to study (fig. 6) the upstream/downstream pattern of metal contents after normalization to organic carbon.

The Zn-POC relation falls into two linear segments. The first corresponds to a constant ratio of 200 $\mu g\ g^{-1}$ Zn/% POC from the estuary entry until about PK67. The second decreases regularly (200-65 $\mu g\ g^{-1}$ Zn/% POC) from PK67 up to the inlet. The constancy of the upstream ratio indicates that the decrease (about 70% for Zn) can be attributed to the organic carbon decrease. Conversely, from PK67 up to the inlet, the ratio decrease seems to be the cause of a mixing process of 2 types of organic matter differently loaded in Zn; a mixing process which has already been determined by the ratio $^{13}C/^{12}C$.[21]

Pb and Cu data are less informative and therefore confirm[20] the weaker relationship between these metals and POC. For Pb, 2 linear

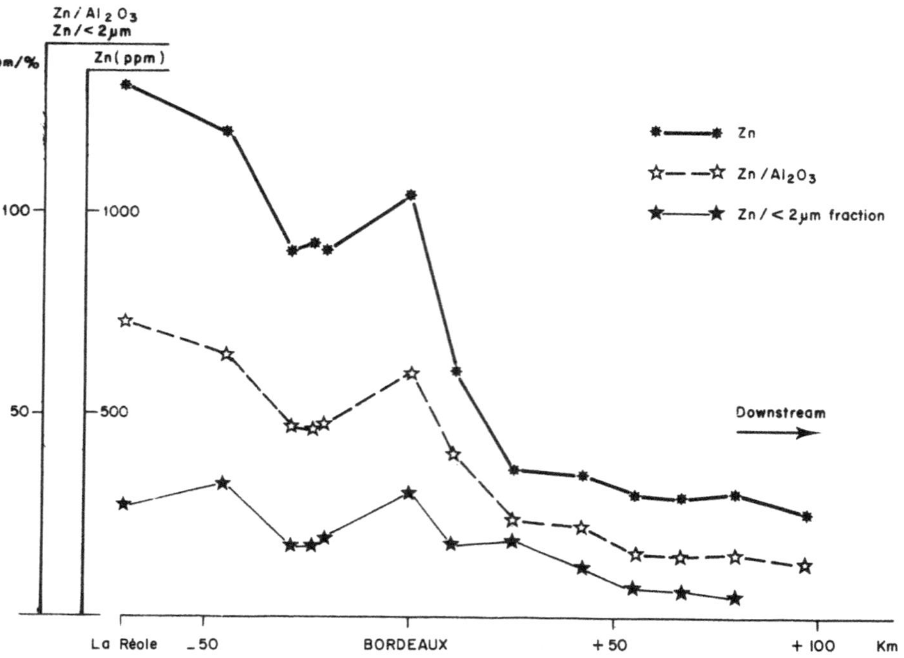

Fig. 5. Lithological impact : Zn as an example in the March 1979 sampling.

segments seem to exist (La Réole and PK52; PK52 - inlet) while the Cu ratio decreases quite irregularly. The upstream/downstream decrease in both elements is not exclusively associated with the POC decrease in any part of the estuary.

Possible Impact of pH and Salinity

In figure 7, examples shown represent the majority of the situations observed during this study.

The metal content decrease, with respect to salinity, takes place mostly upstream of the saline front. During low water, the 0.5 °/oo salinity range goes no further than Bordeaux (PK 0) where 50% of the total decrease has already occurred for Zn, Pb and Cu.

In the upstream zone (La Réole - Ambés) pH variation lies between 7.55 - 8.10. Although limited, this variation is still situated within a range where slight changes may cause important variations in metal adsorption[22]. However, desorption increases when pH decreases[2], and it is interesting to note that pH in the Gironde area decreases in the upstream direction.

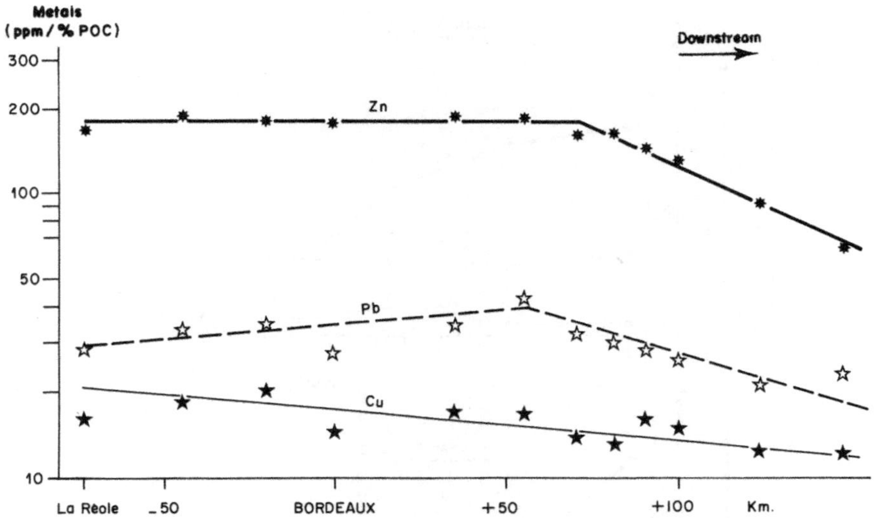

Fig. 6. Longitudinal evolution of the mean ratio:

$$\frac{\text{metals}}{\text{Particulate organic carbon}}$$

In summary, the variations of the two important physico-chemical parameters (i.e. salinity and pH) cannot fully account for the decrease in Zn, Cu and Pb contents, observed in the upper part of the estuary. Furthermore, it must be remembered that this decrease is not related to salinity. Nevertheless, salinity has been, in many estuaries, considered a cause of change in metal contents in particulate form.

Impact of Mixing

According to several sedimentological studies[7,8,10] the Gironde Estuary differs from the North Sea estuaries in that it is not fed by suspended minerals originating in the ocean. Marine microplanktonic species do, however integrate S.M. at the inlet, thus contributing to the enrichment of suspensions with organic phase up to PK65. The marine suspended matter at the inlet is always slightly richer in metal contents than S.M. sampled further upstream, and completely free of marine plankton productivity. Thus, upstream of PK65 the upstream/downstream decrease of metal contents cannot be attributed to mixing processes of metal-poor marine particles, as previously described by Müller and Förstner[23] or Duinker and Nolting[24].

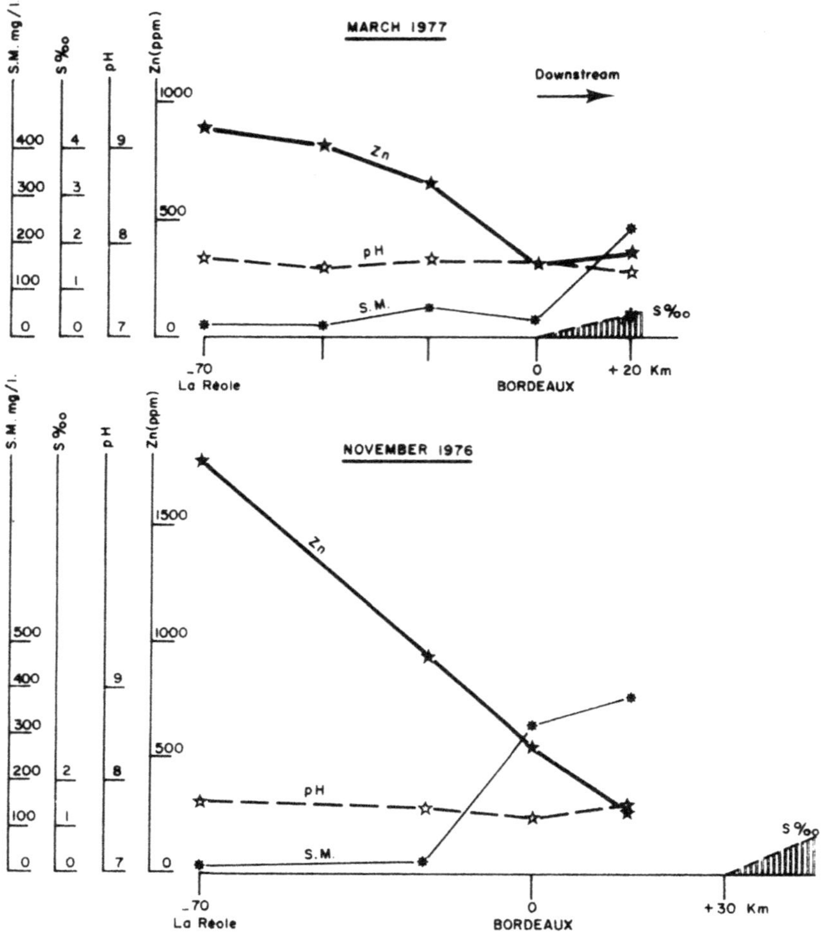

Fig. 7. Decrease physico-chemical conditions of particulate zinc in the upper part of the estuary.

Conversely, the important disproportion between turbidity in the river waters (some tenths of mg l^{-1} in La Réole) and turbidity in the central part of the estuary (in the order of g or 10 g l^{-1}) must be caused by mixing of fluvial particles with the upstream estuarine stock.

During low water (fig. 8a) the estuarine turbidity maximum is pushed towards an upstream extreme and may reach Langoiran. Thus, the decrease of metal contents between La Réole and Langoiran may be mostly attributed to the 2 types of particles pushed towards the upstream direction. Independent evidence of this has been shown in experiments using radioactive isotope tracers[25].

During flooding (fig. 8b) turbidity maximum is forced downstream. Upstream water turbidity is low except during high fluvial discharge, at which time POC and metal contents are maximum at the estuary entry[19]. This coincides with most of the upstream/downstream decreases of metals and POC. In the absence of a turbidity maximum, it appears that these decreases are not attributable to a mixing process of fluviatile particles with impoverished particles from the central estuary. Detailed turbidity analyses have, in fact, shown that a slight increase in S.M. exists almost always between La Réole and Langoiran. This increase which is not representative of the turbidity maximum (sensu stricto) seems to be caused by the tides and also to a great extent by resuspension of material deposited at previous low water period. This material, already impoverished in metals, continues to lose metals after settling. Indeed, in the whole estuary, bottom sediment metal contents appear lower than those of S.M. The metal content of the latter are at most equal to that content measured in the estuarine turbidity maximum. Hence, (fig. 9) the theoretical line indicating the mixing process of x% of fluviatile material (mean Zn content 892 $\mu g\ g^{-1}$) with y% of estuarine material (mean Zn content: 236 $\mu g\ g^{-1}$). In figure 9 decreasing values betwen La Réole and Langoiran, are measured on the basis of the turbidity change reported during the mixing process in this region. The plotting of real value points along the straight line (theoretically decreasing) is regular implying that the decrease reported between La Réole and Langoiran may be the product of the mixing process.

DISCUSSION AND CONCLUSION

Most of the decrease of the metal contents associated with S.M. in the Gironde Estuary takes place (fig. 10) in the upstream part of the estuary, between the dynamic tide (La Réole) and saline intrusion limits (Bordeaux). The decrease is at its maximum during flooding and followed by a corresponding POC decrease.

The decrease in metals and carbon within this upstream zone appears to be caused by mixing of newly arrived fluvial particles with older impoverished estuarine particles. The absence of suspended marine inputs in the central estuary, (upstream of PK65) shows that the decrease in contents cannot be due to a mixing process with marine particles of lower metal content, as has been suggested for the North Sea estuaries[23,24]. These impoverishment processes do not occur in the upper part of the estuary (i.e. La Réole and Langoiran) since the mixing of fluvial and estuarine stocks by itself can explain the observed decreases in a zone where

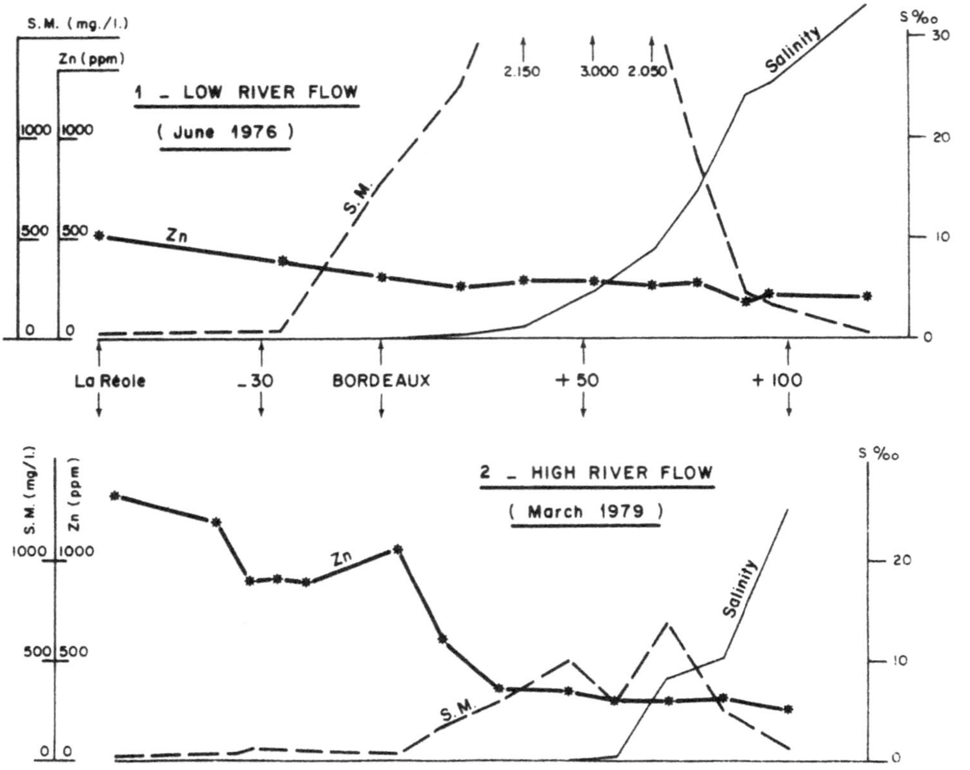

Fig. 8. Upstream/downstream profiles.

salinity and pH remain constant. Impoverishment of fluvial particles does take place in the centre of the estuary and is the result of physico-chemical variations (salinity changes from 0‰ to 10-25‰ depending on season).

Thus, if impoverishment does not occur upstream where the most notable decreases in particulate metal content are reported it implies that DECREASE and IMPOVERISHMENT are two different phenomena which must not be confused. Furthermore, they are not only geographically distinct, especially during flooding but also of a different nature. DECREASE is the result of physical mechanism; the mixing of newly arrived fluviatile particles with older suspended particles. IMPOVERISHEMENT is the result of a chemical process which ends by consuming organic matter and solubilising associated metal. This is shown by the co-related increase in dissolved metal contents and by the results obtained from the double mud tagging' experiment using radio-active isotope tracers (Gironde Estuary)[27].

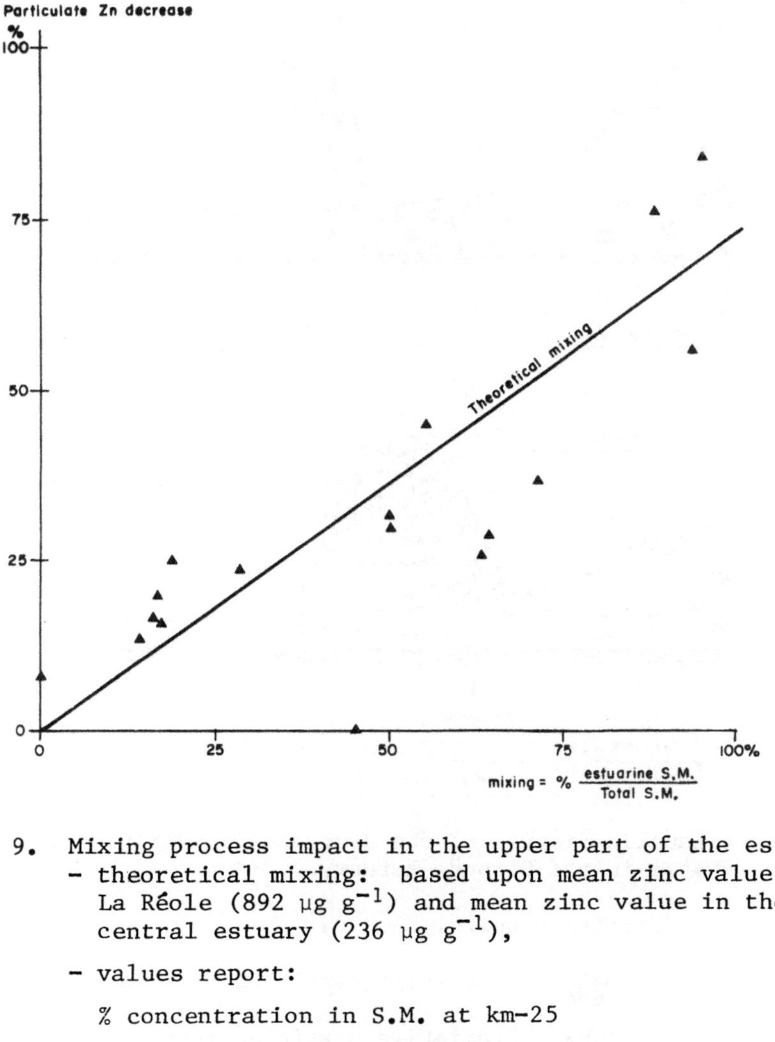

Fig. 9. Mixing process impact in the upper part of the estuary.
- theoretical mixing: based upon mean zinc value at La Réole (892 μg g^{-1}) and mean zinc value in the central estuary (236 μg g^{-1}),
- values report:

% concentration in S.M. at km-25

mixing: $\frac{\text{S.M. concentration at La Réole}}{\text{S.M. concentration at km-25}}$

Zn decrease: $\frac{\text{\% zinc content (La Réole - km-25)}}{\text{zinc content at La Réole}}$

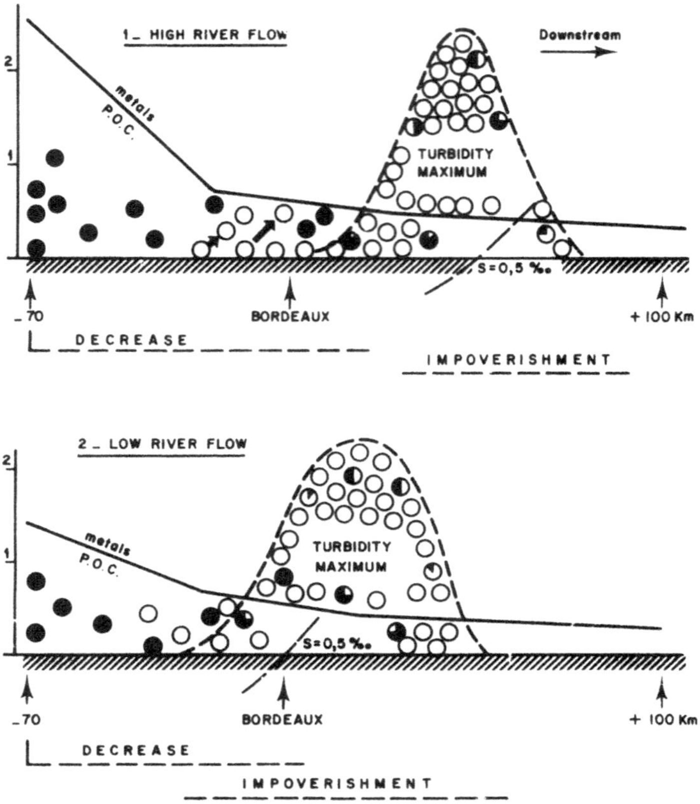

Fig. 10. Schematic diagram of the decrease-impoverishment system in the Gironde Estuary.

1-2: arbitrary units.

 fluviatile particles rich in POC and metal.

 fluviatile particles less rich in POC and metal.

 POC and metal particles at different stages of impoverishment.

 estuarine particles totally impoverished.

Both the geographical dissociation between DECREASE and IMPOVERISHMENT and the different nature of the 2 phenomena means that classical salinity/decrease diagrams cannot be applied in the Gironde Estuary. Given the results obtained, the classical methods would overestimate the role played by estuarine zones with low salinity range (<5 $^{o}/oo$) in terms of the impoverishment in metals and carbon of fluviatile particles. These upstream estuarine zones with low salinity range witness most of the decrease in particulate matter content and could have, in fact, a very minor role to play in solubilization and impoverishment.

ACKNOWLEDGEMENTS

This investigation has been carried out in the framework of the 'Centre de Recherche sur l'Environnment Sédimentaire et Structural des Domaines Marine' (Associated Laboratory n°197 with Centre de la Recherche Scientifique - C.N.R.S.). It has been supported by an "Action Thématique Programmée" - A.T.P. deed "Océanographie Chimique" from the C.N.R.S. We acknowledge the contribution of Y. Lapaquellerie and Ph. Pedemay who completed all analyses in the study.

REFERENCES

1. Burton, J.D., P.S. Liss, 1976: Estuarine Chemistry, Academic Press, London, 228 p.
2. Förstner U., G.T.W. Wittmann, 1979: Metal pollution in the aquatic environment, Springer-Verlag, Berlin and Heidelberg, 486 p., 102 fig., 94 tabl.
3. Etcheber, H., J.M. Jouanneau, 1980: Comparison of the different methods for the recovery of suspended matter from estuarine waters: deposition, filtration and centrifugation: consequences for the determination of some heavy metals. Estuar.Coast.Mar.Sc. 2, 701-707.
4. Etcheber H., 1979: Répartition et comportement du Zn, Pb, Cu, Ni dans l'estuaire de la Gironde. Bull. Inst. Géol. Bassin d'Aquitaine, Bordeaux, no.25, 125-147, 19 fig., 9 tabl.
5. Johnson, M.J., 1949: A rapid micromethod for estimation of non volatile organic matter. J. Biol. Chem., Baltimore, no.181, p. 707-711.
6. Lapaquellerie Y., 1975: Application de la spectrométrie de fluorescence X en géologie marine. Thèse Université, Bordeaux I, no. 96, 98 p., 51 fig., 24 tabl.
7. Allen, G.P., 1972: Etude des processus sédimentaires dans l'estuaire de la Gironde. Thèse Doct. Etat, Univ. Bordeaux I, no. 353, 314 p., 134 fig.

8. Allen, G.P., Sauzay, G., P. Castaing, J.M. Jouanneau, 1976: Transport and deposition of suspended sediment in the Gironde estuary, in: Estuarine Processes, edited by M. Wiley, Acad. Press, New York, London, vol. II, 63-82.
9. Jouanneau, J.M., 1979: Evaluation du volume et de la masse de matières en suspension dans le systeme bouchon vaseux - crème de vase de la Gironde. Bull. Inst. Geol. Bassin d'Aquitaine, Bordeaux, 25, 111-120.
10. Allen, G.P., P. Castaing, A. Klingebiel, 1973: Suspended sediment transport and deposition in the Gironde estuary and adjacent shelf. Mem. Inst. Geol. Bassin d'Aquitaine, Bordeaux, 7, p. 27-36.
11. Allen, G.P., P. Castaing, J.M. Jouanneau, A. Klingebiel, 1974: Les processus de charriage a l'embouchure de la Gironde. Mem. Inst. Geol. Bassin d'Aquitaine, Bordeaux, 7, p. 191-206.
12. Legigan, P., P. Castaing, 1981: La penetration des sables marins dans les estuaires: cas de la Gironde. C.R. Acad. Sc. Paris, t. 292, II, p. 207-212.
13. Bourg, A., H. Etcheber, J.M. Jouanneau, 1979: Mobilisation des metaux traces associés aux matières en suspension dans le complexe fluvio-estuarien de la Gironde. Revue de Biologie et Ecologie méditerranéenne, t. VI, no. 3.4, 161-166, 5 figs.
14. Cauwet, G., F. Elbaz, C. Jeandel, J.M. Jouanneau, Y. Lapaquellerie, J.M. Martin, A. Thomas, 1980: Comportement géochimique des éléments stables et radioactifs dans d'estuaire de la Gironde en période de crue. Bull. Inst. Geol. Bassin d'Aquitaine, Bordeaux, no. 27, 5-33, 12 fig., 11 tabl.
15. Martin, J.M., M. Meybeck, 1979: Elemental mass-balance of material carried by major world rivers. Marine Chemistry Amsterdam, 7, 173-206.
16. Michard, A.G., A. Courmoul, 1978: La sédimentation liasique dans les Causses: contrôle des minéralisations Zn-Pb associées au Lotharingien. Bull. du B.R.G.M., Orléans (II), 2, 57-120, 32 fig.
17. Latouche, C., P. Bertrand, H. Etcheber, J.M. Jouanneau, 1981: Comportement de quelques métaux lourds dans les milieux de transition eaux douces/eaux salées : cas de l'estuaire de la Gironde. Symposium A.S.F., Paris janvier 1981, sous presse dans Mem. Soc. Geol. Franc.
18. Jouanneau, J.M., 1979: Diminution du carbone organique particulaire en zone fluvio-estuarienne. Le cas de la Gironde. C.R. Acad. Sc., Paris, 288 D, 375-378.
19. Etcheber H., J.M. Jouanneau, 1980: Cycles saisonniers des apports en zinc et carbone organique associés aux matières en suspension du système Garronne-Gironde: enseignements et conséquences. C.R. Acad. Sc. Paris, t. 290, 735-738.

20. Bertrand, P., 1979: Etude expérimentale, par fractionnement, de la répartition géochimique du zinc dans les matières en suspension estuariennes: Cas de la Gironde. J. Rech. Oceanog. Paris, vol. IV, no. 4, 39-51, 9 fig.
21. Fontugne, M., J.M. Jouanneau, 1981: La composition isotopique du carbone organique des matières en suspension dans l'estuaire de la Gironde. Application à l'étude du plomb et du zinc particulaire. C.R. Acad. Sc. Paris, in press.
22. Bourg, A., 1979: Effect of ligands at the solid-solution interface upon the speciation of heavy metals in aquatic systems. Proc. Internat. Conf. Heavy metals in the environment, London, 446-449.
23. Müller, G., U. Förstner, 1975: Heavy metals in sediments of the Rhine and Elbe Estuaries: Mobilisation or mixing effect? Environ. Geol. I, 33-39.
24. Duinker, J.C., R.F. Nolting, 1976: Distribution model for particulate trace metals in the Rhine estuary, southern Bight and Dutch Wadden Sea. Neth. J. Sea Res. Texel, 10, 71-102.
25. Castaing, P., J.M. Jouanneau, 1979: Temps de résidence des eaux et des suspensions dans l'estuaire de la Gironde. J. Rech. Oceanogr. Paris, vol. IV, no. 2, p. 41-52.
26. Latouche, C., J.M. Jouanneau, 1981: Evolution des oligo-éléments métalliques liés à la phase particulaire en suspension à l'interface eau douce - eau marine: cas du système fluvio-estuarien girondin. Oceanis, Paris, vol. 6, fasc. 6, 621-636.
27. Etcheber, H., J.M. Jouanneau, C. Latouche, P. Azoeuf, A. Caillot, R. Hoslin, 1980: L'expérience "Double Marquage de Vase en Gironde", Contribution à la connaissance du devenir d'une pollution métallique en estuaire. Oceanologica Acta, Paris, vol. 3, no. 4, p. 477-486.

THE SIGNIFICANCE OF THE RIVER INPUT OF CHEMICAL ELEMENTS TO THE OCEAN

Jean-Marie Martin[+] and Michael Whitfield[++]

+Laboratoire de Geologie
E.N.S., 46 rue d'Ulm
75230 PARIS CEDEX, 05, France

++Marine Biological Association of the U.K.
The Laboratory, Citadel Hill
Plymouth, PL1 2PB, Great Britain

ABSTRACT

The objectives of this paper are to review the role of rivers as a pathway of chemical elements from the land to the ocean and to assess the significance of river input of pollutants to oceanic chemistry.

The major importance of river suspended matter (R.S.M.) on the transport of chemical elements to the ocean is underlined. The relationship between river water/river suspended sediment partition coefficient and the electronegativity function QYO has been reassessed. A comparison between theoretical erosion and actual fluxes of material carried by rivers shows that for most chemical elements these two figures are similar. Additional fluxes are observed in rivers for some elements like Sb, Zn and Pb. This discrepancy is discussed in terms of steady state and non-steady state erosional processes.

During estuarine mixing the discharge of riverine elements to the ocean is drastically modified. More than 90% of the R.S.M. settles with its associated colloidal material produced when river water mixes with sea water. Consequently, due to the strong association of chemical elements with R.S.M., only a small percentage of the continental material will reach the sea. However, the comparison of R.S.M. with deep-sea clay composition emphasizes the prime influence of river input on oceanic sediment composition over a long term period for most elements with the exception of Mn, Co and Cu.

With regards to the dissolved phase, for the elements which exhibit a conservative behaviour during estuarine mixing ("accumulated" elements), corresponding to large mean oceanic residence time (MORT), the influence of anthropogenic discharge on the global ocean will not be readily noticeable although significant changes may be observed in the coastal zone.

For those elements which show "unchanged" or "depleted" concentrations in the ocean, the influence of industrial contamination cannot be significantly observed in the open ocean, because of their rapid removal in the estuarine and coastal zone.

Finally river input of chemical elements to the ocean is compared to atmospheric and volcanic sources of material.

INTRODUCTION

The annual river water discharge to the ocean system is estimated to be about 37,400 km^3 (1), which is 2.7 x 10^{-3}% of the total volume of sea water in the world ocean. This river water flow carries approximately 15 x 10^{15}g of particulate matter and 4 x 10^{15}g of dissolved salts per year[2]. This dissolved and particulate material is considered to be 10 times greater than that introduced by glaciers (Antarctica included) and about 100 times greater than the dust fall-out[3]. The sediment load, water and salt discharges of some major rivers of the world are listed in Table 1.

In order to assess the chemical significance of this flux of material to the ocean system, the following questions must be answered: What is the chemical composition of dissolved and particulate matter carried by rivers? What is the fate of this material within the mixing zone between river and sea water? What is the ultimate quantity and composition of material discharged to the ocean system? How far can the industrial society influence these fluxes and processes? What are the alternative sources of chemical elements for the ocean system?

During the past decades several workshops and books, and a large number of scientific papers, have been devoted to these problems. The present discussion is an attempt to briefly summarize the present state of knowledge in this field.

Table 1 Water, salt and solid discharges of major world rivers

Rank	River	Annual Discharge Water $10^9 m^3$	Salt $10^6 t$	Solid $10^6 t$	References
1	Amazon	5676	296	930	Gibbs[4], Meade et al.[5]
2	Zaire (Congo)	1325	44	110	Peeters[7], Hubert and Martin (unpublished) Van Bennekom and Helder[8], Symoens[6]
3	Orinoco	1070*	38	86	Bueno Romero (unpublished)
4	Chang Jiang (Yang Tse Kiang)	690*	131	280	Martin (unpublished) Yellow River Basin Commission (unpublished)
5	Brahmapoutra	605*	85	720	Handa[9,10]
6	Mississippi	554*	120	310	Clarke[11], USGS[12]
7	Yenissei	554*	(65)	13	Alekin, Brazhnikova[13]
8	Lena	514*	(70)	15	Alekin, Brazhnikova[13]
9	Parana	470*	32	130	de Petris[14]
10	Mekong	470*	46	345	Meybeck-Carbonnel[15]
15	Ganges	366*	61	1450	Handa[9,10]
28	Hoang Ho	47*	-	1780	Yellow River Basin Commission (unpublished)
TOTAL		37400	3800	15500	

* After World Register of rivers discharging to the Ocean, UNESCO Division of Water Sciences.[16]

THE RIVER END-MEMBER

The world-wide composition of the major dissolved elements in river water has been documented since the pioneer work of Clark[11], by Alekin and Brazhnikova[13], Livingstone[17] and Meybeck[18]. Up-to-date world average concentrations of the major ions in river water, which take into account unpublished values for the Yang Tse Kiang river, are given in Table 2. The world average values do not significantly differ from Clarke's[11], and no further investigations are needed to improve this figure. A first assessment of the range of chemical composition of the rivers flowing into each ocean shows differences as large as 100%.

Table 2 Major element composition of rivers flowing to various oceans and world average.

Element	Atlantic	Indian	Arctic	Pacific	World
SiO_2	9.9	14.7	5.1	11.7	10.7
Ca^{++}	10.5	21.6	16.1	13.9	13.3
Mg^{++}	2.5	5.4	1.3	3.6	3.1
Na^{+}	4.2	8.5	8.8	5.2	5.3
K^{+}	1.4	2.5	1.2	1.2	1.5
Cl^{-}	5.7	6.8	11.8	5.1	6.0
SO_4^{-}	7.7	7.9	15.9	9.2	8.7
HCO_3^{-}	37	94.9	63.5	55.4	51.7
Σions	69	140.2	118.6	93.6	90.9
T.D.S.	78.9	154.9	123.7	105.3	101.6

mg. l^{-1}, T.D.S. = Total Dissolved Solids

Table 3 Discharge of nutrients to the world ocean. (10^{12} mol y^{-1})

	C	$N^{(b)}$	$P^{(b)}$
Dissolved	39.7 [a]	-	0.14 [b]
Particulate	15 [c]	-	0.55 [d]
Total	54.7	5.3(1.0) [e]	0.69(0.15) [e]

a) Degens,[19] ; b) Van Bennekom, and Salomon,[20] ; c) Meybeck,[21] ; d) Martin and Meybeck[2]; e) Natural river transport (b).

An estimate of nutrient discharge to the world ocean is given in Table 3. This must be considered as a first approximation and there is an urgent need for time-series nutrient data in the major world rivers. It has been suggested, for example, that man's influence has increased the river input of nitrogen 3 or 4 times and phosphorus about five-fold[20].

With regard to dissolved trace metals, almost nothing is known and according to Edmond[22], "the experience gained in sea water analyses, where the criterion of 'oceanographic consistency' can be applied, is that trace metal concentrations reported before 1973 are almost entirely invalid. It is reasonable to expect that a similar situation holds for the fresh water." Such data as are available are summarized in Table 4.

The chemical composition of particulate trace and major elements has been recently summarized by Gordeev and Lisitzin[23] and Martin and Meybeck[2]. Their findings, as well as some additional data from the authors, are briefly summarized below.

As a first approximation, the chemical composition of river suspended matter is very close to that of surficial rocks (Table 4). However, owing to the importance of chemical weathering and the resulting transport of dissolved material, the suspended matter is relatively enriched in elements such as the lanthanides, aluminium, iron and titanium (see below). The use of an average earth rock composition may therefore significantly change mass-balance computations.

A useful parameter is the distribution (or partition) coefficient K_Y which is the ratio between the mean concentration in the water ($\mu g.l^{-1}$ or $\mu g.kg^{-1}$ dissolved solids) and in the solid phase ($\mu g\ kg^{-1}$). This parameter describes the ease with which the elements are transferred from the solid phase to the solution phase. Despite the great difficulty in assessing the true mean concentration of the elements in river water, order of magnitude assessments can be made of K_Y for most of the elements (Table 5). On a logarithmic scale, it makes little difference whether the crustal rock or the riverine suspended matter is used as the reference phase (compare Figure 1 with Figure 3 of Turner et al.[25]).

The order of magnitude of the elemental partition functions can be simply summarized by relating them to an energy function (Q_{YO}, Figure 1) which depends on the electronegativity (or electron attracting capability) of the individual elements[26,27,28,29]. For river water (Figure 1), the majority of the elements (48 in all) fall on a single sequence in which the degree of affinity for the solid phase decreases with increasing electronegativity on passing from the lanthanide elements to the halides[25]. Only iron falls significantly below this correlation. On the other hand, the alkali and alkaline earth elements form a separate correlation which lies above, but parallel to, the main sequence. Because of their relatively low charge densities the concentrations of these elements are controlled by weak ion-exchange reactions so that they are significantly enriched in the water column relative to elements in the main sequence (such as the lanthanides) which have a similar electronegativity[30]. The concentrations of the main sequence of elements in river water are probably controlled by adsorption and desorption at detrital particle surfaces rather than by simple solubility reactions and it is likely that solid state chemistry rather than solution chemistry is the dominant factor regulating the relative concentrations of the elements.

Table 4. Elemental composition of river dissolved and particulate matter, continental rocks and soils, ocean water and sediments (after Martin and Meybeck[2])

	CONTINENT		RIVERS		OCEAN	
	Rock $\mu g\ g^{-1}$	Soils $\mu g\ g^{-1}$	Dissolved $\mu g\ l^{-1}$	Particulate $\mu g\ g^{-1}$	Water $\mu g\ l^{-1}$	Deep sea clays $\mu g\ g^{-1}$
Ag	0.07	0.05	0.3	0.07	0.04	0.1
Al	69,300	71,000	50	94,000	0.5	95,000
As	7.9	6	1.7	5	1.5	13
Au	0.01	0.001	0.002	0.05	0.004	0.003
B	65	10	18	70	4,440	220
Ba	445	500	60	600	20	1,500
Br	4	10	20	5	67,000	100
Ca	45,000	15,000	13,300	21,500	412,000	10,000
Cd	0.2	0.35	0.02	(1)	0.01	0.23
Ce	86	50	0.08	95	0.001	100
Co	13	8	0.2	20	0.05	55
Cr	71	70	1	100	0.3	100
Cs	3.6	4	0.035	6	0.4	5
Cu	32	30	1.5	100	0.1	200
Er	3.7	2	0.004	(3)	0.0008	2.7
Eu	1.2	1	0.001	1.5	0.0001	1.5
Fe	35,900	40,000	40	48,000	2	60,000
Ga	16	20	0.09	25	0.03	20
Gd	6.5	4	0.008	(5)	0.0007	7.8
Hf	5	-	0.01	6	0.007	4.5
Ho	1.6	0.6	0.001	(1)	0.0002	1
K	24,400	14,000	1,500	20,000	380,000	28,000
La	41	40	0.05	45	0.003	45
Li	42	25	12	25	180	45
Lu	0.45	0.4	0.001	0.5	0.0002	0.5
Mg	16,400	5,000	3,100	11,800	$1.29x10^6$	18,000
Mn	720	1,000	8.2	1,050	0.2	6,000
Mo	1.7	1.2	0.5	3	10	8
Na	14,200	5,000	5,300	7,100	$1.077x10^7$	20,000
Nd	37	35	0.04	35	0.003	40
Ni	49	50	0.5	90	0.2	200
P	610	800	115	1,150	60	1,400
Pb	16	35	0.1	100	0.003	200
Pr	9.6	-	0.007	(8)	0.0006	9
Rb	112	150	1.5	100	120	110
Sb	0.9	1	1	2.5	0.24	0.8
Sc	10.3	7	0.004	18	0.0006	20
Si	275,000	330,000	5,000	285,000	2,000	283,000
Sm	7.1	4.5	0.008	7	0.0005	7.0
Sr	278	250	60	150	8,000	250
Ta	0.8	2	< 0.002	1.25	0.002	1.0
Tb	1.05	0.7	0.001	1.0	0.0001	1.0
Th	9.3	9	0.1	14	0.01	10
Ti	3,800	5,000	10	5,600	1	5,700
Tm	0.5	0.6	0.001	(0.4)	0.0002	0.4
U	3	2	0.24	3	3.2	2.0
V	97	90	1	170	2.5	150
Y	33	40	-	30	0.0013	32
Yb	3.5	-	0.004	3.5	0.0008	3
Zn	127	90	30	250	0.1	120

Concentrations have been updated for dissolved Cd, Cu, Ni in rivers, (Boyle et al., this volume), dissolved U, (Figueres et al.)[24], particulate Pb and Zn (unpublished data). Dissolved Al, As, Ca, Cu, Ni, Pb and Zn in ocean water according to Boyle et al; Bruland and Franks; Burton et al.; Olafsson; Schaule and Patterson (this volume); Mart et al. (personnal communication).

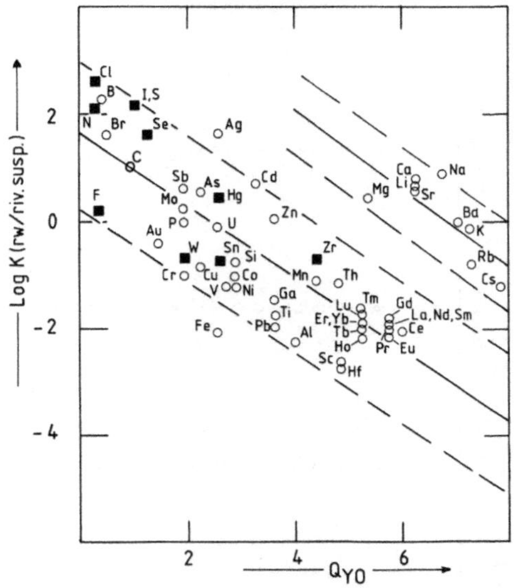

Fig. 1 The relationship between the river water/river suspended matter partition coefficient and the electronegativity function (Q_{YO}) for the elements. Solid squares indicate that values for the river water/crustal rock partition coefficient have been used because of the lack of data for river suspended matter. The data are taken from Table 4. The equations for the straight lines are: (a) for the main sequence log K_Y+ 1.57 - 0.67 Q_{YO} (r = 0.80, n = 48); and (b) for the alkali and alkaline earth metals log K_Y = 5.00 - 0.73 Q_{YO} (r = 0.75, n = 9). The dashed lines show a spread of $\pm$ two standard deviations.

Organic pollutants, such as organochlorine compounds, DDT, aromatic hydrocarbons and herbicides are similarly adsorbed onto particulates, with log K_Y values[31] ranging from 1.58 - 0.42 (see Table 5). Adsorption may result either in a greater stability of these compounds or in a more rapid breakdown according to the nature of the particle. A herbicide such as paraquat, once adsorbed onto particulate matter, loses its herbicidal properties[32,33].

An interesting finding has emerged from the comparison between the total observed elemental fluxes (dissolved + particulate) in the world rivers with theoretical values (Figure 2) derived from the elemental content in the average parent rock and the total quantity of weathered material[2].

The theoretical fluxes Φ_{th} have been computed on the basis of the ratio between the contents of a given conservative element in the suspended matter and in the surficial rock. Aluminium, which

Table 5. Characteristics of the elements[a]

Element	Log K_Y[b]				Mean oceanic residence time (years)	Electronegativity (X_Y)	Classification		
	river/rock	river/susp	sea/rock	sea/deep sea clay			Chemical[c]	Oceanographic[c]	Biological[e]
Ag	1.62	1.62	-1.79	-1.95	4.9×10^3	1.9	b	B	N
Al	-2.15	-2.28	-6.69	-6.83	3.7×10^2	1.5	a,sh	A	?
As	0.32	0.52	-2.27	-2.49	3.2×10^4	2.0	h	B	ET
Au	0.29	-0.41	-1.95	-1.43	7.3×10^4	2.3(I)	b(I)	B	N
						2.9(III)	h(III)		
B	0.43	0.40	-0.25	0.28	0.9×10^7	2.0	h	C	E
Ba	0.12	-0.01	-2.90	-3.43	1.2×10^4	0.85	a	B	N
Br	1.69	1.59	2.67	1.28	1.2×10^8	2.8	an	C	?
C	3.05	1.03	0.60	-	8.0×10^4	2.5	h	B	E
Ca	0.50	0.78	-0.59	0.06	1.1×10^6	1.0	a	C	E
Cd	-0.01	0.70	-2.85	-2.91	1.8×10^4	1.7	b'	B	T
Ce	-2.04	-2.08	-6.48	-6.55	4.6×10^2	1.05	a	A	N
Cl	2.59	-	3.61	-	1.1×10^8	3.0	an	C	E
Co	-0.82	-1.01	-3.96	-4.59	9.2×10^3	1.8	a'	B	ET
Cr	-0.86	-1.01	-3.92	-4.07	1.1×10^4	1.6(III)	a(III)	B	?
						2.1(VI)	h(VI)		
Cs	-1.02	-1.24	-2.50	-2.65	4.2×10^5	0.7	a	C	N
Cu	0.34	0.83	4.05	-4.85	2.4×10^3	1.9(I)	b'(I)	A	ET
						2.0(II)	a'(II)		
Er	-1.98	-1.89	-5.22	-5.08	7.3×10^3	1.2	a	B	N
Eu	-2.09	-2.16	-5.63	-5.73	3.7×10^3	1.1	a	A	N
F	0.19	-	-1.23	-	4.8×10^5	4.0	an	C	E
Fe	-1.96	-2.09	-5.80	-6.03	1.8×10^3	1.8(II)	a'(II)	A	E
						1.9(III)	a,sh(III)		
Ga	-1.26	-1.45	-4.28	-4.37	1.2×10^4	1.6	a,sh	B	N
Gd	-1.92	-1.81	-5.52	-5.60	3.2×10^3	1.1	a	A	N
Hf					2.5×10^4	1.3			
Hg	0.43		-1.98		1.8×10^4	1.9	b	B	T
Ho	-2.21	-2.01	-5.45	-5.25	7.3×10^3	1.2	a	B	N
I	2.14		0.53		3.1×10^5	2.5(-I)	an(-I)	C	E
							h(+5)		
K	-0.22	-0.13	-0.36	-0.42	9.2×10^6	0.8	a	C	E
La	-1.92	-1.96	-5.69	-5.73	2.2×10^3	1.1	a	A	N
Li	0.45	0.67	-0.92	-0.95	5.5×10^5	1.0	a	C	N
Lu	-1.66	-1.77	-4.90	-4.95	7.3×10^3	1.2	a	B	N

Mg	0.27	0.41	0.35	0.31	1.5×10^7	1.2	a	C	E
Mn	-0.95	-1.12	-5.11	-6.00	8.9×10^2	1.4(II) 1.5(IV)	a'(II) h(IV)	A	ET
Mo	0.46	0.21	-0.78	-1.45	7.3×10^5	2.1	h	C	ET
N	2.09	-	-0.15	-	7.3×10^4	3.0(V)	h(V)	B	E
Na	0.56	0.87	1.33	1.18	7.4×10^7	0.9	a	C	E
Nd	-1.98	-1.95	-5.64	-5.67	2.7×10^3	1.1	a	A	N
Ni	0.99	-1.26	-3.94	-4.55	1.4×10^4	1.8	a'	B	?
P	0.27	-0.01	-2.56	-2.92	1.9×10^4	2.1	h	B	E
Pb	-1.21	-2.00	-5.28	-6.07	1.1×10^3	1.6	b'	A	T
Pr	-2.15	-2.07	-5.75	-5.73	3.1×10^3	1.1	a	A	N
Rb	-0.88	-0.83	-1.52	-1.51	2.9×10^6	0.8	a	C	?
S	2.14	-	1.99	-	1.1×10^7	2.5	h	C	E
Sb	1.04	0.59	-2.12	-2.07	8.8×10^3	2.1	h	B	N
Sc	-2.42	-2.66	-5.78	-6.07	5.5×10^3	1.3	a	B	N
Se	1.59	-	-0.95	-	3.7×10^4	2.4	h	B	ET
Si	-0.72	-0.76	-3.69	-3.70	1.4×10^4	1.8	h	B	E
Sm	-1.96	-1.95	-5.70	-5.70	2.3×10^3	1.1	a	A	N
Sn	-0.71	-	-3.85	-	9.2×10^3	1.9	h	B	?
Sr	0.32	0.59	-0.09	-0.04	4.9×10^6	1.0	a	C	?
Tb	-2.03	-2.01	-5.57	-5.55	3.7×10^3	1.2	a	A	N
Th	-0.98	-1.16	-4.52	-4.55	3.7×10^3	1.3	a,sh	A	N
Ti	-1.59	-1.76	-5.13	-5.31	3.7×10^3	1.6		A	?
Tm	-1.71	-1.62	-4.75	-4.85	7.3×10^3	1.2	a	B	N
U	-0.10	-1.10	-1.52	-1.35	4.9×10^5	1.9	a,sh(IV)	C	?
V	-1.00	-1.24	-3.14	-3.33	9.2×10^4	1.85	h(VI)	B	ET
W	-0.71	-	-2.73	-	1.2×10^5	2.1	h	B	N
Yb	-1.95	-1.95	-5.19	-5.12	7.3×10^3	1.2	a	B	N
Zn	0.35	0.07	-4.65	-4.63	1.2×10^2	1.6	a'	A	ET
Zr	-0.75	-	-5.29	-	3.7×10^2	1.4	d,sh	A	N

[a] Data taken from ref.[2,25,36]. [b] Solid phase concentrations expressed in μg kg^{-1}, solution phase concentration expressed in μg kg^{-1} dissolved solids. To convert to μg l^{-1} subtract 4.58 units for river water and 1.45 units for sea water. [c] a = (a) - type cation, b = (b) - type cation, a' = (a) - type cation, b' = (b)' - type cation, h = fully hydrolysed, sh = strongly hydrolysed, an = anionic, Roman numerals indicate the oxidation state of the element. [d] Symbols refer to vertical regions in Figure 4. A = depleted, $R_o < 0.1$, B = unchanged, $0.1 < R_o < 10$, C = accumulated $R_o > 10$ ref.36. [e] E = essential, ET = essential trace element, N = not required, T = toxic, ? = status uncertain.

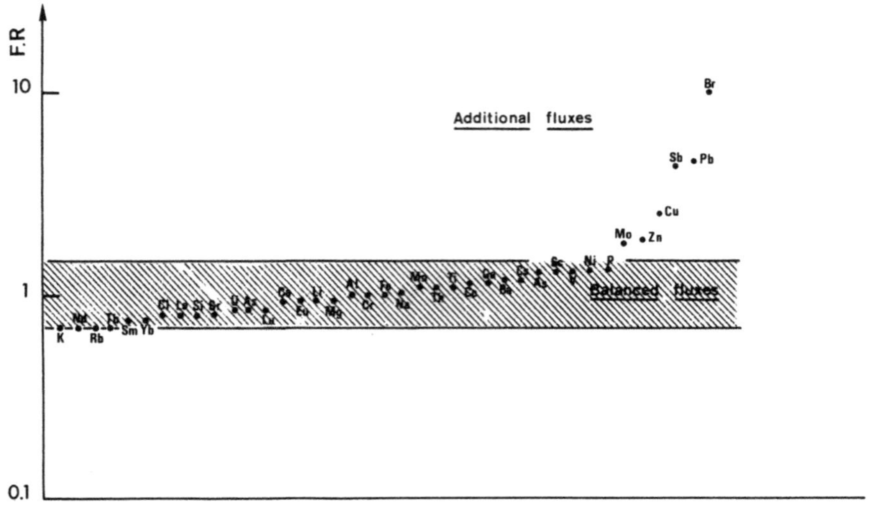

Fig. 2 Chemical elements ranked according to their flux ratio FR. (FR is the ratio between the actual flux - dissolved + particulate - measured in rivers and the theoretical one computed from the elemental content in the average parent rock and the total quantity of weathered material).

is commonly measured, is poorly soluble and, as yet, not affected by pollution, is usually chosen for this purpose. This choice assumes that Al has reached a steady state at the continental surface, i.e., is not at present being accumulated by the soil layer, and is derived only from land erosion. For one given element x the theoretical flux to the ocean is:

$$\Phi_{th} = [x]_{fr} M_{pm} [Al]_{pm} / [Al]_{fr}$$

where $[x]_{fr}$ = average content of element x in the surficial fresh rock, $[Al]_{pm}$ = average Al content in river particulate matter, $[Al]_{fr}$ = average Al content in the surficial fresh rock, M_{pm} = annual river particulate discharge to the ocean.

The total suspended load M_{pm} is estimated to be 15,000 x 10^6 $t \cdot y^{-1}$ and the ratio $[Al]_{pm}/[Al]_{fr}$ is 9.40/6.93 = 1.36. Therefore, the total amount of fresh rock (M_{fr}) from which the river dissolved and particulate products are derived should be around 21,000 x 10^6 $t \cdot y^{-1}$. The difference between M_{fr} and M_{pm} - around 5,500 x 10^6 $t \cdot y^{-1}$ - should be due to the total dissolved transport of material deriving from chemical denudation M_d which is roughly around 2,500 x 10^6 t y^{-1} if that part of the HCO_3^- carried by rivers originating from atmospheric CO_2 is removed. There is therefore an important discrepancy between the M_d derived from the Al-content ratio and the M_d derived the direct measurement of river transport. This may be due to accumulation of the coarser detrital material within the basins, to poor estimation of present-day river suspended load, or to a non-steady-state system, i.e., the river suspended material does not correspond to present chemical denudation, but derived from older soil layers. Finally, it could be due to differential weathering of various types of rocks. It is not possible at the present stage of this study to state which processes are relevant; therefore we will not consider the absolute fluxes, which are still questionable, but solely the flux-ratio (FR) between the theoretical flux Φ_{thx} and the observed flux Φ_{obsx}.

$$\Phi_{obsx} = [x]_{sol}Q + [x]_{pm}M_{pm}$$

where $[x]_{sol}$ = average dissolved content of element x in rivers, Q = annual river water discharge to the ocean (37 400 $km^3 . y^{-1}$ according to Baumgartner and Reichel[7]). Thus the flux ratio is:

$$FR = ([x]_{sol}Q + [x]_{pm}M_{pm})\ ([x]_{fr}M_{pm} \cdot [Al]_{pm}/[Al]_{fr})^{-1}$$

This comparison shows that observed and theoretical fluxes are equivalent ($0.7 < FR < 1.5$) for most elements (e.g lanthanides, alkaline metals and alkaline earths, vanadium, cobalt, iron ...); so that the total amount carried by rivers is equivalent to the theoretical estimate of the quantity weathered. If a steady-state model is assumed, additional fluxes ($FR > 1.5$) are required for some elements namely bromine, copper, molybdenum, antimony, zinc, lead and possibly phosphorus and nickel. It is striking to find that most of these elements are also enriched in aerosols and marine suspended matter. It is important to look for the origin of these enrichments which may correspond either to world-wide pollution or to a natural process such as the atmospheric recycling of marine aerosols or volcanic dust.

These additional fluxes may also be partly explained by a geochemical fractionation in the soil profile which would enrich the top soil layer that supplies the river particulate load through mechanical erosion leaving the remaining soil profile depleted in these elements with respect to Al. This would correspond to a non-steady state weathering. Finally, the excess fluxes may also be partly explained by differential erosion processes: river suspended matter originates mainly from mountainous areas where, on the global scale, the occurrence of elements such as Cu, Pb and Zn is higher[36,37,38]. Therefore, the excess transport in rivers could be an artefact due to the under-estimation of the average composition of surficial rocks effectively exposed to erosion.

In conclusion, one must keep in mind that, in addition to the uncertainty of many of the data listed in Tables 1 to 5, there are important natural geographical fluctuations (e.g. for the major elements, Table 2) from one ocean to the other. Moreover, the river discharge of material to the ocean has probably drastically changed during geological time. As an example, Duplessy[39] has shown that river discharge to the Bay of Bengal decreased by 30-50% during the two last glacial periods; similar situations may be found for other large rivers. The influence of these geographical and historical variations upon ocean chemistry must be carefully examined.

THE ESTUARINE ENVIRONMENT

Introduction

An estuary provides a particularly varied chemical environment[40,41]. The estuarine zone is defined by strong chemical gradients since it is a meeting place where river water mixes with sea water of much higher ionic strength. However, some elements, such as transition metals, and nutrients are much more abundant in river water than in sea water. The nutrients favour a strong primary production of organic matter which represents a potential reservoir of chemical energy. In addition, the small depths enhance the influence of thermal and mechanical energy as compared to the open ocean. This available energy promotes various physico-chemical reactions, often mediated by biological processes. The intensity of these processes will vary with time and their relative importance in fixing the composition of the water will depend on the state of the tide and on the magnitude of the river flow.

These difficulties are compounded by the settling or organic-rich particulate matter and the subsequent formation of anoxic interstitial waters. These waters, with their own distinctive chemistry, are mixed into the estuary from time to time by the action of the tides. Residence times within the estuary are short, varying from hours to weeks depending upon the location within the estuary and on the flux of water through the system.

In recent years, the major chemical reactions occurring in the estuarine environment have been identified. Although a wide range of trace and major elements has been studied, the general understanding of trace-metal behaviour in estuaries is still rudimentary. The behaviour of many elements appears to vary from one estuary to the next and can often show distinct seasonal, or even diurnal variations within a particular estuary[42]. It appears that, even if the processes themselves remain the same from estuary to estuary, we cannot predict the rates at which the various reactions should occur because of our almost total ignorance of their kinetics and mechanisms. In particular, the course of heterogeneous reactions involving both dissolved and solid phases cannot be predicted, owing to our ignorance of the speciation of the dissolved elements and because of the presence of organic coatings on the particles. Since estuaries are characterized by high concentrations of particulates in contrast with the open ocean this is one of the central problems in estuarine chemistry.

A major barrier to our attempts to quantify estuarine removal and release is the lack of information on the solution chemistry of the elements. The biological and geochemical reactivity of a particular element will be strongly influenced by its chemical speciation, i.e., by the molecular groupings it is associated with under particular environmental conditions.

Chemical Speciation of the Elements

Within the estuarine zone, the chemical speciation of the dissolved components is influenced by changes in the salinity, ionic strength and pH accompanying the mixing of fresh water and sea water. Some progress has recently been made in rationalizing equilibrium models for the inorganic speciation of the elements in natural waters[43,44,45]. Water is a ubiquitous ligand so that the dominant process affecting the speciation of the elements is hydrolysis. A simple electrostatic theory[45,46] can be used to delineate the fully hydrolysed, strongly hydrolysed and weakly hydrolysed elements (Table 5). The structure of the hydrolysed species can be predicted from simple chemical rules[45]. Cations with a high charge to radius ratio are fully hydrolysed in natural waters (Table 5) and their speciation is influenced by shifts in the pH and, to a lesser extent, by interactions with the major cations and anions in solution. The speciation of the fully hydrolysed elements has received relatively little attention and only in a handful of cases (C, S, B, P) can useful speciation models be prepared. Such speciation studies are hampered by the fact that there is, as yet, no thermodynamically valid pH scale that can be used under estuarine conditions.

The speciation of the cationic elements that are not fully hydrolysed is best described in terms of the balance struck between covalent and electrostatic bonding during complex formation. Several procedures can be used for estimating the degree of covalency involved in the bonding[43,45,47] but they all result in broadly similar classifications. (a)-type cations (Table 5) tend to form electrostatically bound complexes and bind preferentially with donors from the first row of the periodic table (e.g. N, O, F). Most of these cations are relatively weakly complexed both in fresh water and in sea water. The speciation of the divalent and trivalent elements of this group (mainly the alkaline earth and lanthanide element) tends to be dominated by the formation of carbonate complexes and the stability of these complexes increases with the charge of the central cation. Little shift in speciation is likely to be experienced by these elements in the estuarine zone[44,46]. More significant changes are expected for the strongly hydrolysed (a)-type cations (e.g. Cr(III), Sc(III), Al(III), Be(II)) since their speciation will be strongly dependent on the pH and on the fluoride concentration. It is unlikely that the speciation of the mono and divalent (a)-type cations will be strongly influenced by the presence of organic matter.

(b)-type cations (Table 5) tend to form covalently bound complexes and they bind preferentially with donors from the second and subsequent rows of the periodic table (e.g. Cl, S, P). In sea water they form strong chlorocomplexes and, in fresh water, strong hydroxy and carbonate complexes. Considerable changes in speciation are therefore expected in the estuarine zone. These changes may even result in an alteration in the stable oxidation state (e.g. for Au and Tl) and can be further complicated by the formation of strong organic complexes in the fresh water zone (e.g. for Hg).

Intermediate between the extremes of (a)-type and (b)-type behaviour are the borderline (a)'- and (b)'-cations. The (a)'-cations are mainly found in the first long row of the periodic table (the "nutrient row") and are of considerable biological significance. They have complicated speciation patterns because they can form stable complexes with a wide range of donor atoms. The (b)'-cations show somewhat simpler speciation patterns since they tend to form stronger chloro-complexes and will not be so readily complexed by naturally occurring organic ligands. This simple scheme for the classification of chemical speciation, based on the interplay between electrostatic and covalent bonding during complex formation, is consistent with the biological roles of the individual elements[47,48].

As the picture for inorganic speciation becomes clearer more emphasis is being placed on the interactions of trace elements with organic matter and with the particulate phase. Only limited progress has been made in the isolation and characterization of natural organic ligands. Studies of metal-organic interactions[49] have concentrated, in the main, either on artificial ligands that may be present as pollutants (e.g. NTA, EDTA) or on the rather amorphous assemblage of relatively weak ligands provided by the humic and fulvic acid fractions. Only in the case of mercury, iron and copper are such complexes likely to remain important during the transition to the marine environment. It is likely, however, that highly specific and kinetically stable complexes are produced by the metabolism of marine organisms. In addition a number of elements (notably Hg, Sn, As, Sb, Se) are converted into volatile alkyl compounds by microbial action in anoxic systems.

Recently, some progress has been made in quantifying the adsorption of trace metals onto particle surfaces. The particulate matter can be treated as an additional "ligand" and conditional complexation constants can be estimated for pure mineral surfaces [50,51] and for particles coated with organic matter [52,53]. The concept has even been extended to include planktonic organisms as effective ligands [54,55] and particle-solution interactions have been shown to play a significant role in the control of copper concentrations in the estuarine zone [56].

When interactions with organic components or with particulate matter are significant the chemistry of the system is also likely to be strongly influenced by the rates of chemical processes. This arises either because the biota control the cycling of the elements and produce metastable species or because interactions with the particulate matter are inherently slow. In these circumstances, the equilibrium thermodynamic modelling approach is no longer adequate and it is necessary to develop practical methods to determine directly the most important chemical forms (or groups of chemical species [57]). Such operational procedures are still at an early stage of development and it is essential that particular attention should be paid to providing a clear definition of the fractions that they are able to detect and to investigating the relevance of such fractions to the biological and geochemical behaviour of the elements. Recent studies [58,59] have gone some way towards providing a more accurate assessment of the significance of the various forms sensed by electrochemical methods and a simple analogue has been described which clarifies the relationship between the biological and electrochemical uptake of trace metals [30]. Further progress will be required both in the development of realistic speciation models and in the definition of useful operational procedures before we are able properly to assess the significance of the changes in solution chemistry that accompany the mixing of fresh water and sea water. Until we can achieve such a detailed understanding of the mechanisms involved we must rely on studies of the effects of estuarine mixing on the composition of the particulate and dissolved phases to provide some indication of the major processes at work.

Changes in the Composition of Particulate Matter

Of the initial 15×10^{15} $g \cdot y^{-1}$ of terrigenous material transported by the world rivers, probably less than 10% penetrates into the pelagic area [60]. The overwhelming part of river suspended sediment and attached chemical elements settles at the river-ocean interface. In a very few instances (e.g. Zaire and Ganges-Brahmaputra) suspended sediment and bed-load will reach the deep ocean via a canyon. This also applies when the river mouth is located near the shelf edge (e.g. Mississippi). The composition of the material that can penetrate into the open ocean may become drastically changed in the process. This problem can be partly deciphered by comparison of the composition of river particulate matter and deep-sea clays (Figure 3). Although such a comparison is obviously biased by several factors, such as differential settling, dissolution during settling, remobilization at the benthic boundary layer, etc...., on the whole, there is a fairly good agreement between the composition of these solid phases. This is especially noteworthy for lanthanides, scandium, tantalum, rubudium, chromium, iron, gallium, titanium, silicon and aluminium. These correspond to the "lattice held" elements defined by Chester and Aston [61]. Some elements are in excess in river particulate matter; e.g. antimony, zinc and calcium. This implies that a dissolution or a desorption occurs somewhere in the marine cycle of these elements. If the answer is obvious for calcium, the situation is less clear for zinc. One can look for a desorption within the estuary, or a release at the benthic boundary layer. The final category corresponds to those elements whose concentrations are enhanced in deep-sea clays: cobalt, copper, manganese and nickel. Beyond additional fluxes from the mid-ocean ridge and atmospheric fall-out, one might look for estuarine and nearshore processes such as diffusion of manganese as manganous ions from organic-rich sediment, and subsequent reprecipitation and transport as manganese dioxide colloids in the water column. The enrichment of Na in deep-sea clays is intriguing especially if one considers its ability to diffuse from the sediment to the overlying water during diagenesis [62].

On a local scale, the similarity in chemical composition of river suspended sediments and deep-sea clays with respect to most elements does not imply that these elements are unreactive. As an example the lanthanides and iron are removed from solution within the estuarine zone, but, due to their low K_Y values (Table 5), the particulate concentration is not seriously affected. Only the leachable fraction will be modified. Finally, one must stress that a given element may be mobilized from the sediment in one area and conversely deposited.

Changes in the Composition of the Solution Phase

It is more difficult to assess the situation with regard to the dissolved phase. The most common approach is by comparison of the actual distribution of a given element with that predicted from the theoretical dilution curve of river water in sea water, the chlorinity being considered as a conservative index of mixing.

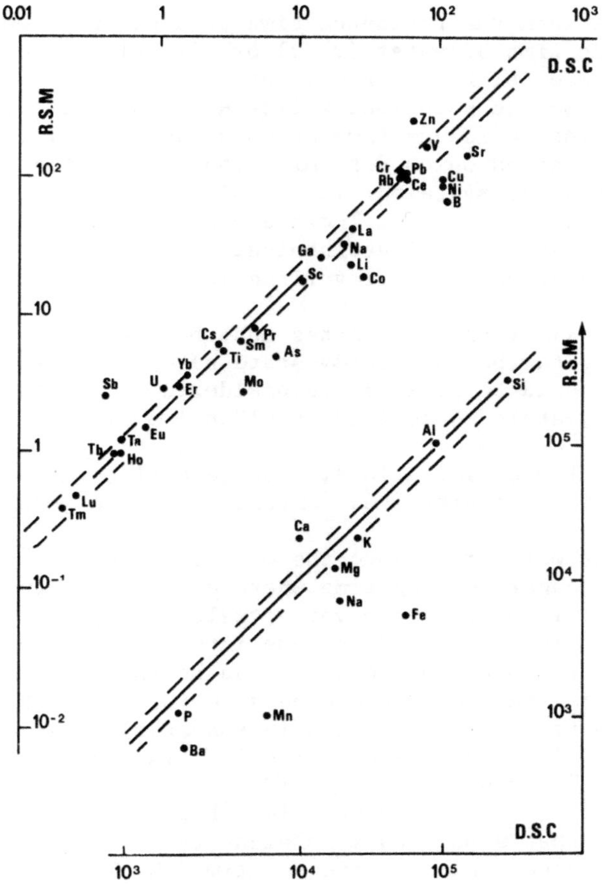

Fig. 3 Comparison of concentrations (ppm) of chemical elements in D.S.C. (deep-sea clays) and R.S.M. (river suspended matter).

Curvilinear relationships are indicative of addition or removal processes, depending upon the sense of curvature. However the comparison is often misleading because of problems related to the definition of correct end-members, the role of the tributaries or bank flow in the saline part of the estuaries, and the occurrence of several water masses of different ages along the estuary. In some cases, the use of tritium and stable isotopes such as oxygen-18 may help unravel these various problems[63,64].

Another approach to the problem of determining the fate of dissolved constitutents in estuaries has been proposed by Martin et al.[32] and Elderfield[41]. They applied the concept of relative residence time (R), defined by Stumm and Morgan[65] for lakes, i.e.

the ratio between the residence time of a given element (t_x) and the residence time of water (t_{H_2O}) or "flushing time". For conservative elements, R = 1; R < 1 means that the chemical element is removed from the solution, whilst R > 1 suggests an internal recycling. The residence time of water and material are probably important variables which determine the fate of the chemical elements in estuarine systems. It is quite obvious that long residence times occurring in well-mixed estuaries will enhance the influence of estuarine biogeochemical processes, as compared to the short residence times prevailing in stratified estuaries.

However all these approaches must be used with a great caution owing to the non-steady-state nature of estuarine systems. The problem could be possibly reconsidered within the concept of "quasi-steady-state" described by O'Kane[66,67].

Despite these limitations, some general comments can be made on the behaviour of dissolved components in the estuarine zone.

As river water runs into the oceans, it enters a reservoir with a much longer holding time where an element can accumulate up to the age limit of the reservoir itself (10^8 years) unless its concentration is controlled at some intermediate level by interactions with particulate matter. It can be shown[34] that the relative residence times of the elements in the ocean (R_o) are given by the ratio (mean concentration in seawater)/(mean concentration in river water). The elements dissolved in sea water may be divided into "accumulated" ($R_o > 10$), "unchanged" ($10 > R_o > 0.1$) and "depleted" ($R_o < 0.1$) categories (Fig. 4, Table 5). The average time that a particular element remains in the oceans is known as its mean oceanic residence time (MORT, Table 5). The "depleted" elements (Table 5) have MORT values which are less than the time required for a single circulation of the oceans ($\sim$ 1500 years). These elements are therefore rapidly deposited and their estuarine chemistry is likely to be controlled by particulate removal. This is usually due to a coagulation-flocculation process involving humic acids or hydrous iron oxide colloids with which they may be associated. The rate of removal will depend, among other things, on the affinity for inorganic ligands. The "accumulated" elements have MORT values in excess of two hundred circulations (Table 5). They are able to accumulate in the world oceans and their concentrations retain remarkably constant ratios to one another despite local changes in the total salt content. Changes in the relative concentrations of even the most biologically active elements of this category (e.g. Ca, C) do not exceed more than a few percent. The "accumulated" elements will in general exhibit conservative mixing behaviour in estuaries. The mean concentrations of the "unchanged" elements (Table 5) are not altered dramatically on transfer from the rivers to the oceans. However, many elements belonging to this category are actively cycled by the biota and will consequently show marked spatial and temporal variations in concentration. Depending on the balance struck between the processes active in a particular estuary these elements may exhibit conservative behaviour (e.g. As, Si, Zn, Ni, N), removal (Zn, Si, Ni) or addition (Ba, As, P, Zn). The controlling reactions might be biological utilisation (e.g. N, P), ion exchange (e.g. Ba), variations in redox chemistry (e.g. As), or adsorption (e.g. Si).

Man's Influence on Estuarine Discharges

Human influence can modify the concentration and fluxes of chemical elements to the ocean, as well as the speciation and geochemical processes. Most usually, man's influence increases the flux of dissolved and suspended matter discharged by rivers. The increase of dissolved components can be significant and the Rhine provides a typical example of contamination due to the discharge of chloride. The concentration of chloride has increased by one order of magnitude since the beginning of this century[68]. The discharge of chloride to the ocean by the Rhine now equals that of the Amazon river. Similarly, Van Bennekom and Salomons (1979) estimated that man's influence has multiplied by 3-4 and 5 the global annual discharges of nitrogen and phosphorus to the oceans respectively (Table 3).

Despite the lack of global studies, and the generally poor level of chemical analyses performed in many national monitoring exercises, it is obvious that the toxic metals have reached very high levels in various rivers. Owing to the low K_Y values of many of these elements (Table 5), the effect of pollution is mainly observed in the particulate phase with the particles settling in the estuarine and nearshore areas. Once they are deposited, the major problem is to know whether or not they will remain there indefinitely. If they do, the environmental damage will be quite small, except to filter feeders (such as oysters and mussels). A particular concern is that the decay of organic matter may set free the trace metals in the surface sediments, which can be resuspended by tides, storms and dredging. It is quite possible that heavy metals released in this way could damage estuarine and nearshore aquatic life[44]. Intensive studies are needed to determine the release rates of heavy metals from estuarine sediments. For those elements that are irreversibly bound to particulate matter, the pollutants discharged by rivers cannot be readily observed in the open ocean. Only those elements which can be readily remobilized from the sediments could have a noticeable effect on the open ocean.

At the moment, it is likely that estuarine processes counteract the discharge of pollutants to the ocean system. As an example, despite the noticeable increase of river input of pollutants to the Baltic Sea, Magnusson and Westerlund[69] noticed that the ranges of dissolved concentrations for some heavy metals such as Pb, Cu, Zn are not significantly higher than those measured in the Indian Ocean. Consequently, severe damage to the fish stock has not yet been observed[70]. However, the concentrations of heavy metals in sediments both nearshore and in the Baltic proper are significantly higher than in the remote areas[71].

PCB's (polychlorinated biphenyls) have been in use for half a century and they have been discovered to cause cancer in experiments on laboratory animals. In the early 1970's, approximately 440,000 pounds of PCB's were discharged to the Hudson River from the General Electric manufacturing plants. They are now trapped in river and estuarine fine-grained sediments and pose a threat to 150,000 upstate New Yorkers who drink the river water as well as to commercial fisheries in the Hudson's estuary[72,73]. Another example is given by the dumping of 5.4×10^6 l of the highly toxic insecticide "Kepone" in the James River, which was stopped in 1975. In a similar manner to the PCB's pollution of the Hudson estuary, this insecticide will affect the aquatic life and the biogeochemical cycle in the James estuary for many years, even though the source of pollution has been eliminated[73].

If the common practice of deforestation and over-grazing has generally increased the riverine solid discharge, there are a few instances in which it has been reduced by damming[74]. This can also affect the dissolved load because of carbonate precipitation and salt retention in soils. The examples of the Indus and Nile are well known. The most detrimental effect is then the coastal zone erosion and the drastic decrease in the productivity of fisheries.

Finally, human influence may severely affect the speciation of chemical elements and hence affect biogeochemical processes. The increasing quantity of nutrients discharged to riverine and estuarine water will enhance nearshore productivity, as observed in the Northern Adriatic Sea, and is at first beneficial. This initial effect is followed by a degradation of the organic matter produced, which leads to an anoxic environment. From a chemical point of view, this process will enhance the dissolution of some elements such as manganese, as observed by Duinker et al.[75] in the Scheldt estuary. Organic contaminants such as detergents can also modify the electrical charge of particulate and hence influence their ion exchange properties[76].

Another example is provided by the geochemical behaviour of uranium. During the processing of calcium fluorophosphate used to produce phosphoric acid and fertilizers, a significant amount of orthophosphate compounds which are naturally enriched in uranium, is released to the aquatic environment. Contrary to the naturally occurring uranium, which is typically conservative in estuaries[24], this additional uranium is quickly removed from the solution during estuarine mixing as observed by Kaufman (unpublished) and Martin et al.[77].

THE RESPONSE OF THE OCEANS TO CHANGES IN COASTAL INPUTS

The mean oceanic residence time (MORT) value indicates the average time that a particular element remains in the ocean before it is removed by incorporation into the sediment. The MORT values are directly related to the K_Y values which describe the partitioning of the elements between the crustal rocks and the sea water[25] and between deep-sea clays and sea water (Fig. 4). The more effectively the elements are incorporated into the solid

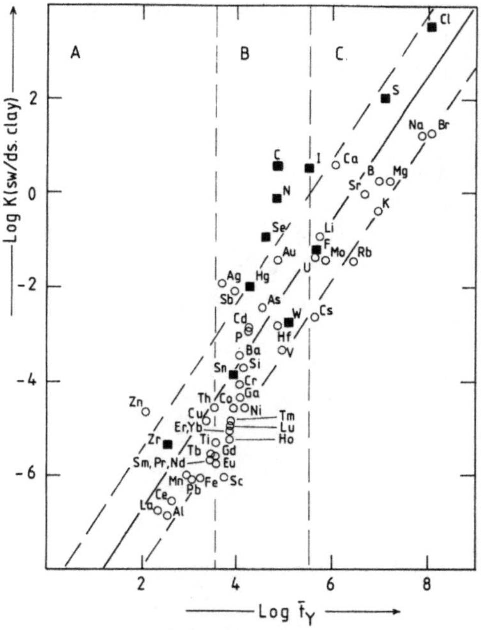

Fig. 4 The relationship between the sea water/deep-sea clay partition coefficient and the mean oceanic residence time. Solid squares indicate that values for the sea water/crustal rock partition coefficient have been used because of lack of data for deep-sea clays. The data are taken from Table 4. The equation for the straight line is: $\log K = -9.86 + 0.73 \log \bar{t}_Y$ ($n = 57$, $r = 0.73$). The dashed lines show a spread of $\pm$ two standard deviations. The diagram is divided vertically into three regions: A, "depleted" elements ($R_o < 0.1$), B, "unchanged" elements ($0.1 < R_o < 10$) and C, " accumulated" elements ($R_o > 10$).

phase, the shorter will be their residence times within the oceans. The relationship that is observed is consistent with the concept of a global ocean at steady state with material added from the world's rivers being approximately counterbalanced by material lost to the sediment. The K_Y values in sea water are directly related to the electronegativities of the elements (Fig. 5). The individual MORT values can therefore be calculated quite simply from the electronegativities of the elements[28,46]. Consequently, the response of the global mean composition of the oceans to changes in the riverine input over geological time can also be estimated from the electronegativities of the elements[34].

In general, for a change in the characteristics of the input to have a significant effect on the mean oceanic concentration of

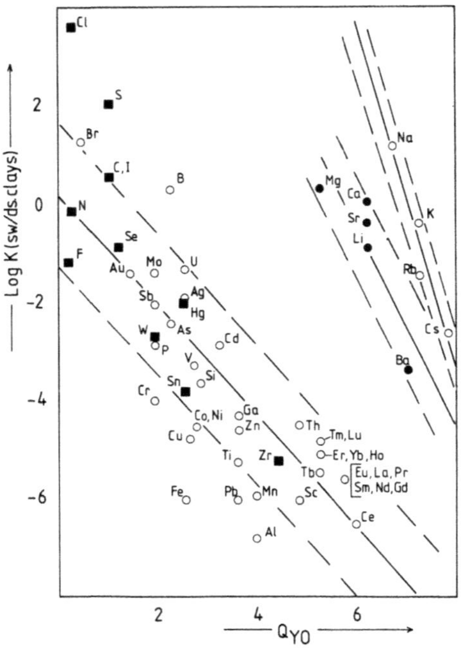

Fig. 5 The relationship between the sea water/deep-sea clay partition coefficient and the electronegativity function (Q_{YO}). Solid squares indicate that values for the sea water/crustal rock partition coefficient have been used because of lack of data for deep-sea clays. The equations for the straight lines are: (a) for the main sequence log K = 0.17 - 1.14 Q_{YO} (r = 0.82, n = 49); (b) for the alkaline earth metals (solid circles) log K = 11.94 - 2.05 Q_{YO} (r = 0.82, n = 5); and (c) for the alkali metals log K = 24.99 - 3.54 Q_{YO} (r = 0.96, n = 4). The dashed lines show a spread of $\pm$ two standard deviations.

a particular element it must persist for periods in excess of the MORT value. Since these values span periods from two hundred years to eight hundred million years (Table 5), it is clear that, with the possible exception of the "depleted" elements (Table 5), anthropogenic inputs are unlikely to have a lasting effect on the global sea water composition. However, since the "depleted" elements exhibit such a strong affinity for the solid phase, it is unlikely that high solution concentrations will be sustained in the estuarine zone. The long term effects of man's discharges on the composition of the global ocean are therefore likely to be small.

The MORT concept is only strictly applicable to the global ocean on time scales in excess of five or ten thousand years. It is therefore more relevant to problems related to the long term geological control of sea water composition than to problems related to anthropogenic inputs. However, the MORT values themselves, and their relationship to elemental electronegativities and to ocean-rock partition coefficients, can provide useful indications of the possible fates of the individual elements. Coastal waters tend to carry a higher particulate load than open ocean waters and, because of their proximity to nutrient sources and their continuous agitation by wind and tide, they tend to be biologically productive. Even in the open oceans the active recycling of the elements produces effective residence times in the wind-mixed surface layers (down to 100 m) ranging from only three or four years for the "depleted" elements[48] to forty years or more (equivalent to the water turnover time) for the "accumulated" elements. Much shorter residence times would be expected in the coastal ocean indicating that anthropogenic inputs are likely to have a significant effect on processes associated with the biological cycling of the elements in this region. The uptake of the elements during the formation of biogenic particles appears to follow a similar pattern to that revealed by the MORT concept in the global oceans since the partition coefficients for phytoplankton, zooplankton and faecal pellets are also related in a simple fashion to the MORT values of the elements[48,78,79]. Little work has been done on the development of quantitative models for the dispersion and removal of components injected into the coastal zone. Here too, a modified version of the residence time concept would be extremely useful. Linear diffusion models[80] suggest that it would be difficult to quantify chemical or biological removal rates by simply measuring coastal concentration profiles (even if the tracer residence time is only of the order of six months) since physical dispersion is likely to exert a dominant influence. More direct measurements are required of the rate of uptake of trace components onto suspended matter (living or dead) and of the rates of transfer of the elements across the air-water and sediment-water interfaces. Special attention should be given to the "depleted" and "unchanged" elements since these are the elements that are most actively recycled by the biota and are most sensitive to short-term fluctuations in the input. It is essential that chemical studies of this kind should be accompanied by physical studies of water movements in the coastal zone and by biological studies of the dynamics of the coastal ecosystems.

COMPARISON OF RIVER INPUT OF MATERIAL TO THE OCEAN WITH OTHER PATHWAYS

Mean trace element concentrations in atmospheric particulate matter over the ocean are generally at least one order of magnitude higher than in river suspended sediment. Although the gross atmospheric flux may be a more important source of trace metals to the coastal zone than is the dissolved input by local rivers (Table 6), the flux of riverborne material makes the dominant contribution on the global scale.

Conversely, recycling of trace metals and radionuclides from the estuarine and coastal environment to the continent must be assessed. Plutonium recycling which was first documented by Fraizier, Masson and Guary[84] is now carefully investigated in the vicinity of Windscale[85] and La Hague[86].

Since the pioneer work of Böstrom and Peterson[87] in the early 60's, hydrothermal activity at oceanic ridge crests has been recognized as a feature common to each fast-spreading oceanic ridge. Penetration of sea water to the depth of magma chamber has been demonstrated by McCulloch et al.[88] and Gregory and Taylor[89] but the implications to the ocean chemistry are still poorly

Table 6 Ratio of atmospheric flux to river input of dissolved trace elements in some selected areas.

Element	South Atlantic Bight (a)	New York Bight (b)	North Sea (c)
Arsenic	2.1	1	1.7
Cadmium	2.7	3.1	1.1
Copper	1.9	-	1.9
Iron	5.8	6.4	1.7
Manganese	0.6	-	0.8
Mercury	22	-	≃ 2.1
Nickel	1.7	-	≃ 1.3
Lead	9.5	20	6.8
Zinc	2.3	3.1	1.9

(a) Windom[81], (b) Duce et al.[82], (c) Cambray et al.[83]

known. On the whole, Edmond et al.[90] have shown that problems in the geochemical balance of Mg and SO_4^{--} can be accounted for by their deposition in the ridge where they precipitate as sepiolite and calcium sulphate respectively. Calcium sulphate is subsequently reduced by Fe^{2+} in the rocks to give a sulphide dominated system. The high acidity concomitant with sepiolite formation leads to hydrogen metasomatism of the basalts and the release of potassium, calcium, manganese, etc...[91].

This volcanic input has been quantified by prorating the chemical input to heat and He^3 fluxes. It has been shown that Ca excess in the sedimentary column, over that supplied by the rivers, is nicely accounted for by hydrothermal activity as are Li, Rb and Mn deposited in deep-sea clays[90].

Table 7 summarizes these main findings and compares the hydrothermal and river inputs of some chemical elements. Owing to the variable rate of crustal production over geological time[93], it is hardly possible to assess a steady-state in the ocean. However, according to these authors, the ocean would be at least as volcanogenic as it is fluvial. Other sources such as diffusion of chemical elements from sediments during diagenesis which can be considered as an indirect flux as compared to the net addition from the atmosphere and the mid-oceanic ridges are not considered here. However, any assessment of the overall chemical functioning of the ocean might consider these sources which can be of major importance for such components as Na^+, Ca^{++}, HCO_3^-[62].

Table 7 Comparison of hydrothermal (H)+ and river (R)++ dissolved fluxes of some chemical elements to the ocean system (mol y^{-1}).

	H^+	R^{++}	H/R
Li	$95\text{-}160 \times 10^9$	64×10^9	1.5 - 2.5
K	1.25×10^{12}	1.4×10^{12}	0.9
Rb	$1.7\text{-}2.7 \times 10^9$	660×10^6	2.6 - 4.1
Ba	$2.45\text{-}6 \times 10^9$	1.6×10^9	0.15 - 0.4
Si	3.1×10^{12}	7.3×10^{12}	0.4
Mn	$60\text{-}180 \times 10^9$	5.5×10^9	11 - 33

These fluxes do not include the slow weathering reactions that results in an extensive oxidation of the sea floor basalt. If one assumes that diffusion is the only transport mechanism, and that alteration cannot affect more than 100 m., a significant flux of Si (0.31×10^{12}) Mn (3.1×10^9) and Fe (0.10×10^{12}) must be added to those given above. (Maynard[92]).
+Edmond et al.,[90]; ++ this paper.

CONCLUSIONS

For the elements which exhibit conservative behaviour in estuaries and which have large MORT values ("accumulated" elements, Figure 4) the river inputs to the ocean system are readily calculated from the global mean dissolved loads of the world's rivers. In order to have a significant effect on the mean oceanic concentration of these elements any changes induced in the riverine input must persist for hundreds of thousands to hundreds of millions of years. The influence of anthropogenic discharges of these elements on the global ocean is therefore likely to be negligible although significant changes may be observed in the coastal ocean.

The flux of the other riverine elements to the ocean is however drastically modified at the river-ocean interface. More than 90% of the river particulate matter settles in the estuarine and coastal zone. A considerable fraction of the dissolved trace constitutents will also bind to the colloidal material (e.g. humic material and hydrous ferric oxides) produced when river water and sea water mix. Only a few elements appear to be released from the particulate to the dissolved phase. Consequently, due to the strong association of most chemical elements to the particulate phase, only a small percentage of the weathered material will eventually reach the open ocean.

It would appear that the mean oceanic concentrations of the "unchanged" and "depleted" elements will not be significantly influenced by anthropogenic inputs because of their rapid removal in the estuarine and coastal zone. However, although many of the reactions which control the removal and recycling of these elements in the inshore zone have been identified, we remain ignorant of their rates and mechanisms. In particular, little is known about the relative importance of the recycling of trace elements after they have been deposited in coastal sediments.

When talking about man's influence on the oceans, it is misleading to consider only mean oceanic concentrations. Man's influence is likely to be most strongly felt in regions where the mean residence times are relatively short (e.g. coastal ocean, surface mixed layer) and where biological processes and chemical water quality are most intimately interlinked. While air-borne inputs have been shown to exert a significant influence on surface concentrations of components such as lead both inshore and in the deep oceans the influence of riverine inputs will be largely confined to the coastal zone. Much more emphasis should be placed on the influence of chemical, physical and biological processes on the recycling of trace elements in this region.

REFERENCES

1. Baumgartner, A. and E. Reichel, 1975: "The World Water Balance". Elsevier, Amsterdam, 179 pp.
2. Martin, J.M. and M. Meybeck, 1979: Elemental mass-balance of material carried by world major rivers. Marine Chem., 7, 173-206.
3. Goldberg, E.D., 1972: Man's role in the major sediment cycle. In: "The changing Chemistry of the Oceans", D. Dyrssen and D. Jagner, eds. Wiley-Interscience, New York, pp. 267-288.
4. Gibbs, R.J., 1972: Water chemistry of the Amazon river. Geochim. Cosmochim. Acta., 36, 1061-1068.
5. Meade, R.H., C.F. Nordin, W.F. Curtis, F.M. Costa-Rodriguez, C.M. Dovale and J.M. Edmond, 1979: Sediment loads in the Amazon river. Nature, 278, 161-163.
6. Symoens, J.J., 1968: La minéralisation des eaux naturelles, Hydrobiological survey of Lake Bang Wenlu and Luapula river basin. Cercle Hydrobiologique de Bruxelles, 2, 1-199.
7. Peters, J.J., 1978: Discharge and sand transport in the braided zone of the Zaire estuary. Neth. J. Sea Res., vol. 12 (3/4), 273-292.
8. Bennekom, A.J. Van, G.W. Berger, W. Helder, R.T.P. De Vries, 1978: Nutrient distribution in the Zaire estuary and river plume. Neth J. Sea Res., vol. 12, (3/4), 296-323.
9. Handa, B.K., 1972: Geochemistry of the Ganga river water. Indian Geohydrology, 8, 71-78.
10. Handa, B.K., 1973: Geochemistry of Indian river waters. Int. Symp. on "Recent Researches in Geochemistry", Patna, India.
11. Clarke, F.W., 1924: "Data of Geochemistry", 5th ed., U.S. Geol. Survey Bull., 770.
12. U.S. Geological Survey, 1966-1967: Quality of surface waters of the United States, Water Supply Papers.
13. Alekin, O.A. and L.V., Brazhnikova, 1960: A contribution on runoff of dissolved substances on the world's continental surface (in Russian). Gidrochim. Mat., 32, 12-34.
14. Depetris, P.J., 1976: Hydrochemistry of the Parana river. Limnol. Oceanogr., 21, 736-739.
15. Meybeck, M. and J.P. Carbonnel, 1975: Chemical transport by the Mekong river. Nature, 255, 134-136.
16. UNESCO, 1979: World Register of Rivers discharging to the oceans. Division of Water Sciences, Techn. Paper.
17. Livingstone, D.A., 1963: Chemical composition of rivers and lakes. Data of Geochemistry, 6th ed., U.S. Geol. Survey Prof. Paper, 440-G, G1-G64.
18. Meybeck, M., 1979: Concentration des eaux fluviales en éléments majeurs et apports en solution aux océans. Rev. Géol. Dyn. Géogr. Phys., 21, 215-246.
19. Degens, E.T., 1981: Carbon transport in world rivers. SCOPE Newsletter, 12, Jan. 1981.
20. Bennekom, A.J. Van and W. Salomons, 1980: "Pathways of nutrients and organic matter from land to ocean through rivers". Proc. SCOR/ACMR/ECOR/IAHS UNESCO/CMG/IABO/

IAPSO workshop on River Input to Ocean Systems, Rome, 26-30 March 1979. J.M. Martin, J.D. Burton, D. Eisma, eds. UNEP/UNESCO, 1980, pp. 33-51.

21. Meybeck, M., 1981: River transport of organic carbon to the ocean. In: "Flux of organic Carbon by Rivers to the Oceans". U.S. Department of Energy, N.T.I.S., Springfield (CONF-8009140), pp. 219-269.
22. Edmond, J.M., 1980: "Pathways of nutrients and organic matter from land to ocean through rivers". Proc. SCOR/ACMR/ECOR/IAHS UNESCO/CMG/IABO/IAPSO workshop on River Input to Ocean Systems, Rome, 26-30 March 1979. J.M. Martin, J.D. Burton, D. Eisma, eds. UNEP/UNESCO, 1980, pp. 31-32.
23. Gordeev, V.V. and A.P. Lisitzin, 1978: Average chemical composition of suspended solids in world rivers and river particulate inputs to oceans, (in Russian). Dokl. Aka. nauk. S.S.S.R., 238, 225-228.
24. Figueres, G., J.M. Martin and A.J. Thomas, 1982: Comportement géochimique de l'uranium dissous dans l'estuaire du Zaïre. Ré-évaluation du bilan global de l'uranium dans l'ocean. Ocean Acta., vol 5, 2, 161-167.
25. Turner, D.R., A.G. Dickson and M. Whitfield, 1980: Water-rock partition coefficients and the composition of natural waters, a reassessment. Marine Chem., 9, 211-218.
26. Zhang, Z.B. and L.S. Liu, 1978a: A Φ (z/l,x) rule of inorganic ion exchange reactions in sea water and its applications. Ocean Selections, 1, 52-71 (in Chinese).
27. Zhang, Z.B. and L.S. Liu, 1978b: A study of the theory of the liquid-solid distribution of the elements in sea water. The theory of distribution equilibria of minor elements on hydrous oxide in sea water. Oceanol. Limnol. Sinica, 9, 151-157 (in Chinese).
28. Zhang, Z.B., L.S. Liu and N. Chen, 1979: A Φ (z/l,x) rule of chemical processes in oceans and its applications. VII. The transport of elements in oceans and the screening loss parameter. Oceanol. Limnol. Sinica, 10, 214-229 (in Chinese).
29. Whitfield, M. and D.R. Turner, 1979a: Water-rock partition coefficients and the composition of sea water and river water. Nature, 278, 132-137.
30. Whitfield, M. and D.R. Turner, 1979b: Critical assessment of the relationship between biological, thermodynamical and electrochemical availability. In: "Chemical Modelling in Aqueous Systems", E.A. Jenne, ed. Am. Chem. Soc., Washington, D.C., pp. 657-680.
31. Odum, W.E., G.M. Woodwell and C.F. Wurster, 1969: DDT residues absorbed from organic detritus by fiddler crabs. Science, 1964, 576-577.
32. Martin, J.M., M. Meybeck, F. Salvadori, and A.J. Thomas, 1976: Pollution chimique des estuaries: état actuel des connaissances, juin 1974. Rapp. Scient. Techn. C.N.E.X.O., 22, 283 pp.
33. Eisma, D., 1981: Suspended Matter as a Carrier for Pollutants in Estuaries and the Sea. In: "Marine Environmental Pollution", R.A. Geyer, ed. vol. 2, Elsevier, Amsterdam.

34. Whitfield, M., 1979: The mean oceanic residence time (MORT) concept, a rationalisation. Marine Chem., 8, 101-123.
35. Egami, F., 1974: Minor elements and evolution. J. Mol. Evol., 4, 113-120.
36. Wilson, H.D.B. and B. Laznicka, 1972: Copper belts, lead belts and copper-lead lines of the world. Proc. 24th Int. Geol. Congress, Montreal, section 4, pp. 37-51.
37. Rona, P.A., 1977: Plate tectonics energy and mineral resources: basic research leading to pay-off. EOS Trans. A.G.U., 58, 629-639.
38. Gibbs, R.J., 1977: Transport phases of transition metals in the Amazon and Yukon rivers. Geol. Soc. Am. Bull., 38, 6, 829-843.
39. Duplessy, J.C., 1980: Correlation of continental precipitations with the marine Quaternary isotopic stratigraphy. Proc. 26th Int. Geol. Congress, Paris, 7-17 July 1980, B 2, abstract, 650.
40. Burton, J.D. and P.S. Liss, 1976: "Estuarine Chemistry". Academic Press, New York.
41. Elderfield, H., 1978: Chemical variability in estuaries. In: " Biogeochemistry of Estuarine Sediments". Unesco, Paris, pp. 171-178.
42. Bewers, J.M. and P.A. Yeats, 1980: Behaviour of trace metals during estuarine mixing. Proc. SCOR/ACMR/ECOR/IAHS UNESCO/CMG/IABO/IAPSO workshop on River Input to Ocean Systems, Rome, 26-30 March 1979. J.M. Martin, J.D. Burton, D. Eisma, eds. UNEP/UNESCO, 1980, pp. 103-115.
43. Ahrland, S., 1975: metal complexes present in sea water. In: "The Nature of Sea Water", E.D. Goldberg, ed. Dahlem Konferenzen, Berlin, pp. 219-244.
44. Dyrssen, D. and M. Wedborg, 1980: Major and minor elements, chemical speciation in estuarine waters. In: "Chemistry and Biogeochemistry of Estuaries", E. Olausson and I. Cato, eds. J. Wiley, New York, pp. 71-119.
45. Turner, D.R., M. Whitfield and A.G. Dickson, 1981: The equilibrium speciation of dissolved components in fresh water and sea water at 25^0C and 1 atmosphere pressure. Geochim. Cosmochim. Acta., 45, 855-881.
46. Whitfield, M., D.R. Turner and A.G. Dickson, 1980: Speciation of dissolved constitutents in estuaries. Proc. SCOR/ACMR/ECOR/IAHS UNESCO/CMG/IABO/IAPSO workshop on River Input to Ocean Systems, Rome, 26-30 March 1979. J.M. Martin, J.D. Burton, D. Eisma, eds. UNEP/UNESCO, 1980, pp. 132-148.
47. Nieboer, E. and D.H.S. Richardson, 1980: The replacement of the non-descript term "heavy metals" by a biologically and chemically significant classification of metal ions. Environ. Poll., series B, 1, 3-26.
48. Whitfield, M., 1981: World ocean, mechanism or machination? Interdisciplinary Science Reviews, 6, 12-35.
49. Mantoura, R.F.C., 1981: Organo-metallic interactions in natural waters. In: "Marine Organic Chemistry", E.K. Duursma and R. Dawson, eds. Elsevier, Amsterdam.
50. Vuceta, J. and J.J. Morgan, 1978: Chemical modelling of trace metals in fresh waters: role of complexation and adsorption. Env. Sci. Technol., 12, 1302-1309.
51. Davies, J.A. and J.O. Leckie, 1979: Speciation of adsorbed ions at the oxide-water interface. In: "Chemical

Modelling in Aqueous Systems", E.A. Jenne, ed. Am. Chem. Soc., Washington, D.C., pp. 299-319.
52. Davies, J.A. and J.O. Leckie, 1978: Effect of adsorbed complexing ligands on trace metal uptake by hydrous oxides. Environ. Sci. Tech., 12, 1309-1315.
53. Bourg, A.C.M., 1979: Effect of ligands on the solid-solution interface upon the speciation of heavy metals in aquatic systems. Proc. Internat. Conf., Heavy Metals in the Environment, CEP consultants, Edinburgh, 446-449.
54. Baccini, P., H. Hohl and T. Bundi, 1978: Phenomenology and modelling of heavy metals distribution in lakes. Verh. Internat. Verein. Limnol., 20, 1971-1975.
55. Balistrieri, L., P.G. Brewer and J.W. Murray, 1981: Scavenging residence times of trace metals and surface chemistry of sinking particles in the deep ocean. Deep-Sea Res., 28, no. 2A, Feb. 1981.
56. Bourg, A.C.M., H. Etcheber and J.M. Jouanneau, 1979: Mobilisation des métaux traces associés aux matières en suspension dans le complexe fluvio-estuarien. Biologie-écologie méditerranéenne, 6, 161-166.
57. Burton, J.D., 1979: Physico-chemical limitations in experimental investigations. Phil. Trans. Roy. Soc. London, B 286, 443-456.
58. Turner, D.R. and M. Whitfield, 1979a: The reversible electrodeposition of trace-metal ions from multi-ligand systems. 1. Theory. J. Electroanal. Chem.. 103, 43-60.
59. Shuman, M.S. and L.C. Michael, 1978: Application of the rotated disk electrode to measurement of copper complex dissociation rate constants in marine coastal samples. Env. Sci. Technol., 12, 1069-1072.
60. Judson, S., 1968: Erosion of the land (what's happening to our continents?). Amer. Scientist, 56, 514-516.
61. Chester, R. and S.E. Aston, 1976: The Geochemistry of Deep-Sea Sediments. In: "Chemical Oceanography", J.P. Riley and R. Chester, eds. 2nd ed., vol. 6, Academic Press, London, pp. 281-390.
62. Sayles, F.L., 1979: The composition and diagenesis of interstitial solutions. I: Fluxes across the sea water-sediments interface in the Atlantic ocean. Geochim Cosmochim. Acta., 43, 527-545.
63. Meybeck, M., P. Hubert, J.M. Martin and P. Olive, 1970: Etude par le tritium du mélange des eaux en milieu lacustre et estuarien: application au lac de Genève et à la Gironde. In: "Isotope Hydrology 1970", I.A.E.A., Vienna, pp. 523-541.
64. Martin, J.M. and R. Letolle, 1979: Oxygen-18 in estuaries. Nature, 282, 292-294.
65. Stumm, W. and J.J. Morgan, 1970: Aquatic chemistry. New York: Wiley Interscience, pp. 425-428.
66. O'Kane, J.P., 1978: An estuarine water-quality model in an oscillating constant-volume reference frame. Proc. R. Irish Acad., series A, 78, 99-118.
67. O'Kane, J.P., 1980: "Estuarine Water Quality Management, with Moving Element Models and Optimization Techniques". Pitman, London.
68. Anon., 1973: "Uitgave Vereniging Milieudefensie". Amsterdam.

69. Magnusson, B. and S. Westerlund, 1980: The determination of Cd, Cu, Fe, Ni, Pb and Zn in the Baltic sea water. Marine Chem., 8, 231-244.
70. Dyrssen, D., 1980: Estuarine chemical processes in the Baltic Sea. Comm. Swedish Soc. for Maritime Research in Kalmar, April 26-27, 1979.
71. Olausson, E., O. Gustafsson, T. Melin and R. Svensson, 1977: The current level of heavy metals pollution and eutrophication in the Baltic proper. Medd. fråh Maringeologiska Laboratoriet, Göteborg, 9, 1-28.
72. Simpson, H.J., R.F. Bopp, B.L. Deck and C.R. Olsen, 1980: Fluxes and behaviour of reactive materials in urban estuaries. Proc. SCOR/ACMR/ECOR/IAHS/UNESCO/CMG/IABO/IAPSO workshop on River Input to Ocean Systems, Rome, 26-30 March 1979. J.M. Martin, J.D. Burton, D. Eisma, eds. UNEP/UNESCO, 1980, pp. 343-351.
73. Anon., 1980: "The Global 2000 Report to the President, entering the twenty-first century". vol. 2, U.S. Gov. Printing Office, Washington, D.C., pp. 305-306.
74. Meade, R.H., 1980: Man's influence on the discharge of fresh water, dissolved material and sediments by rivers to the Atlantic coastal zone of the United States. Proc. SCOR/ACMR/ECOR/IAHS/UNESCO/CMG/IABO/IAPSO workshop on River Input to Ocean Systems, Rome, 26-30 March 1979. J.M. Martin, J.D. Burton, D. Eisma, eds. UNEP/UNESCO, 1980, pp. 309-310.
75. Duinker, J.C., R. Wollast and G. Billen, 1979: Manganese in the Rhine and Scheldt estuaries, Part 2: Geochemical cycling. Est. Coast. Mar. Sci., 9, 727-738.
76. Martin, J.M., J. Jednacak and V. Pravdic, 1971: The physico-chemical aspects of trace elements behaviour in estuarine environments. Thalassia Jugosl., 7, 619-637.
77. Martin, J.M., V. Nijampurkar and F. Salvadori, 1978: Uranium and thorium isotope behaviour in estuarine systems. In: "Biogeochemistry of Estuarine Sediments". Unesco, Paris, pp. 111-127.
78. Cherry, R.D., J.J.W. Higgo and S.W. Fowler, 1978: Zooplankton faecal pellets and element residence times in the oceans. Nature, (London), 274, 246-248.
79. Yamamoto, T., Y. Otsuka, M. Okazaki and K. Okamoto, 1980: A method of data analysis on the distribution of chemical elements in the biosphere. In: "Analytical Techniques in Environmental Chemistry", J. Albaiges, ed. Pergamon Press, Oxford, pp. 401-408.
80. Brewer, P.G. and D.W. Spencer, 1975: Minor element models in coastal waters. In: "Marine Chemistry of the Coastal Environment", T.M. Church, ed. Am. Chem. Soc., Washington, D.C., pp. 80-96.
81. Windom, H.L., 1980: Comparison of atmospheric and riverine transport of trace elements to the continental shelf environment. Proc. SCOR/ACMR/ECOR/IAHS/UNESCO/CMG/IABO/IAPSO workshop on River Input to Ocean Systems, Rome, 26-30 March 1979. J.M. Martin, J.D. Burton, D. Eisma, eds. UNEP/UNESCO, 1980, pp. 360-369.
82. Duce, R.A., G.L. Hoffman, B. Ray, I.S. Fletcher, G.T. Wallace, J.L. Fasching, S.R. Piotrowicz, P.R. Walsh, E.J. Hoffman, J.M. Miller and J.L. Heffter, 1976:

Trace metals in the marine atmosphere: sources and fluxes. In: "Marine Pollutant Transfer", H.L. Windom and R.A. Duce, eds. Lexington Books, Lexington, Mass. USA, 77-119.

83. Cambray, R.S., D.F. Jeffries and G. Topping, 1975: An estimate of the input of atmospheric trace elements into the North Sea and the Clyde Sea. (1972-1973), United Kingdom Atomic Energy Authority Harwell Report. AERE - R 7733, 30 pp.

84. Fraizier, A., M. Masson and J.C. Guary, 1977: Recherches préliminaires sur le rôle des aérosols le transfert de certains radioéléments du milieu marin au milieu terrestre. J. Rech. Atmos., 11, 49-60.

85. Cambray, R.S. and J.D. Eakins, 1980: Studies of environmental radioactivity in Cumbria. Part 1: Concentrations of plutonium and caesium-137 in environmental samples from West Cumbria and a possible maritime effect. A.E.R.E. Harwell Report R 9807, July 1980, 20 p., H.M.S.O., London.

86. Martin, J.M., A.J. Thomas and C. Jeandel, 1981: Transfert atmosphérique des radionucléides artificiels de la mer vers le continent. Oceanol. Acta., 4, no. 3, 263-266.

87. Bostrom, K. and M.N.A. Peterson, 1966: Precipitates from hydrothermal exhalations on the East Pacific Rise. Econ. Geol., 61, 1258-1265.

88. McCulloch, M.T., R.T. Gregory, G.J. Wasserburg and M.A.J. Taylor, 1981: Sm-Nd, Rb-Sr and $^{18}O/^{16}O$ isotopic systematics in an oceanic crustal section: evidence from the Semail ophiolite. Earth Planet. Sci. Lett., (in press).

89. Gregory, R.T. and H.P. Taylor, Jr., 1981: An oxygen isotope profile in a section of Cretaceous oceanic crust, Semail ophiolite, Oman: evidence for $\delta^{18}O$-buffering of the oceans by deep (>5 km) sea-water hydrothermal circulation at mid-ocean ridges. Earth Planet. Sci. Lett., (in press).

90. Edmond, J.M., C. Measures, R. McDuff, L.H. Chan, R. Collier, B. Grant, L.I. Gordon and J.B. Corliss, 1979: Ridge-crest hydrothermal activity and the balances of the major and minor elements in the ocean. Earth Planet. Sci. Lett., 46, 1-18.

91. Edmond, J.M., 1980: Ridge crest hot springs. EOS Trans. A.G.U., 21, 129-131.

92. Maynard, J.B., 1981: Chemical mass-balance between rivers and oceans. Proc. Symp. on "Chemical Cycles in the Evolution of the Earth", Yellow Springs, Ohio, Sept. 1980 (in press).

93. Hays, J.D. and W.C. Pitman, III, 1973: Lithospheric plate motion, sea level changes and climatic and ecological consequences. Nature, 246, 18-22.

AIR-SEA EXCHANGE OF MERCURY

William F. Fitzgerald, G.A. Gill and A.D. Hewitt

Department of Marine Sciences
and Marine Sciences Institute
The University of Connecticut
Groton, Connecticut 06340 U.S.A.

ABSTRACT

Investigations of Hg in the near surface marine atmosphere in rainwater and in seawater have been conducted in open ocean regions of the tropical Pacific and the northwest Atlantic Oceans. Extensive studies of Hg in the atmosphere at the Sea-Air Exchange Program (SEAREX) tower facility at the Enewetak Atoll, Marshall Islands, were completed during 1979 while the southern hemisphere counterpart will be conducted at American Samoa in 1981. During 1979-1980 complementary work was conducted in the coastal marine environment of Long Island Sound. The distribution and chemical composition of atmospheric Hg have been examined using both Au and Ag amalgamation as selective trapping agents. Mercury analyses were conducted by a two-stage Au amalgamation flameless atomic absorption technique (4% precision @ 0.5 ng). Mercury determinations in seawater and in rainwater were made by the Au amalgamation procedure after reduction and aeration.

Most of the near surface atmospheric Hg species over both the open ocean and coastal regions studied are in the vapor phase (>99%). Similar concentrations of Hg were found in the air over the open ocean sites in the northern hemisphere ($\sim$1.5 ng m^{-3}) while smaller concentrations of Hg were observed in the southern hemisphere ($\sim$1.0 ng m^{-3}). Increases in gaseous Hg concentrations in the atmosphere, suggestive of Hg evasion from the sea surface, were observed in the central equatorial Pacific Ocean. At Long Island Sound the gaseous Hg concentrations are about twice the concentrations from oligotrophic oceanic areas, and a significant amount of Hg is in the organic form. The concentrations of Hg in open ocean rains are quite low (∿10 pmol l^{-1}) and reactive Hg concentrations in surface waters are correspondingly small (∿3 pmol l^{-1}).

These observations and the cycling of Hg between the atmosphere and ocean satisfy an air-sea exchange model which treats atmospheric Hg as a trace gas.

INTRODUCTION

Mercury is one of the most frequently measured trace constitutents in nature. During the past decade or so, there has been an extraordinary number of papers published which were concerned with methodologies and techniques for measuring Hg in natural samples. Correspondingly, many investigations concerned with Hg in the environment have been conducted[1,2]. Accurate and sufficient observations in critical parts of the global Hg cycle, however, are still lacking. Inconsistencies, resulting from the paucity of Hg determinations of high quality, are evident in reservoir contents, flux predictions, residence time estimates, and assessments of anthropogenic emissions[3-7].

The atmospheric cycle of Hg over the oceans is a major feature of the global cycle of Hg where information is very limited. It has been predicted that (1) as a consequence of the high volatility of Hg and many of its compounds, there must be a considerable flux of Hg from natural geological and anthropogenic sources to the atmosphere and (2) atmospheric processes must significantly affect the global distribution of Hg and provide a means for Hg transfer between the continents and oceans[1,8-10]. Despite this postulated importance of atmospheric Hg mobilization and exchange with the marine environment, the quantities and distribution of Hg in the marine atmosphere and the fluxes associated with rainfall and dry deposition are not well known. Moreover, the chemical forms of Hg in the atmosphere, the major trajectories and the importance of air-sea exchange, particularly for gas phase species, remain to be determined.

Thus, as part of our more general interests and efforts concerned with the global cycling of Hg and its behavior and fate in the marine environment, we have been investigating the atmospheric cycle of Hg over the oceans and the importance of the vapor phase in regulating Hg exchange at the sea surface. Much of this work has been conducted within the SEAREX (Sea-Air Exchange) Program which is a multidisciplinary, coordinated research endeavor examining the sources, fluxes and air-sea exchange behavior of a variety of trace metals and organic substances. Here, the principal results from our investigations of Hg in the marine atmosphere will be summarized, and modeled in a preliminary manner. In addi tion, unknowns and inconsistencies will be highlighted and suggestions made for future work.

Study Location

Investigations of Hg in the near surface marine atmosphere and in sea water have been conducted in open ocean regions of the tropical Pacific and northwest Atlantic Oceans. Extensive studies of Hg in the atmosphere at Enewetak Atoll, Marshall Islands, were completed during April to August 1979 while the southern hemisphere counterpart is presently underway at Tutuila Island, American Samoa. The Pacific tradewind studies at the islands have been complemented by a recent cruise (R/V T.G. Thompson, October 1-28, 1980) between Hawaii and Tahiti. During the expedition, the Hg distribution in both the atmosphere and surface ocean water was examined in transition from the northern to the southern hemisphere. In July (1979), we carried out similar air and sea investigations as part of the SEAREX program in the northwest Atlantic Ocean. While this Atlantic work was conducted the Enewetak wet season experiment was in progress and thus simultaneous measurements of Hg in the air from both oceans were obtained. The locations of the various studies are indicated in Figure 1.

Potential sources of atmospheric Hg have been studied. For example, we have been investigating our local marine atmosphere and waters of Long Island Sound (Fig. 1) for the past several years [11,12]. A number of experiments to examine the sea surface as a source of particulate and gaseous Hg were carried out on the northwest Atlantic cruise. We have conducted experiments at Kilauea, Hawaii, and at Mt. Etna, Sicily, designed to assess volcanic emissions of Hg, Pb and other metals to the atmosphere[13].

Sampling Techniques and Analytical Procedures

Atmospheric Hg. The field technique which has been developed, tested, and used for the collection of particulate and vapor phase Hg in both the coastal and remote marine atmosphere has been recently described [14]. Briefly, volatile atmospheric Hg species are collected by amalgamation on gold-coated glass bead columns. Gold effectively amalgamates and traps all the major volatile Hg species found in the near ground atmosphere[15]. Atmospheric particulate Hg species are collected on a Gelman Type A-E 47 mm diameter glass-fiber filter which has a pore size of approximately 0.3 μm. Thus, volatile Hg species are operationally defined by passage through this filter.

Mercury analyses are conducted, using a two-stage gold amalgamation gas train with detection of the eluting Hg^o by flameless

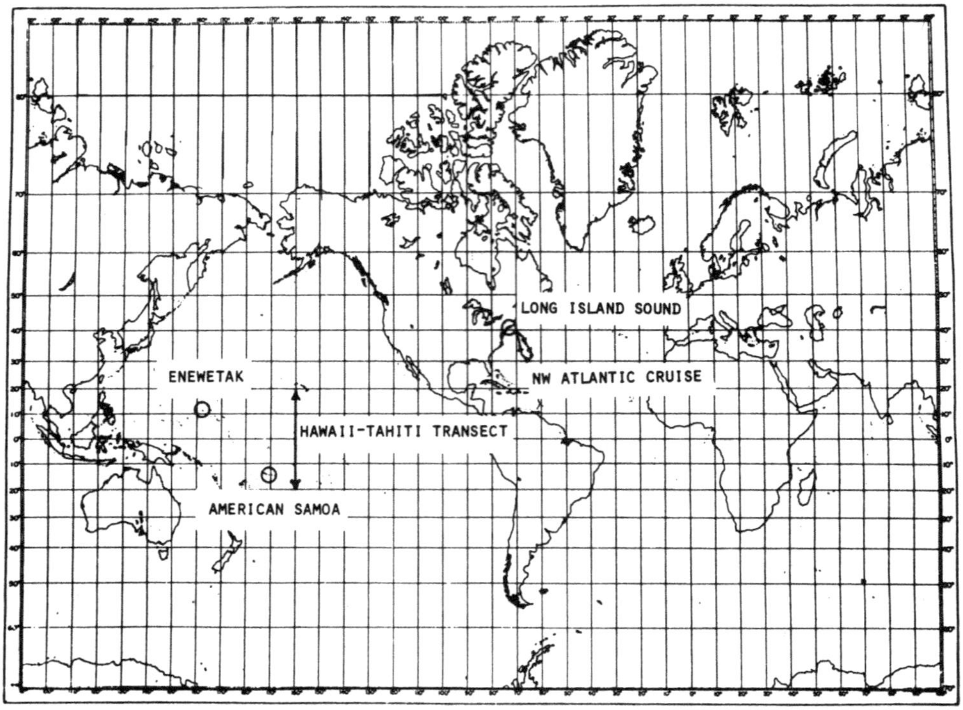

Fig. 1 Locations: Recent investigations of the Hg distribution in the near surface marine atmosphere.

atomic absorption or by d.c. (direct current) plasma emission spectroscopy. The Hg collected on a gilded glass bead tube in the field is transferred by controlling heating to a standardized analytical gold-coated glass bead column using Hg-free air or He as the carrier gas. Following this step, the Hg is eluted from the analytical column by controlled heating, and the gas phase absorption or emission of elemental Hg is determined at 253.7 nm. A standard curve is prepared for the analytical column using known injections of Hg vapor. The coefficient of variation for the determination of 1.5 ng Hg is 1.7% and 0.06 ng of Hg can be confidently measured.

Mercury associated with a particulate matter collection is volatilized by pyrolysis and collected on a gilded glass bead column. Following this step, the Hg is measured in a manner identical to the volatile Hg determination. It is estimated that, in this study, the overall variability associated with the collection, volume determination and analysis of gaseous mercury at 1.0 ng m^{-3} in the marine atmosphere, was about 20%.

Mercury Determinations in Rainwater and Seawater. The mercury analyses in rainwater and in sea water are conducted using a two-stage, gold-amalgamation modification of the $SnCl_2$ reduction-aeration flameless atomic absorption procedure we formerly used.[16,17] A gold-collection column takes the place of the cold-trap preconcentration stage. A significant improvement in precision and sensitivity has been achieved with the two-step amalgamation technique. For example, using a 1 liter sample, the precision of analysis reported as a coefficient of variation for the determination of 1.0 ng Hg is 10% and 0.1 ng Hg l^{-1} can be determined. Analytical details associated with this method can be found in Fitzgerald and Gill[14], and Gill[18].

The procedures used to determine mercury fluxes associated with rainfall are a refinement of an earlier methodology[17]. The rain collection apparatus consists of a 38 cm diameter glass funnel supported in an aluminum holder. The funnel feeds directly into a 1 or 2 liter Teflon storage bottle protected by a glass shroud. The funnels and containers are exhaustively and progressively cleaned at 25°C and at 55°C respectively with various grades of concentrated and dilute nitric acid from reagent to high purity subboiling distilled quality ($\leq$4 ng Hg l^{-1}). The total cleaning process requires a period of 10 days[19]. The sample containers are preacidified with subboiling distilled nitric acid to yield a pH of ~1.0 in the rain collection. In addition 1 ml of $AuCl_4^-$ (30 ppm) is added per liter to ensure that no mercury loss would occur during storage[20]. The rain collector and Teflon bottles are stored before and after collection, doubly bagged in acid-washed polyethylene bags to prevent any deleterious contacts.

The procedures used for seawater collection and sample processing for mercury determinations are, in general, an updated and improved version[21] of our earlier procedures[22]. Briefly, seawater is preserved and surface samples collected in acid-cleaned Teflon bottles (1 and 2 liter), preacidified with high purity concentrated nitric acid (Hg$\leq$0.004 ng ml^{-1}) to yield a 0.5% acid solution (pH 1.2). Open ocean surface waters were carefully sampled from a rubber workboat about 0.8-1.5 km away from the research vessel. Surface samples were obtained by hand (using arm-length polyethylene gloves) off the bow, while the workboat is gently driven or rowed forward into the wind.

Field Sampling. At Enewetak Atoll, (12°N, 162°E), the samplers were mounted atop an 18 m aluminum tower on a small uninhabited island. Sampling was controlled automatically and occurred only if the wind speed exceeded 5 miles per hour, condensation nuclei counts were <500 cm^{-3}, the wind direction was from the open ocean (75° to 135°), and it was not raining. This controlled procedure ensured that air representative of the marine environment would be collected. Uncontrolled collections were also made at ground level (1.5 m) at Enewetak and at Long Island Sound. Sampling at sea is usually conducted when the ship is underway. Collections were made from a 6 meter tower mounted on the bow and by extending the sampling apparatus 2-3 meters forward of the bow using an aluminum mast. While the sampling sector can be controlled with respect to the ship at sea, we generally have depended on the forward motion of the vessel to ensure collection of uncontaminated marine air.

Scientific Results and Interpretations

Atmospheric Hg Distribution. Most of the near surface atmospheric Hg species in the open ocean and coastal regions studied are in the vapor phase (>99%). This physical speciation has been established beyond a reasonable doubt[14,23]. A quantitative indication of the partitioning observed is presented in Table 1. These data agree quite favorably with recent work in the Atlantic Ocean by Slemr et al.[24]. The mean and standard deviation for the gaseous Hg distribution are similar for the open ocean areas studied. Mercury in the vapor form is characterized by a relatively stable concentration in the marine atmosphere. Moreover, some of the variation noted is due to analytical imprecision associated with sample handling, collection and analysis, rather than natural processes. This is evident, for example, at our Long Island Sound campus where the air has a fluctuating continental component, yet the mean and standard deviation for ca. 100 collections over a year period was 2.9 ± 0.5 ng m^{-3} (Table 1). That is, the relative standard deviation is smaller for measurements at Long Island Sound than at the open ocean sites.

It should be noted that many of the deficiencies and misunderstandings concerning the global cycling of Hg can be attributed principally to sampling errors and sampling artifacts. As marine trace metal geochemists are acutely aware, studies of trace constituents in the environment, particularly the oceanic regime, must be built on a foundation that emphasizes careful collection and analytical technique. Our overall analytical precision has been improved and the sampling "noise" reduced by collecting Hg from larger volumes of air.

The open ocean atmospheric particulate Hg fraction does not appear to display much variation. However, the concentrations are quite small (pg m^{-3}), the sampling period is long (14-21 days) and the number of collections is limited. Therefore, it would be difficult to observe variations. Inshore, particulate Hg will show significant variations (Table 1). Such fluctuations are not unexpected and reflect the local source dependence for particulate Hg and the short residence time of particles in the atmosphere.

Table 1. Hg Distribution in the Marine Atmosphere at Searex Sampling Locations

Collection Site	No. of Samples	Gaseous Hg Distribution: Range ng m^{-3}	Gaseous Hg Distribution: Mean & Std. Dev. ng m^{-3}	Rel. Std. Dev.	Particulate Hg Distribution: No. of Samples	Particulate Hg Distribution: pg m^{-3}
Enewetak, Dry Season (April 27–May 21, 1979)	27	0.8 - 2.9	1.6 ± 0.6	0.38	2	0.4 and 0.7
Enewetak, Wet Season (June 28–August 6, 1979)	67	1.1 - 3.2	1.7 ± 0.5	0.29	2	0.5 and 2
Northwest Atlantic Ocean (July 10–24, 1979)	7	1.0 - 1.9	1.6 ± 0.4	0.25	1	0.7
Avery Pt., Groton Ct., Long Island Sound (1979–1980)	108	1.6 - 7.2	2.9 ± 0.5	0.17	18	20 ± 20 (8–60 range)

Our results indicate that Hg should be treated (or modeled) as a trace atmospheric gas with a small particulate component. Gaseous Hg displays relatively small variations in open ocean regions and even in the coastal zone as evidenced by the good agreement between the gaseous Hg concentrations at Enewetak during both seasons and the northwest Atlantic Ocean. This behavior illustrates the well-mixed nature of Hg over the ocean and suggests a long residence time in the atmosphere. Further, the coastal and open ocean determinations put upper and lower limits on the expected range for Hg in the marine atmosphere of the northern hemisphere (Table 1).

Mercury Speciation Studies. The isolation and quantification of the various Hg species that comprise the vapor phase in remote, continental, and oceanic atmospheres may provide a very useful means of tracing certain parts of the Hg cycle in the marine environment. Therefore, we have carried out many experiments designed to separate vapor phase Hg into four fractions according to the Braman and Johnson[15] selective absorbent "stack" scheme. Valuable information and insight has been gained using a sequence of a Ag collection column followed by a Au collection column.* In this arrangement the Ag collects elemental Hg and most likely any other inorganic forms while passing the organic Hg fraction to be trapped on Au.

At Enewetak, in five experiments during the wet season, we found that essentially all the vapor phase Hg was trapped by the Ag absorber. While the analytical uncertainty in these experiments was about $\pm$ 20%, it does appear that for this open ocean region the gas phase consists principally of inorganic Hg of which elemental Hg is probably the major component. In contrast, at our inshore Long Island Sound campus, a significant amount of the gaseous Hg is in the organic form (Table 2).

Mercury in Rainwater. The accurate determination of Hg in rainwater has been a most demanding task. Two satisfactory collections were made during July in the Enewetak wet season. The concentrations of Hg in these rains were small (1.7 and 2.3 ng kg^{-1}). The concentration of particulate Hg measured in July was 2 pg m^{-3}. If we use 2 ng kg^{-1} as an average concentration for rain at Enewetak during July 1979 and the corresponding particulate Hg concentration (2 pg m^{-3}) then a washout ratio (W) reasonably close to other scavenged material is obtained (Table 3). The agreement

*Contrary to the authors' claim, the other absorbents, selective for Hg (II) type compounds and methyl Hg (II) type species do not work under field conditions. This deficiency has been confirmed by others (E. Crecelius, C. Brosset, personal communication, 1980).

Table 2 Gaseous Hg Species in the Coastal Marine Environment: Long Island Sound

Total Gaseous Hg (n = 108)	2.9 ± 0.5 ng m^{-3}
Organo-Hg Species** (n = 13)	27 to 82% of the total Hg (average = 50%)

** Forms of Hg which do not amalgamate with Ag

Table 3 Washout Ratios of metals measured at Enewetak

Scavenged Material	Washout Ratio (W)*
Hg	1200
Pb	480 (CIT)
Pb	600 (URI)
Mn	250 (URI)
Cu	1000 (URI)

$$*W = \frac{\text{amt kg}^{-1}_{\text{precipitation}} \times 1.2 \text{ kg m}^{-3}}{\text{amt m}^{-3}_{\text{air}}}$$

CIT = C.C. Patterson and URI = R.A. Duce, personal communication, 1981.

with other elements is satisfactory considering the limited number of samples. Moreover, the washout ratio suggests that Hg in rainwater consists principally of Hg associated with particles. That is, particulate Hg is behaving in a manner physically similar to other trace particulate metals in the atmosphere that are scavenged by precipitation.

This behavior is geochemically reasonable and expected if the principal components of the gaseous Hg are not very soluble in water (e.g., elemental Hg). Precipitation may be the major removal process for atmospheric Hg and rainfall fluxes must be accurately determined. For example, Nishimura[25] reported Hg in precipitation with a mean and standard deviation of 39 ± 21 ng kg^{-1} for the north Pacific westerlies. These concentrations appear too high and should be checked.

Preliminary Model for the Hg Cycle over the Oceans

The data obtained for Hg thus far are consistent with the hypothesis that the major source of atmospheric Hg is continental. The forms of Hg emitted to the air are principally in the gas phase which is composed of elemental Hg, organo-Hg species and perhaps other inorganic Hg vapors. There are a variety of natural and anthropogenic sources for these various forms of Hg. Our open ocean results indicate that most of the Hg is in the gas phase which probably consists principally of Hg^o. The particulate fraction is about 1000 times smaller at 0.4-2 pg m^{-3} (Table 1).

The Enewetak rainwater results are in agreement with this fractionation of atmospheric Hg. Elemental Hg is very insoluble in both fresh and salt water. For example, the Hg^o concentration for rainwater in equilibrium with an atmosphere containing 1.5 ng m^{-3} at 20°C would be 5×10^{-12} g l^{-1} [26]. Therefore, we would predict that rain would reflect primarily the particulate Hg that is washed out of the atmosphere. The reasonable correspondence of the washout ratio for Hg with other metals (Table 3) suggests an atmospheric composition of principally Hg^o gas and particulate Hg.

By treating Hg as a trace atmospheric gas, a residence time can be estimated using the empirical relationship $\tau_{Hg} \cdot \delta = 0.14$ year established by Junge[27] for trace gases. The residence time for Hg is signified by τ_{Hg}, while δ is the relative standard deviation of the gaseous Hg measurements. A τ_{Hg} between 134-204 days is obtained for measurements at Enewetak and the northwest Atlantic Ocean. A τ_{Hg} based on our Long Island Sound investigations is even longer at 300 days. As noted, we suspect that the smaller estimates of τ_{Hg} for the open ocean are caused in part by analytical variations rather than natural effects. In any event τ_{Hg} between 134-300 days is a significant departure from prevailing steady state box models which predict a τ_{Hg} of 11-13 days[6].

The long residence time for Hg in the atmosphere has additional implications with respect to natural and anthropogenic Hg emissions that are globally mobilized. Our data suggest that such emissions may be 10 to 20 times smaller than assumed[1]. A reduction of such a factor would make volcanic emissions considerably more important. Earlier we suggested that the world-wide volcanic output of Hg based on extrapolation of our measurements at Kilauea, Hawaii would be negligible (<0.1%) compared to anthropogenic emissions[13]. Our recent work in the Mt. Etna plume yielded a Hg/S ratio that was 10 times the Kilauea measurement, and a correspondingly increased Hg flux based on the volcanic input rate of S to the global atmosphere.

Our information concerning the behavior of Hg over the oceans is limited. We suggest, from the speciation experiments and the

Table 4 Hg Fluxes to the Sea Surface During July 1979, at Enewetak Atoll, Marshall Islands

Total Hg burden m^{-2} at Enewetak (1.7 ng m^{-3} for a scaled atmosphere of 6 km)	10,200 ng m^{-2}
Residence time based on the Junge[27] empirical relationship for trace gases ($\tau_{Hg} \cdot \delta$ = 0.14 yr)	175 days
Removal rate = total Hg burden/τ_{Hg}	58 ng m^{-2} d^{-1}
Rain flux to the sea surface based on average concentration of 2 ng Hg l^{-1} (July 1979) and the meterological mean rainfall for July (0.15 m)	10.0 ng m^{-2} d^{-1}
Gas phase transfer from the atmosphere to the oceans.*	∿ 12.0 ng m^{-2} d^{-1}
Dry deposition flux*	∿ 2.0 ng m^{-2} d^{-1}
Total Hg flux	∿ 24 ng m^{-2} d^{-1}

*Estimated (see text)

residence time, that elemental Hg is the predominant form and it is slowly removed from the atmosphere. Also, it appears that the major removal routes are in precipitation, and gas exchange at the sea surface with dry deposition playing a lesser role. The supporting evidence for these fluxes is presented in Table 4.

A removal rate for Hg of 58 ng m^{-2} d^{-1} was determined for the Enewetak atmosphere during July of 1979. This flux was derived using the trace gas residence time for Hg[27] and the assumption that the near surface concentrations of Hg observed can be applied to the entire air column.

The rainflux of 10 ng Hg m^{-2} d^{-1} based on the measurements at Enewetak accounts for part of the predicted flux. The gas phase transfer of Hg from the atmosphere to the oceans can be estimated using the Liss and Slater[28] model for the flux of gases across the air-sea interface. A flux of 12 ng m^{-2} d^{-1} is obtained assuming that the diffusivity of gaseous Hg is similar to radon and the sea surface is a perfect sink. A small depositional flux of 2 ng m^{-2} d^{-1} is obtained if we assume that a transfer velocity of 1 cm s^{-1} can be used with the observed particulate Hg of 2 pg m^{-3}.

The total removal flux of 24 ng m^{-2} d^{-1} agrees surprisingly well with the predicted value based on the Junge relationship. As we have indicated, we think the predicted air to sea transfer of

Hg is actually too high due to the analytical errors associated with the measurements. The order of the removal mechanisms is still tentative and further testing is currently underway in Samoa.

The analysis presented above predicts that the gas phase transfer of Hg from the air to the water is a significant flux. Such vapor exchange at the sea surface is a two-way process. Correspondingly, several of our experiments indicate that the sea is probably a minor source of Hg to the atmosphere. Year long studies at our Long Island Sound sampling station (Table 1) show no evidence of Hg evasion or invasion at the sea surface. Few seasonal variations were noted and the day-night collections generally gave Hg concentrations that agreed with one another within the analytical error (5-10%). At Enewetak the variations observed for gaseous Hg in samples collected simultaneously at three tower levels (ground, middle, top) were within the experimental error (Table 5).

Table 5 Gaseous Hg Distribution. Multilevel Study: Enewetak Atoll, Marshall Island. July 24 - August 6, 1979.

		Hg Concentration	
Sampling Elev., Meters	No. of Samples	Range, ng m^{-3}	Mean and Std. Dev. ng m^{-3}
1.5	8	1.2-2.2	1.5 ± 0.4
10	8	1.2-2.2	1.5 ± 0.3
20	8	0.9-1.8	1.4 ± 0.4

Also, diurnal collections yielded the same average amounts of Hg per unit volume of air (Table 6).

Table 6 Gaseous Hg Distribution. Diurnal Study: Enewetak Atoll, Marshall Island. June 28 - August 6, 1979.

Collection Period	No. of Samples	Range, ng m^{-3}	Mean and Std. Dev. ng m^{-3}
Day	32	1.1-3.2	1.9 ± 0.6
Night	35	0.9-2.4	1.5 ± 0.3
Day and Night	67	1.1-3.2	1.7 ± 0.5

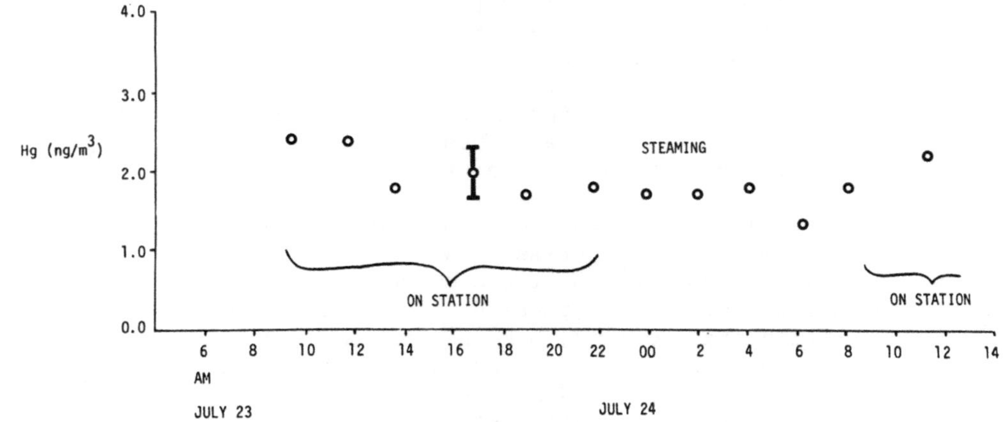

Fig. 2 Vapor phase Hg: Bow sampling time study, R/V. Endeavor 039 - July, 1979. (Sample volumes between 0.88-0.14 m^3 and minimum error is indicated).

During the northwest Atlantic cruise no variation in gaseous Hg other than analytical deviations was observed at bow level during a 24 hour sampling period (Fig. 2). We used the Bubble Interfacial Microlayer Sampler[29] to remove volatile Hg species by degassing sea water under natural conditions. This exploratory experiment indicated that the volatile Hg fraction in surface seawater of the Sargasso Sea must be at least 100x smaller (<0.05 pmol l^{-1}) than the readily reducible species (reactive Hg).

The increase in nearshore atmospheric Hg concentrations and the presence of Hg in organic forms argues further that the continental sources of Hg predominate over oceanic sources. This conclusion is quite different from certain prevailing models[1,3] which predict oceanic emission fluxes of Hg to be very significant (1-2 times the anthropogenic Hg fluxes) and major exchange of gaseous Hg at the sea surface. We expect, however, that in rich biological regions of the marine environment, such as the Peru upwelling, there will be evasion of volatile forms of Hg. We further predict that these sources of Hg may be important locally but not on a global scale.

Our working hypothesis regarding the composition of Hg in the atmosphere, its distribution, the major sources, fluxes to and from the sea surface is limited in terms of mechanisms. For example, we suggest that elemental Hg is slowly removed from the atmosphere, but we are not sure how this is accomplished. Is it conversion to particles following oxidation, is it by gas exchange at the sea surface or another mechanism yet to be identified? We hypothesize that organic Hg forms found in coastal marine air are relatively unstable reacting along pathways that may eventually result in oxidation to particulate species and removal by rain and dry deposition. Again, the gas phase exchange must be evaluated. The elemental Hg hypothesis can be tested by either measuring the elemental Hg present in the atmosphere directly or in sea water in equilibrium with the atmosphere. These are difficult but tractable experiments which we will attempt to conduct.

If the sea is not a significant source of gaseous Hg as we suggest, then meridional experiments will show that the concentrations of Hg in the northern hemisphere are greater than in the southern hemisphere. Such a pattern will reflect the larger number of continental sources, both natural and anthropogenic that are in the northern hemisphere. Indeed, we have preliminary evidence of an interhemispheric difference in the gaseous Hg distribution. During a recent Hawaii-Tahiti cruise along the NORPAX track at 160°W [30] we found the concentrations of Hg in the air from open ocean regions (>10°N) in the northern hemisphere were similar (~1.5 ng m^{-3}), to the observations at Enewetak (Table 1), while smaller concentrations (~1.0 ng m^{-3}) were observed for gaseous Hg in the southern hemisphere (>10°S). This pattern is analogous to carbon monoxide, a trace gas whose primary sources are continental and include both natural and anthropogenic processes [31].

The results from this recent cruise are plotted against latitude in Figure 3. The gaseous Hg distribution differs from the CO

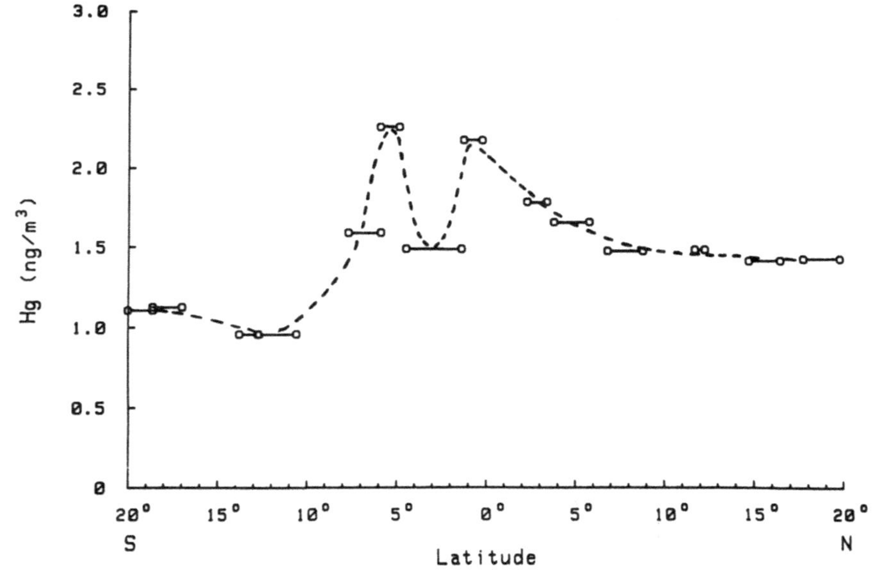

Fig. 3 Vapor phase Hg: Hawaii-Tahiti transect (160°W), R/V T.G. Thompson - Oct. 1 - 21, 1980.

pattern in the equatorial region, where a significant increase in the vapor phase Hg is observed between 10°N and 10°S. A relatively smooth transition from higher concentrations of the northern hemisphere air to the smaller concentration of the southern hemisphere is not observed. While it should be stressed that the maximum is characterized by only 5 measurements, there is ancillary evidence that suggests that evasion of Hg species from the equatorial Pacific Ocean would not be an anamolous process.

This region is characterized hydrographically by a relatively stable thermal structure (>year) with strong but variable zonal water movement to the east and west. The north and south equatorial currents flowing westward are intersected by the eastward-moving equatorial countercurrent that straddles the equator[30]. These upper layers are characterized by divergences, convergences, and enhanced biological activity that are distributed over roughly the same latitude range as the Hg fluctuations. Further, the boundary layer atmospheric O_3 concentrations in these regions are variable.

Routhier et al.[32] measured low levels of O_3 near the latitude range 13°S to 2°N at about the same longitude during 1978 studies in the Gametag program. They suggest that photochemical destruction of O_3 may be an important factor in creating this trough. If this destruction process is significant, it is not unreasonable to suggest interaction with volatile species evading from the sea surface as a sink. The enhanced Hg concentrations in the gas phase are a reflection of the evasion process.

The final speculation is to suggest that organo-Hg species such as dimethyl Hg are being produced by biological processes and evading from the sea surface of the central equatorial Pacific Ocean. It is further suggested that these organo Hg compounds are superimposed on a background of elemental Hg, which itself decreases from the northern to the southern hemisphere. In this hypothesis the organo Hg species must have a short residence time compared to elemental Hg, and may not play an important role in the global Hg cycle.

Surface seawater in equilibrium with the atmosphere should contain the appropriate concentrations of the various vapor phase species. However, the predicted concentrations are much smaller than the reactive Hg determined by the conventional analysis of seawater. For example, from Sanemasa's[26] solubility studies surface seawater in equilibrium with an atmosphere containing 1.5 ng Hg^{o} m^{-3} of elemental Hg at 20°C should have a concentrration of 0.022 pmol l^{-1} (4.5 pg l^{-1}). Reactive Hg determined in surface seawater collected during the Hawaii-Tahiti experiment yielded a concentration of 2.5 ± 1.0pmol l^{-1} (0.5 ± 0.2 ng l^{-1}) that was essentially independent of latitude. Therefore, to obtain sufficient quantities of gaseous Hg^{o} and other volatile Hg species for a precise analysis it will be necessary to purge 100 to 200 l of seawater. Determinations of the gaseous Hg species in surface waters will provide the supporting information to properly close the air-sea exchange cycle of Hg.

Our method of analysis in this equatorial region did not distinguish between the Hg species making up the gas phase. Only total gaseous Hg was determined. Much information concerning the role and nature of air-sea exchange processes in the marine geochemical cycle of Hg would be obtained if the chemical species making up the vapor phase of Hg could be identified and quantified. Both the atmospheric and surface sea water should be investigated. Technically, this should be possible and the work should focus initially on biological productive regions in the marine environment such as the equatorial Pacific, Peru upwelling and inshore areas such as Long Island Sound.

ACKNOWLEDGEMENTS

The authors are grateful for the assistance of the Captain and crew from the R/V Endeavor (U. of Rhode Island) and from the R/V T.G. Thompson (U. of Washington). Shipboard experiments depended on the efforts of K. Bruland and R. Franks, U. of California at Santa Cruz, and A. Ng, California Institute of Technology. M. DeGruy and staff from the Mid-Pacific Marine Laboratory (U. of Hawaii) at Enewetak, provided much physical aid and logistical support to our field program. Additional support for the Enewetak experiments was provided by the Dept. of Energy, Honolulu, Holmes and Narver, Inc., and the University of Hawaii at Manoa. As always, we are indebted to our colleagues in the SEAREX Program for field support.

This work was supported by the National Science Foundation's Office for the International Decade of Ocean Exploration: Sea-Air Exchange Program (SEAREX) Grant Nos. OCE-77-13071 and OCE-77-13072.

REFERENCES

1. National Academy of Sciences (NAS), 1978: An assessment of mercury in the environment. Washington, D.C., p. 185.
2. Nriagu, J.O., 1979: Editor, The Biogeochemistry of Mercury in the Environment. Elsevier/North Holland Biomedical Press, New York. p. 696.
3. Lantzy, R.J. and F.T. MacKenzie, 1978: Atmospheric trace metals: global cycles and assessment of man's impact. Geochim. et Cosmochim. Acta., 43, 511-525.

4. Appelquist, H., K.O. Jensen, T. Sevel and S. Hammer, 1978: Mercury in the Greenland ice sheet. Nature, 273, 657-659.
5. Fitzgerald, W.F., 1979: Distribution of mercury in natural waters. In: "The Biogeochemistry of Mercury in the Environment", edited by J.O. Nriagu. Elsevier/North Holland Biomedical Press, Amsterdam, Chapter 7, 161-173.
6. Andren, A.W. and J.O. Nriagu, 1979: The global cycle of mercury. In: "The Biogeochemistry of Mercury in the Environment", edited by J.O. Nriagu. Elsevier/North Holland Biomedical Press, Amsterdam, Chapter 1, 1-21.
7. McLean, R.A.N., M.O. Farkas and D.M. Findlay, 1980: Determination of mercury in natural waters: Sampling and analysis problems. In: "Polluted Rain", edited by T.Y. Toribara, M.W. Miller and P.E. Morrow. Plenum Publishing Corp., pp. 151-173.
8. Weiss, H.G., M. Koide and E.D. Goldberg, 1971: Mercury in a Greenland ice sheet: Evidence of recent input by man. Science, 174, 692-694.
9. Mackenzie, F.T. and R. Wollast, 1977: Sedimentary cycling models of global processes. In: "The Sea, vol. 6, Marine Modeling", edited by E.D. Goldberg, J.N. McCave, J.J. O'Brien and J.H. Steele. John Wiley, New York, Chapter 19, pp. 765-777.
10. Desaedeleer, G. and E.D. Goldberg, 1978: Rock volatility - Some initial experiments. Geochem. J., 12, 75-79.
11. Gill, G.A. and W.F. Fitzgerald, 1979: "Mercury geochemistry of Long Island Sound: Analytical and field study." Symposium on Trace Elements in the Hydrosphere, 177th Meeting of the American Chemical Society, Honolulu, Hawaii, April 2-7, 1979. Paper #103 in Environmental Chemistry Section, Program and Abstracts.
12. Fitzgerald, W.F., A.D. Hewitt, G.A. Gill and R.M. Ferguson, 1979: Gaseous and particulate mercury in the coastal marine atmosphere: Air/sea exchange considerations. Program and abstracts, 9th Northeast Regional Meeting American Chemical Society, Syracuse, N.Y., October 2-5, Envir. #3.
13. Unni, S., W.F. Fitzgerald, D.Settle, G. Gill, B. Ray, C. Patterson and R. Duce, 1978: "The impact of volcanic emissions on the global atmospheric cycles of sulfur, mercury and lead". Fall Meeting, American Geophysical Union, San Francisco, Dec. 4-8, 1978, Abstract: EOS 59, 1223.
14. Fitzgerald, W.F. and G.A. Gill, 1979: Subnanogram determination of mercury by two-stage gold amalgamation applied to atmospheric analysis. Anal. Chem., 51, 1714-1720.
15. Braman, R.S. and D.L. Johnson, 1974: Selective absorption tubes and emission technique for determination of

ambient forms of mercury in air. Environ. Sci. Tech., 8, 996-1003.

16. Fitzgerald, W.F., W.B. Lyons and C.D. Hunt, 1974: Coldtrap preconcentration method for the determination of mercury in seawater and in other natural materials. Anal. Chem., 46, 1882-1885.
17. Fogg, T.R. and W.F. Fitzgerald, 1979: Mercury in southern New England coastal rains. J. Geophys. Res., 84, C11, 6987-6989.
18. Gill, G.A., 1980: On the geochemistry of mercury in Long Island Sound: An analytical and field study. Master's Thesis, Univ. of Connecticut, p. 199.
19. Patterson, C.C. and D. Settle, 1976: The reduction of orders of magnitude errors in lead analyses of biological materials and natural waters by evaluating and controlling the extent and sources of industrial lead contamination introduced during sample collecting and analysis. In: "Accuracy in Trace Analysis: Sampling, Sample Handling and Analysis", P.D. Lafleur (ed.). NBS, Special Publication 422, Washington, D.C., Vol. I, pp. 321-335.
20. Moody, J.R., P.J. Paulsen, T.C. Rains and H.L. Rook, 1976: The preparation and certification of trace mercury in water standard reference materials. In: "Accuracy in Trace Analysis: Sampling, Sample Handling and Analysis", P.D. Lafleur (ed.). NBS, Special Publication 422, Washington, D.C., Vol. I, pp. 267-275.
21. Fitzgerald, W.F. and G.A. Gill, 1981: A procedure for avoiding mercury contamination during open ocean sampling. In preparation.
22. Fitzgerald, W.F. and W.B. Lyons, 1975: Mercury concentrations in open ocean waters: Sampling procedure. Limnol. Oceanogr., 20, 468-471.
23. Fitzgerald, W.F., A.D. Hewitt and G.A. Gill, 1979: Global cycling of mercury in the marine atmosphere and its exchange with the ocean. IAPSO Program and abstracts for the XVII General Assembly of International Association for the Physical Sciences of the Ocean and the International Union of Geodesy and Geophysics, Canberra, Australia, Dec. 2-15, 1979, PS-9 #5, p. 83.
24. Slemr, F., W. Seiler and G. Schuster, 1981: Latitudinal distribution of mercury over the Atlantic Ocean. J. of Geophys. Res., 86, C2, 1159-1166.
25. Nishimura, M., 1979: Determination of mercury in the aquatic environment and its global movement. Abstracts of Papers ACS/CSJ Chemical Congress, INOR 275, Honolulu, Hawaii, April.
26. Sanemasa, I, 1975: The solubility of elemental mercury vapor in water. Bull. Chem. Soc. Jap., 48, 1795-1798.

27. Junge, C.E., 1974: Residence time and variability of tropospheric trace gases. Tellus, 26, 477-488.
28. Liss, P.S. and P.G. Slater, 1974: Flux of gases across the air-sea interface. Nature, 247, 181-184.
29. Fasching, J.L., R.A. Courant, R.A. Duce and S.R. Piotrowicz, 1974: "A new surface microlayer sampler utilizing the bubble microtome". J. Rech. Atmos., 8, 649-652.
30. Wyrtki, K., E. Firing, D. Halpern, R. Knox, G.J. McNally, W.C. Patzert, E.D. Stroup, B.A. Taft and R. Williams, 1981: The Hawaii to Tahiti shuttle experiment. Science, 211, 22-28.
31. Seiler, W., 1974: The cycle of atmospheric CO. Tellus, 26, 116-135.
32. Routhier, F., R. Dennett, D.D. Davis, A. Wartburg, P. Haagenson and A.S. Delany, 1980: Free tropospheric and boundary-layer airborne measurements of ozone over the latitude range of 58°S to 70°N. J. Geophys. Res., 85, 7307-7321.

CONVERSION TABLE

Since 1 ng Hg = 5 p mol, conversion from 1 ng l^{-1}, 1 ng m^{-3}, 1 ng kg^{-1} is respectively 5 p mol l^{-1}, 5 p mol m^{-3}, 5 p mol kg^{-1}, and 1 pg m^{-3} = 5 f mol m^{-3}.

SEPARATION OF COPPER AND NICKEL BY LOW TEMPERATURE PROCESSES

Gary Klinkhammer

Department of Earth Sciences
The University of Leeds
Leeds LS2 9JT, England

ABSTRACT

When copper and nickel enter estuaries they exhibit distinct geochemical behaviours. Copper tends to become associated with solids while nickel remains dissolved. Nickel is thus free to enter the oceans while copper accumulates in estuarine or near-shore sediments. Early diagenesis in pelagic sediments acts to reverse these estuarine effects. Copper scavenged from sea water is remobilized at the sea-sediment boundary and ninety percent of this copper is recycled to sea water. On the other hand, twenty-five per cent of all the nickel deposited in pelagic sediments is associated with authigenic manganese oxides. Since most pelagic sediments are oxidative near the interface, this gleaned nickel is effectively removed from sea water. These results demonstrate that copper and nickel are fractionated continuously from initial weathering to final burial in sediments.

INTRODUCTION

The observation that copper and nickel are closely associated in marine deposits would indicate that these elements have similar goechemical cycles. However, there are several lines of evidence which suggest that these metals should fractionate from one another at low temperatures. (i) Laboratory studies and speciation calculations[1] show that copper-organic interactions are more important than nickel-organic interactions in natural waters. (ii) Leaching experiments performed on manganese nodule powders[2] are consistent with the presence of copper on adsorption sites but nickel in the lattice of manganese oxides. (iii) Seawater data[3,4,5] reveal that nickel possesses a classical nutrient-depth profile while the copper distribution is influenced by a bottom source and intense deep-water scavenging. (iv) Modeling of oceanic data[6] maintains that the scavenging rate constant of copper is 14 times larger than the corresponding constant for nickel.

Given this contradictory evidence, do copper and nickel separate at low temperatures? The data in this chapter reveal that this fractionation does occur and is prominent at several points during their geochemical cycles.

ANALYTICAL METHODS

Metal and phsophate samples were collected from the Hudson River estuary in October, 1975 during a fall flooding period. Freshwater discharge from the Hudson drainage basin (Fig. 1) ranges between 250-1200 m^3s^{-1}. Discharge during our 1975 collection was 770 m^3s^{-1}. Phosphate samples were stored on ice immediately after collection and analyzed two days later. Trace metal samples were filtered through acid-cleaned 0.4 μm Millipore filters within 8 hours after collection. The "dissolved" fraction was acidified to pH 2 with ultra-pure HCl immediately after filtering. Dissolved nickel and copper were preconcentrated with Chelex-100 and measured by flameless atomic absorption. The filters were low-temperature ashed and dissolved in HF and HNO_3. Particulate copper and nickel were measured by flameless AA and particulate aluminium by neutron activation. Complete analytical details for data from the Hudson are given elsehwere[7].

Pore water samples from site C (Fig. 4) were collected by centrifugation in a cold van. Centrifuged samples were filtered through 0.4 μm Nuclepore filters. Metal aliquots were acidified to pH 2. Manganese was measured by direct injection into a graphite furnace and by AA after extraction with quinolinol in chloroform[8]. Dissolved copper and nickel were coprecipitated with ammonium pyrrolidine dithiocarbamate using cobalt as a carrier. The precipitate was dissolved in 3 N HNO_3 and the metal concentrations were determined by AA. More complete details are given in another paper[9]. The metal results shown in Fig. 5 are average ($\pm$ 1σ) concentrations from subcores taken from two box cores. The bottom water concentrations of copper and nickel agree well with those values reported by Mangum and Edmond[10].

HUDSON RIVER ESTUARY

The Hudson River flows through New York City and its environs (Fig. 1). The largest anthropogenic input into the estuary is New York/New Jersey sewage which can be 20% of the freshwater supply

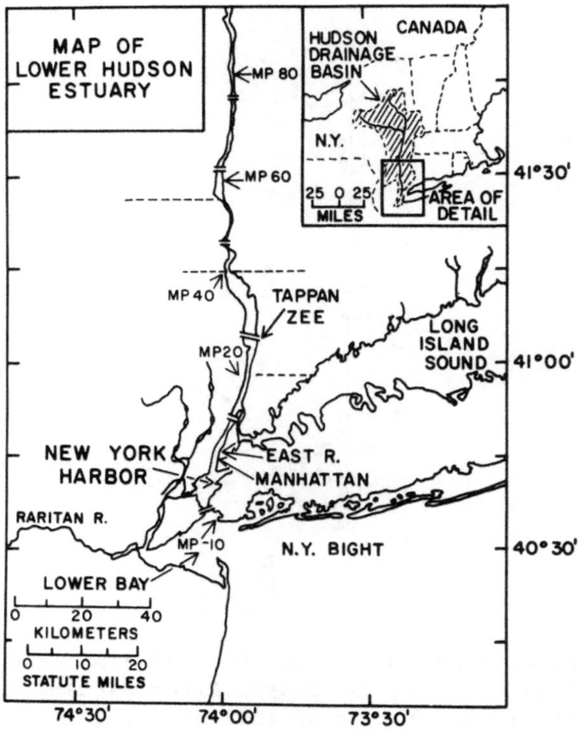

Fig. 1 The Hudson River Estuary. Mile-point (MP) designations refer to statute miles from the southern tip of Manhattan.

during period of low riverine flow. The most diagnostic tracer for this sewage is phosphate. Fig. 2 includes a typical phosphate transect. Simpson et al.[11] have shown that this bowed distribution is maintained by conservative mixing of three components: Hudson River water, New York Bight water and sewage.

Given the magnitude of the anthropogenic source, one would expect any dissolved species having a relatively high concentration in sewage to display a distribution similar to phosphate. Yet dissolved copper and nickel in the Hudson (Fig. 2) have distinctly different distributions. Nickel is distributed much like phosphate with the highest levels occurring in New York Harbor (salinity 10-20 $^{o}/oo$). In fact, the nickel anomaly we observe is the exact effect expected from sewage discharge assuming anthropogenic nickel remains dissolved in the estuary . On the other hand, dissolved copper levels in the Harbor are lower than expected. There are two possible explanations for the differences between these dissolved metal distributions. (i) The concentration of copper in sewage is lower than that of nickel.

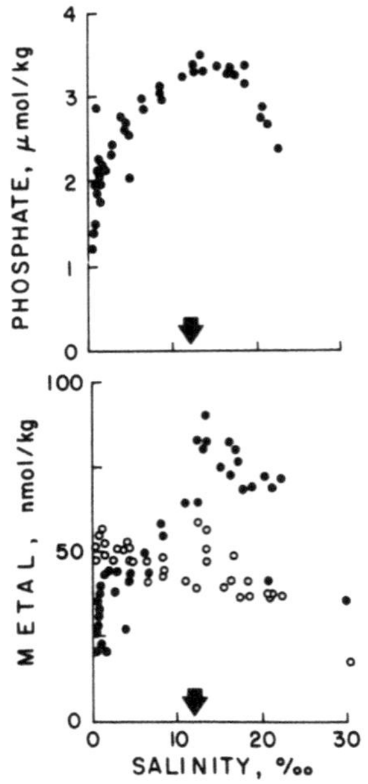

Fig. 2 Phosphate, dissolved copper (○) and dissolved nickel (●) versus salinity in the Hudson River estuary during October, 1975. The arrow indicates the salinity of New York Harbor.

(ii) Anthropogenic nickel is dissolved but anthropogenic copper is associated with particulates.

The first possibility can be eliminated since the total concentrations of copper and nickel in the sewage end-member (Table 1) are nearly identical[12,13]. In fact, modeling confirms that anthropogenic discharges of nickel and copper (Table 1) are equally important in the mass-balance of these elements[14].

Apparently, anthropogenic copper is associated with particulates either initially or within hours after being discharged. This conclusion is supported by metal-aluminium ratios of the suspended material filtered from these samples (Fig. 3). Of the 40 analyzed, only 2 low-salinity samples had copper-aluminium ratios within the range of normal crustal material[15]. The remaining copper ratios were 2-5 times higher than expected from weathering. This observation supports adsorption of anthropogenic copper[7] or discharge of anthropogenic copper as insoluble sulfides[16]. The high copper ratios at low salinities must be maintained by anthropogenic sources upstream of the estuary. Conversely, only 2

Table 1

TOTAL CONCENTRATIONS IN NEW YORK SEWAGE

Copper	1900 nmol kg^{-1}
Nickel	1700 nmol kg^{-1}

COMPONENTS OF THE TOTAL METAL FLUX INTO THE ESTUARY*

October, 1975	Copper	Nickel
River	63	38
Bight	21	34
Sewage	115	102
Total	199	174
Per Cent from Sewage	58%	59%

* mmol s^{-1}

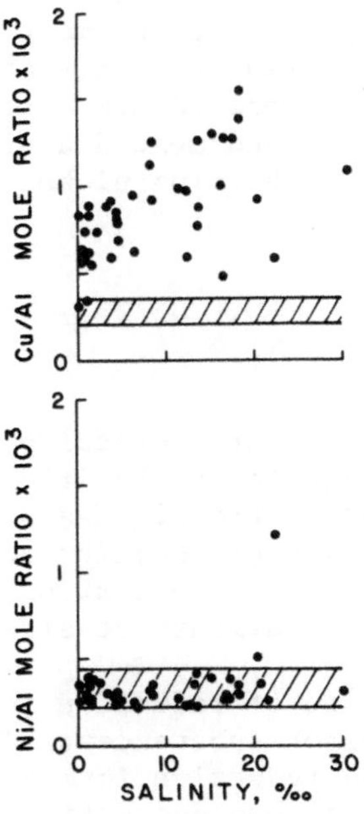

Fig. 3 Cu-Al and Ni-Al mole ratios of suspended material in the Hudson River estuary versus salinity. The hatched areas are ranges expected for normal crustal material.[15]

samples had a nickel-aluminium ratio higher than expected from weathering of normal crustal material. This observation is consistent with both the presence and the magnitude of the dissolved nickel anomaly (Fig. 2).

The association of anthropogenic copper with suspended particulates has global significance. Martin and Meybeck[17] developed a mass-balance with data collected from several major rivers. Based on element-aluminium ratios and the elemental contents of riverine suspended matter, they calculated that several elements (Br, Cu, Mo, Sb and Zn) are unbalanced. That is, these elements are being discharged with riverine particulates at a greater rate than can be accounted for by weathering of average country rock. Martin and Meybeck suggested three possible explanations for this imbalance: 1) worldwide pollution, 2) a non-steady-state condition between weathering and transport or 3) additional natural sources. The results from the Hudson reveal that anthropogenic copper does associate with suspended material and thus pollution may account for the "excess" particulate copper being supplied to the oceans. In fact, the 2.8 enrichment factor required to account for the imbalance calculated by Martin and Meybeck is within the range observed for Hudson particulates (Fig. 3).

While increases in the supply of anthropogenic nickel should increase the seawater concentration, most anthropogenic copper will be deposited in estuarine or nearshore sediments. However, the evidence from pelagic pore waters discussed in the next section reveals that early sediment diagenesis tends to reverse these estuarine effects by burying nickel and remobilizing copper.

PELAGIC PORE WATERS

Introduction

Our laboratory has measured dissolved pore water metal concentrations at four equatorial Pacific sites (Fig. 4). The results from sites M and H were discussed previously[9] in terms of organic diagenesis and mineral equalibria. Of these four sites, the pore waters at site S are the most oxidizing. In terms of reductive intensity, the sediments at site C are intermediate.

A qualitative measure of reductive intensity is revealed in profiles of a few indicator substances. The following trends are suggestive of increasing reductive intensity in pelagic pore waters[18]: (i) increasing then decreasing NO_3^-, (ii) increasing Mn^{2+}, and (iii) increasing Fe^{2+}. The nitrate concentration in

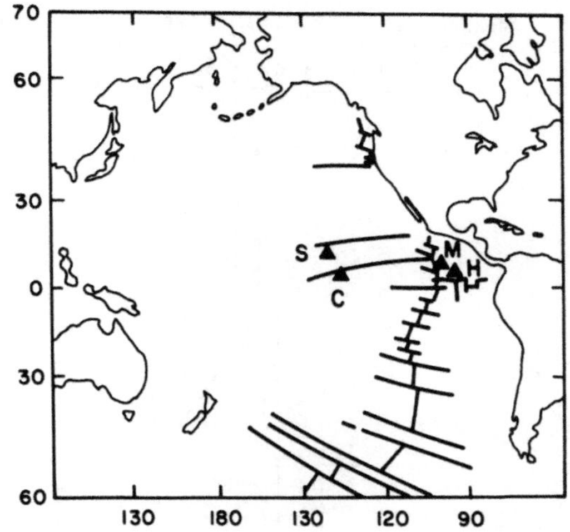

Fig. 4 Pacific pore water study sites.

surficial pore waters at site C is already 10 μmol kg^{-1} above bottom water. Nitrate reaches a maximum of about 51 μmol kg^{-1} at 8 cm but is still 40 μmol kg^{-1} at 30 cm. Mn^{2+} begins to increase below 10 cm (Fig. 5) but dissolved Fe^{2+} is <30 nmol kg^{-1} in the top 30 cm. The profiles of these indicators demonstrate that the pore waters at site C are moderately reducing increasing from oxic-conditions at the interface to suboxic intensity below 15 cm.

In general, the concentration-depth profiles of dissolved copper and nickel at site C (Fig. 5) are similar to results from sites M and H where suboxic conditions were also encountered in the top 30 cm. Nickel concentrations in the oxic-zone are only slightly higher than bottom water but they increase precipitously in the manganese reduction zone. Copper is highest at the sea-sediment interface and decreases systematically deeper in the sediment. While we do not completely understand what maintains these distributions, the observations by themselves carry several important geochemical implications.

Nickel

Nickel behaviour in these pore waters is characterized by two diagenetic zones: the oxic-zone (<15 cm) and the manganese

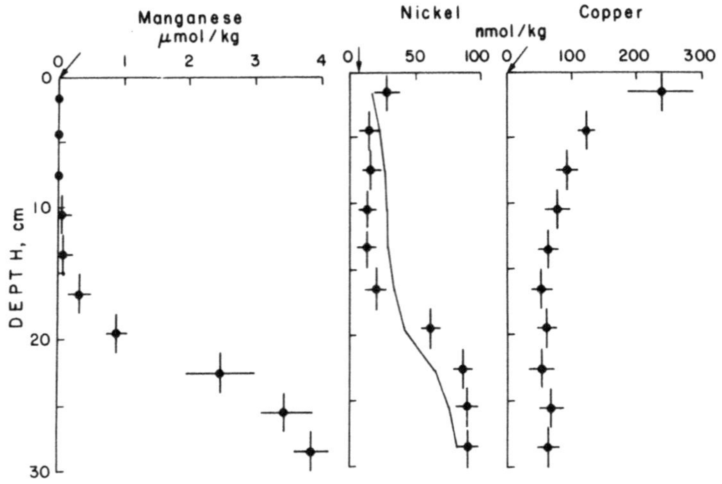

Fig. 5 Dissolved manganese, nickel and copper in pore waters from site C versus depth in the sediment. The arrows are bottom water concentrations of Mn (0.5 nmol kg^{-1}), Ni (6.5 nmol kg^{-1}), and Cu (6.0 nmol kg^{-1}). The curve superimposed on the nickel profile was derived from equation (1).

reduction zone (>15 cm). The simplest model consistent with this zonal distribution involves the degradation of biogenic debris and the reduction of manganese oxides. This stoichiometric model is represented by the following expression.

$$[Ni]_i = \left(\frac{Ni}{NO_3}\right) \frac{D_{NO_3}}{D_{Ni}} [NO3_i - NO3_{i-1}] + \left(\frac{Ni}{Si}\right) \frac{D_{Si}}{D_{Ni}} [Si_i - Si_{i-1}]$$

$$+ \left(\frac{Ni}{Mn}\right) \frac{D_{Mn}}{D_{Ni}} [Mn_i - Mn_{i-1}] + [Ni]_{i-1} \qquad (1)$$

Using this equation the concentration of nickel in interval i can be predicted from coincident changes in nitrate, silica and manganese. The model assumes that dissolved nickel has a simple proportionate relationship with nutrient and manganese levels, being modified only by differences in the rates of diffusion. The first two terms account for the association of nickel with biogenic material. Sclater et al.[3] were the first to point out that nickel in sea water mimics nitrate in the surface ocean and silica in the deep ocean. Equation (1) assumes that these relationships are maintained in pore waters. The D's are molecular diffusion coefficients[19,20]. The ratios of coefficients used for this exercise are $D_{NO_3}/D_{Ni} = 3.0$; $D_{Si}/D_{Ni} = 1.5$ and $D_{Mn}/D_{Ni} = 1.0$. The Ni/NO_3 ratio observed is 6.9×10^{-5} and the Ni/Si regeneration ratio is 3.3×10^{-5} from seawater results[3,5]. The third term in equation (1) is only significant in the manganese reduction zone. $[Mn_i - Mn_i\ \ ]$ is the observed difference in dissolved manganese and Ni/Mn is the ratio in reactive manganese oxides. From previous pore water observations[9], this ratio is about 1.4×10^{-2}.

The dissolved nickel profile predicted from equation (1) is the curve in Fig. 5. While the model predicts the general shape of the nickel profile there are some inconsistencies in absolute concentration. The model does not explain the surface maximum. If real, the maximum at the sediment interface may be related to high productivity in the euphotic zone in this part of the ocean[21] resulting in increased scavenging from the water column. From 5-8 cm the data agree with predicted results within analytical uncertainties. From 8-15 cm the data are lower than predicted values. Apparently, removal of nickel onto solids in the manganese oxidation zone is more important than diagenetic remobilization. The regeneration of nickel in the manganese reduction zone at site C is somewhat greater than previously observed.

Clearly, nickel regeneration in pelagic sediments is accomplished mainly by reduction of manganese oxides. But are these oxides major carriers of nickel to pelagic sediments? The Ni/Mn mole ratio of reducible manganese oxides[9] is about 1.4×10^{-2}. The accumulation rate of total manganese in pelagic sediments is 15 μmol cm^{-2} 10^{-3} yr^{-1} (22). Since at least 50% of this manganese is reducible[18,23], authigenic manganese phases carry about 10 μmol Ni cm^{-2} 10^{-3} yr^{-1} to marine sediments. The total accumulation rate of nickel in these sediments is 40-70 μmol cm^{-2} 10^{-3} yr^{-1} [3] (24). Apparently, as much as 25% of the nickel accumulating in pelagic sediments is associated with authigenic manganese oxides.

Copper

Remobilization in surficial pelagic sediments is the most important source of the copper in the deep ocean[4,9,25]. The dissolved copper maximum of 237 ± 49 nmol kg^{-1} measured at site C is 5 times greater than corresponding levels observed at the three other sites (Fig. 4). Presumably, increased scavenging from the water column related to higher productivity[21] in the euphotic zone accounts for this difference.

Three distinct trends are clear from the distribution of interstitial copper with depth (Fig. 5). Below the surface maximum, copper decreases exponentially to 16 cm. In the bottom of the core (the manganese reduction zone) copper concentrations are constant at about 60 nmol kg^{-1}.

Assuming a linear concentration gradient and a diffusion coefficient of 1.4 x $10^{-6} cm^2\ s^{-1}$ [9] the copper spike at the interface would deliver at least 6.6 μmol $cm^{-2}\ 10^{-3}\ yr^{-1}$ to overlying sea water. This flux is 35 times larger than the accumulation rate of copper in these sediments [25]. This observation suggests that more than 90 per cent of this regenerated copper is returned to sea water. Assuming steady-state, regeneration on this scale requires a large reservoir of solid, reactive copper. As the following calculation demonstrates this reservoir is not the bulk sediment. The average copper content of the top 2 cm of sediment at site C is 1.1 mmol kg^{-1} [25]. Assuming a bulk dry density per wet volume of 0.7 g cm^{-3}, each cm^3 contains only 770 nmoles of copper. The minimum flux calculated above would deplete this entire inventory of copper in 120 years. This result presents a paradox since this much sediment takes 4,200 years to accumulate [25]. This apparent discrepancy must mean that these sediments are coated with a thin layer of copper-rich material. The following discussion is an attempt to constrain the thickness and copper content of this "enriched layer".

A lower limit for the copper content of the enriched layer, $[Cu]_f$, can be estimated with the following equation.

$$[Cu]_f = \frac{F}{B \times S} \tag{2}$$

F is the minimum flux estimate consistent with these data, 6.6 μ mol Cu $cm^{-2}\ 10^{-3}\ yr^{-1}$. B is the bulk dry density per wet volume assumed to be 0.7 g cm^{-3} and S is the sedimentation rate, about 0.24 cm $10^{-3}\ yr^{-1}$ [25]. The <u>minimum</u> copper content of this layer necessary to maintain the dissolved copper profile is 39 mmol kg^{-1}. This requirement must conform to the measured total copper concentration of the top 2 cm, 1.1 mmol kg^{-1}. Equation (3) is the mass-balance for this 2 cm slice.

$$\frac{(2-\alpha)}{2}[Cu]_s + \frac{\alpha}{2}[Cu]_f = 1.1 \text{ mmol kg}^{-1} \tag{3}$$

$[Cu]_s$ is the copper content below the enriched layer. $[Cu]_f$ is the concentration in the copper-rich section and α is the thickness of this layer in cm. Equation (3) reduces to equation (4).

$$\alpha = \frac{2.2 - 2[Cu]_s}{[Cu]_f - [Cu]_s} \tag{4}$$

Since $[Cu] \ll [Cu]$, equation (4) can be reduced to the following expression.

$$\alpha \simeq \frac{2.2 - 2[Cu]_s}{[Cu]_f} \tag{5}$$

Given a minimum estimate for $[Cu]_f$ of 39 mmol kg^{-1}, the enriched layer must be less than 0.06 cm thick.

While only rough estimates, these calculations still reveal several interesting facts concerning copper regeneration. (i) The remobilization process must occur very near (<1 cm) the interface. (ii) The "effective" interface must be extremely fluid since >90% of the copper remobilized leaks into the overlying water column. (iii) The sediment interface must be covered by a thin veneer of copper-rich fluff.

CONCLUSIONS

Copper and nickel are partitioned into different phases in both estuarine and marine systems. In estuaries, anthropogenic copper tends to be particulate while nickel remains dissolved. The net result is that nickel discharged from estuaries enters the open ocean freely while most anthropogenic copper is deposited in coastal sediments.

Early diagenesis in marine sediments tends to reverse these estuarine effects. Copper is released in the boundary layer of pelagic sediments and more than 90% of this regenerated copper fluxes to overlying sea water. On the other hand, nickel is stripped from sea water by authigenic, reducible manganese oxides. This nickel is released in the manganese reduction zone and is either captured in the oxidation zone or diffuses into sea water. Since manganese reduction occurs well below the bioturbated layer in most pelagic sediments, most gleaned nickel is effectively removed from sea water.

REFERENCES

1. Mantoura, R.F.C., A. Dickson and J.P. Riley, 1978: The complexation of metals with humic materials in natural waters. Est. Coast. Mar. Sci., 6, 387-408.
2. Takematsu, N., 1979: Incorporation of minor transition metals into marine manganese nodules. J. Oceanogr. Soc. Jpn., 35, 191-198.
3. Sclater, F.R., E. Boyle and J.M. Edmond, 1976: On the marine geochemistry of nickel. Earth Planet. Sci. Lett., 31, 119-128.
4. Boyle, E.A., F.R. Sclater and J.M. Edmond, 1977: The distribution of dissolved copper in the Pacific. Earth Planet. Sci. Lett., 37, 38-54.
5. Bruland, K.W., 1980: Oceanographic distributions of cadmium, zinc, nickel, and copper in the North Pacific. Earth Planet. Sci. Lett., 47, 176-198.

6. Brewer, P.G. and W.M. Hao 1979: Oceanic elemental scavenging. In: "Chemical Modeling in Aqueous Systems", E.A. Jenne, ed., American Chemical Society, 261-274.
7. Klinkhammer, G., 1977: "The Distribution and Partitioning of Some Trace Metals in the Hudson River Estuary". Msts. Thesis, University of Rhode Island. 125pp.
8. Klinkhammer, G., 1980: Determination of manganese in seawater by flameless atomic absorption spectrometry after pre-concentration with 8-hydroxyquinoline in chloroform. Anal. Chem., 52, 117-120.
9. Klinkhammer, G., 1980: Early diagenesis in sediments from the eastern equatorial Pacific, II. Pore water metal results. Earth Planet. Sci. Lett., 49, 81-101.
10. Mangum, B.J. and J.M. Edmond, 1979: Trace metal profiles from MANOP sites M, H, C, and S. EOS, 60, 858.
11. Simpson, H.J., D.E. Hammond, B.L. Deck and S.C. Williams, 1975: Nutrient budgets in the Hudson River Estuary. In: "Marine Chemistry in the Coastal Environment", T.M. Church, ed., American Chemical Society Symposium Series, No. 18, 616-635.
12. Interstate Sanitation Commission, 1972: Report of the ISC of New York, New Jersey and Connecticut.
13. Klein, L.A., M. Long, N. Nash and S.L. Kinschner, 1974: Sources of metals in New York City wastewater. Report of the New York Water Pollution Control Association.
14. Klinkhammer, G.P. and M.L. Bender, 1981: Trace metal distributions in the Hudson River estuary. Estuar. Coast. Shelf Sci., 12, 629-643.
15. Turekian, K.K. and K.H. Wedepohl, 1981: Distribution of the elements in some major units of the earth's crust. Bull. Geol. Soc. Amer., 72, 175-191.
16. Morel, M.M., J.C. Westall, C.R. O'Mella and J.J. Morgan, 1975: Fate of trace metals in Los Angeles County waste water discharge. Environ. Sci. Tech., 9, 756-761.
17. Martin, J-M. and M. Meybeck, 1979: Elemental mass-balance of material carried by major world rivers. Mar. Chem., 7, 173-206.
18. Froelich, P.N., G.P. Klinkhammer, M.L. Bender, N.A. Luedtke, G.R. Heath, D. Cullen, P. Dauphin, D. Hammond, B. Hartman and V. Maynard, 1979: Early oxidation of organic matter in pelagic sediments of the eastern equatorial Atlantic: suboxic diagenesis. Geochim. Cosmochim, Acta., 43, 1075-1090.
19. Li, Y-H. and S. Gregory, 1974: Diffusion of ions in sea water and in deep-sea sediments. Geochim. Cosmochim. Acta., 38, 703-714.
20. Wollast, R. and R.M. Garrels, 1971: Diffusion coefficient of silica in seawater. Nature Phys. Sci., 229, 94.
21. van Andel, T.H., G.R. Heath and T.C. Moore, Jr., 1975: Cenozoic History and Paleo-oceanography of the Central

Equatorial Pacific Ocean. Geol. Soc. American Mem., 143, 134pp.

22. Bender, M.L., G.P. Klinkhammer and D.W. Spencer, 1977: Manganese in seawater and the marine manganese balance. Deep-Sea Res., 24, 799-812.
23. Sundby, B., N. Silverberg and R. Chesselet, 1981: Pathways to manganese in an open estuarine system. Geochim. Cosmochim. Acta., 45, 293-307.
24. Bostrom, K., T. Kraemer and S. Gartner, 1973: Provenance and accumulation rates of opaline silica, Al, Ti, Fe, Mn, Cu, Ni and Co in Pacific pelagic sediments. Chem. Geol., 11, 123-148.
25. Callender, E. and C.J. Bowser, 1980: Manganese and copper geochemistry of interstitial fluids from manganese nodule-rich pelagic sediments of the northeastern equatorial Pacific Ocean. Amer. J. Sci., 280, 1063-1096.

THE FATE OF PARTICLES AND PARTICLE-REACTIVE TRACE METALS IN COASTAL WATERS: RADIOISOTOPE STUDIES IN MICROCOSMS

Peter H. Santschi, Dennis M. Adler and Michael Amdurer

Lamont-Doherty Geological Observatory of
Columbia University
Palisades, New York 10964
U.S.A.

ABSTRACT

Many trace metals rapidly adsorb to suspended particles when introduced to the water column of coastal marine environments. The transport of these particles and "particle-reactive" trace metals from the water column to the sediments was studied using radiotracer techniques in large microcosm tanks simulating Narragansett Bay.

Transfer rates of radioactive trace metals and plastic particles ("tracer microshperes") to the sediments were much greater in mid-summer than in mid-winter. This coincided with an order-of-magnitude higher particle flux through the water column due to higher sediment resuspension rates in the summer. Laboratory experiments using settling cylinders showed that interaction between the water column and bottom sediments greatly accelerates the deposition velocities of particles. Experiments using a zooplankton cage and sediment traps in the microcosm tanks indicated further that in shallow marine environments zooplankton filter feeding is a much less important transfer process than filter-feeding by benthic organisms which may greatly increase the removal rates of particles and particle-reactive trace metals from the water column during the summer.

INTRODUCTION

It is imperative for us to understand the pathways and rates by which trace metal pollutants, released to coastal waters in soluble or particulate forms, are distributed and retained within the local ecosystem if we are to manage and protect these environments properly. Many trace metals are "particle-reactive"; due to adsorption of hydrolytic species, they tend to partition to suspended solids in seawater. Hence they move through coastal waters, where suspended particles create high surface areas for adsorption, primarily associated with particles. New particles are continually generated through the resuspension of bottom sediment, by coagulation of colloidal material, and by the growth of plankton and bacteria. Sinking material may leave the water column by falling on the bottom while aggregation can increase the settling rate. Furthermore, particles can be filtered and thus be removed from the water column by zooplankton and benthic suspension feeders. These mechanisms of production and removal of suspended particles are sufficiently complicated that it is difficult to study them in the field, let alone in the laboratory.

One way to overcome the complexity and uncertainty in studies of the natural environment is to carry out controlled experiments in large scale synthetic ecosystems (microcosms). Such systems allow manipulation of various physical and chemical parameters of interest while retaining the basic structure and function of the aquatic system under study. The greatest advantage in using microcosms for natural system studies lies in the ability to obtain an accurate material balance. The Marine Ecosystems Research Laboratory (MERL) at Narragansett Bay, Rhode Island, U.S.A., consists of a series of twelve tanks, each 13 m^3 in volume and 5.5 m deep, which simulate the major features of nearby Narragansett Bay. The tanks contain bay water and sediment. They are mixed by a mechanical plunger in a 2 hours on/4 hours off cycle intended to mimic the turbulence and rates of sediment resuspension in Narragansett Bay[1-4]. Nutrients and chlorophyll levels, as well as phytoplankton, zooplankton and benthic organism population densities, species composition and rates of bioturbation are comparable in the bay and tanks. The seasonal cycle observed in the microcosms tanks effectively replicates that found in the bay, even when run in batch rather than flow-through mode[2,4,5] Results from studies in MERL microcosms should also apply to other shallow and well mixed embayments with relatively low tidal velocities such as Buzzards Bay or Long Island Sound.

Radiotracer techniques provide a convenient perturbation- and contamination-free method to trace the movement of substances through natural water systems. Monitoring the decrease in water column activity of radioisotopes added to the MERL tanks, along with their physico-chemical forms in the water column and behavior in the sediments, has been used to assess the rates and mechanisms

Table 1 Comparison of the seasonal removal behavior and distribution coefficients of different groups of elements in the MERL tanks.

Groups of elements	Half-removal times (days) Winter	Summer	Distribution coefficients*) Winter	Summer
Particle-reactive (Fe, Hg, Sn, Cr(III), Pb, Po, Th, Pa, Pu, Am, Be)	10-20	2-3	$(3\text{-}8)10^5$	$(0.5\text{-}2)10^5$
Recyclable (Mn, Co)	30-50	5-30	$1\text{-}5)10^3$	$(0.6\text{-}1)10^3$
Biologically active (Zn, Cd, Se, As)	60-100	10-50	$10^2 - 10^3$	$10^2 - 10^3$
Quasi-conservative (Ba, Ra, Cs, Sb)	150-600	80-150	$10 - 10^2$	$10 - 10^2$

*) Distribution coefficient = activity on particles (dpm g^{-1})/ activity in filtered water (dpm cm^{-3}).

of removal from the water column and mobility in the sediment of the following trace elements: Be, Na, V, Cr, Mn, Fe, Co, Zn, As, Se, Cd, Sn, Sb, Cs, Ba, Hg, Pb, Po, Ra, Th, Pa, Pu, Am[4,6,7] (and unpublished data). Removal and degradation pathways of several polycyclic aromatic hydrocarbons[8,9] have also been studied.

The tracers could be classified in four general groups on the basis of their half-removal times from the water column and chemical forms (Table 1)[10]. The first group, denoted "particle-reactive", is the subject of this paper.

In our experiments we expect the tracer microspheres to emulate the rapid transport of natural particles, and the particle-associated elements to follow the microspheres. It should be noted that particle-reactivity does not imply complete association with particulate material in the water column, but rather the tendency to adsorb on particle surfaces, if given enough time to equilibrate. Particle-reactive elements were removed from the water column with half-removal times of 2-20 days (Table 1)[10]. The other elements investigated were more soluble and therefore had slower removal rates from the water column.

Two natural radionuclides of thorium (^{234}Th, ^{228}Th) are regarded as reasonable analogues to other "particle-reactive" elements[4]. For example, the rate of Th loss from the water column resembles that of Fe, suggesting that iron hydroxide coatings on particles[11] may be a common carrier phase. The thorium isotopes ^{234}Th and ^{228}Th are generated in soluble form by decay of their parents ^{238}U and ^{228}Ra, respectively, both quasi-conservative in seawater. Thorium isotopes therefore, monitor the fate of other elements which are initially added to the water in a soluble form but quickly leave the water column by association with particles. Thorium removal rates can easily be studied by analyzing seawater for its $^{238}U/^{234}Th$ or $^{228}Ra/^{228}Th$ ratio.

An extensive data base of oceanic Th, Ra and U measurements is available on a worldwide scale[12-18]. Our purpose then is to establish a relationship between the removal behavior of Th and that of other trace metals, in order to use Th in the ocean to predict the rates of removal of other trace metals, which are often pollutants in the marine environment. It should be noted, however, that radionuclides are only indicators of the behavior of their stable counterparts of similar physico-chemical form[7,19]. Our approach has been to compare the removal behavior of natural ^{234}Th in Narragansett Bay to that of ^{228}Th, which was added as a spike to the MERL tanks. Both the $^{234}Th/^{238}U$ method (continuous input) in the bay and the ^{228}Th method (spike input) in the MERL tanks gave comparable results. We also compared both methods in the same tank in order to increase the credibility of both the Th method and of the use of microcosm tanks as a tool to study inputs into coastal waters of trace metals added in soluble form.

During the warm summer months, the deposition of Th isotopes and the other particle-reactive trace metals occurred more rapidly than in the winter[4,20]. This result was reproducible two summers in a row in nine tanks, and on duplicate spikes in the winter (see Figure 1, where Fe is used as an example). We report here results from several experiments designed to elucidate the processes which are responsible for this seasonal difference. Even though the particular fraction of these elements and thus their distribution coefficients were higher in the winter than in the summer, removal from the water column was much faster in the summer (Table 1). Something other than differing rates of adsorption onto suspended particles must control the seasonal variation in the transport of particle-associated trace metals. Three such mechanisms have been investigated: (1) the filtering of suspended particles and associated trace elements by organisms such as zooplankton or benthic suspension feeders; (2) aggregation of particles followed by faster settling; (3) impaction of particles i.e. convection or advection and entrainment of particles into the laminar sublayer, and/or adsorption of trace elements at the sediment-water interface. Removal processes of particles and particle-reactive elements which are controlled from the water column were studied by comparing removal rates of radioactively tagged plastic tracer microspheres in zooplankton cages in MERL tanks, and in laboratory

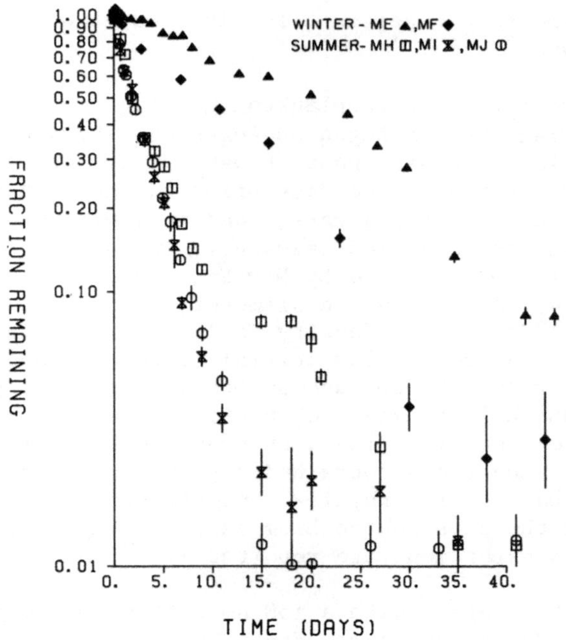

Fig. 1 Removal of ^{59}Fe from the water column of MERL tanks ME (Jan. - Feb. 1979), MF (Feb. - Mar. 1979), MH (anoxic sediments, Aug. - Sept. 1979), MI (Aug. - Sept. 1979), MJ (Aug. - Sept. 1979).

settling cylinders. Benthic removal processes were investigated by comparing the total fluxes from the water column to the tank sediments with the flux which was measured in sediment traps. The settling behavior of natural and artificial particles in the MERL tanks might be complicated by the changing mixing regime, oscillating between 2 hours turbulent and 4 hours stagnant conditions to simulate tidal action. However, the effects of turbulence on the settling velocity of particles can be expected, at the prevailing conditions (the maximum velocity near the plunger is 15 cm sec^{-1}, the Reynolds number for a 15 μm particle is thus <1.5) not to decrease it significantly [see review of this subject in [21]].

EXPERIMENTAL

Th isotopes were measured by alpha (^{228}Th) and beta (^{234}Th) counting of chemically purified fractions of water and sediment samples[4,20]. The test experiment for Th removal was carried out on June 2-3, 1980.

An experiment to test zooplankton grazing as a removal mechanism for trace metals began on August 9, 1979, at the annual temperature maximum for Narragansett Bay, 26°C. The tanks were filled with bay water only two days prior to addition of the radioisotopes. Radioisotopes were added in ionic form in a dilution of one liter of acidified seawater. Two sizes of "tracer micropsheres", 15 μm (supplied by New England Nuclear Company) and 3 m (supplied by 3M Company) in diameter, were used, tagged with ^{141}Ce, ^{57}Co or ^{46}Sc. Their density is 1.3 g cm^{-3}. These artificial particles attained maximum initial concentrations in the well-mixed tank water of 3200 and 50,000 microspheres per liter, respectively, which increased the initial number of suspended particles of the same size class, as measured by Coulter Counter, by at most 50%. The tanks contained a plankton assemblage similar to that of the bay. Chlorophyll-a concentrations in tank MJ were 1.3 μg l^{-1}, and the zooplankton biomass was 33 mg m^{-3}. Details of this zooplankton experiment are reported in [22].

A zooplankton cage[23] with a 150 μm netting was placed for 30 minutes in tank MJ, a few hours after the isotopes were added. The fecal pellets of the trapped zooplankton were collected in a chamber below, subsequently concentrated further by reverse flow[24,25] and in a vertical tube by differential settling. Aliquots of fecal pellets were then enumerate by microscope and analyzed for radioactivity with a Ge(Li) detector for simultaneous detection of the gamma emitting radioisotopes. The experiment showed that the copepods were producing one fecal pellet every 10 minutes. Tank water and wall material samples were routinely sampled and gamma-counted in order to determine removal rates from the water column and adsorption rates on tank walls. Removal from the water column in the first few days appeared to be first order. From the slope of a plot of the logarithm of the concentration versus time, the removal rate constant (λ^{Total}, Table 3) or the half-removal time ($t_{1/2}^{Total} = (\ln 2)/\lambda^{Total}$), can be determined (Figure 2).

Cylindrical sediment traps (with a cross sectional area of 12.5 cm^2 and a height of 12.5 cm) were deployed daily in every experiment to measure particle fluxes. Sediment traps with the given dimensions have been shown to give reasonable estimates of vertical particle fluxes[21,26,27]. New sediments had been transplanted into the tanks several months (tanks MJ, MH, ME) to 1.5 years (tank MI) prior to the spike additions.

Experiments in settling cylinders were carried out in October 1979 and June 1980, at the same time as experiments in MERL tanks. The cylinders are 70 cm tall and 14.5 cm in diameter, containing a

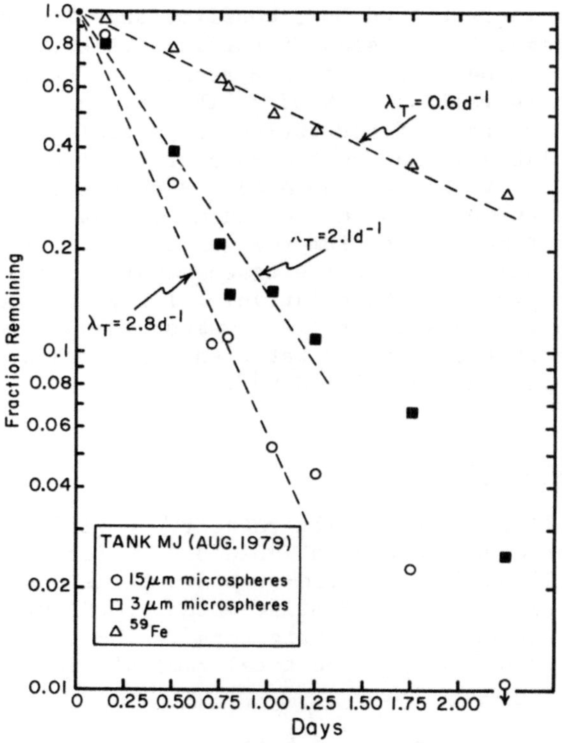

Fig. 2 Removal of tracer microspheres and ^{59}Fe from the water column of tank MJ on August 9-10, 1979.

10 cm^2 paddle (oriented horizontally at 10 cm below the surface) slowly stirring the water at 6 rpm. 15 μm and 3 μm tracer microspheres were added simultaneously to water in the MERL tanks and in the settling cylinders. Water samples were subsequently taken at close intervals to follow removal from the water column.

RESULTS

A similarity in half-removal times of thorium and other particle-reactive trace metals (Fe, Cr(III), Hg, Sn, Pa, Pu, Am) is demonstrated in Table 1. Table 2 shows the half-removal times of ^{234}Th from the water column ($t_{1/2}^{Total}$) in the MERL tanks and in Narragansett Bay and compares these to results obtained from a simultaneous ^{228}Th spike to tank 15. Agreement between half-removal times of natural ^{234}Th and spiked ^{228}Th, and between half-removal times in tanks and bay, confirms that Th can be used to extrapolate removal processes of particle-reactive trace metals in MERL tanks to those occurring in Narragansett Bay. The removal rate of ^{234}Th as well as that of ^{228}Th from the water column is dependent on the rate of adsorption on particles (λ^{ads} = (ln2)/ $t_{1/2}^{ads}$, see Table 2) and on the rate of settling of the particles themselves (λ^{P} = V $\frac{t}{h}$, see Table 4). Both rates are generally faster in the summer than in the winter months[4,20]. The clear seasonal difference in thorium removal rates is reflected also in that of other particle-reactive trace metals. Iron, for example, associated completely with relatively slowly sinking particles in the winter on time scales of 1 - 2 weeks, whereas in the summer, a constant distribution between water and faster sinking particles was obtained in a matter of 1 day[10].

Zooplankton grazing

Fecal pellet incorporation rates (λ^{F} = (ln2)/$t_{1/2}^{F}$) are calculated for each isotope in Table 3. λ^{F} for tracer microspheres as well as for particle-reactive trace metals were found to be only 0.01-0.02 d^{-1}. Similar values (0.02-0.04 d^{-1}) were also found by other MERL investigators[5]. The ratio of the zooplankton filtering rate to the total removal rate of particles and

Table 2 Direct comparison of half-removal times from the water column ($t_{1/2}^{Total}$) or onto particles ($t_{1/2}^{ads}$) for Th in Narragansett Bay (^{234}Th) and MERL tanks (from natural ^{234}Th and artificial ^{228}Th spike) on June 6, 1980 (water temperature = 16.5°C)

System	Particulate matter concentration (mg l^{-1})	$t_{1/2}^{Total}$* (days)	$t_{1/2}^{ads}$** (days)
Narragansett Bay (^{234}Th)	4.1 ± 1.2	4.2 ± 0.3	0.6 ± 0.1
Average of 4 control tanks (^{234}Th)	3.8 ± 0.6	3.0 ± 0.25	1.1 ± 0.2
Tank 15 (^{228}Th spike)	2.9 ± 0.3	3.5 ± 0.4***	-

* $t_{1/2}^{Total} = t_{Th} \cdot R'(1-R')^{-1}$; R' = $^{234}Th/^{238}U$ ratio in unfiltered water; t_{Th} = half life of ^{234}Th.

** $t_{1/2}^{ads} = t_{Th} \circ R(1-R)^{-1}$; R = $^{234}Th/^{238}U$ ratio in filtered water.

*** Here, $t_{1/2}^{Total}$ = (ln2) • $(\lambda^{Total})^{-1}$ was determined from the first order removal rate of ^{228}Th from the water column.

Table 3 Half-removal times for incorporation of radiotracers into fecal pellets ($t_{1/2}^F = (\ln 2)/\lambda^F$) and comparison with the half-removal time from the water column ($t_{1/2}^{Total} = (\ln 2)/\lambda^{Total}$) in MERL tank MJ on August 9, 1979.

	Microspheres 15 μm	Microspheres 3 μm	Trace elements (Cr(III), Sn, Hg, Pa, Fe)
$t_{1/2}^F$ (days)*	30	53	70 - 170
$t_{1/2}^{Total}$ (days)	0.25	0.33	0.7 - 1.4
$t_{1/2}^F/t_{1/2}^{Total}$	120	160	50 - 240

*Incorporation rates of radioisotopes into fecal pellets, (λ_F) were calculated for each isotope [22] using:

$$\lambda_F = Ap \times D \times N/A_T,$$

where Ap = fecal pellet activity (cpm/pellet), D = defecation rate (pellets/animal/day), N = zooplankton population (animals/tank), A_T = total activity in tank (cpm/tank).

particle-reactive trace metals ($\lambda^F/\lambda^{Total} = t_{1/2}^{Total}/t_{1/2}^F$) was only 0.005-0.02. This leads us to conclude that in these tanks, and by extrapolation in Narragansett Bay and other shallow bays, zooplankton grazing does not play a major role in the rapid transfer of trace metals from the water column in the summer.

Tracer microspheres as analogues of natural particles

There may be some doubt as to whether the tracer microspheres do indeed simulate the removal of the natural assemblage of particles. We, therefore, compared the calculated transfer velocities of the tracer microspheres ($V^t = h \cdot \lambda^p$, in md^{-1}), with h = height and λ^p = first order removal rate constant) and the deposition velocity of natural particles (V^d = S/pmc, with S = particle flux, measured in sediment traps, and pmc = particulate matter concentration in the water column) for all experiments in Table 4. If the tracer microspheres settle according to Stokes' law, V^t would increase by a factor of two from winter to summer due to changes in temperature, and therefore, viscosity of the water. Both V^t and V^d, however, increased over more than an order of magnitude from the cold to the warm season. Furthermore, if the natural or plastic particles fall independently through the water column, without any coagulation, or if they are removed by diffe-

Table 4 Comparison of transfer velocities (V^t) of tracer microspheres ($V^t = \lambda^P \cdot h$; λ^P = first order removal rate constant, h = height) with deposition velocities of natural particles V^d caught in sediment traps ($V^d = S \cdot pmc^{-1}$; S = particle flux and pmc = particulate matter concentration.)

Date	Tank	S (mg cm^{-2} d^{-1})	pmc (mg l^{-1})	V^d (m d^{-1})	V^t (m d^{-1}) 15μm	3μm
3/23/78	MA	-	-	-	5.2 ± 0.5	-
6/23 - 8/1/78	MC	5.4 ± 0.6	3.4 ± 0.3	15.7 ± 2.2	9.9 ± 1	-
1/23 - 4/5/79	ME	0.9 ± 0.1	4.3 ± 0.3	2.1 ± 0.3	1.1 ± 0.2	-
2/13 - 3/26/79	MF	0.5 ± 0.07	2.2 ± 0.3	2.5 ± 0.4	-	-
8/9 - 10/9/79	MH	0.8 ± 0.06	2.8 ± 0.3	2.9 ± 0.4	4.3 ± 0.5	1.5 ± 0.2
8/9 - 8/13/79	MI	1.6 ± 0.3	2.2 ± 0.1	7.0 ± 1.6	3.9 ± 0.4	1.1 ± 0.1
8/9 - 9/10/79	MJ	5.7 ± 0.3	1.6 ± 0.2	35 ± 4	12.4 ± 1.2	11.6 ± 1.2
10/9 - 10/16/79	MK	-	-	-	5.8 ± 0.6	-
10/23 - 10/30/79	ML(I)	-	-	-	5.2 ± 0.5	-
6/3 - 6/10/80	ML(II)	4.8 ± 0.1	2.9 ± 0.3	16.4 ± 1.8	19 ± 2	-
6/3 - 6/10/80	MM	12.0 ± 0.9	5.1 ± 0.5	23.5 ± 2.9	-	-

rent mechanisms, then V^t and V^d would not be expected to agree. V^t and V^d values, do however, generally agree to within a factor of two or better, indicating that the plastic particles do indeed trace the movement of natural particles to some degree.

Settling cylinder studies

Since zooplankton grazing is not efficient enough in removing particle-reactive trace metals from the water column, some other aggregation mechanism may be operating on trace metal-laden particles in the water column. In order to examine this, we compared the transfer velocities of tracer microspheres measured in settling cylinders to those measured simultaneously in the tanks. The results of this experiment are listed in Table 5. They indicate clearly that experiments simulating processes in the water column without including the underlying sediments do not adequately reproduce particle removal. While the 15 μm tracer microspheres settled in the settling cylinders at a rate predicted by Stokes' law, they disappeared in the MERL tanks at a rate which was eight times faster (Table 5). The 3 μm particles may have settled slightly faster than predicted by Stokes' law in the settling cylinders, but much less than the 100 fold increase in transfer velocity found in tank MJ. It therefore seems likely that the removal of particles in the MERL tanks is not regulated in the water column but rather is controlled by some benthic processes.

Table 5 Transfer velocities V^t for tracer microspheres (ρ = 1.3 g cm^{-3}) in MERL tanks and in settling cylinders.

Date	System	Water temp (°C)	V^t (m d^{-1})* 15μm	3μm
08/79	Tank MJ	26	13	12
10/79	Tank ML(I)	16	5.2	--
	Settling cylinder	16	--	0.1
06/80	Tank ML(II)	16.5	19	--
	Settling cylinder	16.5	2.5	0.4

* $V^t = \lambda^p \cdot h$, with λ^p = first order removal rate constant and h = height. V^t expected from Stokes' law is 3.1 and 2.5 m d^{-1} for 15μm particles, 0.12 and 0.10 m d^{-1} for 3μm particles at 26 and 16°C, respectively.

Sediment trap studies

Higher sediment resuspension rates during the summer (Table 4) lead to dramatic increase of particle fluxes through the water column. Our experiments show that during the summer the standing stock of particles in the water column is replaced eight times a day by resuspension and deposition, while in the winter only once every two days (Table 4). The increased transfer velocity of microspheres might then be caused by a piggy-back mechanism[28] whereby smaller particles are being scavenged by larger ones with a faster settling velocity. If, accordingly, greater aggregation of particles in the water column results from differential settling then element and particle fluxes in sediment traps should reflect the loss rate from the tank water. If, however, the removal of particles and particle-associated elements occurs at the sediment-water interface (e.g., by benthic suspension feeders, by direct boundary entrainment or adsorption as water is convected to the bottom), the loss rate from the water column would be greater than the "catch" in a sediment trap.

By comparing the trap inventory with the loss rate from the tank, one should be able to distinguiish between these two possibilities, provided the sediment traps catch 100% of the particles falling past them and biodegradation of the trap material is negligible. Release of particle-reactive trace metals from particles settled in sediment traps due to oxidation of organic matter was negligible (a few % only) in tests during different times in the experiment. We tried to minimize this possibility further by limiting the exposure time to 1-2 days. Also, the organic matter content of suspended particles is generally not more than 20%[29]. The first condition can be expected to be met as the aspect ratio (height/diameter) of our cylindrical traps is close to the ideal value as shown in flume tests in the laboratory[26], in the ocean[27] and in lakes[21].

Table 6 lists the ratio (R) of gross sedimentation flux, determined with sediment traps, to overall flux from the water column as measured from the concentration decrease on the first day of the experiment. At the beginning of an experiment little of the radioisotope activity has reached the sediments; thus the capture by sediment traps of resuspended tagged sediments has to be minimal. R is therefore a measure of the trap efficiency. Our best estimate of the trap collection efficiency comes from tank ME (winter, 1979), where removal rates were slow and the benthos was dormant ($R = 1.2 \pm 0.1$). Other tanks with low benthic activity agree with this value. R increases in all tanks after the first day owing to trapping of resuspended tagged sediments (Figure 3). The increase occurs over several days. This is reasonable for the case of eddy diffusional mixing in the sediments. Based on average values of sediment mixing ($D = 0.3 \times 10^{-7}$ and 3×10^{-7} $cm^2\ s^{-1}$

Table 6 Ratio of trap inventory/tank loss (R) on the first day of the experiment.

(A) R in tanks with non-active benthos for determination of trap efficiency:

Tank	Date	Tracer microspheres 15μm	Tracer microspheres 3μm	Particle-reactive trace metals (Fe, Cr(III), Hg, Sn)
ME	1/23/79	1.2	-	1.3
MF	2/13/79	-	-	1.0
MH	8/9/80	1.5	0.9	1.2

(B) R in tanks with active benthos (summer) for determination of removal mechanism:

Tank	Date	Tracer microspheres 15μm	Tracer microspheres 3μm	Particle-reactive trace metals (Fe, Cr(III), Hg, Sn)
MC	6/23/78	0.45	-	0.3
MJ	8/9/79	0.43	0.13	0.13
MI	8/9/79	0.99	0.13	0.6 - 0.9

for winter and summer, respectively) and resuspension rates (S = 0.5 and 5 mg cm^{-2} d^{-1} for winter and summer, respectively) in MERL tanks and in Narragansett Bay [30], the vast majority of radioactively-tagged particles that settle to the sediment surface as a layer of at least 15μm thickness (Z), which is the size of the tracer micro- spheres, are mixed into lower sediment layers and thus being diluted before they can be resuspended. (The transfer coefficient or probability for mixing, D/Z is 1-2 orders of magnitude higher than that for resuspension S/δ, with δ = density of the surface sediments ∿0.3g(dry)/cm^3(wet)). The reservoir of resuspendable particles at the sediment surface would thus contain a mixture of tagged and uncontaminated particles. During the summer months, this ratio (R) was significantly less than one for the microspheres (particularly for the smaller 3 μm size) in normal, non-anoxic tanks on the first day of the experiment, and for several radioisotopes for the first week or two (Table 6 and Figure 3). An efficient removal mechanism operating at or near the sediment surface is therefore implied. The rates of filter feeding were estimated in the MERL tanks during the summer months of 1977 [5]. The total filtering capacity of 0.75-1.37 d^{-1} included 0.03 ± 0.01 d^{-1} for zooplankton and 0.13 d^{-1} for wall fouling

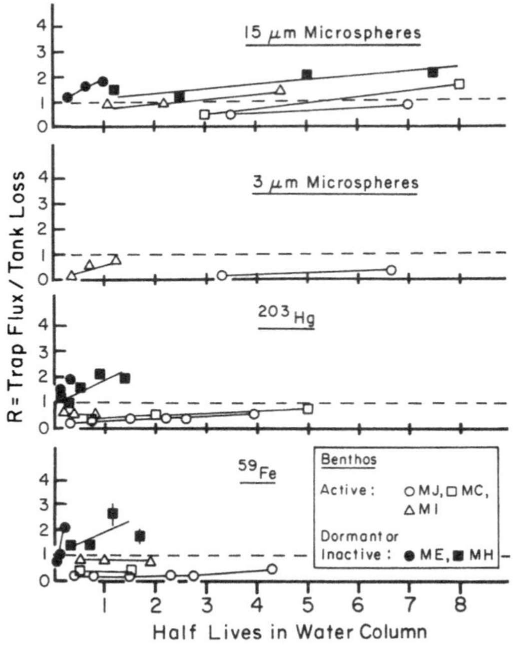

Fig. 3 The ratio (R) of the flux caught in sediment traps to the total removal flux is plotted as a function of time (in units of half-removal times from the water column) for microspheres, ^{203}Hg and ^{59}Fe in different MERL tanks.

organisms, 0.3 d^{-1} for benthic suspension feeders and of 0.6 ± 0.3 d^{-1} for tunicates (which are common in MERL tanks and Narragansett Bay).

A rate constant λ^{Total} ($=\ln 2/t^{Total}$) of 2.8 d^{-1} was calculated for the removal of the 15 μm tracer microspheres during the first days of the experiment in tank MJ (see Table 3) and 0.8 d^{-1} in experiment MI. The difference could be due to different isolation times of bottom sediments from Narragansett Bay (tank MJ: 3 months; tank MI: 1.5 years) which could affect the benthos. For the 3 μm tracer microspheres, the rates were 2.1 and 0.3 d^{-1}, respectively, for both experiments. The rates of disappearance of 3 μm microspheres from these tanks is thus close to the range of estimates of filtering capacities of a MERL tank, while that of the 15 μm size microspheres is somewhat higher. An alternative to filter feeding is entrapment of suspended particles in the laminar sublayer as water is advecting to the bottom. However, one would not expect a large seasonal variation in this removal mechanism. The filter feeding mechanism is therefore preferred.

DISCUSSION

Lowman et al.[31], review the works of a number of investigators reporting on the filtering capacity of benthic biota. Ratios of "biodeposition" (filtering of particles followed by excretion of feces and pseudo feces) to inorganic sedimentation of up to 7 were reported for oyster beds in laboratory experiments[32]. It is thought that mollusks, barnacles and tunicates are even more efficient in removing phytoplankton and detritus from the water by filtering. However, this filtering is almost always size selective. Filter feeders frequently prefer smaller particles (i.e., 2 - 5 μm size) over larger ones, and often organic rich over inorganic ones[32,33]. Our results, which indicate small ratios of trap inventory to tank loss for 3 μm tracer microspheres and particle-reactive elements (and larger R's for 15 μm ones) are consistent with such a mechanism.

The summer tanks MH, MI and MJ, which were spiked on the same day, showed different sediment resuspension and filtering rates, yet exhibited similar removal rates of particle-reactive elements (2 - 8 days half-removal times). It appears therefore that different mechanisms, which compete with one another, exist in the summer for the fast removal of particle-associated elements from the water column. For example, the lack of a benthic filtering capacity in tank MH, which had anoxic sediments at the start of the experiment, was most likely compensated by the rapid adsorption onto the more abundant suspended particles[10] followed by aggregation and settling.

SUMMARY AND CONCLUSIONS

The most important mechanism of transport of particles and particle-reactive trace elements in shallow marine environments is the cycle of resuspension of surface sediments, which provides ample surface area for adsorption, followed by fast removal of the particles either by benthic filter feeding, aggregation/sedimentation and/or impaction at the sediment surface after water is advected to the bottom. It appears that during the summer, processes which introduce and remove particles from the water column are accelerated, as is the rate of adsorption and removal of particle-reactive trace metals such as Th, Fe, Hg, Sn and others. These metals disappear from the water column rapidly by whichever mechanism is most efficient. In winter particle fluxes, rates of adsorption and removal rates of these elements are lowest.

Removal rates of particles are often enhanced by "biosedimentation" caused by filter feeders. In the open ocean, zooplankton are important processors of particles[31,34-38]. However, in shallow marine environments where there is no physical decoupling between surface and bottom waters, their filtering activities appear inconsequential in comparison to those of benthic organisms which may considerably accelerate the removal of particles and particle-reactive trace metals. Benthic organisms might even control the removal of the latter in summer in shallow coastal environments, thus being at greatest risk to be affected by metal pollutants. They might therefore be easily identifiable indicators for pollution[39].

ACKNOWLEDGEMENTS

We are grateful for stimulating discussions with Y.-H. Li and W.S. Broecker, and to J. Bell, K. Hinga, S. Griffiths, B. Evans and K. Kapuchynski for help with sampling and analysis. Also, we wish to thank the staff of MERL for it cooperation.

This work was carried out under contracts from the Environmental Protection Agency (806072020) and the Department of Energy (EY-76-S-02-2185). This is Lamont-Doherty Geological Observatory of Columbia University contribution number 3245.

REFERENCES

1. Pilson, M.E.Q., C.A. Oviatt, G.A. Vargo and S.L. Vargo, 1979: Replicability of MERL microcosms: Initial observations. In: F.S. Jadoff, ed., "Advances in Marine Environmental Research", Proceedings of a Symposium, June 1977, U.S. Environmental Protection Agency, Narragansett, R.I., EPA-600/9-79-05, 359-381.
2. Pilson, M.E.Q., C.A. Oviatt and S.W. Nixon, 1980: Annual nutrient cycles in a marine microcosm. In: J.P. Giesy, ed., "Microcosms in Ecological Research", DOE Symposium Series, Augusta, Georgia, Nov. 8-10, 1978. CONF-781101, National Technical Information Service, 753-778.
3. Nixon, S.W., D. Alonso, M.E.Q. Pilson and B.A. Buckley, 1980: Turbulent mixing in aquatic microcosms. In: J.P. Giesy, ed., "Microcosms in Ecological Research", DOE Symposium Series, Augusta, Georgia, Nov. 8-10, 1978. CONF- 781101, National Technical Information Service, 818-849.
4. Santschi, P.H., D.M. Adler, M. Amdurer, Y.-H. Li and J. Bell, 1980: Thorium isotopes as analogues for "particle-reactive" pollutants in coastal marine environments. Earth Planet. Sci. Lett., 47, 327-335.

5. Elmgren, R., J.F. Grassle, J.P. Grassle, D.R. Heinle, G. Langlois, S.L. Vargo and G.A. Vargo, 1980: Trophic interactions in experimental marine ecosystems perturbed by oil. In: J.P. Giesy, ed., "Microcosms in Ecological Research", DOE Symposium Series, Augusta, Georgia, Nov. 8-10, 1978. CONF-781101, National Technical Information Service, 779-800.

6. Adler, D.M., M. Amdurer and P.H. Santschi, 1980: Metal tracers in two marine microcosms: Sensitivity to scale and configuration. In: J.P. Giesy, ed., "Microcosms in Ecological Research", DOE Symposium Series, Augusta, Georgia, Nov. 8-10, 1978. CONF-781101, National Technical Information Service, 348-368.

7. Amdurer, M., D.M. Adler and P.H. Santschi, 1981: The use of radiotracers in studies of trace metal behavior in microcosms: Advantages and limitations, In: G.D. Grice, ed., "Marine Mesocosms: Biological and Chemical Research in Experimental Ecosystems", Springer Verlag, 81-95.

8. Hinga, K.R., R.F. Lee, J.W. Farrington, M.E.Q. Pilson, K. Tjessem and A.C. Davis, 1980: Biogeochemistry of benzanthracene in an enclosed marine ecosystem. Environ. Sci. Technol., 14, 1136-1143.

9. Lee, R.F., K. Hinga and G. Almquist, 1981: Fate of radiolabeled polycyclic aromatic hydrocarbons and pentachlorophenol in enclosed marine ecosystems. In: G.D. Grice, ed., "Marine Mesocosms: Biological and Chemical Research in Experimental Ecosystems", Springer Verlag, 123-136.

10. Amdurer, M. D.M. Adler and P.H. Santschi, 1982: Studies of the chemical forms of trace elements in sea water using radiotracers. This volume.

11. Aston, S.R. and R. Chester, 1973: The influence of suspended particles on the precipitation of iron in natural waters. Estuarine Coastal Mar. Sci., 1, 225-231.

12. Moore, W.S. and W.M. Sackett, 1964: Uranium and Thorium series inequilibrium in sea water. J. Geophys. Res., 69, 5401-5405.

13. Bhat, S.G., S. Krishnaswami, D. Lal, Rama and W.S. Moore, 1969: ^{234}Th/^{238}U ratios in the ocean. Earth Planet. Sci. Lett., 5, 483-491.

14. Broecker, W.S., A. Kaufman and R.M. Trier, 1973: The residence time of thorium in surface seawater and its implications regarding the fate of reactive pollutants. Earth Planet. Sci. Lett., 20, 35-44.

15. Knauss, K.G., T.L. Ku and W.S. Moore, 1978: Radium and thorium isotopes in the surface waters of the East Pacific and coastal southern California. Earth Planet. Sci. Lett., 39, 235-249.

16. Li, Y.-H., H.W. Feely and P.H. Santschi, 1979: ^{228}Th-^{228}Ra radioactive disequilibrium in the New York Bight and

its implications to coastal pollution. Earth Planet. Sci. Lett., 42, 13-26.

17. Li, Y.-H., H.W. Feely and J.R. Toggweiler, 1980: ^{228}Ra and ^{228}Th concentrations in GEOSECS Atlantic surface waters. Deep-Sea Res., 27A, 545-555.
18. Feely, H.W., G.W. Kipphut, R.M. Trier and C. Kent, 1980: ^{228}Ra and ^{228}Th in coastal waters. Estuar. Coast. Mar. Sci., 11, 179-206.
19. Duursma, E.K. and D. Eisma, 1973: Theoretical, experimental and field studies concerning reactions of radioisotopes with sediments and suspended particles of the sea. Part C: Applications to field studies. Neth. J. Sea Res., 6, 256-324.
20. Santschi, P.H., Y.-H. Li and J. Bell, 1979: Natural radionuclides in the water of Narragansett Bay. Earth Planet. Sci. Lett., 45, 201-213.
21. Bloesch, J. and N.M. Burns, 1980: A critical review of sedimentation trap technique. Schweiz. Z. Hydrol., 42/1, 15-55.
22. Adler, D.M., P.H. Santschi, M. Chervin, K. Hinga and M. Amdurer, 1982: The importance of zooplankton grazing to the removal of metals from shallow estuaries. Submitted for publication.
23. LaRosa, J., 1976: A simple system for recovering zooplanktonic fecal pellets in quantity. Deep-Sea Res., 23, 995-997.
24. Dodson, A.N. and W.H. Thomas, 1964: Concentrating plankton in a gentle fashion. Limnol. Oceanogr., 9, 455-456.
25. Hinga, K.R., P.G. Davis and J. McN. Sieburth, 1979: Enclosed chambers for the convenient reverse flow concentration and selective filtrations of particles. Limnol. Oceanogr., 24(3), 536-540.
26. Gardner, W.D., 1980a: Sediment trap dynamics and calibration: a laboratory evaluation. J. Mar. Res.. 38(1), 17-39.
27. Gardner, W.D., 1980b: Field assessment of sediment traps. J. Mar. Res., 38(1), 41-52.
28. Lal, D., 1980: Comments on some aspects of particulate transport in the oceans. Earth Planet. Sci. Lett., 49, 520-527.
29. Hunt, C.: Graduate School of Oceanography, University of Rhode Island, Kingston, R.I., 02881, U.S.A., Personal communication.
30. Santschi, P.H., S. Carson and Y.-H. Li, 1981: Natural radionuclides as tracers for geochemical processes in MERL mesocosms and Narragansett Bay, In: G.D. Grice, ed., "Marine Mesocosms: Biological and Chemical Research in Experimental Ecosystems", Springer Verlag, 97-110.
31. Lowman, F.G., T.R. Rice and F.A. Richards, 1971: Accumulation and redistribution of radionuclides by marine organisms. In: "National Academy of Sciences, Panel on Radioactivity in the Marine Environment, NRC:, 161-199.

32. Haven, D.S. and R. Morales-Alamo, 1966: Aspects of biodeposition by oysters and other invertebrate filter feeders. Limnol. Oceanogr., 11, 487-498.
33. Jorgensen, C. and E.D. Goldberg, 1953: Particle filtration in some ascidians and lamellibranchs. Biol. Bull., 105, 477-489.
34. Osterberg, C., A.G. Carey and H. Curl, 1963: Acceleration of sinking rates of radionuclides in the ocean. Nature, 200, 1276-1277.
35. Higgo, J.J.W., R.D. Cherry, M. Heyraud and S.W. Fowler, 1977: Rapid removal of plutonium from the oceanic surface layer by zooplankton fecal pellets. Nature, London, 266, 623-624.
36. Beasley, T.M., M. Heyraud, J.J.W. Higgo, R.D. Cherry and S.W. Fowler, 1978: ^{210}Po and ^{210}Pb in zooplankton fecal pellets. Marine Biology, 44, 325-328.
37. Higgo, J.J.W., R.D. Cherry, M. Heyraud, S.W. Fowler and T.M. Beasley, 1980: Vertical oceanic transport of alpha-radioactive nuclides by zooplankton fecal pellets. Natural Radiation Environment III, CONF-780422, Vol. I, NTIS, U.S. Dept. of Energy, 502-513.
38. Fowler, S.W., G. Benayoun and L.F. Small, 1981: Experimental studies on feeding, growth and assimilation in a Mediterranean Euphausiid. Thalassia Jugosl., 7, 47-55.
39. Turekian, K.K., J.K. Cochran, L.K. Benninger and R.C. Aller, 1980: The sources and sinks of nuclides in Long Island Sound. Advances in Geophysics, 22, 129-164.

TRACE METALS IN A LANDLOCKED INTERMITTENTLY ANOXIC BASIN

Michael J. Scoullos

Department of Chemistry, Inorganic Chemistry Labs
University of Athens, 13 A Navarinou St.
Athens 144, Greece

ABSTRACT

Dissolved and particulate trace metals were studied in seawater samples collected regularly from all depths of the almost totally enclosed Gulf of Elefsis (in the northern corner of the Saronikos Gulf, near Athens) polluted by significant amounts of metals through industrial discharges, sewage and ships. The highest metal levels - several times higher than those of the Aegean Sea and similar to those found in known polluted regions elsewhere - were found repeatedly at the N. and N.E. sections of the Gulf due mainly to direct inputs and runoff.

As a result of its hydrography and the prevailing climatological and meteorological conditions, the Gulf is highly stratified during the summer. In August in particular, no oxygen is present at the lower stratum of the water column and the sediments of the deepest section of the Gulf. It is noteworthy that even within the short summer period and the relatively restricted area where the anoxic conditions prevail, several interesting phenomena take place involving decrease of the concentration of the soluble species of certain metals in favour of their particulate forms (for Cu, Zn, Pb and Cd); removal by precipitation of most of them; dissolution and reprecipitation of other metals (Fe, Mn) either in the anoxic zone or at the oxic-anoxic interface. A mobilization of Fe from the surface (top 5-10 cm) of the poorly compacted sea bottom sediment column, into the seawater (in spite of a general inverse tendency for the rest, oxic part of the Gulf) was observed.

INTRODUCTION

Coastal areas provide a variety of environments with fundamentally different conditions, capable either to "trap" in a restricted area considerable amounts of metals or to affect drastically their speciation.

A rigid classification of the coastal systems is difficult, since these are highly dependent on the prevailing hydrographic conditions. Undoubtedly one of the most interesting groups is the one which includes landlocked bays, gulfs and fjords. Several of those are connected to an open or adjacent sea by narrow channels and in some cases water circulation is further restricted by shallow sills. Since the hydrographic (and chemical) regimes of these areas are usually influenced only to a minor extent by the circulation patterns of the adjacent sea, they can be classified in a broad group subdivided in a number of categories. Grasshoff[1] distinguished between enclosed basins located in humid or arid zones and fjords where water exchange with the open ocean is restricted by one or more sills at the mouth.

In the present paper an attempt is made to summarize some of the results on trace metals obtained from our studies in an almost enclosed bay, the Gulf (or Bay) of Elefsis, Greece, which, in spite of combining a great number of the characteristics of the aforementioned categories, cannot be classified properly to any of them. The significance of this area and that of the processes taking place therein, can be appreciated by taking into account: its peculiar hydrographic characteristics resulting in the development of intermittently anoxic conditions in the bottom; its proximity to the industrial area of Elefsis, the city of Athens, the port of Piraeus and the Keratsini Bay where the Athens sewage outfalls are located.

The Area Studied

The Gulf of Elefsis is located in the northern part of the Saronikos Gulf. It is a small (ca 68 km^2) and shallow (max depth 33 m) embayment (Fig. 1) connected to the Saronikos Gulf and the rest of the Aegean Sea by two narrow channels.

The Gulf of Elefsis receives considerable loads of metal and organic matter through the industrial effluents from some 40 big industries located in its eastern part, near the town and the Port of Elefsis. It has already been studied for a number of chemical, biological and physical parameters[2-9]. Its circulation pattern[9] is thermohaline, mainly of a clockwise direction, restricted by shallow sills at the channels. However, inversion in the circulation pattern and wind driven currents often bring into the Gulf

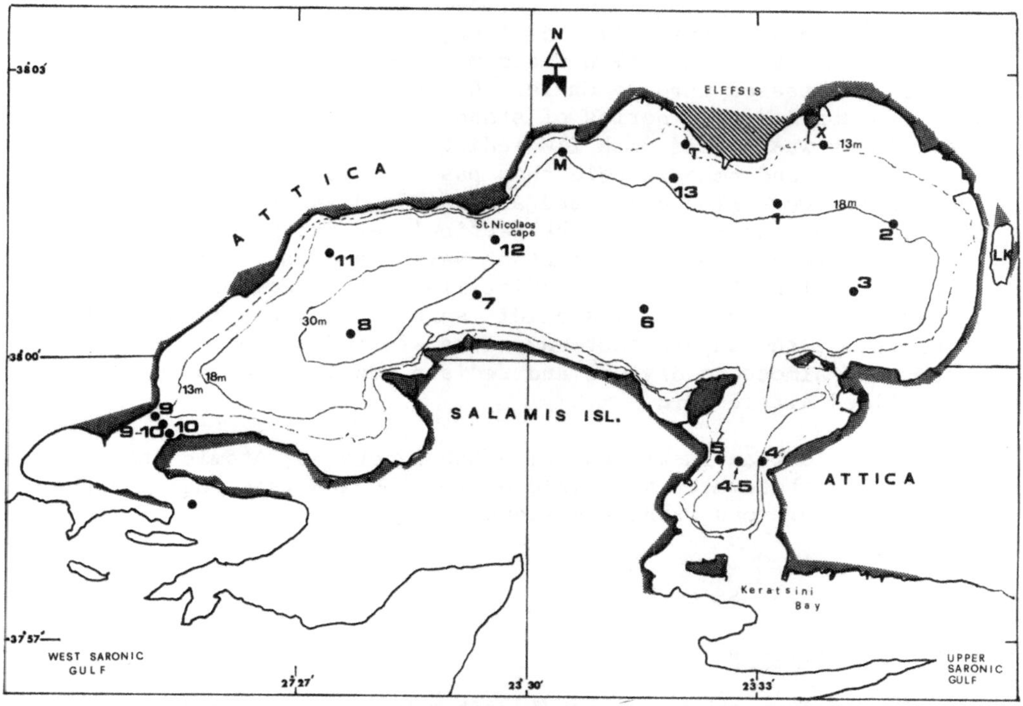

Fig. 1 The bathymetric map of the Gulf of Elefsis showing the grid of stations worked between February 1977 and February 1978.

water from the area of the Athens sewage outfalls, in Keratsini. In a number of instances diverse movement was also observed in the channels. Although the water column is well mixed in winter, it is highly stratified and partly anoxic at the bottom layers of the deeper section during summer.

SAMPLING AND ANALYSIS

Figure 1 shows the permanent stations (number 1-13) and some of the periodical ones (M, T, X) which were sampled over a one year period on the following dates: 17.2.1977, 17.3.1977, 14.4. 1977, 28.5.1977, 5.7.1977, 13.8.1977, 25.9.1977, 16.11.1977, 11.1. 1978 and 25.2.1978 using a plastic-coated vessel with a Diesel motor.

Sea water samples were collected at standard depths of 0 ($\sim$ 0.20), 10, 20 and 30 m using 1.25 l and 7.3 l I.O.S. polypropylene sampling bottles and plastic coated steel wire.

Known volumes (usually ~ 5 l) of sea water samples were filtered through 0.45 μm Nuflow-Oxoid membrane filters and the filtrates were passed through Chelex-100 resin[10]. The same treatment was also applied to a series of standards. The metal retained by the columns was eluted with 2 M redistilled nitric acid. Hydrochloric acid and ammonia were then passed through the columns and the eluates were also collected and analyzed in order to check the efficiency of the columns. The particulates retained on the filters were dried at 50°C to constant weight and they were leached at ca 90°C with 25 ml of 2 M redistilled nitric acid in PTFE containers for 6-12 hours. The acidic solution was centrifuged at 3000 r.p.m., the supernatent was retransferred to the PTFE beaker, evaporated almost to dryness and redissolved in 25 ml of 0.1 M redistilled nitric acid.

A double beam Instrumentation Laboratory 351 Atomic Absorption Spectrophotometer was employed for the analysis, using either an air or a nitrous oxide-acetylene flame.

RESULTS AND DISCUSSION

Area Distributions

Selected monthly surface distributions of dissolved and particulate metals are given in Figure 2. From the distributions it is apparent that quite often the highest concentrations were found at the northern coast (Attica) and mainly in the NE section of the Gulf, where shipyards, steel and iron works, cement, food, chemical and electroplating factories etc. are located.

High concentrations were also found repeatedly at the eastern channel which connects the Gulf of Elefsis to the Keratsini Bay, the rest of the Saronikos Gulf and the Aegean Sea. This is in agreement with the circulation pattern proposed[9], according to which a water mass, rich in metals, originated from the Athens sewage outfall enters the Gulf under favourable meterological conditions, through the eastern channel. Sewage outfalls are responsible in general for high loads of metals, particularly when the road washout is discharged together with industrial effluents and/or domestic raw sewage, as is the case in the Keratsini Bay.

The western part of the Gulf, receiving through the western channel relatively "unimpacted" water from the western Saronikos, has in general lower concentrations and therefore may be considered as a "control" area. However, since this part is also the deepest, the overall picture is different from that of the eastern one. Although the surface concentrations might be low there, the bottom waters are often enriched in metals. Sometimes, under favourable circulation, stimulated by the action of underwater

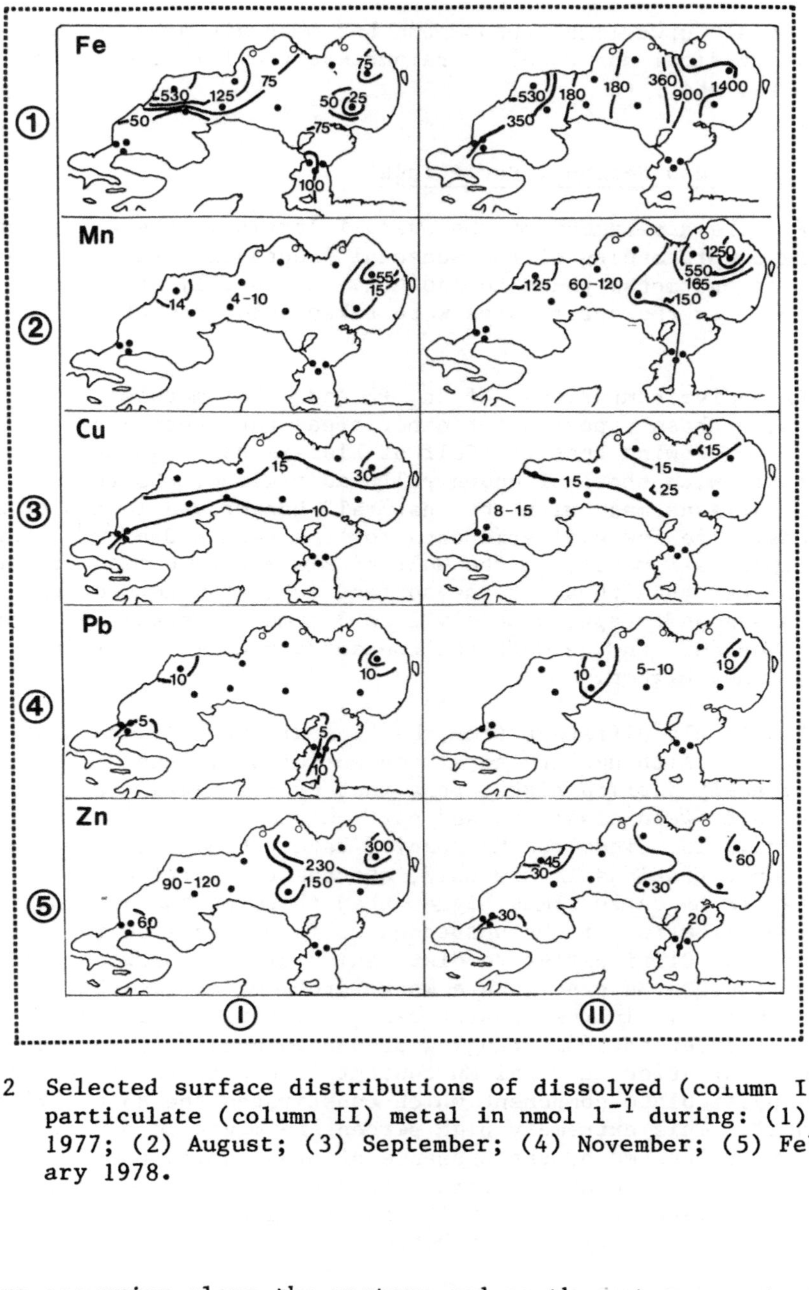

Fig. 2 Selected surface distributions of dissolved (column I) and particulate (column II) metal in nmol l^{-1} during: (1) May 1977; (2) August; (3) September; (4) November; (5) February 1978.

springs occurring along the western and north-western coast, the bottom sea water reaches the surface layers giving, occasionally, patches of higher metal concentration. These phenomena were observed mainly at stations 11 and 12 during several cruises over the studied period[3]. Additional support to this unusual small scale local "upwelling" phenomenon has been given by the patterns of most physical and chemical parameters (salinity, density, nutrients, etc.)[3,5,7].

Comparisons and Seasonal Variations

For the assessment of the overall levels in the area studied and the understanding of the seasonal fluctuation and/or transformation of one metal phase to another, the monthly averaged concentrations over the entire area were taken into account, (see Table 1).

Extensive comparisons of Zn, Pb and other metals' concentrations with those reported for other areas have been given elsewhere[3,7,8] showing that the Gulf of Elefsis has high metal levels comparable with those of known polluted regions. Useful are also the comparisons made with the "natural" background levels of the same sea. The few available data for the Aegean Sea[11-13] show that almost all metals in the Gulf of Elefsis have mean concentrations higher than those of the surface coastal waters of the Eastern Aegean by approximately 2 to 3 times. Certainly, near the "point sources" in the Gulf the concentrations were, often, even twenty times higher.

Distinctly different from the general trend is the distribution of Mn. Although its total concentration in the Gulf is approximately 5 times higher than that of the Aegean Sea, the distribution between dissolved and particulate species is totally different. The dissolved Mn species have, throughout the year, lower concentrations in the Gulf, whereas the particulate Mn species are by ca 20 times higher than those of the Aegean. Anomalous contribution of the suspended solid Mn phase to the total Mn content, and difficulties to distinguish clearly between soluble and particulate Mn species are known problems of the chemistry of this element[14]. In the present case, however, dissolved Mn has very low concentrations, ranging around 20 nmol l^{-1}, so that the seasonal variation in total Mn content is due to the fluctuation of the particulate component which constitutes the 85 to 97% of the total. This extremely high percentage might be partly attributed or influenced by the presence of other solid phases, such as uncompacted sedimentary material, iron hydroxides and organic matter. In fact, Fe in the Gulf has also a high load of particulate species, usually ranging between 82 to 88% of the total.

The high particulate Mn and Fe concentrations are thought to be connected with the redox reactions occuring mainly in the seasonally anoxic part of the Gulf and in the oxic-anoxic interface. This hypothesis is supported by the observation that only these metals show their concentration maxima of the year during the

Table 1. Monthly averaged concentrations of trace metals (Dissolved, Particulate and Total): in the Gulf of Elefsis (in nmol l^{-1}).

Month	Cu			Fe			Mn			Zn			Pb		
	Dis.	Part.	Total	Dis.	Part	Total	Dis.	Part	Total	Dis.	Part	Total	Dis.	Part.	Total
March 1977	35.6	17.6	53.2	120.0	775.3	895.3	25.9	156.6	182.5	374.8	32.1	406.9	9.2	3.4	12.6
April 1977	39.3	17.3	56.6	173.7	465.5	639.2	25.9	145.6	171.5	318.2	33.6	351.8	18.2	8.2	27.0
May 1977	44.8	50.4	95.2	139.7	619.5	759.2	23.7	274.9	298.4	405.4	61.2	466.6	16.9	9.6	26.5
July 1977	30.7	18.9	49.6	128.9	852.3	981.2	10.9	198.4	209.3	446.7	41.3	488.0	12.5	5.6	18.1
August 1977	22.8	14.6	37.4	157.6	1142.3	1299.9	24.6	393.2	417.8	212.6	39.8	252.4	10.1	9.4	19.5
September 1977	22.8	16.2	39.0	148.6	1192.5	1341.1	14.6	504.3	518.9	182.0	48.9	230.9	9.6	7.7	17.3
November 1977	14.6	10.2	24.8	134.3	685.8	820.1	10.9	384.1	395.0	168.3	30.6	198.9	7.2	7.7	14.9
January 1978	16.5	9.6	26.1	150.4	823.6	974.0	11.8	143.8	155.6	272.3	26.0	298.3	14.0	5.5	19.5
February 1978	16.2	9.0	25.2	91.3	508.5	599.8	16.4	134.7	151.1	142.3	26.0	168.3	9.2	8.2	17.4
Mean annual	23.8	14.3	38.1	137.2	784.2	922.1	18.2	258.5	276.7	279.9	36.7	316.6	12.1	7.2	19.3
Aegean Sea coast	11.0	5.5	16.5	96.7	225.6	322.3	57.5	10.0	67.5	85.7	12.2	97.9	5.8	2.1	7.9

August-September period when all the other trace metals show an invariability or even a decrease. All the other metals, had their seasonal peaks during spring, and secondary maxima in January (when Fe had a secondary maximum too) attributable to land wash-out, run-off and effective vertical mixing which redistributes uncompacted muds throughout the water column.

The observed decrease in dissolved and total metal during late spring and summer might be explained by the known accumulation by various organisms, other biological activities and removal of metals to the sediments by formation and/or adsorption on particles.

A Closer Look at the Seasonal Anoxic Regime

Since the Gulf of Elefsis is a basin of maximum depth of 33 m, connected to the rest of the Saronikos by channels having sills shallower than even the depth of its thermocline, the lower part of the stratified water column is "blocked" into the Gulf for a period of, at least, several months. During August the degree of stratification reached a maximum. The uppermost layer (having a temperature of $\sim 27^{o}C$ and very high salinity) extending to a depth of $\sim$15 m, was underlain by an intermediate layer at a depth of 15 to 25 m and by a third layer having a temperature of less than $16^{o}C$, which extended to the bottom. A similar water layering was also present during summer 1973, 1974, 1975 and 1978. The development of the pycnocline and the stratification in the Gulf is clearly shown in Figure 3.

The very high degree of stratification and the decay of the organic material, which falls from the upper photic zone, causes the complete oxygen depletion of the bottom layers at greater depths (25-33 m). This results, in turn, a dramatic increase of ammonia (see Fig. 4) and even in H_2S formation in the very bottom waters of the deeper part of the Gulf, which undoubtedly influence the metal distributions.

The integrated metal concentrations (see also Table 1) showed a decrease for most of the metals, (such as dissolved Cu, Zn, Pb, Cd and particulate Cu, whereas no appreciable changes were noticed for particulate Zn, Cd and Ni; Manganese and Fe show a moderate increase in the concentrations (in nmol l^{-1}) of their dissolved forms and a further considerable one of their particulate ones. Similarly particulate Pb shows a uniform increase throughout the Gulf with no appreciable net increase in the mean total Pb in the Gulf, indicating a transformation of the dissolved to the particulate form. Considering the percentage of the metal in the particulate form, it is apparent that for some metals (eg. Fe, Mn and Cu) it was kept constant throughout summer, whereas for the rest it

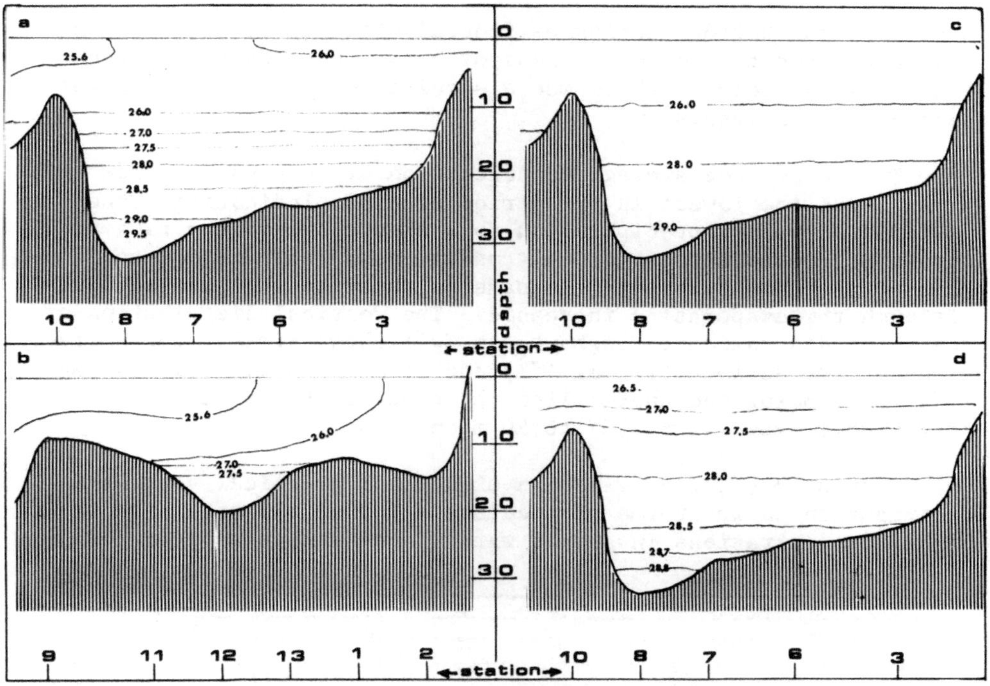

Fig. 3 Stratification in the Gulf of Elefsis during summer: σt profiles for the southern part (a:1977, c:1974, d:1975) and the northern part (b:1977).

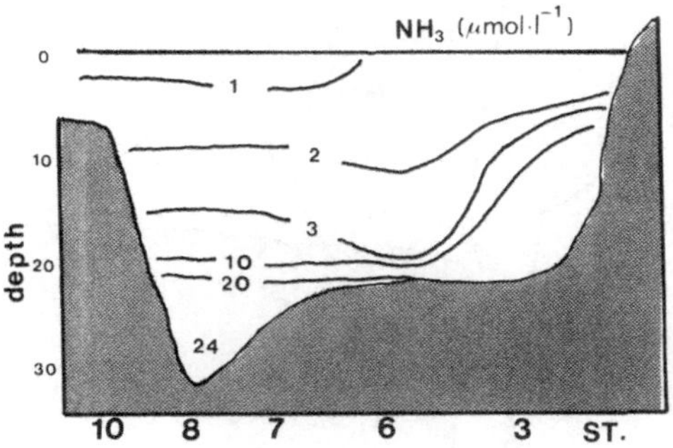

Fig. 4 Cross section of the southern part of the Gulf of Elefsis, showing the vertical distribution of ammonia during August 1977.

increased, a fact which may indicate that metals were removed from the water mass by prior formation of particulates and/or adsorption on suspended solids, which, due to the lack of mixing, sink slowly to the bottom, acting as flocculants. Thus, although in absolute terms the concentrations of both dissolved and particulate forms decreased, there was a simultaneous enrichment of the particles with metals.

The integrated average concentration of suspended solids (2.8 mg l^{-1}) was the lowest in the period studied, indicating rather a maximum removal by precipitation than a decrease in the rate of input, since the meteorological conditions, precipitation and run-off were approximately the same as those during June and July although the evaporation increased. The vertical distribution of the suspended solids throughout the Gulf shows similar concentrations at the surface (2.6 mg l^{-1}) and at a depth of 10 m (2.7 mg l^{-1}), but, below the thermocline the concentrations were considerably higher, viz 4.1 mg l^{-1} at 20 m and 4.5 mg l^{-1} at 30 m.

Station 3 (Fig. 5) gives an example of a typical vertical distribution in which the particulate forms of all metals occur at higher concentrations in deeper water, and their dissolved forms

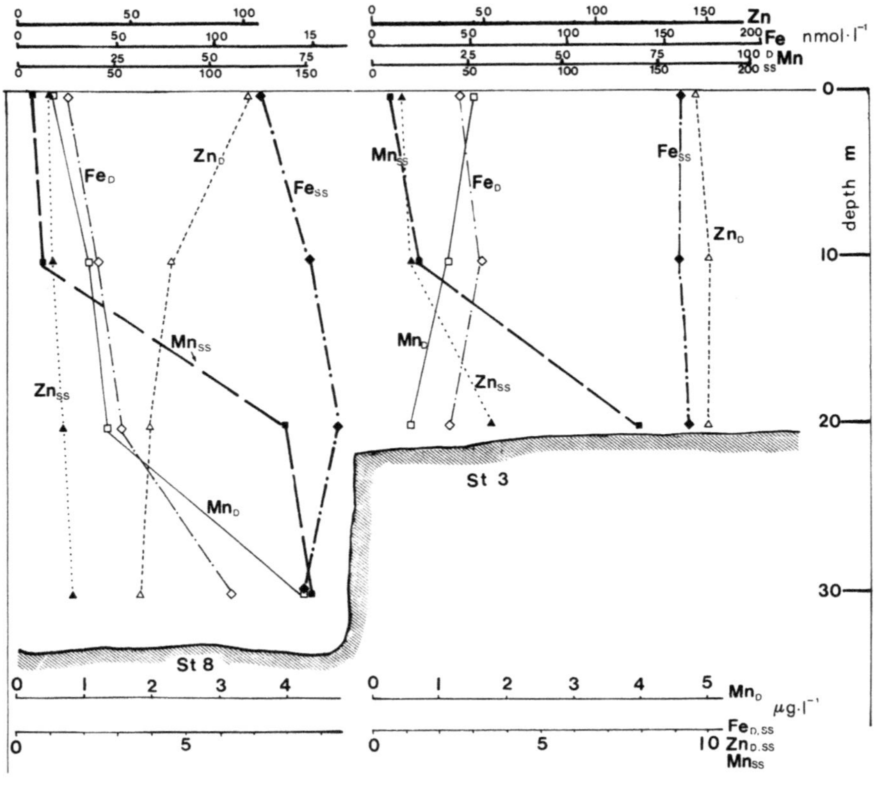

Fig. 5 Typical vertical distributions of Mn, Fe and Zn in selected stations during August 1977.

either do not vary with depth or are present at slightly lower concentrations. This example is typical of a great number of localities, with depths of 20 m or less, which have an anoxic layer only at the very bottom of the water column and/or in the underlying sediments. In this case only Mn(IV), present mainly as MnO_2 in the sediments, is reduced to dissolved Mn(II) which emanates from the sediment and moves slowly upwards. However, because of the presence of appreciable concentrations of oxygen, even below the thermocline, the Mn liberated from the sediments, quickly re-precipitates an MnO_2 at the interface of the oxic and anoxic layers which is very close to the bottom. Thus, in fact this mechanism acts as a chemical process analogous to the physical resuspension of the bottom sediments. Zinc exhibits a similar vertical distribution and this is thought to be mainly attributable not to a similar redox reaction, but rather to the ability of Zn to be easily adsorbed on newly formed oxides [17]. The example of Station 3 is not representative of the entire area and actually the metal profiles of other stations differ slightly from one to another. These differences are due mainly to variations in the topography, the proximity of each station to point sources and the hydrography. Thus at Station 8, representing the deepest part of the Gulf (33 m depth), both dissolved Mn and Fe usually occur at significantly higher concentrations at 30 m, presumably as a result of their dissolution from the bottom deposits. This hypothesis is strongly supported by the vertical distribution of Fe in the sediment cores taken from this area during summer 1978 [3,18], which had distinctly lower iron concentration on the surface than in 20 cm depth, or more (Fig. 6), in spite of the inverse tendency observed in all the other mini-cores throughout the Gulf.

Dissolution of sinking to the bottom Fe particles and formation of soluble iron species (eg. $Fe_3(PO_4)_2$, Fe complexes with organics etc.) is also possible and in agreement with the observed profiles. If any formation of Fe sulphides was taking place in the near bottom layer, this did not balance the particulate Fe dissolution. At this station the O_2-depleted zone was wider and the reprecipitation of MnO_2 took place at the oxic-anoxic water interface, at a depth of 15-20 m, giving rise to a prominent particulate Mn maximum at 20 m. Particulate iron also showed a similar, but less prominent, peak at the same depth. Spencer and Brewer [19] have presented some profiles of trace elements in the Black Sea according to which dissolved Cu and Zn show a decrease below the zero oxygen zone, whereas Ni and Co exhibit an opposite tendency. Although in the Gulf of Elefsis the anoxic layer in the sea water is very narrow and seasonal-and therefore significantly different from that of the Black Sea-the vertical distributions of the above mentioned elements have many similarities in the two regions. The attribution, however, of minor peaks of a certain metal profile to redox reactions must be made with reserve, since anomalies in the vertical distributions might also be produced by

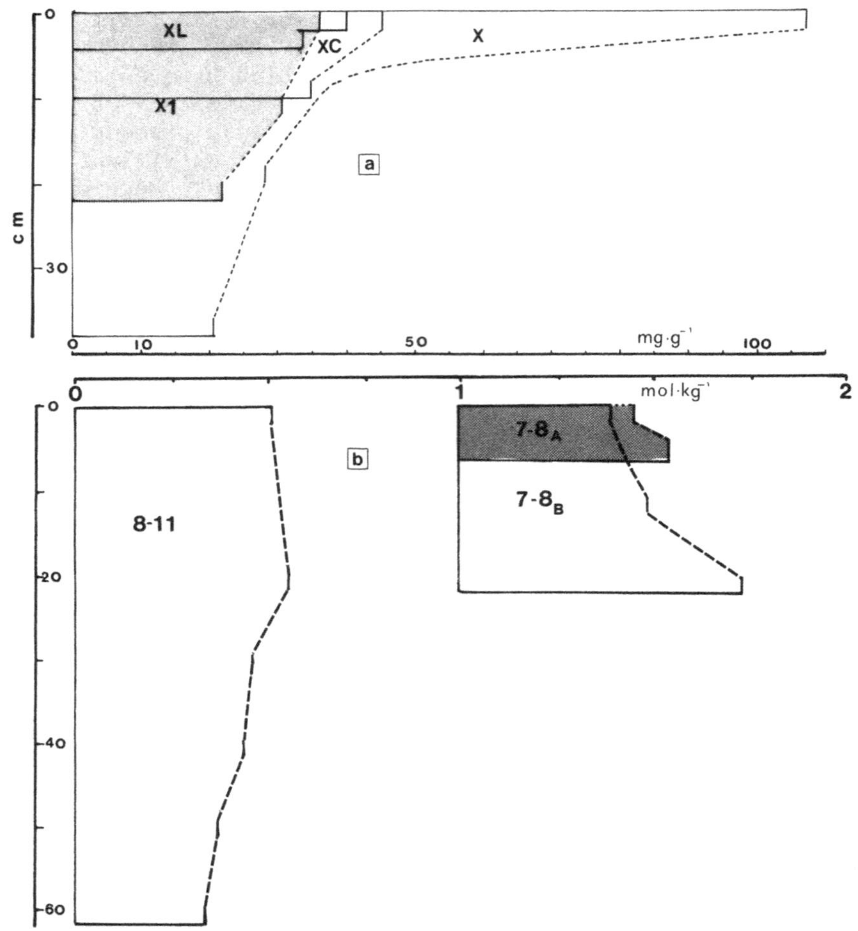

Fig. 6 Vertical distribution of total iron in selected mini-cores of sea bottom sediments of the Gulf of Elefsis taken from its N.E., oxic part, near the industrial site (a); and from the anoxic part (b).

density gradients where particulates are often trapped. Thermoclines with temperature changes of as little as 0.1°C can become centres of major concentrations of suspended matter[20] and consequently of metals.

CONCLUSIONS

The small Gulf of Elefsis is of particular interest because of its hydrographic characteristics and the high loads of metals it receives from the industries and (under favourable meteorological conditions) the sewage outfall area.

The Gulf has trace metals levels similar to those of known polluted regions in other parts of the world and several times higher than those of the open Aegean Sea. Individual concentrations are occasionally extremely high, particularly in the NE section which is directly affected by the industrial discharges and the activities in the port of Elefsis. Biological activity and short period changes in the wind direction are reponsible for dispersing the metals throughout the Gulf.

In the summer, highly stratified waters of its deeper section, oxygen depletion and development of anoxic conditions influence markedly the vertical distribution patterns of most metals.

Iron and Mn show their annual maxima during this period, when both elements seem to emanate from the sediments and reprecipitate in the interface of the oxic-anoxic layers. This phenomenon influences probably the speciation of these elements for the rest of the year. The percentage contribution made by the particulate phases of the above mentioned metals to their total concentration was over 85%. Although most particulate Fe is thought to be authigenic, the good linear correlation between particulate Fe and the magnetic properties of the suspended solids indicates that magnetically active Fe particles which are generated in the industrial site, at the NE coast of the Gulf, constitute a stable component of the particulate Fe population in the studied area.

Most other metals had also higher than usual percentages of particulate species. However more than 85% of Zn was present in dissolved species. Decrease of their soluble forms in favour of the particulate ones and considerable removal through formation of new particles, adsorption and/or coprecipitation, was noticed in the anoxic layer.

For a better understanding of the trace metal hydrochemistry in the area, it will be necessary to couple the dissolved and particulate metal approach to that of the other elements and to the results yielded from the sediments[3,18]. New approaches must be made by use of sediment traps[21], anodic voltammetry[22] and consideration of the metal-humic and fulvic acid complexes[23], useful structural information of which can be produced by comparing them to "model" humic metal systems[24-26].

A careful examination of the sulfur species (bisulfide, polysulfide, thiosulfate, sulfite) using Ag-Ag_2S membrane electrode in combination with $Hg^{(II)-(I)}$ titrations[27] and the study of their possible complexes in the anoxic part of the Gulf will be one of the necessary future steps. Finally the use of radiotracers, isotope ratios, etc. as tracers[28] and the application of the magnetic measurements[4,15,16], will provide useful new tools for the identification of natural and industrial sources, transport and mobilization mechanisms in the Gulf of Elefsis and any other similar coastal system.

ACKNOWLEDGEMENTS

A great part of this work has been done in the Department of Oceanography, the University of Liverpool, where the author worked as a University Fellow. The author would like to thank Professor J.P. Riley for his valuable advice and help.

Thanks are also due to the Director of these Laboratories, Professor D. Katakis and the Faculty of Science, The University of Athens, for the leaves of absence and the facilities provided; Mr. L. Bahas and Miss F. Sakellariadou for their technical assistance; the Ministry of Mercantile Marine - Port Authorities of Elefsis for vessels and crews provided; the Special Marine Sciences Panel, NATO; the Academy of Athens and the Elliniki Etairia (Hellenic Society) for grants.

REFERENCES

1. Grasshoff, K., 1975: The hydrochemistry of landlocked basins and fjords. In: "Chemical Oceanography", Vol. 2, Chapter 15, pp 455-497, J.P. Riley and G. Skirrow, eds. Academic Press, London.
2. Scoullos, M., 1973: Some effects of the pollution on the marine life. Intern. Conference on the Marine Parks of the Mediterranean. Castellabate-Salerno, Italy.
3. Scoullos, M., 1979: Chemical studies of the Gulf of Elefsis, Greece. Ph.D. Thesis, Dept. of Oceanography, The Univ. of Liverpool, Liverpool.
4. Scoullos, M., F. Oldfield and R. Thompson, 1979: Magnetic monitoring of particulate pollution in the Gulf of Elefsis, Greece. Mar. Pollut. Bull., 10, 287-291.
5. Scoullos, M., in press: Oceanographic studies of the Gulf of Elefsis, Greece. Proceedings of the 1st Greek-U.S. working Conf. on Oceanography related to environmental Pollution, Aegina Isl., Greece, 1980.
6. Scoullos, M., 1980: Dissolved and particulate zinc in a polluted Mediterranean bay. Vth ICSEM/UNEP Workshop on Pollution on the Mediterranean, Cagliari, Sardenia Isl., Italy, 483-488.
7. Scoullos, M., 1981: Zinc in seawater and sediments of the Gulf of Elefsis, Greece. Water, Air Soil Pollution, 16, 2, 187-207.

8. Scoullos, M., in press: Dissolved and particulate lead in a polluted coastal environment. Proceedings of the Intern. Symposium on the "Applications of fluid mechanics and heat transfer to energy and environmental problems; Patras, Greece, 1981.
9. Scoullos, M. and J.P. Riley, 1978: Water circulation in the Gulf of Elefsis, Greece. Thalassia, Jugoslavica, 14, 357-370.
10. Riley, J.P. and D. Taylor, 1968: Chelating resins for the concentration of trace elements from seawater and their analytical use in conjunction with atomic absorbtion spectrophotometry. Anal. Chim. Acta., 40, 479-485.
11. Fukai, R. and L. Huynh-Ngoc, 1976: Trace metals in Mediterranean Sea waters. Mar. Pollut. Bull., 7, 9-12.
12. Scoullos, M. and M. Dassenakis, 1980: Dissolved and particulate zinc, copper and lead in surface waters of the Aegean Sea. Report to the Ministry of Coordination, Athens.
13. Scoullos, M., in press: Preliminary report on the trace metal levels of coastal and offshore waters of the Aegean Sea. Proceedings of the Scientific Meeting on Environmental Pollution in the Mediterranean Region, MESAEP, Athens, 1981.
14. Brewer, P.B., 1975: Minor elements in sea water. In: "Chemical Oceanography", Vol. 1, Chapter 7, pp 415-496, J.P. Riley and G. Skirrow, eds. Academic Press, London.
15. Oldfield, F., R. Thompson, M. Scoullos and K. Tolonen, 1979: Magnetic monitoring of particulate pollution. Geophys. J.R. Astr. Soc., 57, 279.
16. Oldfield, F. and M. Scoullos, 1980: Magnetic monitoring of particulate pollution, II. Proceedings of the Balkan Chem. Days, Intern. Conf., Publ. of the Ass. of Greek Chemists, Athens, 124-125.
17. Davis, J.A. and J.O. Leckie, 1978: Effect of adsorbed complexing ligands on trace metal uptake by hydrous oxides. Env. Sci. and Techn., 12, 1309-1315.
18. Scoullos, M., unpublished data.
19. Spencer, D.W. and P. Brewer, 1975: Vertical advection diffusion and redox potentials as controls on the distribution of manganese and other trace metals dissolved in waters of the Black Sea. J. Geophys. Res., 76, 5877-5892.
20. Drake, D.E. and D.S. Gorsline, 1973: Distribution and transport of suspended particulate matter in Hueneme Redondo, NewPort and la Jolla submarine canyons, California. Geological Society of America, Bulletin, 84, (12), 3949-3968.
21. Martin, J.H., in press: Measurement of the flux of particles from surface waters: the use of particle interceptor traps in trace element cycling studies. Proceedings of

the 1st Greek-U.S. working Conf. on Oceanography related to Environmental Pollution, Aegina Isl., Greece, 1980.
22. Nürnberg, H.W., P. Valenta, L. Mart, B. Raspor and L. Sipos, 1976: Applications of polarography and voltammetry to marine and aquatic chemistry II. The polarographic approach to the determination and speciation of toxic trace metals in the marine environment. Z. Anal. Chem., 282, 357-367.
23. Mantoura, R.F.C., 1976: Humic compounds in natural waters and their complexation with metals. Ph.D. Thesis, Dept. of Oceanography, The Univ. of Liverpool, Liverpool.
24. Scoullos, M., 1976: Synthesis and study of complexes of artificial humic acids. Use of them for nitrogen fixation-reduction. D.Sc. Thesis, Dept. of Chemistry, Faculty of Science, University of Athens, Athens.
25. Scoullos, M. and D. Katakis, 1978: Studies on the chemistry of humic acids, I: complexes of Fe (II), Fe (III) and Cu (II) with toluoquinone and oligomeric derivatives. Chimika Chronika, New Series, 7, 65-74.
26. Scoullos, M. and D. Katakis, 1979: Studies on the chemistry of humic acids, II: interaction of Fe (III) and Cu (II) with oxygen containing polymeric ligands. Chimika Chronika, New Series, 3, 169-180.
27. Boulègue, J., J-P. Ciabrini, C. Fouillac, G. Michard and G. Ouzounian, 1979: Field titrations of dissolved sulfur species in anoxic environments - geochemistry of Puzzichello waters (Corsica, France). Chemical Geology, 25, 19-29.
28. Stukas, V.J. and C.S. Wong, 1981: Stable lead isotopes as a tracer in coastal waters. Science, 211, 1424-1427.

THORIUM ISOTOPE DISTRIBUTIONS IN THE EASTERN EQUATORIAL PACIFIC

Michael P. Bacon and Robert F. Anderson

Department of Chemistry
Woods Hole Oceanographic Institution
Woods Hole, Massachusetts 02543
U.S.A.

ABSTRACT

Measurements of the radiogenic thorium isotopes ^{234}Th, ^{230}Th, and ^{228}Th in seawater show that all of the isotopes occur mainly in the dissolved form (< 1.0 μm). The fraction of the Th found in the particulate form increases as the radioactive half-life of the isotope increases. We conclude that chemical scavenging of Th in the deep sea occurs by a reversible exchange process and that the rate of exchange is fast compared with the rate of removal of the particulate matter. As a result the suspended particles are nearly equilibrated with the surrounding water with respect to exchange of Th isotopes at their surfaces. This suggests that chemical equilibrium principles may be applicable to the problem of trace metal scavenging in the deep sea.

INTRODUCTION

Thorium is an especially important trace metal in marine geochemical studies because of the existence of its several radioisotopes having different half-lives (24.1d, 75.2 x 10^3 yr, and 1.91 yr for ^{234}Th, ^{230}Th, and ^{228}Th, respectively). Measurements of $^{234}Th/^{238}U$, $^{230}Th/^{234}U$, and $^{228}Th/^{228}Ra$ radioactive disequilibria in seawater[1,2,3] show that Th is removed from the water column rapidly, its residence time being less than one year in surface waters and less than 100 years in deep waters. It is generally believed that removal occurs by adsorption on particles that sink to the seafloor. This process was referred to by Goldberg[4] as "scavenging", and its importance in controlling the trace metal composition of seawater has been widely recognized[5,6,7].

We determined ^{234}Th, ^{230}Th, and ^{228}Th concentrations, in both dissolved and particulate forms, in seawater samples from the eastern equatorial Pacific. The results indicate that the Th isotopes in the deep ocean are continuously exchanged between seawater and particle surfaces. The estimated rate of exchange is fast compared to the removal rate of the particulate matter, suggesting that the particle surfaces are nearly in equilibrium with respect to the exchange of metals with seawater. A full account of this investigation is being published separately[8]. Here we give a summary of the work and discuss its implications regarding the geochemistry of trace metals in the ocean.

METHODS

Because of the large volumes of water that were required, we used an in situ sampling procedure. Submersible, battery-powered pumping systems[9,10] were used to force the water first through filters (62-μm mesh Nitex followed by 1.0-μm pore-size Nuclepore), then through an adsorber cartridge packed with Nitex netting that was coated with manganese dioxide to scavenge the dissolved Th isotopes, and finally through a flow meter to record the volume of water that was filtered. Natural ^{234}Th served as the tracer for monitoring the efficiency of the adsorber cartridges. Our analytical methods were based on standard radiochemical and nuclear counting techniques[11].

RESULTS

Seventeen seawater samples from five stations in the Panama and Guatemala Basins (Figure 1) were collected with the submersible sampling systems. All of the samples came from depths greater than 1000 m. A full report of the analytical data will be published separately[8]. On the average 4% of the ^{234}Th, 15% of the ^{228}Th and 17% of the ^{230}Th were found in the particulate form; i.e., the percentage increases with increasing radioactive half-life. However, the percentages varied considerably from sample to sample and were found to be strongly dependent on total suspended matter concentration. Figure 2 shows the vertical profiles of dissolved and particulate ^{230}Th.

DISCUSSION

In studying radioactive disequilibrium in seawater, radiochemists usually interpret the data in terms of a scavenging model that assumes irreversible binding of the daughter atoms to particle surfaces followed by slow sinking of the mass of particles to the sea floor[10,12,13]. Surface chemists, on the other hand,

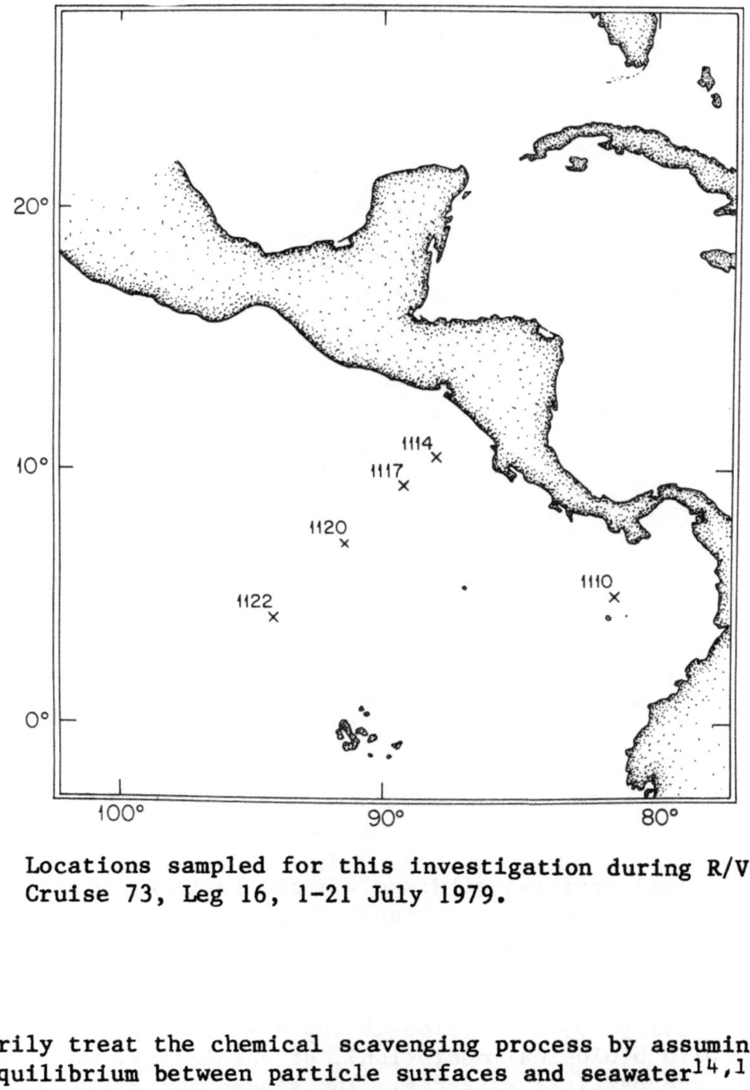

Fig. 1 Locations sampled for this investigation during R/V KNORR Cruise 73, Leg 16, 1-21 July 1979.

ordinarily treat the chemical scavenging process by assuming chemical equilibrium between particle surfaces and seawater[14,15,16]. This implies that reversible exchange occurs. Clearly the two approaches to the problem are not compatible. Our Th isotope results give strong evidence that the latter approach is more reasonable.

One of the immediate discrepancies that became apparent as we completed our first analyses was that, given the amounts of particulate ^{234}Th found, there is too small a fraction of the ^{230}Th in particulate form to be accounted for by an irreversible uptake model. The observed distributions of the different isotopes can be reconciled, however, if an additional removal mechanism for the particulate Th is allowed. We considered that this may occur

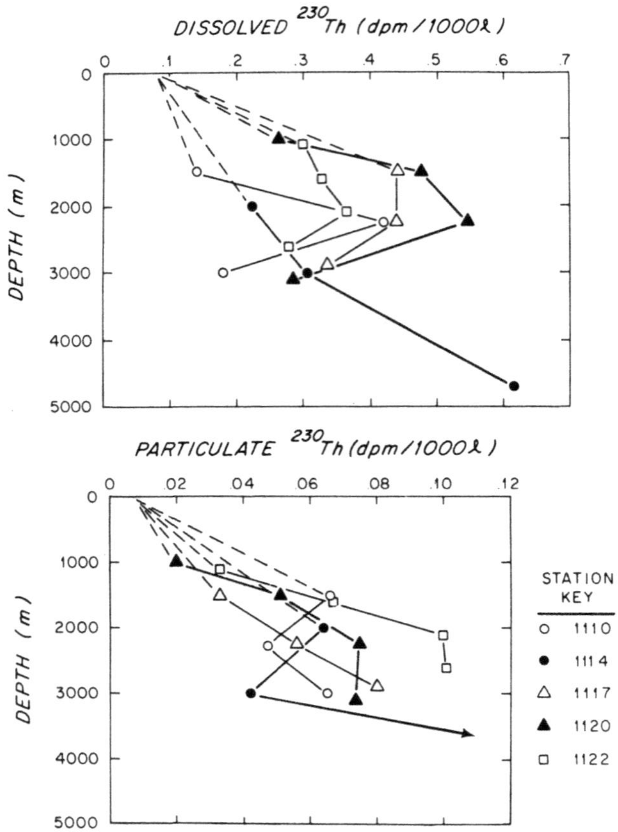

Fig. 2 Vertical profiles of dissolved and particulate ^{230}Th at KNORR 73 stations. Extrapolations to the surface are based on other data[10,30]. 1 dpm = 0.0167 Bq.

either physically, by conversion to rapidly sinking aggregates (possibly with biological mediation), or chemically, by desorption of Th from the particle surfaces. The scheme for production and removal of ^{230}Th from the water column can be summarized as follows:

$$\text{P(dissolved)} \xrightarrow{\lambda} \text{Th(dissolved)} \underset{k_{-1}}{\overset{k_1}{\rightleftarrows}} \text{Th(particulate)} \xrightarrow{k_2} \text{Th(fast particle)} \qquad (1)$$
$$\downarrow S$$

where adsorption, desorption and aggregation are represented by the first-order rate constants k_1, k_{-1}, and k_2. Dissolved Th is supplied to the system by radioactive decay (λ) of the parent nuclide P, and the fine particulate matter settles at a mean speed

S. Radioactive decay causes additional losses of dissolved and particulate Th, its relative importance depending on the half-life of the isotope. The designation "particulate Th" identifies the Th associated with particles collected by filtration, the bulk of which are $\leq 30\ \mu m$ in diameter and are assumed to sink at velocities less than a few hundred meters per year[17,18].

If suspended particles have a long enough residence time in the water column (k_2 and S small), then for any radioisotope that is taken up from solution a steady state is reached in which losses by decay, desorption and aggregation balance the gain by uptake:

$$\frac{dC_p}{dt} = 0 = k_1 C_d - (\lambda + k_{-1} + k_2)\, C_p \tag{2}$$

where C_d and C_p are the concentration of dissolved and particulate Th. Thus, the partitioning of Th between dissolved and particulate forms depends on the half-life of the isotope:

$$\frac{C_p}{C_d} = \frac{k_1}{\lambda + k_{-1} + k_2} \tag{3}$$

Measurements of C_p/C_d for two different isotopes in the same samples allow k_1 and the sum $k_{-1} + k_2$ to be determined. The average values of C_p/C_d for ^{234}Th and ^{230}Th computed from our data set gave $k_1 = 0.52\ yr^{-1}$ and $k_{-1} + k_2 = 2.6\ yr^{-1}$, though individual values of k_1 depended strongly on total particle concentrations (Figure 3).

Consideration of the vertical distributions of the Th isotopes for different values of k_2 showed that only for $k_2 < 0.1\ yr^{-1}$ could the large increases of dissolved and particulate ^{230}Th with depth (Figure 2) be explained. Thus, the most probable pathway for loss of particulate Th is desorption ($k_{-1} \gg k_2$), and many cycles of adsorption and desorption occur before the particles are finally removed from the system. For the long-lived isotope ^{230}Th, we have the further simplification that radioactive decay is negligible, and Equation 3 reduces to $C_p/C_d = k_1/k_{-1}$. Thus, the partitioning is equivalent to an equilibrium distribution with the equilibrium constant $K = k_1/k_{-1}$. The partitioning of the short-lived ^{234}Th, on the other hand, is held at a point removed from equilibrium because of the fast turnover by radioactive decay ($C_p/C_d = k_1/(\lambda + k_{-1})$). Measurement of both isotopes in the same samples, therefore, allows both the kinetics and the equilibrium state of the system to be determined. Since $\lambda = 0$ for stable elements, the implication is that, to a good approximation, the particulate matter suspended in the deep sea may exist in a state

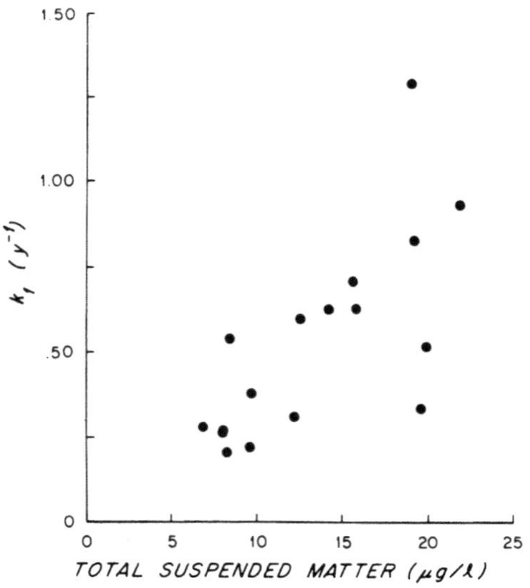

Fig. 3 Plot showing a strong positive correlation between the adsorption rate coefficient for Th (k_1) and total suspended matter concentration.

of equilibrium with respect to exchange of trace metals by adsorption/desorption reactions at the particle surfaces.

According to the equilibrium models [14 15 16], the residence time of any element with respect to removal from the ocean by scavenging is controlled by the equilibrium partitioning of the element between dissolved and adsorbed forms and by the residence time of the particulate matter:

$$\frac{1}{\tau_{Me}} = \frac{1}{\tau_p} \cdot \frac{[Me]_p}{[Me]_t} \qquad (4)$$

where τ_{Me} is the metal residence time, τ_p is the residence time of the particulate matter, and $[Me]_p$ and $[Me]_t$ are the particulate (adsorbed) and total metal concentrations. It is possible to derive expressions that allow the quantity $[Me]_p/[Me]_t$ to be calculated from an appropriate set of equilibrium constants[16]. In principle such an approach would allow estimation of oceanic residence times without the need to measure oceanic concentrations. However, such calculations require that all interactions of the metal ion with ligands in solution as well as with particle surfaces be evaluated.

An alternative procedure is to determine $[Me]_p/[Me]_t$ by direct measurement. In Figure 4 we compare such measurements for Mn, Cu, Pb, Th and Pa with independent estimates of the elemental residence times. The residence times of Pb, Th and Pa are based on radioactive disequilibrium measurements; those of Mn and Cu are based on the fitting of advection-diffusion-model curves to measured vertical profiles of total metal in the water column. For Pb and Pa, which are known to be preferentially removed at ocean boundaries[11,13], values of τ_{Me} representative of midocean areas were selected. According to Equation 4, the data points should lie on a straight line whose slope is inversely proportional to the particle residence time. Despite the fact that the data were drawn from widely separated geographical areas, a linear relationship does appear to be defined by the data. Clearly a more exact treatment is warranted, but the available data suggest that the removal of all five metals from the deep ocean may be controlled by a single population of particles having a residence time of about 5-10 yr in the water column. D.W. Spencer and P.G. Brewer (unpublished manuscript), on the basis of a model describing the concentrations and fluxes of particulate Al, also give a residence time of about 10 yr for the residence time of fine particulate matter in the deep ocean.

In a recent important work on chemical scavenging in the deep ocean, Balistrieri et al.[16] assumed the much shorter particle

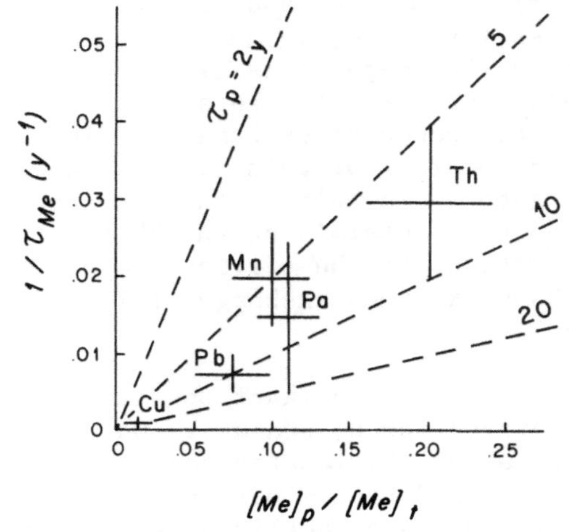

Fig. 4 Plot of removal rate versus fraction of metal in particulate form for five different metals in seawater. The data suggest control of the removal by particles having a five- to ten-year residence time in the water column. Sources of the data are given in reference 8.

residence time of 0.365 yr. We believe this value is unreasonably short. If applied to a particle concentration of 10 μg l^{-1} over a 3800 m water column, it yields a mass flux of about 10 mg cm^{-2} yr^{-1} a value higher than any of those reported from deep-sea sediment-trap deployments. Their value was taken from Brewer et al.[19], who based the estimate on the flux of ^{228}Th measured in sediment traps. Most of the ^{228}Th flux, however, is derived from the upper ocean, where the highest concentrations of ^{228}Ra occur. Since this coincides with the zone of highest biological activity, where particulate matter is efficiently packaged into rapidly sinking aggregates by zooplankton grazing, it is not surprising that most of the ^{228}Th transport should be controlled by particles having relatively high sinking velocities. Several studies have suggested the importance of zooplankton fecal pellets in transporting radionuclides out of the surface ocean[20,21,22,23], and it is now well established that the mass flux of material sedimenting through the water column is dominated by large, rapidly sinking aggregates[17,24,25,26,27]. It does not necessarily follow, however, that chemical reactions at depth in the ocean should be controlled by the same population of particles. Adsorption reactions are governed by particle surface area (e.g., Figure 3), and the vast bulk of the available particles are in the fine, slowly settling fractions[17,18].

We believe that a complete description of chemical processes operating in the deep waters of the ocean must consider not only the effects of the large-particle flux but also the origin and fate of the fine particulate matter. It is likely that the removal of this material (and the adsorbed metals) to the sediments is a complex process involving, in addition to Stokesian settling, interactions with the large-particle flux. Quite possibly there is biological mediation. Recent papers[28,29] contain speculation on the probable significance of exchange of fine particles between the "free" form and the form of being incorporated in large, fast-sinking aggregates. We believe that it should be possible to examine this exchange by determining the Th isotopes in the fine and coarse size fractions of the particulate matter in much the same way that we have examined exchange of Th between dissolved and (fine) particulate form.

We are still unsure of the exact physical significance of the numerical values derived for k_1 and k_{-1}. Our measurements yield estimates of the rates of Th transfer across a somewhat arbitrary boundary, which is defined by the pore size (1.0 μm) of the filters that are used. The rate coefficients that we derive may represent a complex sum of several steps leading to this transfer. Presumably we would infer faster adsorption rates if finer material were included in the particulate fraction. Further work is needed to determine how sensitive the results are to the choice of filter pore size.

If Th atoms are continually absorbed and desorbed as particles fall through the water column, then it follows that dissolved as well as particulate ^{230}Th concentrations should increase with depth. This indeed is what is observed, not only in our profiles (Figure 2), but also in other recent data[30,31]. In fact, the four concentration profiles of Nozaki et al.[31] from the western North Pacific all show, within the uncertainties of the data, linear increases all the way from surface to bottom, exactly as predicted by a reversible exchange model with $k_2 \ll k_{-1}$. Our profiles from the eastern tropical Pacific differ from those of Nozaki et al. and from the model predictions in showing mid-depth maxima and sharp concentration decreases toward the bottom in all but one case (Station 1114). We believe this feature can be explained by an accelerated net uptake of ^{230}Th at the seafloor similar to that shown by ^{210}Pb distributions[13,32]. Such a boundary effect is not included in the models, all of which assume that scavenging is controlled entirely by processes in the water column. The controlling mechanism at the boundary is not known, but the case has been argued that hemipelagic sediments, in which re-mobilized Mn is oxidized and precipitates near the water/sediment interface, act as especially efficient sinks for reactive metals scavenged from the bottom water[33,34]. The fact that our study area is underlain by hemipelagic deposits whereas the area studied by Nozaki et al. is underlain by well oxidized pelagic clay could very well account for the differences between the two sets of profiles.

Thus, we can identify two principal removal pathways for reactive metals scavenged in the deep sea: (1) adsorption and vertical transport by the particle flux and (2) horizontal transport by the circulation and mixing to sinks at ocean boundaries. Those elements that are most strongly bound to the particles (such as Th) are transported mainly by the vertical particulate flux. Elements that are less strongly bound (such as Pb and Pa) undergo considerable horizontal redistribution and may accumulate preferentially in areas such as continental margins, where high particle fluxes and redox cycling of Mn occur[11].

ACKNOWLEDGEMENTS

This research was supported by grants from the National Science foundation (OCE78-26318 and OCE78-25724), a Cottrell Research Grant from the Research Corporation, and a graduate fellowship from the Woods Hole Oceanographic Institution. WHOI Contribution #4918.

REFERENCES

1. Bhat, S.G., S. Krishnaswamy, D. Lal, Rama and W.S. Moore, 1969: Th ^{234}/U^{238}ratios in the ocean. Earth Planet. Sci. Lett., 5, 483-491.
2. Moore, W.S. and W.M. Sackett, 1964: Uranium and thorium series inequilibrium in sea water. J. Geophys. Res., 69, 5401-5405.
3. Moore, W.S., 1969: Measurement of Ra^{228} and Th^{228} in sea water. J. Geophys. Res., 74, 694-704.
4. Goldberg, E.D., 1954: Marine geochemistry 1. Chemical scavengers of the sea. J. Geol., 62, 249-265.
5. Goldschmidt, V.M., 1954: Geochemistry. Clarendon Press, Oxford, 730 pp.
6. Krauskopf, K.B., 1956: Factors controlling the concentration of thirteen rare metals in sea water. Geochim. Cosmochim. Acta., 9, 1-32B.
7. Turekian, K.K., 1977: The fate of metals in the oceans. Geochim. Cosmochim. Acta., 41, 1139-1144.
8. Bacon, M.P. and R.F. Anderson, in press: Distribution of thorium isotopes between dissolved and particulate forms in the deep sea. J. Geophys. Res.
9. Spencer, D.W. and P.L. Sachs, 1970: Some aspects of the distribution, chemistry and mineralogy of suspended matter in the Gulf of Maine. Mar. Geol., 9, 117-136.
10. Krishnaswami, S., D. Lal, B.L.K. Somayajulu, R.F. Weiss and H. Craig, 1976: Large volume in situ filtration of deep Pacific waters: mineralogical and radioisotope studies. Earth Planet. Sci. Lett., 32, 420-429.
11. Anderson, R.F.: The marine geochemistry of thorium and protactinium, Ph.D. dissertation, Massachusetts Institute of Technology/Woods Hole Oceanographic Institution, WHOI-81-1, 287 pp.
12. Craig, H., S. Krishnaswami and B.L.K. Somayajulu, 1973: ^{210}Pb - ^{226}Ra: radioactive disequilibrium in the deep sea. Earth Planet. Sci. Lett., 17, 295-305.
13. Bacon, M.P., D.W. Spencer and P.G. Brewer, 1976: ^{210}Pb/ ^{226}Ra and ^{210}Po/^{210}Pb disequilibria in seawater and suspended particulate matter. Earth Planet. Sci. Lett., 32, 277-296.
14. Schindler, P.W., 1975: Removal of trace metals from the oceans: a zero order model. Thal. Yugoslavica, 11, 101-111.
15. Brewer, P.G. and W.M. Hao, 1979: Oceanic elemental scavenging. In: "Chemical Modeling in Aqueous Systems", E.A. Jenne, ed., ACS Symposium Series, No. 93, pp 261-274, American Chemical Society, Washington, DC.
16. Balistrieri, L., P.G. Brewer and J.W. Murray, 1981: Scavenging residence times of trace metals and surface chemistry of sinking particles in the deep ocean. Deep-Sea Res., 28A, 101-121.

17. McCave, I.N., 1975: Vertical flux of particles in the ocean. Deep-Sea Res., 22, 491-502.
18. Lambert, C.E., C. Jehanno, N. Silverberg, J.C. Brun-Cottan and R. Chesselet, 1981: Log-normal distributions of suspended particles in the open ocean. J. Mar. Res., 39, 77-98.
19. Brewer, P.G., Y. Nozaki, D.W. Spencer and A.P. Fleer, 1980: Sediment trap experiments in the deep North Atlantic: isotopic and elemental fluxes. J. Mar. Res., 38, 703-728.
20. Osterberg, C., A.G. Carey, Jr. and H. Curl, Jr., 1963: Acceleration of sinking rates of radionuclides in the ocean. Nature, 200, 1276-1277.
21. Cherry, R.D., S.W. Fowler, T.M. Beasley and M. Heyraud, 1975: Polonium-210: its vertical oceanic transport by zooplankton metabolic activity. Mar. Chem., 3, 105-110.
22. Higgo, J.J.W., R.D. Cherry, M. Heyraud and S.W. Fowler, 1977: Rapid removal of plutonium from the oceanic surface layer by zooplankton fecal pellets. Nature, 266, 623-624.
23. Beasley, T.M. M. Heyraud, J.J.W. Higgo, R.D. Cherry and S.W. Fowler, 1978: ^{210}Po and ^{210}Pb in zooplankton fecal pellets. Mar. Biol., 44, 325-328.
24. Bishop, J.K.B., J.M. Edmond, D.R. Ketten, M.P. Bacon and W.B. Silker, 1977: The chemistry, biology and vertical flux of particulate matter from the upper 400 m of the equatorial Atlantic Ocean. Deep-Sea Res., 24, 511-548.
25. Honjo, S., 1978: Sedimentation of material in the Sargasso Sea at a 5367 m deep station. J. Mar. Res., 36, 469-492.
26. Honjo, S., 1980: Material fluxes and modes of sedimentation in the mesopelagic and bathypelagic zones. J. Mar. Res., 38, 53-97.
27. Deuser, W.G. and E.H. Ross, 1980: Seasonal change in the flux of organic carbon to the deep Sargasso Sea. Nature, 283, 364-365.
28. Lal, D., 1980: Comments on some aspects of particulate transport in the oceans. Earth Planet. Sci. Lett., 49, 520-527.
29. Tsunogai, S. and M. Minagawa, 1978: Settling model for the removal of insoluble chemical elements in seawater. Geochem. J., 12, 47-56.
30. Moore, W.S., 1981: The thorium isotope content of ocean water. Earth Planet. Sci. Lett., 53, 419-426.
31. Nozaki, Y., Y. Horibe and H. Tsubota, 1981: The water column distributions of thorium isotopes in the western North Pacific. Earth Planet. Sci. Lett., 54, 203-216.
32. Nozaki, Y., K.K. Turekian and K. von Damm, 1980: Pb in GEOSECS water profiles from the North Pacific. Earth Planet. Sci. Lett., 49, 393-400.

33. Bacon, M.P., P.G. Brewer, D.W. Spencer, J.W. Murray and J. Goddard, 1980: Lead-210, polonium-210, manganese and iron in the Cariaco Trench. Deep-Sea Res., 27A, 119-135.
34. Spencer, D.W., M.P. Bacon and P.G. Brewer, 1981: Models of the distribution of ^{210}Pb in a section across the north equatorial Atlantic Ocean. J. Mar. Res., 39, 119-138.

ASPECTS OF THE SURFACE DISTRIBUTIONS OF COPPER, NICKEL, CADMIUM, AND LEAD IN THE NORTH ATLANTIC AND NORTH PACIFIC

Edward Boyle and Sarah Huested

Massachusetts Institute of Technology
Department of Earth and Planetary Sciences
Cambridge, Mass. 02139
U.S.A.

ABSTRACT

The concentrations of copper, nickel, and cadmium have been determined on over 400 surface samples; 10 samples were also analyzed for lead. Nickel and cadmium are closely associated with nutrient distributions, being low (about 2 nmol kg^{-1} for Ni and less than 20 pmol kg^{-1} for Cd) in the central gyres and elevated to 3-3.5 nmol kg^{-1} (Ni) and 40-80 pmol kg^{-1} (Cd) in the equatorial eastern Pacific and in the cool eastern boundary currents. There are large regional variations in their element-nutrient correlations, however. Copper is slightly elevated in cool eastern boundary waters (1.3 nmol kg^{-1}) relative to North Pacific and South Atlantic central gyres, but the Sargasso Sea has higher copper concentrations (1.4 nmol kg^{-1}) than the other central gyres. Copper concentrations in coastal waters of the Gulf of Panama are elevated to 3-4 nmol kg^{-1}; this coastal water advects seaward and creates 'events' of high - Cu waters in the oceanic interior. Lead concentrations in the North Atlantic (100-250 pmol kg^{-1}) are higher than in the North Pacific.

INTRODUCTION

Recent work on the distribution of several trace elements throughout the ocean[1-13] have elucidated the major processes controlling their distributions. It is clear that most of the trace metals exist at much lower concentrations than reported in the earlier literature and also that most are dominated by biological uptake in shallow waters and regeneration at depth from the falling biogenic debris. But trace metal distributions do show distinct differences from the nutrient distributions when examined in detail; this paper will outline some of these differences in surface waters and comment on the possible mechanisms. The data presented here are discussed in greater detail in Boyle, Huested, and Jones [14].

ANALYSIS AND PRECAUTIONS

Since most old trace element data has been discredited as much too high, it is clear that trace element sampling and analysis require more care than nutrient analysis. The response of many laboratories to this problem has been to adopt time-consuming strict ultra-clean room techniques. Our own work, conducted using considerably milder precautions, proves that the extreme approach is not necessary for the elements which we have analyzed: Cu, Ni, Cd, and Pb. The extra expense and time required by the ultra-clean room methods may serve as a hindrance.

Our surface sampling is undertaken by extending a pole (to which a plexiglas bottle holder is attached) over the side of the ship while it is steaming at two knots. The bottles (previously acid leached for one day at 60°C) are capped immediately and kept covered with plastic gloves. They are acidified within a week of collection by adding 2x vycor-distilled 6N HCl to a final pH of 1.8. Our samples stored in this way from a few hours to several years show no changes in the trace metal concentrations with time. 35g samples are analyzed for Cu, Ni, and Cd by a modification of the Co-APDC coprecipitation method of Boyle and Edmond [16]. For Pb, 100 ml samples are analyzed by using Hg^{++} as the coprecipitant, since mercury volatilizes and allows for much lower background absorption. In the lab, sample handling is undertaken in a class 100 laminar flow clean bench; the filters used in this device is the same as used for the (250x more expensive) clean room techniques. The preconcentrated samples are then analyzed by graphite furnace atomic absorption; typical signals for lead are illustrated in Figure 1.

Manganese was determined on a few samples from the Panama Basin by direct injection GFAAS using NH_4NO_3 and ascorbic acid [15] as matrix modifiers.

Precisions are estimated as Cu: $\pm$ 0.1 nmol kg^{-1} or $\pm$5%, whichever is larger; Ni: $\pm$ 0.2 or 5%; Cd: $\pm$ 0.005 or 10%; Pb: $\pm$ 9% at the 10 ng kg^{-1} level; Mn: $\pm$ 1 nmol kg^{-1}.

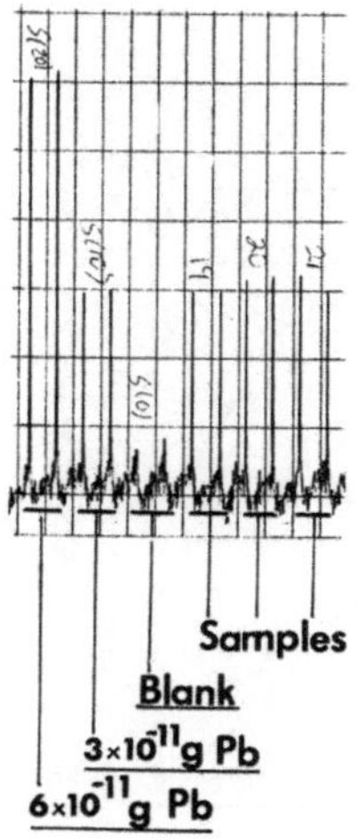

Fig. 1 Lead analysis by flameless AAS. Injection volume was 20 μl; the absolute amounts of lead in this volume are indicated.

DISCUSSION

The copper, nickel, and cadmium data (Figs. 2-14) are from Boyle et al.[14] More detailed discussion of the data is presented there; we will summarize the main points here. The manganese and lead data (Figures 15 and 16) are new.

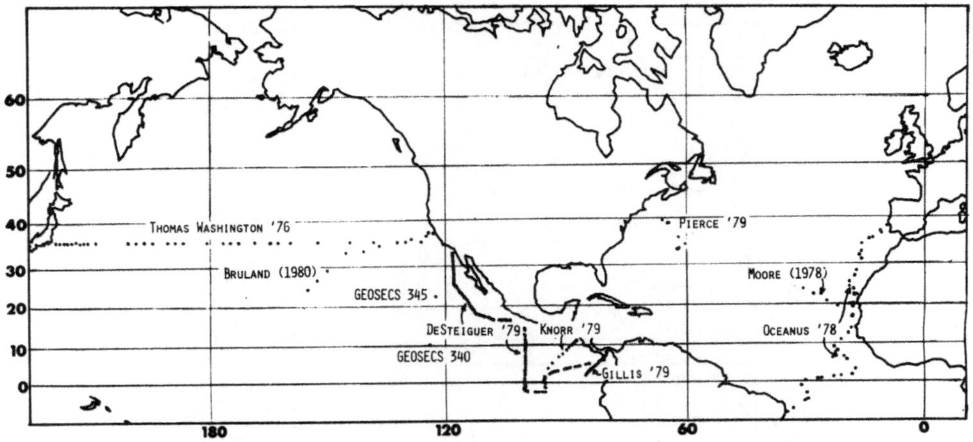

Fig. 2 Surface water sample locations from this study and from Moore (1978) and Bruland (1980).

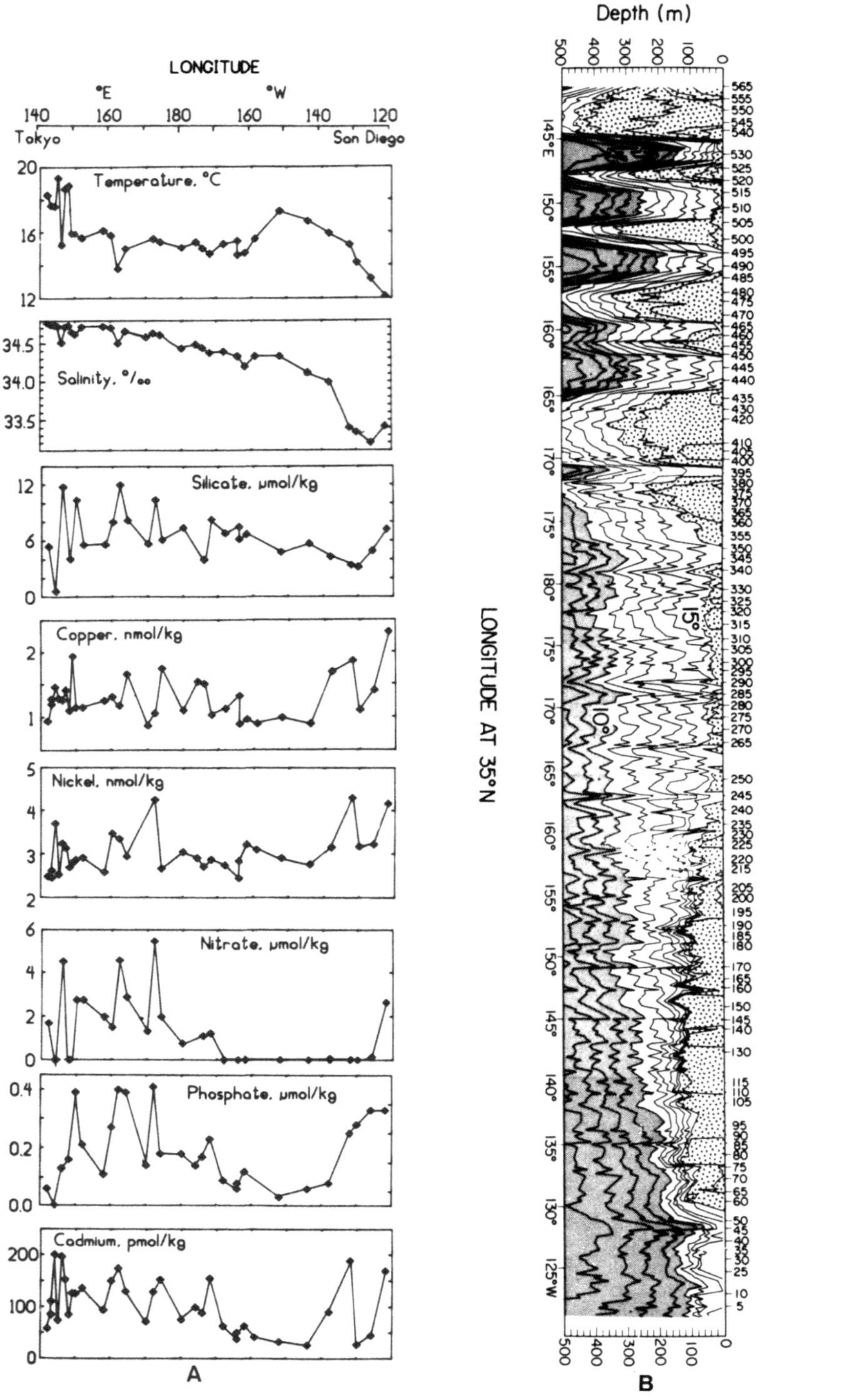

Fig. 3 a) Data from Thomas Washington 1976. b) Transpacific temperature section from this cruise (Kenyon, 1978).

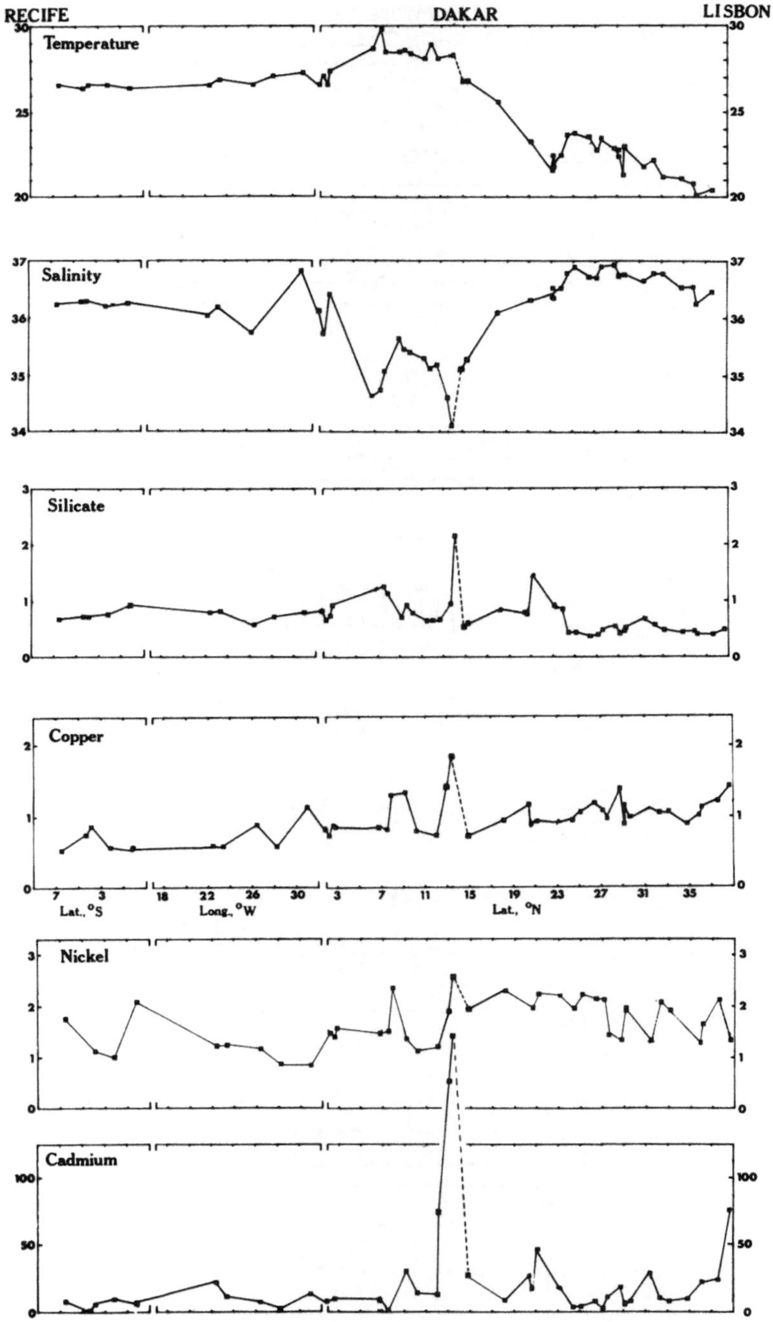

Fig. 4 Data from Oceanus 1978; average concentrations plotted.

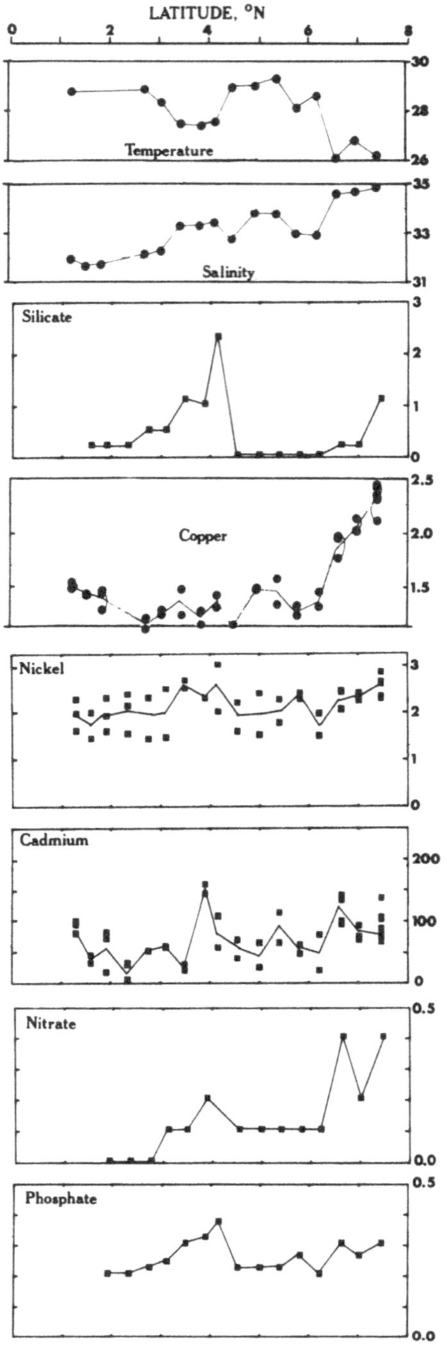

Fig. 5 Data from Gillis 1979; individual analyses plotted.

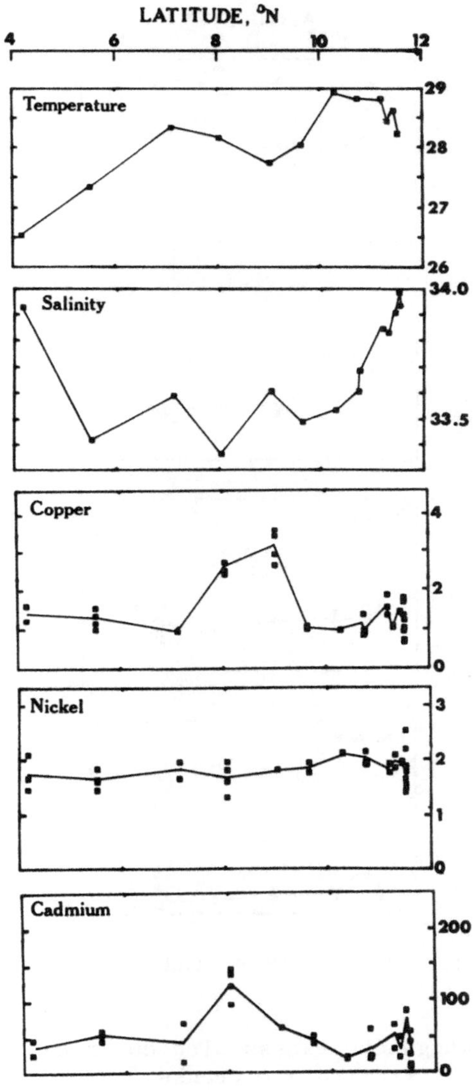

Fig. 6 Data from Knorr 1979; individual analyses plotted.

The major signal in copper is an enrichment (relative to open-ocean concentrations of one to two nmol kg^{-1} in the Gulf of Panama (Figures 5 and 8). Enrichments are also seen in waters at 8° in the Guatemala Basin (Figure 6) and in the waters north of the Gulf Stream (Figure 7). We believe that a possible source of this excess copper is a flux of diagenetically remobilized copper from out of continental shelf sediments. This interpretation is

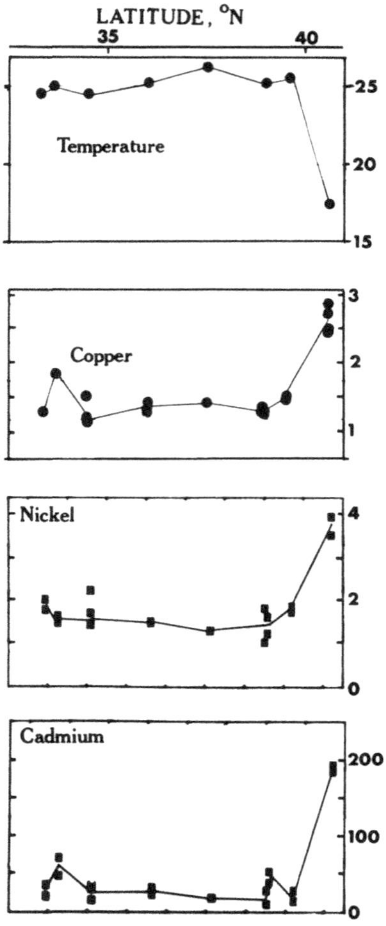

Fig. 7 Data from Pierce 1979; individual analyses plotted.

supported by the manganese excess also observed in the Gulf of Panama (Figure 15). Bruland and Franks (this volume) have interpreted the excess in the waters north of the Gulf Stream as the sum of dissolved riverine sources and continental shelf diagenesis. More detailed studies of trace element profiles over continental shelves are needed to prove the source, however.

The nickel distribution in the surface waters of the ocean is very similar to phosphate, with a slope of 10^{-3} moles Ni/mole P (Figure 9b). However, nickel is not depleted to zero, the lowest concentrations being about 2 nmol kg^{-1}. In profile, the nickel distribution resembles a linear combination of phosphate and silicate; however the slope of the Ni: Si covariance is lower in high

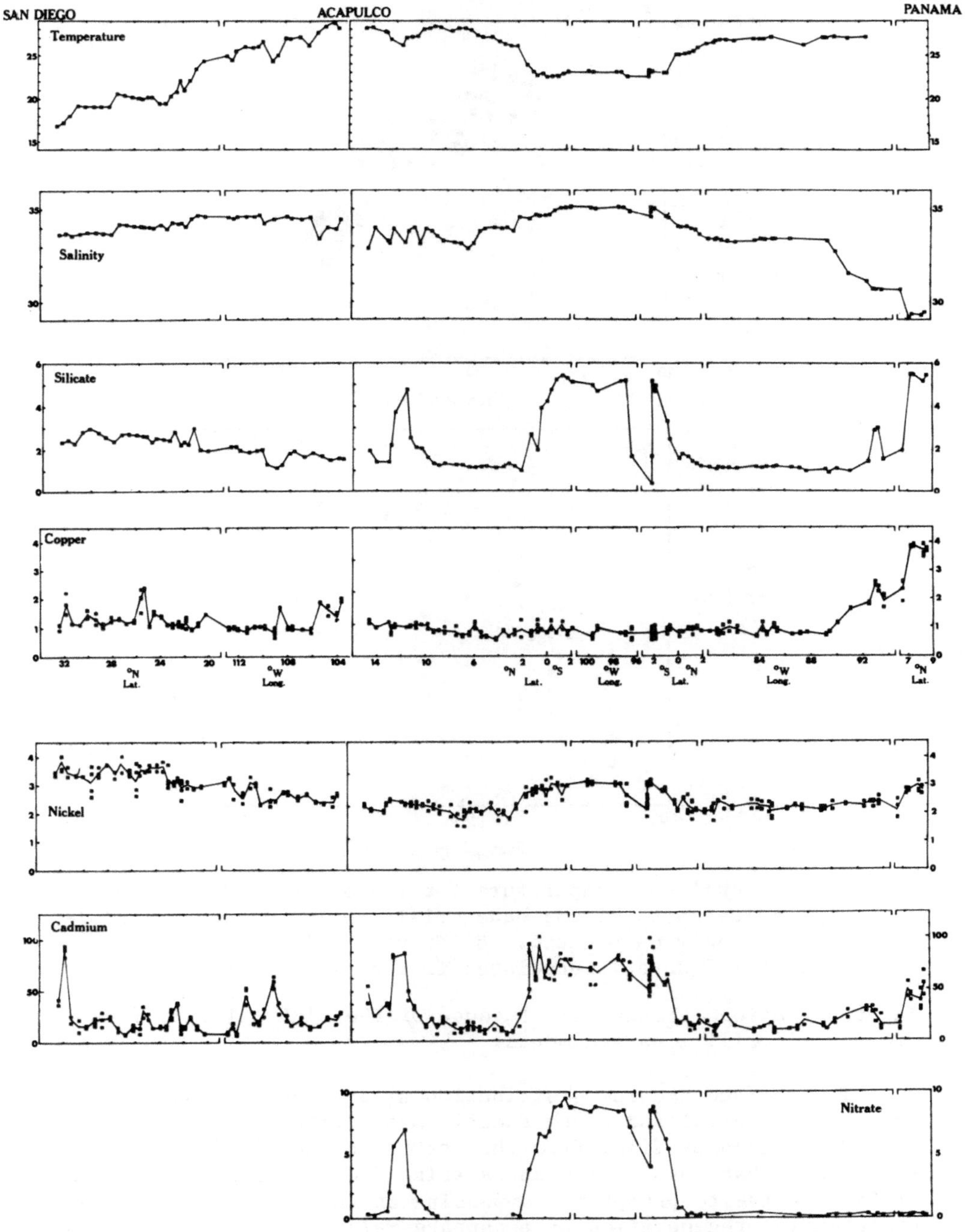

Fig. 8 Data from DeSteiguer 1979; individual analyses plotted.

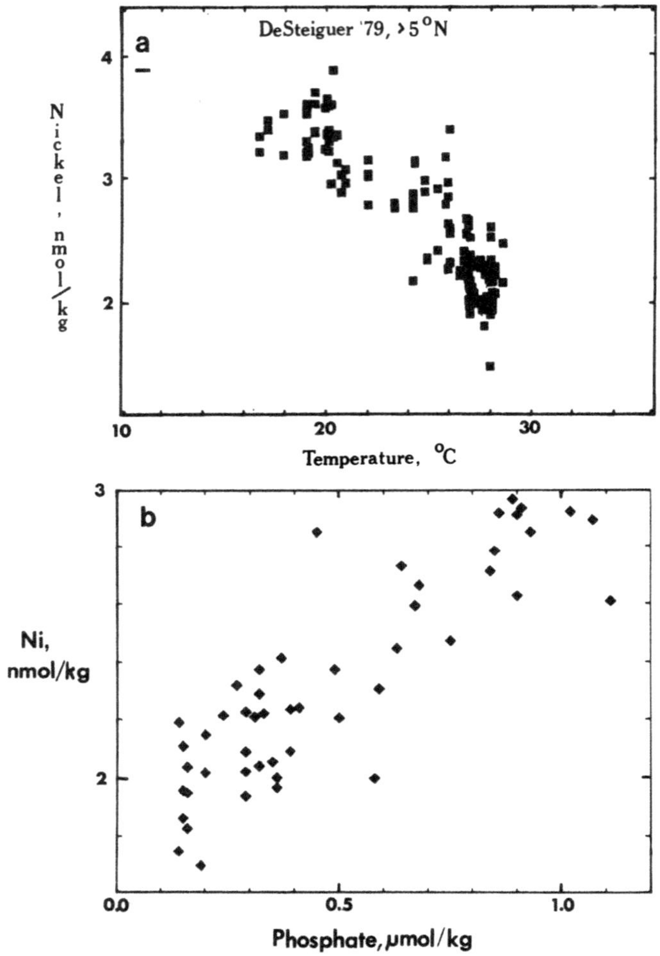

Fig. 9 a) Nickel vs. temperature for DeSteiguer 1979 surface samples north of the equatorial upwelling zone. Individual analyses plotted. b) Ni vs. P, leg 1, DeSteiguer 1979. P data interpolated from cruise report.

latitude regions (Figure 11) because of the high silicate flux from out of siliceous sediments.

The surface cadmium distribution also resembles phosphate. But the equatorial and north Pacific show different Cd-P relations than deep ocean profiles from the Northeast Pacific (Figure 12). We believe that these differences arise from differential uptake of Cd relative to phosphorus depending on the nutrient regime and preferential regeneration of phosphate relative to cadmium in the upper thermocline. Large 'anomalies' in the Cd relative to P may be common in the shallow waters of the ocean.

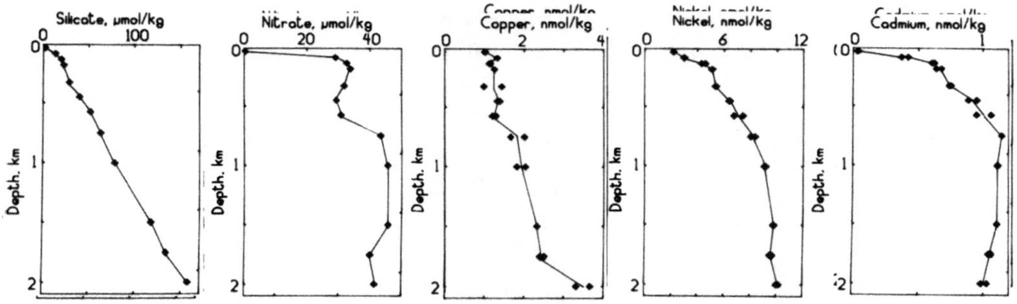

Fig. 10 Profiles from DeSteiguer 1979 station at 100°N, 100°W. Individual analyses plotted.

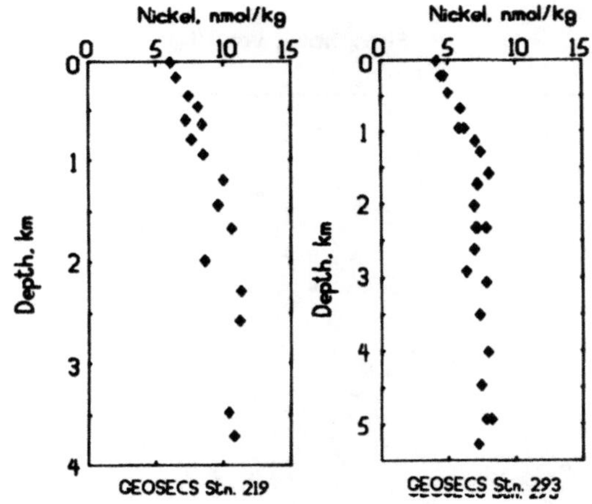

Fig. 11 Nickel profiles from Pacific GEOSECS stations in the Bering Sea (219) and Circumpolar Current (293).

Lead concentrations in the North Atlantic are about a factor of two higher than reported in the Pacific (see Schaule and Patterson[13] and this volume), indicating a greater flux of aerosol Pb to this region. Our numbers have not yet been corrected for recovery efficiency (probably about 80%), which would increase them slightly and make them compatible in general with those of Schaule and Patterson (this volume). But even with the correction, our samples north of the Gulf Stream (collected in August, 1980) are definitely lower than those of Schaule and Patterson (this volume) suggesting that there may be temporal variability in the surface lead concentrations in this area.

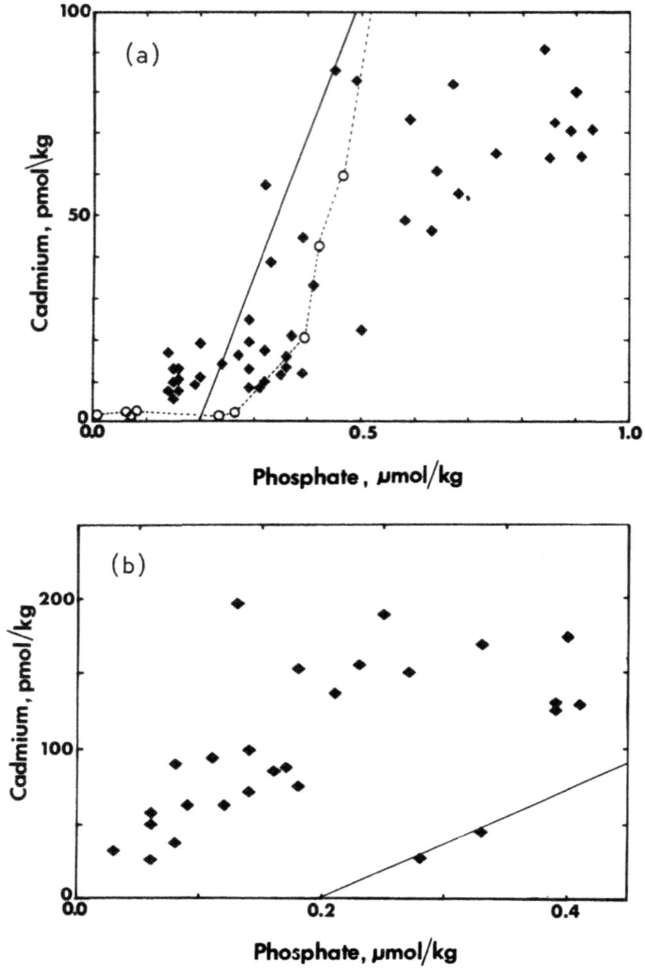

Fig. 12 Cd vs. P; a) Equatorial Pacific (P data interpolated from cruise data report). b) North Pacific.

If we define the fractionations of trace elements (X) relative to phosphorus (P) by primary producers as

$$\alpha = \frac{(X/P)_{PP}}{(X/P)_{water}}$$

and the subsequent fractionation of this primary production to fecal material as

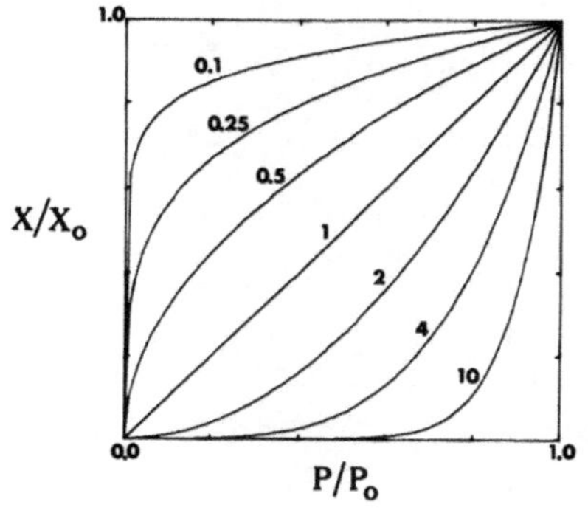

Fig. 13 Plot of trace element concentration (X/X_O) vs. phosphate (P/P_O) for closed system removal conforming to equation (3) for various values of $\alpha\beta$.

$$\beta = \frac{(X/P)_{FM}}{(X/P)_{PP}}$$

we can examine the systematics of trace element uptake by organisms in the surface waters of the ocean. If $\alpha\ \beta$ = constant, then the removal of the trace elements and phosphorus from upwelled water will be as illustrated in Figure 13. In reality, $\alpha\ \beta$ will vary with the nutrient regime; from surface water data we have made estimates of the variation of $\alpha\ \beta$ with phosphate and element/phosphate (Figure 14). Apparently organisms discriminate increasingly against trace elements in low-nutrient regimes. This effect may have a major impact on the steady-state concentrations of trace elements in oligotrophic restricted basins such as the Mediterranean (see Spivack, Huested, and Boyle, this volume).

REFERENCES

1. Andreae, M.O., 1979: Arsenic speciation in seawater and interstitial waters: The influence of biological-chemical interaction on the Chemistry of a trace element. Limnol. Oceanogr., 24, 440-452.
2. Bender, M.L. and C.L. Gagner, 1976: Dissolved copper, nickel and cadmium in the Sargasso Sea. J. Mar. Res., 34, 327-339.
3. Boyle, E.A., R. Sclater and J.M. Edmond, 1976: On the marine geochemistry of cadmium. Nature, 263, 42-44.
4. Boyle, E.A., F.R. Sclater and J.M. Edmond, 1977: The distribution of dissolved copper in the Pacific. Earth Planet. Sci. Lett., 37, 38-54.

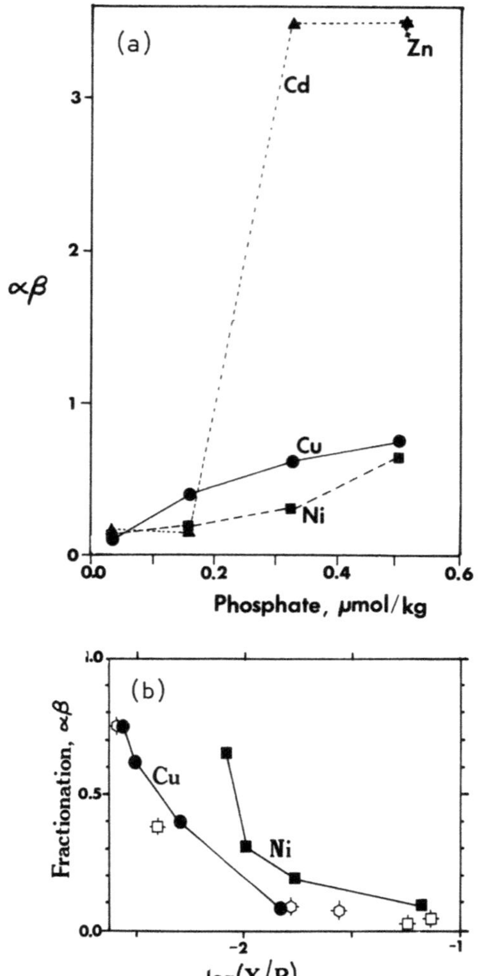

Fig. 14 a) Estimates of the fractionation factor αβ vs. surface phosphate concentrations. b) Estimates αβ vs. the element (X)/P ration in surface waters, derived from surface water data (Bruland 1980).

5. Bruland, K.W., 1980: Oceanographic Distributions of Cadmium, zinc, nickel and copper in the north Pacific. Earth Planet. Sci. Lett., 47, 176-198.
6. Chan, L., D. Drummond, J.M. Edmond and B. Grant, 1977: On the barium data from the Atlantic GEOSECS Expedition. Deep Sea Res., 24, 613-649.

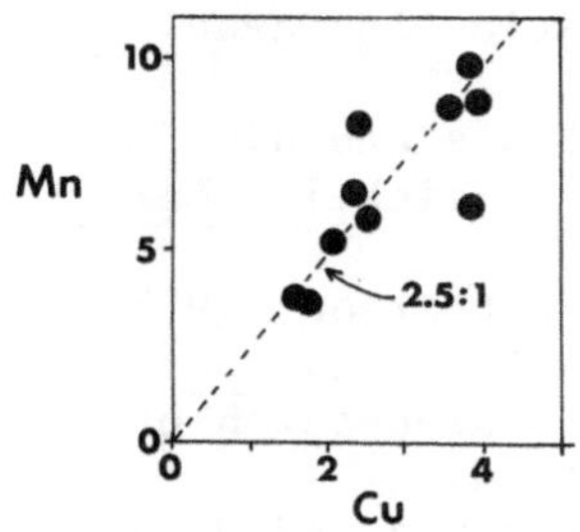

Fig. 15 Manganese vs. copper, surface samples from the Gulf of Panama. Units: nmol/kg.

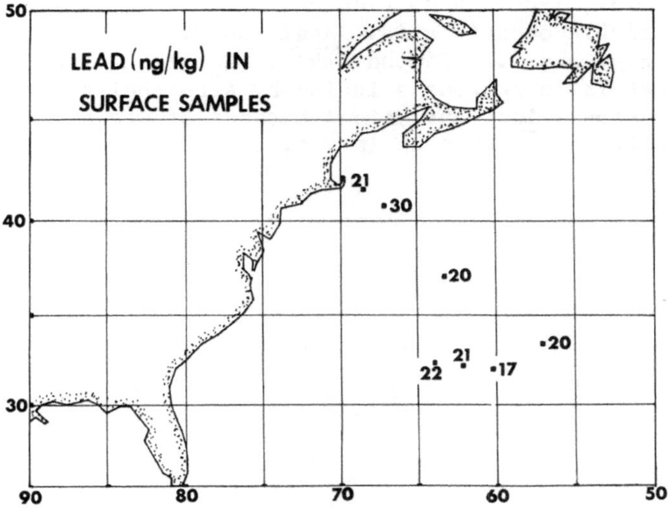

Fig. 16 Lead in the Western North Atlantic, August 1980.

7. Cranston, R.E. and J.W. Murray, 1978: The determination of chromium species in natural waters. Anal. Chim. Acta., 99, 275-282.
8. Landing, W.M. and K.W. Bruland, 1980: Manganese in the North Pacific. Earth Planet. Sci. Lett., 49, 45-56.
9. Martin, J.H., K.W. Bruland and W.W. Broenkow, 1976: Cadmium transport in the California Current. In: "Marine Pollutant Transfer", H. Windom and R. Duce, eds. Lexington Books, Lexington, pp. 159-184.
10. Measures, C.I., R.E. McDuff and J.M. Edmond, 1980: Selenium redox chemistry at GEOSECS I. Earth and Planetary Science Letters, in press.

11. Moore, R.M., 1978: The distribution of dissolved copper in the eastern Atlantic Ocean. Earth Planet. Sci. Lett., 41, 461-468.

12. Sclater, F.R., E. Boyle and J.M. Edmond, 1976: On the marine geochemistry of nickel. Earth Planet. Sci. Lett., 31, 119-128.

13. Schaule, B. and C. Patterson, 1977: The Occurence of lead in the Northeast Pacific. In: "Proceedings of an international experts discussion of lead", M. Branica, ed. Pergamon, Oxford.

14. Boyle, E.A., S.S. Huested and S. Jones, 1981: On the distribution of Cu, Ni and Cd in the surface waters of the North Atlantic and North Pacific Ocean. J. Geophys. Res., 86, C10, 9844-58.

15. Hydes, D.J., 1980: Reduction of Matrix effects with a soluble organic acid. Anal. Chem., 52, 959-963.

16. Boyle, E.A. and J.M. Edmond, 1975: Determination of trace metals in aqueous solution by APDC chelate co-precipitation. In: "Analytical Methods of Oceanography", T.R. Gibb, ed. Advances in Chemistry Series #147, pp 44-55.

MN, NI, CU, ZN AND CD IN THE WESTERN NORTH ATLANTIC

Kenneth W. Bruland and Robert P. Franks

Center for Coastal Marine Studies
University of California
Santa Cruz, CA 95064
U.S.A.

ABSTRACT

The concentrations of Mn, Ni, Cu, Zn and Cd have been determined on surface and deep water samples from the western North Atlantic. The results from a single vertical profile are compared to published results from the North Pacific and interpreted with respect to the hydrographic characteristics of both oceans. Cd, Zn and Ni have nutrient-type distributions in both oceans. They are depleted in surface waters, increase rapidly across the thermocline, then increase or decrease only slightly with depth. The North Atlantic deep waters at depths of 1 to 3 km have average concentrations of Cd, Zn and Ni equal to 0.29, 1.5 and 5.7 nmol kg $^{-1}$, respectively; values substantially lower than their corresponding values in the North Pacific at similar depths of 0.94, 8.2 and 10.4 nmol kg $^{-1}$. Cu concentrations increase gradually with depth in both oceans, with a North Atlantic deep water (1 to 3 km) average value of 1.7 nmol kg $^{-1}$ relative to 2.7 nmol kg $^{-1}$ at similar depths in the North Pacific. Mn concentrations decrease with depth through the thermocline with deep North Atlantic values on the order of 0.6 nmol kg $^{-1}$.

The Atlantic surface samples comprise a transect from the shelf waters off New England to the Sargasso Sea southeast of Bermuda. Metal concentrations are higher in the continental shelf and slope waters with the Gulf Stream separating these from the lower oceanic values. The shelf water vs. open ocean concentrations are: Mn, 21 vs. 2.4 nmol kg^{-1}; Ni, 5.9 vs. 2.3 nmol kg^{-1}; Cu, 4.0 vs. 1.2 nmol kg $^{-1}$; Zn, 2.4 vs 0.06 nmol kg $^{-1}$; and Cd, 200 vs. 2 pmol kg^{-1}. The high shelf water values appear to result from an external, presumably continental source. By comparison, increased levels of Ni, Cu, Zn and Cd in the coastal waters off central California have been interpreted to result primarily from the upwelling of nutrient rich waters. Mn and Cu concentrations are significantly higher in the Sargasso Sea than in surface waters of the North Pacific central gyre: 2.4 compared to 1.0 nmol kg^{-1} for Mn and 1.2 vs. 0.5 nmol kg^{-1} for Cu. Nickel concentrations are not significantly different: 2.1 to 2.4 nmol kg^{-1} in both oceans. Zn concentrations are depleted to approximately 0.06 nmol kg^{-1} in both oceans, while Cd is depleted in both regions to values close to 2 pmol kg^{-1}.

INTRODUCTION

The majority of the oceanographically consistent dissolved trace metal data for Mn, Ni, Cu, Zn and Cd in seawater have been reported in the last five years. Bruland[1] has summarized the oceanographic distributions of Cd, Zn, Ni and Cu in the North Pacific, while Boyle et al.[2] have described the surface distributions of Cu, Ni and Cd in the Northern Atlantic and Pacific. These elements are involved in a biogeochemical cycle involving their net removal from surface waters via sinking biological debris and subsequent regeneration at depth. This involvement in the internal biogeochemical cycles leads to strong correlations between Cd and phosphate, and Zn and silicate, while Ni is best correlated with a combination of phosphate and silicate. The correlation of these trace metals with the various nutrients appears to be primarily a function of the regeneration depth of the metal, with Cd being regenerated at shallow depths simultaneously with the oxidation of organic matter, while zinc has a deeper regeneration depth similar to biogenic opal or calcium carbonate. The distribution of Cu is complicated by significant in-situ scavenging in the deep water column[1,3]. Although a substantial amount of Cu is regenerated from sinking biological debris, in particular at the sediment water interface, in-situ deep water scavenging (with a scavenging rate constant of ca. 10^{-3} yr^{-1}) causes substantial deviation from a nutrient type profile. The vertical distribution of Mn in the North Pacific[4,5] exhibits evidence of substantial deep water scavenging (with a rate constant on the order of 10^{-2} yr^{-1}). Because of this high rate of scavenging, the distribution of Mn in the North Pacific is primarily governed by external inputs which lead to maxima in the surface waters and above active spreading centers[5].

The surface distribution of a trace element or species is to a large extent controlled by its dominant input source. Examples of source terms important to surface waters include: 1) vertical mixing with deeper waters by upwelling or convection; 2) atmospheric fallout or washout; 3) continental input from rivers and/or shelf sediments; and 4) in-situ production. Nutrient species such as nitrate, phosphate and silicate, we well as nutrient-type trace metals such as Ni, Zn and Cd have surface distributions influenced primarily by the rate of vertical mixing[1,2,6,7]. They exhibit depleted levels in subtropical gyres and enhanced levels in areas of coastal upwelling, divergence and subpolar fronts. Pb-210 and common Pb are species whose surface distributions are dominated by atmospheric input[8,9]. These species, in contrast to the nutrient-type elements, have elevated levels within the subtropical gyres and lower values at gyre boundaries as a result of intensified scavenging at the more productive boundaries. An example of a species for which continental input is the dominant source is Ra-228: a radionuclide whose 5.6 yr half-life has proven useful for studying horizontal mixing processes[10,11]. An external, continental input has also been shown to have a major influence on the surface distribution of Mn in the North Pacific[4]. Examples for which in-situ production provides the major source are the radionuclides Th-234 and Th-230 (formed by the decay of their uranium parents), as well as species such as ammonia, chromium (III), and arsenite which are produced by biologically-mediated reduction processes within the euphotic zone[12,13].

This paper presents dissolved Mn, Ni, Cu, Zn and Cd results from a single vertical profile in the western North Atlantic and examines the extent to which these data follow the patterns and relations observed in the North Pacific vertical profiles. Additionally, trace metal data from a surface water transect from the Sargasso Sea to Narragansett Bay is used to provide insight into the relative importance of the various factors affecting the surface water distributions.

SAMPLING AND ANALYTICAL PROCEDURES

Seawater samples were collected in July 1979 during a cruise into the western North Atlantic. A horizontal profile of surface waters and one vertical profile were collected from the R.V. Endeavor in cooperation with the SEAREX program, and, in particular, with Dr. C.C. Patterson's research group at the California Institute of Technology. The vertical profile consisted of unfiltered samples collected with the CIT deep-water, common lead sampler[9]. Unfiltered surface seawater samples were collected from a small inflatable raft moving upwind and more than 200 m away from the research vessel. All samples were collected and stored in conventional polyethylene bottles and acidified with 4 ml of 6N quartz-distilled HCl per liter of seawater. The results obtained will be defined as total dissolved metals.

Ni, Cu, Zn and Cd were determined by flameless atomic absorption spectrometry (AAS) following a 250:1 preconcentration using a dithiocarbamate extraction method. A detailed description of the analytical methods is presented by Bruland[1] and Bruland et al.[14]. Mn was also determined by flameless AAS using the preconcentration method presented by Landing and Bruland[4]. Nutrient samples were immediately frozen until analysis using the methods of Strickland and Parsons[15] for phosphate and silicate. Salinity was measured using an inductive salinometer, while oxygen was determined immediately after sample collection[15].

During the analysis of the samples it became apparent that all samples deeper than 3000 m had been contaminated by a faulty seal in the CIT sampler. These samples were characterized by high levels of nitrate, iron, nickel and copper, indicating leakage of acidified water from the expansion cylinder into the sampling bag. These samples were excluded from further consideration.

RESULTS AND DISCUSSION

Vertical Distributions

Figure 1 presents the salinity, oxygen, phosphate and silicate data for the vertical profile taken at station 4 (34°15'N, 66°17'W) located within the Sargasso Sea approximately 200 km northwest of Bermuda. The profiles are consistent with those

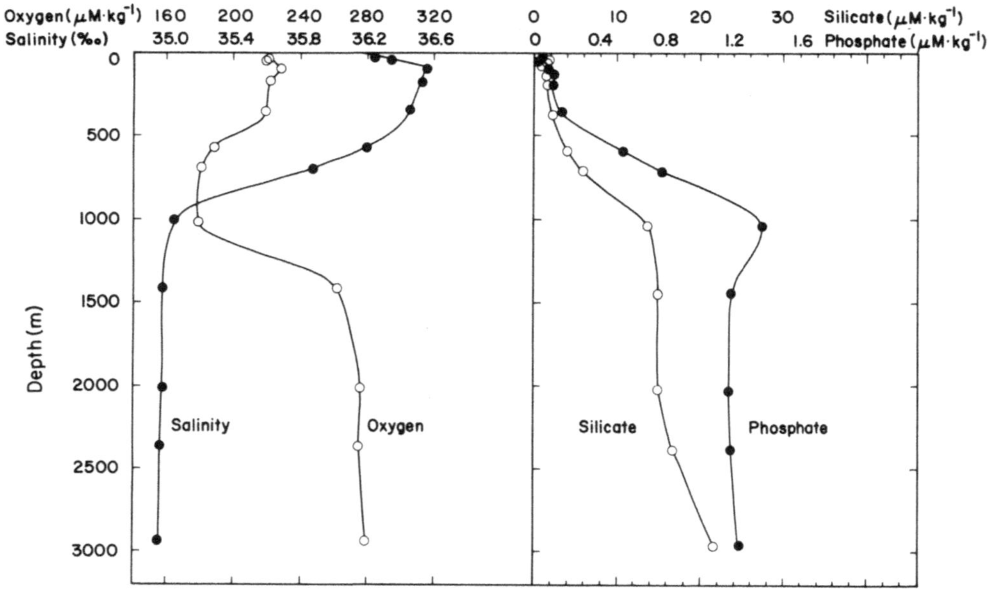

Fig. 1 Hydrographic parameters determined at Station 4. 34°15'N, 66°17'W.

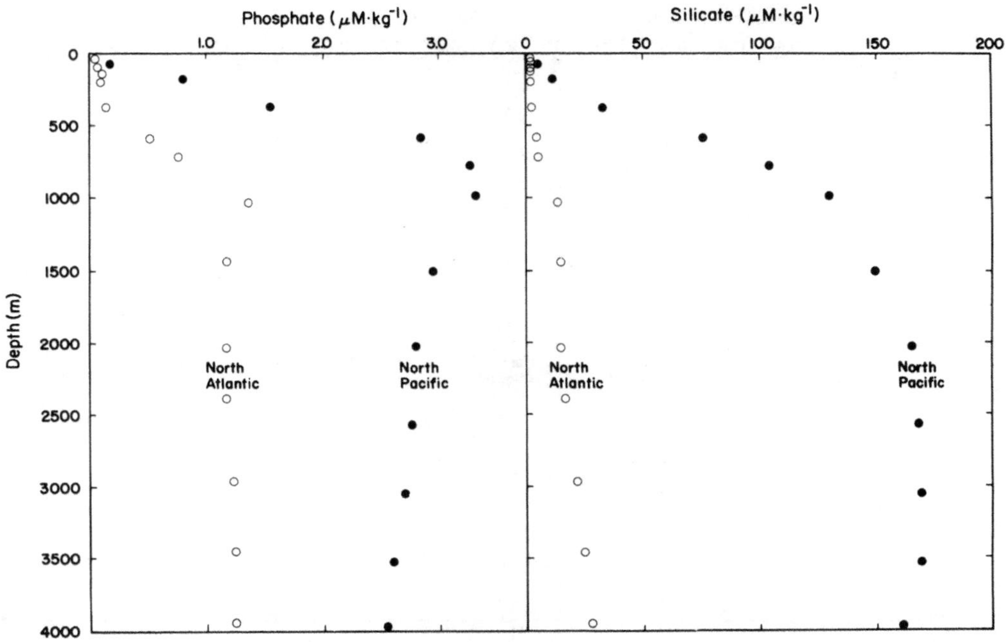

Fig. 2 Phosphate and silicate profiles from Station 4. North Pacific data is from Bruland[1], Station 17, H-77, 32°41'N, 145°00'W.

obtained in the same area during the GEOSECS expedition[16]. Figure 2 compares the phosphate and silicate profiles with profiles from the eastern North Pacific (32°41'N, 145°00'W, stn. 17, H-77; Bruland[1]). Most of the variation in the oceanic vertical distributions of these nutrients can be explained by the interaction of their biogeochemical cycles with the general circulation of the deep ocean. The young deep waters of the North Atlantic are markedly nutrient poor relative to the older North Pacific deep waters with average concentrations of phosphate and silicate approximately 2.4 and 10 times greater in the deep Pacific. The greater fractionation of silicate between the two oceans results primarily from its deeper regeneration depth[17]. The important question of the extent to which the various nutrient-type trace metals show similar inter-ocean patterns is addressed below.

Figures 3 and 4 present the total dissolved trace metal vertical profiles in the North Atlantic and North Pacific. The North Pacific results are from stn. 17, H-77 (Cd, Zn, Ni and Cu from Bruland[1]; Mn from Landing and Bruland[4]; and Pb from Schaule and

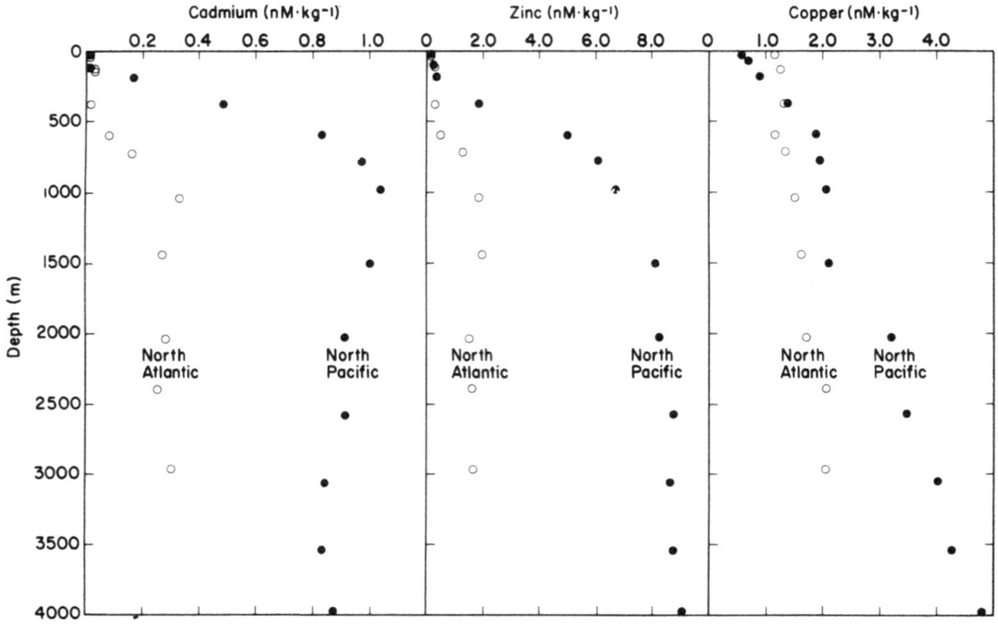

Fig. 3 Total dissolved trace metal profiles from the North Atlantic, Station 4, and North Pacific, Station 17, H-77. (Pacific data from Bruland[1]).

Patterson[9]). Cadmium is depleted in the surface waters of both oceans to values less than or equal to 2 pmol kg^{-1}; this is less than 1% of deep water Cd concentrations. The maximum Cd levels correspond to the phosphate maxima: 0.33 nmol Cd kg^{-1} in the North Atlantic and 1.1 nmol Cd kg^{-1} in the North Pacific. The shape of the North Atlantic Cd profile is similar to that of phosphate, but the deep water Cd concentrations average 15% less than would be predicted from the phosphate distribution and the Cd:phosphate relationship found in the North Pacific[1].

Zinc is also strongly depleted in the surface waters of both oceans and its concentration increases with depth similar to silicate, although a slight mid-depth maximum near the phosphate maximum is evident in the North Atlantic profile. The average concentration at depths of 1 to 3 km in the North Atlantic is 1.7 nmol kg^{-1}, a value only 1/5 of that observed at similar depths in the North Pacific. However, these deep North Atlantic zinc values are approximately twice what would be predicted based upon the Zn:silicate relationship found in the North Pacific[1].

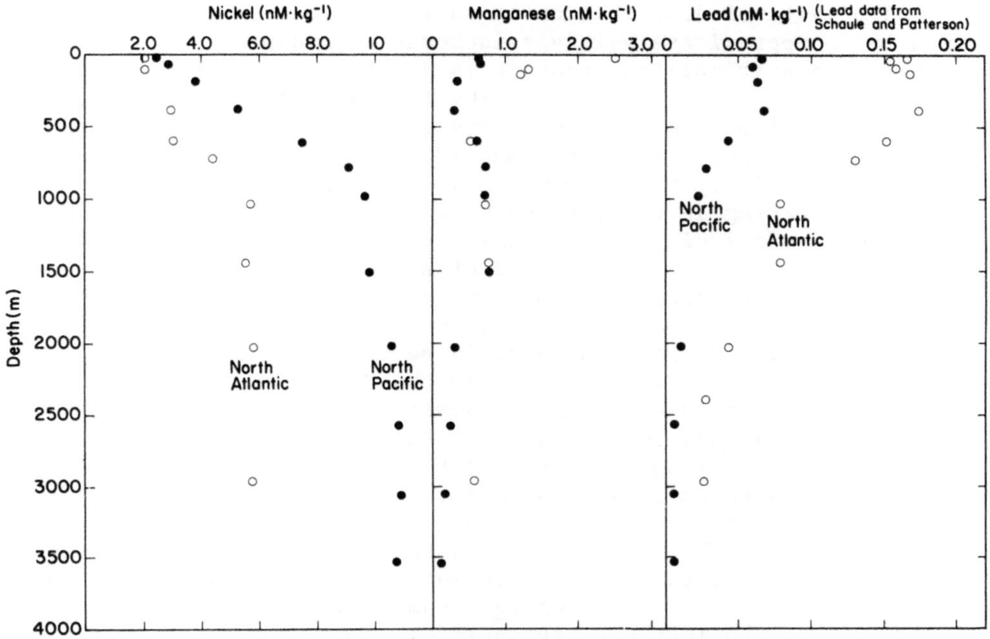

Fig. 4 Total dissolved trace metal profiles from the North Atlantic, Station 4, and North Pacific, Station 17, H-77. (Pacific Ni data from Bruland[1]; Pacific Mn data from Landing and Bruland[4]; Pacific Pb data from Schaule and Patterson[9]; Atlantic Pb data from Schaule and Patterson[18])

The nickel concentration in the surface waters of both oceans ranges from 2.0 to 2.5 nmol kg^{-1}; this is 1/3 to 1/5 of the deep water values. The average concentration at depths of 1-3 km in the North Atlantic is 5.7 nmol kg^{-1}, compared to 10.4 nmol kg^{-1} found at similar depths in the North Pacific.[1] The nickel:nutrient relationship observed in the North Pacific would predict a deep North Atlantic value only 80% of that observed.

Thus, based on the limited data presented here, the three nutrient-type trace metals (Cd, Zn and Ni) have vertical distributions that, similar to the North Pacific profiles, are the result of surface removal and regeneration at depth. In both oceans these nutrient-type metals exhibit vertical profiles that follow one or more of the nutrients, presumably as the result of similar depths of regeneration. However, the individual ratios of metal to nutrient(s) are noticeably different in the two ocean basins.

The vertical distribution of Cu in both oceans is characterized by an approximately linear increase with depth. This deviation from a nutrient-type profile is due to in-situ scavenging within the deep waters [1,3]. Cu, like the nutrient-type elements, is scavenged from the surface waters and regenerated in the deep waters, most likely at the sediment:seawater interface. However, where the nutrient-type elements have their major particulate removal occurring in the surface waters and show no discernible evidence of particulate scavenging within the deep and intermediate waters, the distribution of dissolved Cu, on the other hand, shows evidence of deep water scavenging (with a first order removal rate in the Pacific of ca. 10^{-3} yr^{-1}). This corresponds to a mean-life with respect to particle scavenging and removal of roughly 1000 yrs for dissolved Cu in the deep sea, a value on the same time scale as deep ocean circulation[17]. Although this in-situ scavenging does result in a substantial portion of the deep water Cu being scavenged, enough of the particulate Cu is regenerated back into the deep waters as a result of early diagenesis at the sediment-water interface to cause a net increase in the Cu concentration as the deep water traverses from the Atlantic to the Pacific. The average Cu concentration at depths of 1 to 3 km in the North Atlantic is 1.8 nmol kg^{-1} compared to 2.7 nmol kg^{-1} at equivalent depths in the North Pacific. Thus, the distribution of dissolved Cu is also affected by its involvement in an internal biogeochemical cycle.

The vertical distributions of Mn and Pb (the Pb data are from Schaule and Patterson[9,18]) in both oceans are characterized by surface concentrations which are greater than the deep water concentrations. Higher concentrations at all depths are found in the North Atlantic relative to the North Pacific. These patterns are the result of the high degree of scavenging of Mn and Pb throughout the water column. Deep water residence times ranging from 50 to 300 years have been estimated for these metals[5,9,19,20,21]. There does appear to be some regeneration of the surface-sequestered metals within the major thermocline for both Mn (evidenced by slight Mn maxima associated with the oxygen minimum[4,5]) and Pb (evidenced by the deep penetration depth of Pb and Pb-210[9,21]). However, this is superimposed on a removal throughout the deep water column coupled with little or no regeneration from bottom sediments. As a consequence of this high degree of scavenging in the deep water and the lack of a substantial recycling of either element back into the water column, their distributions are dominated by external inputs. In both the North Atlantic and the North Pacific the atmospheric and/or riverine input of Mn and Pb to the surface waters are the major external inputs. The stable Pb profiles of Schaule and Patterson are further complicated by the recent increase in atmospheric Pb due to anthropogenic emissions, resulting in a transient non-steady state distribution. The Mn profiles can be complicated when influenced by suboxic conditions, or by hydrothermal solutions which may result in localized elevated concentrations of dissolved and particulate Mn[5,19,22,23].

Surface Distributions

Figure 5 is a map of the sampling area showing the location of the stations closest to shore. Stations 3 through 9 have salinity and nutrient values indicating they are within the Sargasso Sea water mass. Table 1 is a compilation of average Sargasso Sea surface water trace metal concentrations from this study and two recent studies by other investigators. Average pacific Ocean values are also included. An assessment of the accuracy of the data reported here is a difficult task. Bruland et al.[14] required agreement between various sampling methods and independent sample concentration techniques. Boyle et al.[3] have stated that "the primary criteria must be interlaboratory agreement and the oceanographic consistency of the data themselves..."

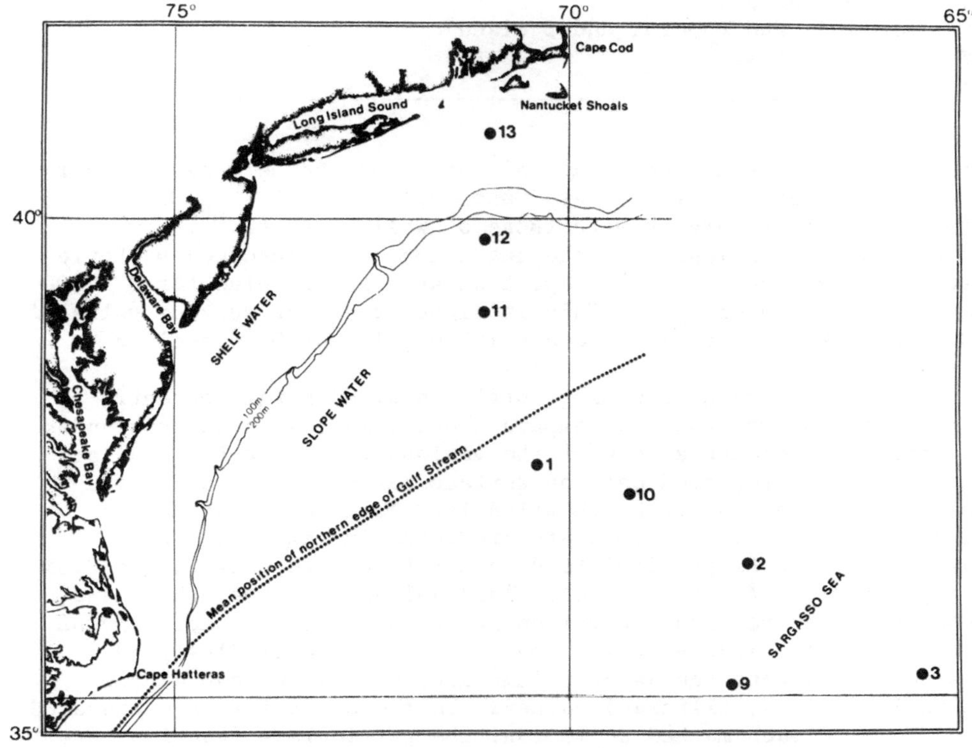

Fig. 5 Location of surface stations. Only stations closest to shore are shown. Map after Csanady[24].

Table 1 Average Metal Concentration of Open Ocean Surface Waters in nmol kg^{-1}.

	Sargasso Sea			North Pacific Gyre	Sargasso Sea Suspended Particulates
Element	A	B	C	D	E
Mn	2.3±.2	--	--	1.0	--
Ni	2.3±.1	2.0±.1	2.3±.2	2.1	.018±.008
Cu	1.2±.1	1.4±.2	1.0±.1	0.5	.028±.006
Zn	.06±.02	--	.30±.05	.07	≤.016
Cd	.002	≤.016	≤.003	.0014	.0009±.0001

A - This paper, Stations 3 - 9
B - Boyle et al.[2]
C - Wallace et al.[25]
D - Bruland[1] and Landing and Bruland[4]
E - Wallace et al.[25]

It can be seen that the Cu and Ni data reported here agree favorably with that of Boyle et al.[2] and Wallace et al.[25]. Our Cd results agree favorably with those of Wallace et al.[25], however these values are less than the detection limit reported by Boyle et al.[2]. Wallace et al.[25] report an average Zn value for 0 - 100 m of 0.3±0.05 nmol kg^{-1}. This is higher than our surface value of 0.06 nmol kg^{-1}, but close to our 100 m value of 0.26 nmol kg^{-1}.

Since we are determining total dissolved metal concentrations, it is important to compare these values with the suspended particulate concentrations of the various trace metals. Wallace et al.[25] also reported data on surface water suspended particulates collected in the North Atlantic. The average values of their Sargasso Sea stations are presented in Table 1. Their average Cu and Ni particulate concentrations are only 2, and 0.8%, respectively, of our total dissolved values. On the other hand, the Cd and Zn particulate values are 45 and 27%, respectively of the total dissolved values. Thus, Cu and Ni total dissolved values throughout the water column are indistinguishable from dissolved (i.e., filtered) values. In the surface waters where Cd and Zn show such marked depletion, the particulate fractions are significant; however, in the deep waters, the particulate fractions become negligible. Alternatively, surface suspended particulate Mn represents a negligible fraction of the total dissolved metal and, as Landing and Bruland[4] have demonstrated, deep water particulate Mn may comprise as much as 60 to 80% of the total dissolved Mn.

One would expect the trace metal concentrations in the surface waters of the subtropical gyres of the North Atlantic and North Pacific to reflect any differences in the magnitude of the source terms in the two regimes. The North Atlantic has a large river input relative to the North Pacific[26] and is also believed to have a higher atmospheric delivery rate of many trace metals (Duce, personal communication). Concentrations of Mn, Cu and Pb in North Atlantic surface waters are all more than twice those in the North Pacific. The lack of a substantial concentration gradient across the upper portion of the thermocline for these elements means that vertical mixing with subsurface waters has little or no effect on the surface water concentrations. In contrast, the nutrient-type trace metals Cd, Zn and Ni have essentially identical concentrations, respectively, in both regions. Thus, even though the rate of vertical mixing within the subtropical gyres may be quite slow, it can still be the dominant source term. Bruland[1] has presented evidence from the North Pacific to support this.

The rest of the stations sampled on the horizontal transect were located in the shelf and slope water of the eastern U.S. continental shelf. These waters are separated from the warm, high salinity Sargasso Sea water by the Gulf Stream. The northern edge of the Gulf Stream is bounded by the 36°/oo isohaline. The salinities of the surface samples range from 36.5^o/oo at stn. 6 in the Sargasso Sea to 30.2^o/oo at stn. 13 in the shelf waters where riverine freshening has occurred. This nearshore freshening can be thought to occur uniformly along the coast, since river plumes mix with the shelf water within a relatively narrow coastal boundary layer[24]. According to NOAA satellite maps of the study region concurrent to our sampling, the stations can be placed accordingly: stn. 1, 1/2° north of Gulf Stream Edge (G.S.E.); stn. 2, ~1° south of G.S.E.; stn. 3-9, Sargasso Sea water; stn. 10, within the Gulf Stream; stn. 11, ~1.5^o north of G.S.E., but on the outer edge of a warm core eddy; stn. 12, slope water; stn. 13, shelf water. The Gulf Stream was located to the south of the mean annual position shown in Fig. 5 during both transects across it. Because of the meandering of the Gulf Stream and the existence of warm core eddies in the slope region, salinity may be used as a better indicator than distance from shore for placing the stations within an appropriate water mass.

The surface water data can thus be plotted against salinity to visually examine the variations in concentrations from Sargasso Sea surface water to the shelf surface waters (Figures 6 and 7). The cluster of data points where salinity is greater than 36°/oo

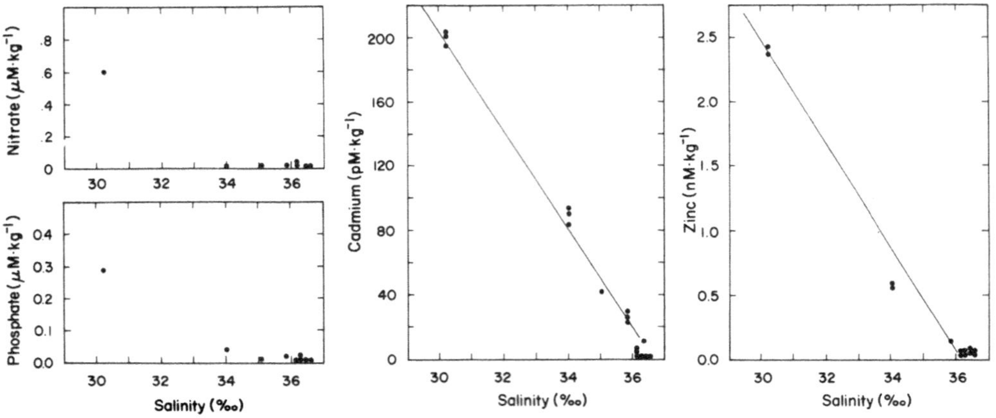

Fig. 6 Variations of elements with salinity along the surface transect. The Sargasso Sea surface waters are the cluster of points with salinities greater than 36°/oo.

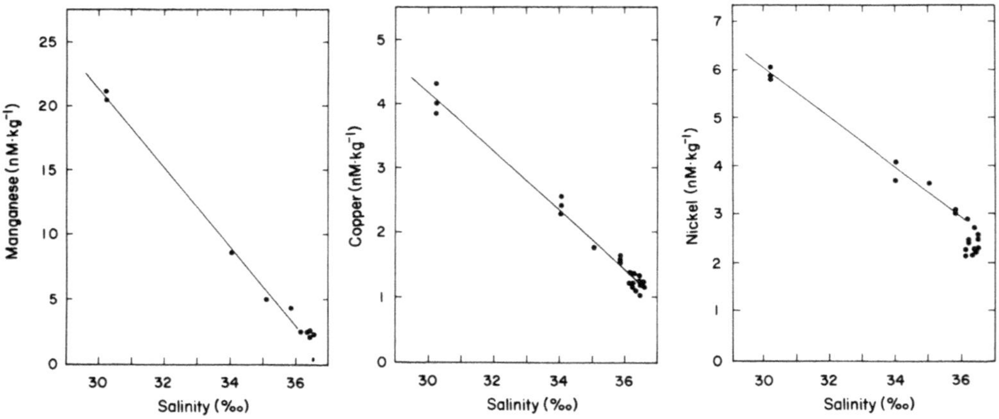

Fig. 7 Variations of elements with salinity along the surface transect. The Sargasso Sea surface waters are the cluster of points with salinities greater than 36°/oo.

represents Sargasso Sea surface water. In a simplistic sense the transect can be considered to cross the mixing zone of two end members, the high salinity water of the Sargasso Sea and the low salinity coastal shelf waters. The data for Cu, Ni, Mn and Cd appear to fall along a salinity mixing line exhibiting conservative behavior. The nutrients phosphate, nitrate and silicate show non-linear behavior and only exhibit significantly higher values at the shelf station. These nutrients are ostensibly substantially removed by biological activity during the mixing process. Zn seems to show a hint of removal during mixing; however, there is only one data point which falls below the line of conservative mixing. To evaluate the results of removal during mixing, a model of the horizontal diffusional mixing of water masses on a 200 km scale was constructed using an apparent horizontal eddy diffusivity for this scale length derived from Okubo[27] and 3 different first-order scavenging rate constants; 10 yr^{-1}, 1 yr^{-1} and 0.1 yr^{-1} (Figure 8). It can be seen that with a scavenging rate constant less than 1 yr^{-1} (or a mean-life greater than 1 yr), a nearly-linear mixing curve is obtained. However, for a rate constant of 10 yr^{-1} (or a 0.1 yr mean life), as might apply to the nutrient species, concave-up curvature would exist. Admittedly, this is a very simplistic exercise fraught with assumptions, but it does suggest that if the metals have surface water residence times greater than 1 yr, then the results of Figures 6 and 7 at least would be consistent.

The end member values for Mn, Ni, Cu, Zn and Cd are given in Table 2 along with the absolute changes in metal concentration. These changes in metal concentration during the mixing of Sargasso Sea and shelf water can be used to derive ratios of Me/ Cu. These "signature ratios" would be distinctive of a certain water mass until removal (or input) processes altered it. Undoubtedly, these ratios would vary both spatially and temporally with changes in the sources of the various metals. In this case the values

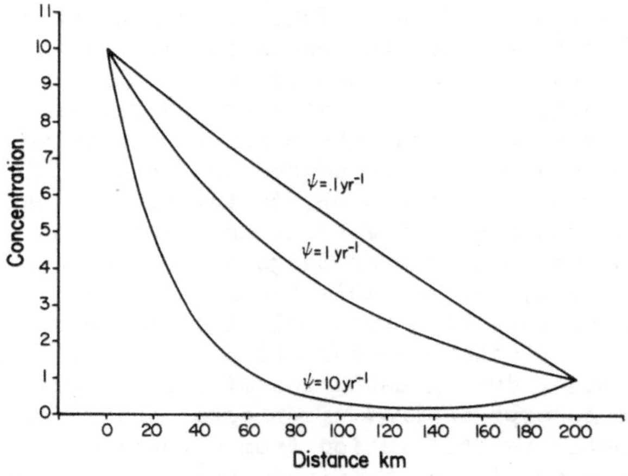

Fig. 8 Theoretical plot of concentration versus distance for a two end-member, vertical diffusion model. Ψ is the first-order removal rate constant. Scale length = 200 km. $K_a = 2.6 \times 10^6\ cm^2\ s^{-1}$ (after Okubo[27]).

Table 2 Sargasso Sea, Shelf Water and Extrapolated Zero Salinity Trace Metal Concentrations (nmol kg^{-1}).

Element	Sargasso Sea End Member	Shelf Water	Δ Me	Δ Me/Δ Cu	Extrapolated Zero Salinity End Member	Ratio of End Members
Mn	2.3	21	19	6.8	114	50
Ni	2.3	5.9	3.6	1.3	21	9
Cu	1.2	4.0	2.8	--	19	16
Zn	0.06	2.4	2.3	0.82	16	270
Cd	0.002	0.20	0.20	0.071	1.1	550

found (relative to Cu) are 1.3 for Ni, 0.071 for Cd, 0.82 for Zn, and 6.8 for Mn. Boyle and Huested[28] reported a Δ Mn/Δ Cu value from the Panama Basin of 2.5.

Whereas wind induced coastal upwelling has been invoked to explain the higher continental boundary values for Cd, Zn, Ni and Cu in the eastern North Pacific off central California[1], a similar argument cannot be used to explain the elevated values in the slope and shelf waters of the eastern U.S. Instead, the source appears to be from riverine input and/or shelf sediments. If the results in Figures 6 and 7 are extrapolated to zero salinity, this would give an estimate of the effective concentration of river water that makes it through the estuarine mixing zone to the shelf waters. These results are presented in Table 2. For each of the trace metals studied, the zero salinity end member concentration is substantially greater than the Sargasso Sea value as indicated by the ratio of end members presented in Table 2. These vary from a ratio of 9 for Ni to a ratio of 550 for Cd. It can be seen that even the shelf waters can have metal concentrations up to two orders of magnitude higher than Sargasso Sea surface waters. There are relatively few reliable river water analyses with which to compare the extrapolated zero salinity end members. Table 3 presents some river water concentrations along with estimates of trace metal removal during estuarine mixing. However, other processes such as desorption from river-borne suspended particulate matter, diagenetic remobilization from estuarine or coastal sediments, or anthropogenic inputs within the estuarine or coastal mixing zone can all affect the final apparent river end member concentration. Although these processes can potentially result in a wide range of apparent river end members, the zero salinity end member concentrations (Table 2) fall within or close to the range in the river data from Table 3.

Table 3

Element	River Concentration nmol kg $^{-1}$	Estimate of Removal During Estuarine Mixing
Mn	130 - Average, Turekian[34]	25% Sholkovitz[33]
	180 - Mississippi, Trefry and Presley[30]	3% Sholkovitz and Copland[32]
Ni	5 - Amazon, Boyle et al.[36]	40% Sholkovitz[33]
	26 - Mississippi, Trefry and Presley[30]	15% Sholkovitz and Copland[32]
Cu	24 - Amazon, Boyle et al.[36]	0 - 25% Boyle et al.[36]
	32 - Mississippi, Trefry and Presley[30]	40% Sholkovitz[33]
	15-54 Average, Boyle[31]	
Zn	No reliable data available	No reliable data available
Cd	$\leq$0.1 - Amazon, Boyle et al.[36]	5% Sholkovitz[33]
	0.9 - Mississippi, Trefry and Presley[30]	3% Sholkovitz and Copland[32]

Conclusions

Vertical profiles of Cd, Zn and Ni show nutrient-type distributions in the North Atlantic with average concentrations in the deep water (1 to 3 km) of approximately 1/3, 1/5 and 1/2 those observed at similar depths in the deep North Pacific[1], although these North Atlantic deep water concentrations are significantly different that what would have been predicted based on the nutrient-metal relationships observed in the North Pacific. This fractionation between the two oceans is a result of the biogeochemical cycles at these trace metals. This internal cycle involves the sequestering from surface waters and regeneration at depth via "the great particle conspiracy"[29] superimposed on the deep water circulation pattern of the oceans. The surface water concentrations of these nutrient-type trace metals are virtually identical in both the North Atlantic and North Pacific gyres. The similarity further emphasizes the dominant role of the internal cycles in the distributions of these elements.

Cu is also enriched in the deep North Pacific by a factor of 1.6 relative to the North Atlantic; however, it deviates from a nutrient-type distribution as a result of deep water scavenging. In contrast to the nutrient-type elements, the lack of a strong concentration gradient across the thermocline reduces the importance of vertical mixing in controlling the surface water concentration. The higher surface values observed in the North Atlantic relative to the North Pacific gyre are presumably caused by higher continental and/or atmospheric sources to the North Atlantic.

Mn has a similar distribution in both oceans with higher surface water than deep water concentrations. The rapid removal of Mn throughout the water column results in its distribution being strongly influenced by external sources. The surface values observed in the North Atlantic are 2.3 times greater than those in the North Pacific gyre, indicating a greater source of Mn to the North Atlantic.

The surface concentrations of all the metals increased greatly from the Sargasso Sea to slope and shelf waters. These increases appear to result from mixing of Sargasso Sea water (high salinity, low trace metal content) with a coastal boundary layer (low salinity, high trace metal content). If this mixing is extrapolated to a hypothetical zero salinity end member, the resulting trace metal concentrations appear reasonable with respect to estimates of effective river concentrations after passage through an estuarine mixing zone.

ACKNOWLEDGEMENTS

We are indebted to G. Gill and A. Ng for their assistance in the collection of the samples. We are grateful to B. Schaule and C.C. Patterson for access to samples from their deep water sampler, and for the use of their Pb data. Thanks to K. Coale, W. Landing and J. Cowen for their editorial assistance. This research was supported by the National Science Foundation, Grant No. OCE 79-19928.

REFERENCES

1. Bruland, K.W., 1980: Oceanographic distributions of cadmium, zinc, nickel and copper in the North Pacific. Earth Planet. Sci. Lett., 47, 176-198.
2. Boyle, E.A., S.S. Huested and S.P. Jones, 1981: On the distribution of Cu, Ni and Cd in the surface waters of the North Atlantic and North Pacific oceans. J. Geophys. Res., 86, 8048-8066.
3. Boyle, E.A., F.R. Sclater and J.M. Edmond, 1977: The distribution of dissolved copper in the Pacific. Earth Planet. Sci. Lett., 37, 38-54.
4. Landing, W.M. and K.W. Bruland, 1980: Manganese in the North Pacific. Earth Planet. Sci. Lett., 49, 45-56.
5. Klinkhammer, G.P. and M.L. Bender, 1980: The distribution of manganese in the Pacific Ocean. Earth Planet. Sci. Lett., 46, 361-384.

6. Sclater, F.R., E. Boyle and J.M. Edmond, 1976: On the marine geochemistry of nickel. Earth Planet. Sci. Lett., 31, 119-128.
7. Bruland, K.W., G.A. Knauer and J.H. Martin, 1978: Cadmium in northeast Pacific waters. Limnol. Oceanogr., 23, 618-625.
8. Nozaki, Y., J. Thomson and K.K. Turekian, 1976: The distribution of Pb-210 and Po-210 in the surface waters of the Pacific Ocean. Earth Planet. Sci. Lett., 32, 304-312.
9. Schaule, B.K. and C.C. Patterson, 1981a: Lead concentrations in the Northeast Pacific: Evidence for global anthropogenic perturbation. Earth Planet. Sci. Lett., (in press).
10. Kaufmann, A., R.M. Trier and W.S. Broecker, 1973: Distribution of Ra-228 in the world ocean. J. Geophys. Res., 78, 8827-8848.
11. Knauss, K.G., T. Ku and W.S. Moore, 1978: Radium and thorium isotopes in the surface waters of the East Pacific and coastal southern California. Earth Planet. Sci. Lett., 39, 235-249.
12. Andreae, M.O., 1978: Distribution and speciation of arsenic in natural waters and some marine algae. Deep-Sea Res., 25, 391-402.
13. Cranston, R.E. and J.W. Murray, 1978: The determination of chromium species in natural waters. Anal. Chim. Acta., 99, 275-282.
14. Bruland, K.W., R.P. Franks, G.A. Knauer and J.H. Martin, 1979: Sampling and analytical methods for the determination of copper, cadmium, zinc and nickel at the nanogram per liter level in sea water. Anal. Chim. Acta., 105, 233-245.
15. Strickland, J.D.H. and T.R. Parsons, 1972: A practical handbook of seawater analysis, 2nd ed. Fisheries Research Board of Canada, Ottawa, 310 pp.
16. Bainbridge, A.E., 1980: GEOSECS Atlantic expedition. Volume 2, Sections and Profiles. Super. of Documents, U.S. Government Printing Office, Washington, D.C.
17. Broecker, W.S., 1974: Chemical Oceanography. Harcourt Brace Jovanovich, New York, N.Y., 214 pp.
18. Schaule, B.K. and C.C. Patterson, 1981: Lead in the North Atlantic. In: "Trace Metals in Seawater" (C.S. Wong et al., ed.). NATO Series, Plenum Press, New York.
19. Weiss, R.F., 1977: Hydrothermal manganese in the deep sea: scavenging residence time and Mn/He-3 relationships. Earth Planet. Sci. Lett., 37, 257-262.
20. Craig, H., S. Krishnaswami and B.L.K. Somayajulu, 1973: Pb-210/Ra-226: Radioactive disequilibria in the deep sea. Earth Planet. Sci. Lett., 17, 295-305.

21. Nozaki, Y., K.K. Turekian and K. VanDamm, 1980: Pb-210 in GEOSECS water profiles from the North Pacific. Earth Planet. Sci. Lett., 49, 393-400.
22. Klinkhammer, G.P., M.L. Bender and R.F. Weiss, 1977: Hydrothermal manganese in the Galapagos Rift. Nature, 269, 319-320.
23. Bolger, G.W., P.R. Betzer and V.V. Gordeev, 1978: Hydrothermally-derived manganese suspended over the Galapagos spreading center. Deep-Sea Res., 25, 721-733.
24. Csandy, G.T., 1981: Circulation in the coastal ocean, Part 2. EOS, 62, 41-43.
25. Wallace, G.T., O.M. Mahoney, R. Dulmage, F. Storti and N. Dudek, 1981: First-order removal of particulate aluminum in oceanic surface layers. Nature, 293, 729-731.
26. Goldberg, E.D., 1975: Marine pollution. In: "Chemical Oceanography", 2nd ed., Vol. 3. Riley and Skirrow (eds.), Chapter 17.
27. Okubo, A., 1971: Oceanic diffusion diagrams. Deep-Sea Res., 18, 789-802.
28. Boyle, E.A. and S. Huested, 1981: Aspects of the surface distribution of copper, nickel, cadmium and lead in the North Atlantic and North Pacific. In: "Trace Metals in Seawater", (C.S. Wong et al., eds.). NATO Series, Plenum Press, New York.
29. Turekian, K.K., 1977: The fate of metals in the ocean. Geochim. Cosmochim. Acta., 41, 1139-1144.
30. Trefry, J.H. and B.J. Presley, 1976: Heavy metal transport from the Mississippi River to the Gulf of Mexico. In: "Marine Pollutant Transfer", H.L. Windom and R.A. Duce (eds.). Lexington Books, Lexington, Mass., Chapter 3.
31. Boyle, E.A., 1979: Copper in natural waters. In: "Copper in the Environment", J.O. Nriagu (ed.). John Wiley and Sons, Chapter 4.
32. Sholkovitz, E.R. and D. Copland, 1981: The coagulation, solubility and adsorption properties of Fe, Mn, Cu, Ni, Cd, Co and humic acids in a river water. Geochim. Cosmochim. Acta., 45, 181-189.
33. Sholkovitz, E.R., 1978: The flocculation of dissolved Fe, Mn, Al, Cu, Ni, Co and Cd during estuarine mixing. Earth Planet. Sci. Lett., 41, 77-86.
34. Turekian, K.K., 1971: Rivers, tributaries and estuaries. In: "Impingement of Man on the Oceans", D.W. Hood (ed.). Wiley, New York, Chapter 2.
35. Boyle, E.A., F. Sclater and J.M. Edmond, 1976: On the marine geochemistry of cadmium. Nature, 263, 42-44.
36. Boyle, E.A., S.S. Huested, B. Grant and J.M. Edmond, (in press): The chemical mass balance of the Amazon plume II: copper, nickel and cadmium.

APPENDIX: NORTH ATLANTIC RESULTS

North Atlantic Vertical Profile (34° 06'N, 66° 07'W)

Depth (m)	Phosphate $\mu M\ kg^{-1}$	Silicate $\mu M\ kg^{-1}$	Mn $nmol\ kg^{-1}$	Ni $nmol\ kg^{-1}$	Cu $nmol\ kg^{-1}$	Zn $nmol\ kg^{-1}$	Cd $pmol\ kg^{-1}$
Surface	0.03	0.8	2.5	1.94, 2.15	1.21, 1.10	0.04	2
100	0.08	1.1	1.3	1.94, 2.13	--	0.26	8
136	0.11	1.4	1.2	--	1.26	--	11
375	0.16	2.2	--	2.88, 3.00	1.28	0.30	18
597	0.53	4.0	0.50	2.96, 3.07	1.15	0.48	85
715	0.77	5.8	--	4.43, 4.36	1.34	1.29	165
1036	1.37	13.6	0.71	5.70	1.54, 1.50	1.86	330
1441	1.18	14.9	0.76	5.51	1.62	1.99	272, 266
2030	1.17	14.9	--	5.78	1.70	1.51	280
2386	1.18	16.7	--	--	2.06	1.59	252
2962	1.23	21.5	0.56	5.87, 5.61	2.03, 2.03	1.64, 1.56	296, 307

North Atlantic Surface Raft Samples

Station #	Latitude °N	Longitude °W	Salinity ‰	Phosphate $\mu mol\ kg^{-1}$	Nitrate $\mu mol\ kg^{-1}$	Silicate $\mu mol\ kg^{-1}$	Mn $nmol\ kg^{-1}$	Ni $nmol\ kg^{-1}$	Cu $nmol\ kg^{-1}$	Zn $nmol\ kg^{-1}$	Cd $pmol\ kg^{-1}$
1	37° 27'	70° 22'	35.072	≤.01	.02	0.9	5.0	3.65	1.76	--	42
2	36° 24'	67° 42'	36.418	≤.01	.03	1.7	2.42	2.72	1.34	--	11
3	35° 14'	65° 30'	36.335	≤.01	.04	0.7	2.36	2.15	1.09	--	2
4	34° 02'	66° 04'	36.161	.03	.01	0.8	2.50	2.15	1.21	--	2
5	31° 29'	66° 01'	36.409	≤.01	.01	1.0	2.08	2.28	1.24	--	(13)
6	30° 12'	66° 37'	36.539	≤.01	.01	1.0	2.30	2.59,2.32, 2.50	1.24,1.16, 1.24	.04,.07	2,≤3
7	29° 22'	65° 57'	36.212	≤.01	.02	1.1	--	2.47,2.47, 2.42	1.13,1.21, 1.21	.04,.05	2,≤3
8	32° 11'	65° 21'	36.476	≤.01	.01	1.0	--	2.21	1.02,1.18	.08,.10	≤3
9	35° 06'	67° 55'	36.250	≤.01	.02	1.1	--	2.45	1.38,1.36	.04,.07	≤3
10	37° 09'	69° 14'	36.173	≤.01	.04	1.2	--	2.88,2.31	1.38	.04,.07	4,4,5
11	39° 03'	71° 01'	35.849	.02	.02	1.1	4.3	3.03,3.08, 3.03	1.59,1.62, 1.53	.15,.15	26,27,30, 24
12	39° 48'	71° 01'	34.023	.04	.01	1.2	8.6	3.72,4.12, 3.71	2.41,2.55, 2.28	.60,.58	93,84,91,
13	40° 55'	70° 56'	30.234	.29	.60	3.0	20.5,21.2	5.81,6.05, 5.84	4.33,4.00, 3.82	2.43,2.37	202,195,201

SOME RECENT MEASUREMENTS OF TRACE METALS IN ATLANTIC OCEAN WATERS

J.D. Burton, W.A.Maher and P.J. Statham

Department of Oceanography
The University
Southampton, U.K.

ABSTRACT

Measurements of dissolved arsenic have been made on samples from vertical profiles at six stations in the Cape Basin. After initial reduction of arsenic (V) to arsenic (III), arsine was generated and trapped in a solution containing potassium iodide and iodine. The concentrates were analysed using a zinc reductor column in conjunction with electrothermal atomic absorption spectrophotometry. Concentrations showed only small variations with depth, the average concentration for 17 samples collected above 110 m being 19.9 (range 17-23) nmol l^{-1} and that for 42 samples from deeper waters, up to 4770 m, being 21.1 (range 17-28) nmol l^{-1}. Overall, the concentrations of arsenic were not significantly correlated with those of phosphate.

A method using extraction of metal dithiocarbamate complexes has been used, in conjunction with electrothermal atomic absorption spectrophotometry, to determine dissolved manganese at four stations in the Nares Abyssal Plain region. Surface concentrations of manganese were about 3 nmol l^{-1}; there was a rapid decrease in concentration over the uppermost several hundred metres and in deeper water concentrations mostly varied about 0.5 nmol l^{-1}. Concentrations of cadmium measured at these stations showed increases with depth which were closely related to those of phosphate, the deep water values fitting well with predictions based on comparable relationships in the Pacific Ocean.

INTRODUCTION

The major features in the distributions of a number of trace metals in open ocean waters have become much clearer as a result of the more accurate measurements made since the mid 1970s. Most of the studies on nickel, copper, zinc and cadmium are cited in a recent paper on these elements by Bruland[1]. As has been pointed out by Measures et al.[2], there is particular interest, from the standpoint of understanding the cycles of trace metals, in obtaining more information about elements with distributions influenced by cycling between different oxidation states and those whose transport is not strongly coupled with micronutrients. Among such elements, arsenic[3,4], selenium[2,5], and manganese[6-9] have been recently investigated.

In this paper an account is given of work carried out, during recent cruises in the Atlantic Ocean, on arsenic and manganese, with respect to both analytical aspects and findings on distributions. Some observations on cadmium are also briefly discussed.

SAMPLES AND METHODS

Samples of sea water were collected during two RRS Discovery Cruises. Analyses for arsenic were made on Cruise 99 (January, 1979) to the Cape Basin and for manganese on Cruise 108 (February-March, 1980) to the Nares Abyssal Plain region; locations are listed in Table 1 for those stations specifically referred to in this paper. The samples for analysis of metals were obtained using either 30 litre Niskin or 7.5 litre NIO bottles.

At each station, in addition to data for temperature and salinity, profiles were obtained for phosphate and silicon, which

Table 1 Details of Stations.

Station number	Position	Bottom Depth (m)
Cape Basin		
9944	36°10.3'S, 9°0.5'E	5017
9948	34°19.8'S, 16°19.2'E	3869
Nares Abyssal Plain		
10163	23°41.3'N, 59°40.8'W	5852
10164	26°12.7'N, 60°22.5'W	6140
10165	23°44.3'N, 61°29.0'W	5825
10169	23°43.5'N, 65°11.7'W	5760

were determined using the conditions described by Strickland and Parsons[10], and dissolved organic carbon (DOC) using the photo-oxidative method of Collins and Williams[11], with minor modifications as described by Gershey et al.[12]. The samples for measurement of DOC were filtered through Whatman GF/F filters, which had been heated previously at 500°C, and the filtrates immediately frozen for return to the shore laboratory for analysis.

The analytical procedures given below were designed to enable the initial stages of separation and concentration to be carried out on board ship as rapidly as possible after sample collection, giving stable concentrates for return to the shore laboratory and final determination. The method for initial concentration of arsenic was designed to be readily compatible with subsequent determination of the element by arsine generation and measurement using electrothermal atomic absorption spectrophotometry. The method used for the concentration of manganese was a solvent extraction method based on the multi-element method of Danielsson et al.[13]. The conditions were substantially modified to allow the determination of manganese and iron on the same sample; they are also suitable for the quantitative recovery of a number of other metals. Measurements of iron were made on samples from Cruise 108 but were invalidated by contamination problems. Fuller details of the development of the procedures will be published elsewhere.

Arsenic

Samples from 1000 m and above were filtered through acid-washed 0.45 μm Sartorius membrane filters. Analyses on samples from depths below 1000 m were made on unfiltered water.

Aliquots of 50 ml were placed in a round bottomed flask, fitted with a modified Dreschel head and an injection syringe in a side arm. Twenty ml of concentrated hydrochloric acid, 1 ml of 1M ascorbic acid solution and 1 ml of 1M potassium iodide solution were added. The solution was stood for 30 min. to allow reduction of As(V) to As(III), which was necessary to ensure quantitative recovery of inorganic arsenic as arsine under the conditions used in the subsequent step. With nitrogen passing through the flask at a flow rate of 150 ml min^{-1}, 0.5 ml of 8% w/v sodium borohydride solution was added from the syringe. The arsine evolved was trapped in 2 ml of a solution containing 0.7% w/v potassium iodide, and excess iodine, over a period of 3 min.

The concentrates were subsequently analysed for arsenic using a Varian-Techtron AA5 atomic absorption spectrophotometer fitted with a Perkin-Elmer HGA 72 carbon furnace, linked to a zinc reductor column for the generation of arsine (see Fig. 1). A continuous stream of argon was allowed to flow with the column

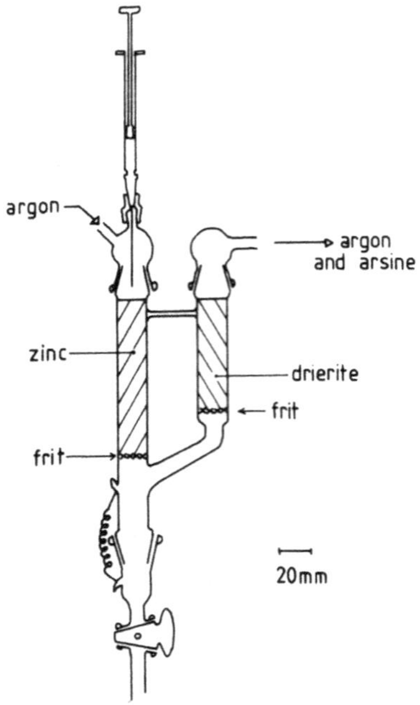

Fig. 1 Zinc reductor column for generation of arsine in determination of arsenic by electrothermal atomic absorption spectrophotometry.

connected into the inert gas line between the HGA 72 control unit and the inlet to the furnace. Calcium sulphate (10-20 mesh) was used as an adsorbent to prevent water vapour entering the carbon furnace. The carbon tube used was of 10 mm i.d. and had a single centrally located inlet hole.

Preliminary reduction of the As(V) present in the concentrates to As(III) was again necessary to obtain quantitative recovery of arsenic as arsine. To the concentrate, 50 µl of 1M ascorbic acid solution was added followed by 0.5 ml of concentrated hydrochloric acid. After standing for 45 min, an aliquot of 0.8 ml was injected into the zinc reductor column, with argon flowing at a rate of 750 ml min^{-1} and the atomic absorption of the arsenic formed by decomposition in the heated carbon tube (ca. 1700℃) was measured at 193.7 nm, with background correction. The procedure was calibrated by making known additions of a standard solution of sodium arsenate to samples of sea water and carrying the spiked samples through the entire procedure, commencing on board ship. During a series of determinations, As(III) standards were injected periodically into the reductor column to allow for any small changes in detector response. Under the conditions used there was a linear relationship of absorbance and concentration over the range 5-70 ng As ml^{-1} of injected solution.

The recovery of arsenic in the above procedure was checked by adding replicate known amounts of As(V) to aliquots of sea water which had been stripped of arsenic. The results were as follows:

Arsenic concentration (nmol l^{-1})	13.3	66.7
Recovery (%)	96	98
Coefficient of variation (%)	3.1	3.7

Replicate analyses commenced during Cruise 99 on a sample containing 21.1 nmol As l^{-1}, gave a coefficient of variation of 3.7%. Because of the ubiquitous presence of arsenic in reagents such as hydrochloric acid and sodium borohydride, blanks have been found to be high (in some cases, about 6.7 nmol l^{-1}) but with a low standard deviation, corresponding to a detection limit (defined as three times the standard deviation of the blank) of 2.0 nmol l^{-1}.

A wide range of elements was tested for interfering effects; the only significant interferences found were at concentrations much higher than those encountered in sea water. No significant difference in the results was found when a sample of sea water was analyzed in the way described and also by the same procedure but using the method of standard additions.

Manganese

The use of 8-hydroxyquinoline as a complexing agent for separation and concentration of manganese from sea water, using solvent extraction, has been favoured in recent work[6-8]. Since an objective in the present work was to develop a procedure for the simultaneous extraction of manganese and iron, the use of this reagent for extraction of iron was examined using ^{59}iron as a tracer. Conditions were established which gave quantitative extraction of both metals into oxine/chloroform, but recoveries of iron in a back extraction step, using nitric acid, were low and variable.

As an alternative, a modification of the procedure of Danielsson et al.[13], to include manganese, was examined. In order to obtain satisfactory recovery of manganese, it was necessary to increase the concentration of the combined ammonium pyrrolidine-dithiocarbamate (APDC)/diethylammonium diethyldithiocarbamate (DDTC) to at least 0.1% $^{w}/v$ of each compound, in the sea water sample. Recoveries were independent of pH in the range 6-9. A study also of the optimum conditions for back extraction from the solvent (Freon TF, i.e. 1,1,2-trichloro-1,2,2-trifluoroethane) led to the adoption of the method outlined below.

Samples for the analysis of manganese were pressure filtered through 0.4 μm Nuclepore filters. To 350 ml of filtrate, 20 ml of an aqueous solution of the complexing agents (2% $^{w}/v$ in both APDC and DDTC) were added, and the solution extracted first with 35 ml and then with 20 ml of freon for 6 min. The combined extracts were shaken with 100 μl of concentrated nitric acid for 30 sec. After standing for 5 min, 5 ml of distilled water was added, and the solution shaken for 30 sec. The aqueous phase was separated and combined with that from a further back extraction using the same procedure. The combined aqueous solutions were returned to the shore laboratory and manganese determined by electrothermal atomic absorption spectrophotometry, using a Pye-Unicam SP9 instrument.

Separations were carried out in a laminar flow hood. The separating funnels were of fluorinated alkene and the concentrates were stored. in high density polyethylene bottles. Method blanks obtained with shipboard use of the procedure were about 7 ng; they have subsequently been reduced to about 1.5 ng. The limit of detection for the determination reported here, estimated as three times the standard deviation of the blank, was about 0.3 nmol l^{-1}.

The conditions established for the determination of manganese have been shown to be suitable also for the determinations of cadmium and copper, as well as iron. Further work is needed, however, to reduce blanks and improve detection limits for these elements.

RESULTS AND DISCUSSION

Arsenic

The vertical profiles of dissolved arsenic at six stations in the Cape Basin, with water depths from 1 to 4.8 km, showed relatively small variations in concentration with depth. Two of these profiles are shown in Fig. 2a and the corresponding profiles of phosphate and silicon are shown in Fig. 2b; the data are listed in Table 2. Overall, there was no significant correlation of arsenic with phosphate. A small but significant difference in concentration of arsenic between near-surface and deeper waters was, however, found. For waters above 110 m the mean concentration for 17 samples was 19.9 nmol l^{-1} with a range of 17-23 nmol l^{-1}, whereas for deeper waters below 110 m the corresponding values for 42 samples were 21.1 and 17-28 nmol l^{-1}, respectively.

The minor influence of transport of arsenic into deeper waters by biogenous particulate material is indicated also by a comparison of these values with those found by Andreae[4] in the northeastern Pacific Ocean. His results show an average concen-

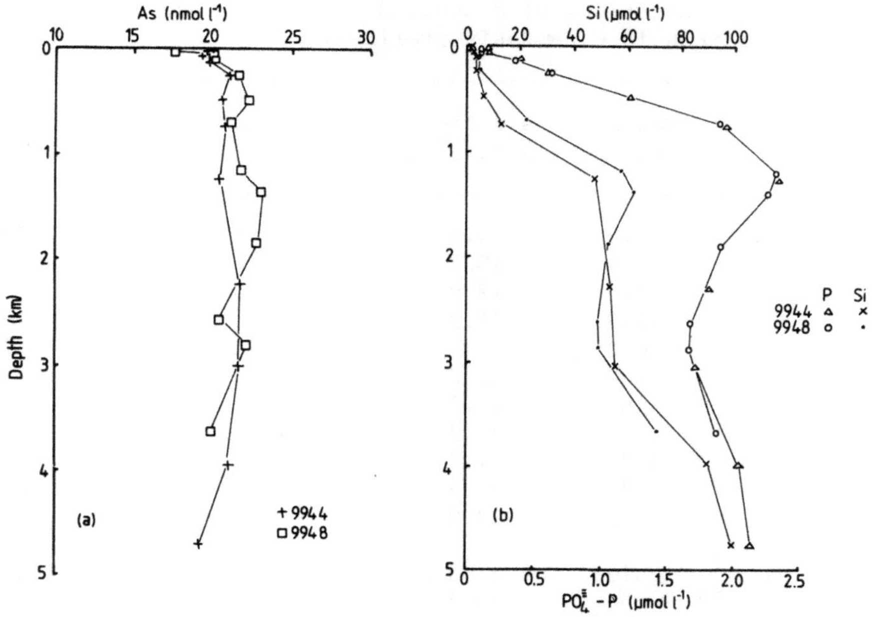

Fig. 2 Vertical profiles at Discovery stations 9944 and 9948 in the Cape Basin: (a) Dissolved inorganic arsenic, (b) Phosphate and silicon. Positions of stations are given in Table 1.

tration for water below 1 km of 24 $\pm$ 1 nmol l^{-1}. There thus appears to be only a small increase in the concentration of dissolved arsenic in deep Pacific waters above that in deep Atlantic waters, in contrast with the marked changes in phosphate and the other micronutrients. The major influence of biological processes on the behaviour of arsenic appears to be cycling between oxidation states in the euphotic zone, with little vertical transport, relative to the reservoir concentration. In this respect the behaviour of arsenic contrasts markedly with that of selenium[2,5,14].

Manganese

The vertical profile of dissolved manganese at station 10169 in the Nares Abyssal Plain region is shown in Fig. 3; the distributions of other oceanographic variables are also shown in this figure together with those for phosphate, silicon and DOC. The

Table 2 Concentrations of dissolved arsenic, phosphate and silicon in Cape Basin profiles.

	Depth (m)	Arsenic (nmol l^{-1})	Phosphate (μmol l^{-1})	Silicon (μmol l^{-1})
Station 9944	10	19.8	0.15	1.80
	50	19.5	0.16	1.86
	110	19.9	0.40	3.03
	245	21.2	0.61	3.10
	490	20.7	1.22	6.83
	750	21.0	1.83	13.1
	1260	20.7	2.34	48.6
	2281	21.9	1.81	53.4
	3021	21.9	1.71	55.8
	3973	21.5	2.04	90.6
	4772	19.8	2.13	99.9
Station 9948	10	17.6	0.11	1.07
	50	20.3	0.10	1.12
	105	20.3	0.37	3.42
	243	21.8	0.64	4.03
	490	22.4	a	a
	720	21.2	1.90	22.1
	1190	22.0	2.31	57.9
	1393	23.4	2.26	62.2
	1883	23.1	1.90	52.6
	2619	20.8	1.67	49.0
	2853	22.4	1.67	49.5
	3664	20.3	1.89	71.5

a no data

data are listed in Table 3. Similar profiles were obtained at stations 10163-5. The concentrations of manganese show a pronounced decrease in the upper 500 m, from surface values averaging about 3 nmol l^{-1}. Concentrations at 500 m and below mostly varied about 0.5 nmol l^{-1}, these concentrations being too close to the detection limit to warrant any analysis of apparent trends. Klinkhammer and Bender[7] found subsurface maxima in concentrations of total dissolvable manganese, associated with concentrations of dissolved oxygen in the oxygen minimum layer of less than 100 μmol l^{-1}. Concentrations of dissolved oxygen in the oxygen minima in the Nares Abyssal Plain region are significantly higher than this.

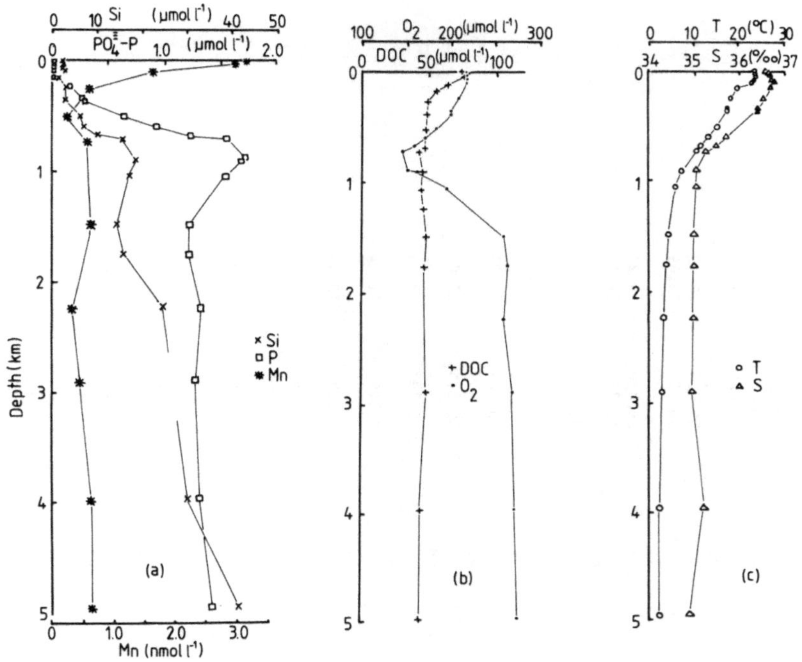

Fig. 3 Vertical profiles at Discovery station 10169 in the Nares Abyssal Plain region: (a) Dissolved manganese, phosphate and silicon, (b) Dissolved oxygen and dissolved organic carbon (DOC), (c) Temperature and salinity. Values for dissolved manganese at depths of 524 m and 2240 m were below the limit of detection as defined in the text. Positions of stations are given in Table 1.

The data are consistent with an input of manganese to surrace waters, and scavenging of the element in deeper waters, as discussed and evaluated by other workers[6-9]. The possible roles of eolian inputs and the advection of manganese from coastal regions, where inputs can arise not only from rivers but also from the release of Mn(II) from reduced sediments, remain to be clarified. Measurements in the upper few hundred metres on suitably chosen sections should provide useful information in this regard. The possible role of biological cycling of manganese in the euphotic zone, which could act as a reductive pump for the element, also requires investigation.

Table 3. Concentrations of manganese in the profile at station 10169, with other chemical and hydrographic data

Depth (m)	Manganese (nmol l^{-1})	Cadmium (nmol l^{-1})	Phosphate (μmol l^{-1})	Silicon (μmol l^{-1})	Oxygen (μmol l^{-1})	Dissolved organic carbon (μmol l^{-1})	Salinity ($^{o}/oo$)	Temperature (°C)
1			b	2.5	216		36.600	23.61
10	3.17					75		
25			0.01	2.2	220		36.654	23.52
40			b	2.4	217		36.670	23.49
50	2.97					76		
88			b	2.3	217		36.730	23.23
110			b	2.6	217		36.812	22.78
120	1.64					64		
178			0.02	a	a		36.703	19.66
188						56		
253			0.16	3.1	207		36.539	18.22
263	0.60					49		
370			0.28	3.7	200		36.409	17.35
388						48		
514			0.64	6.1	182		36.035	15.05
524	0.25	0.14				48		
605			0.93	7.0	170		35.737	13.14
615		0.25				48		
683			1.23	10.3	158		35.492	11.34
693		0.29				47		
722			1.55	15.5	146		35.273	10.49
732	0.58	0.40				43		
900			1.72	a	152		35.052	6.90
903			1.71	18.6	162		35.060	7.21
913		0.48				45		
1055			1.54	17.2	196		35.050	5.98
1065		0.39				44		
1481			1.22	14.3	262		35.025	4.33
1491	0.66	0.35				48		
1750			1.24	15.8	265		35.004	3.79
1760		0.27				47		
2230			1.32	24.8	261		34.989	3.30
2240	0.33	0.29				47		
2885			1.27	a	271		34.942	2.77
2895	0.46	1.60[c]				48		
3949			1.31	30.4	274		34.913	2.35
3959	0.66	0.33				44		
4950			1.42	41.4	267		34.888	2.23
4960	0.67	0.70[c]				43		

a no data b not detected c suspected contamination

Cadmium

Analyses of cadmium were carried out on the concentrates from samples collected in the Nares Abyssal Plain region. As discussed above, the conditions used in the analytical method were optimized primarily for the determination of manganese and the detection limit obtained for cadmium (0.04 nmol l^{-1}, estimated as three times the standard deviation of the blank) was too high to permit precise measurements in surface waters. Nevertheless, the profiles of cadmium were closely related to those of phosphate, as observed in the Pacific[1,15,16] and Indian[17] Oceans. The overall regression equation obtained from data for all the stations gives a value of 0.27 nmol Cd l^{-1} for water containing 1 μmol PO_4^{3-} - P l^{-1}, while the relationship found by Bruland[1] for North Pacific Ocean waters predicts a value of 0.28 nmol Cd l^{-1}. The data thus support the view that a common relationship between cadmium and phosphorus applies through the deep waters of the major ocean basins. Concentrations of dissolved cadmium for samples below 500 m at station 10169 are shown in Table 3.

ACKNOWLEDGEMENTS

The authors thank the officers, crew and scientific complement of RRS Discovery on Cruises 99 and 108 for their assistance. They are especially grateful to Dr. S.E. Calvert and Dr. F. Culkin and Mr. M.J. McCartney who provided some of the oceanographic measurements. The work was supported by the Natural Environment Research Council.

REFERENCES

1. Bruland, K.W., 1980: Oceanographic distributions of cadmium, zinc, nickel, and copper in the North Pacific. Earth Planet. Sci. Lett., 47, 176-198.
2. Measures, C.I., R.E. McDuff and J.M. Edmond, 1980: Selenium redox chemistry at GEOSECS I re-occupation. Earth Planet. Sci. Lett., 49, 102-108.
3. Andreae, M.O., 1978: Distribution and speciation of arsenic in natural waters and some marine algae. Deep-Sea Res., 25, 391-402.
4. Andreae, M.O., 1979: Arsenic speciation in seawater and interstitial waters: The influence of biological-chemical interactions on the chemistry of a trace element. Limnol. Oceanogr., 24, 440-452.
5. Measures, C.I. and J.D. Burton, 1980: The vertical distribution and oxidation states of dissolved selenium in the northeast Atlantic Ocean and their relationship to biological processes. Earth Planet. Sci. Lett., 46, 385-396.

6. Bender, M.L., G.P. Klinkhammer and D.W. Spencer, 1977: Manganese in seawater and the marine manganese balance. Deep-Sea Res., 24, 799-812.
7. Klinkhammer, G.P. and M.L. Bender, 1980: The distribution of manganese in the Pacific Ocean. Earth Planet. Sci. Lett., 46, 361-384.
8. Landing, W.M. and K.W. Bruland, 1980: Manganese in the North Pacific. Earth Planet. Sci. Lett., 49, 45-56.
9. Martin, J.H. and G.A. Knauer, 1980: Manganese cycling in northeast Pacific waters. Earth Planet. Sci. Lett., 51, 266-274.
10. Strickland, J.D.H. and T.R. Parsons, 1972: "A Practical Handbook of Sea Water Analysis" Bull. Fish Res. Bd Can., 167, Second ed., 310 pp.
11. Collins, K.J. and P.J. leB. Williams, 1977: An automated photochemical method for the determination of dissolved organic carbon in sea and estuarine waters. Mar. Chem., 5, 123-141.
12. Gershey, R.M., M.D. Mackinnon, P.J. leB. Williams and R.M. Moore, 1979: Comparison of three oxidation methods used for the analysis of the dissolved organic carbon in seawater. Mar. Chem., 7, 289-306.
13. Danielsson, L.-G., B. Magnusson and S. Westerlund, 1978: An improved metal extraction procedure for the determination of trace metals in sea water by atomic absorption spectrometry with electrothermal atomization. Anal. Chim. Acta., 98, 47-57.
14. Burton, J.D., W.A. Maher, C.I. Measures and P.J. Statham, in press: Aspects of the distribution and chemical form of selenium and arsenic in ocean waters and marine organisms. Thalassia Jugosl.
15. Boyle, E.A., F. Sclater and J.M. Edmond, 1976: On the marine geochemistry of cadmium. Nature, 263, 42-44.
16. Bruland, K.W., G.A. Knauer and J.H. Martin, 1978: Cadmium in northeast Pacific waters. Limnol. Oceanogr., 23, 618-625.
17. Danielsson, L.-G., 1980: Cadmium, cobalt, copper, iron, lead, nickel and zinc in Indian Ocean water. Mar. Chem., 8, 199-215.

DETERMINATION OF THE RARE EARTH ELEMENTS IN SEA WATER

H. Elderfield and M.J. Greaves

Department of Earth Sciences
The University of Leeds
Leeds LS2 9JT, U.K.

ABSTRACT

A method is described for the mass spectrometric isotope dilution analysis of rare earth elements in sea water. The REE are concentrated from sea water by coprecipitation with ferric hydroxide and separated from other elements and into groups for analysis by anion exchange using mixed solvents. Results for synthetic mixtures and standards show that the method is accurate and precise to $\sim \pm 1\%$; and blanks are low (e.g. 10^{-12} moles La and 10^{-14} moles Eu). The method has been applied to the determination of nine REE in a variety of oceanographic samples. Results for N. Atlantic ocean water below the mixed layer are (in 10^{-12} mol kg^{-1}) 13.0 La, 16.8 Ce, 12.8 Nd, 2.67 Sm, 0.644 Eu, 3.41 Gd, 4.78 Dy, 4.07 Er and 3.55 Yb, with an enrichment of REE in deep ocean water, by $\sim$ 2 x for the light REE and $\sim$ 1.3 x for the heavy REE.

INTRODUCTION

The rare earth elements (REE) have occupied an important role in marine geochemical research, particularly as used in the format of REE abundance patterns to describe the geochemical pathways in marine sedimentation and authigenesis[1]. Such patterns in different marine phases have led to the recognition that some fractionation of the REE does take place in the oceans despite the chemical similarities of this group of elements[2,3]. The geochemical processes responsible for this fractionation are not well understood and this prompted us to undertake a study of the marine geochemistry of the REE. Some results for marine sediment components have been described already[4,5,6]. In this report we outline the analytical basis of this research and, specifically, describe the mass spectrometric isotope dilution method which we have developed for the determination of REE in sea water and other natural waters. Variations for REE analysis of sediments and for Nd isotopic studies also are mentioned. Some results are given of the REE analysis of a variety of marine phases.

THE RARE EARTH ELEMENTS

The rare earths or lanthanides are the group of 14 elements (atomic nos. 58-71) in the periodic table following lanthanum, but La itself is usually included in the REE group. The REE can be split into two sub-groups: the light REE (Ce to Gd) with parallel electronic spins in the 4f shell (La can be added) and the heavy REE (Tb to Lu) with opposite spins in the 4f shell.

The naturally-occurring REE isotopes are shown in Fig. 1. Ten of the REE are poly-isotopic (La, Ce, Nd, Sm, Eu, Gd, Dy, Er, Yb and Lu) and these are the elements to which the stable-isotope dilution technique is applicable. Pr, Tb, Ho and Tm are mono-isotopic and Pm has no naturally-occurring isotope. All the REE exhibit artificial radioactivity mostly of short half-life which decays by β+/electron capture. Natural radioactivity has been identified in La, Nd, Sm, Gd and Lu; as a consequence, Ce, Nd, Sm and Yb have radiogenic isotopes. The decay schemes are:

$$ {}^{147}_{62}\mathrm{Sm} \xrightarrow{\alpha} {}^{143}_{60}\mathrm{Nd} $$

$$ {}^{152}_{64}\mathrm{Gd} \xrightarrow{\alpha} {}^{148}_{62}\mathrm{Sm} \xrightarrow{\alpha} {}^{144}_{60}\mathrm{Nd} \xrightarrow{\alpha} {}^{140}_{58}\mathrm{Ce} $$

$$ {}^{149}_{62}\mathrm{Sm} \xrightarrow{\alpha} {}^{145}_{60}\mathrm{Nd} $$

$$ {}^{138}_{56}\mathrm{Ba} \xleftarrow{EC} {}^{138}_{57}\mathrm{La} \xrightarrow{\beta} {}^{138}_{58}\mathrm{Ce} $$

$$ {}^{176}_{70}\mathrm{Yb} \xleftarrow{EC} {}^{176}_{71}\mathrm{La} \xrightarrow{\beta} {}^{176}_{72}\mathrm{Hf} $$

Of particular interest is the α decay of ^{147}Sm ($t_{1/2}$ = 1.06 x 10^{11}y.) which, coupled with natural fractionations of Sm/Nd, allows the application of Nd isotopes as an indicator of the sources of the REE in marine phases[4,7,8,9]. (The β decay of ^{176}Lu, $t_{1/2}$ = 3 x 10^{10}y., has a similar potential but has not yet been exploited in marine studies). The half life of ^{144}Nd is sufficiently long ($\sim$ 5 x 10^{15}y.) for the ^{143}Nd/^{144}Nd ratio, normalized to a natural ^{146}Nd/^{144}Nd ratio of 0.7219, to be the parameter used. Nd isotopic compositions may be obtained by a variation of the method given below.

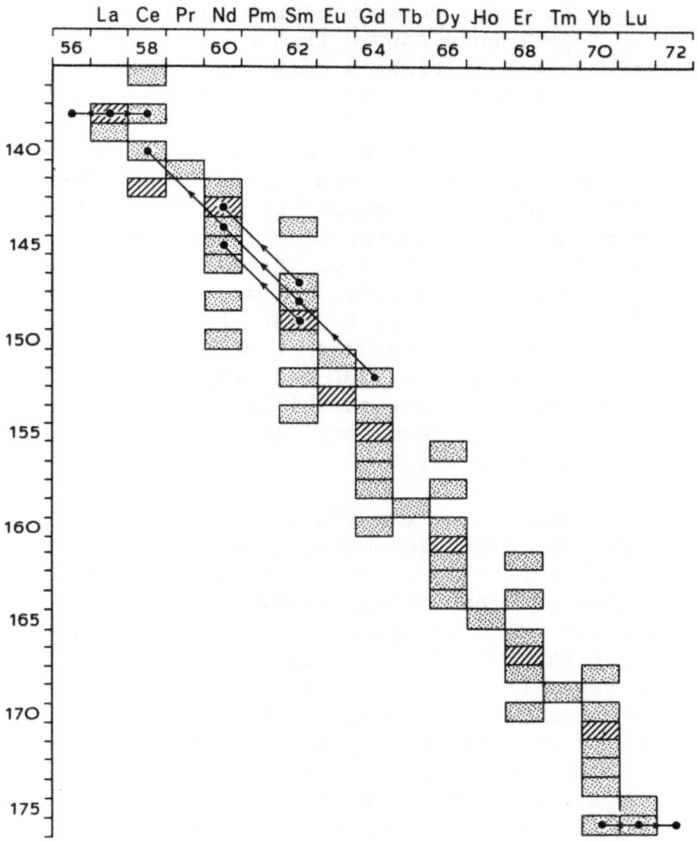

Fig. 1 REE isotopes. The isotopes of each element marked with dark shading are the spike-enriched isotopes. Lines show the sequences and arrows the directions of natural-radioactive REE decays.

EXPERIMENTAL

The basic procedure involves the equilibration of a large-volume sea water sample with a mixture of REE spikes, concentration of the REE from the water sample, chemical separation of the REE from other elements and mass spectrometric analysis.

It is essential that sample and spikes are equilibrated. If the REE are not complexed, equilibration should be almost instantaneous and thorough mixing is all that is required. Normally, equilibration is achieved during the storage period between sample collection and transport of the samples to the laboratory for analysis. However, under certain circumstances equilibration may not be rapid. Experiments showed that equilibration times for river- and estuary-waters rich in dissolved organics ("humics") were of the order of 100 hours (rate constants $\sim 3 \times 10^{-6}$ sec^{-1}).

The REE were concentrated by coprecipitation with ferric hydroxide, a procedure used for the separation of REE in the neutron-activation method of Høgdahl et al.[10]. Instead of using ammonia gas for the precipitation of Fe- and REE-hydroxides, NH_3 solution in water was used which was found easier to manipulate and which could easily be purified. Tracer experiments, using ^{139}Ce and ^{153}Gd, were performed to test the recovery of the coprecipitation stage and it was found that almost quantitative removal of REE from seawater could be achieved (Fig. 2). In addition, it was observed that much smaller quantities of ferric chloride solution ($\ll$ 0.5 mg Fe kg^{-1} of seawater) were required than were used by Høgdahl[11] (15 mg Fe kg^{-1}).

Chemical separation of the REE was carried out using a modification of the two-column ion-exchange procedure described by Hooker[12] and Hooker et al.[13]. This procedure is based upon the technique of combined ion-exchange/solvent extraction used by Korkisch and Arrhenius[14] to separate the REE from marine sediments and by various workers[15,16,17] to separate the REE one from another. The first stage of the separation uses a glacial CH_3COOH /HNO_3 mixture on an anion-exchange resin to separate the REE as a group from Fe, Ba and sea-salt cations. Elution-volume calibrations were performed using radioactive tracers of the REE and ^{133}Ba, with atomic-absorption or flame-emission analysis of Fe, Na, K, Ca and Mg (Fig. 3). The removal of Ba is very important because of interference with La, Eu and Sm (Table 2). In some instances ion beams of $^{138}Ba^+$ and $^{138}BaO^+$ were observed because of carry-over of Ba from the 1st columns due to incomplete elution

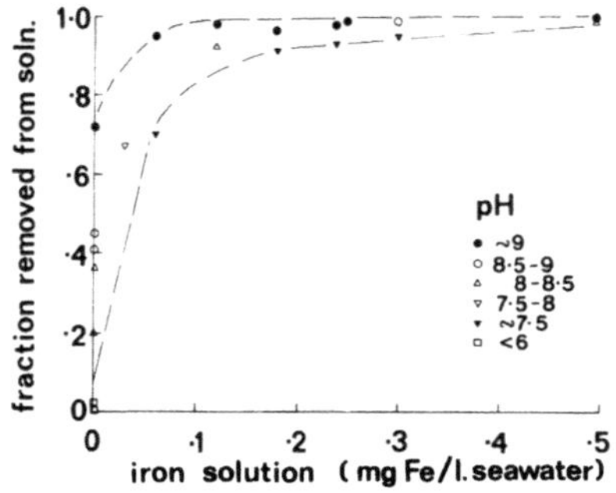

Fig. 2 Results of tracer experiments showing removal of ^{139}Ce from spiked seawater by iron hydroxide coprecipitation as a function of pH and amount of ferric solution added.

when the column is overloaded and which is removed on stripping off the REE (Fig. 3). As shown in Fig. 3 any Ba added to the 2nd columns is eluted at the start of the "light REE fraction" (see below). To ensure Ba removal the sample can be put through the 1st column again. Another difficulty which can arise is overloading because of large amounts of sea salts present from poor washing of the hydroxide precipitate. This can be solved by a preliminary pass through a cation column loaded with HCl.

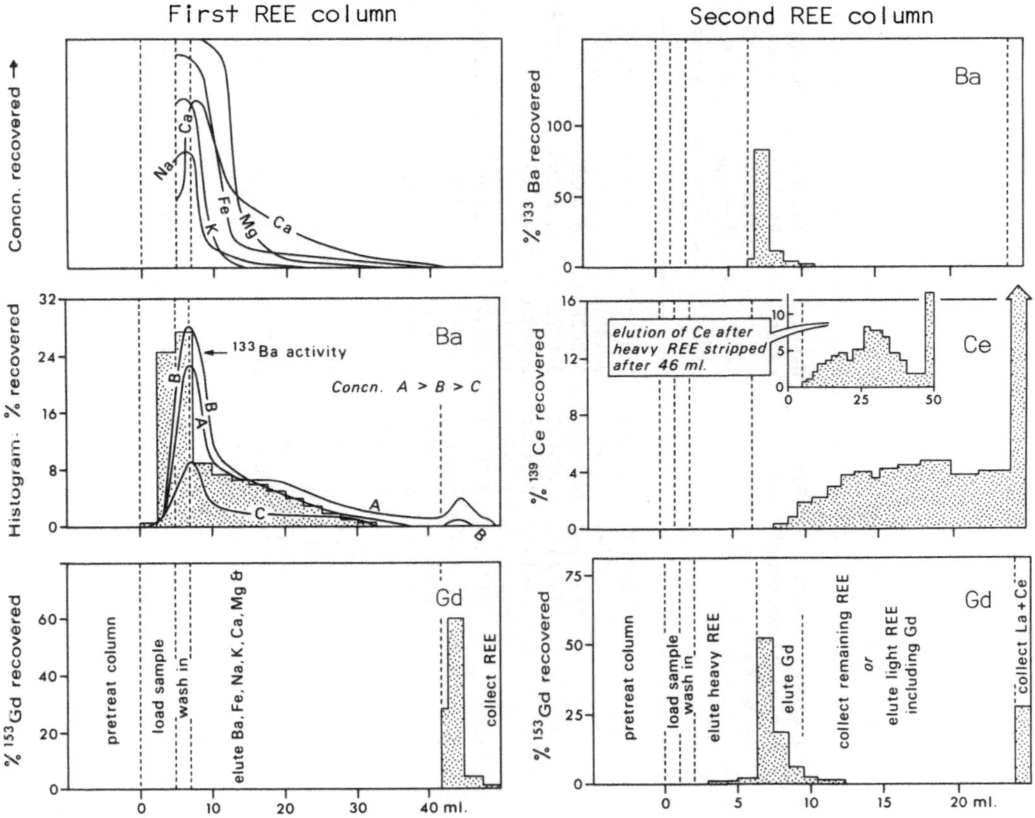

Fig. 3 Elution curves of the REE and interfering elements for anion-exchange separations using CH_3COOH/HNO_3 (1st columns) and $MeOH/HNO_3$ (2nd columns) mixtures.

Table 1. Measured Isotopic Ratios and Concentrations for REE Spikes L2 and H2

Element	Isotope enriched in spike	Ratio measured	Spike ratio	Natural ratio	Natural abundance numerator isotope (as fraction)	Spike concn. µ moles g^{-1} of denominator isotope	Atomic mass of natural element	ppm in chondrite	ppm in shale
La	138	138/139	0.07242	0.00089079	0.00089	0.001237	138.91	0.329	41.0
Ce	142	142/140	4.858	0.12511	0.1107	0.001008	140.12	0.865	83.0
		*138/140	0.001685	0.002825	0.0025				
Nd	143	145/143	0.00632	0.67925	0.08293	0.002832	144.24	0.63	38.0
		146/143	0.01616	1.40826	0.1719				
		**142/143	0.0301	2.22756	0.2717				
Sm	149	147/149	0.00354	1.0854	0.1497	0.001031	150.4	0.203	7.50
		152/149	0.00686	1.9348	0.2671				
Eu	153	151/153	0.00693	0.916	0.4782	0.0004543	151.96	0.077	1.61
Gd	155	156/155	0.00123	1.3897	0.2047	0.001045	157.25	0.276	6.35
		157/155	0.00278	1.0645	0.1568				
Dy	161	162/161	0.05196	1.35222	0.2553	0.002134	162.50	0.343	5.50
		163/161	0.00546	1.32256	0.2497				
Er	167	166/167	0.0319	1.45641	0.3341	0.001594	167.26	0.225	3.75
		168/167	0.0553	1.18003	0.2707				
Yb	171	172/171	0.09163	1.52481	0.2182	0.001652	173.04	0.22	3.53
		174/171	0.01636	2.22502	0.3184				

* used for interference correction on La; ** used for interference correction on Ce

The second stage of the separation uses methanol/HNO_3 mixtures on an anion-exchange column to separate the REE into groups for analysis. Routinely, three fractions were collected by elution or stripping: a heavy REE fraction (Yb, Er, Dy), a light REE fraction (Gd, Eu, Sm, Nd) and a Ce plus La fraction. In practice, the separations tend to be relatively broad for some REE and these elements appear in other than their notational fractions. For example, Ce has a broad elution curve (Fig. 3) and also occurs in the light REE fraction; and Nd invariably is present in the Ce plus La fraction. The separations are partly for convenience in mass spectrometry but also minimize the important isobaric interferences (Table 2). As an alternative to the above scheme, a three-fraction separation producing a heavy REE fraction (as above), a Gd fraction and a light REE fraction of Eu, Sm, Nd, Ce and La has sometimes been used. Ionization of Gd^+ (or GdO^+) has proven difficult when small sample loads are associated with imperfect separations and running of Gd in a separate fraction can help.

Mass spectrometry was performed using a VG Micromass 30 mass spectrometer. Some early work on solids was carried out using an AEI MS5 instrument. Initially, the samples were run with single-filament ionization (using a Ta filament) producing oxide-ion (MO^+) species of La, Ce, Nd and Gd and M^+ species for Sm, Eu and the heavy REE fraction. As the work progressed it was decided to obtain triple-filament ionization (using mixed Re/Ta filaments) to give M^+ species for all the REE. However, when Gd is run separately it is often loaded on a single filament and run as GdO^+. Also, the Ce plus La fraction may be loaded on a single filament with ascorbic acid and colloidal carbon added to obtain M^+ emission. Otherwise samples are loaded with water or H_3PO_4. The observed sequence of metal-ion emission with increasing filament temperature was (Nd) → Ce → La; Eu → Sm → Nd → Gd; Yb → Dy → Er.

SEAWATER ANALYSIS

Reagents and Cleaning Procedures

Analar grade hydrochloric, nitric and glacial-acetic acids, 0.88 ammonia solution and methanol were purified using sub-boiling quartz-finger stills. The HCl was given a preliminary boiling distillation in glass. Water was purified firstly by passing Leeds tap water through a single boiling distillation in glass followed by mixed-bed ion exchange; and, secondly, purifying this distilled-deionized water with a sub-boiling quartz-finger still. Ferric chloride solution was prepared by dissolving Johnson and Matthey Specpure Fe powder in purified HCl and passing through a cation-exchange resin.

Table 2 Isotopic ratios used in mass spectrometry and possible isobaric interferences

Element	Ratios used in isotope-dilution analysis	Masses measured	
		Atomic number	REE isotopes and possible interferences[§]
La	138/139	137	Ba
		138	La + Ba* + Ce*
		139	La
		140	Ce
Ce	142/140	140	Ce
		142	Ce + Nd*
		143	Nd
Nd	145/143, 146/143	143	Nd
		145	Nd
		146	Nd (+ BaO)
Sm	147/149, 152/149	147	Sm
		149	Sm
		152	Sm (+ BaO + CeO + Gd)
Eu	151/153	151	Eu (+ BaO)
		153	Eu (+ BaO)
		154	(BaO + Sm + Gd)
Gd	156/155, 157/155	155	Gd (+ LaO)
		156	Gd (+ Dy + CeO)
		157	Gd (+ PrO)
	172/171, 173/171	171	GdO (+ LaO_2)
		172	GdO (+ CeO + DyO + Yb)
		173	GdO (+ PrO + Yb)
Dy	162/161, 163/161	161	Dy (+ NdO)
		162	Dy (+ NdO + Er)
		163	Dy (+ SmO)
Er	166/167, 168/167	166	Er (+ NdO + SmO)
		167	Er (+ EuO)
		168	Er (+ CeO_2 + SmO + Gd + Yb)
Yb	172/171, 174/171	171	Yb (+ LaO_2 + GdO)
		172	Yb (+ CeO_2 + GdO + DyO)
		174	Yb (+ CeO_2 + GdO + DyO)

§ Interference correction applied for elements marked with asterisk; intereferences in parantheses rarely observed because of differences in emission temperatures and/or chemical separation.

Polypropylene, teflon and glass ware, tubing, filter rings etc. were subjected to rigorous cleaning procedures using hot HNO_3

$/H_2O$, or $HNO_3/HCl/H_2O$ sequences. Nucleopore filters were cleaned by soaking in dilute HCl. All critical manipulations (filter handling, column chemistry) were performed on a laminar-flow clean bench and evaporations were carried out inside teflon hoods swept by filtered air and under infrared lamps.

Spike Solutions

Two mixed spike solutions L2 (containing La, Ce, Nd, Sm, Eu and Gd) and H2 (containing Dy, Er and Yb) were prepared from weighed amounts of stock solutions of enriched isotopes and stored in FEP dropper bottles. The isotopic compositions of the spikes were measured and the spikes were calibrated against gravimetrically-standardized REE solutions prepared from ignited specpure oxides (Table 1). Typically, $\sim$ 0.5000 g L2 (20 drops) and $\sim$ 0.2000 g H2 (8 drops) were used for spiking a $\sim$ 50 kg sea water sample. The weighed spikes were made up in FEP bottles to which 200 g 6M HCl was added. For river waters (having higher REE concentrations) similar spike weights were used for a $\sim$ 10 kg sample.

Ion-Exchange Columns

Bio-Rad AG1X8 resin (200-400 mesh) was packed in a slurry in water into two quartz columns, the "1st column" of 1 cm diameter and the "2nd column" of 0.5 cm diameter, each to give a resin bed of $\sim$ 3 cm depth. The columns in use each consist of a $\sim$ 30 ml-volume bowl above the $\sim$ 10 cm height column plugged with a polypropylene frit above the taper. The resins were treated before use with successive bowlfuls of 10% HNO_3 and water and the procedure repeated.

Procedure

Spiking. On board ship, the pre-weighed acidified spike solution was added, with mixing, to the 0.4 μm Nucleopore-filtered sea water sample of 30-50 kg held in a polypropylene storage bottle. Addition of acid at this stage lowers the sample pH to $\sim$ 2 which stops REE adsorption on the container walls. In the laboratory, the sample weight was recorded.

Equilibration. The mixture of sample plus spikes was allowed to equilibrate. This was achieved in the period of months between sample collection/spiking and arrival of the water at the laboratory. If more rapid equilibrium is required it is suggested that the sample is vigorously stirred for a minimum of one week. If difficulties of equilibration are predicted, pass the sample through a pH cycle with storage under both acid and alkaline conditions and with vigorous mixing throughout.

Coprecipitation/Filtration. A ferric chloride solution was added to the sample to give 0.5 mg Fe per kg of sea water. After mixing overnight approx. 100 g of ∿ 11M ammonia solution was added, with vigorous stirring, to give a pH of 8.3 ± 0.5. After allowing to stand for several days the ferric hydroxide precipitate was recovered by transferring the stirred suspension, via an in-line filter rig containing a Nucleopore filter and using a peristaltic pump, between two 60-litre polypropylene storage bottles. Initially, the suspension was transferred from its original storage bottle through a 12 μm filter, was returned through a 5 μm filter and finally filtered through a 0.4 μm filter.

Preliminary Chemistry. Each filter was rinsed with water to remove sea salts and then placed in a PTFE beaker containing 0.5 to 1.0M HCl in order to dissolve the precipitate. The filters were removed and discarded and the chloride solution evaporated to dryness. (If a preliminary cation-column separation is used the sample is then dissolved in 2 ml 1.75M HCl for loading onto a column of Bio-Rad AG50X8 resin (200-400 mesh) and the REE fraction collected and evaporated to dryness). The evaporated salts were converted to nitrates by the addition of conc. HNO_3 followed by evaporation. The sample was next taken up in 5-6 ml. of 90% CH_3COOH/10% 5M HNO_3 solution. If any particles are undissolved the sample should be centrifuged at this stage.

1st REE Column. The column was pretreated with 5-10 ml. of 90% CH_3COOH/10% 5M HNO_3. The 5 ml. sample in 90% CH_3COOH/10% 5M HNO_3 was loaded, washed in with 2 ml. of 90% CH_3COOH/10% 5M HNO_3 and the column eluted with 35 ml. of 90% CH_3COOH/10% 5M HNO_3. The REE were stripped from the column with 10 ml. of 0.05M HNO_3 which was collected in a PTFE beaker. This solution was evaporated and taken up in 1 ml of 90% MeOH/10% 5.25M HNO_3.

2nd REE Column. The column was pretreated with 3-5 ml. of 90% MeOH/10% 5.25M HNO_3. The 1 ml. sample in 90% MeOH/10% 5.25M HNO_3 was loaded and washed in with 1 ml. of 90% MeOH/10% 5.25M HNO_3. Either the heavy REE fraction (Yb, Er, Dy) was collected by elution with 4.5 ml. of 90% MeOH/10% 5.25M HNO_3, the light REE fraction (Gd, Eu, Sm, Nd) was collected by elution of 17 ml. of 90% MeOH/10% 0.01M HNO_3 and the Ce + La fraction was collected by stripping with 1 ml. of 1M HNO_3; or the heavy REE fraction was collected as above, the Gd fraction was collected by elution with 3 ml. of 90% MeOH/10% 0.01M HNO_3 and the remaining REE (Eu, Sm, Nd, Ce, La) were collected by stripping with 1 ml. of 1M HNO_3. The fractions were collected in PTFE beakers and the contents evaporated to dryness.

Mass Spectrometry. The samples were dissolved in ~ 2 μl of water and each fraction loaded onto the Ta side-filament of mixed Re/Ta triple filaments. Ion beams were measured with a Daly-type scintillation detector or else a Faraday cup collector when beams grew to $\sim 10^{-12}$A. The isotopic masses measured (apart from zeros) are given in Table 2 and REE concentrations were obtained by insertion of the measured ratios into a computer programme which applies interference corrections where appropriate. Two isotopic ratios were measured for all REE except Ce, La and Eu, each ratio yielding an elemental concentration. Ce was corrected for any Nd interference and La for any Ce and Ba (Table 2). Any Ba interference on Eu was monitored by measuring $^{154}BaO/^{153}Eu$ but not corrected for. This is usually very small using triple-filament ionization with the above ratio typically 0.01 or less.

SEDIMENT ANALYSIS

The analysis of REE in solid is routine in the Leeds Isotope and Oceanography laboratories. The Hooker[12,13] type column chemistry used is identical to that given above. The decomposition and preliminary-chemistry steps are given below. The method uses HF which was purified using a two bottle PTFE still.

The sample of ~ 150 mg plus spikes were weighed accurately into a PTFE bomb. Approx. 0.2-0.3 g of spikes L1 and H1 were used which are 10X more concentrated than L2 and H2 (Table 1). 8 ml. HF and 1 ml. conc. HNO_3 were added and the mixture heated overnight at 130°C. The solution was evaporated to dryness, 2 ml. conc. HNO_3 added, evaporated and the process repeated if necessary to oxidize white fluorides. 8 ml. 6M HCl was added and the solution evaporated, 2 ml. conc. HNO_3 added, evaporated and the process repeated. Following this decomposition and equilibration of sample and spike, 5-6 ml. of 75% CH_3COOH/25% 5M HNO_3 solution was added and the mixture heated overnight at 130°C in the bomb to leach out the REE. The sample was centrifuged and the column chemistry performed as from stage 5 above except that the 5 ml. sample was loaded on the 1st column in 75% CH_3COOH/25% 5M HNO_3.

NEODYMIUM ISOTOPIC COMPOSITIONS

Nd isotopic compositions have been measured by a method similar to that developed by O'Nions et al.[18] for geological samples. Following passage through the 1st REE column the sample was loaded onto a temperature-controlled column containing 4 ml. of AGIX8 (200-400 mesh) resin and run at 25°C. The resin was pre-treated with 5-10 ml. of 75% MeOH/10% 8M CH_3COOH/10% 5M HNO_3/5% H_2O, the sample loaded in 1 ml. of this mixture, washed in with 2 ml. and eluted with 45 ml. The column temperature was increased to 35°C and the resin was immediately eluted with 20 ml. of 75% MeOH/10% CH_3COOH/5% 5M HNO_3/10% H_2O followed by collection of the Nd fraction with 30 ml of this mixture.

RESULTS

Precision and Accuracy

The precision and accuracy of the procedure was evaluated by analyzing synthetic REE solutions and international rock standards. Two solutions were prepared: Solution A, a synthetic mixture of the REE in the approximate relative amounts expected in seawater but $\sim 10^6$ times greater in concentration; and Solution B, prepared by dilution of soln. A such that 1 ml. of Solution B contained the equivalent of the REE contents of a 50 kg seawater sample. The results of analyzing these solutions show good agreement both between different sets of analyses and between the

Table 3 REE in Synthetic Mixtures

	La	Ce	Nd	Sm	Eu	Gd	Dy	Er	Yb
Solution A:									
Added ($\mu g\ ml^{-1}$):	14.3	10.9	10.9	2.60	0.482	2.43	1.77	1.85	2.13
Found ($\mu g\ ml^{-1}$):									
14.5.79	14.4	10.7	11.0	2.59	0.491	2.37	-	-	2.08
27.6.70	-	10.8	10.5	2.59	0.495	2.39	1.77	1.85	2.14
19.2.80	14.07	10.86	11.2	2.62	0.481	2.35	1.79	1.95	2.24
23.5.80	14.1	11.1	11.1	2.61	0.480	-	1.75	1.86	2.14
14.1.81	14.33	10.8	10.94	2.59	0.499	2.30	1.75	1.87	2.17
29.1.81	-	10.92	10.62	2.57	0.490	2.41	1.74	1.86	2.163
20.2.80	(13.01)	10.95	11.06	2.59	0.488	2.379	1.765	1.851	2.135
mean	14.2	10.9	10.9	2.59	0.486	2.37	1.76	1.87	2.15
± 1 std. devn.	0.14	0.13	0.26	0.02	0.006	0.038	0.018	0.038	0.048
Solution B:									
Added ($ng\ g^{-1}$):	146	111	111	26.5	4.91	24.7	18.0	18.8	21.7
Found ($ng\ g^{-1}$):	-	110	110	26.6	4.91	25.2	17.8	18.5	22.2

gravimetric and analyzed contents (Table 3), the overall agreement being usually to better than 1% in most cases with better results in the later work as measurements of the spike-isotopic ratios were refined.

The results obtained on the USGS standard rock BCR-1 are shown in Table 4. Despite the inclusion of some early results the precision is good, $\sim 2 \pm 1\%$, and it is considered that current precision is $\sim 1\%$ or better. There is a reasonable body of published data for BCR-1 and the values fall within the literature values which themselves have a range of about 10%. Precision has yet to be tested at natural levels.

Blanks

REE contents of reagents have been monitored using ND contents and the results of reagent-blank determinations are given in Table 5. Total blanks are listed in Table 6 together with the range in sample/blank ratios for the seawaters which have been analyzed. These ratios, at worst $\sim 10^2$, represent a considerable improvement over neutron-activation techinques[10,19] where blanks are unavoidably high[10] (sample/blank ~ 10 but sometimes < 1) because of the use of radioactive tracers.

Table 4 REE in USGS Standard Rock BCR-1 (ppm)

	La	Ce	Nd	Sm	Eu	Gd	Dy	Er	Yb
10.10.77	23.9	-	28.9	6.41	1.92	6.45	-	-	-
31.10.77	23.0	52.6	28.9	6.32	1.83	6.40	6.04	3.50	3.07
6.12.77	24.0	53.9	29.2	6.64	1.94	6.82	6.15	3.54	3.46
7.3.78	24.2	54.6	29.1	6.61	1.89	6.61	6.32	3.70	3.40
8.5.78	-	53.8	29.1	6.67	1.98	6.79	6.41	3.76	3.41
1.3.79	-	-	28.8	6.7	2.0	6.83	6.41	3.72	3.36
19.3.80	(28.0)	-	28.8	6.54	1.85	6.72	6.43	3.71	3.34
14.1.81	-	53.43	28.94	6.57	1.95	6.53	6.40	3.89	3.56
mean	23.8	53.7	29.0	6.56	1.92	6.64	6.31	3.69	3.37
± 1 std. devn.	0.5	0.7	0.1	0.13	0.06	0.17	0.15	0.13	0.15
range in literature values[13]									
min.	23.5	51.5	27.7	6.41	1.89	6.43	6.15	3.51	3.27
max.	26.5	54.9	30.5	7.4	2.2	8.02	6.55	3.71	3.68

Table 5 Reagent Blanks

Reagent/treatment	Nd (10^{-15} moles ml^{-1})
HCl (B.D.H. - Analar)	
(a) 6M by boiling glass distillation	360; 500; 24.3; 347
(b) 6M by sub-boiling quartz redistillation of (a)	1.3; 3.1; 4.2; 13.2; 18.7; 31.2; 9.0; 2.4; 14.6; 9.0
(c) 2.5M prepd. from (b)	8.3; 6.9
(d) 1.0M prepd. from (b)	2.4; 1.6
HNO_3(B.D.H. - Analar)	
(a) untreated	1530; 1460
(b) sub-boiling quartz distillation of (a)	2.63; 1.59; 43.7; 27.0
(c) 1.0M prepd. from (b)	9.0
NH_3 solution (B.D.H. - Analar)	
(a) untreated	125
(b) sub-boiling quartz - or two bottle PTFE - distillation of (a)	4.85; 3.67; 3.74; 6.24; 4.85; 5.06; 0.83; 1.80; 4.16; 5.06; 12.9; 6.52
CH_3COOH (B.D.H. - Analar)	
(a) untreated	16.6
(b) sub-boiling quartz distillation of (a)	2.91; 1.59; 4.09; 2.15
Methanol (B.D.H. - Analar)	
(a) untreated	17.3
(b) sub-boiling quartz distillation of (a)	4.16; 2.36
Water (Leeds tap water)	
(a) boiling glass distillation and mixed-bed ion exchange	0.31; 16.6; 1.39; 2.22
(b) sub-boiling quartz redistillation of (a)	3.12; 1.94; 2.63; 6.10; 1.87; 3.26; 1.94
HF (B.D.H. - Analar 40%)	
(a) untreated	277
(b) two bottle PTFE distillation of (a)	18.7

Table 6 Total REE column blanks (10^{-12} moles)

	La	Ce	Nd	Sm	Eu	Gd	Dy	Er	Yb
14.5.79	-	-	2.08	0.093	0.112	0.394	0.031	0.000	1.040
5.2.80	-	2.64	2.50	0.000	0.000	0.064	0.086	0.000	0.087
30.9.80	2.09	2.43	0.17	0.015	0.010	-	0.098	0.042	0.040
5.11.80	1.01	1.78	0.52	0.160	0.000	0.305	0.178	0.161	0.116
9.2.81	0.72	7.64	1.28	0.465	0.023	0.801	0.160	0.102	0.014
sample/blank max.	$2x10^3$	$7x10^2$	$7x10^3$	$5x10^5$	$9x10^4$	$6x10^3$	$1x10^4$	$5x10^5$	$2x10^4$
sample/blank min.	310	110	256	287	288	213	1340	1260	171

Table 7 REE contents of Marine Phases*

	La	Ce	Nd	Sm	Eu	Gd	Dy	Er	Yb
river water (S.W. Scotland)	842	1630	1010	200	43.6	205	174	106	87.8
sea water (S.W. Scotland)	83.9	148	107	24.6	6.45	26.7	21.5	14.8	13.2
N. Atlantic Ocean Water (100 m)	13.0	16.8	12.8	2.67	0.644	3.41	4.78	4.07	3.55
N. Atlantic Ocean water (2500 m)	29.4	26.1	24.9	4.75	0.895	7.19	6.11	5.09	4.79
plankton	0.146	0.238	(0.034)	(0.0064)	0.0034	(0.0186)	0.0080	-	0.0015
foraminiferal calcite	1.28	0.353	1.00	0.20	0.0601	0.29	0.33	0.27	0.28
diatom	5.77	8.21	6.71	1.56	0.404	1.78	1.69	0.998	0.859
deep-sea sediment	65.1	91.0	92.5	22.9	5.69	25.2	23.0	13.4	13.1
ferromanganese nodule	101	421	143	33.9	8.06	33.0	27.6	16.4	15.0
hydrothermal sulphide	0.097	0.218	0.119	0.030	0.0096	0.033	0.036	0.023	-

* waters in 10^{-12} mol kg^{-1}; solids in ppm

REE in Marine Phases

The method has been applied to the analysis of a variety of types of oceanographic samples (Table 7). Shale-normalized distribution patterns are shown in Fig. 4 and the smoothness of the patterns gives a visual appraisal of the relative accuracy of the analyses. The results for biogenic particles may in part reflect the degree to which our rigorous cleaning techniques have been successful. For example, the REE in foraminifera tests may actually be in an adsorbed phosphatic phase rather than on calcium sites due to ion exchange. The sea water results shown here form part of a study in progress on the vertical distribution of REE in the oceans and these preliminary data are of interest in showing enrichment of REE in deep ocean water relative to water at the base of the mixed layer of 2.26 (La), 1.55 (Ce), 1.95 (Nd), 1.78 (Sm), 1.39 (Eu), 2.11 (Gd), 1.28 (Dy), 1.25 (Er) and 1.35 (Yb), a pattern, in general, of greater deep water enrichment for the light REE than for the heavy REE.

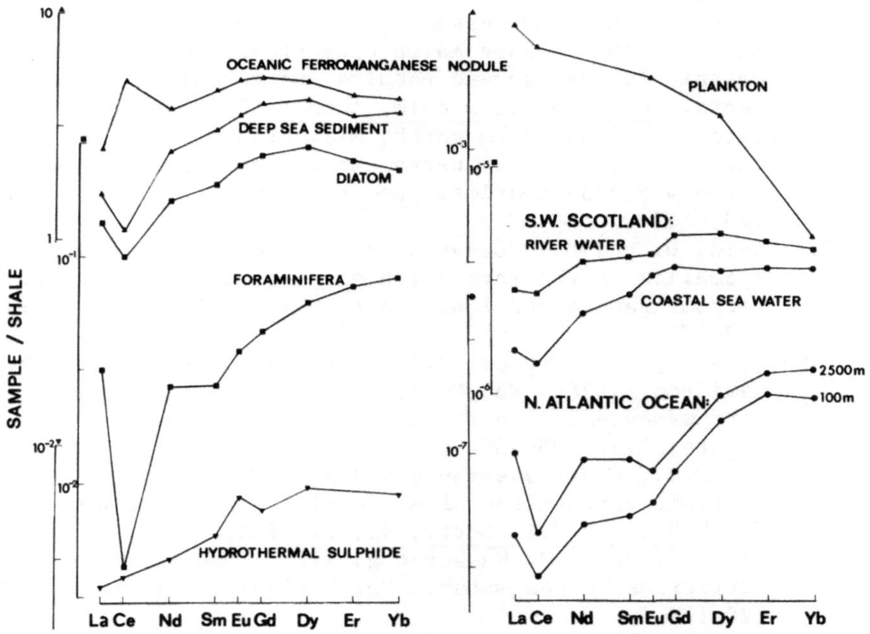

Fig. 4 REE abundance patterns of sea waters, river water, marine particulates and marine sedimentary phases (data from Table 7).

ACKNOWLEDGEMENTS

This work was supported by grant GR3/3125 from the Natural Environment Research Council. It is a pleasure to acknowledge the contributions of several colleagues at Leeds to this work: C.J. Hawkesworth, under whose guidance much of the initial mass spectrometric work was carried out, R.A. Cliff, who has provided invaluable help and advice on mass spectrometry, P. Guise, M. Thirlwell and M.H. Dodson.

REFERENCES

1. Piper, D.Z., 1974: Rare earth elements in the sedimentary cycle: a summary. Chem. Geol., 14, 285-304.
2. Piper, D.Z., 1974: Rare earth elements in ferromanganese nodules and other marine phases. Geochim. Cosmochim. Acta., 38, 1007-1022.
3. Glasby, G.P., 1973: Mechanisms of enirchment of the rarer elements in marine manganese nodules. Marine Chem., 1, 105-125.
4. Elderfield, H., C.J. Hawkesworth, M.J. Greaves and S.E. Calvert, 1981: Rare earth element geochemistry of oceanic ferromanganese nodules and associated sediments. Geochim. Cosmochim. Acta., 45, 513-528.
5. Elderfield, H., C.J. Hawkesworth, M.J. Greaves and S.E. Calvert, 1981: Rare earth element zonation in Pacific ferromanganese nodules. Geochim. Cosmochim. Acta., 45, 1231-1234.
6. Elderfield, H. and M.J. Greaves, 1981: Negative cerium anomalies in the rare earth element patterns of oceanic ferromanganese nodules. Earth Planet. Sci. Lett., 55, 163-170.
7. O'Nions, R.K., S.R. Carter, R.S. Cohen, N.M. Evensen and P.J. Hamilton, 1978: Pb, Nd and Sr isotopes in oceanic ferromanganese deposits and ocean floor basalts. Nature, 273, 435-438.
8. Piepgras, D.J., G.J. Wasserburg and E.J. Dasch, 1979: The isotopic composition of Nd in different ocean masses. Earth Planet. Sci. Lett., 45, 223-236.
9. Piepgras, D.J. and G.J. Wasserburg, 1980: Neodymium isotopic variations in sea water. Earth Planet. Sci. Lett., 50, 128-138.
10. Høgdahl, O.T., S. Melson and V.T. Bowen, 1968: Neutron activation analysis of lanthanide elements in sea water. Adv. Chem. Ser., 73, 308.
11. Høgdahl, O.T., 1966: Distribution of the rare earth elements in sea water. NATO Research Grant No. 203, Semiannual Progress Report No. 2, 43 pp.
12. Hooker, P.J., 1974: B.A. Thesis, Univ. Oxford.

13. Hooker, P.J., R.K. O'Nions and R.J. Pankhurst, 1975: Determination of rare-earth elements in USGS standard rocks by mixed-solvent ion exchange and mass-spectrometric isotope dilution. Chem. Geol., 16, 189-196.
14. Korkisch, J. and G. Arrhenius, 1964: Separation of uranium, thorium and rare-earths in HNO_3 - acetic acid medium. Anal. Chem., 36, 850-854.
15. Faris, J.P. and J.W. Warton, 1962: Anion exchange resin separation of the rare-earths, yytrium and scandium in nitric acid-methanol mixtures. Anal. Chem.34, 1077-1080.
16. Desai, H.B., I.R. Krishnamoorthy and M. Sandar Das, 1964: Determination of scandium, yttrium, samarium and lanthanum in standard silicate rocks, G-1 and W-1 by neutron activation analysis. Talanta 11, 1249-1255.
17. Brunfelt, A.O. and E. Steinnes, 1969: Determination of lutetium, ytterbium and terbium in rocks by neutron activation and mixed solvent anion-exchange chromatography. Analyst. 94, 979-984.
18. O'Nions, R.K., P.J. Hamilton and N.M. Evensen, 1977: Variations in $^{143}Nd/^{144}Nd$ and $^{87}Sr/^{86}Sr$ ratios in oceanic basalts. Earth Planet. Sci. Lett., 34, 13-22.
19. Goldberg, E.D., M. Koide, R.A. Schmitt and R.H. Smith, 1963: Rare-earth distribution in the marine environment. J. Geophys. Res., 68, 4209-4217.

THE CYCLE OF LIVING AND DEAD PARTICULATE ORGANIC MATTER IN THE PELAGIC ENVIRONMENT IN RELATION TO TRACE METALS

George A. Knauer and John H. Martin

Moss Landing Marine Laboratories
Moss Landing, California 95039
U.S.A.

ABSTRACT

The distribution and cycling of many trace elements in the pelagic environment appear to be tightly coupled to biological processes. Therefore, the elucidation of these processes is essential if we are to understand the various interactions between biogenic particles and trace elements. Using particle traps, we have been studying the cycling of total organic material, "living" biomass (as measured by adenosine triphosphate) and various particulate components such as fecal pellets, which represent important potential carriers of many macro-nutrients and trace elements. Our results indicate that most of the macro-nutrients (i.e., C, N and P) are cycled in the upper 200 m of the water column; trace elements such as Cd follow the same pathways. Surface-produced fecal pellets, long thought to be important trace element carriers, appear to be recycled in the upper 1000 m of the water column. However, we have found evidence suggesting that significant amounts of fresh pellets are produced from <u>in situ</u> activity leading to variable fluxes of many elements throughout the water column. The distribution of living organic material as measured in "suspended" particulates (i.e., those collected with conventional water bottles) present the usual classical picture, ranging from > 30% of the total suspended organic carbon at the surface to < 2% at 1500 m. However, the rarer, but rapidly sinking particles collected by our traps, indicate that in the deep ocean (i.e., > 1000 m), the living fraction represents > 400 times more biomass than the suspended material.

This paper presents current information concerning the formation, make-up, distribution and transport of biogenic particles in the belief that this knowledge will provide greater understanding of the role of these particles in the cycling of trace elements.

INTRODUCTION

Chemical processes operating in the oceans are extremely complex, involving the uptake and release of elements from two broad classes of particles: biogenic particles formed in the oceans (consisting principally of phytoplankton, zooplankton, bacteria and their detrital components, organic carbon, opal and carbonates) and particles introduced via riverine and atmospheric input (largely lithogenous). Many elements can be associated with both classes of particles, although some cycles appear to be dominated by one class and not the other. It is the biogeochemical processes associated with the first class of particles that concern us here. The objective of this paper is to present current information concerning the formation, make-up, distribution and transport of biogenic particles in the belief that this knowledge will provide greater understanding of their role in the cycling of trace elements.

In the past, biogenic particles were, for the most part, sampled using standard water bottle casts. Information derived from such studies suggested a strong interaction between many elements and various biological cycles. For example, it has been known for some time that the vertical distributions of the biologically active dissolved nutrients, nitrate and phosphate, are determined by the plankton[1]. These compounds are depleted in surface waters by incorporation into phytoplankton biomass and enriched in waters below the euphotic zone through regeneration processes mediated by bacteria. The distribution and cycling of some trace elements in the pelagic environment (as determined by bottle casts) also appear to be tightly coupled to biological processes. For example, the vertical distribution of dissolved Cd is strongly correlated with the nutrients phosphate and nitrate, being depleted in near-surface waters and enriched in deep water[2,3] Other metals show similar relationships that also suggest that their biogeochemistry is strongly dominated by the cycling of biogenic matter. Nickel has been correlated with both phosphate in surface waters and silicate in deep waters[4,5], while Zn is correlated with silicate[6] (= biogenic opal).

McCave[7] and Lal[8] have indicated that standard sampling techniques involving water bottles primarily collect the small, slow-sinking background material referred to as particulate organic matter (i.e., "suspended" POC, PON, ATP, etc.), and do not adequately sample the large (i.e., > 35 μm) fast-sinking particles. In this respect, recent in situ investigations using SCUBA[9,10] and particle trap collections from deeper water[11-16] indicate that rapidly sinking large particles may be major contributors to total particle flux. Alldredge[9] has estimated that large aggregates may constitute between 20-30% of the total particulate organic carbon, while Shanks and Trent[10] concluded that, in Monterey Bay, California surface waters, > 80% of the carbon flux is from similar aggregates.

Many significant geochemical processes can occur at interfacial boundaries[17] and the presence of this large particle fraction has important ramifications in terms of trace element biogeochemistry in the marine environment. For example, the role of marine bacteria in biogeochemical cycles is becoming more and more evident[18]. Indeed, bacterial participation in a variety of surface-oriented chemical processes involving the oxidation and reduction of metals is apparent[19]. Schweisfeuth et al.[20], working with Mn bacteria, have found that the elements Co, Cu, Zn and Fe were enriched in microbially precipitated Mn oxides.

Although there is relatively little information to date concerning large particle-metal interactions, initial studies indicate the potential importance of this class to metal transport processes[21-24]. However, neither the quality of large particulate biogenic material (e.g., fecal pellets, etc.) nor its quantitative importance (i.e., vertical transport rates) in terms of meso- and bathy-pelagic ocean waters are well understood.

If, as suggested, large particles have not been adequately sampled in the past, and assuming they play a significant role in metal biogeochemistry[23,24], then the qualitative and quantitative evaluation of this fraction is essential if we are to understand the various metal interactions that occur between the biota and the water column and between the biota-water column and the benthos.

Using particle traps (MULTITRAPS), we have been studying the cycling of total organic material, "living" biomass (as measured by adenosine triphosphate) and various particulate components such as fecal pellets, which represent important potential carriers of many macro- and micro-nutrient elements. Our initial results indicate that most of the macronutrients (e.g., nitrogen compounds) are recycled in the upper 200 m[25,26] of the water column (resulting in long residence times); whereas some micronutrients such as Mn have relatively short residence times in surface waters[23]. Surface-produced fecal pellets, long thought to be important trace element carriers[27-30], appear to be recycled in the upper 1000 m of the water column. However, we have found evidence suggesting that significant numbers of pellets are produced from *in situ* zooplankton activity throughout the water column[16], which can affect fluxes of many elements.

The distribution of "living" organic material, as measured in "suspended" particulates (i.e., collected with water bottles), presents the classical picture ranging from > 30% of the total suspended organic carbon at the surface to < 2% at 1500 m. However, our findings indicate that the living fraction associated with the rarer but rapidly sinking large particles collected by traps in the deep ocean (i.e., > 1000 m), represents > 150 times more biomass than that associated with the "suspended" material.

The purpose of this paper is to present and to further discuss our findings concerning large particle flux and their potential importance in overall transport processes.

Fluxes of Carbon, Nitrogen, Fecal Pellets and Living Carbon

Most geochemists recognize the importance of both the phytoplankton and zooplankton in marine biogeochemical cycles, not only because they are able to concentrate large quantities of metals[31], but also because detrital by-products (e.g., exoskeletons, fecal pellets, etc.) produced as a result of normal metabolic activities, represent important mechanisms for metal transport and release throughout the water column. The initial production of organic material which fuels these processes is derived, of course, from unicellular algae living in the illuminated upper water column. It is this primary product that ultimately provides food for all of the deep-living pelagic and benthic organisms. Recent studies by Knauer et al.[13], using particle interceptor traps, indicate that although most of the surface-produced biogenic particles appear to be recycled in the upper 200 m of the water column, significant quantities were nonetheless found at all depths sampled by the traps.

One of the first steps in evaluating the importance of such biogenic particle transport (and associated metal interactions) must include the relationship between primary production and the transport of carbon throughout the water column. This relationship was studied by us during a cruise in California coastal waters[25]. Estimates of primary production were obtained using metal-free collection and processing techniques developed at Moss Landing Marine Laboratories which are fully described in Fitzwater et al.[32] (also see Knauer and Martin, this volume).

Material falling through the water column (i.e., carbon, nitrogen, fecal pellets) was collected in a free-floating trap system (MULTITRAPS; Figure 1). The traps were deployed at depths of 35, 65, 150, 500, 750 and 1500 m on a single line about 80 km off the Monterey Peninsula, California and allowed to drift for 6.17 days.

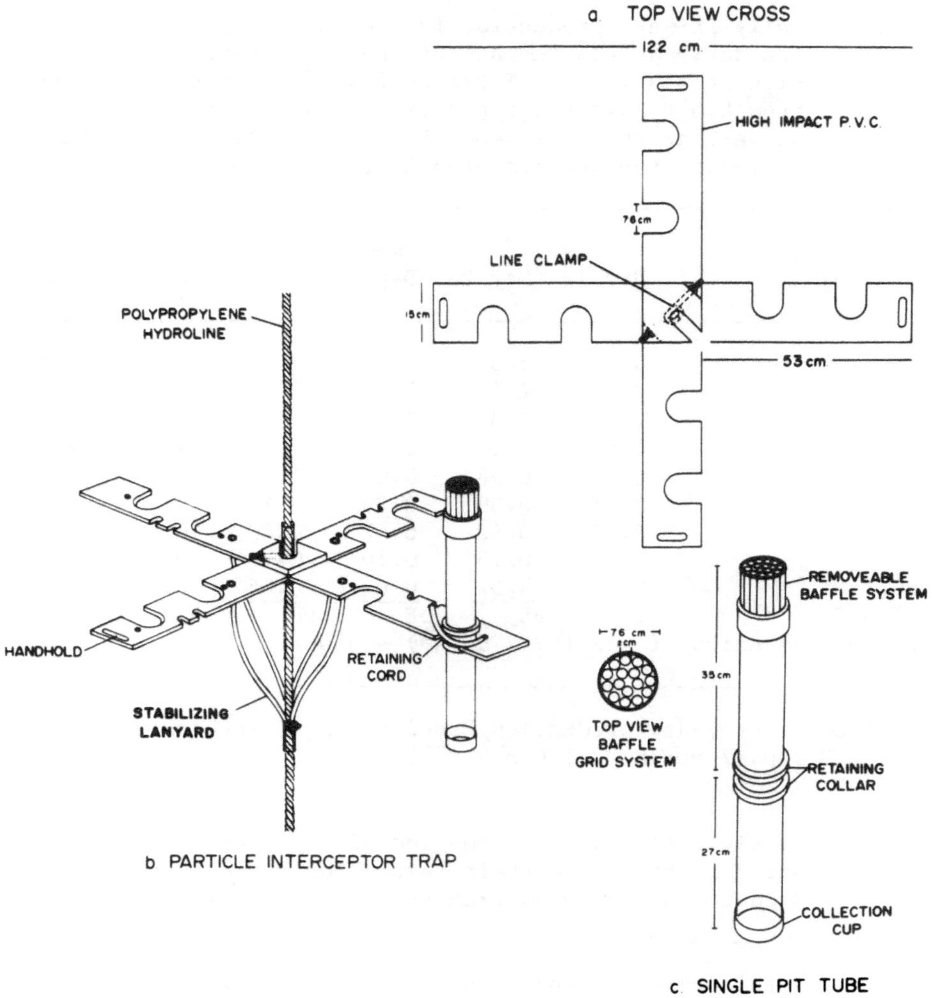

Fig. 1 MULTITRAP collection assembly. Adapted from Knauer, Martin and Bruland[13].

Total daily primary production is presented in Table 1. It is apparent that, since there was very little variation in the daily production rate (i.e., values for days 1-6 range from 40-68 mmol C m^{-2} d^{-1}), comparison of the carbon fixed via primary production with that sinking from the surface would seem appropriate (Table 2). In terms of the total amount of carbon and nitrogen fixed over the 6.17-day trap set (i.e., 350 mmol C), almost 90% appears to have been recycled or converted to the slower-sinking

Table 1 Daily primary production in the water column. Values are derived from the sum of these separate, daily, *in situ* incubations. (Total period of incubations ranged from 8.5 to 9.7 hours per day.) Daily integrated values as well as the total amount of carbon fixed over the 6.17-day trap set are also included.

Depth (m)	Day 1	Day 2	Day 3	Day 4	Day 5	Day 6
			($mmol\ C\ m^{-3}\ d^{-1}$)			
5	3.0	3.0	2.6	2.4	2.2	1.5
10	2.6	2.4	2.1	2.0	1.8	1.3
15	2.0	2.0	1.7	1.7	1.6	1.4
20	1.4	1.4	1.2	1.3	1.1	0.92
25	0.83	0.98	0.82	1.1	0.85	0.67
30	0.61	0.59	0.70	0.58	0.69	0.45
35	0.44	0.52	0.47	0.43	0.52	0.32
50	0.09	0.15	0.10	0.16	0.19	0.09
60	0.03	0.07	0.04	0.05	0.07	0.14
mmoles $C\ m^{-2}d^{-1}$	65	68	58	58	54	40
Total mmoles carbon fixed (6.17 days*) = 350						

*Estimate of 0.17-day production based on day 6 integrated value (i.e., 40 x 0.17 = 6.8 mmol $C\ m^{-2}\ d^{-1}$).

Table 2 Vertical fluxes of carbon and nitrogen collected in particle traps and their associated percents of primary production. Adapted from Knauer and Martin[25].

Depth (m)	Total Organic C and N collected over 6.17 days, m mol m^{-2}: C	N	As percent of primary production: C	N*
35	161	27	46	51
65	78	11.5	22	22
150	41	5.1	12	9.4
500	12.8	1.2	3.7	2.4
750	11.9	1.1	3.4	2.1
1500	37	3.8	10.3	13.4

*Total mmoles fixed in euphotic zone during 6.17-day trap set estimated from the relationship C:N = 6.6 (Redfield et al.[1]) = 350/6.6 = 53 mmol N.

"suspended" particulate organic carbon (POC) and particulate organic nitrogen (PON) pool by 150 m. By 750 m, > 95% of the carbon and nitrogen had been returned to the system.

For illustrative purposes, we have plotted carbon flux (in natural logarithonic scale) vs. depth (Figure 2). This graph clearly indicates at least a two-step regeneration process and perhaps a third (solid and dashed lines) through the depth of the oxygen minimum (500-750 m). In contrast to the minimum flux values observed between 500-700 m, the 1500 m traps revealed an obvious increase in the flux of both carbon and nitrogen. This observed increase is real in a statistical sense. Comparison of the carbon data (t-test) from the 500 and 750 m depth indicated a significant difference ($p < 0.005$) from the 1500 m depth.

We have no conclusive explanation for this increase, although there are a number of possibilities: (1) greater amounts of carbon (and associated materials) were brought in by advective processes (e.g., along isopyncals) from an area of higher productivity. Although this is a possibility, it seems unlikely, since it was the Oceanic-Davidson Current period, which is characterized by low and monotonous production over large areas. (2) Increased carbon and nitrogen were transported by vertically-migrating zooplankton feeding at higher, more productive depths in the water column[33]. This also cannot be ruled out entirely. However, analysis of fecal pellets from 1500 m shows qualitative differences not seen in fecal pellets originating in surface waters. (3) In situ zooplankton populations feeding on background organic

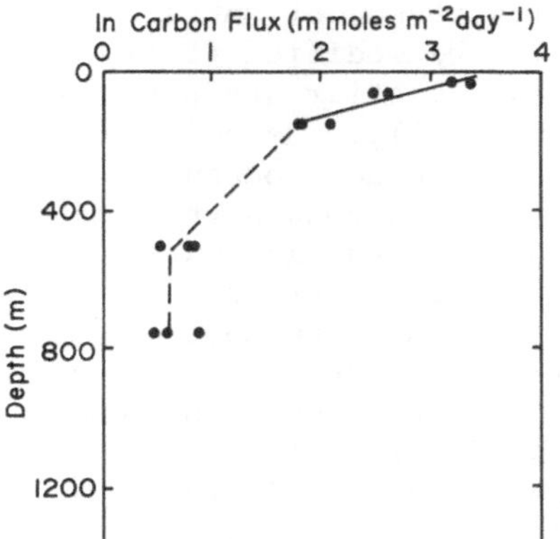

Fig. 2 Ln carbon flux (mmol $m^{-2} d^{-1}$) with depth. From Knauer and Martin[25].

particulate material result in repackaging into larger, faster-sinking particles. The source of such organic particles below the euphotic zone may be microbial in nature, since bacteria and other microheterotrophs can utilize dissolved and "suspended" organic compounds, repackaging them into a more concentrated form of particulate carbon, which can then re-enter the marine food web[34,15]. Indeed, Pomeroy[35] has estimated that the dissolved organic carbon (DOC)-bacteria-POC route may increase the utilizable yield of organic matter from photosynthesis to a value 30% higher than estimated from standard primary productivity measurements.

Although some combination of the above mechanisms probably pertain, we feel that the latter possibility, in situ repackaging, is the prime contributor. This is suggested from zooplankton-fecal pellet data collected during the same trap deployment. During this collection period, fecal pellets were removed from MULTITRAPS, enumerated and separated into size classes. Microscopic examination revealed four primary geometric shapes which were divided into eight subclasses based on their mean diameters (Table 3). The complete procedure in terms of collection, preparation and enumeration, are given in Urrere and Knauer[16]. Upon examination, it became apparent that numerically, this group of particles was important at all depths sampled. Total pellet fluxes decreased from a high of 3.2×10^5 pellets m^{-2} d^{-1} at 35 m to a low of 0.3×10^5 pellets m^{-2} d^{-1} at 500 m, the depth of the oxygen minimum.

However, it is interesting to note that total pellet fluxes increased below this depth, reaching numbers of 0.8×10^5 pellets m^{-2} d^{-1} at 1500 m. If the increased particle fluxes at 1500 m are indeed the result of in situ repackaging, then our attitudes concerning the relative importance of surface-produced particles in transport processes must be modified (i.e., large, rapidly-sinking particles of significant quantity are not produced solely in surface waters). Historically, the most popular mechanism for transporting material to the deep ocean was proposed by Agassiz[36]; i.e., that deep-sea organisms obtain their energy from a "rain" of organic detritus produced in surface waters. The data presented in Table 2 and Figure 2 (i.e., through 750 m) seem to support this idea, indicating that although the majority of surface-produced and relatively labile particulate organic material has been remineralized by the time it reaches the oxygen minimum zone, significant quantities of particulate organic material can still be found at all depths. However, it would be misleading to conclude that the apparent three-step process shown in Figure 2 is the result of a single sinking of surface-produced organic particles. The increased carbon flux at 1500 m argues against such a simplistic interpretation. What is seen is the net result of a dynamic system involving significant transformation of organic material from slow- to fast-sinking particles throughout the water

Table 3 Flux of all fecal pellet size classes collected in MULTITRAPS (numbers m^{-2} d^{-1}) (from Urrere and Knauer[16])

Depth (m)	Fecal Pellet Fluxes: Elliptical Size (mm) 0.05	Elliptical 0.10	Elliptical 0.15	Cylindrical Size (mm) 0.05	Cylindrical 0.10	Cylindrical 0.15	Round	Coiled	Total Average Flux
35	5153	14171	9558	208328	37485	1206	30129	11678	
	7522	16665	11179	202427	52321	1465	28010	10431	
	(6337)*	(15418)	(10368)	(205377)	(44903)	(1330)	(29069)	(11054)	323856
65	5486	4821	5195	174043	30462	831	19615	4945	
	6608	5527	4530	159207	35199	1496	16415	4447	
	(6047)	(5174)	(4862)	(166625)	(32831)	(1164)	(18015)	(4696)	239414
150	13008	3325	332	155301	26431	1164	14919	2577	
	9101	4364	665	129951	30088	1247	12010	2244	
	(11054)	(3844)	(498)	(142626)	(28259)	(1205)	(13465)	(2411)	203360
500	6400	5818	291	12426	5112	291	2203	291	
	7273	5444	665	15085	5860	331	1787	166	
	(6836)	(5631)	(478)	(13756)	(5486)	(312)	(1995)	(229)	34723
750	22441	16124	416	8021	1538	540	1455	125	
	19989	15124	416	9517	1371	125	1205	125	
	(21215)	(15625)	(416)	(8769)	(1455)	(332)	(1330)	(125)	49266
1500	32830	31542	3200	5610	1745	416	1039	0	
	35906	31875	2618	8228	2660	540	1122	0	
	(34368)	(31708)	(2909)	(6919)	(2202)	(478)	(1081)	0	79665

*Numbers in parentheses indicate average values and were calculated from the analyses of two replicate traps/depth.

column. This is based on quantitative information concerning fecal pellet numbers (Table 3) and on qualitative data based on pellet contents (Table 4). Our analysis indicates that five of the eight pellet classes appear to have originated at the ocean surface, in the sense that they show a tendency to decrease exponentially[16]. Furthermore, surface-produced pellets can be used as "tracers" in their descent through the water column in terms of content quality (i.e., if pellets collected from deeper depths contain food items of similar quality and quantity relative to surface pellets of the same size class, then such pellets must have had their origins at the surface).

The 0.10 mm cylindrical and the 0.05 mm elliptical pellet size classes illustrate this point (Figures 3a, b, Table 4). Major food items in the 0.10 mm cylindrical pellets appear reasonably similar over a 750 m depth range, consisting primarily of Coscinodiscus fragments, Pseudoeunotia doliolus and Pseudoeunotia fragments (Table 4). At 1500 m, there is an obvious change in the content quality of this class (= repackaging) with the appearance of numerous "green cells" and concurrent increases in both Coscinodiuscus and Pseudoeunotia fragments. If surface-produced pellets decrease exponentially as they sink (i.e., through 750 m; Figure 3a), then clearly, the majority of surface-produced pellets rarely reach depths below 1500 m (i.e., extrapolation of the

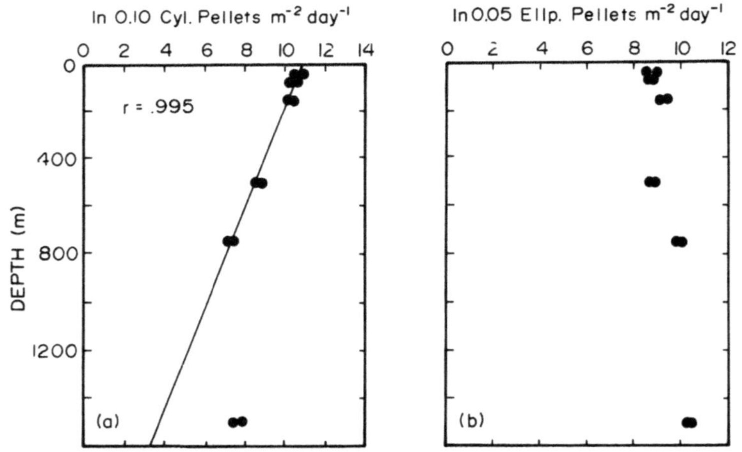

Fig. 3 Ln pellet flux (numbers $m^{-2}d^{-1}$) plotted against depth for the 0.10 mm diameter cylindrical (a) and 0.05 mm diameter elliptical pellet size class (b). The correlation coefficient (r) was computed for the 0.10 mm cylindrical size class over the 35-750 m depth range. Adapted from Urrere and Knauer[16].

Table 4 Analysis of major identifiable food items in all fecal pellet size classes (adapted from Urrere and Knauer[16])

Pellet Size Class	Depth (m)	Navicula sp.	Thalassiosire decipiens	T. decipiens Fragments	Thalassiosira sp.	Coccolithophore	"Green Cells"	Coscinodiscus Fragments	Pseudoeunotia doliolus	Pseudoeunotia Fragments	Nitzschia sp.	Actinoptychus sp.	Prorocentrium micans	Dinoflagellate Fragments	Crustacean Appendages	Distephanus sp.
0.10 Cylin-	35	11*	7	0	0	0	0	117	102	11	10	0	13	14	14	0
drical**	65	5	0	0	0	0	0	107	94	37	1	0	4	2	7	0
	150	5	4	0	0	0	0	196	135	48	0	0	4	3	14	0
	500	8	5	0	0	0	0	130	118	47	1	0	8	0	9	0
	750	9	2	0	0	0	0	84	57	28	0	0	0	1	13	0
	1500	21	8	0	0	0	215	206	95	117	0	0	1	3	6	0
0.05 Ellip-	35	35	15	0	12	100	0	0	0	0	0	3	0	0	0	0
tical	65	12	3	0	0	0	64	3	2	1	1	3	0	0	0	0
	150	11	0	2	2	0	104	9	0	0	0	0	0	0	0	0
	500	1	0	2	0	0	432	61	0	0	0	1	0	0	0	0
	750	1	0	2	0	0	71	11	0	0	0	1	0	0	0	0
	1500	0	0	0	0	0	372	3	0	0	0	0	0	0	0	0

* Number of food items derived from summed analyses of 14 pellets per class per depth.
**Cylindrical pellets standardized to 1 mm length for food item comparisons.
† Boxes emphasize differences.

regression line in Figure 3a indicates that only 0.1% of the surface-produced 0.10 mm pellets can be expected to be found at depths > 1500 m. In contrast to the 0.10 mm cylindrical pellet class, the 0.05 mm elliptical size class shows no tendency to decrease in a clear way with depth (Figure 3b; Table 3), and there is a considerable change in both the quality and quantity of food items consumed, depending on depth (Table 4). Thus, at the 35 m depth, Navicula sp., Thalassiosira decipiens and coccolithophores are the major food items consumed, while at 500 m, they are essentially absent, with "green cells" appearing as the major food source.

Such differences in food quality obviously suggest origins other than the surface for many of the pellets collected in our traps. Thus, what appears to be a relatively smooth three-step decomposition process (Figure 2), with surface-produced material decomposing as it sinks, is actually the result of multiple input throughout the water column derived from (we believe) bacterial and plankton-related processes.

If interactions between microbial populations and the various DOC-POC pools[34,15] are significant in repackaging-type processes throughout the water column, then significant amounts of living biomass must also be present throughout the water column.

On a recent cruise[15] (also off the Monterey Peninsula), another free-floating MULTITRAP array was set at depths of 60, 210 385, 1100 and 1550 m. This deployment was designed to evaluate the association of living biomass with large particle flux. Methodological details concerning the collection, processing and analysis of the trapped particles can be found in Fellows et al.[15]

In this study, two separate classes of particulate materials were considered: (1) "suspended" and relatively abundant, low-density and fine-grained particles collected in water bottles and commonly referred to as particulate organic matter; (2) sediment trap particles which represent the larger and relatively rare particles not normally sampled using water bottles (statistical sense) and which require the use of particle traps for collection.

The suspended particles were analyzed for POC, PON, adenosine triphosphate (ATP) and total adenine nucleotide (= A_T = ATP + ADP + AMP). Trapped particulates were extracted in situ for ATP and A_T in individual collector tubes (Figure 1) filled with two liters of a high-density acid/salt solution (i.e., 0.5 M H_3PO_4/1.5 M NaCl).

The collection and analysis of suspended particulates for POC, PON and ATP are of importance, since classically, these parameters have been used to estimate the qualitative (i.e., C/N ratios) and quantitative (ATP) importance of microbial populations in the water column[37,18]. In addition, the use of these techniques has also provided biological oceanographers with the means for estimating the amount of living carbon (from ATP) relative to suspended POC. In this way, we can then compare suspended living biomass estimates with the amount of living biomass found in the trapped particulates in order to assess their relative importance.

The results of the suspended analyses taken from Fellows et al.[15] are presented in Figures 4 and 5. The POC-PON profiles taken with 5 l Niskin bottles over several days adjacent to our free-floating MULTITRAP array show a classical picture (e.g., Gordon[38]). High values are found in surface waters where particulate primary production is high, followed by a rapid decrease below 200 m. A similar picture is seen in terms of ATP and A_T; high values in surface water, followed by a rapid decrease with depth. (Note that energy charge [EC] values, Figure 5, average approximately 0.7, indicating that the microbial population is viable. No further explanation of EC will be given in this paper; however, a complete explanation can be found in Karl[18]).

Comparisons of the POC profile with the ATP profile indicate that between 30-40% of the total carbon in the euphotic zone (1% light level = 50 m during this deployment) is associated with living biomass. However, with increasing depth, the ratio of living carbon to total POC decreases rapidly to values of < 5% at depths below 200 m, which is also consistent with previously published data[39,40].

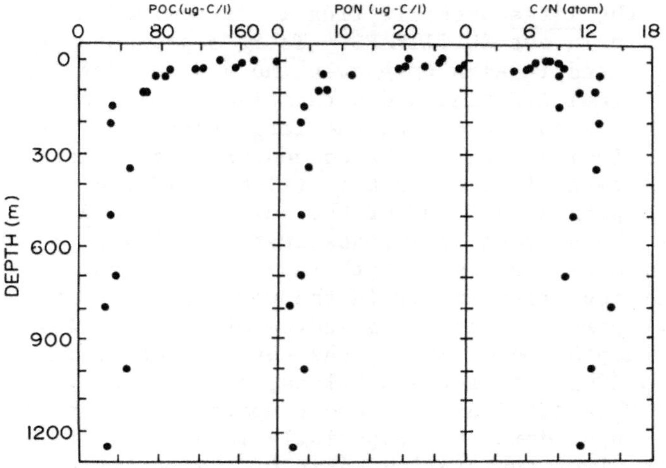

Fig. 4 Water column POC, PON and C:N ratios from "suspended" particulate material. Adapted from Fellows, Karl and Knauer[15].

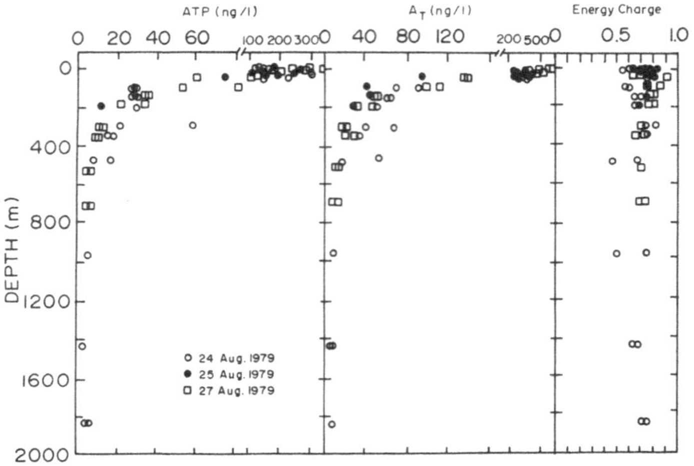

Fig. 5 Vertical distribution of ATP, A_T, and energy charge from water column "suspended" particles. Adapted from Fellows, Karl and Knauer[15].

The point of these comparisons is to re-emphasize the fact that samples taken for suspended particulate organic material and ATP analysis using water bottles generally present a similar picture: surface enrichment and deep depletion, indicating the absence of any significant in situ microbial activity below 200 m. However, a different picture emerges when living biomass associated with the large particle flux is determined using in situ ATP extraction in our MULTITRAPS. Table 5 presents data on the net ATP collected in each trap over the 4.5-day deployment, the downward vertical ATP flux and a relative comparison between ATP concentrations associated with the large particle flux and the "suspended" (from Figure 5) ATP concentrations. There are two important features in these data: (1) Net ATP fluxes decrease to a value of approximately 10% of the surface value at 385 m. Notice that the suspended ATP concentrations (Figure 5) show similar decreases over this depth interval, suggesting that the downward vertical flux of ATP in the upper 385 m of the water column is proportional to the standing stock of living carbon. However, at depths below 385 m, the net flux of living carbon increases to 20-25% of surface values, even at the 1550 m trap depth. (2) When ATP flux data are compared to "suspended" ATP, the former always dominate, especially in deeper waters. (We realize that since the sinking rates of the particles collected at each depth are unknown, a direct comparison between flux [μg ATP m^{-2} d^{-1}] and the suspended concentrations [μg ATP m^{-3}] cannot be made dimensionally.)

Table 5 The relative importance of large-particle ATP (calculated from "swimmer"-corrected ATP flux[15]) to "suspended" ATP*. Adapted from Fellows, Karl and Knauer[15].

Depth (m)	Trap	Net ATP** (µg/trap)	ATP Flux ($\mu g\ m^{-2}\ d^{-1}$)	"Suspended" ATP ($\mu g\ m^{-3}$)	ATP † Flux/ "Suspended"
60	A-1	29.1	1425.1	128	12.9
60	A-2	9.69	552		4.3
210	B-1	--††	--	24	--
210	B-2	1.43	81.5		3.4
210	B-3	--	--		--
385	C-1	0.20	11.4	13.5	0.8
385	C-2	2.42	137.9		10.2
385	C-3	--	--		--
1100	D-1	7.67	437	5	87.4
1100	D-2	7.01	399		80.0
1500	E-1	5.75	328	2	164
1500	E-2	5.50	313		157

* "Suspended" refers to materials trapped in a standard Niskin[R] water bottle.

** Corrected for contribution from "swimmers"[15].

† See Fellows et al.[15] for ATP Flux/"Suspended" calculations in terms of assumptions used for dimensional comparison.

†† Estimates could not be made due to insufficient resolution of the swimmer correction.

In summary, it is becoming apparent that the vertical transport of particulates in the ocean is far more complicated than previously believed. The data presented above strongly support the idea of multiple inputs of particulate material throughout the water column. Many of these features (as determined from the fecal pellet and ATP data) appear to be strongly tied to microorganism-zooplankton processes which can certainly be expected to change both the quality and quantity of organic (and probably inorganic) material as it sinks from the surface and as it is repackaged and transformed at deeper depths.

In the above, we have concentrated our discussion on the complexities involved with the biogenic particles themselves. Obviously, the interactions occurring between these particles and various trace elements are similarly complex. This is exemplified by Mn. In previous studies[23], we found large quantities of alumino-silicates included in trapped fecal pellets and marine snow material. Because of the large amounts of Mn associated with alumino-silicates, it is difficult, if not impossible, to determine how much Mn is associated with the biogenic versus lithogenic fractions. In addition to the Mn in the trapped particulate phase, we also found Mn in the trap salt gradients. This Mn was correlated with organic carbon, which suggests that this part of the Mn may have originally been adsorbed to organic detritus surfaces. In a more recent study (unpublished), we again found Mn associated with the trapped particulate phases mentioned above. We also found evidence that Mn was associated with a third phase, $CaCO_3$. These findings illustrate that the presence or absence of various particulate phases will determine rates at which elements are removed from, or returned to, solution. The basic biological processes described above (particle formation, consumption, decomposition), will in turn determine the rates at which the particles and associated metals will be transported from one layer to another.

Thus, unraveling the exact ways trace elements are transported in the sea is a formidable undertaking. Nevertheless, recent improvements in trace element analytical techniques and the rapid development of particle interceptor traps provide the means necessary for this important task.

REFERENCES

1. Redfield, S.C., B.H. Ketchum and F.A. Richards, 1963: The influence of organisms on the composition of sea water, pp. 26-77. In: Hill, M.N. (ed.), "The Sea", Vol. 2. J. Wiley and Sons, N.Y.
2. Martin, J.H., K.W. Bruland and W.W. Broenkow, 1976: Cadmium transport in the California current, pp. 159-184. In: Windom, H. and R. Duce (eds.). "Marine Pollutant Transfer", D.C. Heath, Lexington, Mass.
3. Bruland, K.W., G.A. Knauer and J.H. Martin, 1978: Cadmium in northeast Pacific waters. Limnol. Oceanogr., 23, 618-625.
4. Sclater, F.R., E.A. Boyle and J.M. Edmond, 1976: On the marine geochemistry of nickel. Earth Planet. Sci. Lett., 31, 119-128.
5. Bruland, K.W., 1980: Oceanographic distributions of cadmium, zinc, nickel and copper in the north Pacific. Earth Planet. Sci. Lett., 47, 76-198.
6. Bruland, K.W., G.A. Knauer and J.H. Martin, 1978: Zinc in northeast Pacific water. Nature, 271, 741-743.
7. McCave, I.A., 1975: Vertical flux of particles in the ocean. Deep-Sea Res., 22, 491-502.
8. Lal, D., 1977: The oceanic microcosm of particles. Science, 198, 997-1009.

9. Alldredge, A.L., 1979: The chemical composition of macroscopic aggregates in two neritic seas. Limnol. Oceanogr., 24, 855-866.
10. Shanks, A.L. and J.D. Trent, 1980: Marine snow: Sinking rates and potential role in vertical flux. Deep-Sea Res., 274, 137-143.
11. Wiebe, P.H., S.H. Boyd and C. Winget, 1976: Particulate matter sinking to the deep-sea floor at 2000 m in Tongue-of-the-Ocean, Bahamas, with a description of a new sedimentation trap. J. Mar. Res., 34, 341-354.
12. Rowe, G.T. and W.D. Gardner, 1979: Sedimentation rates in the slope waters of the northwest Atlantic Ocean measured directly with sediment traps. J. Mar. Res., 37, 581-600.
13. Knauer, G.A., J.H. Martin and K. Bruland, 1979: Fluxes of particulate carbon, nitrogen and phosphorus in the upper water column of the northwest Pacific. Deep-Sea Res., 26, 97-108.
14. Honjo, S., 1980: Material fluxes and modes of sedimentation in the mesopelagic and bathypelagic zones. J. Mar. Res., 38, 53-97.
15. Fellows, D., D.M. Karl and G.A. Knauer, 1981: Vertical distribution and sedimentation of adenosine triphosphate in the upper 1500 meters of the northeast Pacific Ocean. Deep-Sea Res., 28A, 921-936.
16. Urrere, M.A. and G.A. Knauer, 1981: Zooplankton fecal pellet fluxes and vertical transport of particulate organic material in the pelagic environment. J. Plank. Res., 3, 369-382.
17. Hunter, K.A., 1980: Microelectrophoretic properties of natural surface-active organic matter in coastal seawater. Limnol. Oceanogr., 25, 807-822.
18. Karl, D.M., 1980: Cellular nucleotide measurements and applications in microbial ecology. Microbial Rev., 44, 739-796.
19. Nealson, K.H., 1978: The isolation and characterization of marine bacteria which catalyze manganese oxidation, pp. 847-858. In: Krumbein, W. (ed.). "Environmental Biogeochemistry and Geomicrobiology", Ann Arbor, Mich.
20. Schweisfurth, R., D. Eleftheriadis, H. Gundlach, M. Jacobs and W. Jung, 1978: Microbiology of the precipitation of manganese, pp. 923-944. In: Krumbein, W. (ed.). "Environmental Biogeochemistry and Geomicrobiology", Ann Arbor, Mich.
21. Bishop, J.K.B., J.M. Edmond, D.R. Ketten, M.P. Bacon and W.B. Silker, 1977: The chemistry, biology and vertical flux of particulate matter from the upper 400 m of the equatorial Atlantic Ocean. Deep-Sea Res., 24, 511-548.

22. Landing, W.M. and R.A. Feely, 1981: The chemistry and vertical flux of particles in the northeastern Gulf of Alaska. Deep-Sea Res., 28A, 19-37.
23. Martin, J.H. and G.A. Knauer, 1980: Manganese cycling in northeast Pacific waters. Earth Planet. Sci. Lett., 51, 266-274.
24. Knauer, G.A. and J.H. Martin, 1981: Phosphorus-cadmium cycling in northeast Pacific waters. J. Mar. Res., 39, 65-76.
25. Knauer, G.A. and J.H. Martin, 1981: Primary production and carbon-nitrogen fluxes in the upper 1500 meters of the northeast Pacific. Limnol. Oceanogr., 26, 181-186.
26. Eppley, R.W. and B.J. Peterson, 1979: Particulate organic matter flux and planktonic new production in the deep ocean. Nature, 282, 677-680.
27. Osterberg, C.L., A.G. Carey and H. Curl, Jr., 1963: Acceleration of sinking rates of radionuclides in the ocean. Nature, 200, 1276-1277.
28. Boothe, P.N. and G.A. Knauer, 1972: The possible importance of fecal material in the biological amplification of trace and heavy metals. Limnol. Oceanogr., 17, 270-274.
29. Fowler, S.W., 1977: Trace elements in zooplankton particulate products. Nature, 269, 51-53.
30. Higgo, J.J.W., R.D. Cherry, M. Heyraud, S.W. Fowler and T.M. Beasley, 1980: Vertical oceanic transport of alpha-radioactive nuclides by zooplankton fecal pellets, pp. 502-513. In: Gessell, T.F. and W.M. Lowder (eds.). "Natural Radiation Environments." III. International Symp., U.S. Dept. of Energy Symposium Series 51. Technical Information Center.
31. Martin, J.H. and G.A. Knauer, 1973: Elemental composition of plankton. Geochim. Cosmochim. Acta., 37, 1639-1653.
32. Fitzwater, S.E., G.A. Knauer and J.H. Martin, in press: Metal contamination and primary production: Field and laboratory methods of control. Limnol. Oceanogr.
33. Vinogradov, M.E., 1962: Feeding of deep-sea zooplankton. Rapp. P-V. Reun. Cons. Perm. Int. Explor. Mer., 153, 114-120.
34. Fuhrman, J.A., J.W. Ammerman and F. Azam, 1980: Bacterioplankton in the central euphotic zone: Distribution activity and possible relationships with phytoplankton. Mar. Biol., 60, 201-207.
35. Pomeroy, L.R., 1974: The ocean's food web, a changing paradigm. Biosci., 24, 499-504.
36. Agassiz, A., 1888: Three cruises of the United States Coast and Geodetic Survey steamer BLAKE. Vols. 1 and 2. Houghton-Mifflin and Co., N.Y.

37. Karl, D.M. and O. Holm-Hansen, 1978: Methodology and measurement of adenylate energy charge ratios in environmental samples. Mar. Biol., 48, 185-197.
38. Gordon, D.C., 1971: Distribution of particulate organic carbon and nitrogen at an oceanic station in the central Pacific. Deep-Sea Res., 18, 1127-1134.
39. Holm-Hansen, O., 1969: Determination of microbial biomass in ocean profiles. Limnol. Oceanogr., 14, 740-747.
40. Eppley, R.W., W.G. Harrison, S.W. Chisholm and F. Stewart, 1977: Particulate organic matter in surface waters off southern California and its relationship to phytoplankton. J. Mar. Res., 35, 671-696.

TRACE METAL LEVELS IN SEA WATER FROM THE SKAGERRAK AND THE KATTEGAT

Bertil Magnusson and Stig Westerlund

Department of Analytical and Marine Chemistry
Chalmers University of Technology and
University of Göteborg, S-412 96 Göteborg, Sweden

ABSTRACT

Unfiltered, acidified samples from a transection between Sweden and Denmark have been collected. The trace metals were preconcentrated by the freon-extraction method and determined by the graphite furnace technique.

The concentration levels found in Kattegat surface water (0-10 m) were: Cd, 0.2 nmol l^{-1}; Cu, 7.7 nmol l^{-1}; Fe, 36-200 nmol l^{-1}; Ni, 8.7 nmol l^{-1}; Pb, 0.25 nmol l^{-1}; and Zn, 12 nmol l^{-1}. In the bottom water with higher salinity similar concentration levels of Cd, Ni, and Zn were found while the Cu level (3.3 nmol l^{-1}) was lower. The iron and lead levels in the bottom water were considerably enhanced.

INTRODUCTION

Brackish water from the Baltic Sea (net outflow 500 m^3 y^{-1}) flows through the Kattegat and Skagerrak into the North Sea. In the Kattegat, the Baltic Sea water is mixed with bottom water originating from the North Sea. The freshwater discharge into Kattegat is relatively small. This means that the lower salinity of the surface water in this area is mainly a result of outflowing Baltic Sea water.

Several investigations concerning trace metals levels in the eastern and western part of the Baltic[2,3] and the Danish Sounds[4,5] have recently been published. However, for the Kattegat, the Skagerrak and the open North Sea, the trace metal levels are more uncertain.

This small investigation was undertaken to determine the concentration of Cd, Cu, Fe, Ni, Pb and Zn at four stations in northern Kattegat and at one station in the Skagerrak (Fig. 1).

SAMPLING AND ANALYSIS

Samples were collected during a one day cruise with the Swedish Coast Guard (TV 102). Surface water was sampled by hand from a rubber boat. Using arm-length polyethylene gloves, a sampling bottle was lowered into the water, opened, filled and closed below the surface. Samples of deeper water were collected with a 2.5 l teflon-coated GO-FLO sampler (General Oceanics, Miami, Fl.) equipped with teflon valves. The sampler was attached to a stainless steel hydro-wire.

Samples were collected and stored in both teflon and polyethylene bottles. The bottles were acid-cleaned according to an extensive procedure[6] and then soaked in 0.015 M nitric acid for one week, rinsed with ultra-clean water and wrapped in polyethylene bags.

The samples were acidified with 1 ml nitric acid per liter (0.015 M). To allow an acid-leaching of the suspended matter, the samples were stored for five weeks at room temperature before analysis. Trace metal concentrations were determined by graphite furnace AAS after a freon-extraction[7]. By this extraction technique, the trace metals are complexed with a mixture of ammonium pyrrolidine-dithiocarbamate (APDC) and diethylammonium diethylditiocarbamate (DDDC), extracted with freon and back-extracted with nitric acid into a water phase. To obtain lower blank levels, the citrate buffer was replaced with acetate and the extraction carried out with teflon funnels in a clean room.

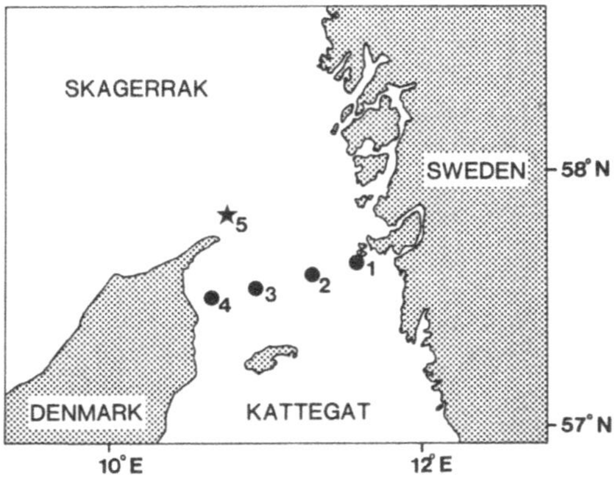

Fig. 1 Sampling stations in the Kattegat and Skagerrak.

Blank determinations were performed by extracting water from a Milli-Q water purification system. The acidified samples (250 ml) were neutralised with the acetate buffer containing an excess of ammonia. Since the evaluated recoveries obtained by standard addition on a few samples were quantitative for all metals (95-101%), the evaluation was made by comparing the extracts with aqueous standard solutions.

RESULTS AND DISCUSSION

The results are presented in Table 1. The difference in salinity in northern Kattegat between the outflowing surface water (the Baltic current) and the incoming bottom water from the Skagerrak is around 9°/oo. The surface water (25°/oo salinity) can be regarded as a mixture of water from the North Sea with water from the Baltic Sea (7-12°/oo).

Cadmium

The concentration of cadmium showed no variation with depth or salinity, and the mean value for all samples was 0.20 nmol l^{-1}. In the Danish Sounds with a slightly lower salinity (20-22°/oo), Rasmussen[4] found 0.21 nmol l^{-1} by two different pre-concentration technqiues, Chelex-100 and freon-extraction in non-acidified samples. Thus a uniform Cd level around 0.20 nmol l^{-1} in sea water from the Kattegat and Skagerrak is observed, and no impact of Cd from the Baltic could be found.

Copper

The upper water mass (0-10 m) in northern Kattegat has a Cu level of 7.7 nmol l^{-1}. Similar Cu levels, 7.9-8.5 nmol l^{-1}, have been found in the Danish Sounds[4] (22°/oo salinity), while the Cu level measured in the Baltic Sea[3] is slightly higher, 9.4-15.7 nmol l^{-1}. A lower Cu level, 3.3 nmol l^{-1}, was found in the inflowing bottom water originating from the North Sea. The Cu values in surface water (0-10 m) are higher than what would be predicted from a mixing of Baltic Sea water (S = 9°/oo, Cu = 13 nmol l^{-1}) with inflowing bottom water from the North Sea. A linear relation is, however, observed from 30 to 35°/oo if Cu is plotted against salinity.

Iron

A large part of the iron in sea water is in a colloidal or particulate form, which mainly will be dissolved during storage

Table 1 Trace metal concentrations in acidified samples from the Kattegat and the Skagerrak

Station	Position and Total Depth	Sampling Depth (m)	Salinity ‰	Cd	Cu	Fe	Ni	Pb	Zn
				nmol l^{-1}					
1	57°39'N	10	31.2	0.21	7.7	65	8.7	0.20	14.4
	11°24'E	70	34.4	0.20	2.9	990	9.0	0.81	11.3
	80 m								
2	57°33'N	Surface	24.9	0.22	8.0	50	9.9	0.24	10.2
	11°11'E	10	30.9	0.20	8.0	29	8.5	0.13	11.3
	40 m								
3	57°32'N	10	31.0	0.20	6.3	180	7.7	0.25	10.6
	10°53'E	35	33.6	0.20	4.1	1200	8.7	0.95	13.9
	50 m								
4	57°31'N	Surface	25.9	0.20	7.4	160	8.7	0.22	10.7
	10°44'E	10	29.6	0.22	8.5	200	9.9	0.30	11.8
	35 m	30	33.8	0.22	5.4	1020	7.3	0.84	15.5
5	57°52'N	Surface	31.9	0.21	5.2	200	6.6	0.27	11.0
	10°47'E	15	33.6	0.20	8.0	110	9.5	0.29	8.0
	110 m	35	34.3	0.20	3.6	300	5.3	0.38	9.8
		100	34.9	0.20	2.9	1110	5.3	0.92	13.3
	Blank level	-	-	0.004	0.2	<2	0.7	0.03	0.3

under acid conditions. In the outflowing surface water the acid-leachable iron concentration ranges from 36-200 nmol l^{-1}, while the inflowing bottom water has a rather constant level around 1070 nmol l^{-1}. The corresponding iron value for Baltic Sea water[3] is around 90 nmol l^{-1}.

Nickel

The nickel concentration found showed no variation with depth in the northern Kattegat (Station 1-4) and had a mean value of 8.7 nmol l^{-1}. At station 5 in the Skagerrak, lower nickel levels were found amounting to 5.3 nmol l^{-1} in the bottom waters. Thus, a decrease in the nickel concentration was observed from the Baltic Sea[3], 10-15 nmol l^{-1} through the Danish Sounds[4], 8.2 nmol l^{-1} and the Kattegat to the Skagerrak.

Lead

A large difference in lead content was found between bottom and surface water. The lead level (mean value 0.25 nmol l^{-1}) in the surface water was comparable with the lowest values given for the Baltic Sea[3]. Furthermore, the results are fairly consistent with the levels found in the North Atlantic surface water[8]. The high lead content in the bottom waters is discussed below.

Zinc

No significant differences in Zn concentration between the different water masses were observed. The mean value of 11.6 nmol l^{-1} found seems rather reasonable in this shallow area. It is also unlikely that an oceanic level[9] will be found in this area for an element like zinc with rather high coastal inputs.

Iron-lead co-variation

The high iron content in the bottom water indicates a high fraction in suspended form, since the solubility of iron in sea water is considerably lower[10]. Iron in suspended form can act as a carrier for other trace metals by adsorption. Of the metals studied, only lead showed a clear co-variation with iron (Fig. 2). According to Rodhe[1], sediments in the North Sea are eroded by the bottom currents and transported into the Kattegat with the inflowing bottom water. Ths high iron and lead contents in the bottom water in northern Kattegat may therefore come from iron and lead present in suspended matter that originates from North Sea sediments. In the Kattegat, the eroded material from the North Sea is

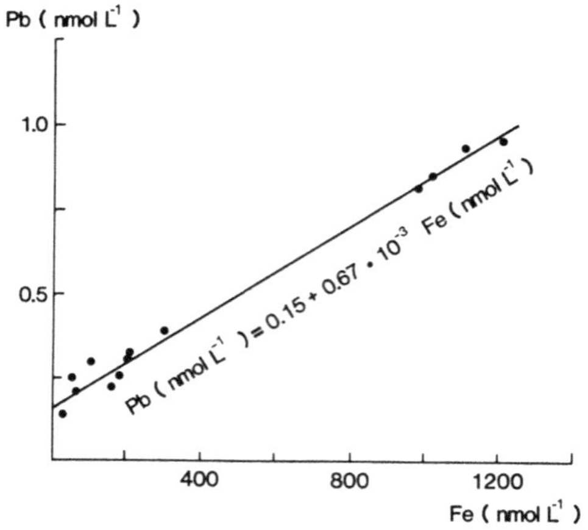

Fig. 2 Iron-lead co-variation. Correlation coefficient R = 0.99.

deposited[1]. Thus, a removal of iron and lead in the bottom waters southwards in Kattegat is probable.

CONCLUSIONS

In the outflowing surface water, a mixture of Baltic and North Sea water, the levels of Cd, Cu, Ni and Zn is rather constant. Fe and Pb show some variation.

In the inflowing bottom water, similar levels of Cd, Ni and Zn were found, while the Cu level was lower (3.3 nmol l). Fe and Pb contents were higher in the bottom water. This increase is explained by a high Fe and Pb content in the suspended matter, originating from North Sea sediments. Variation in this input can explain the variations in Fe and Pb concentrations in the surface water.

ACKNOWLEDGEMENT

The authors wish to thank the Swedish Coast Guard and especially the crew on TV 102. Financial support from the National Swedish Research Council is gratefully acknowledged.

REFERENCES

1. Rodhe, J., 1973: Sediment transport and accumulation at the Skagerrak-Kattegat border. Report No 8, Institute of Oceanography, University of Gotenburg, Sweden.

2. Koroleff, F, in print: Determination of traces of heavy metals in natural waters by AAS after concentration by co-precipitation. Proceedings from the XII conference of the Baltic oceanographers, Leningrad, 14-17 April, 1980.
3. Magnusson, B. and S. Westerlund, 1980: The determination of Cd, Cu, Fe, Ni, Pb and Zn in Baltic Sea water. Mar. Chem., 8, 231-244.
4. Rasmussen, L., 1981: Determination of trace metals in sea water by chelex-100 or solvent extraction techniques and atomic absorbtion spectrometry. Anal. Chim. Acta., 125, 117-130.
5. Schmidt, D., 1980: 14th European Marine Biology Symposium 1979. Helgolander Meeresundersuchungen. 33, 576.
6. Moody, J.R. and R.M. Lindström, 1977: Selection and cleaning of plastic containers for storage of trace element samples. Anal. Chem., 49, 2264-2267.
7. Danielsson, L-G., B. Magnusson and S. Westerlund, 1978: An improved metal extraction procedure for the determination of trace metals in sea water by atomic adsorption spectrometry with electrothermal atomization. Anal. Chim. Acta., 98, 47-57.
8. Schaule, B. and C.C. Patterson, 1981: Lead in the Ocean. This volume.
9. Bruland, K.W., G.A. Knauer and J.H. Martin, 1978: Zinc in the north east Pacific water. Nature, 271, 741-743.
10. Byrne, R.H. and D.R. Kester, 1976: Solubility of hydrous ferric oxide and iron speciation in sea water. Mar. Chem., 4, 255-274.

MERCURY CONCENTRATIONS IN THE NORTH ATLANTIC IN RELATION TO CADMIUM, ALUMINIUM AND OCEANOGRAPHIC PARAMETERS

Jón Ólafsson

Marine Research Institute
Skúlagata 4, Reykjavik
Iceland

ABSTRACT

The water masses investigated are derived from both arctic and temperate regions. The mean reactive mercury concentration in 73 samples is 8.5$\pm$ 3.5 pmol l^{-1} and the mean total mercury concentration in 47 samples is 11$\pm$ 5.0 pmol l^{-1}. The mercury distribution is dissimilar to both that of cadmium and aluminium and it has not a simple correlation with dissolved silicate. The cadmium distribution is strongly related to the nutrients nitrate and phosphate, but the Δ Cd : Δ NO_3 and Δ Cd : Δ PO_4 slope values are significantly lower than have been found in the Pacific. The aluminium concentrations range from 8.1 - 25 nmol l^{-1} and increase generally with depth. At a deep station south of Iceland with Irminger Water at intermediate depths, this water mass had significantly higher mercury concentrations, reactive 17 pmol l^{-1}, than the surface or bottom water. It is suggested that atmospheric input of mercury may have been effective when this water mass formed by deep convective mixing in winter. The East Greenland Current which receives glacial melt water has low mercury concentrations, 7.0$\pm$ 3.5 pmol l^{-1}, insignificantly different from Arctic Intermediate Water, 9.5$\pm$ 3.5 pmol l^{-1}, or Arctic Bottom Water, 11$\pm$ 2.5 pmol l^{-1}. Previous hypotheses of large scale effects of geothermal mercury emanations are examined and it is concluded that the proximity alone of a water body to a site of such an activity is no indication of its mercury concentration.

INTRODUCTION

In recent years significant progress has been made in the study of the oceanic concentration distribution of a number of trace metals other than mercury [1,2]. This has been accomplished through the introduction of sensitive and well controlled analytical methods and by recognition of the crucial importance of the sampling and sample storage steps. With regard to the analysis of mercury in environmental samples a considerable advance in analytical sensitivity was made in the sixties with the introduction of cold vapor atomic absorption. Nevertheless, a recent review [3] shows that since 1971 oceanic mercury concentrations spanning three orders of magnitude have been reported. It furthermore reveals that a considerable proportion of the samples investigated have come from the vicinity of Iceland. The work here presented is a continuation of a previous effort[4] aimed at establishing reliable data on mercury concentrations in North Atlantic water masses. For comparative purposes data were also obtained at one deep station on dissolved cadmium, the distribution of which relates strongly to the labile nutrients[1], and on dissolved aluminium which seems to be controlled largely by dissolution from inorganic solid particles [5].

SAMPLING AND ANALYSIS

The sampling, storage and analytical methods used in this study have recently been evaluated [6,7]. The sampling was carried out on three cruises of R/S Bjarni Saemundsson in the summer of 1980 with sampling stations shown in Fig. 1. The Hydro-Bios water bottles used were modified by replacing internal rubber rings with silicone rubber equivalents. At the commencement of a cruise the water bottles were cleaned by filling with a solution of the detergent Deacon 90. Samples for the analysis of mercury were drawn into 500 ml pyrex vessels and acidified to pH 1 with nitric acid (Merck 457) containing less than 0.05 nmol l^{-1} of mercury impurities. The pyrex bottles were precleaned with both nitric acid and a solution of nitric and hydrofluoric acids (10:1) and subsequently stored up to the time of sampling holding a small volume of nitric acid. Ashore, reactive mercury was determined by cold vapour atomic absorption after preconcentration by amalgamation on gold [8]. A 20 cm long optical cell and a Varian AA6 spectrophotometer were employed. The total mercury concentration was similarly determined following 1 hour U.V. irradiation of a duplicate sample using a 500 W low pressure mercury lamp (Hanovia) and immersion irradiation equipment (Ace Glass Inc.). The precision of the mercury determination assessed by analyzing 19 replicates over a period of 107 days had been found[7] to be $\pm$ 2.0 pmol l^{-1} for a concentration of 12.5 pmol l^{-1}. Unfiltered samples for the analysis of cadmium were collected into 1 l acid washed low

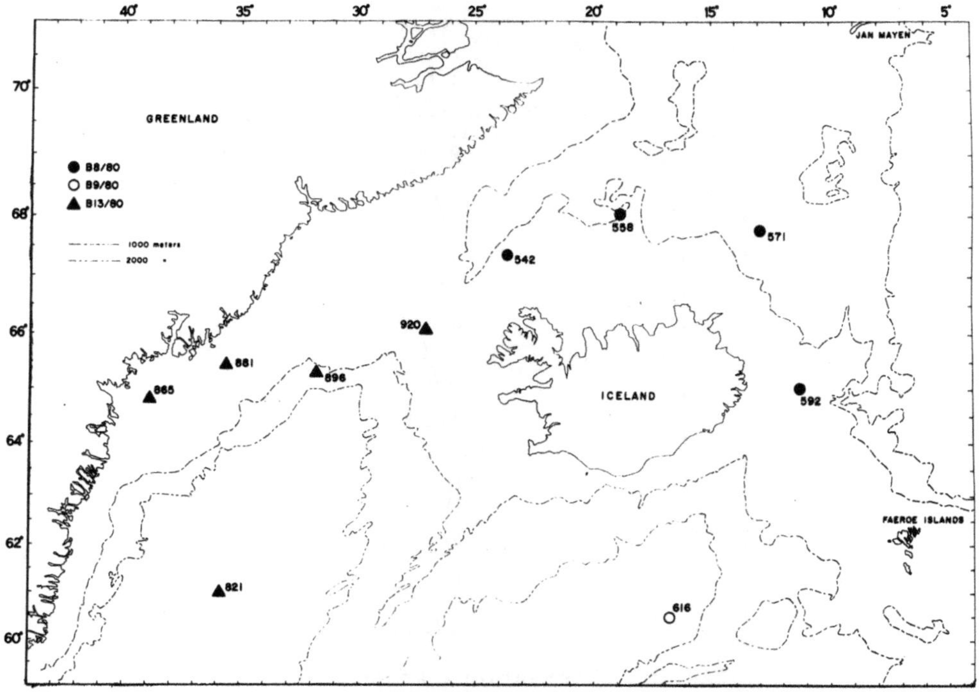

Fig. 1. Locations of sampling stations 1980. Cruise B 8/80: 28th May to 7th June, Cruise B 9/80: 21st June, Cruise B 13/80: 15th August to 26th August.

density polyethylene bottles and acidified to pH 2.2 with hydrochloric acid (Suprapure, Merck). Cadmium was preconcentrated and eluted on board using columns of Chelex-100 resin[10]. The determinations were carried out ashore by atomic absorption using a Varian AA6 instrument fitted with a DC-6 background corrector, CR-90 carbon rod and ASD-35 autosampler. The precision based on replicate determinations is $\pm$ 0.013 nmol l^{-1}. The nitrate and silicate determinations were made on board by colorimetry using an Auto-Analyzer[11]. Determinations of dissolved aluminium in unfiltered samples were also carried out on board within a few hours of sample collection by fluorometric method[12].

RESULTS AND DISCUSSION

The oceanographic conditions in the area investigated are quite complex. Within it are water masses of contrasting temperature and salinity characteristics derived from both arctic and temperate regions, and it is also a region of water mass formation in winter. The water mass definitions and nomenclature used here

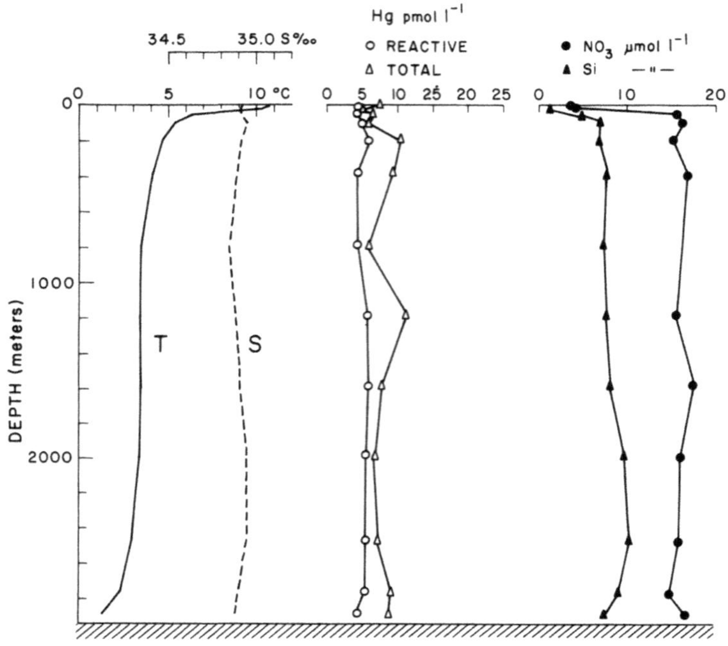

Fig. 3 Observations at station 821.

The aluminium concentrations (8.1-25 nmol l^{-1}) found at station 616 (Fig. 2) are similar or lower than were recently reported for the North Atlantic[5]. However, they do not show a mid-water minimum but increase with depth. The absence of a photic zone concentration minimum and a relatively weak correlation with dissolved silicate (r = 0.86) indicate however, that processes of biological uptake and redissolution with silicate predominant in the Mediterranean[15], may not be controlling the aluminium distribution in the North Atlantic.

Mercury

At station 616 (Fig. 2) the water mass in upper layers down to 1000 m depth is essentially relatively warm and saline Atlantic Water, at intermediate depths Irminger Water, and near the bottom cold water mostly derived from the Iceland-Scotland overflow. It is immediately evident from the concentration profiles (Fig. 2) and also confirmed by statistical tests, that no simple correlations exist between mercury and the dissolved nutrients of nitrate and silicate with r = 0.66 and r = 0.76 respectively. This is contrary to an earlier observation in the Atlantic[16] of a linear

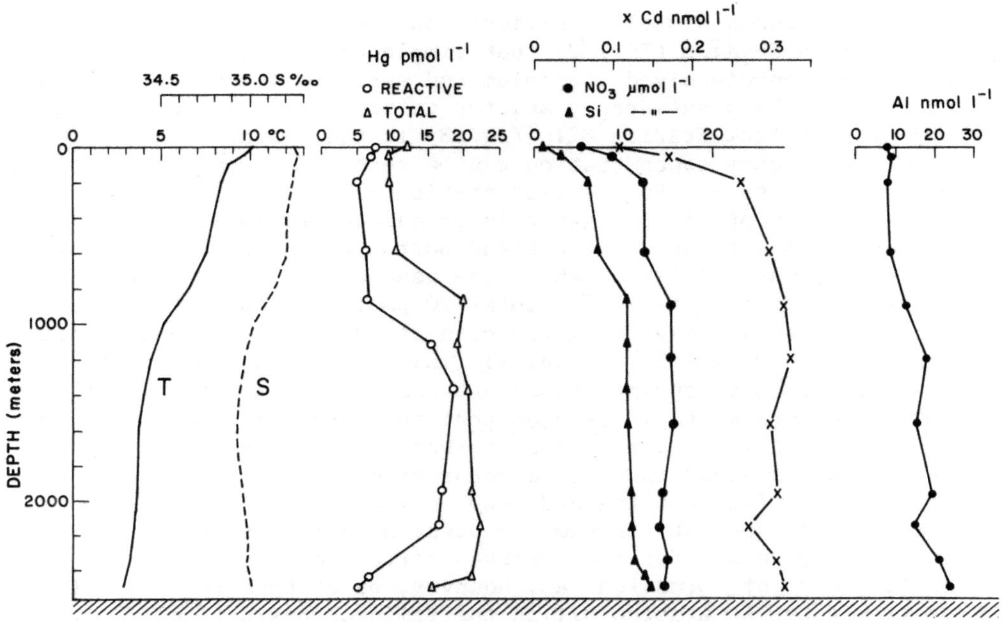

Fig. 2 Observations at station 616.

are according to Stéfansson[13,14]. The results from the deepest stations are presented in Figs. 2 and 3, but from other stations in Table 1.

Cadmium and aluminium

Cadmium at station 616 (Fig. 2) exhibited a strong relation with the nitrate concentration ($r = 0.961$):

$$[Cd] = (0.0197 \pm 0.0019)\ [NO_3] + (0.011 \pm 0.025),$$

and with the phosphate concentration ($r = 0.954$):

$$[Cd] = (0.264 \pm 0.028)\ [PO_4] + (0.016 \pm 0.027),$$

where the concentration units on cadmium are nmol l^{-1} and on the nutrients µmol l^{-1}. The slope values for Δ Cd: Δ NO_3 = 0.0197 and Δ Cd: ΔPO_4 = 0.264 have coefficients of variation of about 10% and they are appreciately lower than values found for the North-Pacific[1] (0.0228 and 0.347, respectively). This is likely due to the few sampling points and the relatively narrow concentration range encountered here.

relation between dissolved silicate and reactive mercury. It is furthermore evident (Fig. 2) that simple correlation does not exist between dissolved aluminium and mercury, r = 0.72. A lack of relation between mercury and the nutrients is also evident in the profile from Station 821 (Fig. 3) in the Irminger Basin since here the mercury concentration can be regarded as uniform from surface to bottom. However, at station 616 (Fig. 2), the mercury concentration profiles suggest intermediate waters with "high" concentrations between surface and bottom waters both with "low" concentrations. A t-test shows the mean of the "high" concentrations (reactive 17 pmol l^{-1}, total 20 pmol l^{-1}) to differ significantly at the 1% level from the mean of the "low" concentrations (reactive: 6.0 pmol l^{-1}, total 11 pmol l^{-1}). When the "high" reactive mercury concentrations are examined in relation to the T-S diagram (Fig. 4) it is seen that they are all in water having T-S characteristics very close to that of Irminger Sea water[14]. This is of interest when it is considered that Irminger Water is formed by surface cooling and deep convective overturn in winter. Atmospheric input at the time of water mass formation may therefore be the reason for the observed mercury distribution. A similar mode of formation can, however, be assumed for the intermediate water at station 821 in the Irminger Basin (Fig. 3), where the mercury concentration was found to be uniform. Atmospheric

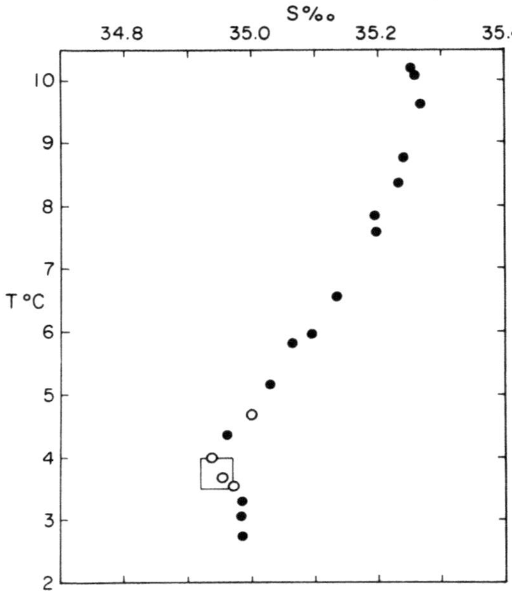

Fig. 4 T-S diagram for station 616. The square marks the boundaries of Irminger Water[14].

transport and input of mercury has then either not been effective in the region of formation of this intermediate water, or its formation has been at a different time from the intermediate water at station 616.

In 1973 reactive mercury concentrations ranging from 5.5 to 40 pmol l^{-1} were found in samples from the East Greenland Current[4]. In Table 1, stations 865, 881 and 920 are in the East Greenland Current as seen from their low salinity surface water, and Polar Water influence was also found at Station 558. The reactive mercury concentration here found in 10 samples of this water type is 7.0 ± 3.5 pmol l^{-1}. Also in Table 1, the reactive mercury in 14 samples of Arctic Intermediate Water is 9.5 ± 3.5 pmol l^{-1} and in 8 samples of Arctic Bottom Water it is 11 ± 2.5 pmol l^{-1}. There are hence no significant differences in the reactive mercury concentrations of these three water masses which predominate in the Iceland Sea.

The overall mean of the reactive mercury concentrations in Figs. 2 and 3 and in Table 1 is 8.5 ± 3.5 pmol l^{-1} (n = 73) and for total mercury the mean is 11 ± 5.0 pmol l^{-1} (n = 47). The difference in total and reactive mercury in 46 samples sets, indicated that 38% of the mercury, or 4.5 pmol l^{-1}, was associated with organic matter and released by photo-oxidation. This percentage is similar to what has been observed in coastal waters[17,18,19] but the concentration of organic mercury is, however, very low and it will perhaps require new analytical approaches to accurately resolve variations in the organic mercury concentrations in oceanic waters.

The results here presented on mercury around and north of Iceland are one to two orders of magnitude lower the earlier data of other workers[20,21,22]. They had found mercury concentrations ranging from 314 to 1810 pmol l^{-1} in the East Greenland Current, from 35 to 424 pmol l^{-1} in Arctic Intermediate Water and from 30 to 1100 pmol l^{-1} in Arctic Bottom Water. They attributed high mercury concentrations to either melt from the Greenland glacier or to volcanic emanations. The Greenland glacier was suggested as a source on account of rather high mercury concentrations reported[23] there in 1971. However, a later study using an intercalibrated neutron activation method revealed mercury concentrations around 50 pmol l^{-1} in Greenland ice[24], thus making it an unlikely source. Although geothermal mobilisation of mercury is well known, the quantity emitted during the 1973 Heimaey eruption was estimated[25] only 7×10^5g, which is little in comparison with e.g. anthropogenic emissions estimated[26] at 10^{10}g Yr^{-1}. Furthermore, significant influences of the Heimaey eruption on sea water mercury concentrations were confined in space and time to the advance of lava into the sea. In the neovolcanic zone of Iceland mercury concentrations less than 50 pmol l^{-1} were found both in Lake Mývatn[27],

Table 1. Mercury Concentrations and Oceanographic Parameters

Station number	Depth M	T ^{o}C	S $^{o}/oo$	Hg pmol l^{-1} Reactive	Total	Silicate $\mu mol\ l^{-1}$	Nitrate $\mu mol\ l^{-1}$	Water mass
542	0	4.38	34.77	11		6.2	11.7	A
	30	4.96	34.94			7.1	12.2	A
	100	2.01	34.78	14		6.7	12.7	AI
	200	0.77	34.83	11		6.9	12.8	AI
	300	0.08	34.89	12		8.5		AI
558	0	0.46	34.29	12		4.1		P+A
	20	0.49	34.40	11		4.4		P+A
	50		34.70	13		5.5		A
	100	-0.10	34.80	11		7.9		AI
	200	-0.15	34.83	7.5		6.0		AI
	392	0.06	34.90	12		7.1		AI
	588	-0.38	34.91	12		8.1		AB
	784	-0.54	34.92	9.5		9.5		AB
571	0	1.30	34.70	12		3.5	7.9	A
	20	0.83	34.71	12		3.9	8.0	A
	50	0.33	34.71	8.0		4.3	8.9	A
	100	-0.32	34.76	12		4.8	10.3	AI
	200	-0.30	34.82	9.5		5.4	11.7	AI
	400	-0.13	34.91	16		7.5	13.2	AB
	767	-0.59	34.91	11		9.4	13.7	AB
	1150	-0.79	34.92	10		10.8	11.7	AB
	1440	-0.88	34.91	13		11.5	12.2	AB
	1730	-0.88	34.92	9.5		12.8	13.5	AB
592	0	4.32	34.69	10		2.5	5.2	
	50	2.26	34.71	10		4.8	11.1	
	177	0.22	34.82	10		6.1	13.4	AI
	355	0.06	34.90	11		7.7	14.7	AI
	470	-0.28	34.90	9.0		9.0	15.2	AB
865	0	5.93	32.96	3.5	16	3.1	0.4	P
	50	4.32	34.56	5.0	22	3.0	8.3	P
	200	3.68	34.77	6.5	9.5	6.3	14.1	
	800	3.44	34.84	5.0	5.5	7.5	14.5	
881	0	3.53	32.01	5.5	13	0.7	0.1	P
	20	1.37	33.78	5.5	10	1.5	3.4	P
	50	3.40	34.42	5.5	9.5	3.0	7.6	P
920	0	3.91	31.83	11	14	1.4	1.2	P
	20	3.04	32.03	5.0	8.0	1.2	0.4	P
	100	3.96	34.54	6.0	8.5	6.6	9.7	P
	200	4.47	34.81	5.5	6.5	7.9	12.7	A
	392	0.66	34.85		7.0	7.8	13.2	AI
	487	0.03	34.90	4.5	6.0	9.3	12.5	AI
	555	0.03	34.90	4.0	14	9.1	13.9	AI
	595	0.04	34.90	6.0	12	8.8	12.1	AI
	640	0.04	34.90	5.0	8.0	10.1	15.4	AI
896	0	10.12	35.07	6.0	11	0.4	2.5	AT
	20	10.07	35.07	10	13	0.4	2.8	AT
	50	7.48	35.05	9.5	12	4.3	12.1	AT
	200	6.72	35.12	12	13	6.5	14.8	AT
	600	4.28	34.94	15	17	8.3	16.5	
	921	2.32	34.81	9.0	11	6.7	13.7	
	1170	0.97	34.83	8.5	9.0	6.9	14.0	OF

Water masses: A = Arctic Water, AI = Arctic Intermediate Water, AB = Arctic Bottom Water, P = Polar Water, AT = Atlantic Water, OF = Iceland-Greenland Overflow Water.

which receives thermal spring-waters, and in the volcanically affected caldera Lake Öskjuvatn[28]. It is important to note, that in the Greenland glacier core study[24] no elevated mercury concentrations were found in strata from the period of the 1783 Laki eruption but this eruption was the largest (12 km^3 of lava) of the last 1000 years in Iceland and it left other conspicuous signs in the Greenland glacier[29]. These considerations lead to the conclusion that the proximity alone of a water body to a site of geothermal activity is no indication of its possible mercury concentration.

Open ocean mercury concentrations of the same order of magnitude but generally somewhat higher than here found, have been reported for the Pacific[30] and the North-Atlantic[16,18]. In an international mercury intercalibration[7] involving surface water collected west of Iceland eleven participants reported mercury concentrations less than 20 pmol l^{-1} and this lends evidence and supports the representativeness of the concentrations here reported as the natural oceanic level. The present results show the mercury distribution to be neither like that of the "nutrient type" trace metals Cd, Zn, Ni, controlled by biological processes nor like aluminium controlled by inorganic processes. It is likely that a detailed understanding of the processes controlling the oceanic mercury distribution will require a closer look at its speciation.

REFERENCES

1. Bruland, K.W., 1980: Oceanographic distributions of cadmium, zinc, nickel and copper in the North Pacific. Earth Planet Sci. Lett., 47, 176-198.
2. Landing, W.M. and K.W. Bruland, 1980: Manganese in the North Pacific. Earth Planet Sci. Lett., 49, 45-56.
3. Fitzgerald, W.F., 1979: Distribution of mercury in natural waters. In, "The biogeochemistry of mercury in the environment", J.O. Nriagu, editor. Elsevier.
4. Ólafsson, Jón, 1975: Mercury in the Atlantic near Iceland. ICES C.M. 1975/E, 34.
5. Hydes, D.J., 1979: Aluminium in sea water: Control by inorganic processes. Science, 205, 1260-1262.
6. Bewers, J.M. and H.L. Windom, in press: Comparison of sampling devices for trace metal determinations in seawater. Mar. Chem.
7. Ólafsson, Jón, in press: An international intercalibration for mercury in sea water. Mar. Chem.
8. Ólafsson, Jón, 1974: Determination of nanogram quantities of mercury in sea water. Anal. Chim. Acta., 68, 207-211.
9. Fitzgerald, W.F. and W.B. Lyons, 1973: Organic mercury compounds in coastal waters. Nature, 242, 452-453.

10. Kingston, H.M., I.L. Barnes, J.J. Brady and T.C. Rains, 1978: Separation of eight transition elements from alkali and alkaline earth elements in estuarine and sea water with chelating resin and their determination by graphite furnace atomic absorption spectrometry. Anal. Chem., 50, 2064-2070.
11. Grasshoff, K., 1970: A simultaneous multiple channel system for nutrient analyses in seawater with analog and digital data record. Technicon Quarterly, 3, 7-17.
12. Hydes, D.J. and P.S. Liss, 1976: Fluorometric method for the determination of low concentration of dissolved aluminium in natural waters. Analyst, 101, 922-931.
13. Stefánsson, Unnsteinn, 1962: North Icelandic Waters. Rit Fiskideildar, 3, 269 pp.
14. Stefánsson, Unnsteinn, 1968: Dissolved nutrients, oxygen and water masses in the Northern Irminger Sea. Deep-Sea Res., 15, 541-575.
15. Caschetto, S. and R. Wollast, 1979: Vertical distribution of dissolved aluminium in the Mediterranean Sea. Mar. Chem., 7, 141-155.
16. Mukherji, P. and D.R. Kester, 1979: Mercury distribution in the Gulf Stream. Science, 204, 64-66.
17. Fitzgerald, W.F., 1976: Mercury studies of sea water and rain: Geochemical flux implications. In: "Marine Pollutant Transfer". H.L. Windom and R.A. Duce, editors. D.C. Heath & Co., Lexington, Mass.
18. Baker, C.W., 1977: Mercury in surface waters of seas around the United Kingdom. Nature, 270, 230-232.
19. Windom, H.L. and F.E. Taylor, 1979: The flux of mercury in the South Atlantic Bight. Deep-Sea Res., 26A, 283-292.
20. Carr, R.A., J.B. Hoover and P. E. Wilkniss, 1972: Cold-vapor atomic absorption analysis for mercury in the Greenland Sea. Deep-Sea Res., 19, 747-752.
21. Gardner, D. and J.P. Riley, 1974: Mercury in the Atlantic around Iceland. J. Cons. Int. Explor. Mer., 35, 202-204.
22. Gardner, D., 1975: Observations on the distribution of dissolved mercury in the ocean. Mar. Pollut. Bull., 6, 43-46.
23. Weiss, V.W., M. Koide and E.D. Goldberg, 1971: Mercury in a Greenland ice sheet: Evidence of recent input by Man. Science, 174, 692-694.
24. Appelquist, H., K.O. Jensen and T. Sevel, 1978: Mercury in the Greenland ice sheet. Nature, 273, 657-659.
25. Ólafsson, Jón, 1975: Volcanic influence on seawater at Heimaey. Nature, 255, 138-141.
26. Andren, A.W. and J.O. Nriagu, 1979: The global cycle of mercury. In, "The biogeochemistry of mercury in the environment". J.O. Nriagu, editor. Elsevier.

27. Ólafsson, Jón, 1979: The chemistry of Lake Mývatn and River Laxá. Oikos, 32, 82-112.

28. Ólafsson, Jón, 1980: Temperature structure and water chemistry of the caldera Lake Öskjuvatn, Iceland. Limnol. Oceanogr., 25, 779-788.

29. Hammer, C.U., 1977: Past volcanism revealed by Greenland ice sheet impurities. Nature, 270, 482-486.

30. Matsunaga, K., M. Nishimura and S. Konishi, 1975: Mercury in the Kuroshio and Oyashio regions and the Japan Sea. Nature, 258, 224-225.

PERTURBATIONS OF THE NATURAL LEAD DEPTH PROFILE IN THE SARGASSO SEA BY INDUSTRIAL LEAD

Bernhard K. Schaule and Clair C. Patterson

Division of Geological and Planetary Sciences
California Institute of Technology
Pasadena, California 91125
U.S.A.

ABSTRACT

Concentrations of dissolved lead were determined by isotope dilution mass spectrometry in 17 seawater samples of a depth profile collected under strict contamination control at a station in the northern Sargasso Sea, 250 km northwest of Bermuda. Dissolved lead concentrations were about 170 pmol$\cdot$kg^{-1} from the surface down to 400 m, then declined sharply to relatively constant values of 25 pmol$\cdot$kg^{-1} below 2000 m depth. This large concentration enrichment in the upper part of the water column is concordant with the distribution observed earlier in the central northeast Pacific, and it confirms the unique character of the occurrence of lead in open ocean water columns: the depth profile for lead differs markedly from those for most other trace metals, which show concentration depletions in surface waters instead.

Lead concentrations at the Sargasso Sea station are about 3-fold larger in surface and thermocline waters, and about 5-fold larger in deep waters than in the northeast Pacific which can be explained as resulting from an eolian input flux of industrial lead to the North Atlantic at least 3-fold higher than in the North Pacific.[1] The residence times of lead in the different portions of the water column are similar to those estimated for the North Pacific . As a consequence of the short residence times in the deep ocean, the vertical lead profile is probably significantly different from steady state due to recent anthropogenic inputs. Geographic distributions of lead contrast with those for most other trace metals by showing lower concentrations in North Pacific deep waters compared to those in the North Atlantic. This difference is a result of the fact that the residence times of most trace metals are longer than the transit times of the deep water masses, while the residence time of lead is shorter than those transit times.

INTRODUCTION

A careful and reliable study of the occurrence of lead in the Northeast Pacific and of historic changes in the eolian flux of lead to the oceans has shown that lead is unique among trace metals in the marine environment because its residence time in the water column is short, which makes the distribution of lead respond quickly to input changes, and that the profile is perturbed because industrial lead inputs to the marine lead cycle have increased significantly beyond natural inputs[1].

The Pb profile in the open North Pacific shows a 15-fold concentration enrichment in the upper part of the water column compared to the lower part. This is contrary to all known trace metal distributions in the North Pacific gyre, which display more or less pronounced "nutrient-type" depletions or, in a few cases, slight enrichments in the upper part of the water column. Oceanic data, backed by a number of other independent observations, suggest the present vertical profile of lead is greatly disturbed and not in steady state. Since characteristics of the increase of industrial inputs to the open ocean as a function of time and the nature of lead redistribution processes in the water column are only approximately known, the model which describes the vertical distribution of lead in the northeast Pacific[1,2] can provide only approximate residence times for lead in the deep ocean (about 80 years) and in the thermocline (more than 20 years).

Marine chemical conditions and anthropogenic inputs in the Sargasso Sea are similar in some respects and different in others in comparison with those in the North Pacific. This study of lead in the Sargasso Sea therefore provides a test of views concerning chemical properties and anthropogenic fluxes of lead that were derived from the earlier North Pacific study.

SAMPLING AND ANALYTICAL METHODS

Samples were collected and analyzed under strictly controlled contamination conditions outlined earlier[3]. Surface samples were taken from a moving raft by hand-dipping bottles at a distance up-current and up-wind from the research vessel. Deep samples were taken using the protected deep water sampler built at Caltech[4] which allowed Pb concentrations of 1 part per trillion, or 5 pmol$\cdot$ kg^{-1} to be sampled in the Northeast Pacific[1]. This is achieved by pointing the sampler intake downward and collecting water only while the sampler is being lowered into undisturbed waters, and using ultra-clean sampling units that have been individually cleaned, assembled and sealed under ultra-clean laboratory conditions and which are inserted into the sampler housing for use before each cast[4].

A vertical profile of collections at 17 depths was sampled in July, 1979 during an NSF-SEAREX cruise at station 4 (34° 15'N; 66° 17'W), located ∿250 km to the northwest of Bermuda and 200 km SE of GEOSECS station II in the northern Sargasso Sea. Aliquots of samples collected by the Caltech sampler were analyzed for Cu, Zn, Cd, Ni, Mn and Fe by K. Bruland[5] and for Hg by W. Fitzgerald. Parallel samples obtained by a different collector during the same casts, 20 m above the samples reported here, have been analyzed for ^{210}Pb by K. Turekian. Hydrographic data are reported elsewhere[5].

Analyses for lead were carried out by isotope dilution mass spectrometry using ultra-clean laboratory techniques described previously[1,3,6]. Concentrations reported here represent dissolved lead, as measured by a chloroform-dithizone extraction method, which has been found to give results similar to those obtained by filtration through 0.4 μm-filters[7]. Ultra-clean mass spectrometric isotopic dilution analyses of lead in surface waters taken in a transect between Narragansett, R.I., and Bermuda[8], show that particle lead declines from 98% of total lead to 20% of the total in a narrow strip of coastal waters, and remains less than 10% of the total in the open Atlantic. Our profile concentration values therefore represent total amounts of lead to within about 10%. The overall error of the reported data is estimated at ±10%[1].

RESULTS

Results of lead analyses are listed in Table 1 and are shown in Fig. 1. One sample was contaminated by sampling too close to the ship, another by error in winch manipulation, and three by failures of improperly made polyethylene bag seals which opened under stress at great depths. These failures were disclosed either by the detection of a tracer in the collected samples which had been previously added to the water outside the sample bag but within the sampler piston housing, or by detection of anomalous concentrations of Zn, Ni and Fe in the samples. All of the data, including the five values which should be disregarded, are listed in Table 1.

The most salient features of this vertical profile for lead are that there is an over 6-fold concentration enrichment in the upper part of the water column over deep water concentrations,

Table 1 Dissolved lead concentrations in the Sargasso Sea at 34° 15'N, 66° 17'W, bottom depth 5100 m. Values in [ ] are probably high due to artifact contamination and should be disregarded.*

Sampling Depth (m)	Diss. Pb $pmol.kg^{-1}$	
0.2	160	
22	[240]	
35	150	
100	160	
150	160	
200	[180]	
400	170	
600	150	
750	130	
1000	77	
1470	77	
2070	43	
2440	27	
2980	26	
3470	[87]	
3900	[39]	Probable true
4800	[63]	value <39

* Bruland, personal communication, provided Zn, Fe and Ni data which he believes are anomalous in the 22 m sample (Zn >100-fold excess and Fe >10-fold excess of true values because of collecting too close to ship) and in the 200 m sample (Ni 3-fold, Zn 10-fold and Fe >10-fold excess of true values because of bag failure). The 3470 m sample was collected while moving up due to winch defect, and also contained piston housing tracer due to bag failure. The 3900 m and 4800 m samples contained piston housing tracer due to bag failure.

similar but not quite as strong as that observed in the Northeast Pacific, and that the concentration levels throughout the water column are about 3 to 5-fold higher than those observed in the Northeast Pacific, whereas most other trace metal concentrations are lower in the Atlantic than in the Pacific[5]. These observations show that the occurrence of lead in the oceans is indeed unique among other trace metals.

There is a remarkable congruence between the depth profiles of lead in the Sargasso Sea and the Northeast Pacific, as shown in Fig. 2. Although concentration levels are 2 to 5-fold higher in

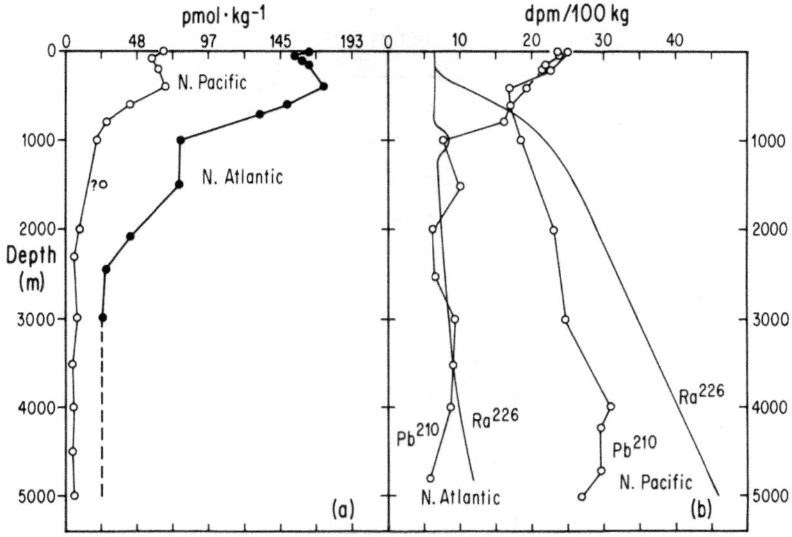

Fig. 1 (a) Vertical profiles of dissolved lead in the North Atlantic at 250 km northwest of Bermuda and in the North Pacific at 1800 km northeast of Hawaii (Pacific data, ref. 1). (b) Vertical profiles of ^{210}Pb and ^{226}Ra at or near these same stations in the North Atlantic and the North Pacific (^{210}Pb Atlantic, unpublished data by K. Turekian; ^{226}Ra Atlantic: GOESECS Station II, ref. 9; ^{210}Pb Pacific: mean values of profiles at station 202, ref. 10, and station 17 [unpublished data, K. Bruland]; ^{226}Ra Pacific, ref. 11).

the Sargasso Sea, the following features appear to be independent of concentrations. In both oceans, there is a sharp decline in concentrations with depth between 400 m and 600 m. Concentrations decline more gradually at greater depths, reaching minimum values at about 2500 m with no significant change below that depth. A modest subsurface maximum ($\sim$ 14% concentration increase in going from about 50 m to 400 m) was observed at both stations that have been studied so far. There is also a slight (8%) drop in concentration in going from the raft-collected samples of surface water to the shallowest samples collected by the deep water sampler at both stations. The significance of these features must await further studies at different stations and times, because at present uncertainties in analytical precision and accuracy of different sampling procedures are not less than 4%.

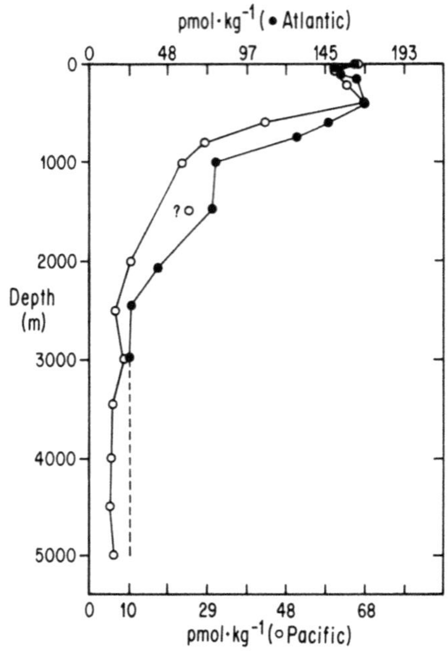

Fig. 2 Overlay of the North Atlantic and North Pacific profiles of dissolved lead using a concentration scale for the Atlantic profile that is reduced by a factor of 2.56 in order to normalize the Atlantic profile to the Pacific profile.

Significant differences between the profiles in the two oceans can also be identified. The Sargasso profile shows a stronger downward penetration of lead below the 400 m depth and a smaller lead concentration enrichment ratio between surface and deep waters. Unfortunately, inaccurate data between 3000 m and the sea floor leave the possibility open that lead concentrations may either decrease of increase slightly in this depth range.

To interpret the depth distribution of lead measured in the Sargasso Sea, consideration must be given to both the 100-fold increase in the flux of industrial lead into the atmosphere which has overwhelmed the natural lead distribution on a global scale [12,13] and the high reactivity and short residence times of lead in the water column[1,14,15].

DISCUSSION

Characteristics of lead inputs to the northern Sargasso Sea

Eolian Source of Lead. Evidence indicating that lead enters the North Pacific mainly from the atmosphere[1] also suggests that this is the principal entry route to the Sargasso Sea. Larger concentrations of lead in surface and thermocline waters relative to those in deep waters in the Sargasso Sea exclude the possibility that upwelling from a long-lived reservoir is the dominant supply route for lead to upper waters. Instead, strong external inputs into surface waters in the form of precipitation and dry deposition from the atmosphere or from ocean margins in the form of continental runoff must be responsible for this near-surface enrichment. A transect between Narragansett, R.I. and Bermuda shows a 20-fold decrease of lead concentrations in surface waters in crossing a narrow strip of coastal and shelf waters toward the Gulf Stream, with little change across the Gulf Stream, followed by a modest increase toward the center of the Sargasso Sea[8]. It appears that most of the fluvial lead is removed over the continental shelves near the coasts by rapidly settling particles which sequester metals like lead[15]. But unlike horizontal surface concentration profiles off California, where low fluvial inputs combined with high productivity induced by upwelling result in low concentrations near the coast with concentration increases towards the open ocean[1], the opposite gradient is observed off the U.S. east coast. This suggests much stronger fluvial inputs of common lead in the east, although the ^{210}Pb gradient is similar to the one in the North Pacific[8]. The observed horizontal gradients would allow for eddy diffusion transport of lead toward the Gulf Stream, but not into the Sargasso Sea. A quantitative estimate of the distance of diffusive transport of lead in the surface waters from the coast towards the open ocean using a ^{228}Ra-derived horizontal eddy diffusion constant of 10^{-6} cm^2 sec^{-1} [16] and a residence time of one year yields a transport limit of only 50 to 100 km, which falls far short of the Sargasso Sea. The pattern of isotopic compositions of lead in the surface waters along the Narrangansett to Bermuda transect[8] shows the presence of recent atmospheric emissions of U.S. industrial lead in a narrow strip of coastal waters ($^{206}Pb/^{207}Pb$ = 1.226), but shows the presence of U.S. industrial lead in Sargasso Sea waters ($^{206}Pb/^{207}Pb$ = 1.199) that had been emitted to the atmosphere a number of years earlier[8]. The $^{206}Pb/^{207}Pb$ ratio in U.S. industrial lead changed from about 1.190 in 1974 to about 1.230 in 1978[17], which permits such an interpretation to be made. The drop of the $Pb/^{210}Pb$ ratio in surface waters between the coast and the Gulf Stream suggests that fluvial lead does not reach the Sargasso Sea[9], but that the dominant input mechanism of lead to the Sargasso Sea must be eolian.

Present Eolian Inputs are Anthropogenic. It has been shown that the input of Pb to the Northeast Pacific has increased at least tenfold over natural levels because of increased emissions of lead from smelters and gasoline exhausts[1]. This conclusion holds for the North Atlantic with only slight modifications.

The historic record of Pb concentrations in Northern Greenland snow layers shows a >200-fold concentration increase[12,18] which is believed to reflect an increase of the Pb dry deposition and precipitation flux by a least a factor of 50[17]. A similar increase of atmospheric lead flux to the oceans probably occurred in the subtropical North Atlantic in recent centuries, because this area is located in the westerlies zone, downwind from central North America, a major source of anthropogenic Pb emissions[1].

This inference is supported by a comparison of present Pb input rates with natural (Pleistocene) rates of dissolved lead output from the North Atlantic water column, as recorded in pelagic sediments. The average sedimentation rate in the North Atlantic is higher than that of the North Pacific[19], resulting in a higher authigenic lead output flux in the North Atlantic. Using a five times higher sedimentation rate, together with measured concentrations of authigenic lead in seven pelagic sediment cores from the western North Atlantic[20,21] and the same measured average ratios between authigenic and total lead for the different sediment types as in the North Pacific[1], a maximum output flux of about 30 ng cm^{-2} yr^{-1} is estimated for the western subtropical North Atlantic. This estimate may have a significant error margin because probabilities of lateral transport and disturbed sedimentation due to re-suspension are higher in the pelagic western North Atlantic than they are in the North Pacific due to strong boundary currents in the vicinity of shelf areas.

The present rate of lead input to the northern Sargasso Sea is estimated by three different methods to be significantly higher than the natural output rate. Using measured lead concentrations in surface water at the profile station and assuming that the residence time of lead in the upper 100 meters is about two years, similar to the North Pacific gyre[22], an input flux of 170 ng cm^{-2} yr^{-1} is found. In a second method, the Pb/^{210}Pb ratio of 390 ng $(dpm)^{-1}$ measured in one rain sample at the profile station[23] is combined with a ^{210}Pb flux of 0.7 dpm cm^{-2} yr^{-1} determined over a period of one year near Bermuda[24] to yield a lead deposition flux by precipitation of 230 ng cm^{-2} yr^{-1}. The total lead input flux is probably 50% higher due to dry deposition[17], and amounts to 330 ng cm^{-2} yr^{-1}. In a third method, the measured Pb/^{210}Pb ratio in the surface water at a neighboring station 100 km west of Bermuda[8] of 270 ng $(dpm)^{-1}$ is combined with the above estimate of the ^{210}Pb input rate to yield a Pb flux of 190 ng cm^{-2} yr^{-1}. These three estimates show that present eolian inputs of industrial lead are about one order of magnitude higher than the estimated output flux of authigenic lead from the North Atlantic during prehistoric times.

Lead Input Fluxes to the North Atlantic exceed those to the North Pacific. The above present-day flux of 170 to 330 ng cm^{-2} yr^{-1} to the North Atlantic is about 3 to 5-fold higher than the present-day input to the northeast Pacific, which ranges from 30 to 70 ng cm^{-2} yr^{-1}, with 60 ng cm^{-2} yr^{-1} being most likely[1]. This trend agrees with differences in air concentrations which are about 0.6 ng Pb m^{-3} in the mid-North Pacific[1] and about 2.4 ng Pb m^{-3} in the mid North Atlantic[25,26]. These higher lead concentrations and fluxes in and to the North Atlantic agree with a model which considers the geographic distribution of the atmospheric occurrences of Pb and ^{210}Pb in relation to their sources[1,27]. The large mid-latitude portions of both the North Pacific and North Atlantic are supplied with Pb and ^{210}Pb by westerlies from continental areas that lie upwind. ^{210}Pb concentrations in westerlies leaving North America are significantly smaller than in westerlies leaving the Asian continent because the latter has a much larger surface area, so the input flux of ^{210}Pb to the North Atlantic at mid-latitudes should be smaller than it is to the North Pacific[1]. The situation is quite the opposite for common lead. Since virtually all common lead in the atmosphere is industrial, lead concentrations in westerlies leaving North America are larger than in westerlies leaving Asia because of larger industrial emissions in North America.

Despite its simplicity, this explanation of increased Pb/^{210}Pb ratios in the atmosphere, precipitation[23] and surface seawater in the North Atlantic[8] is probably correct. On a mass inventory basis, natural inputs of common lead from soil dust and vegetation[28] cannot cause the observed input differences between the two oceans in the absence of anthropogenic inputs of common lead.

Comparison of Lead with other Trace Metals

Manganese. Vertical profiles of manganese are somewhat similar to the unnatural and perturbed vertical profiles of industrial lead, but it is not yet clear whether the undisturbed natural profiles of these two metals were similar in earlier times. The surface maximum in manganese may be caused by fluvial inputs, as is demonstrated by a diminution of the magnitude of this surface maximum with increasing distance from land. In the northeast Pacific the surface maximum nearly disappears[29], while in the Sargasso Sea at the lead profile station the surface maximum is about 2.4 times higher than concentrations at intermediate depths[5]. This residual mangenese enrichment in the Atlantic surface waters may be due to either a shorter distance between the station and land or to higher fluvial inputs of manganese in the North Atlantic [5,19]. The manganese profile has been described as exhibiting a nutrient-type distribution in the water column at large distances from shore, but exhibiting a strong fluvial component in surface waters near shore[30]. No evidence for such inputs governing the oceanic occurrence of lead in the Sargasso Sea can be seen, as discussed in the previous section, but since the present distribution of lead in the water column is disturbed by anthropogenic inputs, factors which control the oceanic occurrence of manganese may also have controlled the natural distribution of lead in the water column in earlier times, as is described in some model calculations[1,2].

Other Trace Metals. There is a great difference between the geographic distribution patterns of lead and most other trace metals in the oceans. Concentrations of lead in the deep Atlantic are larger than in the deep Pacific, whereas the majority of trace metals show lower concentrations in the deep North Atlantic than in the deep North Pacific [16]. Since the latter metals exhibit depletions in the upper part of the water column somewhat similar to nutrients, they might be expected to mimic the known distribution of nutrients. Increased nutrient concentrations in deep ocean waters result from the biological uptake of nutrients in surface waters followed by downward transport within settling particles and then release to the water column at depth. Geographically, nutrient concentrations increase in deep waters which travel from the North Atlantic to the North Pacific and this effect is ascribed largely to a continuous supply of nutrients to deep waters by settling particles along the trajectory of the waters [16,31]. This is consistent with the fact that the residence times of these nutrient tracers are much longer than travel times of water masses from one ocean basin to another. In a similar fashion Cd, Cu, Zn [6] and Ba [31] exhibit 2 to 4-fold increases in concentrations between the depths of the Sargasso Sea and the northeast Pacific. The largest concentration increases correlate with those metals and nutrients which are regenerated at greatest depths [16,31].

Common Lead and ^{210}Pb. Common lead also participates in biological uptake in surface waters, leading to downward transport with settling particles (probably in fecal pellets[33]), followed by both release and scavenging at depth by sedimenting particles and/or removal by processes at the sediment/water interface[1,34,35] The larger concentrations of lead in North Atlantic deep waters compared to lower concentrations in those of the North Pacific, although opposite to trends for other trace metals, nevertheless can be understood in terms of both its high reactivity and short residence time in the water column, and its recently increased eolian input flux to the North Atlantic from anthropogenic sources as outlined above.

Qualitative information on the removal rate constants of lead can be derived from studies of the natural radionuclide ^{210}Pb in the water column. The two types of lead found in the water column are supplied through similar mechanisms in the surface waters but through different mechanisms in the deep waters.

The vertical distribution of ^{210}Pb at the station sampled for common lead in the Atlantic [K. Turekian, unpublished data] (Fig. 1) shows about a 6-fold excess of ^{210}Pb over ^{226}Ra in the surface waters, decreasing almost linearly to zero near 1000 m. This distribution agrees with that of earlier observations in oligotrophic waters in the North Pacific[11]. The excess ^{210}Pb is the result of atmospheric ^{210}Pb inputs and downward transport through particles, which take up ^{210}Pb in the mixed layer and release more ^{210}Pb during transit through the thermocline than they scavenge.

The depth of vertical penetrations of this ^{210}Pb excess and its standing crop is probably mainly limited by the 22-year radioactive half-life, and it is further influenced by in situ production of ^{210}Pb through radioactive decay of ^{226}Ra. Since both of these processes are irrelevant for the distribution of common lead in the water column, the observed similarity of enrichments between common lead and ^{210}Pb in the upper part of the water column does not permit simplistic conclusions. Furthermore, the recent input increases of industrial lead to the ocean has created a non-steady state condition for the distribution of common lead, which does not apply to ^{210}Pb[1].

At depths below 1500 m, excluding the bottom waters, the ^{210}Pb activity in the northwestern Atlantic shows a deficit relative to ^{226}Ra that is smaller than the deficit observed in the northeast Pacific[10], in the southern part of the Labrador Sea[35], or in the tropical North Atlantic[36]. A conventional box model interpretation of these deficits

$$\tau = \tau^* \cdot \frac{A_{210}}{A_{226} - A_{210}} \qquad (\tau^* = 32 \text{ yrs})$$

indicates a longer residence time in the deep waters of the Sargasso Sea (170 years) than in these marginal parts of the North Atlantic (24 years and 20 to 90 years respectively) and longer than at the central northeast Pacific station 17, where the common lead distribution was studied (115 years[1]) but within the range of values determined at other stations within the Pacific, ranging up to 460 years in the central North Pacific[31]. This decrease of removal efficiencies (or increase of residence times) towards the center of the Sargasso Sea is consistent with a similar trend in the North Pacific[10].

The observed geographic variations of ^{210}Pb deficiencies in the deep water column is believed to reflect differences in effects of removal mechanisms, particularly enhanced removal processes in the productive areas and/or resulting from intense contact with the sea floor[10,36]. They may, however, also be influenced by relative input differences of ^{210}Pb derived from the upper part of the water column. For example, the rate of ^{210}Pb input to the deep ocean that is released by settling particles would have an impact on the profile in the North Atlantic that is three times as great as it would on the profile in the North Pacific, because the standing crop of ^{226}Ra is smaller by that factor (Fig. 1). This input rate has been generally disregarded. It may, however, be because of this effect that the observed ^{210}Pb deficiency in the deep Sargasso Sea is less than in the northeast Pacific. The processes involved are crucial for the interpretation of the common lead profile, since it is exactly these addition mechanisms that determine the rate at which increasing anthropogenic inputs are reflected in increasing concentrations in the water column.

The actual residence times of dissolved radioactive or common lead that are dependent upon chemical removal processes from the water column are shorter than the above values because the process of regeneration of dissolved lead from settling particles at depth is not included in the $^{210}Pb/^{226}Ra$ disequilibrium model. This is essential for the supply of common lead, although it has been ignored for ^{210}Pb because it is produced by decay of ^{226}Ra.

A simplified 3-box model of the distribution of both the common lead and ^{210}Pb (with ^{226}Ra) in the northeast Pacific which includes considerations of these processes and increases of industrial lead inputs with time yields a residence time for common lead of about 80 years in the deep waters[1], instead of the 115 years estimated without considering these effects, and at least 20 years in the 100 to 900 m thermocline. Assuming that similar correction factors apply in the northern Sargasso Sea, the residence time of lead in northwestern Atlantic deep waters probably lies between 100 and 150 years. It follows that concentrations in these waters are presently increasing in response to large increases of lead input during past decades, even more slowly than in the central northeast Pacific, and have certainly not reached steady state. On the other hand, in surface waters where the residence time for lead is about 2 years[22], steady-state conditions prevail, so the present larger industrial lead input flux to the North Atlantic compared to that to the North Pacific is recorded directly in the higher lead concentrations in surface waters of the North Atlantic.

This is confirmed by the observed vertical profile of lead in the Sargasso Sea for the following reasons. (1) The trend in the depth distribution matches that observed and modeled in the North Pacific, reflecting essentially the same recent input history in both oceans and the same geochemical processes of uptake and release of dissolved lead in the water column. (2) Concentration levels are about 2.5 times higher in the upper part of the water column than in the North Pacific (Fig. 2), which represents a very satisfactory agreement with an estimated 3 to 5-fold higher eolian input flux to the Sargasso Sea.

These observations show that the concentration and distribution of lead in the water column of each ocean basin is largely determined by local input fluxes, and on a global scale is unaffected by input fluxes along the upcurrent paths of these water masses. This characteristic is unlike that for many other trace metals that have longer residence times. This applies to deep waters that move only short distances on a global scale during 100 years. Lead in the upper parts of the water column, where higher lateral velocities prevail, also cannot be transported far because very short lead residence times in these shallow waters quickly transfer lead to slow moving deep waters.

The geographic pattern of isotopic compositions of authigenic lead in pelagic sediments supports this conclusion, since large isotopic variations on a geographic scale are systematically correlated with isotopic compositions of leads contained in nearby continental drainages, except for correlations with hot vents at spreading centers on small local scales [37]. These isotopic patterns and correlations would be erased by mixing and homogenization if the residence times of lead in the water column were long relative to water travel times through the world oceans.

Isotopic compositions of neodymium in sea water and in pelagic sediments indicate geographic variations similar to those of lead, which implies similarly short residence times of Nd in the water column [38]. Preliminary data indicate that concentrations of Nd in the Sargasso Sea probably increase with depth, as may have been true for lead in preindustrial times. Short residence times coupled with geographic variations in isotopic compositions permit specific continental source regions of metals with these properties to be delineated.

Perspectives

The data presented here provide the basis for a general concept regarding the distribution of lead in the ocean. Future research will have to separate the effects of increased anthropogenic input and to characterize geochemical processes controlling oceanic lead distributions.

The impact of anthropogenic lead inputs to the marine environment could be documented and quantified by studying records of lead concentrations and isotopic compositions preserved in coral. Since industrial ore lead can be distinguished from natural silicate lead by means of isotopic compositions, and since the isotopic compositions of reservoirs of industrial lead change with time and are different in various countries, the various sources of lead in seawater and in biological material in both the North Pacific and the North Atlantic can be further identified[39,40]. Temporal changes in isotopic compositions of lead added to marine reservoirs may also be used to study processes of adsorption and re-dissolution through sinking particles that redistribute lead in the water column and that cannot be identified under steady state conditions without such a time marker. Such use of isotopic analyses of lead in material collected in sediment traps would provide a powerful tool, particularly when it is combined with parallel ^{210}Pb analyses.

Since both profiles of lead in the North Pacific and in the North Atlantic show a characteristic drop in concentration between 400 and 600 m depth, which is not repeated in the ^{210}Pb profiles, a discontinuity is indicated that has not been investigated in the past. The processes that are responsible for this change in concentration and influences of horizontal advection need to be identified and worked into a geochemical model that includes ^{210}Pb as well.

It is possible that the peak of lead inputs into the oceans and the atmosphere from North America has been reached, as the production of lead alkyls decreases. Therefore, detailed studies of the marine lead distribution at present levels of lead input may provide an important benchmark for future investigations of the marine lead geochemistry.

ACKNOWLEDGEMENTS

We are grateful to our colleagues who contributed to this study in various ways, in particular to A. Ng and R. Franks, who carried out the sampling aboard R.V. Endeavor and to K. Turekian and K. Bruland, who provided unpublished ^{210}Pb and trace metal data and gave valuable suggestions. This work was supported by NSF Grant OCE 79-00696.

REFERENCES

1. Schaule, B.K. and C.C. Patterson, 1981: Lead concentrations in the northeast Pacific: evidence for global anthropogenic perturbations. Earth and Planetary Science Letters, 54, 97-116.

2. Schaule, B., 1979: Zur ozeanischen Geochemie des Bleis -- Nachweis einer anthropogenen Störung anhand kontaminationskontrollierter Bleikonzentrationsmessungen im offenen Nord-Pazifik. Dissertation, Universität Heidelberg, West Germany, 165 pp.
3. Patterson, C.C. and D.M. Settle, 1976: The reduction of order of magnitude errors in lead analyses of biological materials and natural waters by evaluating and controlling the extent and sources of industrial lead contamination introduced during sample collecting and analysis. In: "Accuracy in Trace Analysis: Sampling, Sample Handling, Analysis", P. LaFleur, ed. National Bureau of Standards Special Publication, 422, 321-351.
4. Schaule, B.K. and C.C. Patterson, 1981: A deep water collection device which provides seawater samples free of lead contamination. Unpublished manuscript.
5. Bruland, K.W. and R.P. Franks: Mn, Ni, Cu, Zn and Cd in the Western North Atlantic (this volume).
6. Participants of the lead in seawater workshop, Interlaboratory lead analyses of standardized samples of seawater, Marine Chemistry 2 (1974) 69-84; - C.C. Patterson, D.M. Settle, J. Bewers, S. Hartling, R. Duce, P. Walsh, W. Fitzgerald, C.D. Hunt, H. Windom, R. Smith, C.S. Wong and P. Berrang, 1976: Comparison determinations of lead by investigators analyzing individual samples of seawater in both their home laboratory and in an isotope dilution standardization laboratory. Mar. Chem., 4, 389-392.
7. Patterson, C.C., D.M. Settle and B. Glover, 1976: Analysis of lead in polluted coastal seawater. Marine Chemistry, 4, 305-319.
8. Ng, A., B. Schaule, K. Bruland and C.C. Patterson, 1981: Concentrations of common lead and ^{210}Pb and isotopic compositions of lead in surface waters of the North Atlantic. Unpublished manuscript.
9. Chung, Y. and H. Craig, 1973: Radium-226 in the Eastern Equatorial Pacific. Earth Planet. Sci. Lett., 17, 306.
10. Nozaki, Y., K.K. Turekian and K. Van Damm, 1980: Pb-210 in GEOSECS water profiles from the North Pacific. Earth Planet. Sci. Lett., 49, 393-400.
11. Chung, Y., 1976: A deep ^{226}Ra maximum in the Northeast Pacific. Earth Planet. Sci. Lett., 32, 249-257.
12. Murozumi, M, T.J. Chow and C.C. Patterson, 1969: Chemical concentrations of pollutant lead aerosols, terrestrial dusts and sea salts in Greenland and Antarctic snow strata. Geochim. Cosmochim. Acta., 33, 1247-1294.
13. D. Settle and C.C. Patterson, 1980: Lead in albacore: guide to lead pollution in Americans. Science, 207, 1167-1176.

14. Robbins, J.A., 1978: Geochemical and geophysical applications of radioactive lead, In: "The Biogeochemistry of Lead in the Environment", J.O. Nriagu, ed., Part A, Elsevier, Amsterdam, 285-394.
15. Turekian, K.K., 1977: The fate of metals in the oceans. Geochim. Cosmochim. Acta., 41, 1139-1144.
16. Broecker, W.S., 1974: Chemical Oceanography, Harcourt Brace Jovanovich, New York, 214 pp.
17. Settle, D.M. and C.C. Patterson, 1981: Summary Report to NSF, Grant OCE 77-14520.
18. Ng, A. and C.C. Patterson, 1981: Natural concentrations of lead in ancient Arctic and Antarctic ice. Geochim. Cosmochim. Acta., 45, 2109-2121.
19. Horn, D.R., M.N. Delach and B.M. Horn, 1974: Physical properties of sedimentary provinces, North Pacific and North Atlantic Oceans, In: "Deep Sea Sediments", A.L. Inderbitzen, ed., Plenum Press, New York, 417-461.
20. Chow, T.J. and C.C. Patterson, 1962: The occurrence and significance of lead isotopes in pelagic sediments. Geochim. Cosmochim. Acta., 26, 263-308.
21. Chow, T.J., J.L. Earl and C.B. Snyder, 1980: Lead isotopes in Atlantic DSDP cores, In: "Isotope marine geochemistry", E.D. Goldberg, Y. Horibe and K. Saruhashi, eds., Uchida Rokakuho Publishing Co., Tokyo, 399-412.
22. Nozaki, Y., J. Thomson and K.K. Turekian, 1976: The distribution of Pb-210 and Po-210 in the surface waters of the Pacific Ocean. Earth Planet. Sci. Lett., 32, 304-312.
23. Settle, D.M., C.C. Patterson, K.K. Turekian and J.K. Cochran, 1980: Common lead and ^{210}Pb concentrations in precipitation at tropical oceanic sites. Jour. Geophys. Res. In press.
24. Benninger, L.K., E.P. Dion and K.K. Turekian, 1981: ^{7}Be and ^{210}Pb precipitation fluxes at New Haven Conn. and Bermuda. In preparation.
25. Ng, A. and C.C. Patterson, 1981: Upper limit contributions of lead from the oceans to the atmosphere determined in seaspray generated by a bubble interfacial microlayer samples. Unpublished manuscript.
26. Duce, R.A., G.L. Horrmann, B.J. Ray, I.S. Fletscher, G.T. Wallace, J.L. Fasching, S.R. Piotrowicz, P.R. Walsh, E.J. Hoffmann, J.M. Miller Heffter, 1976: Trace metals in the marine atmosphere: Sources and fluxes, In: "Marine Pollutant Transfer", H. Windom and R.A. Duce, eds., Heath and Co., Lexington, MA, 77-119.
27. Turekian, K.K., Y. Nozaki and L.K. Benninger, 1977: Geochemistry of atmospheric radon products. Ann. Rev. Earth. Planet. Sci., 5, 227-255.

28. Nriagu, J.O., 1978: Lead in the atmosphere, In: "The biogeochemistry of lead in the environment", J.O. Nriagu, ed., Part A, 137-174.

29. Landing, W.M. and K.W. Bruland, 1980: Manganese in the North Pacific. Earth Planet. Sci. Lett., 49, 45-56.

30. Martin, J.H. and G.A. Knauer, 1980: Manganese cycling in the Northeast Pacific waters. Earth. Planet. Sci. Lett., 51, 266-274.

31. Redfield, A.C., B.H. Ketchum and R.A. Richards, 1963: The influence of organisms on the composition of sea water, In: "The Sea", Vol. II, E.D. Goldberg, ed., 26-77.

32. Chan, L.H., D. Drummond, J.M. Edmond and B. Grant, 1977: On the Ba data from the Atlantic GEOSECS expedition. Deep Sea Research, 24, 613-649.

33. Burnett, M.W. and C.C. Patterson, 1981: Perturbation of natural lead transport in nutrient calcium pathways of marine ecosystems by industrial lead, In: "Isotope Marine Geochemistry", E.D. Goldberg, Y. Horibe and K. Saruhashi, eds., Uchida Rokakuho Publishing Co., Tokyo, 413-438.

34. Craig, H., S. Krishnaswami and B.L.K. Somayajulu, 1973: ^{210}Pb, ^{226}Ra: radioactive disequilibrium in the deep sea. Earth Planet. Sci. Lett., 17, 295-305.

35. Bacon, M.P., D.W. Spencer and P.G. Brewer, 1980: Lead-210 and Polonium-210 as marine geochemical tracers: Review and discussion of results from the Labrador Sea, In: "Natural Radiation Environment III", T.F. Gesell and W.F. Lowder, eds., U.S. Dept. Energy Report CONF-780422, Vol. 1, 473-501.

36. Bacon, M.P., D.W. Spencer and P.G. Brewer, 1976: Pb-210/Ra-226 and Po-210/Pb-210 disequilibria in seawater and suspended particulate matter. Earth Plan. Sci. Lett., 32, 277-296.

37. Unruh, D.M. and M. Tatsumoto: Lead isotopic composition and uranium, thorium and lead concentrations in sediments and basalts from the Nazca Plate. Initial Reports of the Deep Sea Drilling Project (1976), 34, 341-347.

38. Piepgras, D.J. and G.J. Wasserburg, 1980: Neodymium isotopic variations in seawater. Earth Planet. Sci. Letters, 50, 128-138.

39. Ng., A. and C.C. Patterson, 1981: Changes of Pb and Ba with time in California off-shore basin sediments. Submitted to Geochimica et Cosmochimica Acta.

40. Flegal, A.R., B.K. Schaule, D.M. Settle and C.C. Patterson, 1981: Lead Concentrations and stable isotopic compositions in surface waters of the Central Pacific. Abstract, meeting, Trans. Amer. Geophys. Union.

COPPER, NICKEL AND CADMIUM IN THE SURFACE WATERS OF THE MEDITERRANEAN

A.J. Spivack, S.S. Huested and E.A. Boyle

Massachusetts Institute of Technology
Department of Earth and Planetary Sciences
Cambridge, MA. 02139 U.S.A.

ABSTRACT

Copper, nickel, and cadmium have been determined on 50 surface samples throughout the Mediterranean. All of these elements are substantially more concentrated in the surface waters of the Mediterranean than in similar nutrient-depleted waters in the open ocean. A mass-balance model for the Mediterranean allows limits to be placed on the magnitudes of the trace element sources. The actual source or sources cannot be resolved from the data, and it is not clear whether this trace element enrichment is natural or anthropogenic in origin.

INTRODUCTION

Oceanic distributions of copper, nickel and cadmium have been accurately determined and reported[1,2,3,4,5]. The distribution of these elements has been shown to be predominately controlled by biogeochemical cycles within the ocean; as determined by comparing their distributions with the distributions of the nutrients nitrate, phosphate and silicate. This type of analysis does not lead to detailed mechanistic explanations but has indicated that biogenic particles remove these elements from the water column, sink and are remineralized (releasing the elements at depth).

In nutrient-depleted areas where the relative importance of the biogeochemical cycling may be lessened other processes such as sediment fluxes, aeolian fluxes, river and anthropogenic inputs may be discernable. The Mediterranean Sea, which has been described as the most impoverished large body of water known in terms of nutrients may be an area where these processes can be observed. Surface samples for copper, nickel, and cadmium were collected in the Mediterranean in August, 1980, and the results are discussed here.

SAMPLING AND ANALYSIS

The samples were collected during NORDA cruise 1309-80, of the U.S.N.S. Barlett in August and September 1980 in collaboration with Dr. D. Reid of NORDA.

The collection procedure for the metals are described in detail elsewhere[5]. Two unfiltered samples were collected at each location. In addition, samples were occasionally taken for filtration through 0.4 μm Nucleopore filters under vacuum. Within a few days of collection samples were acidified with double vycor-distilled 6N HCl. This process was carried out in a laminar flow clean bench. The samples were analyzed by a modified cobalt-APDC coprecipitation method[5].

RESULTS

Within the data set itself there is a high level of consistency as well as good agreement where other reliable data overlaps[5]. There was no significant difference between filtered and unfiltered samples. The distribution of copper, cadmium and nickel show the same general trends (Figures 1, 2 and 3). Outside of the Strait of Gilbraltar, the concentrations of copper and cadmium are lower than at any site sampled in the Mediterranean. For nickel, a few locations sampled in the sea had about the same concentration found outside of the Strait. However, the bulk of the Mediterranean samples had higher concentrations. Outside of the Strait, copper is below 1.4 nmol kg^{-1} and cadmium is below 25.0 pmol kg^{-1} while all the samples from the Mediterranean were above 1.43 nmol kg^{-1} for copper and 48.0 pmol kg^{-1} for cadmium. Concentrations increase immediately within the Strait and then remain relatively constant through the Alboran and Southern Balearic Seas.

In the central Balearic there is a concommitant rise in concentrations of the three metals. Highest concentrations in the western Mediterranean were found in the Tyrrhenian Sea. The eastern Mediterranean had generally higher concentrations increasing with highest concentrations towards the Straits of Messina.

Figures 1a, 2a, and 3a are plots of metal concentration vs. longitude. The dots represent sites along the more northerly track of the cruise when more than one sample was collected at a

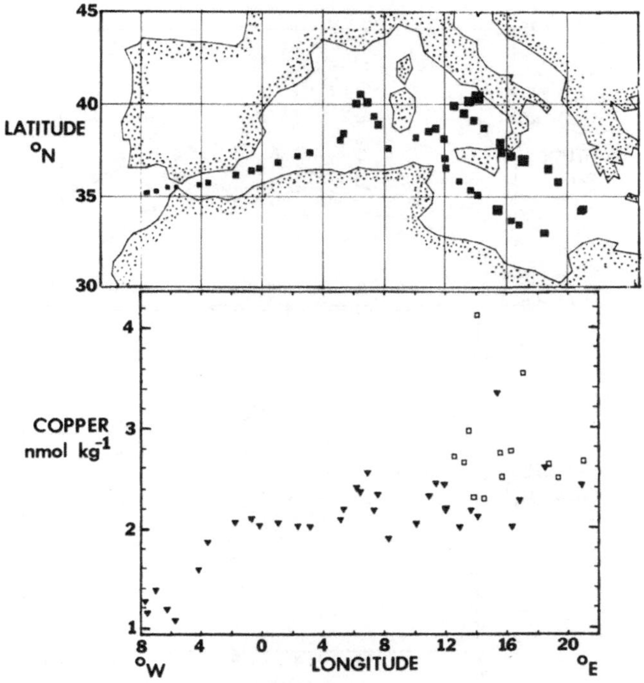

Fig. 1 a) Concentration of copper plotted vs. latitude. b) Relative copper concentrations. Length of symbol perimeter proportional to concentration.

particular longitude. A more illustrative representation of the relative concentration distributions is given in Figures 1b, 2b, and 3b. The area of the symbol is proportional to the square of the concentration.

The distribution of silicate differs from that of the metals. Immediately inside the Strait of Gilbraltar the concentration drops by a factor of 3. However, as with the metals there is a general rise in concentration towards the east.

Salinity does not vary significantly between the western Alboran Sea and the waters directly outside of the Strait of Gibraltar. Salinity increases towards the east as has been noted previously .

DISCUSSION

Recent work on the distribution of copper, nickel and cadmium emphasizes the controlling influence of biogeochemical cycl-

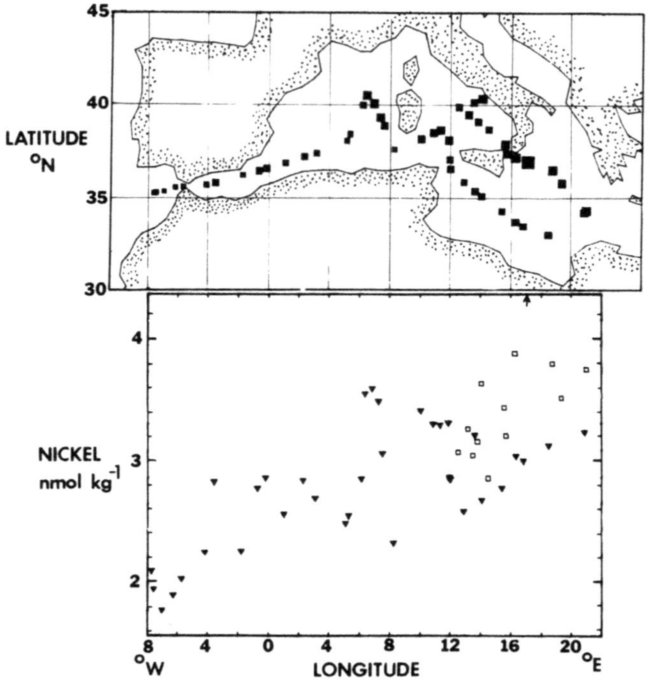

Fig. 2 a) Concentration of nickel plotted vs. latitude. b) Relative nickel concentrations. Length of symbol perimeter proportional to concentration.

ing in the ocean. Only for copper is there evidence supporting the importance of other processes. The relatively high copper levels found in the Gulf of Panama as well as the surface maximum observed at GEOSECS Stn. 340[7] have been reasonably explained by the mobilization of copper from continental shelf sediments.

The biogeochemical cycling results in these metals having nutrient-like distributions. That is, there is a general depletion in surface waters and enrichments with depth. Both nickel and cadmium mirror the distributions of the labile nutrients. Cadmium profiles have been described by linear equations with phosphate or nitrate as the independent variable, while nickel covaries with both phosphate and silicate[4,5]. It has been pointed out that the coefficients of these equations are not globally unique and vary regionally[5]. The relationship of cadmium and nickel to nutrients are notably different in that cadmium is depleted to less than 3 pmol kg^{-1} while nickel has a residual concentration of 2 nmol kg^{-1}, in nutrient depleted waters[4,5].

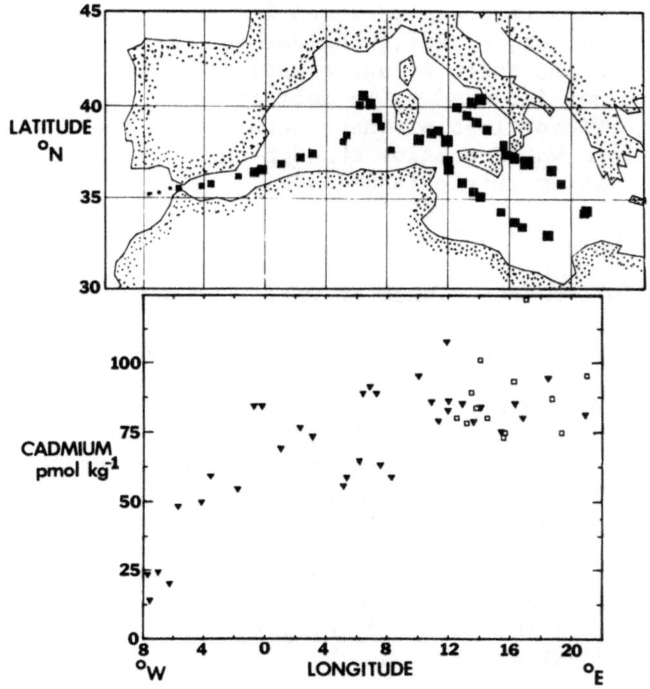

Fig. 3 a) Concentration of cadmium plotted vs. latitude. b) Relative cadmium concentrations. Length of symbol perimeter proportional to concentration.

The distribution of copper is not as simple. Advection - diffusion models with scavenging have been used to model its deep ocean distribution[7]. Concentrations in oligotrophic surface ocean are 0.5 - 1.4 nmol kg^{-1} [4,5].

In comparison, everywhere in the Mediterranean there are higher concentrations of nickel, copper and cadmium than in comparably nutrient-depleted waters in the open ocean. It is not possible to simply model the surface distributions of these metals quantitatively. Clearly, unlike in the ocean, these elements do not just simply reflect the distribution of nutrients.

The water budget and circulation of the Mediterranean have recently been reviewed[6]. The system is dominated by surface entry of water through the Strait of Gibraltar with a subsurface outflow of intermediate water. The outflowing intermediate water is formed from surface waters during winter convective processes predominately in the eastern Mediterranean. This type of circulation has been described as anti-estuarine.

A simple mass balance model can be constructed to estimate the influx of copper, nickel, cadmium, and phosphate from other sources besides the North Atlantic water flowing in at Gibraltar. This model is not intended to be definitive. It is assumed that the only significant loss of these metals from the basin is via the intermediate water outflow through the Strait and that the equation describing this balance is:

$$Q_i C_i - Q_0 C_0 + X = 0$$

where Q_0 and Q_i are the fluxes of water out of and into the basin, respectively, C_0 and C_i are the concentrations of the outflow and inflow and X is the total non-Atlantic input of the metal.

The outflow concentrations were not measured directly. We use estimates based on the concentrations of the eastern Mediterranean surface waters which form the outflowing intermediate water. The values of Q_i and Q_0 are 38 x 10^{15} kg yr^{-1} and 36 x 10^{15} kg yr^{-1}(ref.8). C_i for copper, nickel and cadmium are respectively 1.2 nmol kg^{-1}, 2.0 nmol kg^{-1} and 19 pmol kg^{-1}. The values of C_0 used are 2.6 nmol kg^{-1}, 3.9 nmol kg^{-1} and 78 pmol kg^{-1} for copper, nickel and cadmium respectively. The calculated values of X are, copper, 5 x 10^7 mol yr^{-1}; nickel, 6 x 10^7 mol yr^{-1}; and cadmium 2 x 10^6 mol yr^{-1}.

There are a number of possible sources of these fluxes, both natural and anthropogenic, river transport of weathering products, mobilization from shelf sediments, both rainout and fallout from the atmosphere and wastes, municipal as well as industrial.

As there are no reported data for the various inputs it is impossible to quantify the relative importance of them. However, extreme cases can be considered. For example, if rivers are considered as the sole source of the excess metals then an average river concentration can be estimated for the basin. Using published flow data[8] the average inflow concentrations (including the Black Sea inflow in the estimates) are: copper, 64 nmol l^{-1}; nickel, 84 nmol l^{-1} and cadmium 2.3 nmol l^{-1}. These can be compared to the concentrations found in the Amazon River: copper, 24 nmol kg^{-1}, nickel, 5 nmol l^{-1} and cadmium $\leq$ 0.1 nmol kg^{-1}(ref.9).

Alternatively, if these metals were derived from rain, dividing the yearly excess inputs by the yearly precipitation will give an estimate of the average rain water concentrations. Using published precipitation data[8] these estimates are: copper 150 nmol kg^{-1}, nickel 200 nmol kg^{-1} and cadmium 5.3 nmol kg^{-1}. This atmospheric contribution can also be estimated as an area-normalized flux. The fluxes are: copper 1.8 nmol cm^{-2} yr^{-1}, nickel, 2.3 nmol cm^{-2} yr^{-1} and cadmium 60 pmol cm^{-2} yr^{-1}. These fluxes can be compared to those derived for the North Atlantic[10]. It should be kept in mind that the North Atlantic fluxes are estimates based on model calculations. The ranges are due to the use of different models. The estimated fluxes are: copper 0.29-6.4 nmol cm^{-2} yr^{-1} nickel, 0.03-0.16 nmol cm^{-2} yr^{-1} and cadmium 25 pmol cm^{-2} yr^{-1}.

Another source which should be considered are shelf sediments. It is possible that metals associated with oxide phases on river transported terrigenous material are released under mildly reducing conditions encountered in the shelf sediments. The fluxes of metals normalized to the shelf area[11] are: copper 8.8 nmol cm^{-2} yr^{-1}, nickel 11 nmol cm^{-2} yr^{-1} and cadmium 0.31 nmol cm^{-2} yr^{-1}. The flux of copper from the shelf in the Panama Basin has been estimated roughly at 16 nmol cm^{-2} yr^{-1} (Ref. 5).

These calculations are not meant to suggest that any one of these sources are in fact important. Rather they do show, by comparison with other published data that no single source for the high concentrations observed in the Mediterranean can be determined apriori.

REFERENCES

1. Martin, J.H., K.W. Bruland and W.W. Broenkow, 1976: Cadmium transport in the California Current. In: "Marine Pollutant Transfer", edited by H. Windom and R. Duce. Lexington Books, Lexington, Mass., p. 159-184.
2. Bender, M.L. and C.L. Gagner, 1976: Dissolved copper, nickel and cadmium in the Sargasso Sea. J. Mar. Res., 34, 327-339.
3. Moore, R.M., 1978: The distribution of dissolved copper in the eastern Atlantic Ocean. Earth Planet. Sci. Lett., 41, 461-468.
4. Bruland, K.W., 1980: Oceanographic distributions of cadmium, zinc, nickel and copper in the north Pacific. Earth Planet. Sci. Lett., 47, 176-198.
5. Boyle, E.A., S.S. Huested and S.P. Jones, 1981: On the distribution of copper, nickel and cadmium in the surface waters of the North Atlantic and Pacific Ocean. J. Geophys. Res., 86, C9, 8048-8066.
6. Hopkins, T.S., 1978: Physical Processes in the Mediterranean Basin. In: "Estuarine Transport Processes", ed. by B. Kjerfve. University of South Carolina Press, 269-310.
7. Boyle, E.A., F.R. Sclater and J.M. Edmond, 1977: The distribution of dissolved copper in the Pacific. Earth Planet. Sci. Lett., 37, 38-54.
8. Lacombe, H. and R. Tchernia, 1972: Caracteres hydrologiques et circulation des eaux en Mediterranee. In: "The Mediterranean Sea: A Natural Sedimentation Laboratory",

ed. by D.J. Stanley. Dowden, Hutchinson and Ross Inc., 75-110.

9. Boyle, E.A., S.S. Huested, B. Grant and J.M. Edmond, 1982: The Chemical Mass Balance of the Amazon Plume II: Copper, Nickel, and Cadmium, Deep Sea Research. (in press).

10. Duce, R.A., G.L. Hoffman, B.J. Ray, I.S. Fletcher, C.T. Wallace, J.L. Fasching, S.R. Piotrowicz, P.R. Walsh, E.J. Hoffman, J.M. Miller and J.L. Heffter, 1976: Trace Metals in the Marine Atmosphere: Sources and Fluxes. In: "Marine Pollutant Transfer", ed. by H.L. Windom and R.A. Duce. Lexington Books, 77-121.

11. Menard, H.W. and S.M. Smith, 1966: Hypsometry of ocean Basin Provinces. Journal of Geophysical Research, 71, 4303-4325.

ACCURATE AND PRECISE ANALYSIS OF TRACE LEVELS OF CU, CD, PB, ZN, FE AND NI IN SEA WATER BY ISOTOPE DILUTION MASS SPECTROMETRY

V.J. Stukas and C.S. Wong*

Seakem Oceanography Ltd.
2045 Mills Road
Sidney, B.C. Canada

*Institute of Ocean Sciences
P.O. Box 6000
Sidney, B.C. Canada

ABSTRACT

Accurate and precise determinations of natural levels of trace metals in sea water are highly reliant upon the size and variability of the analytical blank, the method for determining the yield, and, to a lesser extent, the inherent precision of the instrument used. Thermal source, isotope dilution mass spectrometry together with ultra-clean room techniques were successfully used in the determinations of Cu, Cd, Pb, Zn, Ni and Fe in sea water. Multi-element analyses were performed in a single experiment owing to the differing release with filament current for each element. A single Re filament loaded with a substrate of silica gel and phosphoric acid gave high precision (0.1% to 0.5%) for the determination of a single ratio, and allowed low detection limits (from 0.02 fmol Cd to 0.07 pmol Fe). Yields were accurately and uniquely determined in the same sample by the addition of two isotopically enriched spikes, one before, and one after an extraction. Blanks were assessed as the summation of individual contributions determined in separate experiments prior to sample analysis. Simply obtaining one value for a total blank contribution can be misleading and can generate larger errors. Rigorous clean room procedures allowed very low blanks; 0.0002 nmol Cd kg^{-1}, 0.0005 - 0.002 nmol Pb kg^{-1}, 0.02 nmol Cu kg^{-1}, 0.03 nmol Zn kg^{-1} and 0.7 nmol Fe kg^{-1}.

Sea water, collected from relatively low-lead, coastal waters was stored acidified in rigorously cleaned, conventional polyethylene carboys at pH 2.0, pH 1.6 and at pH 1.1. Strict, clean room conditions allowed accurate comparisons to be made between carboys with varying pH's of sample pretreatment and to the stability of the concentration with time. Extraction of sea water stored at pH 1.6 to 2.0 by dithizone and by APDC/DDDC in Freon TF yielded values representing the dissolved fraction. Acidification to pH 1.1 yielded higher Pb and Zn results but appeared insufficient for 'total' results as defined by aqua regia digestion of sea water. Isotopic composition work on Pb extracted or treated in the above manner supports the definition of soluble and total. The precision in the determination of soluble Cd and Zn with time was ca 1%, a level approaching that attained in measuring a single ratio. The Pb results were less precise, 5%, reflecting the low levels encountered, 0.04 nmol kg^{-1}. Work at higher Pb concentrations indicated a much higher precision was possible. The determination of Cu, Ni and Fe were complicated by isobaric interferences. These could be minimized or eliminated by an ion-exchange purification of the final aqueous extract, and by a stabilization of the ratio with time and with filament current. Analysis of comparable amounts of Cd, Pb and Zn in NBS standard 1643a corroborated the accuracy of the IDMS approach/dithizone extraction procedure. Comparison of the sea water results with FAA and ASV methods indicated that for these metals, the IDMS approach was the most precise and the most accurate.

INTRODUCTION

In the analysis of sea water, isotope dilution mass spectrometry (IDMS) offers a more accurate and precise determination than is potentially possible with other conventional techniques, such as flameless atomic absorption (FAA) spectrophotometry or anodic stripping voltimetry (ASV). Instead of using external standards measured in separate experiments, an internal standard, which is an isotopically enriched form of the same element, is added to the sample. Hence, only a ratio of the spike to the common element need be measured. The quantitative recovery necessary for the FAA and ASV techniques is not critical to the IDMS approach. This factor can become quite variable in the extraction of trace metals from the salt laden matrix of sea water. Yield may be isotopically determined in the same experiment, however, by the addition of a second isotopic spike after the extraction has been completed.

An outline of the elements in sea water that may be analyzed by IDMS has been presented by Chow[1]. Most of the subsequent IDMS work has pertained to the analysis of Pb in sea water[8,9,11]. The extension of the technqiue to the analysis of Cu, Cd, Tl and Pb has been made by Murozumi[5] using a thermal source (TS) mass spectrometer; whereas, Mykytiuk et al.[6] employed a spark source (SS) mass spectrometer in the determination of Fe, Cd, Zn, Cu, Ni, Pb and U in sea water. It may be noted that TSMS has two prime advantages over SSMS that may be of particular value in sea water analysis. One is the much higher precision (ca 0.1%) capable of the instrument, a factor important in assessing the stability of the trace metals in sea water with time or the extraction technique itself. Secondly, the ratios to be determined may be measured at various filament currents (temperature) for an internal corroboration. This is of particular importance in detecting isobaric interferences that may lead to spurious ratios, and hence misleading results.

In this study, we have investigated the feasibility of using a TSMS in the isotope dilution analysis of Cu, Cd, Pb, Zn, Ni and Fe in sea water. The approach basically follows that which had been successfully employed by the author in the analysis of Pb in sea water [11]. Herein, great importance was attached to the definition of the blank and the initial clean up schedule. Once the operating parameters had been established by the analysis of pure isotopic spikes, the various components that constitute a blank were identified and minimized where possible using ultra-clean room techniques. The subsequent extractions of reagent water and sea water by dithizone, and by APDC/DDDC ion exchange resins were evaluated for suitability for the IDMS approach, yield and contamination levels. Corroboration of the results obtained on the primary batch of sea water were performed by two independent techniques (ASV, FAA) by three independent labs/personnel.

EXPERIMENTAL: MASS SPECTROSCOPY

A Nuclide, 30 cm, 90° sector, single focussing mass spectrometer was used to determine the isotope ratios. Re was selected over Ta and W filaments as it provided the best sensitivity and the widest range in release of the elements with increasing filament power (Fig. 1, Table 1). This spread in power or temperature at which the elements were thermally ionized indicated that many elements could be analyzed sequentially with little or no burnoff of the next element, an important consideration in performing multi-element analyses. Sensitivity data (Table 1) for Re, with a silica gel/phosphoric acid as a loading substrate, appear to be of the order of 10^{-13}mol to 10^{-16}mol. Although these values are quite acceptable in light of typical seawater concentrations the sensitivity appears to be non linear with the quantity of sample loaded. In later experiments, nanomole sized samples yielded signals one to two orders of magnitude higher than would be predicted from the data in Figure 1. This effect may be related to either the work function of the metal and/or the amount of silica gel available per nmol of element loaded. Precision was also affected

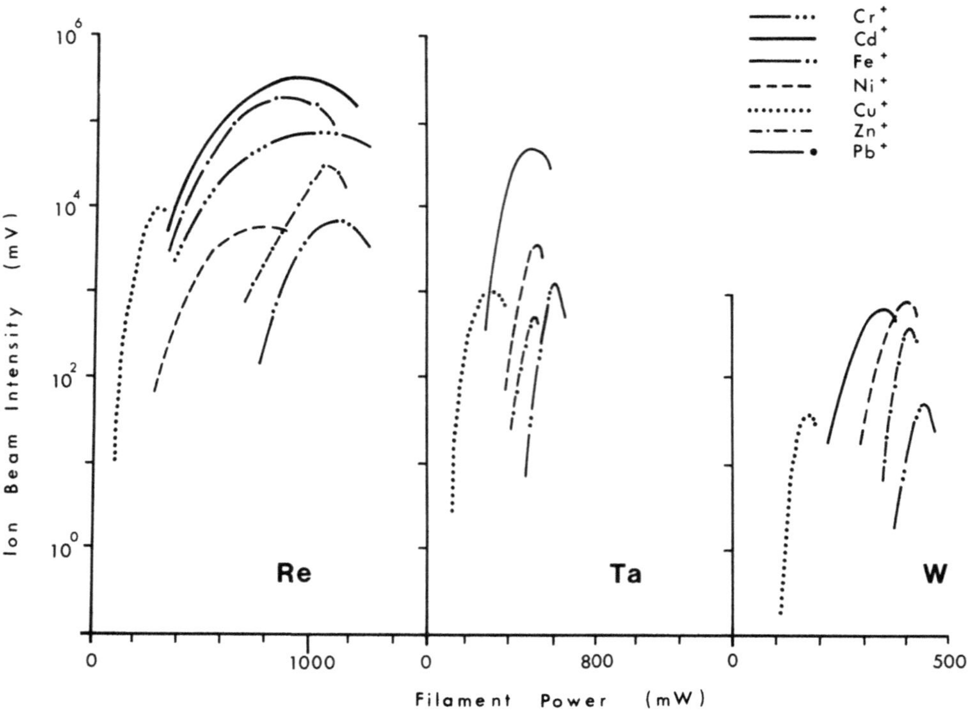

Fig. 1 Release spectra of metals using a TSMS. Note the change in sensitivity with different filament materials. One microgram of each isotope has been loaded. Sensitivity is non linear with respect to the amount of sample loaded. Nanogram-sized loads yield signals much higher than that predicted from the above data.

by the size of load. Nanomole to picomole sized samples, such as those encountered in sea water extractions, allowed a precision of 0.1% to 0.5% to be attained. Precisions of 0.01% to 0.05% could be attained with micromole sized loads.

The accuracy and precision with which the ratios can be determined is in part dependent upon the mass discrimination encountered for that element. In a single filament TSMS, this preferential treatment of the lighter isotope over another heavier one is most likely the result of Rayleigh distillation[4]. Thus, in a sample load with a finite supply, variations in mass discrimination were observed with increasing filament current or temperature. A secondary variation may be related to time and/or the amount of sample available for ionization.

The measurement of mass discrimination was performed initially on Pb, Cu and Cr using the NBS standard reference materials

Table 1 Operating Parameters and Sensitivity Using a Re Filament

Element	Sensitivity (fmol)	Optimum Operational Range (filament current, A)	Optimal Preci sion determined one ratio	Comments On Signal Characteristics
Cu	60 (<20)	1.20-1.40	0.2%	beam is short lived, especially at higher currents
Cd	0.9 (0.02)	1.65-1.80	0.1%	very stable, long lived, intense beam
Pb	0.5 (0.01)	1.65-1.80	0.1-0.01%	very stable, long lived, intense beam
Ni	100	1.75-1.85	0.2%	beam is moderately stable
Zn	20	1.95-2.05	0.2%	although life span is moderate, ratio stays stable
Fe	70	2.1-2.3	0.5%	low intensity

*value in brackets is derived from filament loads in the ng range (i.e. best results).

(isotopic), NBS 981, 976 and 979, respectively. When unspiked sea water was analyzed, mass discrimination was assessed for Cd and Zn by comparing the natural bundance data with values from the CRC handbook. That of Fe and Ni was deduced mass spectrometrically from a known mixture of pure isotopic spikes.

Isotopic spikes themselves were obtained from the Oak Ridge National Laboratory (ORNL). ^{206}Pb, ^{114}Cd, ^{68}Zn, ^{62}Ni, ^{57}Fe and ^{65}Cu were chosen as primary spikes (S_1); whereas, ^{208}Pb, ^{112}Cd, ^{68}Zn, ^{54}Fe and ^{60}Ni were chosen as secondary spikes. Spikes were prepared by digestion of ORNL isotopes in NBS HNO_3 or Seastar Quartz (SQ) HCl, then dilution with quartz distilled water (QD). The concentrations of the multielement spikes were checked periodically against a stock solution.

If there are more than two commonly occurring, stable isotopes, then it becomes possible to add two isotopic spikes: once before the sample is extracted or operated upon, and the second, after the extraction. The first spike (S_1) can be related to the amount of element present, while the ratio of the spikes added may be related to the yield (Y). Blanks incurred during and after the extraction may be weighted according to the yield. Thus:

$$Y = \frac{S_2}{S_1} \qquad \frac{A(n - Rk) - G(m - rk)}{G(e - rd) - A(f - Rd)} \tag{1}$$

and $$X = \frac{Rd - f}{G} S_1 - \frac{n - Rl}{GY} S_2 - \frac{1}{Y} B_2 - B_1 \tag{2}$$

where: $A = c - rb$ (3)
$G = a - Rb$ (4)
S_1 = first spiked added
S_2 = second spike added
B_1 = blank up to and including the extraction
B_2 = blank of components after the extraction

and the isotopic composition of the common element and the spike are as follows:

	S_1	S_2	common
amu 1	d	k	b
amu 2	e	m	c
amu 3	f	n	a

The ratios measured on the mass spectrometer are:

$$r = (\text{amu } 2/\text{amu } 1) \tag{5}$$
$$R = (\text{amu } 3/\text{amu } 1) \tag{6}$$

Thus, the yield of an extraction can be isotopically determined by (1) in the same experiment as is the concentration of that element. Moreover, situations in which there exists a relatively high blank, B_2, or a low yield can now be dealt with. The accuracy and precision of X, in this case, are limited by the precision with which the ratios, r and R, and the blank can be determined. However, the blank must now be known as the sum of various components rather than one, general number. In practice, yields of less than 5-10% lead to unreliable results.

REAGENTS, PURIFICATION AND BLANK LEVELS

Accurate and precise seawater analysis is highly dependent upon the size of the blank involved, owing to the low levels encountered, and, to a lesser extent, the precision with which the blank can be determined. This situation has been classically illustrated by the lead in sea water approach developed by Patterson's group at Cal Tech[8]. Herein, the various components contributing to the blank are identified, measured and minimized if at all possible. Thus, the assignment of a value for a blank is the summation of individual values rather than the measurement of one value, a composite which may not accurately reflect all of the blank components. For example, bottle leaching by acidified sea water or the NH_3 used in neutralization may be omitted in the re-extraction of a seawater sample for blank assessment. Moreover, this component approach is particularly favoured when a yield related formula (2) is used.

In this study Pb levels were used as a guideline for assessing the cleanliness of a reagent. From previous Pb work[11], it appeared that the overall blank level was highly dependent upon the cleanliness of the ambient air and upon the levels found in the water. In fact, these levels appeared to be the limiting factor in the cleanliness of labware and many reagents. Concentration levels in reagents and exposure rates for air were determined by single spike IDMS in independent experiments before sea water samples were attempted.

Class 100 clean room facilities were used for all of the extractions and the sample loading. In addition to the general supply of HEPA filtered air, epoxy painted clean hoods located within the clean rooms further scrubbed the air and provided an ultra clean work area. Access to these clean rooms was via one air lock in which cap and gown attire was donned. Exposure rates, measured in various areas with 30 ml FEP beakers of spiked acidified QD water, were very low (Table 2). The rates, ranging from 1 fmol d^{-1} for Cd to 1 pmol d^{-1} for Fe, were considered insignificant in relation to the other identified sources of contamination.

The water used for producing reagents and leaching labware was provided by a sub boiling, quartz distilled (QD) setup located in the room with the lowest exposure rates. Supply water for the still, located in a 30 l CPE header tank, is fed via a CPE line from a modified Milli Q (MQ) system. Modifications included the replacement of many fittings and tubes with CPE ones, and the substitution of clean Nuclepore filter (0.45 μm)/teflon case for the 'Twin 90' filter (styrene acrylonitrile). Supply water to this clean room, MQ system is provided by a 65 l CPE header tank fed with MQ from a standard Milli Q/RO setup. Values obtained for QD and the clean room MQ system (Table 2) are gratifyingly low. A major reduction in levels may be made, for Pb and Cu at least, by spiking QD water with clean NH_3 to pH 9-10 and letting sit for a month (see below). The low values attained by QD water are necessary for the final leaching of labware and in aqua regia type experiments.

Cleanup of CPE, teflon and quartz (the only materials employed in our labware) is similar to that employed by Patterson and

Table 2 Air Exposure Rates and Blank Levels in Reagents

	Air		Water		Acids	
	Clean Room pmol d^{-1}	Hood pmol d^{-1}	MQ pmol kg^{-1}	QD pmol kg^{-1}	SQ HCl pmol g^{-1}	NBS HNO_3 pmol g^{-1}
Cu	0.6		60	10	0.9	0.8
Ni	2	0.1		70	0.7	1.4
Cd	0.4	0.001	9	0.7	0.02	0.2
Pb	0.4	0.01	5-24	0.2-1.0	0.1	0.2
Zn	6	0.2	500	24	1.1	0.9
Fe	40	1	3000	400	150	40

	NH_3 pmol g^{-1}	$CHCl_3$ pmol g^{-1}	0.001% Dz pmol g^{-1}	1%APDC/DDDC pmol g^{-1}	Chelex 100 pmol ml^{-1}
Cu	2	0.04	0.2	16	2
Ni	8	0.15	0.4	340	
Cd	0.07	0.003	0.03	80	0.09
Pb	0.3	0.005	0.01	8	0.5
Zn	4	0.08	0.15	47	3
Fe	23	3	11	400	100

Settle (1976).[7] CPE containers intended for sea water were given an additional hot soak in acidified QD. Previously cleaned labware that has been exposed to ∿10 ng per element is hot soaked in acidified QD. Soaking at room temperature provides insufficient leaching.

Purified ammonia was produced by passing NH_3 through a series of borosilicate filters and bubbling into a chain of CPE bottles containing MQ and then QD. Further reduction in blank levels is possible by isopiestic distillation of the concentrated HN_4OH using a teflon elbow still operated at room tmeperature. Ammonia was then stored in a clean FEP bottle to allow Cu and Pb to 'plate out' onto the walls (Fig. 2). Providing a fresh, acid cleaned surface increases the rate of adsorption, (a 40% decrease in concentration in 30 days) a situation analogous to an ion pump in a vacuum system. Ni, Zn and Cd appeared to display a slight drop in concentration by this method. Isopiestic distillation appeared to be a more efficient purifying process for Ni, Zn, Cd and Fe. It

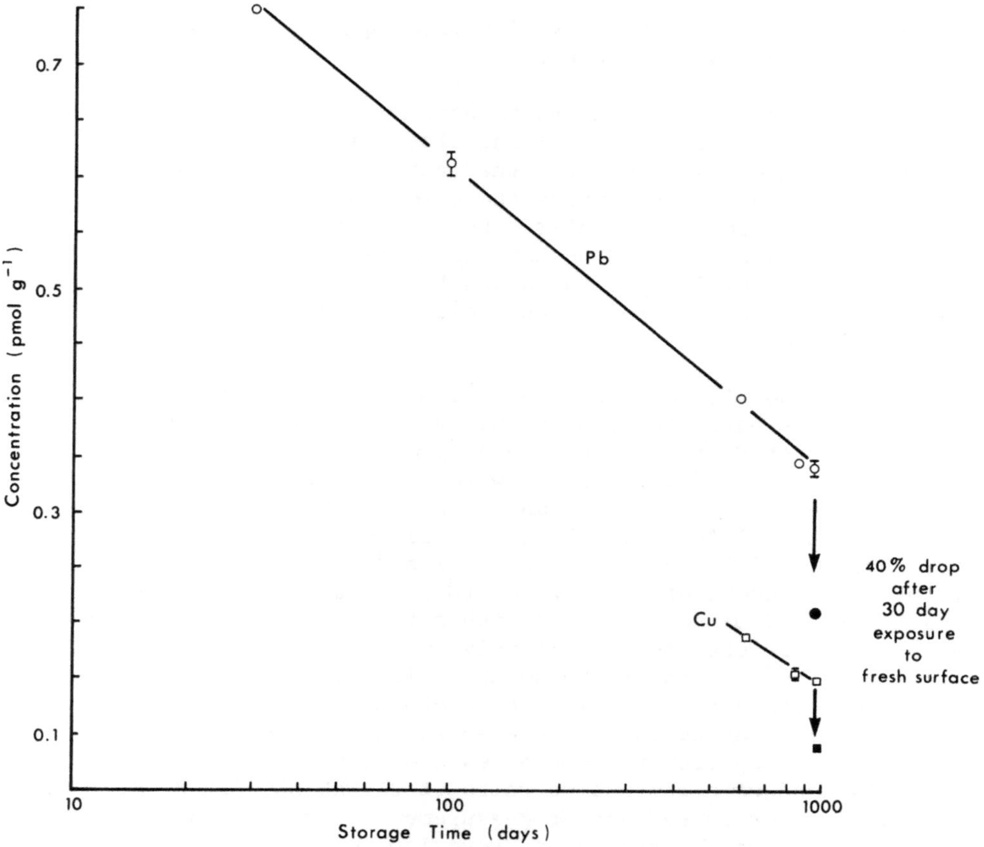

Fig. 2 Decrease of Pb and Cu in concentrated ammonia with time. The sample, maintained in the same clean FEP bottle (open symbols), was subsampled at day 904. After 30 days of storage in a newly cleaned CPE bottle, the concentration of Pb and Cu dropped by 40% (solid symbols). Small to virtually zero decreases were noted in the concentration of Ni, Zn, Cd, Fe and Cr. In QD water stored at pH 9-10, the rate of decrease in concentration for most of the metals is very similar to the higher rate observed in concentrated ammonia.

may be noted that a much larger decrease in concentration with time was noted for these metals in QD water spiked with ammonia.

Acids used were sub boiling distilled versions of reagent grade material. In house distillation of HCl using a quartz still (termed SQ) indicated levels comparable to NBS values for doubly distilled HCl (Table 2). NBS HNO_3 (doubly distilled, quartz)

(Table 2). NBS $HClO_4$ (doubly distilled, quartz) was used for digestion of organic products.

Chloroform was initially produced by extraction with acidified QD water and then sub boiling distillation (quartz or teflon elbow). However, it was found that any residual HCl in the chloroform tended to decompose the chloroform into phosgene upon prolonged standing. A simpler method of repeated pairs (6) of acid extractions and pH 8 washes was found to produce chloroform notably low (fmol levels) in Cd and Pb and acceptable for Fe (Table 2). Moreover, the product did not appear as prone to decomposition.

Dithizone (Dz) appeared to be a major source of both contamination and isobaric interferences. From preliminary studies, it was noted that Cd and Pb were readily extracted by 1-3% HNO_3 washes, while Cu, Ni, Zn and Fe were difficult to clean up without actually destroying the carbamate itself with acidified washes of greater than 5-10% HNO_3. Thus, a novel, non-destructive approach was devised: (a) reagent grade Dz was dissolved in clean chloroform; (b) aliquots of pH 10 water (QD) are used to extract the Dz, leaving behind Dz contaminated with metals; (c) aliquots of clean chloroform are used to repetitively extract/clean up the basic, Dz rich solution; (d) the basic solution is acidified sufficiently to just precipitate Dz; (e) clean chloroform is added to dissolve the precipitate, then decanted and evaporated to dryness; (f) 0.001% (w/v) is made up from the above residue with clean chloroform; (g) a final rinse with successive pairs of 2% NBS HNO_3/pH 8 water is performed prior to use and/or measurement. The Dz produced in the above manner appears stable and very low in trace metals, particularly Cd, Pb, Zn and Cu (Table 2). The 0.001% Dz is stored triple-bagged, in a refrigerator.

Reagent grade ammonium pyrrolidine dithiocarbamate (APDC) and diethylammonium diethyldithiocarbamate (DDDC) were purified by recrystallization and then dissolution to 1% levels in QD water. The solution was extracted repetitively with Freon TF (reagent grade). The Freon was extracted with concentrated NBS HNO_3.

The ion exchange resin, Chelex 100 was purified by successive batch extractions and soaking in 5 M SQ HCl, 5 M NBS HNO_3, and 1% HCl in QD water. The final product was stored as a 5 M HCl/slurry and loaded via a CPE dropper.

Quartz wool, used to stopper the separatory funnels for ion exchange experiments, and silica gel, used for loading the sample onto the filaments, were cleaned by a modified version of the quartz cleanup[7]. Repeated decanting/addition of the acidified QD was employed.

Baker Ultrex-grade, phosphoric acid was sublimed onto a clean, cool quartz surface. A .35 N solution was made up in QD for use in the filament loading process.

To reduce Fe blanks, Re was leached lightly with conc. HNO_3, then leached for a couple of hours with hot 2M HCl. Filaments were then assembled and degassed for three hours at 2.5 A.

The multi-element spikes were evaluated for inter-contamination by loading various weights of solution directly onto the Re ribbon. For most metals, there appears to be a linear relationship between the size of the spike and the contamination level encountered. Typical levels encountered are thus 0.02-0.090 nmol for Fe and fmol range for Cu, Cd, Pb, Ni, and Zn. Contamination has presumably arisen from either the dissolution process itself or from the actual manufacture of the isotopic spikes, themselves. These blanks are generally small with respect to the amounts encountered in sea water analysis.

Sea water was peristaltically pumped from 25 m in late April from the centre of Saanich Inlet, British Columbia. Although a plankton bloom was evident in surface waters, the rinse water was evidently clear. The CPE tubing employed was held at depth by a clean CPE coated weight. A new, short piece of acid washed tygon tubing was used for the portion of the line within the pump body itself. A clean section of CPE tubing delivered the output flow to the two 65 l, CPE carboys stored upwind of the generator/pump unit. The clean carboys had been double bagged. Sea water was stored acidified (pH 2.0, 1.6, 1.1) under clean room conditions.

ANALYTICAL METHODS FOR SEAWATER ANALYSIS

Two basic approaches in the extraction of trace metals from sea water were explored with varying degrees of success. They were: i) 0.001% dithizone in $CHCl_3$; and ii) 1% APDC/DDDC with Freon TF followed by Chelex 100 purification of the aqueous back extraction.

The dithizone approach was investigated initially on account of past familiarity and success with extracting Pb from sea water. In addition, Dz provided what appeared to be the lowest blank levels (Table 2), particularly for Cu, Cd, Pb and Zn. Dz has been used to extract Cu, Cd, Pb, Zn and Ni from sea water[8,10,11]. Patterson et al.[8] had found that simple Dz extraction of untreated sea water gave Pb results identical to sea water filtered with a 0.45 μ Nuclepore filter. Hence, this value was termed 'soluble' and is applied herein as a description of all the trace metals extracted in this manner. Briefly, the soluble metal was extracted as follows: (a) the acidified sea water is subsampled (approx.

800 ml) and stored in a 1 l FEP separatory funnel; (b) a multi-element spike is added and allowed to equilibrate for ca one hour; (c) the sea water is adjusted to pH 8 to 8.5 with a measured amount of concentrated ammonia; (d) a weighed aliquot (approx. 45 g) of 0.001% Dz is added to the sea water and shaken vigorously for two minutes; (e) after 30 minutes, the Dz layer is decanted into a 125 ml FEP separatory funnel and rinsed twice with two 10 ml portions of pH 8 water (neutralized QD); (f) the Dz is then back extracted by shaking with 5 ml of 2-5% (w/v) NBS HNO_3 for two minutes; (g) the aqueous back extract is then rinsed twice with 5 ml portions of clean $CHCl_3$; (h) the second multielement spike is weighed out into a 30 ml FEP beaker and the back extract is added; (i) the sample is evaporated slowly at low heat on a mini hot plate located in the clean hood; (j) the dried residue is redissolved in 2% NBS HNO_3 and loaded onto the Re filament with the silica gel/phosphoric acid approach.

A further refinement of this technique to minimize isobaric interference for Cu is the addition of a 'mini-Chelex' step between step (i) and (j). However, only 1 ml or less of Chelex slurry is loaded and no buffer is used. In addition, step (f) can be modified so that > 5% HNO_3 is used in the back extract for a better yield.

The other major variation of extraction involving Dz is the aqua regia or 'total' approach. Patterson et al.[8] (1976) demonstrated that the evaporation of sea water in the presence of 2-3 ml aqua regia and subsequent redissolution in QD water/extraction by 0.001% Dz (as per c-j above) yielded accurate, total values for Pb. This term, 'total', is used to describe all of the trace metals extracted in this manner. It may be noted that, herein, approx. 400 ml of sea water was evaporated in a 500 ml quartz beaker at low heat under ultra-clean hood conditions.

The use of 1% APDC/DDDC and Freon TF[2] has been applied to the extraction of Cu, Cd, Pb, Zn, Ni, Fe and Co from sea water. The approach was modified somewhat to accommodate the loading operation of the filament. Isotopic spikes were added in a manner similar to the above (b, h). The large unloadable residue left upon evaporation of the back extract was redissolved in pH 8 water and then re-extracted with less than 1 ml of Chelex 100 using the general method of Sturgeon et al.[12]. The evaporated residue was loaded as before. To reduce the size of unloadable residue in later work, the amount of 1% APDC/DDDC was reduced from 3 g to 0.1-0.2 g. While little change was observed in the isotopically determined yields, the residue was sufficiently small to be loaded as before. Isobaric interferences were commensurately smaller.

ISOTOPIC COMPOSITION DATA ON SOLUBLE AND TOTAL LEAD

The differentiation of 'soluble' and 'total' lead as defined by Patterson et al.[8] is in part dependent upon the contamination levels introduced and their control at various stages in performing either the extraction, the filtering or the aqua regia digest itself. It may be argued that in spite of good control on contamination levels and their variability, the higher levels of lead produced by the aqua regia approach may be an artefact. An independent approach provided by the determination of the isotopic composition of lead in 'soluble' versus 'total' experiments is presented here as support of the above differentiation.

Sea water was selected from a fjord location in British Columbia in which there typically is a significant particulate fraction. More important, however, the site was chosen so that the isotopic composition of that lead was substantially different from that determined on lead in the blank, local gasoline or local sea water[11]. It may be noted that lead in local gasoline dominates the blank or local sea water. Thus, any contamination introduced above and beyond that calculated for in the known blank would manifest itself as a mixing of the two end members. This would tend to lower the ratios of the lead in the 'total' sample.

Two separate sea water samples were analyzed for 'soluble' metals and 'total' metals in the prescribed fashion. It may be noted that in these samples in which there appeared to be a 90% particulate fraction (Table 3), the aqua regia blank comprised only 2% of the 'total' sample. The isotopic results (Table 3)

Table 3 Isotopic Composition of Soluble and Total Leads in Sea Water

	Blank, local	Sea Water near a mine with "Young" Pb Soluble A	Total A	Soluble B
$^{206}Pb/^{207}Pb$	1.1476 $\pm$ 0.0011	1.2279 $\pm$ 0.0006	1.2276 $\pm$ 0.0007	1.2296 $\pm$ 0.0009
$^{206}Pb/^{208}Pb$	0.4748 $\pm$ 0.0009	0.4988 $\pm$ 0.0009	0.4993 $\pm$ 0.0008	0.4996 $\pm$ 0.0002
$^{206}Pb/^{204}Pb$	18.02 $\pm$ 0.10	19.138 $\pm$ 0.030	19.148 $\pm$ 0.050	19.148 $\pm$ 0.043

Notes:
1) concentration of soluble lead; 0.0229 $\pm$ 0.0005 nmol kg^{-1}
2) concentration of total lead: 0.231 $\pm$ 0.005 nmol kg^{-1}
3) aqua regia blank; 0.005 $\pm$ 0.001 nmol kg^{-1}
4) natural abundance, soluble blank; 0.0010 $\pm$ 0.0001 nmol kg^{-1}

indicate a concordance of the lead ratios at one or two sigma for the 'soluble' and the 'total' samples. The 'total' results are substantially different from the local gasoline lead and do not give any indication of "excess" lead from unknown sources. Hence, the lead extracted and analyzed by the 'soluble' and 'total' approaches defined above are isotopically homogeneous and are indigenous to that sea water sample.

DISCUSSION OF IDMS SEAWATER DATA

The application of IDMS for the analysis of trace metals in sea water appears very promising. Very precise and stable values have been obtained for Cd, Pb and Zn using one of the simplest extraction procedures, 0.001% dithizone in chloroform (Table 4A, Fig. 3). This method yielding 'soluble' results also yielded the lowest blanks (0.2 pmol, 0.5 pmol and 30 pmol, respectively) of the various methods tried. Isotopically determined yields generally ranged from 80-99%. Acidification of the sea water from pH 2.0 to pH 1.6 (after day 93, Table 3) appeared to have no measurable effect on the concentration data. However, further acidification of a separate aliquot to pH 1.1 produced significant increases in the Pb and Zn values (0.0423 $\pm$ 0.0019 nmol kg^{-1} to 0.0475 $\pm$ 0.0006 nmol kg^{-1}; and 11.18 $\pm$ 0.13 nmol kg^{-1} to 11.55 $\pm$ 0.06 nmol kg^{-1}, respectively) while the Cd concentration remained constant (0.708 $\pm$ 0.007 nmol kg^{-1}, 0.719 $\pm$ 0.005 ng kg^{-1}). Use of APDC/DDDC in Freon TF (Table 4B) provided values concordant with (1 to 2σ) but less precise than those obtained by simple dithizone extraction. Extraction of metals with these carbamates appears entirely equivalent. Analysis of comparable amounts of Cd, Pb and Zn in NBS standard 1643a using Dz extraction yielded results concordant at 1σ. Hence, the IDMS approach for Cd, Pb and Zn, at least, appears to be accurate.

The more rigorous aqua regia digestion of 'total' approach gave yet higher Pb (0.0892 $\pm$ 0.0046 nmol kg^{-1}) and Zn (12.73 $\pm$ 0.43 nmol kg^{-1}) values indicating the presence of a significant particulate fraction (52%, 12% respectively) that was not readily leached or extracted. Cd values appeared to remain constant (0.691 $\pm$ 0.009 nmol kg^{-1}) indicating that all of the Cd available was in a dissolved form. It may be noted that the use of a mini-Chelex step as a final purification of one aqua regia sample not only gave excellent ion beam intensities but also yielded the only single total Cd value in agreement at 1σ with soluble Cd results. The remaining aqua regia samples, concordant at 2σ, appeared to have suppressed ion beam signals.

The total Cd, Pb and Zn results were less precise than soluble values (Table 4C), a condition most probably attributed to the greater complexity of the method and hence, greater blank

Table 4 Concentration of Trace Metals in Sea Water by IDMS

A. Soluble Metals Extracted by Dithizone

Storage Time (days)	Cu	Cd	Pb	Zn	Ni	Fe
50	5.26 ± 0.05	0.726 ± 0.004*	0.0440 ± 0.0006	11.23 ± 0.08	8.09 ± 0.17	14*
	5.21 ± 0.06	0.709 ± 0.002	0.0483 ± 0.0008[c]	11.21 ± 0.05		15*
75	5.07 ± 0.11	0.712 ± 0.002	0.0486 ± 0.0007[c]	11.21 ± 0.05	9.20 ± 0.17	11*
85	5.32 ± 0.05	0.698 ± 0.007	0.0416 ± 0.0011	11.36 ± 0.03		1.4*
93		0.698 ± 0.002	0.0404 ± 0.0012	11.12 ± 0.09		0*
	5.98 ± 0.47*	0.699 ± 0.007	0.0407 ± 0.0016			
151	5.41 ± 0.17	0.709 ± 0.003	0.0397 ± 0.0005	11.17 ± 0.05	10.56 ± 0.26	
209		0.717 ± 0.004	0.0451 ± 0.0006	11.39 ± 0.06	9.88 ± 0.85	
211		0.715 ± 0.005	0.0398 ± 0.0014	11.27 ± 0.15		
294	5.48 ± 0.35	0.706 ± 0.004	0.0421 ± 0.0010	11.23 ± 0.08		64 ± 4
		0.712 ± 0.005	0.0436 ± 0.0007	11.20 ± 0.08		52 ± 11
	5.04 ± 0.39	0.703 ± 0.003	0.0440 ± 0.0007	11.13 ± 0.03		
563		0.713 ± 0.009	0.0415 ± 0.0024			
		0.715 ± 0.008	0.0447 ± 0.0029	11.27 ± 0.23		
Average	5.26 ± 0.09	0.708 ± 0.007	0.0423 ± 0.0019	11.18 ± 0.13	9.43 ± 1.06	58 ± 6

Notes:
Concentration is in nmol kg^{-1}.
pH changed from 2.0 to 1.6 after day 93.
*values are omitted in the calculation of the average.
[c]values are considered to be elevated due to an improperly cleaned separatory funnel.
No hot leaching after use was performed in this early work.

Table 4 (cont'd.)

B. Soluble Metals Extracted by APDC/DDDC in Freon TF

Cu	Cd	Pb	Zn	Ni	Fe
5.26 ± 0.16	0.715 ± 0.004	0.0492 ± 0.0048*	12.24 ± 0.31*	5.16 ± 0.02*	
	0.750 ± 0.018	0.0410 ± 0.0029	11.21 ± 0.14	7.19 ± 0.09	
5.54 ± 0.47		0.0381 ± 0.0043	10.86 ± 0.38	7.85 ± 0.09	
5.71 ± 0.24	0.765 ± 0.027	0.0372 ± 0.0034	11.62 ± 0.61	7.58 ± 0.17	
5.19 ± 0.63				8.18 ± 0.03	
5.43 ± 0.24	0.743 ± 0.026	0.0388 ± 0.0020	11.23 ± 0.38	7.70 ± 0.42	

Notes:
Concentration is expressed in nmol kg^{-1}.
*values omitted in the calculation of the average.

C. Total Metals by Aqua Regia Digest Followed by Dithizone Extraction

Cu	Cd	Pb	Zn	Ni	Fe
	0.687 ± 0.001	0.0874 ± 0.0019	13.15 ± 1.38		62 ± 1
	0.680 ± 0.004	0.0738 ± 0.0001*	12.21 ± 0.08		
5.16 ± 0.06	0.697 ± 0.001	0.0835 ± 0.001	13.15 ± 0.17		
	0.626 ± 0.016*	0.0922 ± 0.0063	12.73 ± 0.21		
5.41 ± 0.13	0.698 ± 0.006	0.0936 ± 0.0014	12.40 ± 0.09		
5.29 ± 0.13	0.691 ± 0.009	0.0892 ± 0.0046	12.73 ± 0.43		62 ± 1

Notes:
Concentration is expressed in nmol kg^{-1}.
*values omitted in the calculation of the average.

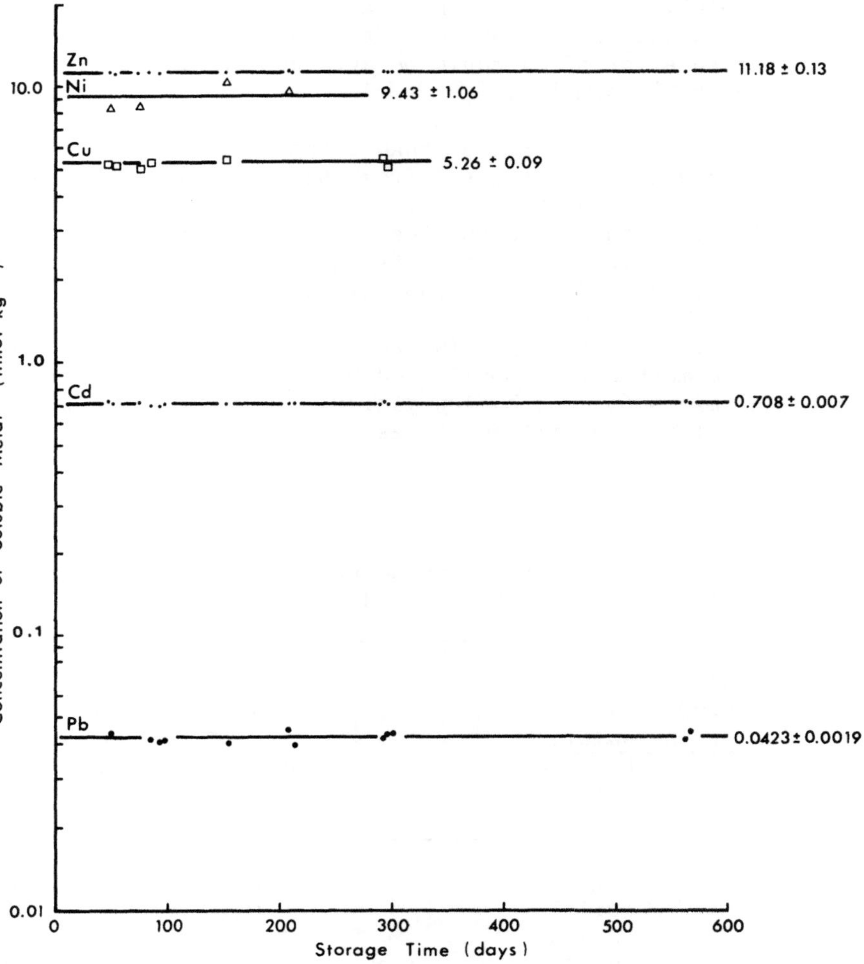

Fig. 3 Concentration of metals stored at pH 2.0 (day 0 - day 93) and at pH 1.6 (day 93 on) in a rigorously cleaned CPE carboy with time. One sigma error bars are symbol size or smaller. Note that even very low level Pb can be reliably stored for a period of a year and a half given the proper conditions.

variability. The precision in the determination of the time averaged, soluble data for Cd and Zn (Table 4A) was ca 1%, a level approaching that attained in the determination of a single ratio or sample. The precision of the soluble Pb ca 5%, reflected the low levels of Pb encountered, 0.04 nmol Pb kg^{-1}, and the relatively larger proportion of the sample that the blank, 0.002 nmol constituted. Soluble Cd and Zn blanks represented only 0.03% and 0.3%, respectively, of the sample. Previous work on Pb[11] had indicated that an error of 1% could be attained in Pb-rich (greater than 0.1 nmol kg^{-1}) samples.

If a constant blank, rather than a summation of predetermined components, were to be subtracted from the Pb and from the Cd data (Table 5), the error in the Pb data doubles (from 0.002 nmol to 0.004 nmol) while that of Cd remains virtually unchanged. The former appears to reflect the magnifying effect of the term (B_2/Y) in equation (2) coupled with a proportionately high (5%) blank.

The measurement of copper in sea water using an organic solvent carbamate method was complicated by isobaric interferences at m/e = 63 and m/e = 65, and by low yields. In the early APDC/DDDC approach, the large quantities of residue left upon evaporation of

Table 5 Comparison of Data using Constant Blank Versus Component Blank Corrections in a Yield Dependent Formula

Cd Blank ~ 0.0002 nmol or 0.03% of the sample		Pb Blank ~ 0.0019 nmol or 5% of the sample	
Component Blank	Constant Blank	Component Blank	Constant Blank
0.709	0.710	0.0440	0.0443
0.712	0.712	0.0416	0.0408
0.698	0.698	0.0404	0.0394
0.698	0.699	0.0407	0.0397
0.699	0.699	0.0397	0.0411
0.709	0.710	0.0451	0.0461
0.717	0.717	0.0398	0.0408
0.715	0.715	0.0421	0.0431
0.706	0.706	0.0436	0.0523
0.712	0.713		
0.703	0.703		
0.707	0.707	0.0419	0.0431
±0.007	±0.007	±0.0020	±0.0041

Note:
a) concentrations expressed in nmol kg^{-1}
b) error limits are one sigma

the back extraction necessitated a mini Chelex purification step. For dithizone extraction perchloric digestion proved to be of little use in eliminating isobaric interferences. Moreover, yield and quantity of carbamate breakdown products varied with the concentration of the nitric used in the back extract. Concentrated nitric allowed 100% yields [10], but also permitted large isobaric interferences at m/e = 63. With 2-3% HNO_3, the level sufficient for the extraction of Cd, Pb and Zn, low yields (ca. 5%) concomitant with a low residue resulted. Interferences were still significant. Use of 5% HNO_3 allowed a balance to be reached between yield and quantity of residue, and a stabilization (to some extent) in the ratio (63/65) m with time and filament current. In these cases, the values obtained for 'soluble' Cu, 5.26 $\pm$ 0.09 nmol kg^{-1} were relatively well defined and stable with time (Figure 3). The precision of this data is modest (approximately 2%) when compared with the potential precision capable of the instrument (approximately 0.2%). These agreed with results, 5.43 $\pm$ 0.24 nmol kg^{-1}, obtained using the APDC/DDDC approach.

While little improvement in signal characteristics was gained by the more rigorous aqua regia approach (Table 4C), a definite improvement occurred when the redissolved residue was treated with a 'mini-Chelex' cleanup step. The eluent from this purification step produced a relatively clean residue that was easily loaded. Signal characteristics for Cu were greatly enhanced and isobaric interferences were virtually non-existent. The values obtained, 5.29 $\pm$ 0.13 nmol kg^{-1}, were concordant at 1σ with the soluble results.

The measurement of Ni and Fe in sea water was hampered by either low yields or by loading problems/isobaric interference. With the dithizone approach, low yields (2-5%) were encountered for Ni, and isobaric interference at m/e 56 and 57, for Fe. Soluble results for Ni and Fe tended to be poorly defined (Table 4A). One aqua regia sample treated with a mini-Chelex purification step did yield a well defined reasonable (see below) Fe result, 62 $\pm$ 1 nmol kg^{-1}. Isobaric interferences appeared to be minimal. The use of dimethylglyoxime or hydroxylamine hydrochloride was considered for improving the yield, but rejected in view of altering the low Cd, Pb and Zn blanks.

The use of APDC/DDDC in Freon TF together with a mini-Chelex step provided much higher Ni yields (over 60%), and hence, better defined soluble results, 7.70 $\pm$ 0.42 nmol kg^{-1}. Although all of the components of this blank have yet to be identified, Ni values appeared much lower than those obtained by dithizone, 9.43 $\pm$ 1.06 nmol kg^{-1}. Isobaric interference was suspected at m/e = 58. Use of ^{61}Ni as S_2 and ^{62}Ni as S_2, with m/e = 60 used to assess the common element may avoid this problem.

INTERCOMPARISON OF IDMS DATA ON SEA WATER FROM C1 WITH FAAS AND ASV RESULTS

To establish the accuracy of a method, it is useful to compare the results obtained by one procedure with those produced by completely independent techniques. In this study, the sea water contained in carboy C1 was used as a standard since the water mass appeared stable in concentration with time for many elements (Table 4, Fig. 3) and the stability was capable of being defined quite precisely. Furthermore, corrobration for Cd, Pb and Zn analysis at low levels was established by the determination of small quantities of NBS 1643a. This data was established by IDMS. Comparisons for the independent techniques were also provided by independent operators. The FAA approach together with the Danielsson technique (APDC/DDDC in Freon TF) was employed by Keith Johnson of I.O.S. using the same clean room facilities. The FAA approach together with an APDC/DDDC extraction or 'column' (i.e. Chelex) technique[12] was applied by R. Sturgeon of the National Research Council. Sea water was shipped in rigorously cleaned[7], 2 l CPE bottles. For ASV analysis, sea water from C1 was shipped to J.A. Page in a 2 l CPE bottle that had been rigorously cleaned, and had previously provided low-level lead results.

The data obtained on C1 (Table 6) is broadly consistent, and in some cases has provided excellent corroboration. Copper as determined by one FAA, 5.35 $\pm$ 0.09 nmol kg^{-1}, is in excellent agreement at 1σ level with IDMS values for total, 5.29 $\pm$ 0.13 nmol kg^{-1}, and soluble, 5.26 $\pm$ 0.09 nmol kg^{-1} Cu. The ASV value, 3.93 $\pm$ 0.60 nmol kg^{-1} is poorly defined and may be considered broadly consistent at 2σ. It is not clear why the second FAA result, 2.52 nmol kg^{-1}, is discrepantly low.

For Cd, the ASV value of 0.649 $\pm$ 0.044 nmol kg^{-1} is reasonably well-defined and in good agreement with the IDMS values for total, 0.691 $\pm$ 0.009 nmol kg^{-1} and soluble, 0.708 $\pm$ 0.007 nmol kg^{-1} Cd. The poorly defined FAA value of 0.498 $\pm$ 0.089 nmol kg^{-1} is just concordant at 2 levels. Since the recovery of Cd is the most variable (85 $\pm$ 14%) of the elements extracted in the Danielsson technique, the yield for these samples may have been consistently low. With a different extraction technique the second FAA result, 0.721 $\pm$ 0.027 nmol kg^{-1}, is well defined and in excellent agreement with the IDMS results.

The lead results are particularly interesting. Considering the possibilities of lead contamination, it is remarkable that the first FAA value of 0.0381 $\pm$ 0.0100 nmol kg^{-1} is in such close agreement with the soluble Pb IDMS value of 0.0423 nmol kg^{-1}. The ASV value is anomalously high and is considered to be the result of contamination, either in Queen's laboratory or during transit of the sample. It is worthwhile to note that both dithizone and

Table 6 Comparison of IDMS Data with FAA and ASV Approached on Sea Water from C1

Method	Cu	Cd	Pb	Zn	Ni	Fe
'Soluble'						
IDMS	5.26 ± 0.09	0.708 ± 0.007	0.0423 ± 0.0019	11.18 ± 0.13	9.43 ± 1.06[d] 7.70 ± 0.42	58 ± 6
'Total'						
IDMS	5.29 ± 0.13	0.691 ± 0.009	0.0892 ± 0.0046	12.73 ± 0.43		62 ± 1
APDC/DDDC with FAA[a]	5.35 ± 0.09	0.498 ± 0.089*	0.0381 ± 0.0100	11.01 ± 0.31	5.79 ± 0.17	50 ± 5
Chelex or APDC/DDDC with FAA[b]	2.52	0.721 ± 0.027	0.087	11.62	10.39 ± 0.51	
ASV[c]	3.93 ± 0.60	0.649 ± 0.044	0.290 ± .034	12.47 ± 1.07		

Notes:
a) FAA results by W.K. Johnson of IOS using the same clean room.
b) FAA results by R. Sturgeon, NRC, Ottawa.
c) ASV results by J. Page, Queen's University, Kingston, Ontario.
d) Dual values resulting from different extraction techniques.
Concentrations are expressed in nmol kg^{-1}, errors quoted are one sigma.
*yields were low and variable (85 ± 14%), error quoted reflects average of FAA data alone.

APDC/DDDC extracted only the soluble or readily leached component of lightly acidified (pH 1.6 - 2.0 $\pm$) sea water. It is not clear whether the difference in sample extraction by NRC allowed concordance with the total results. Conceivably, the use of strong acids in the 'column' technique could leach out more of the total metal. For total Pb results, it appears that the more rigorous aqua regia treatment of the sea water would be necessary.

The Zn results parallel those for Pb. The IDMS soluble Zn, 11.18 $\pm$ 0.13 nmol kg^{-1}, is considered to be equivalent at 1σ to the FAA values of 11.01 $\pm$ 0.31 nmol kg^{-1} and 11.62 nmol kg^{-1} as the sample pretreatment was identical (i.e. acidification to pH 1.6). The intermediate Zn value of 11.48 $\pm$ 0.06 nmol kg^{-1} (pH 1.1 pretreatment) appears to represent additional Zn gained from leaching of the particulate fraction. The relatively poorly defined ASV result (12.47 $\pm$ 1.07 nmol kg^{-1}) does not allow a definitive comparison with the total Zn (12.73 $\pm$ 0.43 nmol kg^{-1}) or soluble Zn (11.18 $\pm$ 0.13 nmol kg^{-1}) values determined by IDMS. The total results appear to be favoured.

Analysis of Fe by FAA (50 $\pm$ 5 nmol kg) yielded results concordant at 1σ with the poorly defined, soluble Fe (58 $\pm$ 6 nmol kg^{-1}) results by IDMS and just concordant at 2σ with the single total IDMS value (62 $\pm$ 1 nmol kg^{-1}).

The comparison of Ni values by the two FAA techniques appears somewhat discrepant (5.79 $\pm$ 0.17 nmol kg^{-1} versus 10.39 $\pm$ 0.51 nmol kg^{-1}) and tend to bracket the dual IDMS values of 9.43 $\pm$ 1.06 nmol kg^{-1} and 7.70 $\pm$ 0.42 nmol kg^{-1}. As difficulties were encountered with either low yields in the dithizone extraction or with isobaric interferences in the APDC/DDDC approach, a definitive statement as to the validity of any value cannot be made. It may be noted that the poorly defined IDMS (9.43 $\pm$ 1.06 nmol kg^{-1}) value is concordant with the higher of the two FAA values, 10.39 $\pm$ 0.51 nmol kg^{-1}. However, it is in the author's experience that low yields and/or isobaric interferences tend to yield values erroneously high. Substitution of ^{61}Ni for the primary spike and the measurement of m/e = 60 as the common element may overcome this problem.

CONCLUSIONS

The combination of IDMS and ultra-clean room techniques for the determination of natural levels of Cu, Cd, Pb and Zn in sea water appears to be a method capable of accuracy and very high precision. Ni and Fe are more difficult to determine but show some promise. Through the use of IDMS, multielement analysis, yields and concentrations can be determined on the same sample, a situation not possible with conventional techniques. Corroboration of the accuracy of the IDMS approach using dithizone extraction for low levels of Cd, Pb and Zn was made by the analysis of small quantities of NBS 1643a. Quantitative extraction does not appear to be a prerequisite for high precision or accuracy. In certain cases, approximate concentrations can be determined with as little as 5% yield. Blanks, predetermined as a sum of component parts, may be weighted according to the yield of a particular step or extraction. This approach can significantly reduce the error in situations of lowered yields and/or relatively high blanks. It should be noted, however, that the method is slow (0.5 to 3 samples per day), expensive and highly dependent upon the experience of the operator. The quality of the results makes IDMS ideal for a "referee" technique.

For IDMS work, soluble trace metals may be extracted from sea water that has been unacidified or previously acidified to pH 2.0 - 1.6 with dithizone in chloroform or with 1% APDC/DDDC in Freon TF. Isobaric interferences, resulting most likely from carbamate breakdown products, may be minimized by an ion exchange purification of the aqueous back extract. Further work is necessary for the routine analysis of Cu, Ni and Fe. For the determination of total metals, an aqua regia digestion of the sea water appeared necessary. Simply increasing the acidification of the sea water appears to produce intermediate results. Isotopic composition work on Pb supports this differentiation initially proposed by Patterson et al.[8]. Sea water may be stored for a period of up to 1 1/2 years at pH 1.6 in rigorously cleaned CPE containers.

The highest precision was obtained in the determination of soluble trace metals. Levels of ca 1% may be obtained for Cd, Zn and Pb (under certain conditions). This level may be approached in the determination of Cu, Ni and Fe ratios. Comparison of the results for Cu, Cd, Pb, Zn and Fe with FAA and ASV appeared to indicate that not only was the IDMS approach accurate, but was also capable of the highest precision.

REFERENCES

1. Chow, T.J., 1968: Isotope dilution analysis of sea water by mass spectrometry. Journal of WPCF, pp. 399-411.
2. Danielsson, L., B. Magnusson and S. Westerlund, 1978: An improved metal extraction procedure for the determination of trace metals in sea water by graphite furnace A.A. Anal. Chim. Acta., 98(1), 47-57.
3. Garner, L.G. and L.P. Dunstan, 1978: Determination of nanogram per gram concentrations of iron by isotope dilution mass spectrometry. In: "Advances in Mass Spectrometry", Vol. 7A, N.R. Daly, ed. Heydon and Sons, London, 481-485.

4. Moore, L.J., E.F. Heald and J.J. Filliben, 1978: An isotope fractionation model for the multiple filament thermal ion source. In: "Advances in Mass Spectrometry", Vol. 7A, N.R. Daly, ed. Heydon and Sons, London, 448-474.

5. Murozumi, M., 1979: Isotope dilution mass spectrometry of copper, cadmium, thallium and lead in marine environments. Presented at ACS/CSJ Chemical Congress, Hawaii.

6. Mykytiuk, A.P., D.S. Russell and R.E. Sturgeon, 1980: Simultaneous determination of iron, cadmium, zinc, copper, nickel, lead and uranium in sea water by stable isotope dilution spark source mass spectrometry. Anal. Chem., 52, 1281-1283.

7. Patterson, C.C. and D.M. Settle, 1976: The reduction of orders of magnitude errors in lead analyses of biological materials and natural waters by evaluating and controlling the extent and sources of industrial lead contamination introduced during sample collecting, handling and analysis. NBS spec. pub. 422, 321-351.

8. Patterson, C.C., D.M. Settle and B. Glover, 1976: Analysis of lead in polluted coastal sea water. Mar. Chem., 4, 305-319.

9. Schaule, B. and C.C. Patterson, 1980: The occurrence of lead in the Northeast Pacific, and the effects of anthropogenic inputs. In: "Lead in the Marine Environment", M. Brancia and Z. Konrad, eds. Pergamon Press, Oxford, 31-43.

10. Smith, R.G., Jr. and H.L. Windom, 1980: A solvent extraction technique for determining nanogram per litre concentrations of cadmium, copper, nickel and zinc in sea water. Anal. Chim. Acta., 113, 39-46.

11. Stukas, V.J. and C.S. Wong, 1981: Stable lead isotopes as a tracer in coastal waters. Science, 211, 1424-1427.

12. Sturgeon, R.E., S.S. Berman, A. Desaulniers and D.S. Russell, 1980: Preconcentration of trace metals from sea water for determination by graphite furnace atomic absorption spectrophotometry. Talanta, 17, 85-94.

STUDIES OF THE CHEMICAL FORMS OF TRACE ELEMENTS IN SEA WATER USING RADIOTRACERS

M. Amdurer, D. Adler and P.H. Santschi

Lamont-Doherty Geological Observatory of
Columbia University
Palisades, New York 10964 U.S.A.

ABSTRACT

Important clues as to the chemical forms, seasonal variations, biogeochemical transformations and removal mechanisms of trace elements in natural waters are revealed by the use of chemical separation procedures. The use of radioisotopes greatly facilitates these studies by offering simple, rapid multielement analysis. We have added a total of 26 radiotracers in various forms to large marine microcosms (13 m^3 tanks, Marine Ecosystems Research Lab, U.R.I.) and monitored their chemical and biological behavior over periods of up to nine months. Our chemical separation scheme consists of serial extraction by 0.4 μm filter, activated charcoal and chelex resin; other techniques include fractionation with XAD resin or ultra-filtration of the "dissolved" fraction. Results indicate that the transition metals and metalloids studied have distinct "fingerprints" which allow the identification of groups of similarly-behaving elements. The distribution of chemical forms varies with season, condition of the water column with respect to biological activity or suspended particulate matter, and the form of the added tracer. The fractionations of some radioactive trace elements in the microcosms are compared with chemical speciations predicted by equilibrium chemical speciation models.

INTRODUCTION

Examining the vertical distribution of trace elements in the ocean yields valuable clues about their chemical behavior and cycling mechanisms in the water column[1,2]. The next step in refining our understanding of the elements' marine chemistry is to examine their chemical speciation. Knowledge of the chemical forms of trace elements in seawater, their relative reactivity and distribution coefficients for uptake on particles, greatly assists the interpretations of observed vertical distributions or removal rates. This information is particularly important for estuarine or coastal environments where, because of rapidly-changing conditions of the water column, steady-state vertical distributions of elements can not be used to understand their reactivities. Furthermore these waters receive large natural or anthropogenic inputs of potentially harmful substances. With their often restricted circulation, it is imperative to know what controls the movement of these chemicals through the system, in order to predict whether they will be concentrated locally and in what form, or if they will be diluted into the ocean[3,4].

Chemical separation studies can help to understand the fate of various trace elements by addressing the following questions: 1) Do the elements associate rapidly with a particulate phase and settle to the bottom to accumulate or be transported with bottom sediments? 2) To what extent, and at what rate, do they become involved in the biological cycle, either a) through complexation with dissolved organic matter, which may keep them in solution, or b) by bioaccumulation in organisms and subsequent transport and recycling with biotic components? 3) Do the trace elements remain in true solution, to be diluted throughout the ocean by mixing? To begin to answer these questions we have studied the chemical forms and removal rates of more than 20 trace metals and metalloids, added in radioactive form to large synthetic ecosystems ("microcosms").

Chemical speciation studies of trace metals in seawater are of two types. One group consists of theoretical, equilibrium models based on known thermodynamic data, derived generally from studies of less complex solutions[3,5,6]. These models do not include kinetic or biological effects and many do not consider adsorption to solid particles. Especially for trace metals, where stability constants with even inorganic ligands are poorly known, the results may be quite variable, depending on the set of thermodynamic data selected[6] (Table 5). The second approach is to determine directly the chemical forms of trace elements by electrochemical techniques or wet-chemical separation schemes[7-10]. The problem with this approach is that the "species" identified are often operationally-defined by the method used, making it difficult to compare the results of various methods or to compare these results to equilibrium predictions. Nevertheless, these "operational" systems may be more useful for predictive purposes.

The studies reported here are of the latter type. Our experiments are not intended to study the chemical speciation or cycling of natural elements in the ecosystem. To do this, the tracers must be fully equilibrated with all of the reactive (i.e. non-matrix) phases of the stable elements. This equilibration may require a time period equal to or greater than the duration of the experiment. There is ample evidence that ^{65}Zn, for instance, may take years to fully equilibrate with the natural organically-complexed fraction of stable Zn[11] and the same may be true for ^{55}Fe[12]. Rather, the radiotracers, added in ionic form, are analogs for the low level input of similarly dissolved pollutant metals, such as may occur in some industrial or municipal effluents.

The use of multiple radiotracers allows us to simultaneously examine the relative behavior of a large number of trace elements with widely-differing chemical properties. By studying the changes in chemical form with time following addition of the tracers, we can infer the reactivity, chemical transformations in the water column, and major removal mechanisms of these elements. These studies provide unique "fingerprints" of each trace element according to the phases in which it occurs. This approach permits the grouping of elements with similar chemical forms and removal patterns; it is analogous to the grouping of trace metals in the ocean according to their similarity to phosphate or silicon distributions[2]. If we understand the pathways of an element through a particular environment, we can then predict the behavior of like elements in similar systems.

This paper presents an overview of the response of a suite of trace elements to the same separation techniques. We discuss the relationship between the various phases in which the elements occur, their speciation based on equilibrium calculations, and their environmental behavior. These observations indicate the basis on which the elements have been grouped.

EXPERIMENTAL

Setting

The Marine Ecosystems Research Laboratory (MERL) of the University of Rhode Island maintains 12 controlled experimental ecosystems ("microcosms"), 3 of which are dedicated to radiotracer studies. The microcosms are outdoor fiberglass tanks, 5.5 m high and 1.8 m in diameter, containing 13 m^3 of water from the adjacent Narragansett Bay and a 30 cm layer of silty-clay bay sediments. The water column is cyclically mixed to simulate tidal currents and the turbulence and sediment resuspension regime in the bay. Tank walls are cleaned twice weekly with a rotary brush to remove fouling organisms. Water in the tanks may be exchanged with Narragansett Bay or the microcosms may be run as a closed (batch) system. Most radiotracer experiments are run in the batch mode. Previous studies have shown that a batch experiment can continue for seven months with no appreciable divergence from flow-through conditions[13].

Over several annual cycles, the MERL microcosms have proved to be good simulations of Narragansett Bay with respect to planktonic and benthic biotic structure, nutrient and other chemical concentrations and cycling, suspended particle concentrations and flux rates through the water column, and sediment mixing processes[13-15]. In addition, the removal rates of ^{228}Th and ^{59}Fe added to the microcosms were equal to those of the natural isotope ^{234}Th in Narragansett Bay measured during the same season (when corrected for the different mean depths of tanks and bay), and they showed similar seasonal variations[16]. This suggests that the processes controlling the removal of these particle-reactive elements from the water column are similar in the tanks and bay, and is a strong indication that the microcosms may be confidently used as analogs of Narragansett Bay for the study of trace metal behavior.

Procedures

Radiotracer experiments generally last from one to four months. In spiking, acid-stabilized microliter quantities of each isotope are added to one liter of seawater, which is then poured into each tank during a mixing cycle. The tank is completely mixed within 15 min[17]. Initial concentrations in the tanks are $7 \times 10^2 - 1.5 \times 10^4$ Bq l^{-1} (becquerel) of each γ-emitter and 0.7 - 3.0 Bq l^{-1} of each α-emitter. These levels permit detection of the radiotracers at <1% of their original concentration in most samples. Tracer measurements are carried out either by gamma-ray spectrometry using a Ge(Li) detector and 4096 channel multichannel analyzer coupled to a microcomputer with spectrum scanning and peak identification programs, or by alpha spectrometry using surface-barrier Si detectors.

Acidified (5 ml conc. HNO_3 per 100 ml) bulk water samples are collected frequently to monitor removal rates of the tracers from the water column. Continuous collection of settling particles is accomplished using sediment traps suspended in the tanks, which are sampled every 1-3 days. Nutrient and chlorophyll-*a* concentrations are measured biweekly.

The standard chemical separation system consists of filtration of a 500 ml bulk water sample through a 0.4 μm Nuclepore filter, followed by sequential flow through 3 ml activated coconut charcoal and Chelex-100 (Na form) columns (flow rate 2 ml min^{-1}). The column effluent is collected and acidified. Other fractionating methods used include XAD resins, Amberlite IRA-938 resin, and ultra-filtration using an Amicon stirred cell under N_2 pressure. In general, duplicate chemical separations are replicable to within 5-10% (relative) for fractions that contain more than 50% of the tracer activity and to within 20% (relative) for fractions that comprise less than 30% of the total activity.

Particulate and Colloidal. The particulate fraction is defined as that portion of the trace element retained on a 0.4 μm Nuclepore filter. Though this is an arbitrary cutoff, it has become a fairly standard definition. The affinity of trace metals for adsorption to particle surfaces, measured by their particle distribution coefficient (K_D), governs the percent of tracer activity found in this phase. K_D could be variable, however, if the trace metals are stabilized in solution, if they are not exposed to particles long enough to reach an equilibrium distribution, or if the adsorptive capacity of the particle surfaces is altered (e.g. by organic or hydroxide coatings).

Some fraction of the "dissolved" species of trace elements (i.e. those which pass through the filter) may be colloidal. Sequential ultra-filtration through membranes with nominal molecular weight cutoffs of 10^5, 10^4, and 10^3 u (unified atomic mass unit) was used to determine the concentration and size spectrum of these colloidal species. Both activated charcoal and Amberlite IRA-938 resin, because of their macroporous structures, can trap a wide range of colloidal particle sizes.

Charcoal and XAD. Activated carbon (a.c.) has long been used to remove non-polar trace organics (such as phenols, dyes, pesticides) from wastewaters. Vaccaro[18] and Kerr and Quinn[19,20] have shown that is can adsorb 70-90% of the dissolved (<0.45 μm) organic carbon from ocean waters, although in coastal water they found that it may remove somewhat less. This suggests that a.c. is less effective for terrestrial organics (such as humic substances), which may be enriched in nearshore waters. XAD-2 resin, however, adsorbs ~90% of both terrestrial[21] and marine humates[22]. Using both a.c. and XAD, combined with ultra-filtration, we hoped to isolate the organically-complexed species of trace metals. As mentioned above, however, a.c. can also adsorb inorganic colloidal matter. Furthermore, though a.c. surfaces are predominantly non-polar, its high specific surface area (up to 2000 $m^2.g^{-1}$) and presence of some surface functional groups (e.g. phenol, carboxyl) could lead to the adsorption of certain trace metals.

For trace metal adsorption, activated carbons are usually modified using strong acid or base to alter the surface charge, or treated with sulfonate or H_2S (to produce surface S-groups) or chelating agents to improve their adsorptive capacity[23]. Adsorption is very pH-dependent, being much more efficient at low pH

(1-5) for metals such as Cr, Hg, Fe and V. Since our a.c. is untreated and is used at the pH of seawater (8.2-8.4), it is likely that it adsorbs mostly uncharged complexes. For Cd, however, optimal adsorption occurs at alkaline pH[23].

The sequence of charcoal and chelex columns is important: placing chelex columns before charcoal (the reverse of the normal fractionation sequence) showed that much of the "charcoal-extractable" fraction of some metals could be extracted by chelex, which removes transition metal cations and their inorganic complexes. At slow flow rates and high chelex:solution ratios, chelex can also strip trace metals bound in organic complexes (Mantoura, personal communication); under the conditions of these experiments, however, the "charcoal" fraction of these elements which was also removed by chelex represented inorganic species. Thus the charcoal fraction of our separation system represents some combination of organically-complexed and colloidal metals, plus dissolved inorganic species of specific elements.

Chelex. Chelex-100 has been used to quantitatively remove and concentrate the inorganic species of many transition-series trace elements from seawater[24]. Since the radiotracers are added in ionic form, if they do not undergo speciation changes in the water column the transition-series elements should remain chelex (or charcoal) extractable.

Effluent. Species found in the effluent must be quite unreactive toward most adsorbing surfaces or complexing agents in seawater, since they have passed through a very strong adsorbent (a.c.) and chelating agent (chelex). Effluent species include alkaline and alkaline earth elements, anions, and some dissolved organic matter (probably more polar and low molecular weight) which is not adsorbed by a.c.

RESULTS AND DISCUSSION

The trace elements we have studied can be divided into four major groups (Table 1), according to their chemical forms and removal rates in the water column, and their behavior in the sediments[28]. Elements within each group move through the system in similar phases, although the specific adsorption/complexation reactions and interactions with biota may be different. In the following pages we will discuss the fractions identified using our separation system in two simultaneous experiments (MH and MJ) during the summer of 1979, and how they relate to the major removal mechanisms of each group of elements (some elements studied in earlier experiments are also discussed). Tank MJ contained normal, healthy sediments but the sediments in tank MH were made anoxic (the bottom water was not aerated for one week by turning

Table 1 Element Groups, Dissolved Speciations and Half-removal Times

Element	Form Added	Probable major species in aerobic seawater	Removal Half-times (days)		Ocean Residence Time (years)*
			summer	winter	
Group I					
Fe	$^{59}FeCl_3$	$Fe(OH)_3$	2-4	9-18	77**-200
Cr(III)	$^{51}CrCl_3$	$Cr(OH)_2^+ \rightarrow Cr(OH)_3 \rightarrow CrO_4^{2-}$	3-5	11	-
Sn	$^{113}SnCl_4$	$SnO(OH)_3^-$ $SnO(OH)_2$	2	11	-
Pa	$^{233}PaCl_5$	(?)$Pa(OH)_5$, $PaO(OH)_3$, $PAO_2(OH)$	3	-	31**
Hg	$^{203}HgCl_2$	$HgCl_4^{2-}$, $HgCl_3^-$, $HgCl_3Br^{2-}$	2-8	21	$8x10^4$
Group II					
Mn	$^{54}MnCl_2$	Mn^{2+}, $MnCl^+$	6-27	28-30	20**; $1x10^4$
Co	$^{58,60}CoCl_2$	Co^{2+}, $CoCl^+$, $CoCO_3$, $CoSO_4$	5-21	40-50	$3x10^4$
Group III					
Se	$H_2{}^{75}SeO_3$	SeO_3^{2-}, SeO_4^{2-}	15-60	-	$2x10^4$
As	$Na_3{}^{74}AsO_4$	$HAsO_4^{2-}$	12-40	-	$5x10^4$
V	$^{48}VCl_3$	$H_2VO_4^-$, HVO_4^{2-}	15	-	$8x10^4$
Zn	$^{65}ZnCl_2$	Zn^{2+}, $ZnCl^+$, $ZnCl_2$, $Zn(OH)_2$, $ZnCO_3$	6-20	60-70	$2x10^4$
Cd	$^{109}CdCl_3$	$CdCl_2$, $CdCl^+$, $CdCl_3^-$	20	>100	$5x10^4$
Group IV					
Cs	$^{134}CsCl$	Cs^+	80-180	~150	$6x10^5$
Ba	$^{133}BaCl_2$	Ba^{2+}	105	~600	$4x10^4$
Ra	$^{228}RaCl_2$	Ra^{2+}	150	150	$4x10^6$
Sb	$^{125}SbCl_3$	$Sb(OH)_6^-$	-	~210	$1x10^4$
Cr(VI)	$Na_2{}^{51}CrO_4$	CrO_4^{2-}	26	-	$6x10^3$

* Brewer[25], Stumm and Brauner[26]. ** Brewer et al.[27].

off the mixer), killing all the macrobenthos. Tank MH also had a large phytoplankton bloom at the time of the experiments. In addition, we will briefly discuss the seasonal contrast in chemical forms and removal rates of Group I elements in similar studies performed during the winter and spring.

GROUP I

Group I comprises the "particle-reactive" elements Fe, Cr(III), Hg, Sn, Pa (and Be, Th, Po, Pb, Pu and Am, which will not be discussed in this paper). These elements are removed from the water column rapidly (Table 1) and do not show significant back-diffusion from the sediments. Chemical separations of these elements are presented in Figures 1 and 2. The rapid attainment of a relatively constant distribution among the chemical forms in each tank (except for Cr, discussed below) but the tremendous difference between the two microcosms is immediately obvious.

Particulate and Colloidal

In all experiments, Group I elements had consistently larger particulate fractions than any other elements. Figures 1 and 2 demonstrate, however, that the percentage of Group I elements on particles may range from <20% to >95%, depending on the element

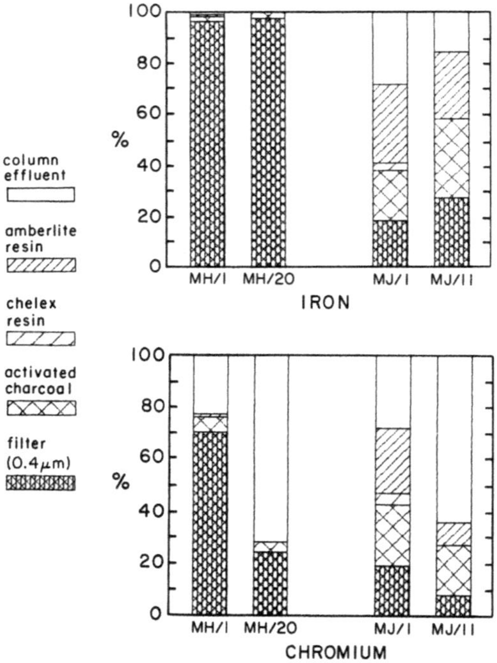

Fig. 1 Results of chemical separation procedures for Fe and Cr. Vertical scale = percent of each fraction in the water column. Tank and day of experiment on horizontal axis (e.g. MJ/11 = tank MJ, day 11). "Amberlite resin" refers to Amberlite IRA-938 resin.

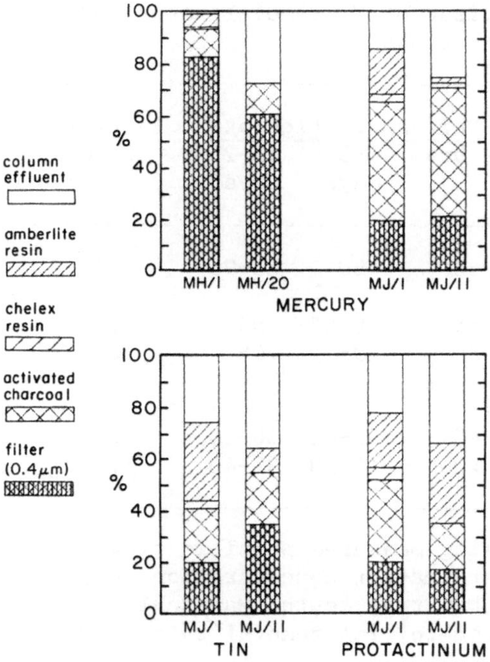

Fig. 2 Results of chemical separation procedures for Hg, Sn and Pa. Vertical scale = percent of each fraction in the water column. Tank and day of experiment on horizontal axis (e.g. MJ/11 = tank MJ, day 11). "Amberlite resin" refers to Amberlite IRA-938 resin.

and the conditions of the microcosm. In general, during the winter when particle residence times in the water column are long and benthic activity is minimal, the particulate fraction of these elements is 80-95% of the total activity (in Table 2 Fe is used to exemplify the behavior of Group I elements). During the spring, the particulate percentage is somewhat lower, and the residence time of both suspended particles and Group I elements in the water column is intermediate between winter and summer. In summer experiments, very short particle and trace metal half-removal times are accompanied by relatively low particulate percentages (20-50%, (Table 2), and a great deal of benthic faunal activity. The MH summer experiment, however, had a long particle residence time and a high filterable percentage of Group I elements, but their residence times in this tank were close to those in MJ.

The major difference in chemical forms of the particle-reactive elements between tanks MH and MJ was the distribution between particulate and colloidal phases. Ultra-filtration of the

Table 2 Comparison of Fe removal rates and percent particulate in MERL tanks

Expt.	Season	Percent particulate 2 hrs	1-3 days	15-25 days	Particle conc (mg l^{-1})	Particle $t_{1/2}$ (days)*	Iron $t_{1/2}$ (days)**
MF	winter-spring	60	80	85-95	2.2	2	9.4
MA	spring	55	60-65	>95	4.5	1.2	5.2
MH	summer, anoxic sediments	80	>95	>95	2.8	1.8	3.5
MC	summer	45	35-45	30-50	3.4	0.2	2.1
MJ	summer	20	15-20	30-40	1.6	0.2	2.6

* Calculated using measured particle fluxes in sediment traps (F) and particulate matter concentration (pmc) as $t_{1/2} = \ln 2 \times pmc \times h/F$, where h = tank depth (Santschi et al., this volume).

** Determined from initial removal rate of Fe from the water column.

"dissolved" fraction of these elements in both tanks indicated that they are predominantly colloidal, with a molecular weight $>10^5$u (Figures 3 and 4). In MH, however, the particulate fraction predominated, while in MJ the colloidal form was most important. Some fraction of this colloidal phase was adsorbed by activated charcoal and by Amberlite IRA-938 resin (Figures 1 and 2), but extraction with XAD resin demonstrated that these colloids were not predominantly humic in nature (Table 3).

Tanks MH and MJ had very different particle size spectra, as determined by Coulter Counter. The total number of particles greater than 2 μm in diameter in the former tank was 20 times greater than the latter, and their surface area was 12 times as large. Furthermore, the residence time of particles in tank MH was about ten times as long as in MJ (Table 2). The long residence time and large number of particles in MH thus allowed the flocculation of the colloidal Group I elements. Tank MJ had the opposite conditions of fewer particles and very short particle residence times (Table 2). A much smaller percentage of the colloids could thus be aggregated into particulate form. In this tank, as in other healthy tanks during the summer, much of the removal of particles from the water column was accomplished by benthic filter feeding, which enhanced particle removal rates (Santschi et al., this volume). This produced short residence

Table 3 Percent of tracers in XAD fraction

Element	Tank MH		Tank MJ	
	Day 2	Day 9	Day 2	Day 9
Fe	1.2	0.7	0.6	0.5
Cr	0.8	2.1	1.9	2.2
Mn	0.0	0.0	0.2	0.0
Co	0.0	0.0	0.0	0.2
Sn	-	-	4.8	3.3
Hg	12.1	10.0	22.6	15.5
Se	11.8	23.1	-	-
As	0.3	0.0	-	-
Zn	1.5	0.0	0.0	0.0
Cd	-	-	2.0	0.7

times of particle-reactive elements during the summer, even though these elements may not have been predominantly in particulate form. During the spring, a less active benthos contributed less to removal of particles from the water column and their subsequent resuspension. Particle residence times were longer and consequently Group I elements were more particle-associated. During the winter and in the summer in tank MH, where the benthos was inactive, Group I elements were predominantly particulate and their removal was by particle settling only.

Organic

In our experiments, trace metal interaction with a.c. showed two distinct patterns. Some of the tracers immediately reached a distribution which did not change appreciably over the course of the experiment (Table 4). Most of these elements (except for Group I) were extractable by chelex if it preceded the charcoal column, and they were not XAD-extractable (Table 3). This behavior suggests that the species removed by a.c. were rapidly-forming inorganic complexes, which were maintained at a constant percentage of the water column concentration throughout the experiment (Table 4). As discussed above, a somewhat greater proportion of Group I elements was adsorbed by charcoal during experiment MJ -- this represents the ultra-filterable, colloidal phase which dominated the speciation in that tank.

In contrast with Fe and Cr, Hg in the charcoal fraction decreased somewhat with time (Table 4), as did the fraction

Table 4 Distribution and probable species in activated charcoal

Element	% in charcoal*	Probable species in charcoal ( ) = % in seawater
1. Constant Distribution		
Fe	<5-10 (30)**	colloidal $Fe(OH)_3$ (95-100)
Cr	0-10 (25)**	colloidal $Cr(OH)_3$ (95-100 before oxidation to CrO_4^{2-})
Co	10-20	$CoCO_3$ (7); $CoSO_4$ (7)
Cd	40-60	$CdCl_2$ (15-50); $CdCl_3^-$ (6-28); $CdCl_4^{2-}$ (0-10); $CdCl_6^{4-}$ (0-9)
Zn	35-50	$Zn(OH)_2$ (<1-50); $ZnCO_3$ (4-15); $ZnSO_4$ (2-16)
2. Decreasing fraction in charcoal		
Hg	30-60 → 15-25	organic; $HgCl_4^{2-}$ (66); $HgCl_3^-$ (12); $HgCl_3Br^{2-}$ (12)
Sn	35 → 10	organic; $SnO(OH)_2$
3. Increasing fraction in charcoal		
Se	<10 → >60	organic; SeO_3^{2-} + SeO_4^{2-} (100)
V	15 → 25	organic; $H_2VO_4^-$ + HVO_4^{2-} (100)
As	0 → 15	organic; $HAsO_4^{2-}$ + $HAsO_3^{2-}$ (100)

* Based on 9 experiments during all seasons, 1978-1980.
** Percent in tank MJ; higher because of large colloidal fraction.

extracted by XAD (Table 3). The low molecular weight (l.m.w.) colloidal fraction also decreased, while the high molecular weight colloidal fraction increased (Figure 3). This behavior suggests that Hg had rapidly associated with an organic phase[29-33], though with time it became less extractable by XAD or charcoal (for instance, it may have been methylated[34], or the organic matter to which it initially adsorbed may have altered with time).

Very little work exists on Sn speciation in the marine environment. Methylation and the involvement of Sn in other biological cycles have been discussed, however[34,35], and a large percentage of dissolved Sn in seawater may by in the form of methylated species (Braman, reported in Zingaro[35]). In these 4 experiments Sn was slightly extractable by XAD (3-5%, Table 3). Although Hg and Sn thus show some organic associations, their speciation and removal rates are dominated by their particle-reactivity in this shallow estuarine environment with high resuspended particle fluxes through the water column.

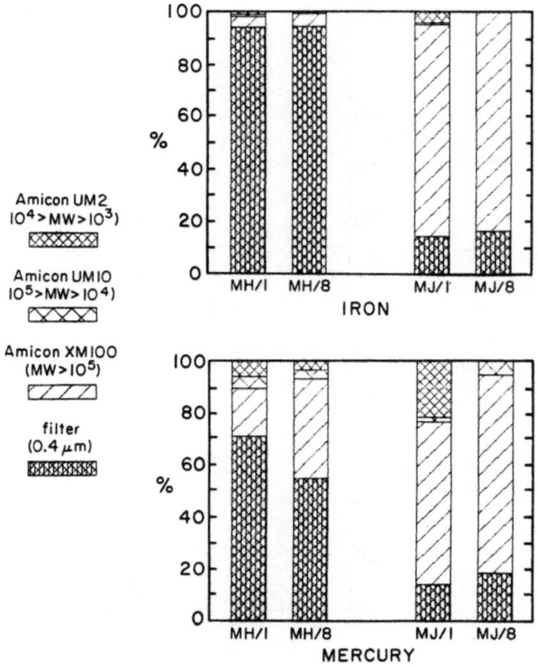

Fig. 3 Colloidal fractions of Fe and Hg determined by ultra-filtration.

Ionic and Effluent

As discussed above, Group I elements undergo an immediate speciation change when added to the microcosms. Spiked as cations in acid solution, they should be extracted by chelex, but even on the first day of the experiments they are essentially unextractable (Figures 1 and 2), indicating that these elements no longer exist as simple cations or chelex-extractable complexes. A variable percentage of Group I elements occurs in the effluent fraction (Figures 1 and 2). Except for Cr, this probably represents colloidal phases which were not adsorbed by a.c. or Amberlite IRA-938 resin.

The residence time of Cr in the water column increased substantially during the first 4-6 weeks of all experiments in which Cr was added as Cr(III). During this period a dramatic increase in the effluent fraction (Figure 1), and concomitant decreases in the percent of filterable (Figure 1) and ultra-filterable (Figure 4) Cr were observed. A simultaneous spike of Cr(III) ($t_{1/2}$ =3-5 days, initially) slowly approached that of chromate ($t_{1/2}$=26 days), and the chemical separation in the Cr(III) tank gradually

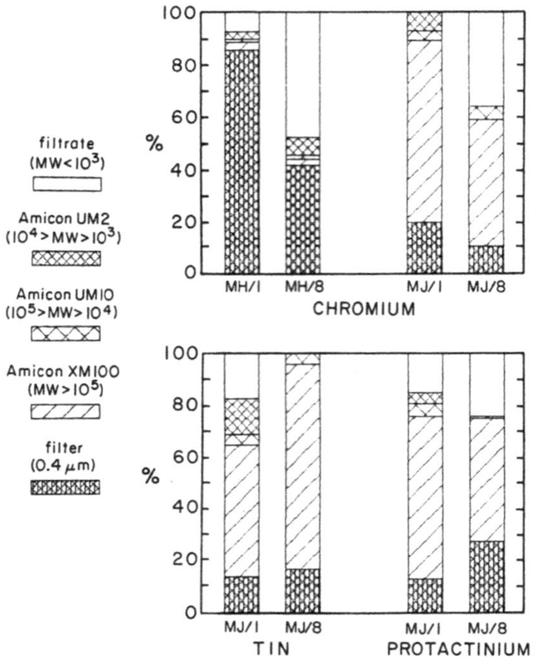

Fig. 4 Colloidal fractions of Cr, Sn and Pa determined by ultra-filtration.

shifted towardthat of Cr(VI) (>95% in the effluent fraction). This suggests that we were observing the oxidation of Cr(III) to chromate, the thermodynamically stable form of Cr in seawater[36]. To test this hypothesis, the method of Cranston and Murray[37] was used to separate these two species. Initially there was no measurable chromate in the Cr(III) tank; after one day, a measurable amount was detected, and after two weeks ~ 90% of the Cr remaining in the water column was in the form of chromate. Modelling the removal of Cr(III) and Cr(VI) with time and the changing ratio of these two species yielded an oxidation half-time of about 14 days (Amdurer, in preparation).

Our data indicate, therefore, that Group I elements, added to the water column as cationic species, rapidly hydrolyze to a colloidal form and can be flocculated onto suspended particles. In these experiments the degree to which these colloids are adsorbed onto particles is related to a combination of a) the kinetics of flocculation compared to particle removal rates; b) the amount of particle surface area available for adsorption. Destabilization and flocculation of colloids has been shown to be an important removal mechanism of trace elements in estuaries[38]. Because of their predominantly colloidal nature, Group I elements should be most affected by these processes.

GROUP II

Manganese and cobalt form the second group of "recycable" elements. Despite fairly rapid removal rates (similar to Group I elements), their removal mechanisms must differ since Mn and Co are far less "particle-reactive". Only very small percentages of Co and Mn (<3%, Figure 5) were associated with particles in the water column; equally small percentages occurred in colloidal fractions. As these elements are neither colloidal nor XAD-extractable (Table 3), there is no evidence that they rapidly become organically-associated. The 10-20% of Co that is found in the charcoal fraction can be extracted by chelex if it precedes the a.c. column, suggesting that the a.c. fraction of Co represents the adsorbed inorganic species listed in Table 5. Thus, on the time scale of our experiments, these elements remain predominantly as chelex-extractable dissolved inorganic species (Figure 5).

There is much evidence, however, that Mn is rapidly cycled between sediments and the overlying water column[39-42]. Manganese and Co are probably removed from the water column by Mn oxyhydroxide precipitation (and Co coprecipitation[43]), both in the water and at the sediment-water interface. Their behavior in the sediments is very sensitive to redox conditions. When mixed down into suboxic sediments (depending on the concentration of organic material in the sediments, temperature, bioturbation, etc., this may be just below the sediment surface), MnO_2 is reduced, releas-

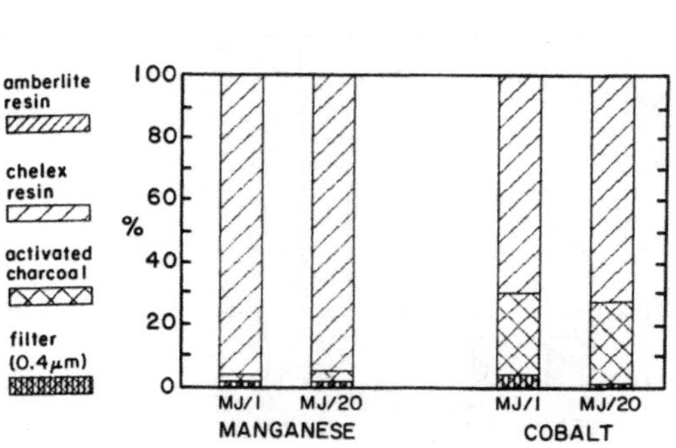

Fig. 5 Results of chemical separation procedures for Mn and Co. Vertical scale = percent of each fraction in the water column. Tank and day of experiment on horizontal axis (e.g. MJ/11 = tank MJ, day 11). "Amberlite resin" refers to Amberlite IRA-938 resin.

ing dissolved Mn^{+2} and Co^{+2} to the pore waters. Since the kinetics of Mn oxidation are relatively slow and pore water exchange in these highly-bioturbated sediments is rapid[44], some of the Mn and Co are "recycled" to the water column rather than being "trapped" in the oxidized sediment layer as they are in the ocean[45]. Thus, both Mn and Co were observed to rapidly diffuse back to the water column after the overlying water had been replaced with tracer-free seawater (our unpublished data).

Mn and Co are the only elements that have a substantial portion of their total inventory on the microcosm walls (up to 40%) at the end of an experiment. Mn-oxidizing bacteria attached to the walls may greatly enhance removal[46]. Although this wall adsorption may represent a strictly inorganic reaction (since the walls provide a large surface area bathed in oxidizing waters) separate adsorption experiments in the laboratory showed no such enhanced adsorption for Mn and Co compared to other elements.

GROUP III

Group III consists of elements whose chemical forms and/or cycling through the system may be more directly involved in biological processes: Se, As, V, Zn and Cd. In this shallow, particle-dominated ecosystem, Zn and Cd are less obviously involved in biotic cycling than in the deep ocean[2]. Cu and Ni, which have not yet been studied, would probably also fall in this group.

Se and As were ultra-filterable, but unlike Group I elements, they occurred predominantly in the l.m.w. fraction (Figure 6), demonstrating that they are not associated with the h.m.w. colloi-

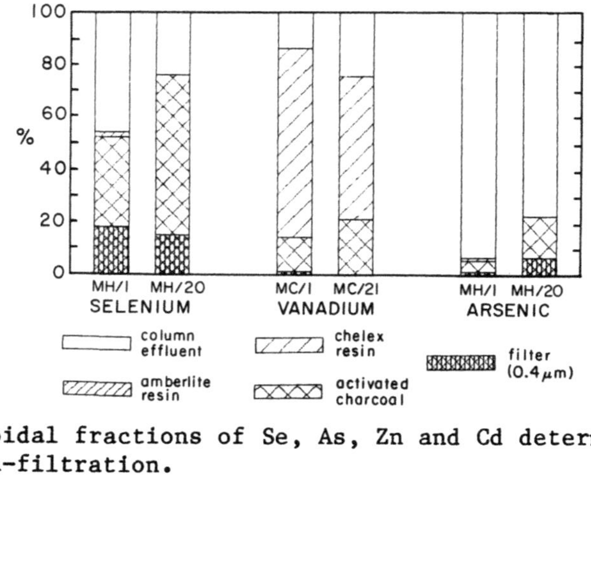

Fig. 6 Colloidal fractions of Se, As, Zn and Cd determined by ultra-filtration.

dal phase of particle-reactive elements. The charcoal fractions of Se, V and As increase with time (Figure 7). Several lines of evidence indicate that at least some part of the charcoal fraction of these elements represents organically-associated species. The large effluent fraction of these three elements contains a mixture of simple anions and organically-associated species not adsorbed by activated charcoal.

The percent of XAD-extractable Se increased with time (Table 3), as did both the charcoal (Figure 7; Table 4) and the l.m.w. colloidal fractions (Figure 6) at the expense of the effluent fraction, which would contain the anionic selenite species that was initially added. These observations suggest that in the water column Se is slowly converted from an inorganic to an organically-associated species. A biological role in the speciation and removal of Se concurs with the observation that the oceanic distribution of Se is dominated by biologiical processes[47], and its importance as a micronutrient for many organisms[35,48,49].

Arsenic was not XAD-extractable and a relatively small fraction was adsorbed by the charcoal column. But a significant proportion of the As, which increased with time, appeared in the l.m.w. colloidal fraction (Figure 6). Arsenic can be complexed by l.m.w. organic matter[50], or be alkylated by marine organisms[51-53]. Our data suggest that As added to the tanks slowly enters a relatively polar (therefore not extractable by XAD or charcoal) "dissolved" organic phase.

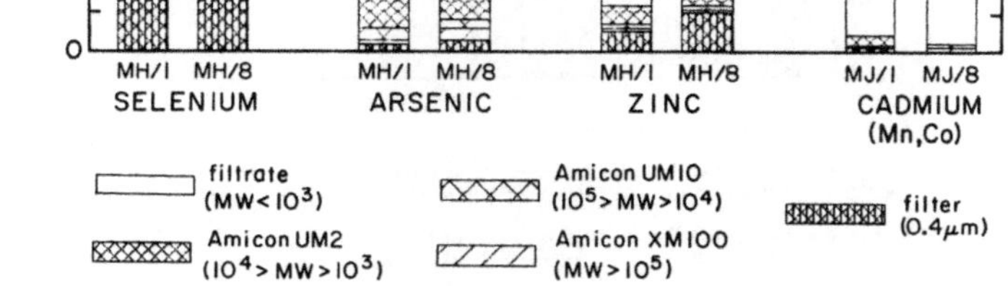

Fig. 7 Results of chemical separation procedures for Se, V and As. Vertical scale = percent of each fraction in the water column. Tank and day of experiment on horizontal axis (e.g. MJ/11 = tank MJ, day 11). "Amberlite resin" refers to Amberlite IRA-938 resin.

Vanadium was not added in the experiments which utilized XAD or ultra-filtration. However, this element is a micronutrient for many protozoa and is an essential part of the oxygen transport system in tunicates[48]; for this reason, and because of its chemical fractionation pattern, it is included in the Group III elements.

As mentioned above, Cd and Zn are included in this group of elements because of their oceanic transport phases[2], though organically-associated fractions were not isolated in these experiments. A small percentage of the Zn, which increased with time, occurred in the l.m.w. colloidal fraction (Figure 6). This may represent slow uptake of Zn into an organic phase; the amount is too small, however, to draw any definite conclusions. A significant fraction of the Zn was filterable in tank MH but not in MJ (Figure 8). In previous experiments (except in one tank which had a large phytoplankton bloom, as did tank MH), Zn was always less than 6% particulate, independent of the percent of Group I elements that was particulate. These observations, combined with ultra-filtration data, suggest that Zn was not associated with the Group I particulate and colloidal phases; instead Zn may associate with particulate organic matter produced in situ. Further study is required, however, since Zn may also be scavenged by Mn oxyhydroxides[43], thus associating it with Group II elements. Zn may also adsorb to siliceous (biogenous or inorganic) particles, or precipitate directly from seawater as $ZnSiO_4$[54].

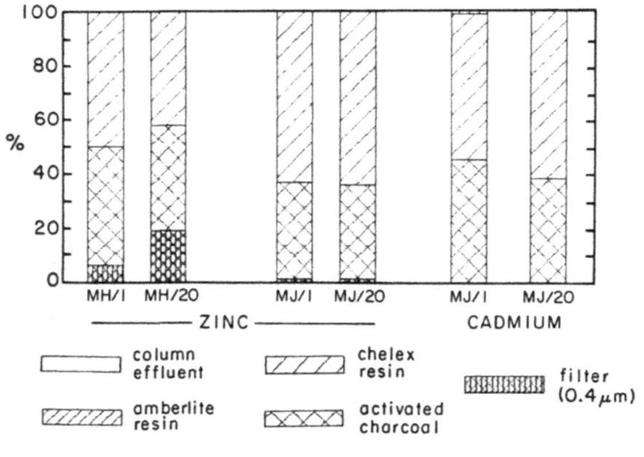

Fig. 8 Results of chemical separation procedures for Zn and Cd. Vertical scale = percent of each fraction in the water column. Tank and day of experiment on horizontal axis (e.g. MJ/11 = tank MJ, day 11). "Amberlite resin" refers to Amberlite IRA-938 resin.

Although some authors have concluded that a large fraction of soluble Zn in seawater is organically-bound, there is considerable disagreement on this topic [55]. The apparently long equilibration time of Zn with organic phases [11], however, would preclude the observation of these forms in our experiments. Both Zn and Cd have large a.c. fractions (Figure 8) which are chelex-extractable when chelex precedes the a.c. It is possible that some of the a.c. fraction of these elements represents organically-complexed species; the immediate attainment of a constant distribution among the fractions in the separation scheme and the absence of XAD-extractable or colloidal fractions, however, mitigate against this conclusion. The inorganic species probably represented by the a.c. fraction are listed in Table 4. Cd and Zn remain, therefore, primarily as inorganic species on the time scale of our experiments.

The half-removal times of "biologically-active" elements are longer than Groups I and II (Table 1). Removal from the water column may take place by incorporation into living or dead organic material which is scavenged by filter-feeding organisms or flocculated by settling particles. Direct adsorption to suspended particles without intermediary organic complexation will also occur. Uptake by phytoplankton during blooms and settling with dead cells may be an effective removal mechanism, but transfer to the sediment in zooplankton fecal pellets has been shown to be unimportant [56] (Santschi et al., this volume).

An adjunct study during these spike experiments investigated enrichment of the radiotracers at the air-sea interface (Amdurer, in preparation). Of all the trace elements added, only Se and As were found to be slightly enriched in the soluble (<0.4 μm) phase at the interface. However, the specific activity of all of the Group III elements and Hg, but not other tracers, was significantly higher in interfacial particles than in particles from the water column below. This was interpreted as an indication that these elements were associated with hydrophobic, dissolved and particulate organic phases which migrate to the interface.

GROUP IV

The fourth group consists of the "quasi-conservative" elements Cs, Ba, Ra and Sb, which remain in solution in ionic form and are removed to the sediments extremely slowly (Table 1). Although biological processes may be important for the removal of some of these elements in the ocean, their time scale is probably too long to observe in the microcosms (or in estuarine environments). An exception to this may be the small particulate fraction of Ba in tank ME (Figure 9), which is most likely related to a large diatom bloom in the tank at that time.

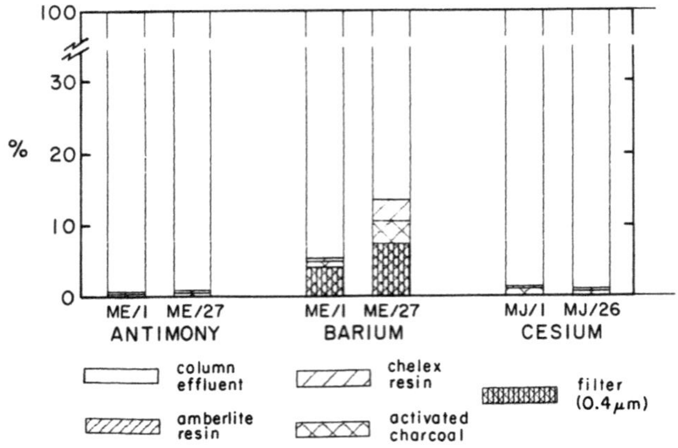

Fig. 9 Results of chemical separation procedures for Sb, Ba and Cs. Vertical scale = percent of each fraction in the water column. Tank and day of experiment on horizontal axis (e.g. MJ/11 = tank MJ, day 11). "Amberlite resin" refers to Amberlite IRA-938 resin.

In general, Group IV elements adsorb very little to suspended particles and they are neither chelex- nor charcoal-extractable; they are thus found in the effluent fraction of our separation system (Figure 9 - note change in scale). These elements are probably slowly incorporated into the sediments by direct adsorption and ion-exchange. Their removal rates thus depend on the rate at which they are exposed to sorption/exchange sites in sediments, i.e. by the rates of pore water irrigation and sediment turnover by benthic macro-fauna.

COMPARISON TO CALCULATED SPECIATIONS

Results of our separation system can be compared to the speciations of dissolved trace metals in seawater calculated by equilibrium modelling. The overwhelming particle-reactivity of Group I metals and the precision of our fractionation methods make a study of the truly dissolved phases of these elements impractical. Table 5 shows that the probable species of Mn, Co, Cd and Zn separated by our system fall within the range of values predicted by equilibrium models. This table demonstrates the wide range in calculated speciations, clearly indicating the problems with this type of approach. As shown above, in our microcosm experiments these elements occur predominantly as soluble inorganic phases. The lack of agreement in the models, though distrubing from a theoretical point of view, would not affect our prediction of the environmental fate of these elements.

Table 5 Comparison of Chemical Separation with Equilibrium Predictions

Element	Species in seawater (%)	Predicted fraction	Total predicted in each fraction	% found in each fraction
Mn [57]	Mn^{2+}(58)	chelex		
	$MnCl^+$ (30)	chelex	90	95-98
	$MnHCO_3^+$ (2)	chelex		
	$MnCl_2$ (2)	charcoal-chelex	9	2-5
	$MnSO_4$ (7)	charcoal-chelex		
Co [58]	Co^{2+} (54)	chelex	85	80-90
	$CoCl^+$ (31)	chelex		
	$CoCO_3$ (7)	charcoal-chelex	14	10-20
	$CoSO_4$ (7)	charcoal-chelex		
Cd [58-60]	Cd^{2+} (0-2)	chelex		40-60
	$CdCl^+$ (29-57)	chelex	29-57	
	$CdCl_2$ (15-50)	charcoal-chelex	15-50	40-60
Zn [5,6,60]	Zn^{2+} (22-40)	chelex		50-65
	$ZnCl^+$ (16-22)	chelex	42-66	
	$ZnCl_2$ (0-7)	charcoal-chelex		
	$Zn(OH)_2$ (<1-50)	charcoal-chelex	30-59	35-50
	$ZnCO_3$ (4-6)	charcoal-chelex		
	$ZnSO_4$ (2-16)	charcoal-chelex		
	ZnOHCl (0-12)	charcoal-chelex		

SUMMARY

Using a suite of radiotracers added to a simulated estuarine ecosystem as analogs, we are studying the chemical behavior and transfer mechanisms of pollutant trace metals in the coastal environment. This paper presents a brief review of the separation techniques that have been employed, and the relationship between the environmental behavior of various trace elements and the phases in which they occur. The power of radiotracer techniques is illustrated by our simultaneous investigation of a large number of trace elements -- this approach has enabled us to divide them into groups of similarly-behaving elements.

ACKNOWLEDGEMENTS

We are grateful for stimulating discussions with Y.-H. Li and W.S. Broecker, and to S. Griffiths, B. Evans, K. Kapuchynski and

Glenn Adler for help with sampling and analysis. Also, we wish to thank the staff of MERL for its cooperation.

The work was carried out under contracts from the Environmental Protection Agency (806072020). This is Lamont-Doherty Geological Observatory contribution number 3249.

REFERENCES

1. Boyle, E.A., F. Sclater and J.M. Edmond, 1977: The distribution of dissolved copper in the Pacific. Earth Planet. Sci. Lett., 37, 38-54.
2. Bruland, K., 1980: Oceanographic distributions of cadmium, zinc, nickel and copper in the North Pacific. Earth Planet. Sci. Lett., 47, 176-198.
3. Morel, F.M.M., J.C. Westall, C.R. O'Melia and J.J. Morgan, 1975: Fate of trace metals in Los Angeles County wastewater discharge. Envir. Sci. Technol., 9, 756-761.
4. Galloway, J.N., 1979: Alteration of trace metal geochemical cycles due to the marine discharge of wastewaters. Geochim. Cosmochim. Acta., 43, 207-218.
5. Long, D.T. and E.E. Angino, 1977: Chemical speciation of Cd, Cu, Pb and Zn in mixed freshwater, seawater and brine solutions. Geochim. Cosmochim. Acta., 41, 1183-1191.
6. Dyrssen, D. and M. Wedborg, 1980: Major and minor elements, chemical speciation in estuarine waters. In: "Chemistry and Biogeochemistry of Estuaries", E. Olausson and I. Cato, eds. John Wiley & Sons, New York, 71-119.
7. Florence, T.M. and G.E. Batley, 1976: Trace metal species in seawater. Talanta, 23, 179-186.
8. Fukai, R. and L. Huynh-Ngoc, 1975: Chemical forms of Zn in seawater. J. Oceanogr. Soc. Japan, 31, 1-13.
9. Cranston, R.E. and J.W. Murray, 1980: Chromium species in the Columbia River and estuary. Limnol. and Oceanog., 25, 1104-1112.
10. Hart, B.T. and S.R.H. Davies, 1981: Trace metal speciation in the freshwater and estuarine regions of the Yarra River, Victoria. Estuar. Coastal Shelf Sci., 12, 353-374.
11. Young, D.R., T-K. Jan and G.P. Hershelman, 1980: Cycling of zinc in the nearshore marine environment. In: "Zinc in the Environment", J.O. Nriagu, ed. Wiley-Interscience, New York., 298-335.
12. Jennings, C.D., 1978: Selective uptake of ^{55}Fe from seawater by zooplankton. Mar. Sci. Comm., 4, 49-58.
13. Pilson, M.E.Q., C.A. Oviatt and S.W. Nixon, 1981: Annual nutrient cycles in a marine microcosm. In: "Microcosms in Ecological Research", J.P. Giesy, ed. DOE Symposium Series, CONF-781101, N.T.I.S. Augusta, Ga., 753-778.

14. Elmgren, R., J.F. Grassle, J.P. Grassle, D.R. Heinle, G. Langlois, S.L. Vargo and G.A. Vargo, 1981: Trophic interactions in exerpimental marine ecosystems perturbed by oil. In: "Microcosms in Ecological Research", J.P. Giesy, ed. DOE Symposium Series, CONF-781101, N.T.I.S. Augusta, Ga., 779-800.
15. Santschi, P.H., S. Carson and Y-H. Li, 1981: Natural radionuclides as tracers for geochemical processes in MERL tanks and Narragansett Bay. In: "Marine Mesocosms: Biological and Chemical Research in Experimental Ecosystems", G. Grice, M. Reeve, eds. Springer-Verlag, 97-109.
16. Santschi, P.H., D. Adler, M. Amdurer, Y-H. Li and J. Bell 1980a: Thorium isotopes as analogues for "particle-reactive" pollutants in coastal marine environments. Earth Planet. Sci. Lett., 47, 327-335.
17. Nixon, S.W., D. Alonso, M.E.Q. Pilson and B.A. Buckley, 1981: Turbulent mixing in aquatic microcosms. In: "Microcosms in Ecological Research", J.P. Giesy, ed. DOE Symposium Series, CONF-781101, N.T.I.S. Augusta, Ga., 818-849.
18. Vaccaro, R.F., 1971: Estimation of adsorbable solutes in seawater with ^{14}C-labeled phenol and activated carbon. Environ. Sci. Tech., 5, 134-138.
19. Kerr, R.A. and J.G. Quinn, 1975: Chemical studies on the dissolved organic matter in seawater. Isolation and fractionation. Deep-Sea Res., 22, 107-116.
20. ----- and -----, 1980: Chemical comparison of dissolved organic matter isolated from different oceanic environments. Mar. Chem., 8, 217-229.
21. Mantoura, R.F.C. and J.P. Riley, 1975: The analytical concentration of humic substances from natural waters. Anal. Chim. Acta., 76, 97-106.
22. Stuermer, D.H. and G.R. Harvey, 1977: The isolation of humic substances and alcohol-soluble organic matter from seawater. Deep-Sea Res., 24, 303-309.
23. Huang, C.G., 1978: Chemical interactions between inorganics and activated charcoal. In: "Carbon Adsorption Handbook", P.N. Cheremisinoff and F. Ellerbusch, eds. Ann Arbor Science, Ann Arbor, MI., 281-329.
24. Bruland, K.W., R.P. Franks, G.A. Knauer and J.H. Martin, 1979: Sampling and analytical methods for the determination of copper, cadmium, zinc and nickel at the nanograms per liter level in seawater. Anal. Chim. Acta., 105, 233-245.
25. Stumm, W. and P.A. Brauner, 1975: Chemical speciation. In: "Chemical Oceanography", J.P. Riley and G. Skirrow, eds. Academic Press, New York, 173-239.
26. Brewer, P.G., 1975: Trace metal speciation. In: "Chemical Oceanography", J.P. Riley and G. Skirrow, eds. Academic Press, New York, 415-496.

27. Brewer, P.G., Y. Nozaki, D.W. Spencer and A.P. Fleer, 1980: Sediment trap experiments in the deep North Atlantic: isotopic and elemental fluxes. J. Mar. Res., 38, 703-728.
28. Santschi, P.H., Y-H. Li and S. Carson, 1980: The fate of trace metals in Narragansett Bay, Rhode Island: Radiotracer experiments in microcosms. Estuar. Coastal Mar. Sci., 10, 635-654.
29. Windom, H.L. and F.E. Taylor, 1979: The flux of mercury in the South Atlantic Bight. Deep-Sea Res., 26, 283-292.
30. Fitzgerald, W.F., 1976: Mercury studies of seawater and rain: geochemical flux implications. In: "Marine Pollutant Transfer", H.L. Windom and R.A. Duce, eds. Heath, New York, 121-134.
31. Lindberg, S.E., A.W. Andren and R.C. Harriss, 1975: Geochemistry of mercury in the estuarine environment. In: "Estuarine Research", E.A. Cronin, ed. Academic Press, New York, 64-107.
32. Crecelius, E.A., M.H. Bothner and R. Carpenter, 1975: Geochemistries of arsenic, antimony, mercury and related elements in the sediments of Puget Sound. Environ. Sci. Technol., 9, 325-333.
33. Loring, D.H. and J.M. Bewers, 1978: Geochemical mass balances for mercury in a Canadian Fjord. Chem. Geol., 22, 309-330.
34. Jernelov, A., 1975: Microbial Alkylation of metals. In: "International Conference on Heavy Metals in the Environment", T.C. Hutchinson, ed., 845-859.
35. Zingaro, R.A., 1979: How certain trace elements behave. Environ. Sci. Technol., 13, 282-287.
36. Elderfield, H., 1970: Chromium speciation in seawater. Earth Planet. Sci. Lett., 9, 10-16.
37. Cranston, R.E. and J.W. Murray, 1978: The determination of chromium species in natural waters. Anal. Chim. Acta., 99, 275-282.
38. Sholkovitz, E.R., 1978: The flocculation of dissolved Fe, Mn, Cu, Ni, Co and Cd during estuarine mixing. Earth Planet. Sci. Lett., 41, 77-86.
39. Graham, W.F., M.L. Bender and G.P. Klinkhammer, 1976: Manganese in Narragansett Bay. Limnol. Oceanog., 21, 665-673.
40. Evans, D.W., N.H. Cutshall, F.A. Cross and D.A. Wolfe, 1977: Manganese cycling in the Newport River Estuary, North Carolina. Estuar. Coastal Mar. Sci., 5, 71-80.
41. Grill, E.V., 1978: The effect of sediment-water exchange on manganese deposition and nodule growth in Jervis Inlet, British Columbia. Geochim Cosmochim. Acta., 42, 485-494.
42. Yeats, P.A., B. Sundby and J.M. Bewers, 1979: Manganese recycling in coastal waters. Mar. Chem., 8, 43-55.

43. Murray, J.W., 1975: The interaction of cobalt with hydrous manganese dioxide. Geochim. Cosmochim. Acta., 39, 635-647.
44. McCaffrey, R.J., A.C. Myers, M. Bender, N. Luedtke, D. Cullen, P. Froelich and G. Klinkhammer, 1981: Benthic fluxes of nutrients and manganese in Narragansett Bay, Rhode Island. Limnol. Oceanog., 25, 31-44.
45. Landing, W.M. and K.W. Bruland, 1980: Manganese in the North Pacific. Earth Planet. Sci. Lett., 49, 45-56.
46. Nealson, K.H. and J. Ford, 1979: Surface enhancement of bacterial manganese oxidation: implications for aquatic environments. Geomicrobiol. J., 2, 21-37.
47. Measures, C.I. and J.D. Burton, 1980: The vertical distribution and oxidation states of dissolved selenium in the northeast Atlantic ocean and their relationship to biological processes. Earth Planet. Sci. Lett., 46, 385-396.
48. Bowen, H.J.M., 1966: Trace Elements in Biochemistry. Academic Press, N.Y., 241 pp.
49. Lakin, H.W., 1973: Selenium in our environment. In: "Trace Elements in the Environment", E.L. Kothny, ed. Am. Chem. Soc., Washington, D.C., 96-111.
50. Waslenchuk, D.G. and H.L. Windom, 1978: Factors controlling the estuarine chemistry of arsenic. Estuar. Coastal Mar. Sci., 7, 455-464.
51. Andreae, M.O., 1979: Arsenic speciation in seawater and interstitial waters: the influence of biological-chemical interactions on the chemistry of a trace element. Limnol. Oceanogr., 24, 440-452.
52. Sanders, J.G. and H.L. Windom, 1980: The uptake and reduction of arsenic species by marine algae. Estuar. Coastal Mar. Sci., 10, 555-568.
53. Johnson, D.L. and R.S. Braman, 1975: The speciation of arsenic and the content of germanium and mercury in members of the pelagic sargassum community. Deep-Sea Res., 22, 503-507.
54. Willey, J.D., 1977: Coprecipitation of zinc with silica in seawater and distilled water. Mar. Chem., 5, 267-290.
55. Florence, T.D., 1980: Speciation of zinc in natural waters. In: "Zinc in the Environment", J.O. Nriagu, ed. Wiley-Interscience, New York, 199-227.
56. Adler, D.M., 1981: Tracer studies in marine microcosms: transport processes near the sediment-water interface. Ph.D. Thesis, Columbia Univ., New York, 346 pp.
57. Crerar, D.A. and H.L. Barnes, 1974: Deposition of deep-sea manganese nodules. Geochim. Cosmochim. Acta., 38, 279-300.
58. Ahrland, S., 1975: Metal complexes present in seawater. In: "The Nature of Seawater", E.D. Goldberg, ed. Dahlem Konferenzen, Berlin, 219-244.

59. Zirino, A. and S. Yamamoto, 1972: A pH-dependent model for the chemical speciation of copper, zinc, cadmium and lead in seawater. Limnol. Oceanog., 17, 661-671.

60. Lu, J.C.S. and K.Y. Chen, 1977: Migration of trace metals in interfaces of seawater and polluted surficial sediments. Environ. Sci. Technol., 11, 174-182.

TRACE METALS (FE, CU, ZN, CD) IN ANOXIC ENVIRONMENTS

Jacques Boulègue

Laboratoire de Géologie Appliquée
Université Pierre et Marie Curie
4, place Jussieu 75230 Paris Cedex 05
France

ABSTRACT

Anoxic environments can be characterized by the appearance of several reduced sulfur species: hydrogen sulfide, polysulfide ions, organic sulfides and thiosulfate. We have studied the possible complexations of Fe, Cu, Zn and Cd ions by polysulfide and thiosulfate ions. Laboratory results show that Cu(I) is strongly complexed by polysulfide (S_4^{2-} and S_5^{2-}) and thiosulfate and that Cu(II) and Cd are complexed by thiosulfate.

The combination of the above data with previous data on complexation with hydrogen sulfide enables us to describe the possible speciations of the above metals in the following systems: H_2S-H_2O, H_2S-S_8-H_2O and H_2S-O_2-H_2O.

These results are employed to explain the concentrations of Fe, Cu, Zn and Cd observed in several anoxic environments (Pavin Lake, Black Sea, Delaware Estuary). A good agreement is obtained between thermodynamic calculations and field measurements.

Recent studies of the sulfur system in reducing conditions in water have been done in the laboratory[1-7] and in the field[8-12]. They have emphasized the importance of the formation of polysulfide ions (HS_i^-; S_i^{2-}; $i > 1$). Polysulfide ions are rapidly produced upon reaction of hydrogen sulfide with elemental sulfur or incomplete oxidation of hydrogen sulfide. Polysulfides, elemental sulfur and thiosulfate are the main intermediary products of the incomplete and complete oxidations of hydrogen sulfide as observed in the laboratory[6,7,13,14]. In reducing environments, one may expect that the variations of concentration of hydrogen sulfide due to environmental changes will bring about changes in the control of metal concentrations by sulfide formation and changes due to possible complexation by sulfide, polysulfides and thiosulfate.

In this paper, we report some results on the speciation of trace metals in anoxic environments and their changes during sulfate reduction.

DESCRIPTION OF REDUCING ENVIRONMENTS

In this paper, we will give a brief survey of some possible processes controlling the solubility of some trace metals (Fe, Cu, Zn, Cd) in different reducing environments with emphasis on marine processes. In several cases, we have chosen the H_2S - Sea Water (S.W.) and the H_2S -S_8 - S.W. systems as representative of sulfur chemistry in reducing environments. This is in agreement with the results of field studies[8-12]. The H_2S - S.W. system is representative of pore waters where bacterial sulfate reduction occurs without secondary oxidation of hydrogen sulfide. The H_2S-S_8-S.W. system corresponds to pore waters where bacterial sulfate reduction and oxidation of hydrogen sulfide occur at the same time. This is typical of redox boundaries in open waters or in recent sediments. The possible changes in the sulfur system and the environmental parameters between these two environments have been described by Boulègue et al.[11,12].

To represent the changes in environmental parameters during bacterial sulfate reduction, we have chosen two models as developed by Gardner[15]. The first one corresponds to quartz sediment, i.e. an unreactive environment, and the second corresponds to carbonate sediments (Figure 1).

The results of these models are used for the description of the changes in the concentrations of dissolved trace elements during bacterial sulfate reduction in the model environments. The initial concentrations of trace metals which have been considered are taken from the work of Lu and Chen[16]. They are (expressed in mol kg^{-1}):

$$5x10^{-9} < [Fe] < 10^{-6} ; 5x10^{-9} < [Cu] < 5x10^{-8} ;$$
$$7x10^{-9} < [Zn] < 1,5x10^{-7} \text{ and } 2x10^{-10} < [Cd] < 10^{-8}.$$

One change was made in Gardner's model to take account of results on the control of the redox potential in reducing environments[5]. In the presence of the H_2S-S_8-S.W. system, the redox potentials were taken as those found at equilibrium[5] rather than as that of

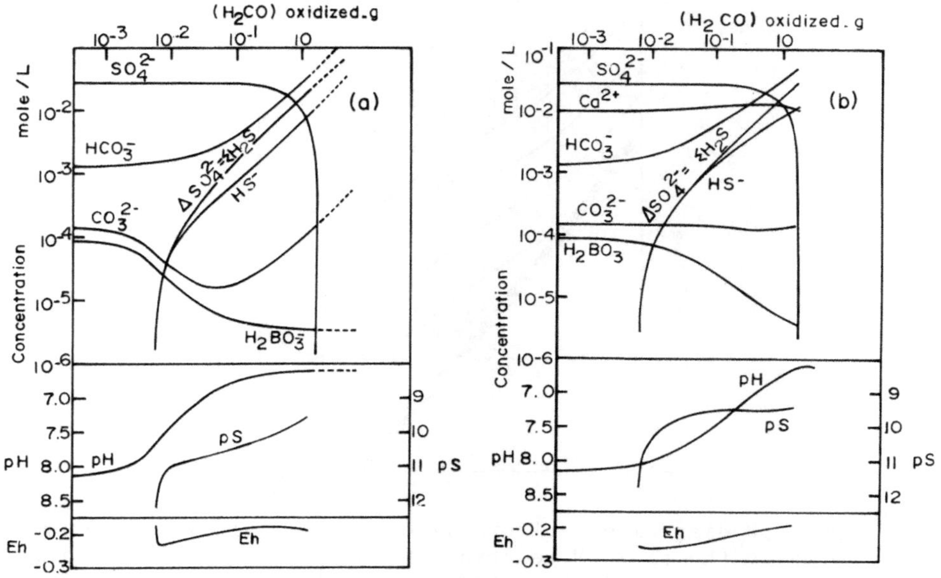

Fig. 1 Computer model of the changes in some properties of sea water during bacterial sulfate reduction. The organic matter which is consumed is assumed to be only carbohydrate $(CH_2O)_n$. Results after Gardner[15] (reprinted with permission of R.L. Gardner and Pergamon Press). (a) Pore water in a pure quartz sediment (non reactive environment). (b) Pore water in a pure carbonate sediment.

the HS^-/SO_4^{2-} redox couple which was used only for the H_2S-S.W. system. These differences are illustrated in Figure 2.

COMPUTATION OF THE SOLUBILITY OF TRACE METALS IN REDUCING ENVIRONMENTS.

The equilibrium of the metallic ion Me^{n+} with its sulfide $MeS_{n/2}$ corresponds to:

$$MeS_{n/2} + (n/2)\ H^+ \rightleftarrows Me^{n+} + (n/2)\ HS^-$$

with the corresponding equilibrium constant

$$K_s^* = (Me^{n+}) \cdot (HS^-)^{n/2} / (H^+)^{n/2} \qquad (1)$$

where the brackets correspond to activity.

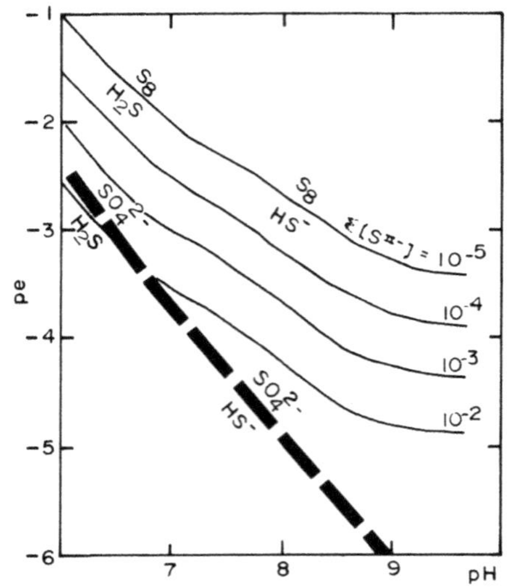

Fig. 2 pe - pH diagram of sulfur systems. The full lines correspond to the pe - pH relation in the H_2S-$S_8(\alpha)$-S.W. system with constant $\Sigma[S^{II-}] = 10^{-x}$ mol kg^{-1}. The large dotted line corresponds to the HS^-/SO_4^{2-} redox couple. From data by Boulègue[2] and Boulègue and Michard[4,5].

In the H_2S-S.W. System, (HS^-) can be expressed[5] as a function of (H^+) and the total concentration of hydrogen sulfide $\Sigma[H_2S]$, and one can rewrite (1):

$$(Me^{n+}) = K_s^* \cdot \{(H^+)/\Sigma[H_2S]\}^{n/2} \cdot \{(1/\gamma HS^-)+(H^+)/(K_{A1}\cdot\gamma H_2S) + K_{A2}/((H^+)\cdot\gamma S^{2-})\}^{n/2} \qquad (2)$$

where γ is the activity coefficient and K_{A1} and K_{A2} are respectively the first and second ionization constant of H_2S.

In the system H_2S-S_8-S.W., (HS^-) can be conveniently expressed as a function of measurable parameters[5]:

(H^+), $\Sigma[S^{II-}] = \Sigma[H_2S] + \Sigma[H_2S_i]$ and

$\chi_S = \Sigma(i - 1)\ \Sigma[H_2S_i]/\Sigma[S^{II-}]$

and one can compute:

$$(Me^{n+}) = K_s^* \cdot \{(H^+)/((1+\chi_S).\Sigma[S^{II-}])\}^{n/2} \cdot (1/\gamma HS^-)+(H^+)/(K_{A1}.\gamma H_2S)+\sum_{i>1}(i.K_{H/i}^*)/(K_i^*.\gamma HS_i^-)+\sum_{i>1}(i.K_i^*)/((H^+).\gamma S_i^{2-})\}^{n/2} \quad (3)$$

where $K_{H/i}^* = (S_i^{2-}).(H^+) / (HS_i^-)$ and $K_i^* = (S_i^{2-}).(H^+)/(HS^-)$.

The calculation of the maximum concentration of Me in the presence of $MeS_{n/2}$ must take into account all possible complexes in the H_2S-S.W. and H_2S-S_8-S.W. systems.

The following complexes with hydrogen sulfide and polysulfide can be found[2,17,18]:

$Me(HS)_k^{n-k}$	with β_k	$= (Me(HS)_k^{n-k})/(Me^{n+}).(HS^-)^k$
$MeS_{n/2}(H_2S)_1^o$	with β_1	$= (MeS_{n/2}(H_2S)_1^o).(H^+)^{n/2}/(Me^{n+}).(HS^-)^{n/2}.(H_2S)^1$
$MeS_{n/2}(HS)_m^{m-}$	with β_m	$= (MeS_{n/2}(HS)_m^{m-}).(H^+)^{n/2}/(Me^{n+}).(HS^-)^{m+n/2}$
$MeS_iS_j^{n-4}$	with β_{ij}	$= (MeS_iS_j^{n-4})/(Me^{n+}).(S_i^{2-}).(S_j^{2-})$
$Me(S_i)_2^{n-4}$	with β_{ii}	$= (Me(S_i)_2^{n-4})/(Me^{n+}).(S_i^{2-})^2$
$Me(HS)S_i^{n-3}$	with β_{iH}	$= (Me(HS)S_i^{n-3})/(Me^{n+}).(HS^-).(S_i^{2-})$

We have checked the possible complexes of Cu(I) and Fe(II) with polysulfide ions by comparing the solubility of Cu_2S, CuS and FeS in H_2S-H_2O and H_2S-S_8-H_2O systems. Significant complexing of iron by sulfide or polysulfide was not detected.

Complexing of Cu(I) by polysulfide was found[2] with $\log\beta_{44}$ = 23 and $\log\beta_{45}$ = 22. These values are in good agreement with similar complexes determined by Cloke[19]. Otherwise, values for sulfide complexing of Cu(II), were taken respectively from Romberg and Barnes[20]. Log K for $Zn(OH)(HS)^o$ was taken to be 16.72, and for $Cd(OH)(HS)^o$ 19.6. $Cd(OH)(HS)^o$ was used rather than $Cd(OH)^+$ after consideration of the data of Gubeli and Sainte Marie[21] and Sainte Marie et al.[22] and check of the consistency of these data (F. Morel, personal communication , 1981).

Thus, the ratio between the saturation concentration of Me, [Me], and the activity (Me^{n+}) can be written:

$$
\begin{aligned}
[Me]/(Me^{n+}) = \{1/\gamma_{Me^{n+}}\} &+ \sum_k \{\beta_k \cdot (HS^-)^k\}/\gamma_{Me(HS)_k^{n-k}} \\
&+ \sum_l \{\beta_l \cdot (HS^-)^{n/2} \cdot (H_2S)^l\}/\{\gamma_{MeS_{n/2}(H_2S)_l^0} \cdot (H^+)^{n/2}\} \\
&+ \sum_m \{\beta_m \cdot (HS^-)^{m+n/2}\}/\{\gamma_{MeS_{n/2}(HS)_m^{m-}} \cdot (H^+)^{n/2}\} \\
&+ \sum_{ij} \{\beta_{ij} \cdot (S_i{}^{2-}) \cdot (S_j{}^{2-})\}/\gamma_{MeS_iS_j^{n-4}} + \\
&\quad \sum_i \{\beta_i \cdot (S_i{}^{2-})^2\}/\gamma_{Me(S_i)_2^{n-4}} \\
&+ \sum_i \{\beta_{iH} \cdot (S_i{}^{2-}) \cdot (HS^-)\}/\gamma_{Me(HS)S_i^{n-3}} + C \qquad (4)
\end{aligned}
$$

Where C is a term related to the complexes of Me^{n+} with ions other than those of the H_2S-S_8-S.W. system, i.e. F^-, Cl^-, OH^-, $SO_4{}^{2-}$, $SO_3{}^{2-}$, $S_2O_3{}^{2-}$, $NH_4{}^+$, $PO_4{}^{3-}$, $HPO_4{}^{2-}$, etc. These complexes can be computed with WATEQ2 program[23]. Thiosulfate and sulfide complexes were taken from Naumov et al.[24].

Equations (2) and (4) and equations (3) and (4) enable computation of the speciations of Me and [Me] in H_2S-S.W. and H_2S-S_8-S.W. systems respectively owing to a modification of WATEQ2 program[2,25]. Data for iron sulfides were taken from Berner[26] and data for CuS, Cu_2S, ZnS and CdS from Naumov et al.[24]. They have been discussed by the present author[2,8,9,12].

We present in the following the results of the computations for different trace metals. The results are given as log[Me]-pH diagrams for a range of $\Sigma[H_2S]$ and $\Sigma[S^{II-}]$ values found in natural reducing environments.

Iron

In the case of iron, we have considered the formation of three possible sulfides: amorphous FeS, greigite Fe_3S_4 and pyrite. Pyrite and greigite were considered to form only in the presence of elemental sulfur and/or polysulfide ions[27,28]. Thus, one can compute for greigite:

$$Fe_3S_4 + 3H^+ \rightleftarrows 3Fe^{2+} + 3\,HS^- + (1/8)S_8 \quad K_S^* = 10^{-12.9}$$

and for pyrite :

$$FeS_2 + H^+ \rightleftarrows Fe^{2+} + HS^- + (1/8)S_8 \qquad K_S^* = 10^{-7.49}$$

The results for iron solubility in the presence of these sulfides are presented in Figures 3 and 4. The initial concentrations of iron in the water submitted to reduction of sulfate are in the range $5 \times 10^{-9} - 10^{-6}$ mol kg^{-1}. One can note that under reducing conditions, it will be difficult to differentiate between mackinawite (FeS) and greigite in the control of iron solubility. Mackinawite will give an excess iron concentration of 10% compared to greigite. Such a small difference will be difficult to detect at low iron concentration.

Copper

In reducing systems the solubility of copper depends on three variables: pH, pe and $\Sigma[H_2S]$ or $\Sigma[S^{II-}]$. Given a set of those variables one can compute which of CuS or Cu_2S should form. However, we have also considered the case when only CuS was forming since experiments[29] have shown that CuS forms even in conditions where Cu_2S should have precipitated. In the H_2S-S_8-S.W. system, only CuS should form. We have also computed the ratio of the solubilities [Cu(I)]/[Cu(II)] (see Table 1). Knowing [Cu(I)] and [Cu(II)] in the presence of Cu_2S and CuS, respectively, one can calculate the total concentration [Cu] = [Cu(I)] + [Cu(II)].

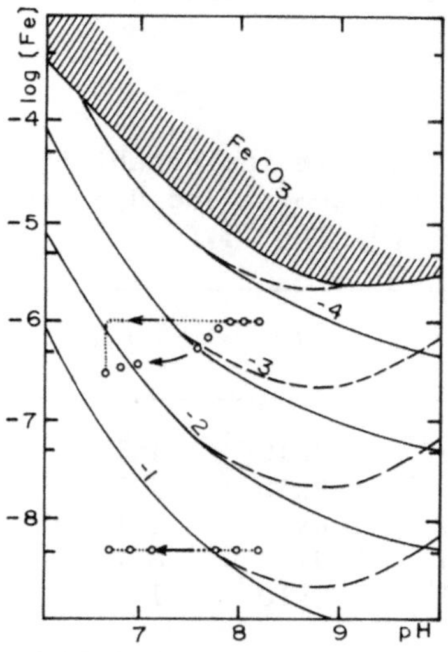

Fig. 3 Solubility diagram of Fe in the presence of amorphous FeS. Results given as log (total molality of iron) versus pH. Full lines: H_2S-S.W. system for $\log\Sigma[H_2S]$ = x. Dashed lines: H_2S-S_8-S.W. system for $\log\Sigma[S^{II-}]$ = x. Changes in iron concentration during bacterial sulfate reduction (see Fig. 1). Dotted line: quartz sediment. Open dots line: carbonate sediment.

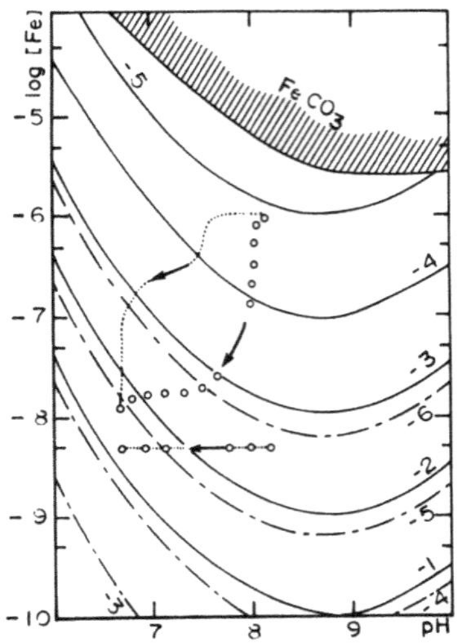

Fig. 4 Solubility diagram of Fe in the presence of Fe_3S_4 and FeS_2 in the H_2S-S_8-S.W. system with $\log\Sigma[S^{II-}] = x$. Results given as log (total molality of iron) versus pH. Full lines: Fe_3S_4. Dashed lines: FeS_2. Changes in iron concentration during bacterial sulfate reduction. Dotted lines: quartz sediment. Open dotted lines: carbonate sediment.

Table 1 Ratio of the equilibrium solubilities [Cu(I)]/[Cu(II)] in the systems H_2S-S.W. and H_2S-S_8-S.W. as a function of pH and $\Sigma[H_2S]$ OR $\Sigma[S^{II-}]$.

pH		6	7	8	9	10
	10^{-1}	1.0×10^{-8}	6.8×10^{-11}	1.7×10^{-11}	1.9×10^{-11}	2.5×10^{-11}
H_2S-S.W.	10^{-2}	2.1×10^{-5}	5.9×10^{-7}	1.7×10^{-7}	1.9×10^{-7}	2.5×10^{-7}
$\Sigma[H_2S]$=	10^{-3}	2.4×10^{-2}	2.6×10^{-3}	1.6×10^{-3}	1.9×10^{-3}	2.5×10^{-3}
(mol	10^{-4}	2.1×10^{1}	3.91	8.28	1.8×10^{1}	2.5×10^{1}
kg^{-1})	10^{-5}	2.4×10^{4}	4.1×10^{3}	1.4×10^{4}	9.5×10^{4}	2.3×10^{5}
	10^{-1}	1.5×10^{-2}	2.65	1.0×10^{2}	4.9×10^{3}	6.5×10^{4}
H_2S-S_8-SW	10^{-2}	1.0×10^{-2}	7.29	3.2×10^{2}	1.5×10^{4}	2.0×10^{5}
$\Sigma[S^{II-}]$=	10^{-3}	3.6×10^{-3}	1.0×10^{1}	9.1×10^{2}	4.9×10^{4}	6.4×10^{5}
(mol	10^{-4}	1.2×10^{-3}	4.81	1.4×10^{3}	1.2×10^{5}	1.8×10^{6}
kg^{-1})	10^{-5}	3.7×10^{-4}	1.60	7.6×10^{2}	1.4×10^{5}	2.5×10^{6}

The results of the solubility calculations are given in Figures 5 and 6. The initial concentrations of Cu considered in the beginning of the reduction of sulfate are in the range 5×10^{-9} - 5×10^{-8} mol kg $^{-1}$.

One can question the applicability of the above results to anoxic environments. Several non-stoichiometric copper sulfides are found and thermodynamic data on them are lacking. Also association of copper and iron are frequent as in chalcopyrite $CuFeS_2$. From the results of Cabri[30], it seems that both iron sulfide (FeS) and copper sulfide (CuS) precipitate as a solid solution which then evolves according to environmental conditions. In this case only a correction for the activity coefficient of CuS in the solid solution should be made. Thus the overall results of the above calculations are valid and may be used in environmental conditions to check copper geochemistry.

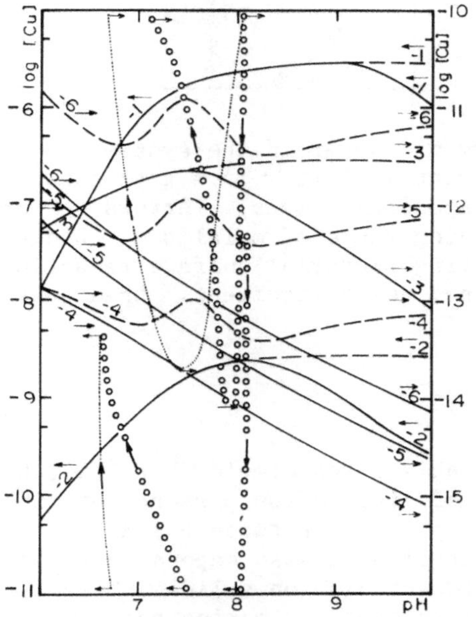

Fig. 5 Solubility of copper in the system H_2S-S.W. versus pH for different log ($\Sigma[H_2S]$) = x. (x is shown as parameter on the corresponding curves). The full lines correspond to the solubility at thermodynamic equilibrium. The dashed lines correspond to the solubility in the presence of CuS. Arrows (→ or ←) indicate the corresponding scale of molality. Changes in solubility of copper during bacterial sulfate reduction (see Figure 1): Dotted line: quartz sediment. Open dots: carbonate sediment.

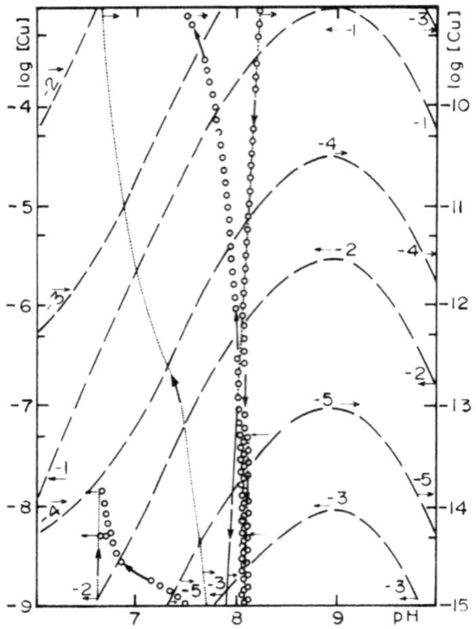

Fig. 6 Solubility of copper in the system H_2S-S_8-S.W. versus pH for different $\log(\Sigma [S^{II-}]) = x$ (x shown as parameter on the corresponding curves). Arrows (→ or ←) indicate the corresponding scale of molality. Changes in solubility of copper during bacterial sulfate reduction (see Figure 1): Dotted line: quartz sediment. Open dots line: carbonate sediment.

Zinc

Zinc has a stable mixed complex[21] $Zn(HS)(OH)^{o}$. As in Figure 7, the solubility of zinc in the presence of zinc sulfide should be constant in the pH range between 6 and 10. It is possible that the solubility control will also depend on the solubility of $ZnSiO_3$ or on the adsorption of zinc on silica. These results are in agreement with observations on marine sediments made by Lu and Chen[16].

Cadmium

Cadmium exhibits a series of stable sulfide complexes $Cd\ (HS)_n^{(n-2)-}$, n = 1,4, and a stable mixed complex $Cd(OH)(HS)^{o}$. The influence of these complexes can be seen in Figure 8.

Interestingly, it appears that the solubility of cadmium should not change by a large factor in reducing environments.

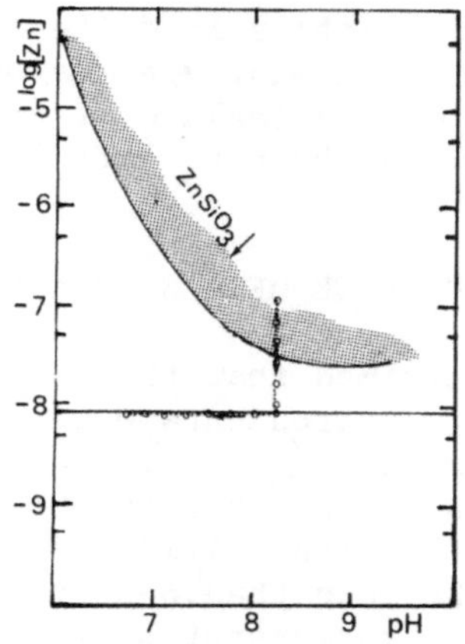

Fig. 7 Solubility of zinc in the presence of zinc sulfide in the H_2S-S_8-S.W. and the H_2S-S.W. systems results given as log (total molality of Zn). Changes in solubility of zinc during bacterial sulfate reduction: Dotted lines: quartz sediment. Open dots lines: carbonate sediment.

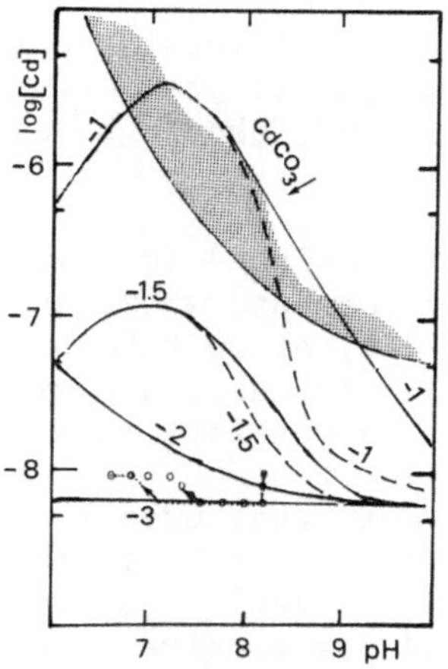

Fig. 8 Solubility of cadmium in the presence of CdS as a function of pH results given as log (total molality of Cd). Full line: System H_2S-S.W. for log ($\Sigma[H_2S]$) = x. Dashed line: System H_2S-S_8-S.W. for log ($\Sigma[S^{II-}]$) = x. Changes in solubility of cadmium during bacterial sulfate reduction: Dotted line: quartz sediment. Open dots line: carbonate sediment. The solubility curve of cadmium carbonate is also shown.

Thus, cadmium can be quite mobile in different types of sediments. In waste disposal areas or in mining areas where reducing conditions are promoted to stop the leakage of trace metals, one may expect that cadmium will not be very efficiently retained.

GEOCHEMICAL BEHAVIOUR OF TRACE METALS IN ANOXIC ENVIRONMENTS.

It has often been claimed that the concentrations of trace metals found in reducing environments were in large supersaturation versus sulfide formation[16,31-33]. One may note that in all these cases FeS was present and that the trace metals (Cu, Cd, Zn) were associated within the iron sulfide. In this case possible supersaturation cannot be more than one order of magnitude[34]. Generally, the computed supersaturation was due to incomplete information on the physico-chemical parameters of the environment, incomplete titration of sulfur speciations and employment of incorrect thermodynamic data. In several careful analyses of reducing environments it has been shown that equilibrium between metal and its sulfide was generally obtained for Cu, Fe, and Cd[2,8,9,11,12,35]. In another instance our calculations showed that in the deep zone of the Black Sea, with $\Sigma[H_2S] \sim 2\text{-}4 \times 10^{-4}$ mol kg^{-1} and pH ~ 7.7, one should expect [Fe] in the range 5×10^{-8} - 2×10^{-7} mol kg^{-1}, at equilibrium with FeS, which compares well with data given by Spencer and Brewer[32].

Thus one should expect that metal sulfides will precipitate according to simple thermodynamic predictions. Solid solution formation, as with FeS and CuS, should be expected in numerous cases.

Phosphate is strongly enriched in some organic rich environments. In lacustrine environments where sulfide production is often limited by availability of sulfate one can observe[36,37] vivianite, $Fe_3(PO_4)_2$. In this case, one will expect a separation of iron from metals which do not form insoluble phosphates.

It has often been argued that organic complexing could explain the computed supersaturation of trace metals versus sulfide precipitation. As was shown by Gardner[38], simple organic molecules cannot be called for to give such an explanation. However, some organo sulfur derivatives of humic and fulvic acids can be very strong binding agents for trace metals such as Cu and Fe as shown by Boulegue et al.[12]. Humic and fulvic acids could also be good complexing agents although in sea water, Na, Ca and Mg compete successfully for binding sites[39].

CONCLUSION

The calculations presented in this paper give an account of the state of knowledge about the speciations and the changes in the concentrations of some trace metals in marine reducing environments. The assessment of these speciations in field studies depends on the availability of accurate knowledge of sulfur chemistry in the samples. Combined studies of the chemistry and geochemistry of sulfur and trace metals are rare and much remains to be done in this field.

REFERENCES

1. Giggenbach, W.F., 1972: Optical spectra and equilibrium distribution of polysulfide ions in aqueous solutions at 20°C. Inorg. Chem., 11, 1201.
2. Boulègue, J., 1978: "Géochimie du soufre dans les milieux réducteurs", Thesis, University of Paris 7, Paris.
3. Boulègue, J., 1978: Electrochemisty of reduced sulfur species in natural waters. I - The H_2O-H_2S system. Geochim Cosmochim. Acta., 42, 1439.
4. Boulègue, J. and G. Michard, 1978: Constantes de formation des ions polysulfurés S_6^{2-}, S_5^{2-} et S_4^{2-} en phase aqueuse. Jour. Fr. Hydrologie, 9, 27.
5. Boulègue, J. and G. Michard, 1979: Sulfur speciations and redox processes in reducing environments. In: "Chemical Modeling in Aqueous Systems", E.A. Jenne, ed. A.C.S. Symp. Ser. 93, Washington, D.C.
6. Chen, K.Y. and J.C. Morris, 1972: Kinetics of oxidation of aqueous sulfides by O_2. Environ. Sci. Technol., 6, 529.
7. Gourmelon, C., J. Boulègue and G. Michard, 1977: Oxydation partielle de l'hydrogène sulfuré en phase aqueuse. C.R. Acad. Sc. Paris, 284(C), 269.
8. Boulègue, J., 1977: Equilibria in a sulfide rich water from Enghien-les-Bains (France). Geochim. Cosmochim. Acta., 41, 1751.
9. Boulègue, J., 1978: Metastable sulfur species and trace metals (Mn, Fe, Cu, Zn, Cd, Pb) in hot brines from the french Dogger. Amer. J. Sci., 278, 1394.
10. Boulègue, J., 1979: Formation des eaux thermales sulfurées des Pyrénées Orientales. Origine du soufre. Géochimie de fer et du cuivre. Jour. Fr. Hydrologie, 10, 91.
11. Boulègue, J., J.P. Ciabrini, C. Fouillac, G. Michard and G. Ouzounizn, 1979: Field titrations of dissolved sulfur species in anoxic environments. Geochemistry of Puzzichello waters (Corsica, France). Chem. Geol., 25, 19.

12. Boulègue, J., C.J. Lord and T.M. Church, in press: Sulfur speciations and associated trace metals (Fe, Cu) in the pore waters of Great Marsh, Delaware. Geochim. Cosmochim. Acta.
13. Cline, J.D. and F.A. Richards, 1969: Oxygenation of hydrogen sulfide in sea water at constant salinity, temperature and pH. Environ. Sc. Technol., 3, 838.
14. Boulègue, J., 1972: Mise en évidence et dosage des polysulfures au cours de l'oxydation de l'hydrogène sulfuré dans l'eau de mer. C.R. Acad. Sci. Paris, 275(C), 1335.
15. Gardner, L.R., 1973: Chemical models for sulfate reduction in closed anaerobic marine environments. Geochim. Cosmochim. Acta., 37, 53.
16. Lu, J.C.S. and K.Y. Chen, 1977: Migration of trace metals in interfaces of seawater and polluted surficial sediments. Environ. Sci. Technol., 11, 174.
17. Sillen, L.G. and A.E. Martell, 1964: "Stability constant of metal ion complexes". Spec. Publ. 17, Chemical Society, London.
18. Sillen, L.G. and A.E. Martell, 1971: "Stability constant of metal ion complexes", suppl. 1. Spec. Publ. 25, Chemical Society, London.
19. Cloke, P.L., 1963: The geologic role of polysulfides. II- The solubility of acanthite and covellite in sodium polysulfide solutions. Geochim. Cosmochim. Acta., 27, 1299.
20. Romberger, S.B. and H.L. Barnes, 1970: Ore solution chemistry, III. Solubility of CuS in sulfide solution. Econ. Geol., 65, 901.
21. Gubeli, A.O. and J. Sainte Marie, 1967: Constantes de stabilité de thiocomplexes de sulfures de métaux; II. Sulfure de zinc. Can. J. Chem., 45, 2101.
22. Sainte Marie, J., A.E. Torma and A.O. Gubeli, 1964: The stability of thiocomplexes and solubility products of metal sulphides; I. Cadmium sulphide. Can. J. Chem., 42, 662.
23. Ball, J.N., E.A. Jenne and D.K. Nordtrom, 1979: WATEQ2- A computerized chemical model for trace and major element speciation and mineral equilibria of natural waters. In: "Chemical Modeling in Aqueous Systems" E.A. Jenne, ed. A.C.S. Symp. Ser. 93, Washington, D.C.
24. Naumov, G.B., B.V. Zhyzenko and I.L. Khodakovskiy, 1974: "Handbook of Thermodynamic Data", I. Barnes and V. Spectz, eds. Natl. Tech. Infor. Serv. Rep. PB-226722, U.S. Dep. Comm., Washington, D.C.
25. Gourmelon, C., 1977: "Simulation des interactions eaux minéraux sulfurés des roches", Thesis, University of Paris 7, Paris.

26. Berner, R.A., 1967: Thermodynamic stability of sedimentary iron sulfides. Amer. J. Sci., 265, 773.
27. Rickard, D.T., 1969: The chemistry of iron sulfides formation at low temperatures. Stockholm Contrib. Geol., 20, 67.
28. Rickard, D.T., 1975: Kinetics and mechanism of pyrite formation at low temperatures. Amer. J. Sci., 275, 636.
29. Rickard, D.T., 1972: Covellite formation in low temperature aqueous solutions. Mineral. Deposita., 7, 180.
30. Cabri, L.J., 1973: New data on phase relations in the Cu-Fe-S system. Econ. Geol., 68, 443.
31. Brooks, R.R., B.J. Presley and I.R. Kaplan, 1968: Trace elements in the interstitial waters of marine sediments. Geochim. Cosmochim. Acta., 32, 397.
32. Spencer, D.W. and P.G. Brewer, 1971: Vertical advection diffusion and redox potentials as controls on the distribution of manganese and other trace metals dissolved in waters of the Black Sea. J. Geophys. Res., 76, 5877.
33. Presley, B.J., R.R. Brooks and I.R. Kaplan, 1967: Manganese and related elements in the interstitial waters of marine sediments. Science, 158, 906.
34. Charlot, G., 1969: "L'analyse qualitative et les reactions en solution", Masson, Paris.
35. Lyons, W.M.B., K.M. Wilson, G.M. Smith and H.E. Gaudette, 1980: Trace metal pore water geochemistry of nearshore Bermuda carbonate sediments. Oceanologica Acta., 3, 363.
36. Nriagu, J.O., 1972: Stability of vivianite and ion-pair formation in the system $Fe_3(PO_4)_2$-H_3PO_4-H_2O. Geochim. Cosmochim. Acta., 36, 454.
37. Emerson, S., 1976: Early diagenesis in anaerobic lake sediments: chemical equilibria in interstitial waters. Geochim. Cosmochim. Acta., 40, 925.
38. Gardner, L.F., 1974: Organic versus inorganic trace metal complexes in sulfidic marine waters - some speculative calculations based on available stability constants. Geochim. Cosmochim. Acta., 38, 1297.
39. Nissenbaum, A. and D.J. Swaine, 1976: Organic matter-metal interactions in recent sediments: the role of humic substances. Geochim. Cosmochim. Acta., 40, 809.

THE BEHAVIOR OF TRACE METALS IN MARINE ANOXIC WATERS: SOLUBILITIES AT THE OXYGEN-HYDROGEN SULFIDE INTERFACE

Steven Emerson*, Lucinda Jacobs*, Brad Tebo†

*School of Oceanography
University of Washington
Seattle, WA 98195 U.S.A.

†Scripps Institution of Oceanography
University of California - San Diego
La Jolla, CA 92093 U.S.A.

ABSTRACT

The predicted equilibrium behavior of trace metals across the oxygen-hydrogen sulfide interface in the environment is dependent upon the metal affinity for solid sulfide and metal sulfide complex formation. Equilibrium solubility estimates for iron, manganese, copper, nickel and cadmium demonstrate a wide range of trace metal behavior and generally reproduce environmentally observed trends, although the data are extremely limited. The absolute magnitude of the metal concentrations in sulfide containing waters is difficult to predict from equilibrium calculations because of uncertainty in the solubility data and the identification of the solid phase which forms.

The kinetics of oxidation of the reduced form of metals which change oxidation states across the oxygen-hydrogen sulfide interface is variable because of the chemical behavior of the ion in water and the role of catalysis by bacteria. Manganese oxidizing bacteria greatly enhance the rate of Mn(II) oxidation in waters containing low oxygen concentrations at the oxygen-hydrogen sulfide interface in Saanich Inlet, an anoxic fjord in British Columbia, Canada.

INTRODUCTION

Dramatic changes occur in the solubility of trace metals at the redox front represented by the oxygen-hydrogen sulfide boundary in natural systems. Regions where these reactions are of potential importance in controlling the aqueous chemistry of metals are: within sediment pore waters; at the surface exit of marine hydrothermal systems; and in closed basin anoxic systems. It is our goal in this study to examine the processes which control metal solubility across the redox front. We focus our investigation on closed basin anoxic systems because they are the least complex and some field trace metal data is available. From the analysis we hope to gain insight into processes controlling metal solubility at other redox boundaries in the environment. The metals discussed here, Mn, Fe, Cu, Ni and Cd, are very different in their behavior in H_2S-containg systems and are representative of the range of trace metal behavior at the oxygen-hydrogen sulfide interface.

Processes which control the concentration of metals at the redox front are categorized in this study into chemical equilibrium and kinetic considerations. Chemical reactions that occur as a result of the reduction of sulfate to sulfides on the anoxic side of the O_2-H_2S boundary are metal sulfide precipitation and the formation of soluble metal sulfide complexes. To a first approximation, the concentration of trace metals in sulfide containing waters can be viewed in terms of thermodynamic equilibrium. In the section on "Equilibrium Considerations" we reconstruct the trace metal solubility across the oxygen-hydrogen sulfide interface using available thermodynamic equilibrium constants for the sulfide containing waters and measured trace metal values for the oxic region.

The region immediately surrounding the oxygen-hydrogen sulfide interface is very poorly poised because of the low concentration of oxidant and reductant species[1,2]. The calculated $p\varepsilon$ changes from a value of (8)-(12) to (-2)-(-4) over a relatively short distance. Equilibrium models, thus, only approximate the true trace metal behavior. Chemical kinetic factors which are important in determining the metal concentration are the varying rates of oxidation and adsorption above the O_2-H_2S interface and the rates of reduction and metal sulfide precipitation and dissolution on the anoxic side. Since the redox front is a zone of a variety of energy yielding reactions, it is also a region of specialized bacterial activity which results in a catalysis of many of the redox reactions. A classic example of the influence of bacterial catalysis on the determination of trace metal behavior in this region is the oxidation of manganese(II). In the discussion of kinetic processes we present the results of an investigation of the rate of manganese(II) oxidation and the role of bacteria in catalyzing the oxidation reaction.

BACKGROUND

In order to eventually understand the processes which control the behavior of trace metals at the redox front one must investigate: (1) the state of our knowledge of the aqueous chemical behavior of metal-ligand complexation and solid formation so that meaningful qualitative generalizations can be made; and (2) the degree to which expected trends are followed by environmental measurements. In this section we present a brief review of trace metal complexation and solid sulfide solubility behavior in aqueous solution, a literature review of trace metal measurements in sulfide containg waters, and background information on the oxygen-hydrogen interface in Saanich Inlet, on which we shall focus much of our discussion.

Metal Sulfide Solubility

The solubility of metals in sulfide containing waters is controlled by the degree to which they form solid sulfides and metal sulfide complexes. Langmuir[3] and Stumm and Morgan[4] review the classification of metal ions in aqueous solutions and the reasons that they tend to form complexes with various ligands. The classification schemes are based on the deformability or polarizability of the electron cloud. Briefly, metal ions which have an inert gas configuration (spherical symmetry and low polarizability, class A or "hard sphere" cations of Schwarzenbach[5]) tend to form strong electrostatic bonds with ligands which also have low polarizability (F, O and N containing complexes). Pearson[6] includes in this category a number of additional metal cations which have a tendency to form ion pairs with the above ligands (Table 1). Metals with an electron configuration of Ni^{o}, Pd^{o} and Pt^{o} are termed class B[5] or "soft spheres." They are characterized by a highly deformable electron cloud, and high polarizability. They form largely covalent bonds with ligands which are also polarizable (for example, I and S containing species). Metal cations of intermediate configuration are the transition metals (class C). These cations have partially filled 3d subshells. The tendency for metals of this category to form complexes is usually explained in terms of the "crystal field theory." In very simplistic terms, protons are progressively added to the nucleus (and electrons to the 3d subshell) in the sequence Sc through Cu. The addition of protons to the nucleus increases the attraction of the inner electrons and decreases the size of the cation. The decreased size increases the ionic potential (valence/radius) and the electronegativity (the power to

Table 1 Classification of metal ions and their tendency to form complexes in aqueous solution (largely from Stumm and Morgan[4]). Underlined ions are those considered in this study.

I. Cations which form largely electrostatic bonds with ligands. Low polarizability, hard spheres.

<u>Class A</u> (Schwarzenbach[5]) Electron configuration is that of noble gases

H^{+}, Li^{+}, Na^{+}, K^{+}
Be^{2+}, Mg^{2+}, Ca^{2+}, Sr^{2+}
Al^{3+}, Sc^{3+}, La^{3+}
Si^{4+}, Ti^{4+}, Zr^{4+}, Th^{4+}

<u>Hard Acids</u> (Pearson[6])

All class A metals plus
Cr^{3+}, Mn^{3+}, Fe^{3+}, Co^{3+}, UO^{2+}, $\underline{Mn}^{2+}$

II. Cations which form largely covalent bonds. Polarizable, soft spheres. Electron number corresponds to Ni^{o}, Pd^{o} and Pt^{o}.

<u>Class B</u> (Schwarzenbach[5])

$\underline{Cu}^{+}$, Ag^{+}, Au^{+}, Te^{+}, Ga^{+}
$\underline{Zn}^{2+}$, $\underline{Cd}^{2+}$, Mg^{2+}, Pb^{2+}, Sn^{2+}
Tl^{3+}, $\underline{Au}^{3+}$, In^{3+}, Bi^{3+}

III. Transition metal cations.

U^{2+}, Cr^{2+}, $\underline{Mn}^{2+}$, $\underline{Fe}^{2+}$, $\underline{Co}^{2+}$, $\underline{Ni}^{2+}$, $\underline{Cu}^{2+}$
Ti^{3+}, U^{3+}, $\underline{Cr}^{3+}$, $\underline{Mn}^{3+}$, $\underline{Fe}^{3+}$, $\underline{Co}^{3+}$

---

attract electrons). The sequence of increasing tendency to form complexes increases with the increases in ionic potential and electronegativity and is called the Irving-Williams order, $Mn^{2+} < Fe^{2+} < Co^{2+} < Ni^{2+} < Cu^{2+} > Zn^{2+}$.

Bisulfide, polysulfides and sulfide are ligands which are polarizable and fall into the "soft sphere" category[4]. Hence, they have a tendency to form strong complexes and insoluble sulfides with class B metals. Figure 1 shows the trend of metal sulfide solubility products for different transition and class B metal ions. Included in this figure are the activity ratios of complexed metal ion to free metal ion for selected soluble sulfide species at a total sulfides concentration of 10^{-3} mol l^{-1} and a pH of 7.4. There are no transition metal-sulfide complexes in the figure because there is little data to suggest that they form to

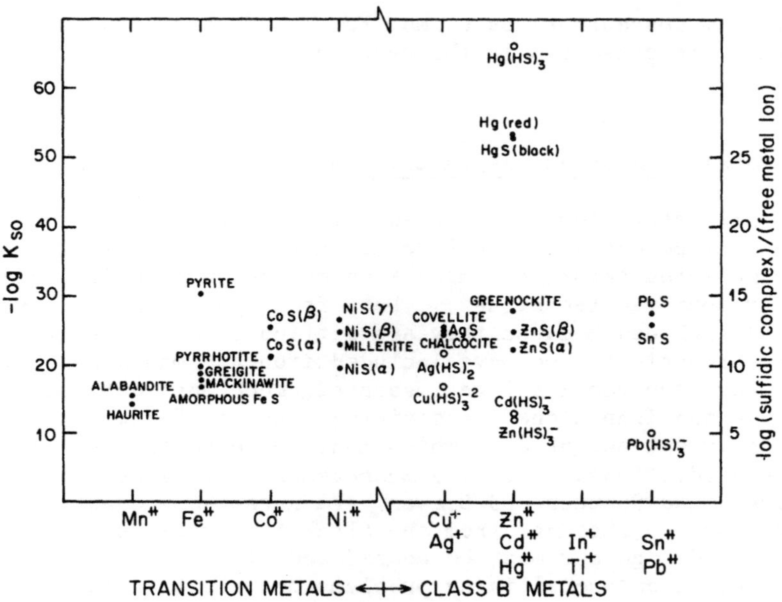

Fig. 1 Metal sulfide solubility products for transition metal and class B cations (closed symbols). Open symbols are the activity ratios of complexed metal ions to free metal ion. The predominant complex species varies; representative complexes were selected to illustrate enhanced solubility. Calculations were made at a total sulfide concentration of 10^{-3}mol $l^{-1}$and pH = 7.4. Equilibrium data are from Smith and Martell[20], Barnes[23], and Table 3. Values of K_{SO} are of the form $(M^{+n})\ (S^{=})n/2$.

any significant extent. Class B metals ions form strong complexes with reduced sulfur species in addition to forming insoluble sulfides.

The trends in Figure 1 imply distinctly different behavior of the transition and class B metal cations in waters which contain reduced sulfur species: transition metal cations should decrease in concentration with increasing total sulfide levels because they form solid sulfides but not strong dissolved sulfide complexes. Class B metal cations should show solubility increases with increasing sulfide concentration because of the strong complexation by reduced sulfide ligands.

The detailed behavior across the oxygen-hydrogen sulfide interface is complicated by the interplay of the metal cation solubility in oxygen containing waters and the varying influence of the solid sulfide and dissolved ligand equilibria as the total sulfide concentration increases. Predictions for the trend in equilibrium concentrations of Mn, Fe, Cu, Ni and Cd across the interface are presented in the section on "Equilibrium Considerations."

Trace Metal Concentrations in Sulfide Containing Waters

A literature survey of recent trace metal measurements in marine sulfide containing waters is presented in Table 2. Low levels measured before the mid 1970s may be in question because of recently improved techniques[7]. Data from hydrothermal waters have not been included because they are difficult to compare with low temperature values. We have included iron and manganese only in cases where another metal was measured; much more data exist for manganese and iron alone. A striking feature of Table 2 is lack of a consistent set of data which could be used to test thermodynamic predictions. Only the measurements of Jacobs and Emerson[8], Boulegue[9], and Spencer and Brewer[1] are detailed enough to use for this purpose. (The data from the Black Sea[1] were determined more than a decade ago and must be considered with caution because of contamination and sensitivity problems prevalent at that time.)

Saanich Inlet

In the following discussion of trace metal solubility in anoxic systems we use, as an example, Saanich Inlet, a fjord on Vancouver Island, British Columbia. We have concentrated our efforts on the behavior of a selected group of trace metals (Mn, Fe, Cu, Ni and Cd) across the oxygen-hydrogen sulfide interface in this intermittently anoxic fjord[2,8,10]. Saanich Inlet has a maximum depth of 220 m and a sill depth of 70 m. A seasonal cycle of anoxia is caused by the isolation of water behind the sill during winter and summer, which is followed by a flushing event in late summer or fall[11]. The general characteristics of the redox conditions in Saanich Inlet during July of 1979 are shown by the profiles of oxygen, hydrogen sulfide, nitrate, ammonia and manganese in Figure 2. The linear temperature-salinity curve (Figure 2) and increase in concentration of ΣH_2S to the bottom indicate that the chemical structure of the water column near the O_2-H_2S interface had reached a period of "quasi steady state" at this time[2]. The metal concentrations reported in Table 2 were determined from samples taken during this period. (For detailed profiles of the metals, the reader is referred to Jacobs and Emerson[8].)

Table 2 Trace metal data from sulfide containing waters. Temperature is (^{o}C) and concentrations are in μ molar for H_2S and nmolar for the metals.

T	PH	H_2S	Cr	Mn	Fe	Co	Ni	Cu	Zn	Ag	Cd	Hg	Pb	Source
6-8	6.9	1,000	19		267	2.9			92					Framvarian Fjord[48]
9	7.6	200 -400		4,200	89	8.0	50	5.0	11					Black Sea[1]
10	7.5	235		20,000	1,400			6.0	46					Beppa Bay, Japan[49]
14	7.4- 7.6	775		390	280			70	75		5		10	Enghien-les-Bains[9]
8	7.4	30	1	7,000	500									Saanich Inlet[2]
10	7.3	110		3,000	300			1-2						Lake Nitinat[13,53]
8	7.4	40		7,000	800		6	2			0.1			Saanich Inlet[8]

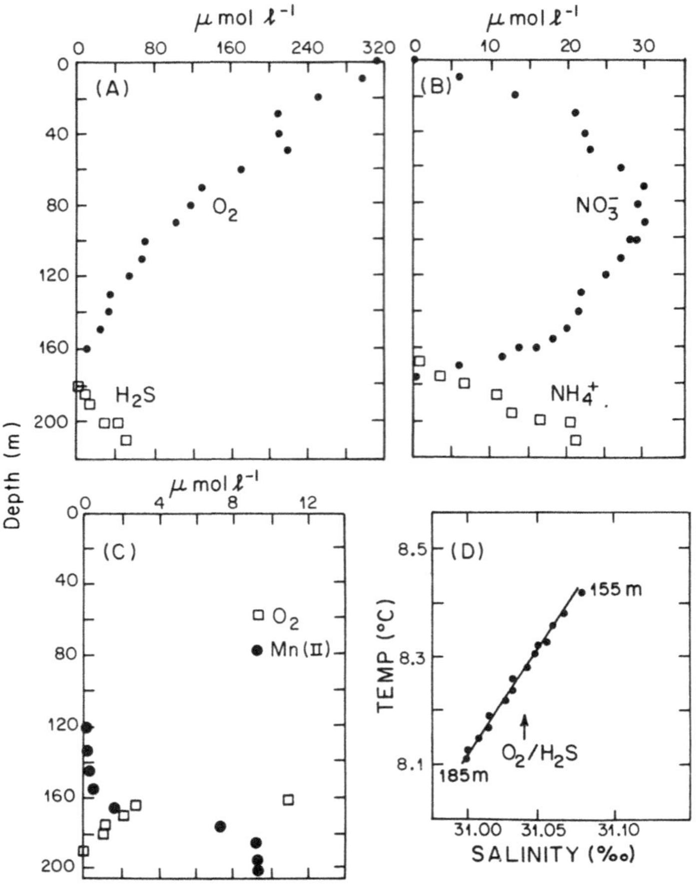

Fig. 2 Profiles of (A) oxygen and ΣH_2S, (B) nitrate and ammonia, and (C) manganese(II) in the water column of Saanich Inlet, July 1979. Figure (D) is a temperature salinity plot for the region 155-185 meters.

EQUILIBRIUM CONSIDERATIONS

A thermodynamic equilibrium approach to the behavior of metals in natural aquatic systems is seldom entirely consistent with observations, but does provide a valuable frame of reference. This simplified approach assumes that the solubility of metals in marine anoxic waters is dependent upon the oxidation state of the metal, complexation by reduced sulfur species, and solid metal sulfide formation. A pure metal sulfide is assumed to be the controlling solid phase. The sulfate-bisulfide couple is used to define the pε, and thus the oxidation state of the metal ion.* The distribution of dissolved sulfur species is calculated at constant pH with a mass balance equation and thermodynamic equilibrium constants. Elemental sulfur is assumed to be a metastable intermediate. The bisulfide ion activity is combined with the metal sulfide solubility product to derive the activity of the free metal ion. From the predicted activity of free metal ion, the activity of the major complexed species can be calculated using thermodynamic stability constants. The total soluble metal concentrations is then predicted from a metal mass balance and activity coefficient estimates.

Complexation of metals by dissolved organic ligands was not included in our calculations because of the results of a recent study of copper-fulvic acid complexation. Lieberman [13] showed that only 20-50% of the copper in oxygen and sulfide containing (300μ M) waters of Lake Nitinat, a fjord in Vancouver, B. C. is bound by fulvic acids extracted from the reducing waters of this anoxic basin. Other metals should be less strongly complexed by organic ligands [14], suggesting the inorganic complexes are of much greater importance even in waters with high (2 mg l^{-1}) dissolved organic matter concentrations. (Dissolved organo sulfur species may represent an exception to the above contention because of their very high affinity for metal complexation, J. Boulegue, unpublished results. Since their concentration in seawater and metal binding constants are poorly known we have not included them in the model.)

In the calculations presented here ligands other then reduced sulfur species are assumed to be at typical seawater concentrations and complexed according to the computer program MINEQL 2 [15]. Ion activity coefficients are based on the Davies extension of the Debye-Huckel Equation [16]. For an ionic strength of seawater (I = 0.7), this results in activity coefficients of 1, 0.7, 0.2 and 0.035 for 0, ± 1, ± 2 and ± 3 species respectively.

Our approach is similar to that of Leckie and Nelson [17]. However, the master variable is formulated as $-\log \frac{(HS^-)}{(SO_4^=)}$ rather than pε, which makes the presentation more sensitive to changes in

---

* Boulegue and Michard [12] have shown that the pε measured by a platinum electrode responds to the sulfide-polysulfide couple rather than the sulfate-sulfide couple. This observation indicates that the polysulfide-sulfide couple is very electroactive and the reaction at the platinum surface is rapid. This result does not necessarily imply that the overall pε of the environment containing sulfur and bisulfide is controlled by this couple.

total reduced sulfur species (ΣH_2S_n; n = 1, 4, 5, 6) and more easily compared with environmental results. A sulfur mass balance:

$$S_T = 10^{-1.55} = [H_2S] + [HS^-] + [S^=] + 4[S_4^=] + 5[S_5^=] + 5[HS_5^-]$$
$$6[S_6^=] + [SO_4^=] + [CaSO_4^o] + [MgSO_4^o] + [NaSO_4^o]$$

where [ ] denotes concentration, constrains total dissolved sulfur to the concentration of sulfate in seawater. The concentration of elemental sulfur is assumed to be negligible in mass balance considerations, as are the concentrations of thiosulfate and sulfite. Reduced sulfur species are assumed to be free from major cation complexation [18].

The equilibrium constants used in the model calculations are presented in Table 3. Constants determined at conditions other than 25°C and I = 0 were corrected to zero ionic strength using activity coefficients derived from the Davies equation. For constants determined at higher ionic strength (>0.7), this correction is less accurate and resultant calculations contain a greater degree of uncertainty (see Table 3). Only the constants of the most important species at seawater conditions and pH = 7.4 are included in the tabulation. The selection of constants for the species distributions of metals was in some cases based on published speciation models; the Davison [19] model was used for ferrous iron. In general, however, the equilibrium constants are taken from the critical review and compilation of Smith and Martell [20]. Polysulfide equilibrium data are from Boulegue and Michard [21] and are in reasonable agreement with the constants determined by Cloke [22]. Bisulfide and polysulfide metal complexation has not been thoroughly investigated and quantified, although bisulfide complexation of class B metals is well known [23]. Naumov [24] reports free energy data for the relatively weak complexes $Fe(HS)_2^0$ and $Fe(HS)_3^-$. These data do not appear in the critical compilation of stability constants and are thus deleted from consideration. The study of Crerar and Barnes [25] indicates that copper-bisulfide complexes involve cuprous rather than cupric copper. The constants derived for cuprous-bisulfide complexes [26] and cuprous-polysulfide complexes [27] were used in these calculations. The stability constant of the cuprous bisulfide complex presented in these calculations is questionable because the original determination was performed at an ionic strength of 2.1, and correction to I = 0 involved an over-extension of the Davies activity coefficient model. The complex was included to indicate qualitative rather then quantitative behavior.

We have deleted from consideration the $Cd(OH)^+$ complex reported by Ste-Marie et al. [28]. The magnitude of this constant

Table 3. Stability constants for the reactions used in the equilibrium model (Figures 3-7)

Equation	log k	($T^{o}C$,I) Conditions	Source
$NiS + H^{+} \rightleftharpoons Ni^{++} + HS^{-}$ (Millerite)	- 9.23	(25°,0)	50
$CdS + H^{+} \rightleftharpoons Cd^{++} + HS^{-}$ (Greenockite)	-14.1	(25°,0)	50
$H^{+} + FeS \rightleftharpoons Fe^{++} + HS^{-}$ (Amorphous)	- 3.05	(25°,0)	Footnote 1
$3H^{+} + Fe_3S_4 \rightleftharpoons 3Fe^{++} + 3HS^{-} + S_0$ (Greigite)	-13.17	(25°,0)	Footnote 1
$H^{+} + FeS \rightleftharpoons Fe^{++} + HS^{-}$ (Mackinawite)	- 3.64	(25°,0)	Footnote 1
$H^{+} + FeS \rightleftharpoons Fe^{++} + HS^{-}$ (Pyrrhotite)	- 5.95	(25°,0)	50
$H^{+} + FeS_2 \rightleftharpoons Fe^{++} + HS^{-} + S^{0}$ (pyrite)	-16.4	(25°,0)	50
$H^{+} + Cu_2S \rightleftharpoons 2Cu^{+} + HS^{-}$ (chalcocite)	-34.7	(25°,0)	50
$\frac{1}{2}H^{+} + CuS \rightleftharpoons Cu^{+} + \frac{1}{2}HS^{-} + \frac{1}{2}S^{0}$ (covellite)	-18.42	(25°,0)	50
$5HS^{-} + Cu_2S + H^{+} \rightleftharpoons 2Cu(HS^{-})_3^{-2}$	2.02±.26	(22°,2.1-4.4)	26
$MnS + H^{+} \rightleftharpoons Mn^{++} + HS^{-}$ (Alabandite)	- 0.40	(25°,0)	50
$MnS_2 + H \rightleftharpoons Mn^{++} + HS^{-} + S_0$ (Haurite)	- 1.60	(25°,0)	50, Footnote 2
$MnCO_3 \rightleftharpoons Mn^{++} + CO_3^{=}$ (Rhodochrosite)	-10.54	(25°,0)	50
$FeCO_3 \rightleftharpoons Fe^{++} + CO_3^{=}$ (Siderite)	-10.50	(25°,0)	50
$NiCO_3 \rightleftharpoons Ni^{++} + CO_3^{=}$	- 6.87	(25°,0)	20
$CdCO_3 \rightleftharpoons Cd^{++} + CO_3^{=}$	-13.74	(25°,0)	20
$CuCO_3 \rightleftharpoons Cu^{++} + CO_3^{=}$	- 9.63	(25°,0)	20
$NiS_{\alpha} \rightleftharpoons Ni^{++} + S^{=}$	-19.4	(25°,0)	20
$NiS_{\beta} \rightleftharpoons Ni^{++} + S^{=}$	-24.9	(25°,0)	20
$NiS\gamma \rightleftharpoons Ni^{++} + S^{=}$	-26.6	(25°,0)	20
$Ni^{++} + SO_4^{=} \rightleftharpoons NiSO_4^{0}$	2.32	(25°,0)	20
$Ni^{++} + Cl^{-} \rightleftharpoons NiCl^{+}$	0.00	(25°,1.0)	20
$Cd^{++} + HS^{-} \rightleftharpoons Cd(HS)^{+}$	7.6	(25°,1.0)	20
$Cd^{++} + 2HS^{-} \rightleftharpoons Cd(HS)_2^{0}$	14.6	(25°,1.0)	20
$Cd^{++} + 3HS^{-} \rightleftharpoons Cd(HS)_3^{-}$	16.5	(25°,1.0)	20
$Cd^{++} + 4HS^{-} \rightleftharpoons Cd(HS)_4^{=}$	18.9	(25°,1.0)	20
$Cd^{++} + Cl^{-} \rightleftharpoons CdCl^{+}$	1.98	(25°,0)	20

Table 3 continued.

Equation	log k	(T°C,I) Conditions	Source
$Cd^{++} + 2Cl^{-} \rightleftharpoons CdCl_2^0$	2.60	(25°,0)	20
$Cu^{+} + S_4^{=} + S_5^{=} \rightleftharpoons CuS_4S_5^{-3}$	21.85	(25°,0)	12,27,50
$Cu^{+} + 2S_4^{=} \rightleftharpoons Cu(S_4)_2^{-3}$	22.81	(25°,0)	12,27,50
$Cd^{+} + 3Cl^{-} \rightleftharpoons CdCl_3^{-}$	2.40	(25°,0)	20
$Cd^{++} + SO_4^{=} \rightleftharpoons CdSO_4^0$	0.95	(25°,0)	20
$Cu^{++} + CO_3^{=} \rightleftharpoons CuCO_3^0$	6.75	(25°,0)	20
$Cu^{++} + OH^{-} \rightleftharpoons CuOH^{+}$	6.3	(25°,0)	20
$Cu^{++} + Cl^{-} \rightleftharpoons CuCl^{+}$	0.4	(25°,0)	20
$Cu^{+} + 2Cl^{-} \rightleftharpoons CuCl_2^{-}$	5.5	(20°,0)	20
$Cu^{+} + 3Cl^{-} \rightleftharpoons CuCl_3^{=}$	5.7	(20°,0)	20
$Fe^{++} + OH^{-} \rightleftharpoons FeOH^{+}$	4.5	(25°,0)	19
$Fe^{++} + SO_4^{=} \rightleftharpoons FeSO_4^0$	2.12	(25°,0)	19
$Fe^{++} + Cl^{-} \rightleftharpoons FeCl^{+}$	0.14	(25°,0)	19
$Mn^{++} + Cl^{-} \rightleftharpoons MnCl^{+}$	0.61	(25°,0)	51
$Mn^{++} + SO_4^{=} \rightleftharpoons MnSO_4^0$	2.26	(25°,0)	20
$S^{=} + H^{+} \rightleftharpoons SH^{-}$	13.9 + 1	(25°,0)	20
$SH^{-} + H^{+} \rightleftharpoons H_2S\ aq$	7.02 $\pm$ 0.4	(25°,0)	20
$H_2S\ (g) \rightleftharpoons H_2S\ aq$	- 0.99	(25°,0)	20
$\frac{1}{4}S_8 + HS^{-} \rightleftharpoons S_3^{=} + H^{+}$	-12.5	(25°,0)	21
$\frac{3}{8}S_8 + HS^{-} \rightleftharpoons S_4^{=} + H^{+}$	- 9.52	(25°,0)	21
$\frac{1}{2}S_8 + HS^{-} \rightleftharpoons S_5^{=} + H^{+}$	- 9.41	(25°,0)	21
$\frac{5}{8}S_8 + HS^{-} \rightleftharpoons S_6^{=} + H^{+}$	- 9.62	(25°,0)	21
$HS_4^{-} \rightleftharpoons S_4^{=} + H^{+}$	- 7.0	(25°,0)	21
$HS_5^{-} \rightleftharpoons S_5^{=} + H^{+}$	- 6.1	(25°,0)	21
Redox Equations	log K (=pε)	Conditions	Source
$\frac{1}{8}SO_4^{=} + \frac{9}{8}H^{+} + e^{-} \rightleftharpoons \frac{1}{8}HS^{-} + \frac{1}{2}H_2O$	4.25	(25°,0)	4
$Cu^{++} + e^{-} \rightleftharpoons Cu^{+}$	2.7	(25°,0)	4
$Fe^{+++} + e^{-}\quad Fe^{++}$	13.2	(25°,0)	4

Sources: 1. Recalculated from the data of Berner[37] using the Davies approximation for activity coefficients and the hydrogen sulfide stability constants of Smith and Martell[20].

2. This solubility product is calculated from free energy data. Aqueous ion free energy is from Robie et al.[50] and haurite free energy was calculated from the entropy and enthalpy data of Mills[52].

implies that the OH^- complex should be of greater importance than the chloro and bisulfide complexes. This is not in accordance with other studies[29]. In addition, cadmium falls among a group of metals in the periodic table which do not form strong OH^- complexes[30]; the complex reported by Ste-Marie et al.[28] is many orders of magnitude greater than would be expected according to this scheme. Since the equilibrium constants for the other cadmium-sulfide complexes reported in Table 3 also come originally from the study of Ste-Marie et al.[28], deletion of the OH^- complex from consideration without substitution of another complexing ligand is not consistent with their experimental data[8]. This represents a potential flaw in the construction of the equilibrium diagram for cadmium. The resolution of this potential inconsistency awaits a critical reinterpretation of the cadmium-sulfide complexation data.

The solubility products of covellite (CuS) and chalcocite (Cu_2S) are written in terms of cuprous ion. For covellite it was necessary to re-derive the solubility product from ΔG^0 data, since most compilations report it in terms of cupric copper. Covellite has an unique crystal structure that is not apparent in its simple stoichiometry. The presence of covalent disulfide pairing and the occupation by copper of cupric and cuprous coordination sites in the crystal indicate that covellite is not a simple cupric sulfide[31]. The consideration of cuprous copper in equilibrium with covellite is probably the most realistic approach, especially in light of the stabilizing effect that extensive complexation by sulfidic ligands has on this oxidation state.

Before presenting the results of the equilibrium calculations and comparing them with the observations, we would like to point out some important caveats to the prediction of equilibrium trace metal concentrations in sulfide containing waters. The metastable persistence of thiosulfate, sulfite and polysulfides near the oxygen-hydrogen sulfide interface at levels greater than those predicted by equilibrium calculations has been observed[9,32]. Elevated concentrations of these species would contribute significantly to the solubility of those class B metals with which they tend to complex, causing significant departures from predicted levels. An additional uncertainty lies in the assumption of the metastable persistance of elemental sulfur, which is an intermediate to the oxidation of sulfide by oxygen[33] and an intermediate in the metabolism of certain chemolithotrophs and photolithotrophs[34]. The assumption that elemental sulfur exists throughout the anoxic zone may be incorrect, especially in regions far removed from the interface[32].

The assumption of pure sulfide solid phase control in the anoxic zone is an additional simplification. Iron sulfides are ubiquitous in anoxic sediments and probably form in anoxic waters[1,35]. Formation of solid solutions of trace metal sulfide and iron sulfide may be an important consideration in evaluating the concentration of the dissolved metal ion[9]. The simplest treatment of solid solution behavior assumes ideality, which means the activity of each end member is equal to its mole fraction. In low temperature, biologically driven anoxic systems, the flux of iron is much greater than any trace metal, which could result in a dilute solid solution in which the trace species has a mole fraction of 10^{-3} or less and iron sulfide has a mole fraction of about 1[9]. This simple ideal solid solution treatment indicates that the solid trace metal sulfide activity would be 10^{-3} or less, which would decrease the free metal ion and all complexed species concentrations below the value they would have in equilibrium with a pure metal sulfide by an equivalent amount. In non-ideal solid solutions this effect may be compensated to some extent by an activity coefficient which is unknown and probably variable. Craig and Scott[36] showed that the activity coefficient of pyrrhotite in spalerite approached a constant value of 2.6 as the mole fraction of pyrrhotite approached zero. If the activity coefficient does not reach levels of 10^2- 10^3 it would not totally compensate for the decreased solubility of the solute sulfide in very diluted solid solution. The metal concentration would, thus, be far lower than that predicted to be at equilibrium with the pure metal sulfide.

A final caveat is that the equilibrium distributions were calculated at a pH of 7.4. In natural systems the pH drops in response to sulfate reduction. Changes in pH will redistribute the reduced sulfur species, especially the polysulfides, which results in changes in metal solubility. The effect of pH on metal solubility is shown in the figures of the predicted equilibrium behavior.

The predicted values of iron, manganese, nickel, cadmium and copper at 25°C are presented in Figures 3 to 7. Solid lines represent individual species activity and the dashed (pH = 7.4) and dotted (pH = 7.8) lines represent the concentration of total soluble metal in seawater. The metals considered in these figures exhibit a wide range of behavior across the redox front. Copper, iron and manganese change oxidation state. Iron and manganese retain their classification of transition metal, but copper changes from a transition metal to a class B metal when it is reduced frrom cupric to cuprous copper. Nickel, a transition metal, and cadmium, a member of the class B metals, do not change oxidation state. Values at the far left of the Cu, Ni, Cd and Mn diagrams represent observed metal concentrations in oxic deep open ocean water[37,38]. The observed concentrations were partitioned using metal mass balance equations, the complex stability constants, Davies activity coefficients approximations, and calculated ligand activities in seawater. The response of a metal to

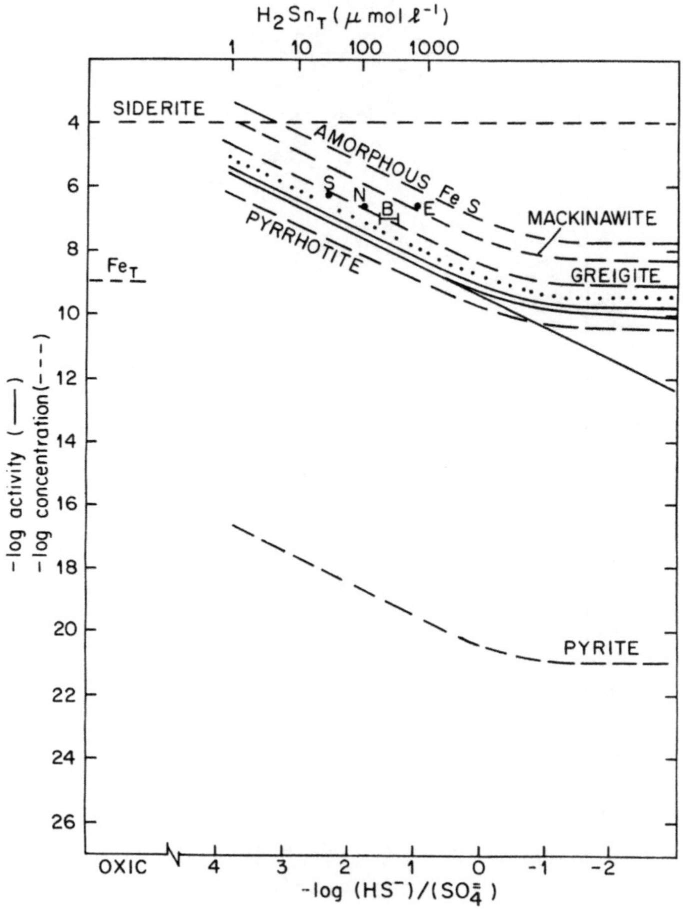

Fig. 3 The activity or concentration of iron in seawater as a function of the $-\log(HS^-)/(SO_4^=)$. Solid lines represent calculated individual species activities (shown for greigite only). The dashed line on the left side of the figure is an estimate of the upper limit of the total dissolved iron concentration, Fe_T, in open ocean oxic seawater. The dashed line in the region of sulfide containing waters represents the concentration of total soluble iron, Fe_T, predicted to be in equilibrium with various iron sulfide mineral phases at a pH of 7.4, and the dotted line represents total iron in equilibrium with greigite at pH = 7.8. The line representing total soluble iron in equilibrium with siderite was calculated assuming a constant total carbonate composition of 2.2 mmolar. The lines are predicted using the thermodynamic data in Table 3. The symbols represent measured values reported in Table 2; (S) is Saanich Inlet, (B) is the Black Sea, (N) is Lake Nitinat, and (E) is from Enghien-les-Bains (see text for further explanation).

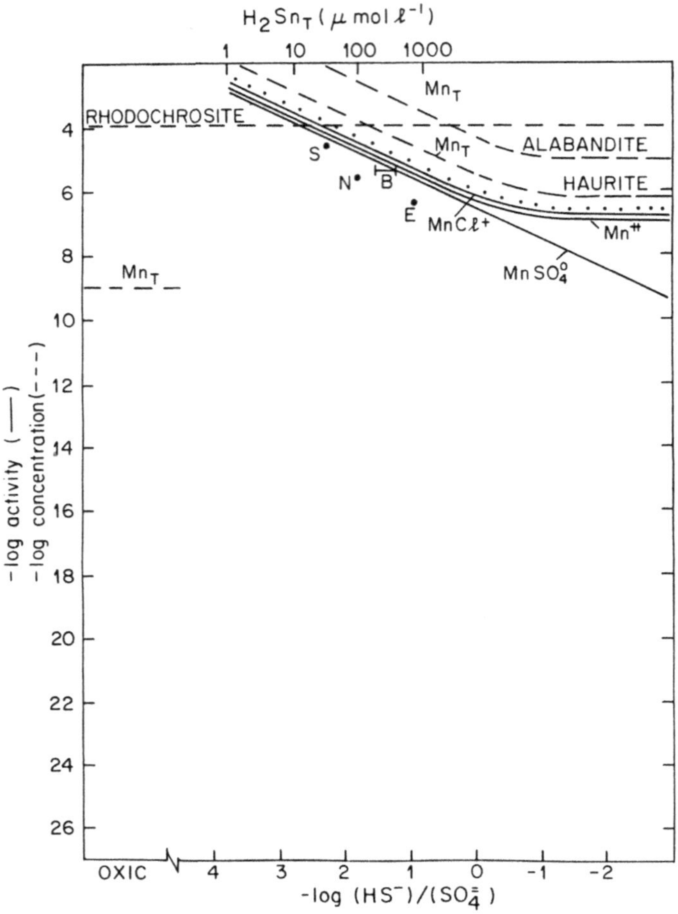

Fig. 4 The activity or concentration of manganese in seawater as a function of the $-\log(HS^-)/(SO_4^=)$. Solid lines represent calculated individual species activities. The dashed line on the left side of the figure represents the total measured value, Mn_T, in deep ocean water (from Martin and Knauer [38]). The dashed line in the region of sulfide containing waters represents the predicted concentration of total metal, Mn_T, in equilibrium with alabandite and haurite at a pH of 7.4, and the dotted line represents Mn_T for haurite at pH = 7.8. Species distributions are shown for haurite at pH = 7.4. The line representing total soluble manganese in equilibrium with rhodochrosite was calculated assuming a constant total carbonate composition of 2.2 mmolar. The equilibrium calculations were made with data from Table 3. Symbols represent measured values reported in Table 2. (S) is Saanich Inlet, (B) is the Black Sea, (N) is Lake Nitinat and (E) is from Enghien-les-Bains (see text for further explanation).

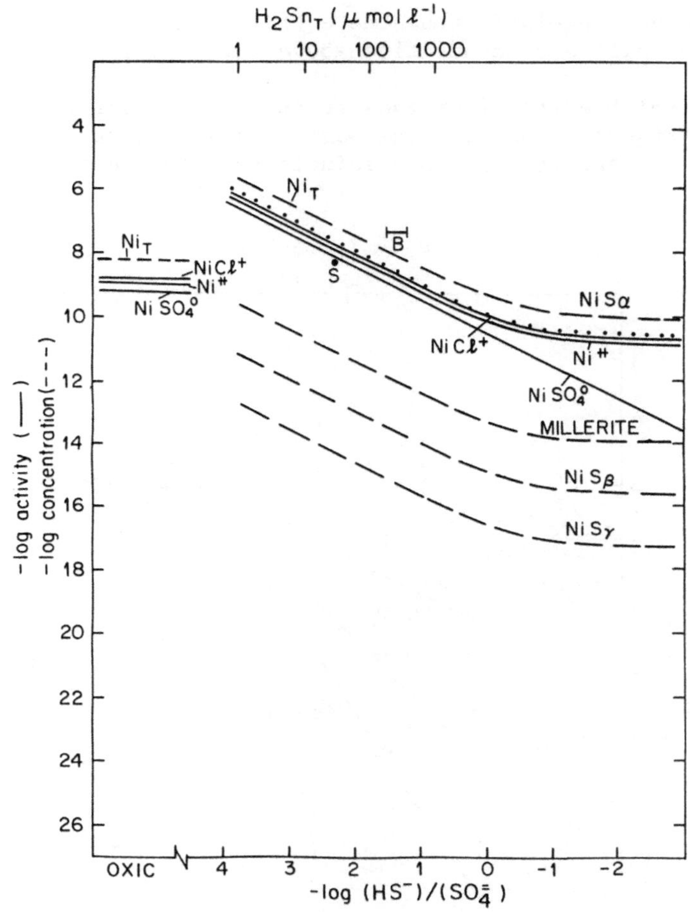

Fig. 5 The activity or concentration of nickel in seawater as a function of the $-\log(HS^-)/(SO_4^=)$. Total nickel concentration (Ni_T, dashed line) in oxic waters is that measured in deep ocean water (Bruland[37]). The lines representing Ni_T values in sulfide containing waters are predicted for equilibrium with NiSα, NiSβ, NiSλ and millerite. Species distribution are shown for NiSα. See the caption of Figure 4 and the text for further explanation of the lines. Symbols are measured values reported in Table 2 (see Figure 3 for symbol explanation).

the redox front is seen as the discontinuity between observed oxic concentrations and predicted anoxic concentrations. In practice, oxygen and hydrogen sulfide coexist in anoxic basins at values slightly below 1 μmolar[2], thus the equilibrium model is only viable at a sulfide concentration above this value.

Total soluble iron increases at the oxygen-hydrogen sulfide interface (Figure 3) because the most likely candidate for the sulfide of ferrous iron is more soluble than the controlling phase

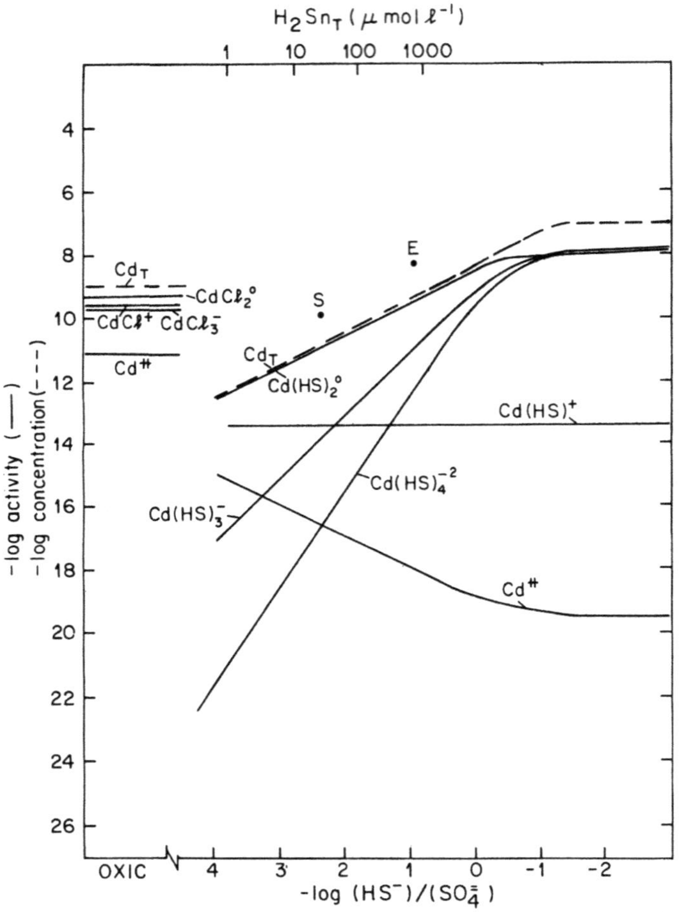

Fig. 6 The activity or concentration of cadmium in seawater as a function of the $-\log(HS^-)/(SO_4^=)$. The total cadmium concentration (Cd_T, dashed line) in oxic waters is that measured in deep ocean water (Bruland[37]). The line representing Cd_T values in sulfide containing waters is predicted for equilibrium with CdS. See the caption of Figure 4 and the text for further explanation of the lines. Symbols are measured values reported in Table 2 (see Figure 3 for symbol explanation).

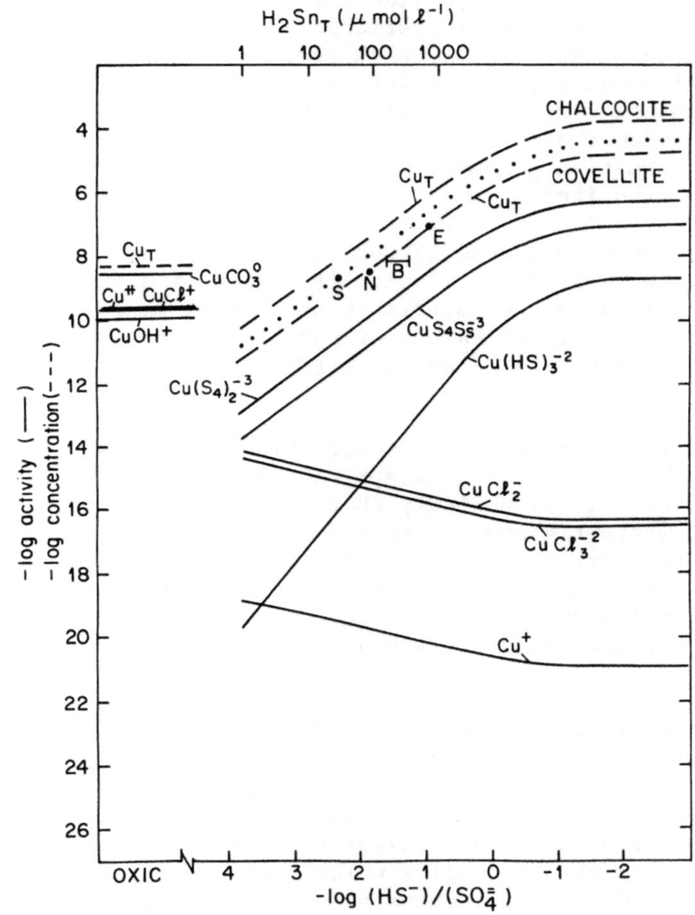

Fig. 7 The activity or concentration of copper in seawater as a function of the $-\log(HS^-)/(SO_4^=)$. The total copper concentration (Cu_T, dashed line) in oxic waters is that measured in deep ocean water (Bruland[37]). Oxic species distributions are calculated at pH = 7.4 and total carbonate = 2.2 mmol/kg. The lines representing Cu_T values in sulfide containing waters are predicted for equilibrium with chalcocite, Cu_2S and covellite, CuS (the dotted line is for covellite at pH = 7.8). Species distributions are shown for covellite. See the caption of Figure 4 and the text for further explanation of the lines. Symbols are measured values reported in Table 2 (see Figure 3 for symbol explanation).

in oxic waters. The dissolved iron concentration responds to increasing levels in total sulfide with a decrease in concentration. This behavior is characteristic of a transition metal that does not complex significantly with reduced sulfide and results in a concentration maximum at low sulfide levels. Several iron sulfide solid phases were considered in the calculations. Although pyrite is favored thermodynamically, kinetic factors result in the preliminary formation of amorphous iron sulfide, greigite, mackinawite or pyrrhotite[39].

Total soluble manganese (Figure 4) also decreases with increasing total sulfide, following the trend of a free metal ion that is controlled by the solubility product of manganese sulfide. Several manganese sulfide values are available in the literature. We present the values for alabandite (MnS) and haurite (MnS_2) in Figure 4.

The predicted behavior of nickel (Figure 5) is similar to iron and manganese, with the exception that nickel does not change oxidation state. It is unlikely that there is a sufficient nickel flux to intermittently anoxic regions in natural systems to support the full extent of the predicted solubility maximum for the most soluble phase, NiSα. Thus, metal sulfide control of the dissolved nickel concentration may not occur until relatively high levels of total sulfide are reached.

Cadmium does not undergo a change in oxidation state, but complexes strongly with bisulfide, resulting in an enhanced cadmium solubility of 3 to 10 orders of magnitude over the predicted free ion activity at the range of total sulfide found in natural waters (1 to 6000 μm). Cadmium solubility (Figure 6) is representative of the behavior of a class B metal. The importance of the stoichiometry of the complex which forms is illustrated by the response of the different cadmium bisulfide complexes [$Cd(HS)_n^{2-n}$, where n = 1 to 4] to increasing bisulfide activity. $Cd(HS)^+$ activity is unchanged with increasing total sulfide because the increase in bisulfide activity is cancelled by the decrease in the activity of the free cadmium ion. As (n) increases, the slope of the line representing species response to bisulfide activity steepens. Since the thermodynamic data indicate that $Cd(HS)_2^0$ is the most important species, total soluble cadmium increases with increasing bisulfide activity. In seawater, soluble cadmium concentrations should go through a minimum at low total sulfide.

Copper (Figure 7) changes oxidation state and, as cuprous copper, is capable of forming strong complexes with chloride, bisulfide and polysulfide ions. At a pε of -4, the ratio of Cu^+ to Cu^{2+} activity is about 10^6, thus cupric copper was not considered in the calculations for anoxic zone. Copper shows behavior similar to cadmium, with a predicted concentration minimum at low total sulfide. Two solid phases were considered, chalcocite and covellite. In the anoxic zone, soluble copper is controlled by copper polysulfide complexes. If the polysulfide ions do not exist at regions far removed from the interface, soluble copper would be controlled by bisulfide or chloride complexes, resulting in a decrease in predicted concentrations.

Comparison of Field Data with Equilibrium Model

Selected field measurements from Table 2 are plotted on the equilibrium diagrams. Four environments were plotted: Saanich Inlet (S), Lake Nitinat (N), Enghien-les-Bains (E), and the Black Sea (B). The environments S, N, and B are close to seawater model conditions however, E has a much lower ionic strength and can be compared directly to equilibrium predictions only under certain conditions.

Iron(II) measurements from the Black Sea and Saanich Inlet follow the generally predicted trends (Figure 3) and fall on the line representing greigite. Data from Enghien-les-Bains cannot be directly compared because model calculations include $FeCl^{+}$ as a major species and ionic strength differences come into play. The measured concentrations are nearly ten orders of magnitude greater than that which would be predicted for pyrite equilibrium indicating the importance of the monosulfide precursors in limiting iron concentrations in these anoxic environments. Manganese(II) measurements from the four environments (Figure 4) follow the predicted trend for equilibrium with a manganese sulfide phase, decreasing roughly parallel to the lines representing metal sulfide equilibrium control. The qualification for Enghien-les-Bains values stated for iron applies to manganese as well. Equilibrium solubility behavior for manganese in sulfide containing waters is not unexpected since MnS crystals have been observed in anoxic marine sediments[40].

The nickel concentration in Saanich Inlet (Figure 5) is below the level which is predicted for NiS equilibrium indicating, perhaps, that the Ni flux to the anoxic region is not sufficiently large during the anoxic period to raise the value to that required for formation of the most soluble phase. The Black Sea data, however, are about five times higher than the predicted value for the most soluble phase, $NiS\alpha$.

Cadmium results from Saanich Inlet and Enghien-les-Bains (Figure 6) indicate close agreement with the predicted solubility trends. The difference in ionic strength between (E) and the model are not critical in this case because the major Cd complex is uncharged. A factor that introduces considerable uncertainty into these distributions is the activity coefficient estimates that must be applied to the original complex stability constants to convert them to the infinite dilution scale (I = 0) from the laboratory conditions (I = 1.0).

A comparison of the observed copper concentrations in Saanich Inlet, the Black Sea, and Enghien-les-Bains with model predictions agrees quite well with the predicted solubility trends indicating the probable importance of polysulfide complexation to the solubility of copper in these environments.

KINETIC CONSIDERATIONS

The degree to which trace metals attain their equilibrium concentrations across the redox front is controlled by the relative rates of chemical reactions and mixing (or transport) of the waters at the interface. If chemical reactions are slow with respect to the rate of mixing, the distributions may be interpreted in terms of end member mixing. This type of analysis appears to hold for some metals in regions of hydrothermal circulation at mid-ocean ridges although the data are limited[41]. If the rates of chemical reactions are relatively rapid, then the distributions predicted by the equilibrium models of Figures 3 through 7 should be approached providing the source of the metal in question is sufficient to support the higher values.

The limiting factor for the approach to chemical equilibrium at the redox front is the degree to which the major dissolved redox species -- oxygen and hydrogen sulfide -- attain chemical equilibrium. Among the examples mentioned here, the degree of approach to chemical equilibrium is greatest in sediment pore waters and least in regions of hydrothermal circulation, with the redox front in anoxic basins intermediate between these two. There is a significant overlap of H_2S and O_2 concentrations at the exit of the Galapagos hydrothermal vents[42]. In anoxic basins, the overlap also exists, but at concentrations below one micromolar (Figure 2).

A study of the distribution of a number of redox species across the O_2-H_2S interface in Saanich Inlet[2], revealed that the rates of reduction of the oxidized form of a redox couple in sulfide containing waters is usually much more rapid than the rate of oxidation of the reduced form in oxic waters. The tendency for non-equilibrium among redox couples was much greater in oxygen containing than in sulfide containing waters. One reason for this is the absence in anoxic waters of photosynthetically driven biological processes (with the exception of photolithotrophic bacteria) which tend to increase the free energy of the environment in oxygen containing waters. Another contribution to disequilibrium in oxic waters is the sluggish nature of oxidation rates. The explanation for the slow oxidation kinetics is uncertain but may be related to the energy barrier which must be exceeded in order to break the bond between the oxygen atoms in the O_2 molecule.

Of the metals discussed here, the greatest concentration gradients across the redox front occur in the distributions of iron and manganese. The oxidation rate of Fe(II) is rapid enough that iron concentrations are unmeasurable by spectrophotometric methods in oxygen containing waters[2]. This observation is consistent with the well known rapid environmental oxidation rate of Fe(II)[4,43]. By contrast, measurable concentrations of Mn(II) coexist with oxygen (Figure 2) and a distinct layer of manganese rich particulates exists above the interface[1,2]. The oxidation rate of Mn(II) is known to be sluggish[44], but may in some cases be catalyzed by surfaces[45] and bacteria[46]. Since the remobilization of Mn(II) in anoxic environments is extensive and its fate in the oxygen containing waters may also influence the fate of other metals, we have undertaken the study of the factors controlling its oxidation rate at the redox front in Saanich Inlet.

The Manganese(II) Oxidation Rate

Using a one-dimensional model, the Mn(II) gradient, information about the turbulent transport at the redox front in Saanich Inlet, Emerson et al.[2] showed that the mean life of manganese with respect to oxidation was on the order of days. Because this rate is rapid with respect to laboratory measurements of the oxidation rate[44] and also with respect to catalysis by particulates[45], one of us (Bradley Tebo) has undertaken the investigation of bacterial catalysis of Mn(II) oxidation in this environment. Details of the bacterial study and Mn oxidation rate estimates from mass balance arguments are presented in Emerson et al.[10].

Figure 8 shows the distribution of oxygen, ΣH_2S, and Mn(II) in Saanich Inlet in September 1980. The potential for bacterial catalysis of Mn(II) oxidation was investigated in the depth range 10-150 m using a radioactive isotope technique that measures binding of manganese to particulates, >0.2 μm. Water samples were incubated with Mn^{54} at 4-10°C and atmosphere oxygen concentrations. Manganese binding to bacteria was differentiated from binding to either organic or inorganic particulates by the addition of buffered formaldehyde (2%) to some of the experiments. Laboratory tests showed that the formaldehyde neither chelates nor otherwise significantly interacts with Mn(II) adsorption to inorganic manganese oxides (R. Rosson, unpublished data). Since formaldehyde generally inhibits bacterial metabolism and growth, the component of binding dependent on the interaction of Mn(II)

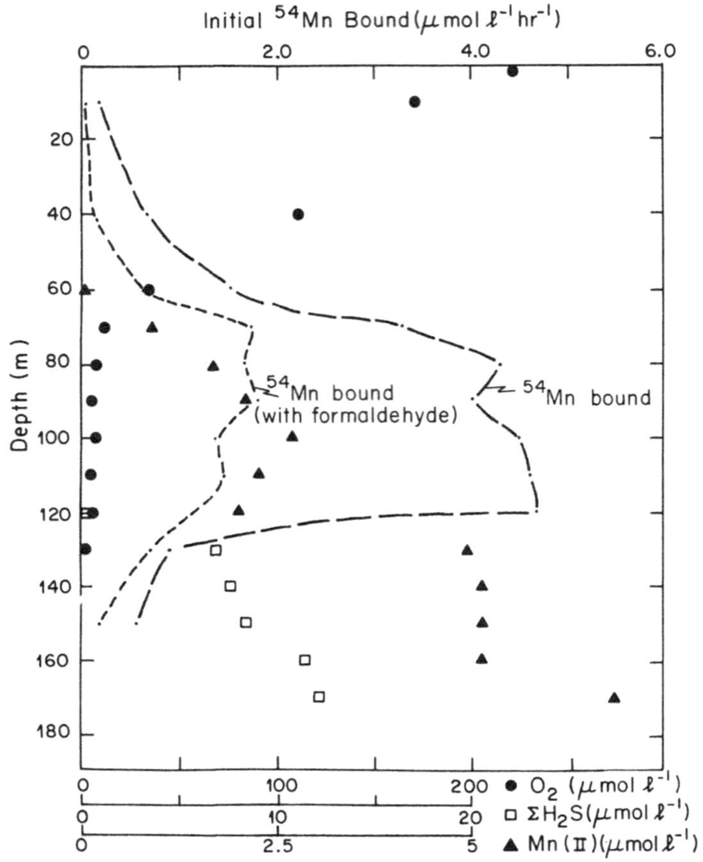

Fig. 8 Depth distribution of oxygen, H_2S, and Mn(II) in Saanich Inlet in September of 1980. Also shown is the depth distribution of the initial Mn^{54}(II) binding rate (0 to 1 hour interval) with and without the addition of formaldehyde (see text and Figure 9).

with bacteria was inhibited in experiments which included the addition of this compound.

The results of time course experiments for samples taken at 100 m (Figure 9) demonstrate that the rate of Mn(II) binding decreased as a function of time. In this figure four different experimental runs are reported: one to which no formaldehyde was added and three to which formaldehyde was added at different

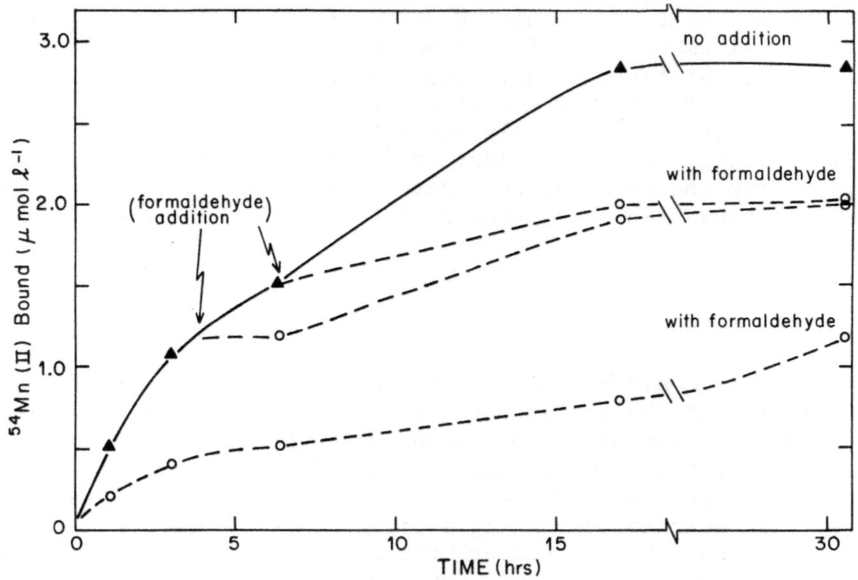

Fig. 9 The time course of Mn^{54}(II) binding on to particulate material from Saanich Inlet. The amount of Mn(II) bound is plotted as a function of time for a sample from 100 meters. The lines represent: A sample to which no formaldehyde was added, a sample to which formaldehyde was added at the outset, and two samples to which formaldehyde was added after the beginning of the experiment (4 and 6.3 hours).

times. Samples poisoned with formaldehyde, regardless of the time of addition, exhibited a substantial decrease in the binding rate of Mn^{54} by the particulate material. The fact that Mn^{54} was not removed from the particulate material by the addition of formaldehyde at times after the beginning of the experiment (4 and 6.3 hours), provides additional evidence that the formaldehyde solution does not chelate the dissolved manganese(II). The slow rate of binding after the formaldehyde addition is probably a result of inorganic adsorption of Mn(II) by the particulate material.

The initial (0-1 hr) Mn(II) binding rates at various depths are plotted in Figure 8. A peak in manganese binding by particulates is located just above the O_2-H_2S interface. In the presence of formaldehyde the Mn^{54} binding rate is much slower. Only a small amount of Mn(II) is removed from solution above 60 m and below 120 m. We believe that these results indicate that there is a substantial amount of Mn(II) binding that can only be attributed to bacterial catalysis. While the results are inconclusive with respect to the relative rates of adsorption and oxidation, the binding process is most likely autocatalytic oxidation with an adsorption step preceding oxidation[44]. Electron micrographs of material from the particulate layer[16] reveal structures which are similar to manganese oxide coated bacteria grown in laboratory studies.

CONCLUSION

Although chemical equilibrium models generally lead to overestimates of the concentrations of dissolved trace metal in oxic seawater, some of the processes which strongly influence metal concentrations in the open ocean, e.g., incorporation into biological cycles and adsorption on particulate oxides, are not as effective in sulfide containing waters. Thermodynamic equilibrium models may; thus, be of greater utility in predicting the solubility of trace metals in anoxic environments. Thermodynamic data for trace metal-sulfide solid and complex formation; however, is limited, and in some cases where it does exist, inconsistent: for some metals (e.g., copper) there is an uncertainty with respect to the oxidation state of the dissolved metal cation in anoxic waters; for metals which form polysulfide complexes (Cu, Hg, and Ag)[47], equilibrium calculations are in doubt because of known metastable existence of the polysulfide complex[9,32]; and, formation of solid solutions of trace metals and iron sulfides is not well understood or even known to be important in solubility considerations. In spite of these caveats, we have shown that the general trends of the solubilities of Mn, Fe, Cu, Ni and Cd in hydrogen sulfide containing waters follow, in many cases, those predicted from thermodynamic and aqueous inorganic chemistry principles.

The redox front immediately surrounding the oxygen-hydrogen sulfide interface represents a very poorly poised chemical system where one would not necessarily expect an equilibrium state to be attained; thus, kinetic considerations are important. The most serious departure from thermodynamic equilibrium appears to be on the oxic side of the interface where the oxidation rate of the more soluble reduced metals is variable and in many cases unknown. In the case of manganese(II), the rate of oxidation is much more rapid than would be predicted in an inorganic system. Bacterial catalysis enhances the rate of oxidation in the low oxygen containing waters at the O_2-H_2S interface in Saanich Inlet.

In order to better understand the processes which control the distribution of trace metals in sulfide containing waters and at the redox front it is essential that further research be done in the following areas: (1) the evaluation of equilibrium constants for metal-sulfide solid and complex formation; (2) the determination of concentrations of dissolved metals and sulfide species across the O_2-H_2S interface in environments with a variety of different sulfide concentrations; (3) the evaluation of trace metal gradients in sediment pore waters and regions of hydrothermal circulation; and (4) the study of the role of bacterial catalysis in enhancing the rate of approach to equilibrium. Continuing effort in these areas of research are prerequisites for the meaningful interpretation of the effect of anoxic environments on the concentration of trace metals in the sea.

REFERENCES

1. Spencer, D.W. and P.B. Brewer, 1971: Vertical advection diffusion and redox potentials as controls on the distribution of manganese and other trace metals dissolved in waters of the Black Sea. J. Geophys. Res., 76, 5877.
2. Emerson, S., R.E. Cranston and P.S. Liss, 1979: Redox species in a reducing fjord: Equilibrium and kinetic considerations. Deep-Sea Res., 26A, 859-878.
3. Langmuir, D., 1979: Techniques of estimating thermodynamic properties for some aqueous complexes of geochemical interest. In: "Chemical Modeling in Aqueous Systems", (Jenne, ed.), ACS Symposium 93, 353-387.
4. Stumm, W. and J.J. Morgan, 1970: Aquatic Chemistry. Wiley-Interscience, New York, 583 pp.
5. Schwarzenbach, G., 1961: The general, selective and specific formation of complexes by metal cations. Adv. Inorg. Radiochem., 3, 257, 265-270.
6. Pearson, R.D. (ed.), 1965: Hard and Soft Bases. Dowden, Hutchinson and Ross, Stroudsburg, Pennsylvania, 480 pp.
7. Boyle, E.A. and J.M. Edmond, 1977: Determination of copper, nickel and cadmium in seawater by APDC chelate coprecipitation and flameless atomic adsorption spectrophotometry. Analytica Chemica Acta., 91, 189.
8. Jacobs, L. and S. Emerson, 1982: Anoxic controls of trace metal solubility in a reducing fjord.
9. Boulegue, J., 1977: Equilibria in a sulfide rich water from Enghien-les-Bains (France). Geochim. et Cosmochim. Acta., 41, 1751-1758.
10. Emerson, S., L. Jacobs, S. Kalhorn, R. Rosson, B. Tebo, and K. Nealson, 1982: Environmental oxidation rate of manganese(II): Bacterial catalysis.
11. Anderson, J.J. and A.H. Devol, 1973: Deep water renewal in Saanich Inlet, an intermittently anoxic basin. Estuarine and Coastal Marine Sci., 1, 1.

12. Boulegue, J. and G. Michard, 1979: Sulfur speciations and redox processes in reducing environments. In: "Chemical Modeling in Aqueous Systems", (Jenne, ed.) ACS Symposium 93, 25-50.

13. Lieberman, S., 1979: Stability of copper complexes with seawater humic substances. Ph.D. Thesis, University of Washington, Seattle, WA.

14. Mantoura, R.F.C., A. Dickson and J.P. Riley, 1978: The complexation of metals with humic materials in natural waters. Estuarine and Coastal Marine Science, 6, 387-408.

15. Nordstrom, D.K., L.N. Plummer, T.M.L. Wigley, T.J. Wolery, J.W. Ball, E.A. Jenne, R.L. Bassett, D.A. Crerar, T.M. Florence, B. Fritz, M. Hoffman, G.R. Holdren, Jr., G.M. Lafon, S.V. Mattigod, R.E. McDuff, F. Morel, M.M. Reddy, G. Sposito and J. Thrailkill, 1979: A comparison of computerized chemical models for equilibrium calculations in aqueous systems. In: "Chemical Modeling in Aqueous Systems", (Jenne, ed.), ACS Symposium 93.

16. Whitfield, M., 1975: Seawater as an electrolyte solution. In: "Chemical Oceanography", Vol 1, (Riley and Chesters, eds.), Academic Press.

17. Leckie, J.O. and M.B. Nelson, 1975: Role of natural and hetrogeneous sulfide systems in controlling the concentration and distribution of heavy metals: Second International Symposium on Environmental Biochemistry, Burlington, Ontario, Canada (unpublished manuscript).

18. Herr, F.L. and G.R. Helz, 1978: On the possibility of bisulfide ion-pairs in natural brines and hydrothermal solutions. Econ. Geo., 73, 73-81.

19. Davison, W., 1979: Soluble inorganic ferrous complexes in natural waters. Geochim. et. Cosmochim. Acta., 43, 1693-1699.

20. Smith, R.M. and A.E. Martell, 1976: Critical Stability Constants, Vol. 4, Inorganic Complexes, Plenum Press, New York, 257 pp.

21. Boulegue, J. and G. Michard, 1978: Costantes de formation des ions polysulfures S_6^{2-}, S_5^{2-} et S_4^{2-} eu phase aqueuse. Journal Francais d'Hydrologie, 9, 27-34.

22. Cloke, P.L., 1963a: The geologic role of polysulfides, Pt. 1: The distribution of ionic species in aqueous sodium polysulfide solutions. Geochim. et Cosmochim. Acta., 27, 1265-1298.

23. Barnes, H.L., 1979: Geochemistry of Hydrothermal Ore Deposits. Wiley-Interscience, New York, 798 pp.

24. Naunov, G.B., B.N. Ryzhenco, I.L. Khodakovsky, 1974: Handbook of Thermodynamic Data (translated from the Russian edition, 1971, by G.J. Solermani) NTIS Report PB-226722, 328 pp.

25. Crerar, D.A. and H.L. Barnes, 1976: Ore solution chemistry, V: solubilities of chalcopyrite and chalcocite assemblages in hydrothermal solution at 200°C to 350°C. Econ. Geol., 71, 722-794.
26. Snellgrove, R.A. and H.L. Barnes, 1979: In: Solubilities of Ore Minerals (H.L. Barnes) In: " Geochemistry of Hydrothermal Ore Deposits", 2nd ed. (Barnes, ed.), Wiley-Interscience.
27. Cloke, P.L., 1963b: The geologic role of polysulfides, Pt. 2: The solubility of acanthite and covellite in sodium polysulfide solutions. Geochim. et. Cosmochim. Acta., 27, 1299-1319.
28. Ste-Marie, J., A.E. Torma and A.O. Gübeli, 1964: The stability of thio-complexes and solubility products of metal sulphides 1, cadmium sulphide. Canadian J. of Chem., 42, 662-668.
29. Baes, D.F. and R.E. Mesmer, 1976: The Hydrogen of Cations, John Wiley Inc., Interscience.
30. Whitfield, M., D.R. Turner and A.G. Dickson, 1981: The equilibrium speciation of dissolved components in fresh water and seawater at 25°C and 1 atmosphere pressure, unpublished manuscript.
31. Wuensch, B.J., 1974: Sulfide crystal chemistry. In: "Sulfide Mineralogy", Vol. 1, (Ribbe, ed.), Mineral.Soc. of Am. Short Courses Notes.
32. Jorgensen, R.B., J.G. Kuenen and Y. Cohen, 1979: Microbial-transformations of sulfur compounds in a stratified lake (Solar Lake, Sinai). Limnol. Oceanogr., 24, 799-822.
33. Chen, K.Y. and S.K. Gupta, 1973: Formation of polysulfides in aqueous solution. Environ. Lett., 4, 187-200.
34. Trudinger, P.A., 1979: The biological sulfur cycle. In: "Biogeochemical Cycling of Mineral-Forming Elements", (Trudinger and Swain, eds.), Elsevier Press, New York.
35. Richards, R.A., 1965: Anoxic basins and fjords. In: "Chemical Oceanography", Vol. 1, (Riley and Skirrow, eds.), Academic Press, New York, 611-641.
36. Craig, J.R. and S.D. Scott, 1974: Sulfide phase equilibrium. In: "Sulfide Mineralogy", Vol. 1, (Ribbe, ed.), Mineral Soc. of Am. Short Course Notes.
37. Bruland, K.W., 1980: Oceanographic distributions of cadmium, zinc, nickel and copper in the North Pacific. Earth and Planet. Sci. Lett., 47, 176-198.
38. Martin, J.H. and G.A. Knauer, 1980: Manganese Cycling in the N.E. Pacific Waters. Earth and Planet. Sci. Lett., 51, 266-274.
39. Berner, R.A., 1967: Thermodynamic stability of sedimentary iron sulfides. Amer. J. Sci., 265, 773-785.
40. Suess, E., 1979: Mineral phases formed in anoxic sediments by microbiol decomposition of organic matter. Geochim. et Cosmochim. Acta., 43, 339-352.

41. Edmond, J.M., C. Measures, R. McDuff, L.H. Chan, P. Collier, B. Grant, L.I. Gordon and J.B. Corliss, 1979a: Ridge crest hydrothermal activity and the balances of the major and minor elements in the ocean: The Galapagos data. Earth and Planet. Sci. Lett., 46, 1-18.
42. Edmond, J.M., C. Measures, B. Manigum, B. Grant, F.R. Sclater, R. Collier, A. Hudson, L.I. Gordon and J.B. Carliss, 1979b: On the formation of metal rich deposits at ridge crests. Earth and Planet. Sci. Lett., 46, 19-30.
43. Murray, J.W. and G. Gill, 1978: The geochemistry of iron in Puget Sound. Geochim. et. Cosmochim. Acta., 42, 9-20.
44. Morgan, J.J., 1964: Chemistry of aqueous manganese II and IV. Ph.D. Thesis, Harvard University, Cambridge, MA. 244 pp.
45. Sung, W. and J.J. Morgan, 1981: Osidative removal of Mn(II) from solution catalyzed by the α-FeOOH (lepidocrocite) surface. Geochim. et Cosmochim. Acta., (in press).
46. Nealson, K., 1978: The isolation and characterization of marine bacteria which catalize manganess oxidation. In: "Environmental Biochemistry and Geomicrobiology 3", (Krumbein, ed.), Ann Arbor Science Publ. 847 pp.
47. Gardner, L.R., 1974: Organic versus inorganic trace metal complexes in sulfidic marine waters - some speculative calculations based on available stability constants. Geochim. et Cosmochim. Acta., 38, 1297-1302.
48. Piper, D., 1971: The distribution of Co, Cr, Cu, Fe, Mn, Ni and Zn in Framvaren, a Norwegian anoxic fjord. Geochim. et Cosmochim. Acta., 35, 531-550.
49. Shiozawa, T., K.Kawana, A. Hoshika, T. Tanimoto and O. Takimura, 1977: Vertical distribution of heavy metals and its seasonal variations in Beppu Bay. J. of the Oceanogr. Soc. of Japan, 33, 350-356.
50. Robie, R.A., B.S. Hemingway and J.R. Fisher, 1978: Thermodynamic properties of minerals and related substances at 298.15 k and 1 bar (10^5 Pascals) pressure and at higher temperatures. U.S. Geologic Survey Bulletin No. 1452, 457 pp.
51. Wagman, D.D., W.H. Evans, V.B. Parker, I. Halow, S.M. Bailey and R.H. Schwinn, 1968: Selected values of chemical thermodynamic properties. Pt. 3, Nat. Bur. Std. Tech. Note, 270-273, 264 pp.
52. Mills, K.C., 1974: Thermodynamic Data for Inorganic Sulphides, Selenides and Telluridge. Butterworths, London, 845 pp.
53. Jacobs, L. and S. Emerson, unpublished data.

VARIATIONS OF DISSOLVED ORGANIC COPPER IN MARINE WATERS

Klaus Kremling, Alfred Wenck
and Christoph Osterroht

Institut für Meereskunde
Kiel, F.R. Germany

ABSTRACT

Dissolved organic copper was isolated from large amounts of seawater during two experiments, one in open Baltic waters during an anchor station in May 1980, and the second in July 1980 at Saanich Inlet (Canada) using a controlled ecosystem enclosure (CEE). Samples, taken in intervals between 4 and 24 h, showed average organic copper fractions in the range of about 5% of total copper. Diurnal fluctuations could be detected in Baltic waters, with minimum and maximum Cu values of around 1 and 0.3 nmol l^{-1}, respectively. Short-term variations of organic copper were also evident in the CEE experiment with a range of 0.5 to 3 nmol l^{-1}. The results strongly indicate the release of copper-organic substances (or complexing agents) by primary producers and short-term variations due to microbial activity.

INTRODUCTION

Numerous studies have indicated that the association of copper with dissolved organic matter (DOM) is significant and that these substances may play a major role in the ecology of phytoplankton[1-7]. In two very recent investigations[8,9] it has been shown by direct chemical analysis that copper-organic substances actually exist in marine waters, although their molecular nature is still waiting for further identification.

This paper presents some observations on the occurrence of dissolved organic copper including the detection of short-term variations in two differrent types of marine waters.

EXPERIMENT

Sample collection and materials

Two experiments were conducted, one in open Baltic waters (57°05' N; 19°26' E) during an anchor station in May 1980, and the second in July 1980 at Saanich Inlet (Vancouver Island, Canada) using a controlled ecosystem enclosure extensively employed and described under the CEPEX project[10].

The apparatus used for collecting the copper-organic substances in both runs has been already described in detail[9]. The seawater enters the system by a flow rate of 250 ml min^{-1} and is in contact only with teflon, polypropylene or glass. The total volume of the apparatus is 600 ml resulting in a residence time of ca 3 min during actual sampling. A pH controller allows the introduction of acid (3 mol l^{-1} HCl has been used in our runs) if a pH below the natural value is wanted for the experiment.

Silanized porous glass packed in columns of 3 cm i.d. and a length of 25 cm was used for the isolation of dissolved organic material (DOM). This adsorbant has a relatively high adsorption capacity, allows the fractioned desorption of accumulated polar and non-polar material, and is nearly free from metal contamination. The porous glass (CPG 10 from Serva, Heidelberg) was pretreated and silanised according to procedures outlined in the literature[11,12]. Prior to sample elution the column was rinsed with 100 ml of distilled water. Adsorbed organic substances were eluted with 150 ml of acetone. The eluates were collected in quartz bottles and stored at 4°C under nitrogen until further analysis.

Blank 'tests' of the system revealed negligible contamination or adsorption[9]. The acetone purified by distillation showed copper concentrations of <1 nmol l^{-1}. Regarding an actual Cu content in the organic fractions of about 0.8 μmol l^{-1}, the blank of the solvent can be neglected. (Normally >100 liters of seawater containing ∿0.8 nmol l^{-1} of 'organic Cu' were collected and concentrated in about 150 ml of organic extracts.)

Analyses

The concentration of dissolved Cu in seawater was determined after a APDC/freon extraction[13]. The coefficient of variation has been found to be around 5%. Analysis of copper in the eluted organic fractions was made directly by using calibration curves of aqueous solutions, showing the same slope as the standard addition curves. Atomic absorption spectrophotometers (Perkin-Elmer,

Models 400 and 503) equipped with background correctors and heated graphite atomizers (HGA 72/2100) were used for the measurements.

The nutrient analysis was performed using the Technicon Auto Analizer system[14], and the analyses of chlorophyll a were made fluorometrically[15].

The amount of dissolved organic material (DOM) reported for the Baltic anchor station was estimated by means of the absorbance at 345 nm; the maximum absorption was found at this wavelength in a recorded UV-spectrum of the extracts. The measurements were performed in 1 cm cells using a Zeiss spectrophotometer (Model PM4) and acetone as reference solvent.

RESULTS

Baltic anchor station

During the cruise conducted with R.V. "Poseidon" into the Baltic DOM has been collected on a 14-days anchor station from depths of about 4 - 6 m. Samples were taken over a 24-h cycle with sampling intervals between 4 and 24 h (corresponding to seawater volumes between 60 and 360 liters). The dissolved organic material was accumulated at different pH values (natural and pH 2.5) following to the findings in our previous experiments[9], which have shown considerable increase of isolated DOM with decreasing pH. This provides, first, much more organic material from the same sample volume and secondly, offers the possibility of gathering additional information on the functional groups associated with the metal.

The results are presented in Figs. 1, 2 and 3. The most striking feature is the short-term variation of the organic copper fraction monitored during the first week of the anchor station (Fig. 1). It follows a distinct pattern showing generally maximum values during the day with minimum concentrations in the night. It is interesting to note that, after 24 h of observation, the values obtained are in the same range as measured the day before, with maximum and minimum values of around 1 and 0.3 nmol l^{-1}, respectively. No variations were shown during the second part of the anchor station when a 24-h sampling interval was performed, showing organic copper concentrations were about 0.4 nmol l^{-1}.

The total copper concentrations were relatively constant over the whole period of sampling, varying between 9.5 and 12.5 nmol l^{-1}. The temperature of the surface layer varied between about 4 and 6°C with minimum values on May 20/21 (see Fig. 2) indicating the penetration of colder waters from lower depths. This was also reflected for nitrate which showed a light increase of concentra-

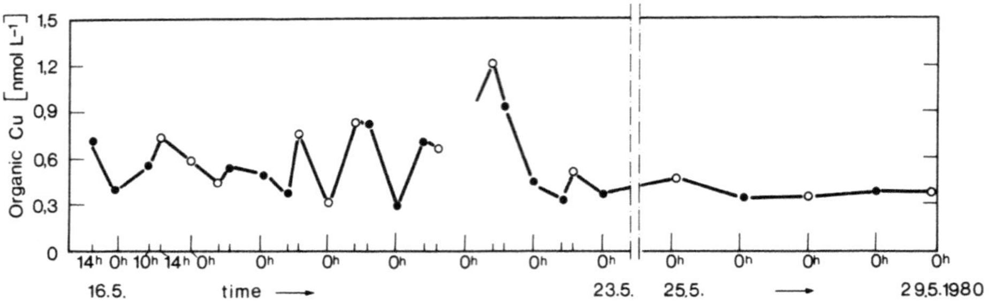

Fig. 1 Diurnal variations in dissolved organic copper monitored at the Baltic anchor station. Isolation of DOM was at natural pH (●) or pH 2.5 (○).

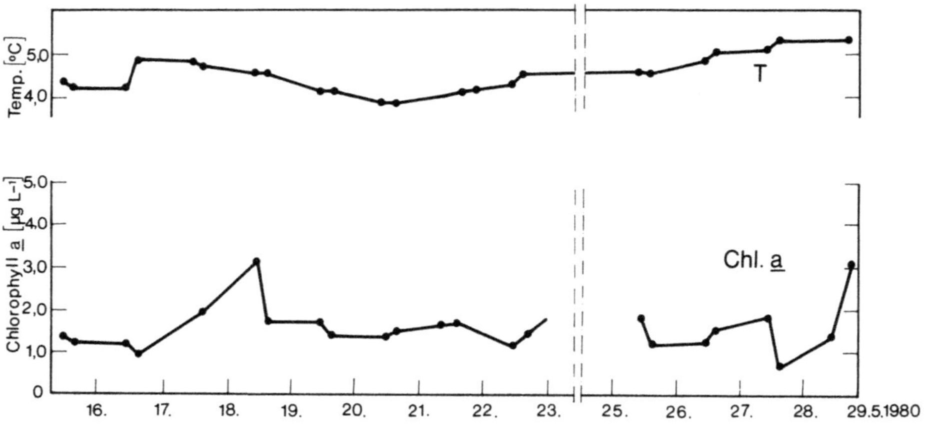

Fig. 2 Variations of temperature and chlorophyll a at the Baltic anchor station (sampling depth 4 - 6 m).

tion during this period. During all the other sampling days the NO_3^- values stayed below 0.1 μmol l^{-1} (phosphate and silicon concentrations values were about 0.3 and 5 μmol l^{-1}, respectively) probably causing the low amount of biomass monitored in the waters (see chlorophyll a values in Fig. 2).

The chlorophyll a and nutrient concentrations did not reflect the pattern of dissolved organic Cu described above. This can also be seen from the measurements of dissolved organic carbon (not reported here) varying between about 0.25 and 0.33 mmol C l^{-1} during the course of the anchor station. This proves that conventional methods are much too insensitive to quantify the excretion of chelating agents during a plankton growth.

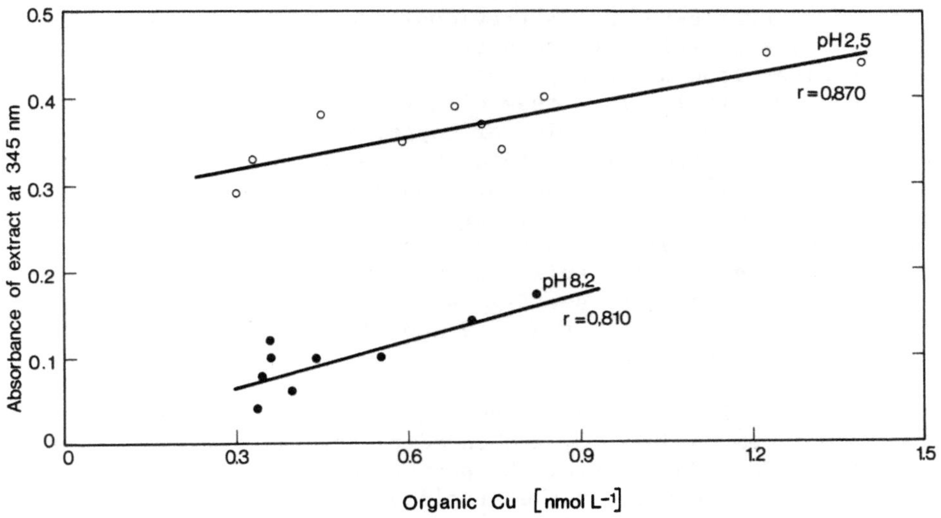

Fig. 3 Relationship between the UV absorbance of isolated DOM and the concentration of organic copper in seawater at natural pH (●) and pH 2.5 (○).

We have shown in our previous paper[9] that a significant correlation exists between the isolated DOM, expressed as absorbance of the extracts at 345 nm, and the concentration of organic copper determined therein. This was confirmed for the anchor station described here (see Fig. 3); only slight differences were found in the copper-complexing capacity (or slope) for the organic substances accumulated at different pH. The much higher absorbance for DOM at pH 2.5 ("background" absorbance corresponds to 0.27 in Fig. 3) is probably contributed by substances such as humic (fulvic) and amino acids. They are represented at relatively high concentrations in Baltic waters and are known to change their physico-chemical behaviour at lower pH with the result that they are more effectively adsorbed. The fraction of organic copper determined under the conditions during this cruise was in the same range (about 5% of total copper) already found during our previous investigations[9].

Controlled ecosystem experiment

In order to support our findings under conditions which guarantee a uniform body of water, a controlled experiment was initiated. Such an experiment should be conducted as closely as possible to field conditions but at the same time recognizing the limitations imposed. The technique adopted was that for the

0.25-scale CEPEX experiments previously described by several authors. The plastic enclosure (Ø 2.44 m), consisting of an acrylic float from which a polyethylene bag ca. 16 m deep was suspended, was filled by a team of divers on July 14, 1980 in Saanich Inlet. About 2 mmoles of copper were added on July 17 to the bag to bring the initial total Cu concentration to approximately 33 nmol l^{-1} (the stock solution had been prepared by diluting $CuCl_2$ in 25 liters of surface seawater). To initiate the experiment and to maintain optimal conditions for phytoplankton growth, the bag was also spiked with nutrients at the beginning of the experiment (July 18, see Fig. 4) resulting in concentrations of about 9 μmol l^{-1} for nitrate, 9 μmol l^{-1} for silicon, and 1 μmol l^{-1} for phosphate.

The phytoplankton population began to grow and reached a biomass maximum on July 20, indicated by the chlorophyll a peak (Fig. 5) on July 20. During the phytoplankton bloom, nutrient concentrations decreased rather rapidly (see Fig. 4). Nitrate most probably became the limiting nutrient, as its concentration decreased to less than 0.1 μmol l^{-1} (all samples were taken from 4 m depth). Shortly after plankton growth was initiated the biological material started to precipitate. This phenomenon is observed in most of the experiments conducted with isolated water masses and is probably caused by a reduction in turbulence and vertical advection in the bag[16]. It seems, however, doubtful if the chlorophyll decline observed on July 21 (see Fig. 5) at 4 m

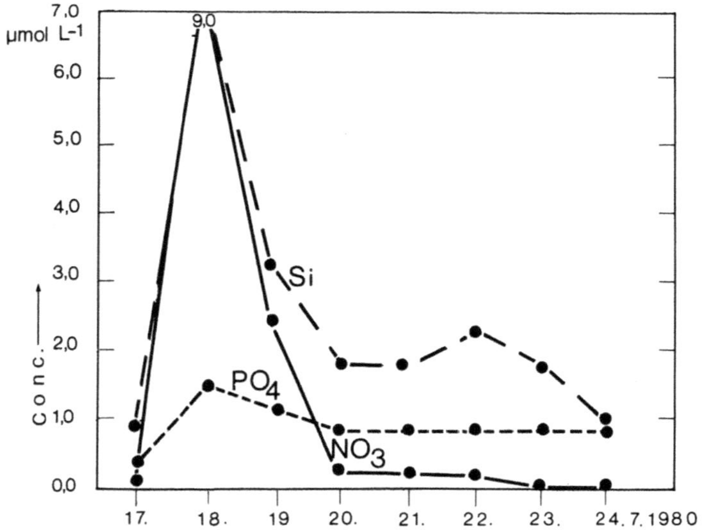

Fig. 4 Changes in concentrations of phosphate, nitrate, and silicon in controlled enclosure at 4 m depth.

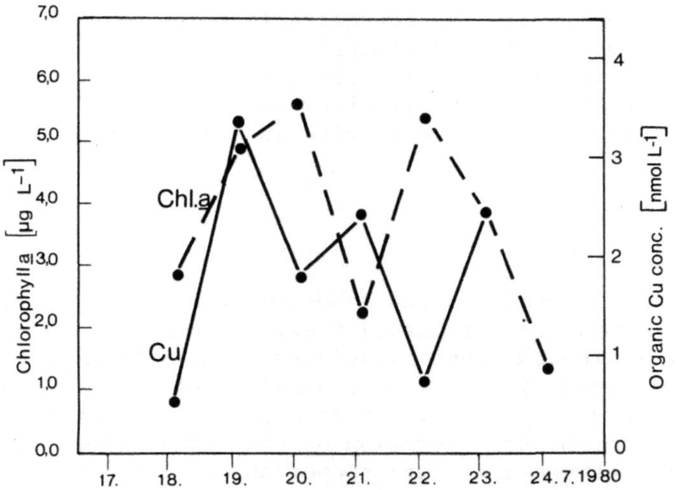

Fig. 5 Variation of chlorophyll a and dissolved organic copper concentrations in seawater during controlled ecosystem experiment.

depth was representative for the water column. It is more likely that the sedimentation actually started at July 23 since a second biomass maximum within only 2 days seems rather improbable.

The composition of the phytoplankton bloom was dominated by diatoms with a dominance (75% of the biomass) of Skeletonema costatum at the beginning of the experiment. The investigation on July 23 showed a decrease of Skeletonema costatum to about 50%, the residual biomass consisting mainly of Chaetoceros, Septocylindrus damicus, and Nitzschia species.

Figure 5 presents the results of the organic copper extraction performed at natural pH. The collection was done daily with sampling intervals between 5 and 10 h corresponding to seawater volumes between about 80 and 160 liters. The most obvious feature of the copper curve is its inverse relationship to the chlorophyll a values. This indicates that the release of copper complexing ligands or the excretion of copper-organic substances is more related to the physiological state of the cells, than just to the cell density. It is important to note that the copper spike (31 nmol l^{-1}) was added one day before the first sampling started on July 18. The pronounced increase of organic copper from about 0.5 to more than 3 mmol l^{-1} on July 19 is, therefore, very probably due to the excretion of chelating agents during phytoplankton growth. The variations monitored in the following days of the experiment were in the same order as described for the anchor station (see Fig. 1). The concentrations of collected organic copper, however, were approximately twice those measured in Baltic waters reflecting the addition of copper to the enclosed water body. The level of total dissolved copper remained relatively constant during the experiment with an average value of 33 ± 3 nmol l^{-1}.

DISCUSSION

The experimental results described here strongly suggest that dissolved organic copper actually exists in marine waters. This confirms our previous investigations [9] and the findings very recently obtained by others with similar techniques [8]. We suggest that these substances originate from primary producers (biosynthesis), either by the excretion of complexed copper or by the interaction between inorganic copper and released complexing agents. It is also conceivable that copper-organic compounds originate from the decompostion of dead organisms. The relatively rapid decrease of the organic copper, however, indicates compounds which are easily metabolized by bacteria and are very likely not identical with high-molecular humic substances (Gocke, personal communication). This is especially evident for the Baltic anchor station where the collection of organic material has been carried out also at rather low pH values. Although the amount of isolated DOM increased by a factor of about 3 (in comparison to the absorbance of extracts collected at natural pH) the concentrations of accumulated organic copper were nearly identical under acidic and natural conditions (see Fig. 1 and 3). These results differ slightly from our previous findings obtained in a highly brackish area of the Baltic (Bothnian Sea) which is known for a high abundance of polyphenolic substances.

The biosynthesis of copper-organic compounds has been investigated by several authors mainly on the basis of biologically motivated culture experiments. Chemical evidence for the excretion of strong copper-complexing agents was recently documented for a number of species of blue-green algae[7]. Although the investigations of these authors were performed in a freshwater medium, there is no major argument against the suggestion that the copper-complexing properties of extracellular metabolites from marine algae are expected to be similar to those from freshwater algae. The culture experiments[7] produced copper-complexing agent concentrations of about 5×10^{-7} mol l^{-1} with stability constants in the range of 10^{8} to 10^{12}. Theoretical approaches[6,17] with compounds such as citrate, histidine, and EDTA, respectively, suggest that organic ligands with similar affinities to copper can markedly affect its chemical state in marine waters by organic complex formation. It must, however, be stressed that chelating compounds are probably not available in the marine environment in micromolar concentrations. The investigation of individual amino acids in Baltic waters[18,19] (stations were located near the anchor station described here) showed generally concentrations of $<10^{-7}$ mol l^{-1} implying that these compounds do not account for complexation properties of seawater. Unfortunately, the state of our knowledge does not allow yet any suggestions on the molecular structure of the isolated substances. In the previous paper, however, we have demonstrated the usefulness of high-performance liquid chromatography for separation of copper-organic substances from the bulk of dissolved organic material[9]. Future studies are directed towards further characterization of these substances by well established physical-chemical procedures.

The CPG 10 column removed only about 20% of the DOM as indicated by DOC analysis. It is, therefore, important to emphasize that not all of the copper-organic substances may be extracted by the technique applied. Repeated extraction on unloaded columns, however, gave no indications of further adsorption indicating a very selective accumulation. It is also necessary to consider possible artifacts in the isolation of the copper-organic material on the CPG 10 column. It is conceivable that copper-organic substances are formed when Cu^{2+} ions or inorganic copper species pass through the organic rich column. It was already argued[8] that this process prerequisites a fast reaction and strong affinity of the organic ligands with copper because of competing Mg^{2+} and Ca^{2+} ions. One would assume that such compounds would have already been formed under natural conditions. It should also be expected that the amount of organically associated copper would increase with the pool of adsorbed organic material. This has never been observed in our experiments, for example when extractions of the same water masses were processed at pH 2.5 (see Fig. 1 and 3). These considerations and the excellent reproducibility of our extractions support the assumption that the copper-organic substances are actually present in the seawater and not formed during the isolation process.

The most exciting, but somewhat surprising finding is the distinct variation documented in dissolved organic copper. This is especially evident during the first week of the anchor station with maximum concentrations during the day, and minimum values generally overnight. The pronounced rhythm, re-establishing the values after 24 h, led us to the suggestion that microorganisms are responsible for the variations. It is interesting to mention that indications for "diurnal" fluctuations in Baltic waters have also been determined recently for dissolved sugars and amino acids[18,19]. These compounds followed fluctuations similar to those described here for organic copper (the ratio of the maximum/minimum concentrations corresponded also to about 3). The causes of this rhythm are unknown. However, the findings indicate very probably that the material cycle is governed by rhythms of microbiological and/or chemical parameters. Although minor inhomogeneities in the water masses have to be considered during the anchor station, we believe that, statistically, such a distinct pattern can hardly be expected if different bodies of water had been sampled.

Another characteristic of the results is that the strong variations of organic copper did not reflect the fluctuations in chlorophyll a or inorganic chemical parameters suggesting that the exudation of the material investigated here is not merely related to the biomass concentration. McKnight and Morel[7] showed that the copper-complexing agents occured mainly during the stationary phase of their batch cultures, indicating the importance of the physiological state of the plankton for the excretion of chelating substances.

Studies on short-term variations of dissolved organic materials are very limited in the literature (see refs. 18 and 19). From our knowledge, variations of metal-organic compounds in marine waters have never been investigated before. One reason is that it is difficult to follow uniform water bodies over sufficient periods of time. The main reason, however, seems to us the lack of adequate analytical systems to measure small, but pronounced differences and to process a frequent sample series.

The results and problems described above stress the need for further joint chemical and microbiological investigations on the short-term dynamics of dissolved copper-organic compounds. We consider these studies as a main assumption in understanding the biological interaction of this element. However, similar studies should also be started for other metals initiating possibly a new field of trace element research in marine waters.

ACKNOWLEDGEMENTS

We are grateful to Dr. C.S. Wong for giving us the opportunity to conduct the controlled experiments in Saanich Inlet. Special thanks go to P. Munro, F. Whitney, and K. Johnson for their great help in preparing and processing the experiment. We also thank Dr. V. Smetacek for identifying the phytoplankton organisms.

This investigation was supported by the German Federal Ministry of Research and Technology (Bundesministerium für Forschung und Technologie).

REFERENCES

1. Slowey, J.F., L.M. Jeffrey and D.W. Hood, 1967: Evidence for organic complexed copper in sea water. Nature, 214, 377-378.
2. Williams, P.M., 1969: The association of copper with dissolved organic matter in seawater. Limnol. Oceanogr., 14, 156-158.
3. Steemann-Nielsen, E. and S. Wium-Andersen, 1970: Copper ions as poison in the sea and freshwater. Mar. Biol., 6, 93-97.
4. Sunda, W. and R.R.C. Guillard, 1976: The relationship between cupric ion activity and the toxicity of copper to phytoplankton. J. Mar. Res., 34, 511-529.
5. Swallow, K.C., J.C. Westall, D.M. McKnight, N.M.L. Morel and F.M.M. Morel, 1978: Potentiometric determination of copper complexation by phytoplankton exudates. Limnol. Oceanogr., 23, 538-542.
6. Jackson, G.A. and J.J. Morgan, 1978: Trace metal-chelator interactions and phytoplankton growth in seawater media: theoretical analysis and comparison with reported observations. Limnol. Oceanogr., 23, 268-282.
7. McKnight, D.M. and F.M.M. Morel, 1979: Release of weak and strong copper-complexing agents by algae. Limnol. Oceanogr., 24, 823-837.
8. Mills, G.L. and J.G. Quinn, 1981: Isolation of dissolved organic matter and copper-organic complexes from estuarine waters using reverse-phase liquid chromatography. Mar. Chem., 10, 93-102.
9. Kremling, K., A. Wenck and C. Osterroht, 1981: Investigations on dissolved copper-organic substances in Baltic waters. Mar. Chem., 10, 209-219.
10. Menzel, D.W. and J. Case, 1977: Concept and design: controlled ecosystem pollution experiment. Bull. mar. Sci., 27, 1-7.
11. Karch, K., 1974: Unpolare stationäre Phasen in der Hochdruckflüssigkeitschromatographie. Thesis, Univ. Saarbrücken.
12. Derenbach, J.B., M. Ehrhardt, C. Osterroht and G. Petrick, 1978: Sampling of dissolved organic material from seawater with reversed-phase techniques. Mar. Chem., 6, 351-364.
13. Danielsson, L.-G., B. Magnusson and S. Westerlund, 1978: An improved metal extraction procedure for the determination of trace metals in sea water by atomic absorption spectrometry with electrothermal atomization. Anal. Chim. Acta, 98, 47-57.
14. Grasshoff, K., 1976: Methods of seawater analysis. Verlag Chemie, Weinheim.

15. Strickland, J.D.H. and T.R. Parsons, 1972: A practical handbook of seawater analysis. Bull. Fish. Res. Bd. Canada, 167, 1-311.
16. Kremling, K., J. Piuze, K. v. Bröckel and C.S. Wong, 1978: Studies on the pathways and effects of cadmium in controlled ecosystem enclosures. Mar. Biol., 48, 1-10.
17. Stumm, W. and P.A. Brauner, 1975: Chemical speciation, in: "Chemical Oceanography", Vol. 1, J.P. Riley and G. Skirrow, ed. Academic Press, London.
18. Meyer-Reil, L.-A., M. Bölter, G. Liebezeit and W. Schramm, 1979: Short-term variations in microbiological and chemical parameters. Mar. Ecol. Prog. Ser., 1, 1-6.
19. Liebezeit, G., 1980: Aminosäuren im marinen Milieu - neuere analytische Methoden und ihre Anwendung. Dissertation, Univ. Kiel.

TRACE METALS SPECIATION IN NEARSHORE ANOXIC AND SUBOXIC PORE WATERS

W. Berry Lyons and W. F. Fitzgerald

Department of Earth Sciences
University of New Hampshire
Durham, New Hampshire 03824 U.S.A.

Marine Sciences Institute
University of Connecticut
Avery Pt., Groton, CT 06340 U.S.A.

ABSTRACT

Sedimentary pore waters were obtained seasonally from two nearshore areas in Long Island Sound, U.S.A. The pore waters were collected, separated from the sediments and analyzed utilizing an inert atmosphere and ultra-clean laboratory techniques in order to minimize any artifacts due to improper sampling and contamination. The pore fluids were analyzed for a variety of major and minor components as well as the trace metals Fe, Mn, Cu, Cd and Ag. The sediments from Branford Harbor were anoxic and bacterially mediated sulfate reduction was the major biogeochemical process taking place. As this process proceeds, SO_4^{2-} is consumed and HS^-, PO_4^{3-}, NH_4^+, titration alkalinity and dissolved organic carbon are produced. Thermodynamic calculations indicate that metal-polysulfide and bisulfide complexes may control the concentrations of dissolved Cu, and Ag in these pore waters, while the concentrations of Fe, Mn and Cd are controlled by other inorganic species. Except for Cu, there is little evidence to suggest that complexation with dissolved organic matter plays a major role in controlling the solubility of these trace metals in the anoxic waters under investigation.

On the other hand, the sediments from Mystic River showed little evidence of sulfate reduction. Enrichments of Fe, Mn, Cu, and Ag did occur in the surface pore waters. These enrichments are through to be due to either the reduction of Mn and Fe oxides or the oxidation of iron monosulfides.

INTRODUCTION

The chemistry of trace metals other than iron and manganese in sedimentary pore fluids is poorly understood and until recently has not been studied. In addition, the majority of recent studies concerning pore water trace metals were conducted without the knowledge of oxidation effects[1-4], or without adequate concern for contamination problems.[5,6]

During early diagenesis trace metals are remobilized from and partitioned to various reservoirs in the sediments. The pathway for these diagenetic changes is undoubtedly the pore fluids of the sediments. Trace metals are solubilized into the pore waters and subsequently removed from the pore fluids as authigenic products.

The primary approach in studying trace metal removal is the assumption that thermodynamic equilibrium is attained between the liquid and solid phases. This approach has been utilized to identify mineral-water equilibrium reactions in the sediments. It has proved to be very successful in the understanding of iron and manganese chemistries in nearshore sediments.[7-12] However, there has been considerable debate concerning whether equilibrium models are applicable to other transition elements.[13,14] For example, Presley et al.[13] and Duchart et al.[15] found that the concentrations of Cu, Zn, Ni, Co and Pb were all higher than what might be expected by simple sulfide precipitation (e.g., the least soluble mineral phase). Yet, because metals other than Fe and Mn do not apparently form authigenic minerals that can be detected by conventional x-ray diffraction techniques, this thermodynamic approach still appears to be the best first approximation to the delineation of sediment-pore water interactions involving trace metals geochemistry. This approach assumed, however, that all the dissolved metal species are known and that the metal concentrations measured are truly "dissolved" and not colloidal in character.

The purpose of this study was: 1) to accurately measure the concentrations of Fe, Mn, Cu, Cd and Ag as well as the various ligands that can form complexes with these metals in the pore waters of nearshore sediments and 2) by the use of thermodynamic calculations to determine the speciation of these metals in the pore water samples.

ENVIRONMENTAL SETTING

Two different areas were chosen as sampling sites. Branford Harbor in the central portion of Long Island Sound (Figure 1), 5 nautical miles east of the entrance of New Haven Harbor, was chosen because it represents an area that has been greatly disturbed by anthropogenic activities. The sampling site is located on the Pawson Park Marsh tidal flat about 0.3 km south of the American Wire Company and approximately 100 m E-SE from the Branford city sewage treatment plant outfall. The site is exposed at mean low water and the temperature of the surface sediments varies from 20.0°C in late August to 2.5°C in mid-February. Throughout the year the majority of the tidal flat is covered with a dense growth of sea lettuce, *Ulva lactuca*, and a large population of the mud snail, *Ilyanassa obsoleta*, grazing on the sediment surface.

The second sampling area is located in the Mystic River estuary, 4 nautical miles due east of the University of Connecticut's Marine Sciences Institute in Groton in eastern Long Island Sound (Figure 1). The Mystic River estuary opens into Fishers Island Sound. The sampling site is located under 3 m of water and was chosen because it is representative of the nearshore sedimentary environment in the eastern part of Long Island Sound.

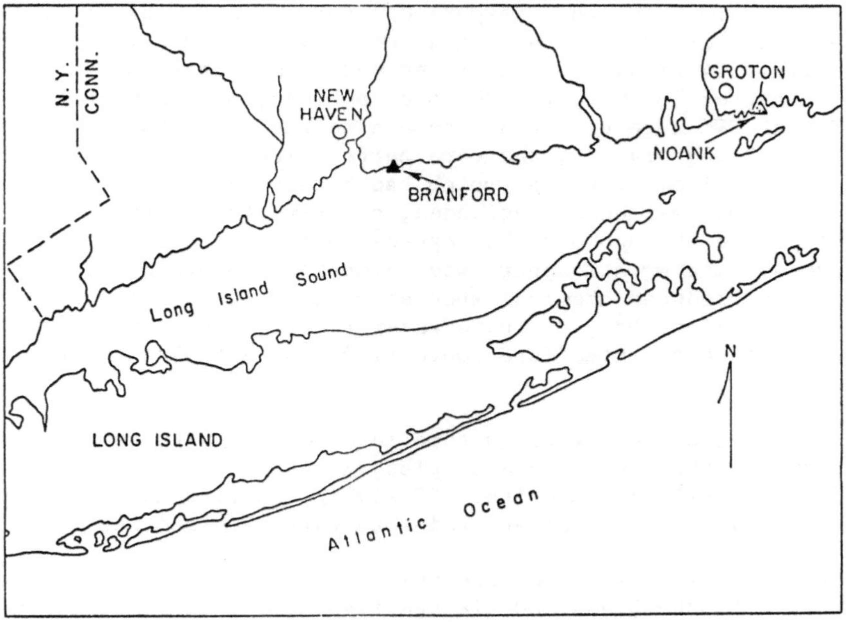

Fig.1 Sampling sites in Long Island Sound

During the summer and early fall the sediment surface is covered by a dense growth of eel grass, Zostera marina. The surface sediment temperatures ranged from 17.5° in mid-September to 4.0°C in mid-February.

There are differences in the seasonal trends of physical oceanographic parameters such as sediment transport in these two regions of Long Island Sound.[16,17] Bohlen[17] suggests that the eastern part of the Sound is more representative of a coastal embayment compared to the more classically defined estuarine character of the central and western Long Island Sound.

The sediments from Branford Harbor are anoxic with sulfate reduction being the most important biogeochemical reaction taking place, while the Mystic River sediments (at least in February) were suboxic with little to no sulfate reduction occurring.[18,19] The mean concentration of organic carbon is 2.34% at the Branford site and 1.85% at the Mystic River sampling area.[19] In general, titration alkalinities, and phosphate increase with depth in the pore waters at Branford Harbor and remain relatively constant at Mystic River.

SAMPLING AND ANALYTICAL PROCEDURES

The methods of sediment collection and pore water extraction are outlined in Lyons and Fitzgerald[18] and Lyons.[19] Briefly, the sediment was collected in acid cleaned polyvinyl chloride (PVC) core barrels (4 cm o.d.). The cores were immediately placed in larger diameter (6 cm i.d.) PVC core carriers, purged with N_2 gas, sealed with PVC core caps and returned to the laboratory. Upon return to the laboratory, the core barrels were placed into a 183 cm x 91 cm x 61 cm glove bag which had been purged with N_2. The cores were then extruded, sectioned, centrifuged and the pore fluid filtered through 0.4μ Nuclepore T.M. filters under the N_2 atmosphere. The N_2 atmosphere was maintained to avoid possible oxidation of reduced chemical species present in the pore waters. While in the glove box, the pore water to be analyzed for trace metals was then decanted into conventional polyethylene (CPE) bottles.

All material that were to come into contact with the trace metal samples (CPE centrifuge bottles, Nuclepore T.M. filters, Satorious T.M. filtration units, CPE storage bottles and Eppendorf T.M. pipette tips) were cleaned according to Patterson and Settle.[5]

The pore water samples were then removed from the glove bag, transported to the Class 100 air quality ultra-clean laboratory at the Marine Sciences Institute, University of Connecticut and acidified to pH 1 with Baker Ultrex T.M. nitric acid. With the exception of the extrusion of the sedimentary material from the cores and the pore water separations, all sample preparations, manipulations and storage of these samples were performed in the clean laboratory. The laboratory provides a uniform air flow which through constant filtration minimizes any particulate contamination. This is of utmost importance in trace metal analysis. Q-water or ultrapure quartz sub-boiling distilled water was used for the preparation of all solutions used in trace metal analysis[6,20] as well as the cleaning of all laboratory ware that was to come into contact with the trace metals samples. In this work much effort and care were taken in the cleaning of sampling equipment and laboratory ware and sample handling to minimize any trace metal contamination.

Iron and manganese were analyzed after dilution with Q-water. In this technique 100 μl of the pre-acidified pore water was pipetted by an Eppendorf T.M. pipette into a 10 ml volumetric Pyrex T.M. flask. To this flask 50 μl of Ultrex T.M. HNO_3 was added and the solution was then brought to volume with Q-Water. Standards were made up by adding various amounts of 1 ppm stock metal solutions, 50 μl of Ultrex T.M. HNO_3 and 100 μl of a sea water solution of approximately the same salinity as the samples to be analyzed. This final addition ensures that the standards are of approximately the same ionic strength and contain the same salts as the samples. The sea water solution was made up by filtering Long Island Sound water at 27 $^{o}/oo$ salinity through a cleaned Satorious filtering unit using a clean 0.4 μ Nuclepore T.M. membrane filter. The filtered water was diluted on a weight basis with Q-water to arrive at the desired salinity. Blanks were made up with 100 μl of the appropriate salt solution and made up to volume with Q-water.

The samples were analyzed by injecting 25 μl aliquots into a HGA 2000 Perkin-Elmer graphite furnace attached to a Jarrell-Ash 82-800 Double Beam Atomic Absorption Spectrophotometer. The signal response was monitored by a Fisher series 200 Recordall recorder, recording at 1.3 cm per minute. The input scale of the recorder was varied with the element being measured. Graphite tubes in the furnace were replaced after 75 to 100 analyses. Metal concentrations were determined by comparing the peak heights of the samples to the standard curve established by the determination of at least five known standards. The detection limits of this technique for 1% absorption were: Fe, 0.9 μmol l^{-1} and Mn, 0.2 μmol l^{-1}. The 33 cm depth pore water sample from Branford Harbor was analyzed in triplicate using this technique. The coefficient of variation was: Fe, $\pm$11% at 6.5 μmol l^{-1} and Mn, $\pm$12% at 11.8 μmol l^{-1}.

Cadmium, copper and silver were determined by an ammonium pyrrolidine dithiocarbamate (APDC) chelation followed by a methyl isobutyl ketone (MIBK) extraction of the metal chelate from the aqueous phase.[21,22] The standard addition method of Kremling and Peterson[23] was used due to the small quantity of sample available. In this technique, a 1.00 ml sample of the pre-acifidied pore water was introduced using an Eppendorf T.M. pipette into a pre-cleaned 2.0 ml quartz glass vial. The pH of the sample was then adjusted to 3.5-4.0 by the addition of from 50 to 125 μl of a 1% ammonium acetate buffer solution. The buffer solution was made up in Q-water and the metals removed by the addition of 500 μl of 2% APDC followed by three extractions with 25.0 ml of chloroform. The buffer was then extracted three more times with 25.0 ml of chloroform after the addition of 200 μl of the 2% APDC solution. Finally, the buffer was extracted with from 2 to 15 ml of double distilled MIBK. This procedure reduced the metal concentrations in the buffer solution to below the detection limits for all the metals studied.

After adjusting the pH of the sample, 100 μl of a 2% APDC solution was added to the quartz vial and the vial capped and stirred for 2 minutes with a Bronwill T.M. autoswitch mixer. The 2% APDC solution was made by adding 0.5 g of APDC to 25 ml of Q-water. After the filtration of this solution through a Nuclepore T.M. 0.4 μ membrane filter, it was extracted four times with chloroform using 25 ml in the first extraction and 10 ml in the following three extractions. The extractions were carried out in a 125 ml Teflon (FEP) separatory funnel and the final solution was stored in a pre-cleaned 125 ml CPE bottle. After its preparation the APDC solution was frozen and used again the following day. On the third day a new solution was prepared.

Finally, 1.00 ml of double distilled MIBK was added to the quartz vial with an Eppendorf pipette. The vial was then capped and mixed for 3 minutes with the Bronwill T.M. autoswitch mixer. Two blanks consisting of Q-water, buffer, APDC and MIBK were analyzed along with two samples and three standard additions for each sample. The concentrations of Cu, Cd and Ag were determined using the standard additions technique of Brewer et al.[22] after the blanks were subtracted from both the samples and the standards.

The detection limits of this technique for 1% absorption were determined to be: Cu, 0.03 μmol l^{-1}; Cd, 2 nmol l^{-1}; Ag, 2 nmol l^{-1}.

Due to the small volume of pore water available for trace metal determination (less than 10 ml) replicate extractions and analysis could not be undertaken. Instead, surface water from the Branford Harbor area was collected in May of 1976 and 1 ml aliquots were analyzed for Cu, Pb, Cd and Ag by the above technique. The coefficients of variation of triplicate analysis were: Cu,

±9% at 0.52 μmol l^{-1}; Cd ±50% at 2 nmol l^{-1}; and Ag, ±50% at 2 nmol l^{-1}. The large coefficients of variation for Cd and Ag are due to the face that their concentrations were near their detection limits.

The detection limit reported here is higher than that reported by Kremling and Peterson [23] for Cu, the only metal mutually studied. These authors report values as low as 0.011 μmol l^{-1} using this technique with a relative standard deviation of ±14% at 0.025 μmol l^{-1}.

Values of ΣH_2S were determined on a sub-set of pore water samples using the technique of Cline [24] which had been modified for pore water by Goldhaber.[25] The coefficient of variation of this method was ±12% at 250 μmol l^{-1}. Dissolved organic carbon was determined using the method of Menzel and Vaccaro [26] with modifications for the higher concentrations expected in pore water by Lindberg [27]. The precision of this measurement was ±5% at 3.33 mmol l^{-1}. Analytical techniques for nutrient and titration alkalinity are outlined in Lyons and Fitzgerald [18] and detailed in Lyons.[19]

TRACE METAL PROFILES

Iron

The dissolved iron data are shown in Tables 1 and 2. The values obtained in this study are similar to those reported recently for other nearshore brackish water/marine environments where prevention of pore water oxidation was taken into account, Troup [8] has observed between 2 to 37 μmol l^{-1} in the top 70 cm of southern Chesapeake Bay pore fluids, with the majority of values between 2-12 μmol l^{-1}. Martens and Goldhaber [28] have observed dissolved iron concentrations between 9-18 μmol l^{-1} in brackish North Carolina pore waters. Murray and Gill [11] have found dissolved iron concentration in Puget Sound pore waters as high as 110 μmol l^{-1} but in sediments where dissolved sulfide is present, the values usually range between 1 and 7 μmol l^{-1}. The highest ferrous iron concentrations observed are usually in the top 10 cm of the sediments.

Troup [8] and Murray and Gill [11] have found increases in pore water iron with depth similar to that observed in the core.

Manganese

The dissolved manganese data are shown in Tables 1 and 2. The concentrations and profiles of Mn^{2+} reported here as similar

Table 1 Trace Metal Pore Water Data from Branford Tidal Flat Sediments

Depth (cm)	Fe $\mu mol\ l^{-1}$	Mn $\mu mol\ l^{-1}$	Cu $\mu mol\ l^{-1}$	Ag $nmol\ l^{-1}$	Cd $nmol\ l^{-1}$
May, 1976					
4	11.5	23.8	0.09	<2	4
12	7.7	14.0	0.08	<2	4
20	6.4	11.8	0.05	<2	4
33	4.8	5.5	<0.03	<2	3
45	4.3	2.7	<0.03	<2	4
63	3.6	0.5	<0.03	<2	3

Table 2 Trace Metal Pore Water Data from Mystic River Estuary Sediments

Depth (cm)	Fe $\mu mol\ l^{-1}$	Mn $\mu mol\ l^{-1}$	Cu $\mu mol\ l^{-1}$	Ag $nmol\ l^{-1}$	Cd $nmol\ l^{-1}$
February, 1976					
3.5	36.5	12.7	0.42	9	2
12	4.1	3.8	0.13	4	<2
23	2.9	1.1	0.08	2	<2
36.5	9.7	2.2	0.06	2	<2

to those reported by other investigators in the recent literature. Holdren et al.[29] have observed concentrations as high as 275 $\mu mol\ l^{-1}$ in Chesapeake Bay. Sanders[30] found values of Mn^{2+} from 1 to 5$\mu mol\ l^{-1}$ in a nearshore marine environment near Morehead City, N.C. with maximum concentrations at 3 cm. Grill[31] has observed similar Mn^{2+} profiles in pore waters from Jarvis Inlet, British Columbia. Murray et al.[32] have reported values from 0.7 to 95 $\mu mol\ l^{-1}$ in pore fluids in Saanich Inlet, B. C., with maximum concentrations at 5 cm depth of sediments.

Copper

The pore water copper values are shown in Tables 1 and 2. The high Cu concentration at 3.5 cm at Mystic River is found in the sediment layer which contained visible quantities of dead eel grass (Zostera marina). It is possible that the dissolved Cu and possibly the Fe, Mn, Ag and Cd found at this depth have come from the in-situ decomposition of the eel grass. Recent work by Elderfield et al.[33], however, indicates that this increase in pore water trace metal values in the surface of nearshore sediments in the winter is more likely to be due to the oxidation and dissolution of metal-rich sulfides. Elderfield et al.[33] have observed pore water enrichments of Cu and Ni in Narragansett Bay and attribute these increases to the above cited mechanism. Aller[12] has observed similar winter enrichments for Fe in central Long Island Sound pore fluids which are produced by the oxidation of iron monosulfide minerals. The oxidation of metal sulfides and the subsequent solubilization of metals has been documented via laboratory experiments.[34]

The Cu pore water values observed in this study are similar to those reported by Bricker and Troup[35] and Heggie and Burrell.[36] Bricker and Troup[35] found that profiles of Cu followed those of the Fe^{2+} pore water profile in Chesapeake Bay pore fluids. Their Cu values ranged from 0.05 to 0.48 $\mu mol\ l^{-1}$. Heggie and Burrell[36] found from 0.02 to 0.16 $\mu mol\ l^{-1}$ in pore water from Resurrection Bay, Alaska. Both of these investigations found maximum Cu pore water values at 4-7 cm depth, similar to the results presented here. Elderfield et al.[33] have found similarly shaped Cu pore water profiles in Narragansett Bay, R.I. Their values range from less than 0.04 to 0.36 $\mu mol\ l^{-1}$. Tramontano[37] has observed values from 0.10 to 0.42 $\mu mol\ l^{-1}$ in the pore waters of Thames River, Conn. sediments. His profiles also show that, in general, the Cu concentration decreases with depth. Recently Lyons et al.[38] have observed in pore water Cu values ranging from 4.7 to 7.9 $nmol\ l^{-1}$ in Devil's Hole, Bermuda carbonate sediments.

Cadmium and Silver

The Cd and Ag pore water data are shown in Tables 1 and 2. At the Branford location Ag^+ is below the detection limits of our method throughout the core. The Ag^+ values at Mystic River are higher with the highest value observed at 3.5 cm. It is possible that this large concentration of Ag^+ has been generated from the degradation of recently deposited eel grass detritus, the reduction of Mn and/or Fe oxides or the oxidation of iron monosulfides.

Cadmium profiles at Branford Harbour increase across the sediment-water interface to a steady-state value (i.e. 4 $nmol\ l^{-1}$)

Table 3 Branford Harbor and Mystic River Pore Water Data

Depth cm	Salinity $^{o}/oo$	Alkalinity meq ℓ^{-1}	PO_4^{-3} mmol 1^{-1}	DOC mmol 1^{-1}	NH_4^+ mmol 1^{-1}	SO_4^{2-} mmol 1^{-1}	ΣH_2S mmol 1^{-1}	pH
BRANFORD								
4	17.2	7.39	0.57	1.2	0.10	13.8	0.010	6.8
12	20.6	11.05	0.73	1.4	0.27	10.6	0.010	6.8
20	-	15.49	0.93	3.0	0.25	10.0	0.065	6.7
33	20.6	18.31	0.82	3.1	0.59	6.4	0.229	7.0
45	21.1	22.01	0.85	3.3	1.01	3.5	0.448	6.9
63	18.2	16.27	0.71	3.5	1.15	4.4	0.558	-
MYSTIC								
3.5	18.2	2.77	0.03	0.8	0.11	24.2	0.002	7.1
12	26.5	1.88	0.03	1.3	0.07	23.6	0.006	6.9
23	28.4	1.43	0.02	2.0	0.07	19.9	0.006	6.6
36.5	34.4	1.88	0.04	3.9	0.06	23.4	0.004	6.4

while at Mystic River they reach a maximum value at 3.5 cm then decrease below detection limits with depth.

In general, the dissolved concentrations of Fe, Mn, Cu, Cd and Ag increase across the sediment-water interface (highest values in top 3-4 cm core sections) and then decrease with depth.

The higher concentrations of Fe, Mn and Cu, in the surface pore waters are due to production or solubilization of these metals from some solid phase. These decreases of concentration with depth suggest that the metals are being removed from solution either by diffusion and/or precipitation of an authigenic mineral phase or absorption onto a solid substrate.[9,29]

Trace Metal Speciation: Branford Harbor

The solution chemistries of the pore water samples upon which the speciation calculations were made are shown in Table 3. Both hand calculations and the MINEQL program were utilized in the speciation calculations.

Irons and manganese. Lyons and Fitzgerald[39] have shown that the pore waters under investigation are all undersaturated with respect to $FeCO_3$ (siderite), $Fe_3(PO_4)_2$ $8H_2O$ (vivianite), $Mn_3(PO_4)_2$ $3H_2O$ (reddingite) and albandite (MnS). Saturation calculations indicated that the ion activity product (IAP) of ferrous sulfide in these pore waters varies from 4.0×10^{-21} to 1.8×10^{-17}.[19,39] These values range from equilibrium with respect to amorphous FeS (1.2×10^{-17}) to 180 undersaturated with respect to Fe_3S_4 (6.3×10^{-19}).

These calculations indicate that both Mn^{2+} and Fe^{2+} pore water concentrations are controlled by the precipitation of the authigenic mineral phases $MnCO_3$ and FeS. The black color of the top 20 - 30 cm of these sediments support the idea that the monosulfide FeS was present in these sediments.[40] Berner[40] has also observed both the monosulfide and pyrite (FeS_2) in the top 10-15 cm of the Branford Harbor sediments.

The major iron species present in these pore waters are $FeCl^+$ and Fe^{2+} (Table 4). It is also possible that organic complexes of Fe^{2+} or Fe^{2+} - Fe^{3+} colloids exist in these samples. Krom and Sholkovitz[41] and Lyons et al.[42] have observed Fe - organic matter associations in anoxic pore waters or nearshore sediments. What importance, if any, these species play in the Fe chemistry of present samples is unknown.

Calculations indicate that the major Mn^{2+} species in these pore fluids are $MnCl^+$, Mn^{2+} and $MnCl_2^{\circ}$ (Table 4). Organic matter

- Mn interactions in nearshore anoxic pore waters appear to be minimal.[41,42]

Metals other than Fe-Mn. Many investigators[13,15,43-46] have observed the apparent lack of simple sulfide solubility in the control of trace metals other than Fe^{2+} and Mn^{2+} in anoxic systems. These authors have suggested that dissolved metal distributions are controlled by: (1) some unknown material -- water equilibrium[43,45], (2) organic matter -- trace metal speciation[13], (3) inorganic speciation[14] and (4) a kinetic hindrance due to organic coatings on nucleation sites[46].

Cadmium. Because of the imprecision of the cadmium pore water data it is difficult to quantify as well as qualify cadmium speciation. However, calculations do indicate that the activity of the species $CdOH^+$ is within the range of the observed values (Table 4). Boulegue[47] has also shown that $CdOH^+$ should be the major cadmium species in anoxic ground waters. At the higher ΣH_2S concentrations $Cd(HS)^+$ may also be an important Cd species.

Silver. The Ag^+ measurements are also of very low precision which makes thermodynamic modeling even more difficult. Calculations indicated, however, that the silver-chloride species are not of importance in these pore waters. Calculations do indicate (Table 4) that the silver bisulfide complex, $AgHS^o$, may be significant in these systems.

Copper. Bouleque[47] has shown through thermodynamic considerations that Cu^{2+} should be reduced to Cu^+ at the pH's and Eh's encountered in anoxic pore waters (e.g., pH = 6-8, Eh = 0.1-0.4v). It is reasonable to conclude that the copper observed in the pore waters under investigation is Cu^+. This is an important consideration because Cu^+ forms strong polysulfide complexes while Cu^{2+} does not.

Table 4 Major trace metal species in Branford Harbor and Mystic River pore waters

METAL	BRANFORD	MYSTIC RIVER
Fe	$FeCl^+$, Fe^{2+}	Fe^{2+}, $FeSO_4^o$
Mn	$MnCl^+$, Mn^{2+}, $MnCl_2{}^o$	Mn^{2+}, $MnCl^+$, $MnCl_2{}^+$, $MnSO_4{}^o$
Cu	$Cu(S_4)_2{}^{3-}$, CuDOM?	Cu^{2+}, $Cu(OH)_2{}^o$, $CuHCO_3{}^+$?
Cd	$Cd(OH)^+$, $CdHS^+$?	Cd(OH), $CdCl^+$?
Ag	$AgHS^o$?	$AgCl_4{}^{2-}$

Gardner [14] and Boulegue [47,48] have demonstrated that Cu^+-polysulfide complexes are of extreme importance in systems rich in reduced sulfur. Although polysulfide was not measured in this study, the polysulfide species (S_4^{2-}) can be determined empirically using the expression derived by Boulegue and Michard [49].

$$a_{S_4^{2-}} = \frac{10^{-9.3}\, a_{HS-}}{a_{H+}}$$

The calculated activities of $Cu(S_4)_2^{3-}$ are four to eight orders of magnitude lower than the observed values of total copper in the top three sections of core. This discrepancy can be explained by three possible mechanisms: 1) enhanced polysulfide levels above theoretical concentrations, 2) organic-Cu associa tions, 3) collodial, rather than dissolved, Cu present in the samples.

Enhanced Polysulfide Values

Boulegue [47] has shown that in the French Dogger waters polysulfide concentrations are found some 100 times higher than predicted theoretical values. In addition, Aizenshtat et al.[50] have measured polysulfide values as high as 150 mmol l^{-1} in blue-green algal mat pore fluid in Solar Lake, Sinai, where the redox conditions fluctuate rapidly on a daily basis. It would seem likely that in environments such as salt marshes, tidal flats and other transition environments that are intermittently exposed to anoxic and toxic conditions, polysulfide concentrations should be found in greater concentrations than their theoretical ones. This "disequilibrium" of the dissolved sulfur system could lead to enhanced soluble metal concentrations as well. Calculations indicate that dissolved polysulfide values of from 1.5 to 3 times greater than those calculated would be needed to explain the Cu^+ concentrations observed in the top part of Branford core. Based on the previous work [47,48] cited above, this is entirely possible.

Cu-Organic Interactions

The large concentration of DOC produced in anoxic pore waters (Table 3) may greatly affect the availability of trace metals to participate in various inorganic reactions. Presley et al. [13] and Elderfield and Hepworth [44] have suggested that organic compounds may also control the mobility of trace metals in anoxic nearshore pore waters. Laboratory experiments by Akiyama [51] have shown that the pigment and lipid fractions (1500-50,000 MW) from the bacterial degradation of dead algae can form water soluble iron-organic associations.

Possibly even more important in this regard are the high molecular weight organic compounds known as "humic substances." Nissenbaum at al.[52], Lindberg and Harris[53], and Krom and Sholkovitz[41] have shown that a high molecular weight fraction of anoxic pore water DOM may play a major role in trace metal chemistry.

Nissenbaum's data (Nissenbaum et al.[52], Nissenbaum and Swaine[54]) indicate that as much as 100% of the Fe^{2+}, Zn^{2+} and Cu^{2+} and from 50 to 80% of the Co^{2+} and Ni^{2+} found in Saanich Inlet pore waters are bound to this high molecular weight organic polymer. Lindberg and Harris[53] have shown that Hg^{2+} is associated with the lowest and the highest molecular weight organic carbon fraction in nearshore Gulf of Mexico pore waters.

If the model developed by Elderfield and Hepworth[24] to examine the importance of DOM speciation of trace metals in anoxic pore waters is utilized to calculate the formation constant of the Cu - organic matter complex needed to explain the dissolved copper concentrations in this study[19], the values range from 10^{12}-10^{16}.

Although Cu^{2+}-dissolved organic matter associates have been suggested in seawater,[55-57] the above calculated values are higher than Cu^{2+}-organic matter association constants observed by other investigators (for example, $K = 10^{8.7}$ for Cu^{2+}-soil fulvic acid[58]; $K = 5x10^{5}$ for Cu^{2+}-freshwater dissolved fulvic acid[59] and $K = 10^{9-10}$ for Cu^{2+}-seawater fulvic acid[56]). In addition, these values have been determined for Cu^{2+}. It is expected that values for the equivalent organic complexes of Cu^{+} would be lower by at least an order of magnitude. However Boulegue et al.[60] have suggested the organo-sulfur ligands such as cysteine and/or penicillamine could be important metal binders in salt marsh pore waters. Very strong complexes are formed by Cu^{+} with these compounds ($K_f = 10^{19}$). As pointed out by Boulegue et al.[60], one might expect Ag^{+} and Cd^{2+} to also form strong organo-sulfur complexes in natural waters. Thus metal-organo-thiol associations may be very important in organic rich, anoxic environments.

Interpretation of metal-DOM interactions in marine sedimentary pore fluids is made difficult by the fragmentary information available on the composition of DOM and on its relative importance in trace metal complexing. Therefore the role of DOM in controlling trace metal solubilities in anoxic marine waters is still unfortunately, debatable.

Presence of Colloidal Copper

Because "dissolved pore water metal concentrations are operationally defined by filtration through a 0.4μ membrane, the possibility of colloidal Cu species cannot be ruled out at this time. Experimental work[46] has shown that at humic material concentrations of 0.4 mmol l^{-1} of carbon inhibition of CuS nucleation occurs and stable CuS sols are formed. Humic material concentrations greater than 0.4 mmol l^{-1} of carbon are commonly observed in nearshore anoxic sediments[61]. Whether this sol stabilization is due to chelation, surface poisoning or steric stabilization by polymeric material is unknown. Horzempa and Helz[46] calculate that the time needed for anoxic waters with 10 nmol l^{-1} concentration of total copper to form colloidal Cus particles of 1000nm is ∿ 100 yr. Therefore, the question of colloidal Cu as a major contribution to "dissolved" Cu in anoxic waters remains unanswered.

Trace Metal Speciation: Mystic River

The nutrient, sulfate and reduced sulfide data from the pore waters of these sediments suggest that this environment should be considered suboxic rather than anoxic. The little reduced sulfur present is probably due to sulfate reduction in anoxic microniches[62] present in these sediments, for sulfate reduction does not appear to be a major process. In addition to this difference, the Mystic River samples are of coarser grain size material and the entire core was grey-olive in color not the characteristic black of estuarine sediments where sulfate reducing is occurring[40]. Recently, Iizumi et al.[63] have shown that in shallow water sediments where eelgrass is abundant, nitrate reduction is the dominant biogeochemical process occcurring. This is true, even though dissolved sulfide was observed in low quantities (i.e. 2-25μM). Their findings support our data which suggests that even though reduced sulfur is present in the pore fluids, sulfate reduction is a secondary process in these sediments.

Previous work on hemipelagic suboxic sediments have shown that Fe, Mn, Ni, Zn and Cu can all be remobilized into the pore fluids.[64-67] This metal remobilization is thought to be due to the reduction of manganese and iron oxide phases within the nitrate reduction biogeochemical zone. It is unclear whether the remobilization of metals into the pore fluids of the surficial sediments at this location is due to the reduction of oxides, the oxidation of the eelgrass detritus or the oxidation of authigenic monosulfides.

If the ΣH_2S data are neglected, metal pore water speciation can be determined as tabulated in Table 4. Again, in these calculations organic matter - metal associations were not considered.

CONCLUSIONS

Due to our lack of knowledge concerning the reactions taking place, the kinetics of these reactions, the various trace metal species being formed and the nature and reactivity of the dissolved organic matter in the pore solutions, it is not surprising that the theoretical approach taken here cannot precisely explain all of the field observations. The anoxic pore fluids it is apparent, however, that with the possible exception of Cu, inorganic reactions and inorganic speciation are the primary controlling factors on the distribution and concentrations of Fe, Mn, Cd and Ag in these samples.

The results outlined in this investigation strongly suggest that bisulfide and polysulfide complexes may be very important in trace metal solubility in nearshore organic rich sediments. As Boulegue[48] has pointed out, in the past these species have been neglected in the study of reducing environments and thus in thermodynamic calculations on these environments. This is especially true in transitional environments where oxidation of sulfide minerals may produce high concentrations of S_x^{2-} ions.

In suboxic environments, metals are remobilized primarily through the reduction of manganese and iron oxides. These metals are complexes inorganically with the major anions of seawater.

ACKNOWLEDGEMENTS

The authors appreciate the help of J. Crowell, S. Hornor, T. Fogg and M. Hines in obtaining the samples. We thank C.D. Hunt for his analytical expertise and H.E. Gaudette for his encouragement. D. Templeton was very helpful with the speciation calculations. We greatly appreciate the discussions of anoxic chemistry with Dr. J. Boulegue. This manuscript greatly benefited from the comments of the editor and an anonymous receiver. This work was initially support by Army Corps of Engineers Grant DACW-33-75-0085 (W.F. Fitzgerald, principal investigator) and was aided by NSF Grants OCE-77-20484 and EAR-79-11194 in the latter stages of the writing of this manuscript.

REFERENCES

1. Bray, J.T., O.P. Bricker and B.N. Troup, 1973: Phosphate in interstitial waters of anoxic sediments: oxidation effects during sampling procedure. Science, 180, 1362.
2. Troup, B.N., O.P. Bricker and J.T. Bray, 1974: Oxidation effect on the analysis of iron in the interstitial water of recent anoxic sediments. Nature, 249, 237.

3. Loder, T.C., W.B. Lyons, S. Murray and H.D. McGuiness, 1978: Silicate in anoxic pore waters and oxidation effects during sampling. Nature, 273, 373.
4. Lyons, W.B., H.E. Gaudette and G.M. Smith, 1979: Pore water sampling in anoxic carbonate sediments: oxidation artifacts. Nature, 227, 48.
5. Patterson, C.C. and D. M. Settle, 1976: The reduction of orders of magnitude errors in lead analyses of biological materials and natural waters by evaluating and controlling the extent and sources of industrial lead contamination introduced during sample collecting and analysis, in, The National Bureau of Standard special publication: Accuracy in Trace Analysis, Proceeding of the 7th Materials Res. Symp. (LaFleur, P., ed.), 321.
6. Brewer, P.G., N. Frew, N. Cutshall, J.J. Wagner, R.A. Duce, P.R. Walsh, G.L. Hoffman, J.W.R. Dutton, W.F. Fitzgerald, C.D. Hunt, D.C. Girvin, R.G. Clem, C.C. Patterson, D. Settle, B. Glover, B.J. Presley, J. Trefrey, H. Windom and R. Smith, 1974: Interlaboratory lead analyses of standardized samples of seawater, Mar. Chem., 2, 69.
7. Calvert, S.E. and N.B. Price, 1972: Diffusion and reaction profiles of dissolved manganese in the pore waters of marine sediments. Earth Planet. Sci. Lettr., 16, 245.
8. Troup, B.N., 1974: The interaction of iron with phosphate, carbonate and sulfide in Chesapeake Bay interstitial waters: A thermodynamic interpretation, The John Hopkins University, Ph.D. Dissertation, 114 pp.
9. Robbins, J.A. and E. Callendar, 1975: Diagenesis of manganese in Lake Michigan sediments. Am. Jour. Sci., 275, 512.
10. Emerson, S., 1976: Early diagenesis in anaerobic lake sediments: chemical equilibria in interstitial waters. Geochim. Cosmochim. Acta., 40, 925.
11. Murray, J.W. and G. Gill, 1978: The geochemistry of iron in Puget Sound, Wash. Geochim, Cosmochim. Acta., 42, 9.
12. Aller, R.C., 1977: The influence of macrobenthos on chemical diagenesis of marine sediments, Ph.D., dissertation, Yale University, 600 pp.
13. Presley, B.J., R. Kolodny, A. Nissenbaum and I.R. Kaplan, 1972: Early diagenesis in a reducing fjord, Saanich Inlet, B.C. - II. Trace element distribution in interstitital water and sediment. Geochim, Cosmochim, Acta., 36, 1073.
14. Gardner, L.R., 1974: Organic versus inorganic trace metal complexes in sulfidic marine waters--some speculative calculations based on available stability constants. Geochim, Cosmochim. Acta., 38, 1297.

15. Duchart, P., S.E. Calvert and N.B. Price, 1973: Distribution of trace metals in the pore waters of shallow water marine sediments. Limnol. Oceanogr., 18, 605.
16. Riley, G.A., Oceanography of Long Island Sound 1952-1954, 1956: I. Introduction; II. Physical Oceanograpy, Bull, Bingham Ocean. Coll., 15, 9.
17. Bohlen, W.F., 1975: An investigation of suspended material concentrations in eastern Long Island Sound. Jour. Geophys. Res., 80, 5089.
18. Lyons, W.B. and W.F. Fitzgerald, 1978: Nutrient production in nearshore tidal flat pore waters: A kinetc study. In Environ. Biogeochem. and Geomicro. (ed. W.E. Krumbein) vol. 1, 237.
19. Lyons, W.B., 1979: Early diagenesis of trace metals in nearshore Long Island Sound sediments. Ph.D. dissertation, University of Connecticut, p. 257.
20. Bewers, J.M., S. Hartling, R. Duce, P. Walsh, W. Fitzgerald, C.D. Hunt, H. Windom, R. Smith, C.S. Wong, P. Berrang, C. Patterson and D. Settle, 1976: Comparison determinations of Pb by investigators analyzing individual samples of seawater in both their home laboratory and in an isotope dilution standardization laboratory. Mar. Chem., 4, 389.
21. Brooks, R.R., B.J. Presley and I.R. Kaplan, 1967: The APDC-MIBK extraction system for the determination of trace elements in saline waters by AAS. Talanta, 14, 809.
22. Brewer, P.G., D. W. Spencer and C.L. Smith, 1969: Determination of trace metals in seawater by atomic absorption spectrophotometry. Am. Soc. Test. Mat., S.T.P., 443, 70.
23. Kremling, K. and H. Peterson, 1974: APDC-MIBK extraction system for the determination of Cu and Fe in 1 cm^3 of seawater by flameless AAS. Anal. Chim. Acta., 70, 35.
24. Cline, J.D., 1969: Spectrophotometric determination of hydrogen sulfide in natural waters. Limnol. Oceanogr., 14, 454.
25. Goldhaber, M.B., 1974: Equilibrium and dynamic aspects of the marine geochemistry of sulfur, Ph.D. thesis, Dept. of Geology, U.C.L.A., 399 pp.
26. Menzel, D.W. and R.F. Vaccaro, 1964: The measurement of dissolved organic carbon and particulate carbon in seawater. Limnol. Oceanogr., 9, 138.
27. Lindberg, S.E., 1973: Hg in interstitial solutions and associated sediments from estuarine areas on the Gulf of Mexico, M.Sc. Thesis, Florida State Univ., 124 pp.
28. Martens, C.S. and M.B. Goldhaber, 1978: Early diagenesis in transitional sedimentary environments of White Oak River estuary, N.C. Limnol. Oceanogr., 23,428.

29. Holdren, G.R., Jr., O.P. Bricker III and G. Matisoff, 1975: A model for the control of dissolved manganese in the interstitial waters of Chesapeake Bay, in, "Marine Chemistry in the Coastal Environmental". ACS Symp. 18, ed., Church, T.M., Am. Chemical Soc., Washington, D. C., 364-381.
30. Sanders, J.G., 1978: The sources of dissolved Mn to Calico Creek, N.C. Est., Coast. Mar. Sci., 6, 231.
31. Grill, E.V., 1978: The effect of sediment-water exchange on Mn deposition and nodule growth in Jervis Inlet, B.C. Geochim. Cosmochim. Acta., 42, 485.
32. Murray, J.W., V. Grundmanis and W.M. Smethie, 1978: Interstitial water chemistry in the sediments of Saanich Inlet. Geochim. Cosmochim. Acta., 42, 1011.
33. Elderfield, H., R.J. McCaffrey, N. Luedtke, M. Bender and V.M. Truesdale, 1981: Chemical diagenesis in Narragansett Bay sediments. Am. Jour. Sci., 281, 1021.
34. Engler, R.M. and W.H. Patrick, 1975: Stability of sulfides of Mn, Fe, Zn, Cu and Hg in flooded and non flooded soil. Soil Sci., 119, 217.
35. Bricker, O.P., III and B.N. Troup., 1975: Sediment-water exchange in Chesapeake Bay, in, "Estuarine Research," vol. 1, ed. by Cronin, L.E., Academic Press, Inc., New York, 3-27.
36. Heggie, D. and D. Burrell, 1975: Depth distribution of copper in interstitial waters and the water column of an Alaskan Fjord. ABSTRACTS, Fall meeting of A.G.U., San Francisco, Calif.
37. Tramontano, J., 1978: The effects of dredging and winter storm activity upon the nutrient and trace metal goechemistry of the waters of the Thames River and adjacent eastern Long Island Sound. M.Sc. Thesis, University of Connecticut, 105 pp.
38. Lyons, W.B., K.M. Wilson, P.B. Armstrong, G.M. Smith, and H.E. Gaudette, 1980: Trace metal pore water geochemistry of carbonate sediments, Bermuda. Ocean. Acta., 3, 363.
39. Lyons, W.B. and W.F. Fitzgerald, 1976: Iron and manganese geochemistry of tidal flat pore waters: A thermodynamic study. Presented at the Annual G.S.A. Meeting, Denver, Colo.
40. Berner, R.A., 1970: Sedimentary pyrite formation. Am. Jour. Sci., 268, 1.
41. Krom. M.D. and E.R. Sholkovitz, 1978: On the association of iron and manganese with organic matter in anoxic marine pore waters. Geochim. Cosmochim. Acta., 42, 607.
42. Lyons, W.B., H.E. Gaudette and P.B. Armstrong, 1979: Evidence for organically associated iron in nearshore pore waters. Nature, 282, 202.

43. Spencer, D.W. and P.G. Brewer, 1971: Vertical advection diffusion and redox potentials as controls on the distribution of manganese and other trace metals dissolved in waters of the Black Sea. Jour. Geophys. Res., 76, 5877.
44. Elderfield, H., and A. Hepworth, 1975: Diagensis, metals and pollution in estuaries. Mar. Pollut. Bull., 6:(6), 85.
45. Helz, G.R. and S.A. Sinex, 1974: Chemical equilibrium in the thermal spring waters of Virginia. Geochim. Cosmochim. Acta., 38, 1807.
46. Horzempa, L.M. and G.R. Helz, 1979: Controls on the stability of sulfide soils: colloidal covellite as example. Geochim. Cosmochim. Acta., 43, 1645.
47. Boulegue, J., 1978: Metastable sulfur species and trace metals in hot brines from the French Dogger. Am. Jour. Sci., 278, 1394.
48. Boulegue, J., 1977: Equilibria in a sulfide rich water from Enghien-les-Bains, France. Geochim. Cosmochim. Acta., 41, 1751.
49. Boulegue, J., and G. Michard, 1979: Sulfur speciation and redox processes in reducing environments: in, "Chemical Modeling in Aqueous Systems", ed. Jenne, E.A., Am. Chem. Soc. Symp. Ser. 93, Washington, D.C., 25-50.
50. Aizenshtat, Z., A. Stoler, H. Nielsen and Y. Cohen: The geochemical sulfur enrichment of recent organic matter by polysulfides in the Solar Lake, in press.
51. Akiyama, T., 1973: Interactions of ferric and ferrous irons and organic matter in water environment. Geochem. Jour., 7, 167.
52. Nissenbaum, A., M.J. Baedecker and I.R. Kaplan, 1972: Studies of dissolved organic matter from interstitial water of a reducing marine fjord. In, "Advances in Organic Geochemistry," H.R. von Gaertner and H. Webner, eds., Pergamon Press, Oxford, pp. 427-440.
53. Lindberg, S.E. and R.C. Harris, 1974: Mercury organic matter associations in estuarine sediments and interstitial water. Environmental Science and Technology, 8, 459.
54. Nissenbaum, A. and D.J. Swaine, 1976: Organic matter-metal interactions in recent sediments: The role of humic substances. Geochim. Cosmochim. Acta., 40, 809.
55. Fitzgerald, W.F., 1969: Study of certain trace metals in seawater using anodic stripping voltammetry. Ph.D. dissertation, WHOI-MIT, 180 pp.
56. Mantoura, R.F.C., A. Dickson and J.P. Riley, 1978: The complexation of metals with humic materials in natural waters. Est. Coast. Mar. Sci., 6, 387.
57. Montgomery, J.R. and R.J. Santiago, 1978: Zn and Cu in particular forms and soluble complexes with inorganic and organic ligands in the Guanajibo River and coastal zone, Puerto Rico. Est. Coast. Mar. Sci., 6, 111.

58. Schnitzer, M. and S.I.M. Skinner, 1966: Organo-metallic interactions in soils: 5. Stability constants of Cu^{2+} Fe^{2+} and Zn^{2+} - fulvic acid complexes. Soil Sci., 102, 361.
59. Shuman, M.S. and G.P. Woodward, 1977: Stability constants of Cu-organic chelates in aquatic samples. Environ. Sci. Tech., 11, 809.
60. Boulegue, J., C.J. Lord III and T.M. Church: Sulfur speciation and associated trace metals in the pore waters of Great Marsh, Delaware, in review.
61. Krom, M.D. and E.R. Sholkovitz, 1977: Nature and reactions of dissolved organic matter in the interstitial waters of marine sediments. Geochim. Cosmochim. Acta., 41, 1565.
62. Jorgensen, B.B., 1977: Bacterial sulfate reduction within reduced microniches of oxidized marine sediments. Mar. Biol., 41, 7.
63. Iizuma, H., A Hattori and C.P. McRoy, 1980: Nitrate and nitrite in interstitial waters of eelgrass beds in relation to the rhizosphere. J. Exp. Mar. Biol. Ecol., 47, 191.
64. Brooks, R.R., B.J. Presley and I.R. Kaplan, 1968: Trace elements in the interstitial waters of marine sediments. Geochim. Cosmochim. Acta., 32, 397.
65. Sawlan, J.J. and J. W. Murray, 1980: Trace metal profiles in pore fluids from hemipelagic and red clay sedimentary provinces. EOS, 61, 274.
66. Lyons, W.B., M.E. Hines, G.M. Smith and A.D. Hewitt, 1980: The biogeochemistry of sediments in two Gulf of Maine basins. Mar. Chem., 9, 307.
67. Boniforti, R., personnel communication.

THE CONTRASTING GEOCHEMISTRY OF MANGANESE AND CHROMIUM IN THE EASTERN TROPICAL PACIFIC OCEAN

James W. Murray, Berry Spell, and Barbara Paul

School of Oceanography
University of Washington
Seattle, Washington 98195
U.S.A.

ABSTRACT

The study of the different oxidation states of trace metals can put limits on the oxidation state of seawater. In addition, the changes in oxidation state can influence input and removal in the ocean. Manganese and chromium form a contrasting pair with opposite tendencies. Under oxidizing conditions Mn(IV) is scavenged as MnO_2 while Cr(VI) is soluble as CrO_4^{2-}. Under reducing conditions Mn(II) is soluble as Mn^{2+} while Cr(III) is removed as $Cr(OH)_3$. Thus the distributions of manganese and chromium can be influenced by the oxygen minimum zone.

These chemical transformations as well as the influence of the continental margin, can be demonstrated using data from the Eastern Tropical Pacific. Vertical profiles of manganese in the Guatemala Basin exhibit pronounced features that all intensify approaching the continental margin. There is a surface maximum (up to 9 nM), a maximum in the oxygen minimum (up to 6 nM) and a strong increase towards the sediments in the bottom 1000 m (up to 9 nM). The surface maximum is probably due to atmospheric or riverine input or remobilization out of shelf sediments. The maximum in the oxygen minimum is due to *in situ* reduction or remobilization out of slope sediments followed by horizontal mixing. The partitioning between dissolved and acid soluble manganese tends to support the latter explanation. The increase in the bottom water is mostly in the acid soluble fraction and increases as the continental margin is approached. The regional distribution favors a sediment rather than a hydrothermal source.

Chromium marine geochemistry is equally complicated. Important features can be seen in data from a line of stations normal to the coast of Baja at 22°N. Along this section the oxygen minimum intensifies to the east. Total chromium is between 3.0 and 4.0 nM at the surface and then decreases sharply to a minimum of less than 2.5 nM at the top of the oxygen minimum. The concentrations then increase to about 5.0 nM in the deep water. In general, the deep profiles are similar to those of silica. Coincident with the minimum in total chromium are high values in Cr(III) and particulate chromium. A logical explanation is that Cr(VI) is reduced to Cr(III), much of which is rapidly adsorbed and scavenged by the vertical particle flux.

INTRODUCTION

As the first order characterization of the distribution of trace metals in seawater has become established, attention has increasingly focused on the mechanisms that control those distributions. It is clear from the shapes of the individual profiles and the correlations of trace elements with nutrient elements in the water column that biological factors play a major role in trace metal cycles (see for example, Cu[1,2]; Ni[3]; Cd[4]; Zn[5]; Ba[6]; Cr[7]; As[8]; Se[9]). In situ adsorption or scavenging influence the distribution of the more reactive elements such as Pb[10,11]; Cu[12]; ^{234}Th, ^{230}Th and ^{228}Th[13]. Hydrothermal circulation of seawater through newly formed ocean crust is a major input of Mn and other trace metals to the ocean[14,15,16].

Several elements can undergo transformation among different oxidation states. Examples of such elements include Mn(II,IV), Cr(III,VI), I(-I,V), Se(IV,VI) and As(III,V). Analyses of these couples combined with classic dissolved oxygen and nitrate determinations put limits on the oxidation state (e.g., pE) of seawater in regions of very low or unmeasurable dissolved oxygen. These redox transformations influence the input and removal and distribution of these elements from seawater.

Another important factor yet to be properly evaluated is the effect of the ocean boundary on trace metal distributions. Few field experiments have been specifically designed to test for horizontal variations. Bacon et al.[11] have suggested that Pb is being scavenged by processes at the edge of the oceans. Bender et al.[17] and Landing and Bruland[18] have proposed that reduction of MnO_2 in the sediments results in a flux of dissolved manganese to seawater that is mixed and advected horizontally into the interior of the ocean. This source may contribute to the manganese maxima at the ocean surface and at the depth of the oxygen minimum.

We have analyzed for manganese and chromium species along two transects of stations in the Eastern Tropical North Pacific. In both regions the oxygen minimum is intense and the effect of oxidation-reduction reactions and the ocean boundary are especially pronounced. The samples we have collected provide an excellent opportunity to compare the seawater chemistry of manganese and chromium.

OXIDATION-REDUCTION CHEMISTRY

Oxidation-reduction reactions in seawater can be identified by the characteristic pE value [$pE = -\log(e^-)$] at which the activity of oxidized and reduced species are equal. Some relevant reduction couples are listed in Table 1 together with the equilibrium constants and characteristics pE values. In addition to the couples for I, Mn, As, Cr, U, Se and Fe we have listed the couples for O_2, NO_3^-, and SO_4^{2-} as these compounds are thought to be the major electron acceptors used by bacteria to facilitate the decomposition of organic matter. The couples are listed in order of decreasing electron free energy levels which corresponds to the equilibrium sequence of reactions occurring during organic matter decomposition.

Table 1 The reduction half reactions for some nutrients and trace metals in seawater. The equilibrium constants have been taken from Stumm and Morgan[19] and Emerson et al.[20]. The pE values [$pE = -\log(e^-)$] have been calculated for pH = 7.4 which is a reasonable pH value for the oxygen minimum zone of the Eastern Tropical Pacific Ocean. We also assume that the activities of oxidized and reduced species are equal.

Equation	pK	pE
$1/4 O_2(g) + H^+ + e^- = 1/2 H_2O$	+20.8	12.8
$1/5 NO_3^- + 6/5 H^+ + e^- = 1/10 N_2(g) = 3/5 H_2O$	+21.1	11.2
$1/6 IO_3^- + H^+ + e^- = 1/6 I^- + 1/2 H_2O$	+18.3	10.5
$1/2 MnO_2(s) = 2H^+ + e^- = 1/2 Mn^{2+} + H_2O$	+21.8	8.9
$1/2 SeO_4^{2-} + H^+ + e^- = 1/2 SeO_3^{2-} + 1/2 H_2O$	+14.8	7.4
$1/3 CrO_4^{2-} + 2 H^+ + e^- = 1/3 Cr(OH)_2^+ + 2/3 H_2O$	+22.0	6.6
$1/2 H_3AsO_4 + H^+ + e^- = 1/2 As(OH)_3 + 1/2 H_2O$	+9.4	2.0
$1/2 UO_2(CO_3)_2^{2-} + e^- = 1/2 UO_2(s) + CO_3^{2-}$	+1.5	-0.5
$1/2 FeOOH(s) + 2 H^+ + e^- = 1/2 Fe^{2+} + H_2O$	+16.5	-2.7
$1/8 SO_4^{2-} + 9/8 H^+ + e^- + 1/2 HS^- + 1/2 H_2O$	+4.2	-3.4

Previous studies of nutrients have shown evidence for denitrification (reduction of NO_3^- to N_2) in the region of the Eastern Tropical Pacific where O_2 is less than 10 μM[21]. There is one report of H_2S in the oxygen minimum zone off Peru[22]. Normally sulfate reduction is not achieved but there is a large gap between sulfate reduction and denitrification and evidence for trace metal reduction could narrow our knowledge about this gap.

Manganese and chromium provide an especially interesting and contrasting pair in this regard. In oxidizing seawater Mn(IV) should be scavenged as $MnO_2(s)$ while oxidized Cr(VI) is soluble as CrO_4^{2-}. In the reduced states Mn(II) is soluble as Mn^{2+} while Cr(III) forms insoluble $Cr(OH)_3(s)$. The differences in the distributions of these two elements should be especially pronounced in the intense oxygen minimum zone of the Eastern Tropical Pacific.

MANGANESE AND CHROMIUM CHEMISTRY

Chromium chemistry has not been previously described in any detail in the oceanographic literature. Chromium (atomic number 24) has an outer shell electron population of $3d^5 4s^1$ and tends to exist in nature in two main oxidation states, Cr(III) and Cr (VI).

Chromium(III) has three 3d electrons. In octahedral coordination these three electrons are in the high spin states and the crystal field stabilization energy is $6/5\ \Delta_0$ which is the largest octahedral site preference energy available to the transition metals[23]. As a result chromium(III) tends to form strong kinetically inert complexes[24]. For example, the rates of substitution of the waters of hydration are extremely slow with a residence time of about two hours[25,26]. The lack of lability of the coordination sphere imposes limitations on the mechanisms of electron transfer reactions.

The solubility of Cr(III) oxides or hydroxides is low. The solubility minimum of solid $Cr(OH)_3(s)$ is about 400 nM at pH 8.5 and is controlled by the intersection of the species $Cr(OH)_2^+$ and $Cr(OH)_4^-$. Another candidate suggested for solubility control is chromite ($FeCr_2O_4$). Hem[27] has proposed that chromite could result from the reaction of Cr(III) with iron(III) hydroxide as follows:

$$Fe(OH)_3(s) + 2Cr(OH)_2^+ + e^- = FeCr_2O_4(s) + 3H_2O + H^+$$

If chromite controls the solubility dissolved Cr(III) should be less than 10^{-10} M at pH 8. This reaction could occur after the adsorption of Cr(III) by $Fe(OH)_3(s)$. Dissolved Cr(III) does have a strong tendency to adsorb onto surfaces. This has been demonstrated by Cranston and Murray[28] using Cr(III) spikes added to Columbia River estuary water. Rapid adsorption has also been demonstrated in river and lake water by Benes and Steinnes[29]. Chromium(VI), on the other hand, does not tend to be strongly adsorbed. Chromium(III) sulfides tend not to form because of the stability of $Cr(OH)_3(s)$[30].

Chromium(VI) tends to form mostly oxo compounds. Because of its large ionic potential it is an extremely strong acid and the soluble tetrahedral chromate ion (CrO_4^{2-}) is the common species. It is known as a strong oxidizing agent and is only stable in oxygenated environments. The reduction of Cr(VI) as CrO_4^{2-} to Cr(III) as $Cr(OH)_2^+$ has been described by Elderfield[31].

$$CrO_4^{2-} + 6H^+ + 2H_2O + 3e^- = Cr(H_2O)_4(OH)_2^+ \qquad \log K = 66.1$$

The equilibrium ratio of dissolved chromium is then:

$$\log \frac{Cr(VI)}{Cr(VI)} = 6\ pH + 3\ pE - 66.1$$

At pH = 3.1 and pE = 12.5 the ratio equals 19.9 and Cr(VI) should predominate. The pE - pH relationships are shown in Figure 1[32]. At pH = 8.0 the transition from Cr(VI) to Cr(III) occurs at pE 6.6.

The kinetics of the oxidation reaction are known to be slow. Earley and Cannon[33] concluded that this is due in part to the differences in coordination between Cr(III) (tetrehedral) and Cr(VI) (octahedral). Using Cr(III) spikes added to Columbia River estuary water we have observed a first order oxidation half life of one month[28]. Other estimates range from 2 to 20 months[34,35]. Emerson et al.[20] modeled the distribution of Cr(III) above the anoxic interface in Saanich Inlet and obtained a residence time for Cr(III) of 6 to 20 days. The complexity of the natural system made it difficult for them to distinguish between scavenging and oxidation of Cr(III).

Manganese chemistry has been extensively reviewed in the past[36,19,37]. Mn(IV) is the stable form under oxidizing conditions and forms a wide range of solid phases with various stoichiometries and structures[38]. There is evidence that hausmannite (Mn_3O_4) may be the first solid phase to form under conditions of rapid oxidation[36,39]. Mn^{2+} has five d electrons and has no crystal field stabilization energy in octahedral coordination. Thus Mn^{2+} tends to form relatively weak complexes and probably exists in seawater primarily as free Mn^{2+}. Both Mn^{2+} and MnO_2 show strong tendencies to be scavenged.

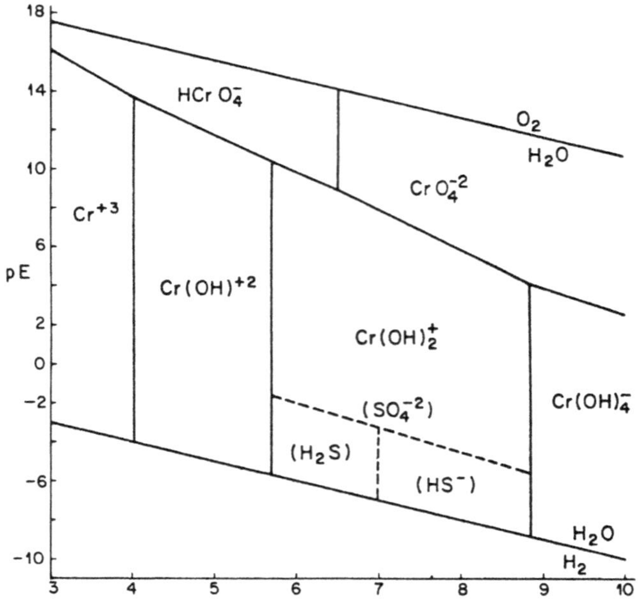

Fig. 1 pE-pH diagram for the major dissolved species of chromium III and VI.

Information on the kinetics and mechanism of $MnO_2(s)$ reduction is sparse. The reaction must be bacterially mediated and $MnO_2(s)$ probably can serve as a terminal electron acceptor during the oxidation of organic matter[40]. The kinetics of oxidation of Mn(II) in clean laboratory solutions are slow in the pH range of 7.0 to 8.0[41,42,37]. It is clear, however, that surfaces such as γ FeOOH[43] and bacteria[20] can speed the oxidation enormously.

SAMPLE COLLECTION AND ANALYSES

The data reported here were obtained on samples collected during two cruises of the R/V T.G. Thompson. Cruise TT 126 to the Guatemala Basin was during February 1978 and samples were collected at four stations (Figure 2). Cruise TT 145 from Hawaii to the coast of Baja California was during October 1979 and samples were collected at five stations along a transect normal to the continental margin (Figure 2). In addition to the trace metal data reported here, samples were analyzed for hydrographic parameters (θ, S, O_2) and nutrients (NO_3^-, NO_2^-, NH_4^+, PO_4^{3-}, SiO_2). All water samples were collected using 30-liter Niskin bottles fitted with

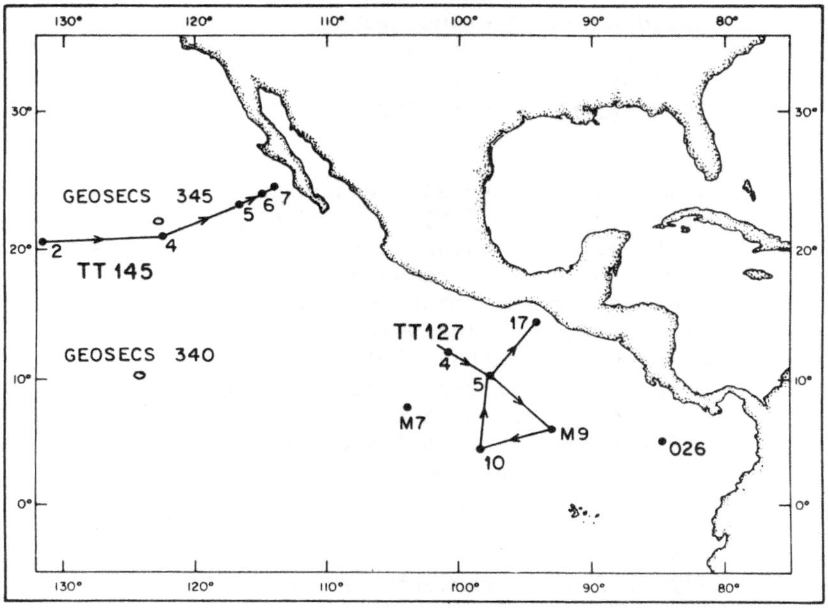

Fig. 2 Cruise tracks and station locations for R/V Thompson cruises TT 127 and TT 145. Also shown are MANOP stations M7 and M9, two GEOSECS stations 340 and 345 and Oceanographer station 0 26.

surgical rubber-tubing closures, hung on stainless steel hydrowire and tripped with brass messengers. All of these analyses as well as our previous work has indicated that no serious contamination results from this approach.

The details of the chromium analytical procedures have been described elsewhere[7,28] so only a brief summary is given here. Immediately after the Niskin bottles were on deck, samples were drawn into 140-ml polyethylene bottles that had been precleaned at $60^{\circ}C$ for four days with 1% distilled hydrochloric acid. Total chromium [Cr(VI) + Cr(III) + Cr_p] was coprecipitated with iron(II) hydroxide and reduced chromium [Cr(III) + Cr_p] was coprecipitated with iron(III) hydroxide. These coprecipitation steps were completed within minutes of sample collection to minimize storage problems. The iron hydroxide precipitates were filtered through 0.4 μm Nuclepore filters and stored in polyethylene vials for later analyses in the lab. Particulate chromium was also obtained by filtering unaltered samples through 0.4 μ filters. In the laboratory the iron hydroxide coprecipitates were dissolved in 6N distilled hydrochloric acid and analyzed by flameless atomic absorption. Based on our previous work[32], the limit of detection is about 0.1 to 0.2 nM. Precision is about 5%. Six closely spaced samples around 3500 m at station 7 give 4.9±0.25 nM and eight replicate analyses from a sample at 2650 m at station 6 gave 5.2 ±0.31 nM.

Water column samples were also analyzed for dissolved and total dissolvable manganese (TDM)[17,44]. The samples for TDM were collected into 500 ml polyethylene bottles immediately after the Niskin bottles were on deck. The same cleaning procedure was used for these bottles as for the chromium bottles. After all nutrient, salinity and oxygen samples were drawn the Niskin bottles were shaken and samples of about 15-liter were pressure filtered through 0.4 μm Nuclepore filters in Teflon filter holders. Samples for dissolved manganese were collected from the filtrate near the end of the filtration run. All samples were brought to pH < 2 with nitric acid free of trace metals and stored in individual zip-lock plastic bags to minimize contamination.

When the samples were returned to the lab the pH was adjusted to approximately pH 8 using concentrated NH_4OH (ultrapure, G. Frederick Smith). Twenty ml of chelating cation exchange resin in the ammonia form (Chelex-100, 100-200 mesh, Bio-Rad) was added to the samples and they were batch extracted on a shaker table for 36 hours. The resin was decanted into columns and the manganese eluted using 2N HNO_3[45]. The eluant was then analyzed by graphite furnace atomic absorption spectrophotometry. Replicate analyses of samples indicate a precision of about 5%.

RESULTS

Guatemala Basin

The Guatemala Basin is located off the coasts of Southern Mexico, Guatemala and El Salvador in the tropical Pacific Ocean. The basin is bounded on the east by the Middle America Trench, on the north by the Tehuantepec Ridge, on the west by the East Pacific Rise and on the south by the Galapagos Rise and Cocos Ridge. The average depth of the center of the Basin is about 4100 m and the depth decreases towards the ridges on the northern, western and southern boundaries. The depth increases to over 6000 m along the eastern boundary in the Middle America Trench.

Hydrography. Few hydrographic data have been previously published on the deep water of the Guatemala Basin. We bring together here our data from four stations on the Basin, the results from MANOP stations M (m7) and H (m9) collected during INDOMED I, data from the northwest corner of the Panama Basin at R/V Oceanographer station 26 and data from GEOSECS station 340.

The location of these stations is shown in Figure 2. GEOSECS 340 is located about 2000 km west of the East Pacific Rise and serves as a representative data set for source water from the main basin of the Central Pacific.

The circulation in the Guatemala Basin at shallow and intermediate depths is complicated. The California and Peru currents enter the area from the north and south respectively. The Equatorial Counter Current and Equatorial Undercurrent are major inputs from the west. Surface water leaves the basin in the north and south Equatorial currents. Large variations of strength and location occur in these currents due to seasonal wind variations. Additional contributions are from upwelling which occurs at the Intratropical Convergence, from the Equatorial Undercurrent and at the continental margin. The abyssal flow into the Guatemala Basin is poorly known. Lonsdale[46,47] has described the abyssal circulation east of the East Pacific Rise. Pacific Deep Water enters the Chile Basin through gaps in the Chile Rise, then proceeds across sills or through trenches into the Peru Basin and then to the Panama Basin. Deep water may also enter the Guatemala Basin from the west via fracture zones in the East Pacific Rise. Mantyla[48] has suggested a sill depth of 3000 m.

The water mass structure can be identified from θ-S characteristics (Figure 3). The warm and fresh surface water is brought in by the California and Peru current with salinities of 33 to 34 ^{o}/oo and temperatures of 15 to 25 ^{o}C. The salinity maximum at 150 to 200 m is due to subtropical subsurface water formed at dry subtropical latitudes by evaporation and sinking of surface water[49,50]. The salinity minimum centered at 800 m reflects Intermediate Water that moves into the region after it is formed by the cooling of low salinity surface waters at high latitudes. Examination of the GEOSECS longitudinal section at 130 ^{o}W indicates that the Intermediate Water can originate in either the Arctic or Antarctic. A clear boundary exists at about 5 ^{o}N. A θ-S plot of our data along isopycnal surfaces shows that the Intermediate Water in the Guatemala Basin is more than 50% of Antarctic orgin.

Pacific Deep Water with a salinity of about 34.670 ^{o}/oo is found below the Intermediate Water. The potential temperature of the deep water in the Guatemala Basin is 1.56 ^{o}C and is adiabatic in the bottom 1000 m (Figure 3). Temperature is the best diagnostic parameter to identify the source of the deep water. The departure from the θ-Z profile for GEOSECS 340 occurs at about 2700 m suggesting that this is the effective sill depth for water entering the Guatemala Basin from the west across the East Pacific Rise. The deep water in the Panama Basin is everywhere too warm to be a source (Figure 3). Mantyla[48] has proposed that the Guatemala Basin receives abyssal water through the Seguieros Fracture Zone at 7 ^{o}N 102 ^{o}W and between the East Pacific Rise and

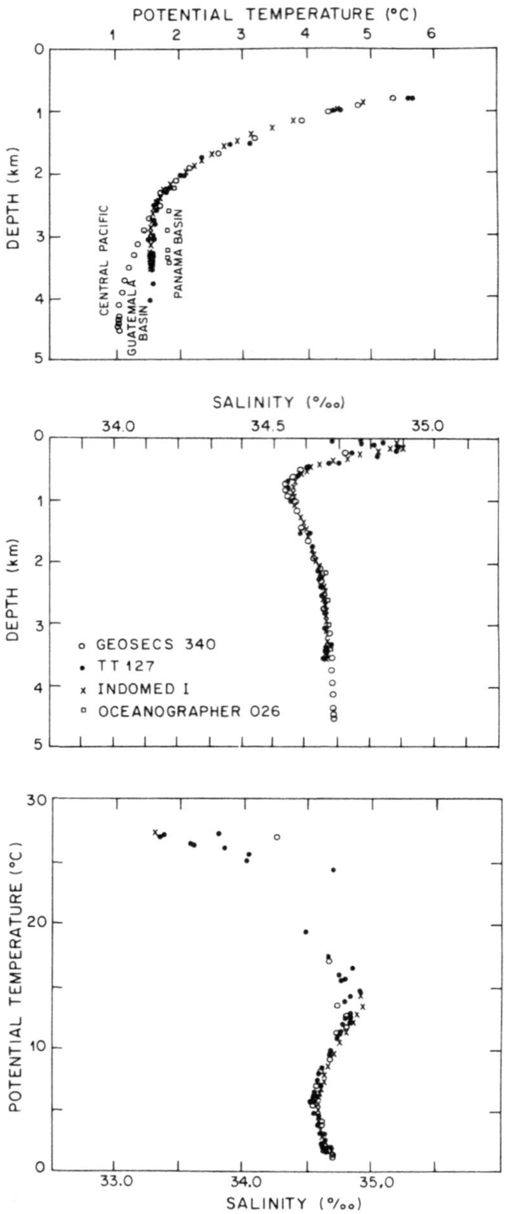

Fig. 3 Hydrographic properties of the Guatemala Basin during TT 127. Top: potential temperature versus depth; Middle: salinity versus depth; Bottom: potential temperature versus salinity. GEOSECS 340 is a reference station from the central Pacific and 0 26 is a reference station from the Northern Panama Basin.

the Cocos Ridge. The double temperature minima proposed by Mantyla was not observed in our data.

Nutrients and oxygen. The nutrient profiles reflect the intense biological activity that takes place in the water column in the Guatemala Basin. The most characteristic feature is the intense minimum in dissolved oxygen that extends from 75 to 1000 m (Figure 4). This low oxygen water lies in a linear θ-S region between the S maximum and the S minimum (Figure 3). At station 17 oxygen is less than 10 μM from 75 to 1000 m. At station 10 in the southwest part of the basin oxygen is never less than 10 μM. The values at stations 4 and 5 and MANOP stations M and H fall inbetween these two extremes. The variations are due to the fact that the low oxygen water extends as a lobe away from the coast producing N-S as well as E-W gradients. The profiles of NO_3 and PO_4 mirror the dissolved oxygen results (Figure 4). A pronounced inflection in the NO_3 profile from 100 to 500 m is due to NO_3 reduction by denitrification. The NO_3/PO_4 ratio in the water column is close to 16/1 with a PO_4 intercept of 0.4 μM. Dissolved silicate increases steadily from low surface values to a broad maximum centered at 2500 m. There are pronounced horizontal gradients with deep water silica increasing from 150 μM at station 10 to 180 μM at station 17 (Figure 4).

Manganese. Our profiles of manganese reveal many characteristic features, some of which have been described before. Together with the MANOP profiles[44], the set of six profiles provides sufficiently detailed coverage to reach stronger conclusions

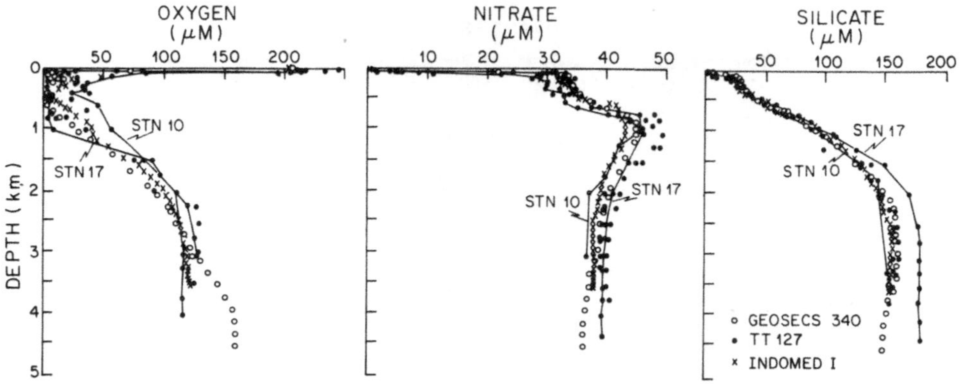

Fig. 4 Nutrient element profiles from the Guatemala Basin. Left: dissolved oxygen versus depth; Middle: nitrate versus depth; Right: dissolved silicate versus depth. R/V Thompson stations 10 and 17 are shown separately to emphasize the horizontal gradients within the Guatemala Basin.

about the manganese sources to the Guatemala Basin. In addition, we report here values for dissolved manganese (<0.4μ) that help clarify the partitioning between oxidized and reduced manganese. Our four profiles are located approximately along a transect off the Central American coast at the Gulf of Tehuanepec (Figure 2). Station 4 is northwest of station 5 and is about equal distance from shore. Station 10 is south of the other stations and out of the core of the water of lowest oxygen concentration.

All profiles of TDM show high surface values of about 2 nM (Figure 5). At all four of the TT 127 stations TDM increases sharply to a shallow maximum between 50 m (stations 5, 10 and 17) and 150 m (station 4). The mixed layer at these stations is thin and the shallow maximum occurs at the base of the mixed layer. At stations 5, 10 and 17 TDM decreases to a minimum of less than 1 nM

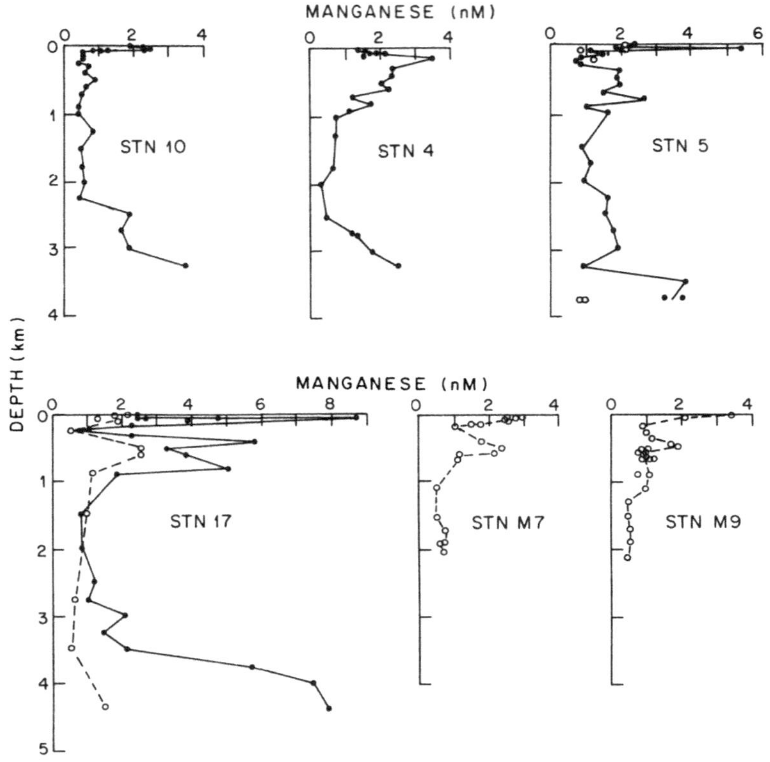

Fig. 5 Manganese profiles from the Guatemala Basin. Solid circles represent total dissolvable manganese and open circles represent dissolved manganese. All concentrations are given in nmole kg^{-1}. Data for MANOP stations M7 and M9 are taken from Klinkhammer and Bender [44].

at about 200 m and then increases to a broad maximum between 300 and 900 m. This broad maximum coincides with the oxygen minimum zone. The magnitude of the TDM maximum in the oxygen minimum increases from station 10 to station 17. The profile at station 4 is different in that TDM increases to a maximum at 125 m and then decreases gradually through the oxygen minimum. The profiles at MANOP stations 7 and 9 are similar to ours with the exception that those profiles show no shallow manganese maximum. They do show high surface values of TDM[44].

We have dissolved manganese values for shallow water from stations 5 and 17. At both locations the shallow manganese maximum at 100 m is composed of reactive particulate manganese. Landing and Bruland[18] also found high values (up to 12 nM) in the near surface waters from five profiles off the coast of California. Horizontal gradients exist in their data and the highest values were found in the surface waters closest to shore in the California Current. They also found that the increases in manganese were in the particulate form. Martin and Knauer[51] have recently reported a pronounced leachable particulate manganese maximum at about 200 m at a station west of ours.

Dissolved values in the oxygen minimum zone are only available from station 17. There we have a maximum in dissolved manganese with values reaching about 2.5 nM. TDM at the same depths are about twice that value indicating that manganese is about equally partitioned between dissolved and particulate forms.

In the mid-water column, from about 1000 to 2500 m, TDM is less than 1 nM. Again horizontal gradients are present. At station 10 which is farthest from the continental margin and out of the most intense oxygen minimum, TDM is about 0.5 nM. At station 17, the mid-water values are about 1.0 nM. The dissolved and TDM values agree well at station 17 suggesting that most of the manganese is dissolved at these depths.

In the bottom water of the Guatemala Basin, TDM increases steadily toward the sediment-water interface. The sill depth across the East Pacific Rise, as determined from our temperature profiles, is at 2700 m and within the basin the water column is adiabatic below that depth. The steady increase toward the bottom suggests a sediment rather than a hydrothermal source. The horizontal gradients in the bottom water support that conclusion. At station 10, TDM increases to about 3.5 nM, at station 5 values reach 3.8 nM and station 17 the highest value is 8 nM. The dissolved manganese values in the bottom water at station 5 and 17 indicate that the increase in TDM is due mainly to particulate manganese.

Baja Transect

During R/V Thompson cruise TT 145 stations were occupied along an east-west transect at about 23°N (Figure 2). The sediments under these stations grade from red clay to hemipelagic. The transect was chosen to look simultaneously at the effects of the oxygen minimum and horizontal distributions away from the continental margin. GEOSECS 345 was chosen as a reference station.

Hydrography and nutrients. The circulation off the coast of Baja is dominated by the California current and the California undercurrent. The hydrographic properties can be summarized from the θ-S plot and profiles of θ and S versus depth (Figure 6). The surface waters are warm and salty due to evaporation. There is a shallow salinity minimum around 100 m. This water has been described in detail by Reid[52] and Kenyon[53] and is formed at the sea surface in the North Pacific. The depth of the salinity minimum increases from east to west but it always lies within the main pycnocline in a region of large vertical stability. A shallow oxygen maximum lies on the average about 20 m above the salinity minimum (Figure 7). All of these features are clearly seen in the TT 145 data. The intensity of the salinity minimum along our transect increases in the onshore direction from about 33.5°/oo at station 7 to 34.0°/oo at station 4. Over this distance the salinity minimum deepens from 50 to 150 m.

Below the salinity minimum is the subsurface salinity maximum centered at about 500 m that is most pronounced at station 7 and only slightly evident at station 4. This reflects the California undercurrent which transports warm salty water poleward along the Baja coast. Wooster and Jones[54] described this feature at 31°N and found the undercurrent to be confined close to the coast with a width of only 20 km. They calculated an average current of 30 cm s^{-1} with a flow of 2 x 10^6 cm^3 s^{-1}. In our data the maximum salinity values were observed between 8° and 11°C. At station 7 where the effects of the current are most clearly seen, the salinity at the maximum reaches 34.55°/oo at 10.5°C. The potential temperature in the deep water decreases smoothly toward the bottom and a small increase of 0.1 to 0.2°C is observed moving inshore from station 4 to 7. Salinity increases towards the bottom where maximum values of 34.68°/oo were observed. Horizontal gradients were less than .005°/oo.

The undercurrent is also characterized by low dissolved oxygen (Figure 7). The oxygen minimum is not as intense here as observed in the Guatemala Basin; however, values less than 10 μM were observed at both stations 4 and 7 below 400 m. Our transect was located along the northern fringe of the most intense part of the oxygen minimum and thus we never saw water that was below the

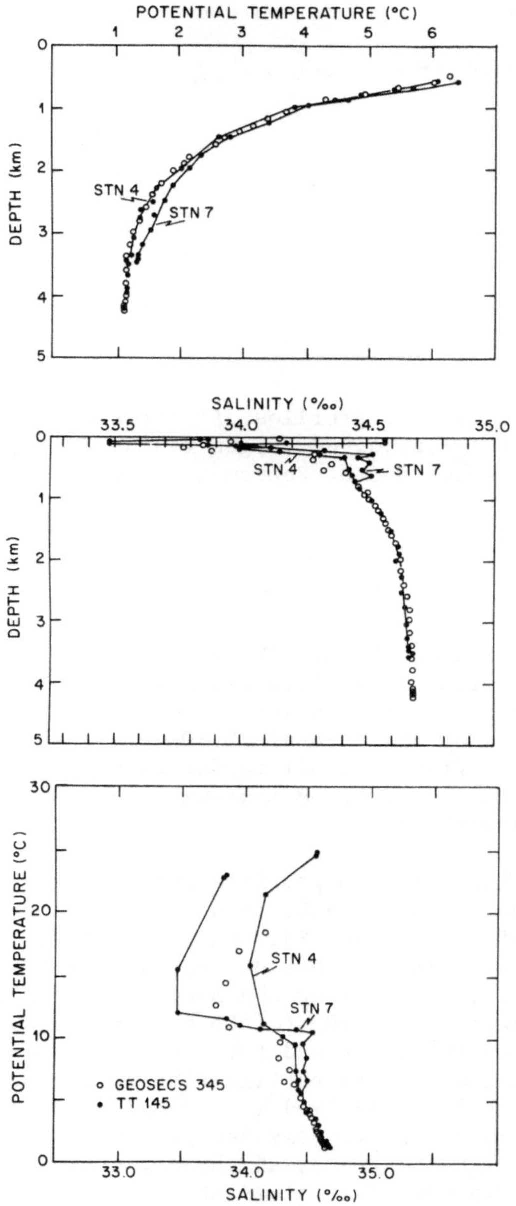

Fig. 6 Hydrographic properties off the coast of Baja during TT 145. Top: potential temperature versus depth; Middle: salinity versus depth; Bottom: potential temperature versus salinity. GEOSECS 345 data is shown in open circles. Only data from R/V Thompson stations 4 and 7 are shown to emphasize horizontal variations. R/V Thompson data is shown in closed circles.

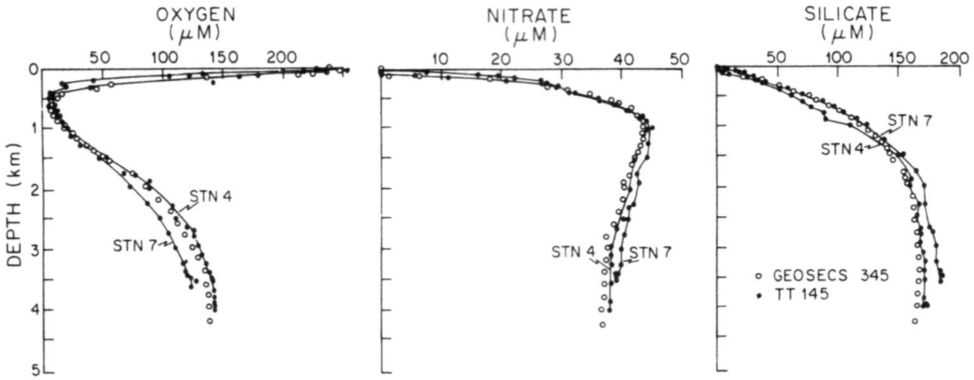

Fig. 7 Nutrients element profiles off the coast of Baja during TT 145. Left: dissolved oxygen versus depth; Middle: nitrate versus depth; Bottom: dissolved silicate versus depth.

detection limit of about 2 μM. The nitrate profiles mirror the dissolved oxygen with the maximum nitrate concentrations located about 500 m deeper than the lowest dissolved oxygens. This may be an artifact caused by the removal of NO_3^- by denitrification at shallower depths. Dissolved silica increases steadily from low surface water values to a broad maximum around 3000 m and then decreases slowly with depth.

Chromium. The chromium profiles are strongly influenced by the oxygen minimum. This can be seen most dramatically in the chromium(VI) profiles (Figure 8). Chromium(VI) at the surface is consistently about 3 nM and then decreases sharply to a minimum that becomes more intense along the section toward the east. The minimum is centered in the 200 to 400 m depth range. At station 2 the Cr(VI) minimum is absent and the profile increases to a broad maximum at about 2 km. This profile mirrors the Si profile which suggests that Cr(VI) is involved in a deep regeneration cycle. A similar conclusion has previously been reached on samples from the Cascadia Basin[7]. At station 2, at the top of the oxygen minimum, Cr(VI) is about 3.6 nM. The values at the minimum at the remaining stations are 2.5 nM at station 4, 2.1 nM at station 5, 1.7 nM at station 6, and 1.1 nM at station 7. Within the 5% error range the deep water concentrations are uniform at about 4.5 nM with no significant horizontal variations. Horizontal gradients were observed in the Cascadia Basin[32]. Only at station 6 is there a clear increase toward the sediments suggesting a sediment input.

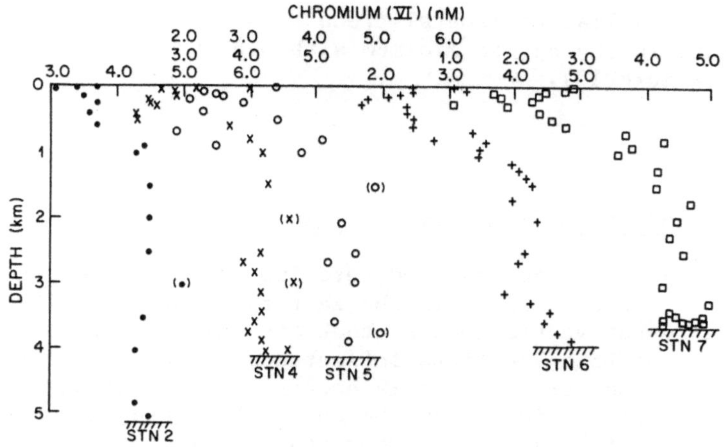

Fig. 8 Profiles of chromium(VI) versus depth at stations 2, 4, 5, 6 and 7 in the Baja transect.

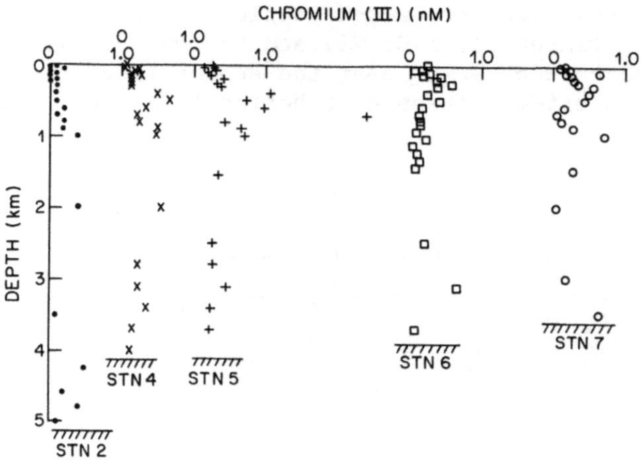

Fig. 9 Profiles of chromium(III) versus depth at stations 2, 4, 5, 6 and 7 in the Baja transect.

Chromium(III) is generally at or below the level of detection (0.2 nM) except at the top of the oxygen minimum where there is a well-defined maximum at all stations except station 2 (Figure 9). The Cr(III) maximum reaches about 0.5 nM at stations 4, 6 and 7. At station 5, the maximum reaches 1.0 nM. While the Cr(VI) minimum intensifies approaching the coast the Cr(III) maximum does not exhibit a clear horizontal trend. Particulate Cr also shows high values in the oxygen minimum although the scatter is somewhat greater (Figure 10).

DISCUSSION

Oxidation-Reduction Chemistry

Although the manganese and chromium profiles presented here are from different regions in the Eastern Tropical Pacific we can made some tentative statements about the oxidation-reduction chemistry. At both locations inflections in the NO_3 profiles indicate that denitrification is occurring. The dissolved manganese maximum at station 17 in the Guatemala Basin and the dissolved chromium(VI) minima at four stations in the Baja transect suggest that there are, at least locally, conditions reducing enough to reduce Mn(IV) and Cr(VI). Thermodynamic calculations suggest that chromium reduction should occur at around pE 6.6. Local conditions are implicated because at the Baja transect stations dissolved oxygen is always greater than about 10 μM. Denitrification is thought to only occur at oxygen concentrations less than 10 μM. Thus the redox couples are not at all internally consistent. Elements like Cr(VI) and Mn(IV) are either reduced *in situ* in microzones or at or near the sediments at the continental margin and the effects translated horizontally into the interior.

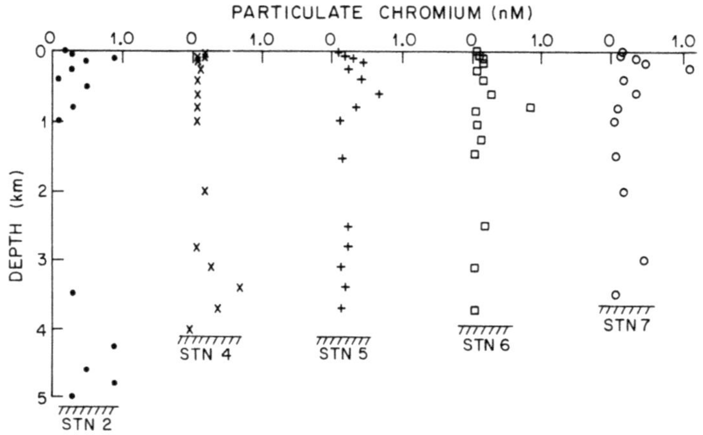

Fig. 10 Profiles of particulate chromium versus depth at stations 2, 4, 5, 6 and 7 in the Baja transect.

Manganese Chemistry

The TDM profiles are characterized by three features. These are the maxima near the surface, in the oxygen minimum and near the bottom. Bender et al.[17] concluded that the sources for Mn in the surface maximum include 1) riverine input; 2) diffusion from shelf sediments; and 3) deposition of atmospheric particulates. Upwelling is probably not a source because Mn concentrations below the surface are less than the surface values. Klinkhammer and Bender[44] favored an atmospheric input; however, Landing and Bruland[18] cast doubt on this hypothesis when they found an inverse relationship between ^{210}Pb and TDM along a surface water transect from the central Pacific to the west coast of North America. Martin and Knauer[55] also ruled out direct riverine and atmospheric inputs and suggested that an indirect input from weathering of clay minerals directly injected into the water or resuspended along the continental slope may be a significant source. Riverine input into the Eastern Tropical Pacific is negligible, and thus we tentatively rule out this source. It is difficult to rule out horizontal transport from the continental shelf but it is difficult to use this source to explain the location of the maxima in the 50 - 150 m depth range. This is the depth range of the chlorophyll maximum and thus the surface maximum in TDM may be due to biological enrichment at the base of the photic zone.

The maximum in the oxygen minimum is composed of both dissolved and particulate manganese. The highest values we observed were 5.5 nM at station 17. Klinkhammer and Bender[44] reported that they consistently detected a Mn maximum when oxygen fell below 100 µM. They found that TDM increased with apparent oxygen utilization (AOU) and nitrate deficit when oxygen was less than 4 µM. It is unclear whether these Mn maxima are the results of reactions taking place <u>in situ</u> or whether they are advective and diffusive features from the continental margin. Klinkhammer and Bender[44] proposed three explanations. 1) Manganese in seawater is in oxidation-reduction equilibrium with hausmannite (Mn_3O_4). 2) Manganese is reduced in nearshore sediments and transported horizontally by advection and diffusion. 3) Manganese is released <u>in situ</u> from organic matter decomposed by O_2 and NO_3. There is evidence for and against all three of these models. Klinkhammer and Bender[44] presented the case for redox equilibria in some detail. From the following reduction reaction

$$Mn_3O_4 + 8H^+ + 2e^- \text{ --- } 3Mn^{2+} + 4H_2O$$

they derived

$$\log(Mn) = 21.17 - 2/3\ pE - 8/3\ pH$$

Using this equation the predicted equilibrium dissolved manganese tends to be much larger than measured, typically reaching values of greater than 10 nM. Landing and Bruland[18] summarized data supporting the third model in which Mn is produced by decomposing organic matter. Recently Martin and Knauer[51] have conducted a detailed study of the distribution and fluxes of manganese at a site between the Clipperton-Clarion fracture zone much further to the west than our stations. The oxygen minimum was still very intense at their location. They observed that the vertical organic carbon flux decreased sharply yet the Mn to organic carbon flux ratio increased. This appears to argue against a net source from decomposing organic matter and suggest that in situ scavenging is more important[56]. Their observations also tend to argue against the increased dissolution of hausmannite at lower pE values because there was no manganese released from particles within the oxygen minimum and dissolved manganese stayed uniformly low. They concluded that horizontal transport from nearshore along isopycnal surfaces is probably the major source.

Our data can be brought to bear on these different possibilities. TDM (Figure 5) is much less than predicted using the hausmannite model. In addition, a significant fraction (50%) of the TDM maximum is in the particulate and not the dissolved phase (station 17). In addition, hausmannite is generally formed by the rapid oxidation of Mn(II)[36,39], a condition which seems unlikely to be fulfilled in the intense oxygen minimum zone of the Eastern Tropical Pacific where dissolved oxygen is less than 2 μM. Based on the observed partitioning between dissolved and particulate forms and the oxidation environment within the oxygen minimum it appears to us that enrichment of Mn(II) is biological. Alternately, inorganic particulate matter is the most consistent explanation. Clearly more measurements of dissolved manganese are needed to substantiate our findings and new measurements of Mn(IV) could also help clear up these questions.

Simple model calculations confirm the feasibility of the argument for horizontal transport. Following Weiss[14] we suggest a simple balance between horizontal diffusion (K_H) and removal of total manganese by scavenging. The removal is described by a first order scavenging rate constant ψ (yr^{-1}). This is an oversimplification for a region in which the T-S characteristics clearly suggest an advective component. More thorough model development will be presented elsewhere. For the case where horizontal advection equals zero and the distribution is at steady state

$$C_x = C_o \exp - \sqrt{\psi / K_H}\, x$$

where C_x is the concentration at a distance x from the coast and C_o

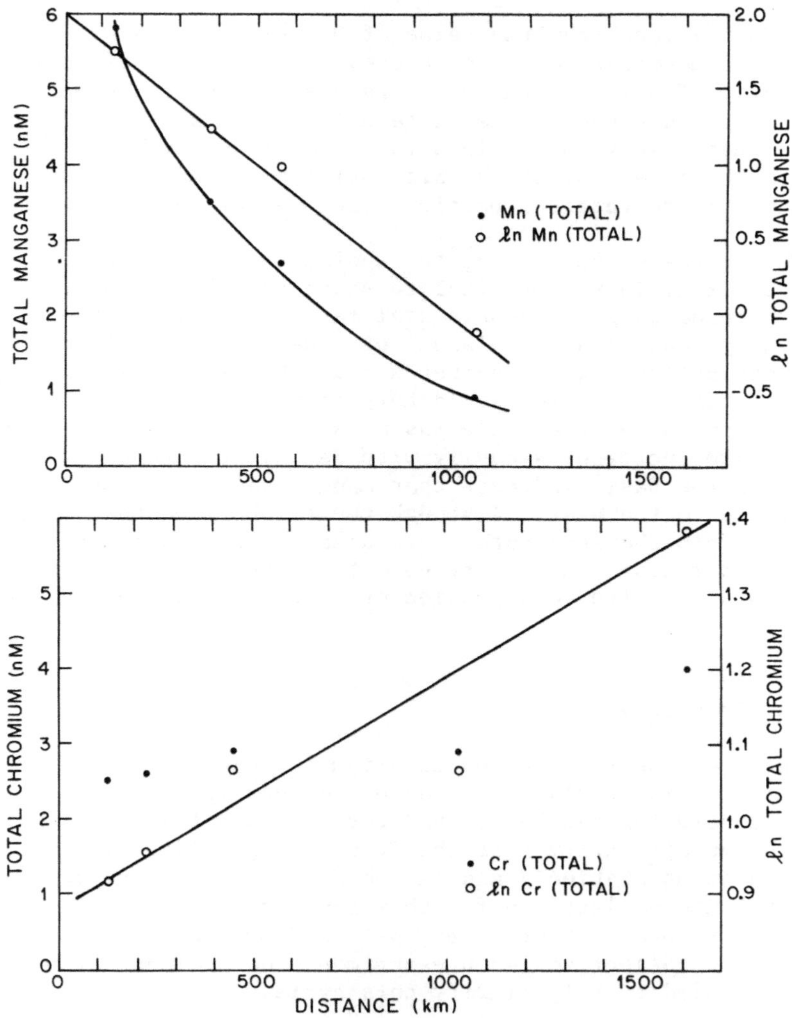

Fig. 11 Total manganese and chromium and ln total manganese and chromium versus distance from the continental shelf break in kilometers.

is the boundary concentration. Using $K_H = 5 \times 10^6\ cm^2\ s^{-1}$ Weiss[14] found that a value of $\psi = 0.02\ yr^{-1}$ (or residence time $\tau = 50$ yr) adequately described the distribution of hydrothermal manganese away from the Galapagos Rift site. For our four stations in the Guatemala Basin we have plotted the maximum value of TDM in the oxygen minimum versus the distance (x km) from the continental shelf break. The exponential decrease is fitted best when $(\sqrt{\psi / K_H})^{-1}$ equals 500 km (Figure 11). This characteristic distance is significantly less than the value of 900 km observed by Weiss[14] for the parameter values given above. Assuming a horizontal eddy diffusivity of 5 x 10^{-6} cm^2 s^{-1}, the scavenging residence time (τ) in the oxygen minimum of the Eastern Tropical Pacific is about 200 years. That this value is less than that found by Weiss[14] is not too disturbing because of the high primary productivity and presumably large vertical particle flux in the Guatemala Basin.

The increase approaching the sediment-water interface is almost entirely in the particulate phase and the increase becomes greater approaching the continental margin. Similar findings were observed by Landing and Bruland[18] off the coast of California. This distribution argues against a hydrothermal origin although ^3He data from this region would help answer this question. The water column in the Guatemala Basin is adiabatic below 2700 m, however, TDM increases steadily with depth. Our pore water studies in the Basin indicate that manganese remobilization intensifies to the east. Although the remobilized Mn is mostly trapped within the sediments, this does produce manganese rich surface sediments. Thus we propose that the near-bottom increase in TDM is due to the resuspension of Mn-rich fine-grained sediments.

Chromium Chemistry

Our previous research on chromium in seawater has indicated that chromium is involved in a deep regeneration cycle[32]. A similar conclusion can be reached for the Baja Transect stations, based on the similarities of the Cr and Si profiles over the entire water at station 2 and in the deep water at the remaining stations. The explanation for this regeneration is unclear as the mechanism of uptake of CrO_4^{2-} by opal or $CaCO_3$ phases has not been established. Selective leach experiments on sediment trap material are needed to help clarify this cycle.

There is one main feature which is superimposed on the deep biological cycle. This is the pronounced minimum in Cr(VI) and maximum in Cr(III) and particulate chromium observed in the oxygen minimum zone. The most straight-forward explanation is that Cr(VI) is reduced to Cr(III) in the low oxygen water. This Cr(III) is rapidly scavenged and thus reduced chromium shows up as both Cr(III) and particulate chromium. The scavenged Cr must be rapidly removed from the water column resulting in a pronounced minimum in total chromium as well. This cycle is consistent with the solution chemistry of chromium which was reviewed earlier. The problem with this explanation is that even though dissolved oxygen is very low in the oxygen minimum off Baja it was still measurable; thus chromium reduction can only be explained by invoking microzones of lower oxidation potential.

An alternate explanation can be proposed that is similar to that presented earlier for manganese. In this case there is a scavenging (or reduction plus scavenging) sink for total chromium at the continental margin and chromium is transported to that sink by horizontal mixing along isopycnal surfaces. Again, as a first approximation we neglect advection and assume the flux of total chromium is onshore. The minimum value in the oxygen minimum is plotted against distance from the shelf break in Figure 11. The scale length for total chromium is 330 km. Again, assuming $K_H = 5 \times 10^6 \ cm^2 \ s^{-1}$ we find that $\psi = 0.0135 \ yr^{-1}$ and the scavenging residence time $\tau = 740$ years. This residence time is significantly slower than calculated earlier for manganese. The difference may arise from the fact that while both oxidation states of manganese can be rapidly scavenged, chromium is probably only scavenged as Cr(III). Cr(VI) as CrO_4^{2-} remains fairly soluble.

CONCLUSIONS

It is clear that chromium and manganese form oxidized and reduced species which have contrasting chemistries. During reduction reactions chromium becomes more insoluble and manganese becomes more soluble. This chemistry can explain the profiles for these trace metals in the low oxygen water of the Eastern Tropical Pacific. The chromium profiles are characterized by a pronounced minimum in Cr(VI) and a maximum in Cr(III) while dissolved manganese [mostly Mn(II)] shows a maximum in the same region.

When measured as a pair these two metals help constrain the oxidation-reduction conditions in the oxygen minimum. Inconsistencies exist, however, that need to be explained. The maxima in Cr(III) and Mn(II) imply that both chromium and manganese can be reduced. At one location, however, measurable (but low) dissolved oxygen was present throughout the oxygen minimum. This implies that either the metals are reduced <u>in situ</u> in reducing microzones or that they are produced at the continental margin, near or at the sediments, and transported horizontally into the interior. In the case of manganese there is a maximum in both dissolved and acid soluble manganese in the oxygen minimum. The simplest explanation is that both peaks correspond to Mn(II). But clearly what is needed is an analytical method to distinguish between Mn(II) and Mn(IV). In addition, a careful study of iron and uranium in the oxygen minimum would extend the oxidation-reduction implications of our study.

ACKNOWLEDGEMENTS

We are especially grateful to the captain and crew of the R/V Thompson and to many people who helped us collect water samples during TT 127 and TT 145. This research was supported by NSF Grant OCE 80-18335, University of Washington Contribution 1250.

REFERENCES

1. Boyle, E. and J.M. Edmond, 1975: Copper in surface waters south of New Zealand. Nature, 253, 107-109.
2. Bender, M.L. and C. Gagner, 1976: Dissolved copper, nickel and cadmium in the Sargasso Sea. J. Mar. Res., 34, 327-339.
3. Sclater, F.R., E.A. Boyle and J.M. Edmond, 1976: On the marine geochemistry of nickel. Earth Planet. Sci. Lett., 31, 119-128.
4. Martin, J.H., K.W. Bruland and W.W. Broenkow, 1976: Cadmium transport in the California current. In: "Marine Pollutant Transfer", (H.L. Windom and R.A. Duce, eds.), D.C. Heath and Co., 159-184.
5. Bruland, K.W., G.A. Knauer and J.H. Martin, 1978: Zinc in northeast Pacific Waters. Nature, 271, 741-743.
6. Chan, L.H., J.M. Edmond, R.F. Stallard, W.S. Broecker, Y.C. Chung, R.F. Weiss and T.L. Ku, 1976: Radium and barium at GEOSECS stations in the Atlantic and Pacific. Earth Planet. Sci. Lett., 32, 35-44.
7. Cranston, R.E. and J.W. Murray, 1978: The determination of chromium species in natural waters. Anal. Chim. Acta., 99, 275-282.
8. Andreae, M.O., 1979: Arsenic speciation in seawater and interstitial waters: the influence of biological-chemical interactions on the chemistry of a trace element. Limnol. Oceanogr., 24, 440-452.
9. Measures, C.I. and J.D. Burton, 1980: The vertical distribution and oxidation states of dissolved selenium in the northeast Atlantic Ocean and their relationships to biological processes. Earth Planet. Sci. Lett., 46, 385-396.
10. Craig, H., S. Krishnaswami and B.L.K. Somayajulu, 1973: ^{210}Pb-^{226}Ra: radioactive disequilibria in the deep sea. Earth Planet. Sci. Lett., 17, 295-305.
11. Bacon, M.P., D.W. Spencer and P.G. Brewer, 1976: $^{210}Pb/^{226}Ra$ and $^{210}Po/^{210}Pb$ disequilibrium in seawater and suspended particulate matter. Earth Planet. Sci. Lett., 31, 119-128.
12. Boyle, E.A., F.R. Sclater and J.M. Edmond, 1977: The distribution of dissolved copper in the Pacific. Earth Planet. Sci. Lett., 37, 38-54.
13. Brewer, P.G., Y. Nozaki, D.W. Spencer and A.P. Fleer, 1980: Sediment trap experiments in the deep North Atlantic. J. Mar. Res., 38, 703-728.

14. Weiss, R.F., 1977: Hydrothermal manganese in the deep sea: scavenging residence time and Mn/^{3}He relationships. Earth Planet. Sci. Lett., 37, 257-262.
15. Klinkhammer, G., M. Bender and R.F. Weiss, 1977: Hydrothermal manganese in the Galapagos Rift. Nature, 269, 319-320.
16. Edmond, J.M., C. Measures, B. Magnum, B. Grant, F.R. Sclater, R. Collier, A. Hudson, L.I. Gordon and J.B. Corliss, 1979: On the formation of metal rich deposits at ridge crests. Earth Planet. Sci. Lett., 46, 19-30.
17. Bender, M.L., G.P. Klinkhammer and D.W. Spencer, 1977: Manganese in seawater and the marine manganese balance. Deep-Sea Res., 24, 799-812.
18. Landing, W.M. and K.W. Bruland, 1980: Manganese in the North Pacific. Earth Planet. Sci. Lett., 49, 45-56.
19. Stumm, W. and J.J. Morgan, 1970: Aquatic Chemistry. Wiley-Interscience, New York, 583 pp.
20. Emerson, S., R.E. Cranston and P.S. Liss, 1979 Redox species in a Reducing fjord: equilibrium and kinetic considerations. Deep-Sea Res., 26, 859-878.
21. Codispoti, L.A. and F.A. Richards, 1976: An analysis of the horizontal regime of denitrification in the Eastern Tropical North Pacific. Limnol. Oceanogr., 21, 379-388.
22. Dugdale, R.C., J.J. Goering, R.T. Barber, R.L. Smith and T.T. Packard, 1977: Denitrification and hydrogen sulfide in the Peru upwelling region during 1976. Deep-Sea Res., 24, 601-608.
23. Burns, R.G., 1970: Mineralogical Application of Crystal Field Theory. University Press, London, 224 pp.
24. Pearson, R.G., 1966: Acids and bases. Science, 151, 172-177.
25. Taube, H., 1970: Electron Transfer Reactions of Complex Ions in Solution. Academic Press, New York, 103 pp.
26. Linck, R.G., 1976: Oxidation-reduction reactions. In: "Survey of Progress in Chemistry", (A.F. Scott, ed.), Vol. 7, Academic Press, 89-147.
27. Hem, J.D., 1977: Reactions of metal ions at surfaces of hydrous iron oxide. Geochim. Cosmochim. Acta., 41, 527-538.
28. Cranston, R.E. and J.W. Murray, 1980: Chromium species in the Columbia River and estuary. Limnol. Oceanogr., 25, 1104-1112.
29. Benes, P. and E. Steinnes, 1975: Migration forms of trace elements in natural waters and the effect of the water storage. Water Research, 9, 741-749.
30. Lingane, J.J., 1966: Analytical Chemistry of Selected Metallic Elements. Reinhold, New York, 143 pp.
31. Elderfield, H., 1970: Chromium speciation in seawater. Earth Planet. Sci. Lett., 9, 10-16.

32. Cranston, R.E., 1979: Chromium species in natural waters. Ph.D. Thesis, University of Washington, 304 pp.
33. Earley, J.E. and R.D. Cannon, 1965: Aqueous chemistry of chromium(III). Trans. Metal Chem., 1, 34-109.
34. Fukai, R. and D. Vas, 1969: Changes in the chemical forms of chromium on the standing of seawater samples. J. Oceanogr. Soc. Japan, 25, 47-49.
35. Schroeder, D.C. and G.F. Lee, 1975: Potential transformations of chromium in natural waters. Water, Air and Soil Pollut., 4, 355-365.
36. Morgan, J.J., 1967: Chemical equilibrium and kinetic properties of manganese in natural waters. In: "Principals and Applications of Water Chemistry", (S.D. Faust and J.V. Hunter, eds.), John Wiley, New York, 561-624.
37. Murray, J.W. and P.G. Brewer, 1977: Mechanisms of removal of manganese, iron and other trace metals from seawater. In: "Marine Manganese Deposits", (G.P. Glasby, ed.), Elsevier, 291-325.
38. Burns, R.G. and V.M. Burns, 1977: Mineralogy. In: "Marine Manganese Deposits", (G.P. Glasby, ed.), Elsevier, 185-248.
39. Stumm, W. and R. Giovanoli, 1976: On the nature of particulate manganese in simulated lake waters. Chemica, 30, 423-425.
40. Ghiorse, W.C. and H.L. Ehrlich, 1976: Electron transport components of the MnO_2 reductase system and the location of the terminal reductase in a marine bacillus. Appl. Microbiol., 31, 977-985.
41. Hem, J.D., 1963: Chemical equilibria and rates of manganese oxidation. U.S. Geol. Survey Water-Supply Paper 1667A, A1-A64.
42. Morgan, J.J., 1964: Chemistry of aqueous manganese(II) and (IV). Ph.D. Thesis, Harvard University, 244 pp.
43. Sung, W. and J.J. Morgan, 1981: Oxidative removal of Mn(II) from solution catalyzed by the γ-FeOOH (lepidocrocite) surface. Geochim. Cosmochim. Acta., 45, 2377-2384.
44. Klinkhammer, G.P. and M.L. Bender, 1980: The distribution of manganese in the Pacific Ocean. Earth Planet. Sci. Lett., 46, 361-384.
45. Kingston, H.M., I.L. Barnes and T.C. Rains, 1978: Separation of eight transition elements from alkali and alkaline earth elements in estuarine and seawater with chelating resin and their determination by graphite furnace atomic absorption spectrometry. Anal. Chem., 50, 2064-2070.
46. Lonsdale, P., 1976: Abyssal circulation of the southeastern Pacific and some geological implications. J. Geophys. Res., 81, 1163-1176.
47. Lonsdale, P., 1977: Inflow of bottom water to the Panama Basin. Deep-Sea Res., 24, 1065-1101.

48. Mantyla, A., 1975: On the potential temperature in the abyssal Pacific Ocean. J. Mar. Res., 33, 341-354.

49. Reid, J.L., 1965: Intermediate waters of the Pacific Ocean. The Johns Hopkins Oceanographic Studies, No. 2. The Johns Hopkins Press, Baltimore, 85 pp.

50. Wyrtki, K., 1967: Circulation and water masses in the eastern tropical Pacific Ocean. Int. J. Oceanol. Limnol., 1, 117-147.

51. Martin, J.H. and G.A. Knauer, in press: Manganese cycling in Northeast Pacific Equatorial Waters. J. Mar. Res.

52. Reid, J.L., 1973: The shallow salinity minima of the Pacific Ocean. Deep-Sea Res., 20, 51-68.

53. Kenyon, K.W., 1978: The shallow salinity minimum of the Eastern North Pacific in Winter. J. Phys. Ocean., 8, 1061-1069.

54. Wooster, W.S. and J.H. Jones, 1970: California Undercurrent off northern Baja California. J. Mar. Res., 28, 235-250.

55. Martin, J.H. and G.A. Knauer, 1980: Manganese cycling in Northeast Pacific waters. Earth Planet. Sci. Lett., 51, 266-274.

56. Balistrieri, L., P.G. Brewer and J.W. Murray, 1981: Scavenging residence times of trace metals and surface chemistry of sinking particles in the deep ocean. Deep-Sea Res., 28, 101-121.

POTENTIALITIES AND APPLICATIONS OF VOLTAMMETRY IN CHEMICAL SPECIATION OF TRACE METALS IN THE SEA

Hans Wolfgang Nürnberg and Pavel Valenta

Institute of Applied Physical Chemistry
Chemistry Department, Nuclear Research Center
(KFA), Juelich, Federal Republic of Germany

ABSTRACT

Substance specificity and general methodological properties with respect to sensitivity and accuracy make advanced modes of polarography and voltammetry a powerful and convenient approach to study at trace levels the speciation of heavy metals as Cd, Pb, Zn, Cu, Hg which are important from the viewpoint of marine ecochemistry as well as of raw materials with respect to their incorporation in manganese nodules.

Detailed informations can be obtained on the speciation behaviour of defined complex species with respect to stability, ligand number, formation kinetics and mechanism and the significance of specific side-effects by salinity components on these parameters. The selection of the experimental procedure depends predominantly on the stability and thus electrochemical reversibility of the studied complex type. In this manner the general pattern of the speciation distribution in the sea with the predominant labile complexes forming inorganic ligands has been determined for Cd and Pb. Systematic studies with well-known defined model ligands of moderate strength, as NTA, have provided for Cd, Pb and Zn the experimentally founded basis for a comprehensive understanding and prognostic conclusions on the parameters governing their speciation as nonlabile species by components of dissolved organic matter (DOM). Moreover, determination of complexation capacities for various heavy metals and speciation-minded voltammetric analysis of heavy metal levels in the sea can provide important global data and informations of diagnostic character on the speciation-capability of sea water from various regions and depths.

INTRODUCTION

The sea is an important stage in the biogeochemical cycle of metals. As in other parts of the environment certain heavy and transition metals, as Hg, Cd, Pb, Cu, Zn have gained, due to their pronounced ecotoxicity[1,2], growing particular significance since several years in marine trace metal chemistry. In general, there is a distinct tendency for those metals to be transported from the land into the sea either immediately by fluvial input and run-off or indirectly via the atmosphere from which they are deposited by dry dust and to a more significant extent by rain and snow[3,11]. There act also in certain parts of the oceans marine metal sources at the sea bottom in connection with marine volcanism and sea floor spreading. Usually the toxic metals brought to the sea have rather substantial residence times in this environmental compartment and the sea has actually properties of a pseudo-sink for them. Nevertheless, the biogeochemical cycle is ultimately closed due to the transfer of toxic metals from the sea back to the terrestrial environment via the marine food and with the sea spray. The situation is schematically depicted in Fig. 1.

Generally the heavy and transition metals have a distinct tendency to be bound to the surface layers of suspended particu-

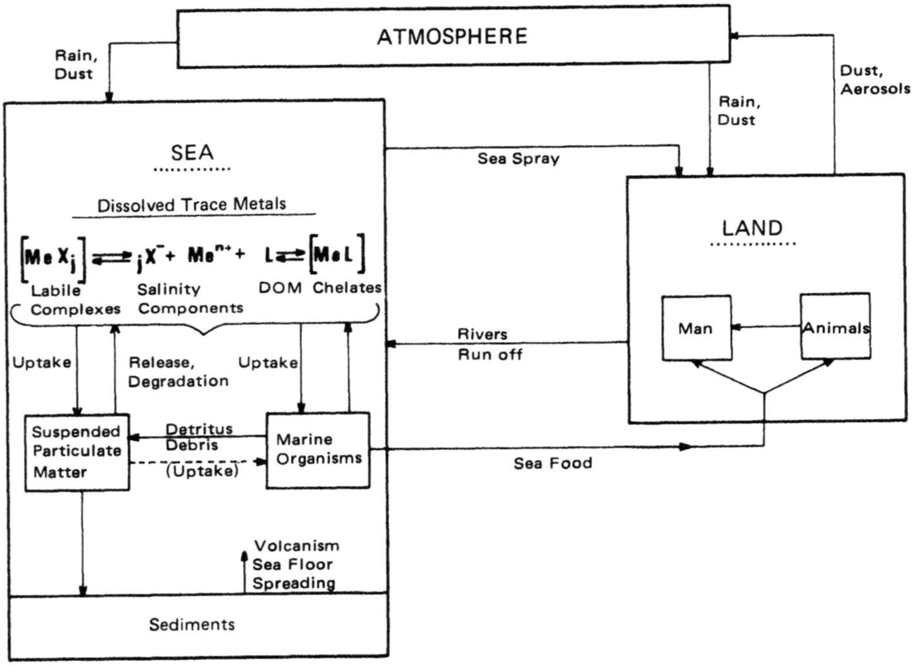

Fig. 1 Biogeochemical cycle of heavy metals

late matter and sediments while the concentrations in the dissolved state are usually at the lower trace level. Thus, in the rather shallow coastal waters the sediments have besides the suspended particulates an important depot function for those toxic trace metals and therefore in the regulation of their levels available in the dissolved state, e.g. if a continuing demand occurs due to metal uptake by marine organisms from the various trophic stages of the food webs. Although the concentrations of the considered toxic metals in the dissolved state are rather low, typically between 10^{-3} and 1 $\mu g\ l^{-1}$, the dissolved state remains very important. In and from this state occurs to a major extent the transfer of the toxic metals to and from the interfaces of suspended particulates, sediments and the marine organisms, usually taking up these metals but sometimes also releasing certain amounts due to biological regulation effects. For these various interactions and ultimately toxic effects the overall concentrations of the respective trace metals in the dissolved state are not without significance but the various chemical species in which the trace metals exist in the sea have a key function.

In this context it has to be noted that the dissolved state is usually defined for metals in the sea and other natural waters in an operational manner by a convention. Thus, the metal content in the filtrate from filtration through a membrane filter with 0.45 μm pore size is counted as dissolved. This includes necessarily all inorganic and organic colloids with a diameter below 0.45 μm. These colloids need particular investigations in the future, because indications exist that their participation in the trace metal speciation in sea water might be significant. There are authors[4] who even assume that with high probability for the speciation of dissolved trace metals by organic material the binding of trace metals by organic colloids is of predominant significance.

The above-mentioned toxic trace metals (Cd, Pb, Zn, Cu, and Hg) and a number of other trace metals of particular interest, e.g., Co, Ni, Se(IV), etc., are readily accessible to voltammetry. Because of its striking determination sensitivity (typical limit 1 ng l^{-1}, or even less), its inherent particularly low accuracy risks, a satisfactory determination rate (due to the possibility of simultaneous determination of several trace metals in one run), and its low cost requirements, voltammetry has in recent years become an important if not the most powerful method in the analysis of the dissolved overall levels of toxic metals in natural waters[5-10]. Moreover, the compactness of the instrumentation creates a potential for very convenient shipboard analysis[11]. Usually differential pulse stripping voltammetry (DPSV) at suitable working electrodes is employed. Only Ni and Co are determined just by conventional or differential pulse voltammetry (DPV) after adsorption of their dimethylglyoxime chelates, achieving in this manner interfacial proconcentration at the surface of the working electrode[12]. While this approach provides high analytical sensitivity and still permits also global speciation investigations for Ni and Co the addition of the chelator dimethylglyoxime prohibits studies of individual defined complex species of these two latter heavy metals.

However, in general the substance specificity of polarography and voltammetry provides also at present the most informative and versatile approach for speciation studies of dissolved toxic trace metals, as Cd, Pb, Zn, Cu, Hg, in natural waters at or at least sufficiently close to their rather low concentration levels[4-8,13,14].

The selection of the respective voltammetric procedure depends on the kind of the speciation type and problem to be studied[8,14]. One has to distinguish between fundamental investigations of defined individual inorganic and organic complexes of the respective trace metal in sea water medium under carefully, adjusted experimental conditions aimed to yield results of general validity for the sea, and studies aimed to explore just the overall speciation capacity including the colloids of sea water originating from various regions. The respective methodological principles of the relevant voltammetric procedures will be outlined in the following together with the presentation and discussion of corresponding typical results from our speciation studies with Cd, Pb and Zn. The obtained experimental data furnish reliable evidence and support for the emerging conclusions and arguments of general significance.

The applied voltammetric procedure for the investigation of defined complex species depends primarily on the stability of the complexes.

SPECIATION BY LABILE COMPLEXES

General Physicochemical and Methodological Aspects

The inorganic ligands X commonly abundant in sea water and significant for trace metal speciation are Cl^-, OH^-, CO_3^{2-}, HCO_3^- and SO_4^{2-}. They form usually labile mononuclear complexes MeX_j[15]. Their complex equilibria are very mobile due to high rate constants as well for formation k_f as for dissociation k_d of the complexes. Therefore these labile complex species MeX_j undergo reversible electrode processes. The same applies[16] also to weak organic complexes MeL_j with certain organic ligands L which are possibly components of dissolved organic matter (DOM). Other DOM components form certainly rather stable complexes MeL_j (viz. next section). It is to be emphasized, however, already here that the significance of the contributions of organic ligands to the trace metal speciation will depend significantly on their concentration level and thus, on the level of DOM in the studied area of the sea. In large parts of the open oceans DOM has rather low concentrations (< 1 mg l^{-1}).

Generally the (consecutive) stability constant K of a complex MeX_j or MeL_j is related to the rate constants according to eq. (1) $K = k_f/k_d$. It is thus obvious that K will be small, if both rate constants are of not too different magnitude as it is the case with weak and therefore labile complexes. In this context k_f denotes the formation rate constant for complex formation from the components of the complex (recombination mechanism) according to

$$Me + X \underset{k_d}{\overset{k_f}{\rightleftharpoons}} MeX \text{ or } Me + L \underset{k_d}{\overset{k_f}{\rightleftharpoons}} MeL$$

(Signs are omitted for simplicity as usual.)

Also for rather stable complexes k_f remains large, having typically the order of 10^8 to 10^{10} l . mol^{-1} s^{-1}, because in a recombination mechanism complex formation is predominantly or fully diffusion controlled[17]. Yet the dissociation rate constant k_d is now small and according to eq. (1) there will exist for different strong complexes according to the "constancy" of diffusion controlled k_f a more or less good reciprocal relationship between the respective stability constants K and the corresponding k_d-values. It has, however, to be emphasized that, as will be pointed out later, different formation mechanisms might be operative in sea water for the formation of trace metal complexes. The measured formation rate constant k_f' refers then to a different complex formation mechanism and can not be inserted into eq. (1). Nevertheless, the outlined fundamental correlations resulting from eq. (1) between the stability constant K and the rate constants k_f and k_d referring to the recombination mechanism remain always valid and relevant for the stability of the complexes.

As the labile complexes MeX_j formed with the above mentioned ligands X abundant in sea water yield reversible responses one can apply for their investigation an approach known since several decades in the application of polarography and voltammetry to complex chemistry[18,19]. It consists in the measurement of the dependence of the reversible half wave potential $E_{1/2}$ on the logarithm of the ligand concentration [X]. The reversible half wave potential is closely related to the redox potential and is thus a thermodynamically well defined value. The half wave potential shift $\Delta E_{1/2}$ depends on the ligand concentration [X] according to eq. (3).

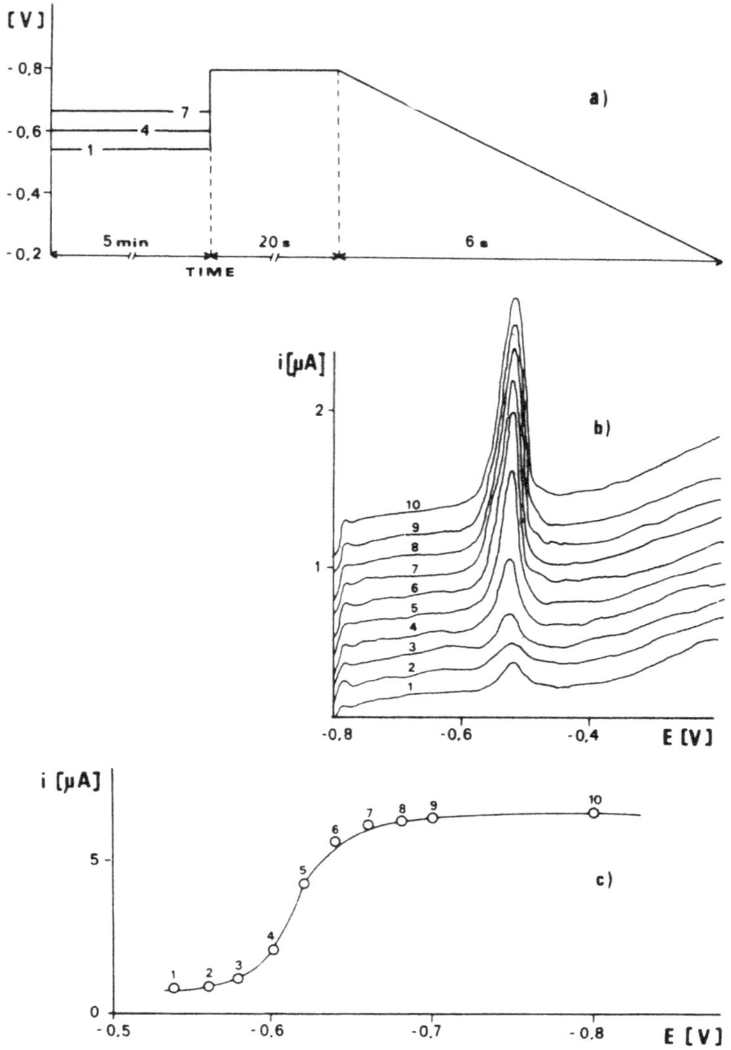

Fig. 2 Construction of pseudo-polarogram from automated ASV-measurement. a) Course and timing of polarising voltage; b) Family of ASV-peaks as function of cathodic deposition potential; and c) Resulting pseudo-polarogram.

$$\Delta E_{1/2} = -\frac{RT}{nF} \ln \sum_{j=o}^{j} \beta_{MeX_j} [X]^j \qquad (3)$$

R is the gas constant, T the temperature in ^oK, n the number of electrons transferred in the electrode process (usually for the considered toxic trace metals n equals to 2), F is the Faraday (96500 C mol^{-1}) and β_{MeX_j} is the overall cumulative stability constant according to

$$\beta_{MeX_j} = \frac{[MeX_j]}{[Me]\ [X]^j} \tag{4}$$

With a certain ligand X frequently a sequence of complexes MeX_j with different ligand numbers j = 1, 2, 3 and corresponding stability constants β_{MeX_j} will be formed at a given ligand concentration [X]. The corresponding consecutive stability constants K_j are defined according to eq. (5) $K_j = \beta_j / \beta - 1$.

Yet throughout this paper the overall cumulative stability constants β_j are used and it is presumed that they are conditional values, i.e. they refer to the ionic strength of the considered medium, e.g. sea water, and contain therefore the general salt effects or in other words the activity coefficients of the reactants. Therefore the β_j are related to the concentrations of the reactants of the complex equilibria and not to their activities which would be for a natural water in principle also a possible but a rather impractical treatment.

With respect to the characterisation of a certain type of labile inorganic complexes MeX_j in sea water the task is now to measure the $\Delta E_{1/2}$ - ln[X]- relationship over a relevant range of [X]. From this dependence (viz. Fig. 3) the stability constants β_{MeX_j}, and the ligand numbers j can be evaluated according to the method of De Ford and Hume[20] and in this manner the various complexes of the type MeX_j are identified from j and their strength is characterised by the corresponding β_{MeX_j}- values.

If those measurements shall be performed at or relatively close to the rather low concentration levels of toxic trace metals prevailing in sea water the classical approach using conventional dc-polarography cannot be applied for sensitivity reasons.

Yet the method of constructing pseudo-polarograms[21] from measurements by conventional anodic stripping voltammetry (ASV), applying a linear potential scan in the stripping stage, permits to work at the 10^{-9} M level for the studied trace metals (viz. Fig. 2). The pseudo-polarogram is constructed from the heights of ASV-peaks corresponding to a series of cathodic deposition potentials E_d (about 10 E_d-values) adjusted in the potential range of the hypothetic dc-polarogram. As those measurements have to be performed over an extended range of the ligand concentration [X] to obtain a $\Delta E_{1/2}$-ln [X]- relationship (viz. Fig. 3) relevant to all consecutive complexes MeX_j of the studied type, which exist in sea water, a larger number of experiments has to be made, as usually for [X] a range of several orders of magnitude has to be investigated. Therefore, the procedure for the recording of the

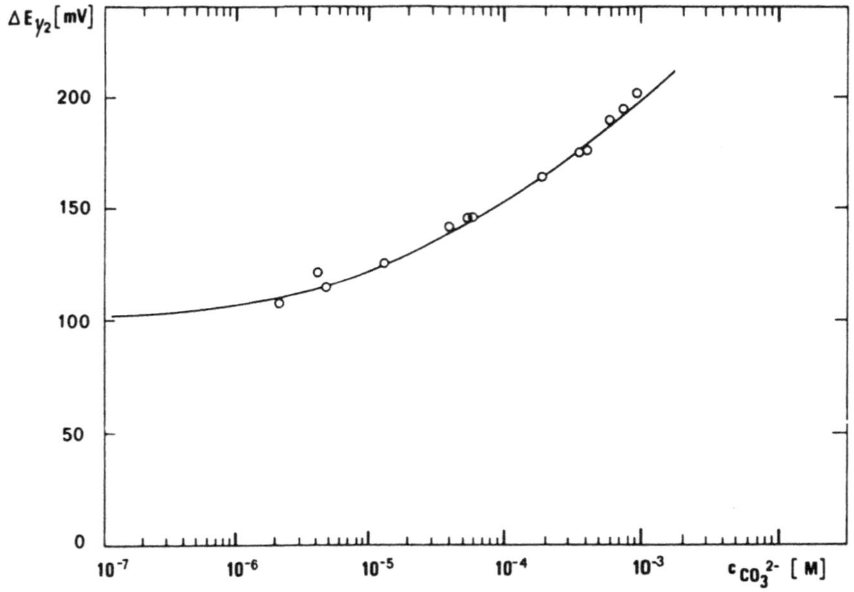

Fig. 3 $\Delta E_{1/2}$ - lg $[CO_3^{2-}]$-relationship for Pb(II) in artificial sea water. Total Pb(II) concentration 6 x 10^{-9}M.

relationship of the ASV peak heights on the E_d-values has been automated[13].

Results

In this manner the role of CO_3^{2-} and HCO_3^- in the complexation of Pb(II) and Cd(II) in sea water could be determined in the 10^{-9} M range employing a mercury film electrode (MFE) on glassy carbon support as working electrode[22]. To avoid interferences from DOM the medium was artificial sea water containing, of course, all salinity components. For Cd(II) the obtained data have been also reproduced in genuine sea water with negligible DOM-levels sampled at the Ligurian coast. The concentration of CO_3^{2-} and HCO_3^- and thus the pH has been adjusted by controlling the CO_2 partial pressure in the inert gas phase over the solution simulating in this matter closely the real situation in nature. In a similar way the data for the formation of chlorocomplexes of both trace metals have been determined. From these results and literature data for the hydroxo complexes of Pb(II) and the sulphato complex of Cd(II) the inorganic species distribution in sea water has been evaluated for both trace metals (Table 1). In this evaluation a noticeable improvement of the percentage values of the inorganic complex species existing in sea water has been achieved by taking into

Table 1 Stability and distribution of the predominant species of Cd(II) and Pb(II) in sea water with and without correction for ion pairing of the major salinity components.

Species	Without ion pairing correction		With ion pairing correction	
	β_j	Distribution (%)	β_j	Distribution (%)
Cd^{2+}	-	2.7	-	1.9
$CdCl^+$	27.25	34.6	32.69	29.1
$CdCl_2$	61.66	36.8	88.74	37.2
$CdCl_3$	91.20	25.6	157	31.0
other CdX_j	-	0.3	-	0.8
Pb^{2+}	-	3.0	-	1.8
$PbCl^+$	8.57	12.0	10.28	8.6
$PbCl_2$	22.24	14.7	32.00	12.6
$PbOH^+$	7.9×10^6	50.0	7.9×10^6	30.0
$Pb(OH)_2$	6.3×10^{10}	0.9	6.3×10^{10}	0.5
$PbCO_3$	4.2×10^5	18.9	1.6×10^6	43.0
$Pb(CO_3)_2^{2-}$	6.25×10^8	3.7	9.1×10^9	3.7

account the influences of ion pairing between the salinity components effective in concentrated electrolytes as sea water[23]. The net results of this study were, that for Cd(II) the three chloro-complexes are present in the sea at rather comparable quantities, with a slight predominance of $CdCl_2$, and account altogether for 97% of the inorganic speciation of this trace metal. Moreover, it could be established, that although the percentages are rather low, there exist carbonato- and bicarbonato complexes for Cd(II) while no bicarbonato complexes of Pb(II) exists in the sea. Pb(II) is most abundant as carbonato-complexes, predominantly as $PbCO_3$ with 43%, followed by $PbOH^+$ while the chloro-complexes of Pb(II) correspond altogether to only 21% (see Table 1). With respect to modelling the experimental results for Pb(II) have further confirmed the relevance of the sea water model of Pytkowicz and Hawley[24] for this trace metal. However, the controversial species distributions resulting from model computations of several other authors (see ref. 22,23) could not be confirmed. This emphasizes impressively the necessity to regard modelling results only as tentative until they have found experimental verification.

Conclusions

The inorganic speciation pattern for Pb(II) and Cd(II) in Table 1 ignores contributions due to chelate formation with strong organic ligands being components of DOM. Nevertheless, it is concluded that the elucidated speciation is for both trace metals valid in the major part of the open sea, at least in the large oceanic regions with low biological productivity. In those regions the DOM level will be too low to contribute significantly to the speciation of most dissolved trace metals, as will be discussed in more detail in the next section. Moreover, also the amount of suspended particulate matter remains rather small in large parts of the open oceans and seas and also in certain coastal waters, e.g. along large parts of the Ligurian and Tyrrhenian coast in the Mediterranean sea[25]. More complicated speciation situations may arise, however, due to elevated levels of DOM and organic colloids in upwelling areas as well as a number of coastal waters and particularly in estuaries where besides increased DOM and colloid levels also increased amounts of suspended particulate matter frequently exist and can effect the speciation.

Labile complexes with organic ligands

An example for a weak complex with organic ligands MeL_j is the Cd-glycinate which has been recently studied at the 6×10^{-7} M level for Cd(II) applying differential pulse polarography (DPP) at the dropping mercury electrode (DME) to measure the dependence of the reversible peak potential on the ligand concentration[16]. Experiments have been performed in artificial sea water and in genuine samples from the Pacific spiked with Cd(II) to the beforementioned concentration. The latter samples had been subjected to UV-irradiation to exclude interference by other DOM components. The results in both media agree and show that up to a glycine concentration of $3{,}6 \times 10^{-3}$ M the adjusted Cd(II)-concentration is chelated to less than 5%. Thus, glycine can be excluded as a natural chelator of Cd(II) in the open sea. This conclusion is confirmed by the results for Zn(II), although this trace metal forms stronger complexes with glycine requiring the usage of a different experimental approach (see next section).

SPECIATION BY STRONG COMPLEXES WITH ORGANIC LIGANDS

General and Methodological Aspects

A number of organic ligands can form rather stable complexes or even chelates MeL_j with trace metals, as Cd, Pb, Zn etc. The reduction of those MeL_j-species requires a considerable overvoltage and the electrode process becomes therefore irreversible.

Then the dependence of the shift of the half wave potential $\Delta E_{1/2}$ or a related peak potential ΔE_p on the ligand concentration [L] provides no more a reliable source for the evaluation of the characteristic complex parameters, ligand number j and stability constant β or K. However, the overvoltage difference between the complexed trace metal amount $[MeL_j]$ and the organic ligand uncomplexed amount of the respective trace metals $\Sigma[MeX_j]$, predominantly existing in sea water as in the previous section treated various labile complexes MeX_j with the salinity components, provides a very informative alternative possibility for studies on this kind of speciation by organics.

The investigation principle is as follows[26,27]. In natural sea water, containing negligible amounts of natural DOM or being prepared artificially, an appropriate overall concentration of the trace metal to be studied is adjusted. The required concentration $[Me]_{tot}$ depends on the applied voltammetric method. One will be inclined to keep $[Me]_{tot}$ always rather small, but while the application of differential pulse polarography (DPP) at a DME required typical $[Me]_{tot}$ levels of 10^{-7} M or more the usage of differential pulse stripping voltammetry (DPASV) in connection with a HMDE permits to lower $[Me]_{tot}$ to the 10^{-9} M level. In this solution increasing concentrations of the studied ligand [L] are adjusted and the reversible response, due to the amount of by the organic ligand uncomplexed remaining trace metal, i.e. the remaining concentration of $\Sigma[MeX_j]$ is measured. It will decrease with increasing ligand concentration [L] (see Fig. 4). In principle, if a polarographic measurement is made, also the increase of the more

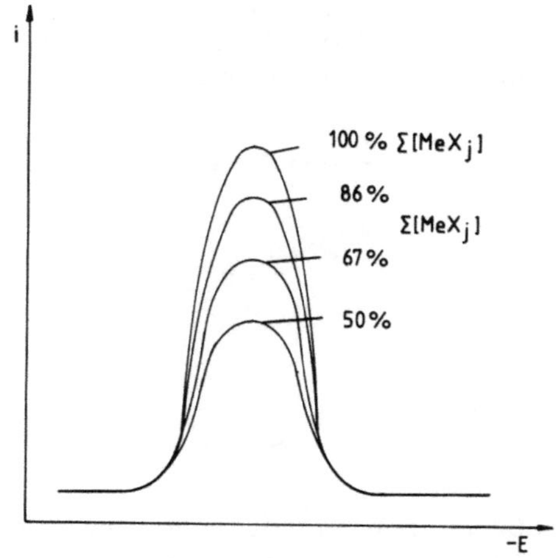

Fig. 4 Schematic behaviour of decreasing voltammetric response corresponding to reversible reduction of $\Sigma[MeX_j]$ according to increasing additions of chelating ligand [L], e.g. NTA.

negative and irreversible response, due to $[MeL_j]$, could be measured. But at the trace level this is less precise than the better defined reversible response corresponding to $\Sigma [MeX_j]$ and if stripping voltammetry, e.g. DPASV, is applied, the involved preconcentration step of Me at the electrode by cathodic deposition makes the evaluation of the reversible response due to $\Sigma [MeX_j]$ mandatory.

In this manner one records a semi-logarithmic relationship between the decrease of unchelated $[MeX_j]$ and log [L] as shown in Fig. 5. This relationship will give an immediate information on the concentration required for the studied chelating or inert complexes forming organic ligand L to achieve in sea water a certain complexation degree of the studied trace metal. Thus, this voltammetric approach provides a versatile and convenient experimental tool to test if a certain organic chelator or inert complexes forming ligand, presumed to be a possible component of DOM, can be with respect to the concentration level of DOM a candidate for a significant contribution to the speciation of the studied trace metal in the sea or in other types of natural waters. The primary major source of DOM in the sea is phytoplankton as the main component of the biological productivity. Thus, the DOM components released by living and dead phytoplankton belong to the following types of organic compounds: carbohydrates, proteins, amino acids, humic acids and nucleic acids. An example[28,32] of a typical DOM composition is given in Table 2.

A prerequisite of fundamental importance, but easily to fulfill is that $[Me]_{tot}$ is kept orders of magnitude lower than the lowest value of [L] adjusted. Then several in principle possible interferences will remain negligible and the measured data correspond in practice to an unperturbed complex equilibrium according to eq. (1). In principle the voltammetric measurement always per-

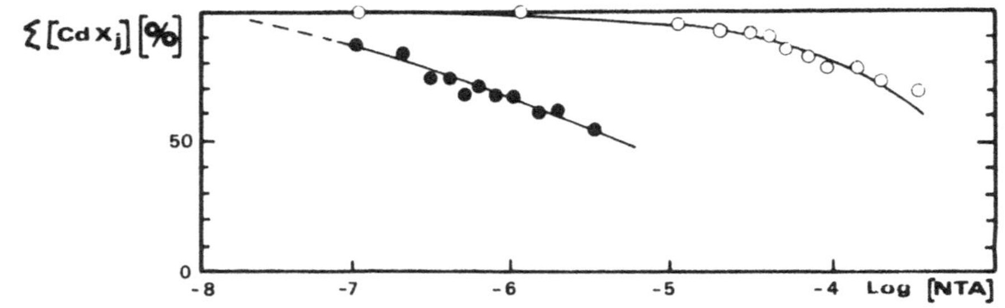

Fig. 5 Relation between percentage of unchelated Cd (present as labile complexes $\Sigma[CdX_j]$ and logarithm of chelating ligand concentration (lg[NTA]). 100% $\Sigma [CdX_j]$ correspond to a total Cd-concentration of 3×10^{-9} M. ● model solution of 0.59 M NaCl, pH 8.5, borate buffer, I 0.7; ○ Adriatic Sea water, pH 7.9.

Table 2 Typical Composition of DOM in the open sea[28,32].

Substance Class	Concentration (mg l^{-1})
Carbohydrates	≤ 0.8
Amino Acids	≤ 0.04
Proteins	≤ 0.04
Humic Acids	≤ 0.11

turbs somewhat the complex equilibrium in the vicinity of the electrode, as the electrode process consumes a minute amount of Σ $[MeX_j]$. This will induce dissociation of MeL_j. The point is to keep these effects negligible by having a sufficient excess of [L] over $[Me]_{tot}$ and by selecting an appropriate voltammetric method with small measuring time, i.e. preferentially a pulse technique[26,34,35] as DPP or at very low $[Me]_{tot}$ -levels DPASV with small cathodic deposition times, e.g. 2 min[27,30,31,39].

There are several independent ways to check that the beforementioned interferences remain negligible. Theoretically one can estimate if the selected voltammetric procedure will permit to fulfill under the given experimental conditions the criteria of Davison [29]. Frequently a more practical approach is to prove experimentally the absence of noticeable perturbations of the chelate equilibrium. A corresponding indicator is that the reversible response due to Σ $[MeX_j]$ is diffusion controlled[26,27]. A further very certain and inherent test is as follows[26,27,30,31]. If the complexation data of all relevant side reactions for the studied chelate MeL_j are taken into account (an example gives Fig. 6) one must be able to evaluate from every point of the relationship in Fig. 5 the common conditional stability constant K_{MeL_j} of the studied chelate species MeL_j. This K_{MeL_j} defined by eq. (1) is the value affected no more by any specific side effects operative in the respective medium, e.g. sea water, but only by the general salt effect due to the existing ionic strength of the medium, i.e. 0.7 M for sea water.

The outlined voltammetric approach has been successfully used under the mentioned preconditions and applying the tests for practically existing chelate equilibrium, for fundamental studies on the chelation of Cd, Pb and Zn with model chelators in sea water and inland waters and to explore in these media the conditions for a contribution to the speciation of those trace metals by potential and known chelating components of DOM.

Extended investigations[26,27,30,31] with the from the viewpoint of complex chemistry well known model chelator NTA, which

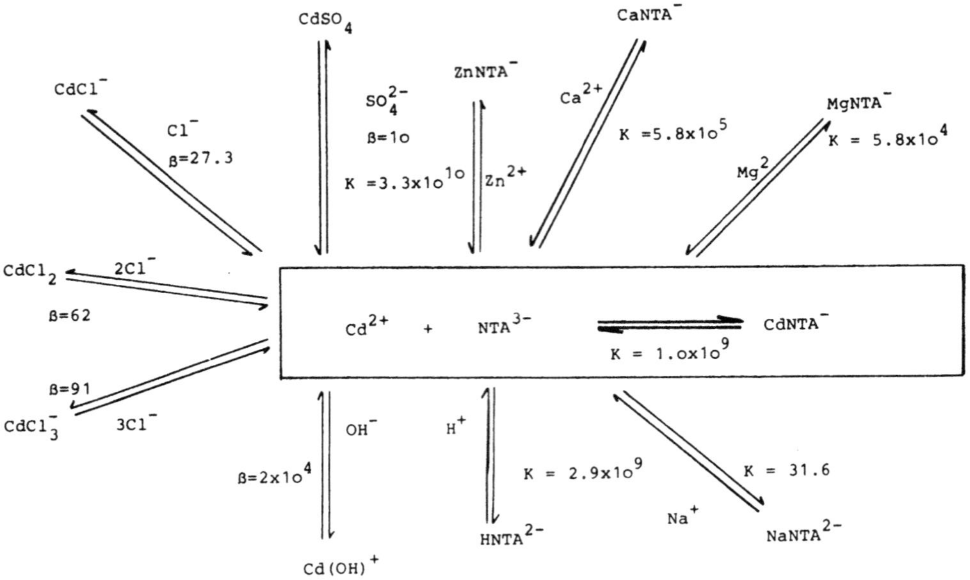

Fig. 6 Relevant side reaction interacting in chelation of Cd(II)-traces with NTA in sea water. The β_j- and K-values of the involved equilibria have been determined experimentally[23,26]. All β_j-values are not corrected for ion pairing of major salinity components.

has been selected, because it is a rather strong but in contrast to EDTA not an extremely strong chelator, have provided the experimental data basis to clarify a number of general aspects of trace metal speciation by organic chelation in sea water and to draw general prognostic conclusions on the requirements for significant contributions of chelating DOM components to the speciation of Cd, Pb and Zn in the sea. Besides measurements in sea water it has been found elucidating with respect to the clarification of the kind and magnitude of the specific side effects of the various salinity components to perform supporting measurements in model solutions of certain salinity components, e.g. 0.01 M $CaCl_2$ etc., at the sea water ionic strength adjusted with 0.59 M NaCl and appropriate addition of $NaClO_4$.

Due to the potential differences between the reversible responses for the studied trace metals and the fact, that always the chelating ligand L was kept in substantial excess, simultaneous measurements with several trace metals could be carried out providing the important experimental evidence that the trace metals do not affect their respective chelation to a noticeable extent.

Results and Conclusions

Systematic studies[26,27,30,31] for Cd, Pb and Zn with the chelator NTA have revealed that the most significant specific side effects on the speciation of those trace metals by organic chelation will be exerted by Ca and Mg. These alkaline earth ions, present in substantial excess in the sea (Ca 0.01 M and Mg 0.0536 M), compete significantly for the organic chelator, although the stability constants of their organic chelates are smaller than those of the trace metal chelates (viz. Table 3). This competitive side effect is the main reason that relatively large amounts of the rather strong chelator NTA (log K 9-11) are required to achieve a moderate chelation degree of 20% for the studied trace metals (viz. Table 3).

It was thus concluded that for chelating DOM components with K-values smaller than NTA correspondingly higher concentrations are to be required if a noticeable contribution to the speciation of the studied trace metals is to occur in the sea.

With the rather conservative assumption that the chelating DOM-components would have only a molecular weight MW of 100 one computes as function of log K from the experimental data in Table 3 the amounts of chelating DOM-components (in mg l^{-1}) required to achieve a 20% chelation degree of the listed heavy metals in the sea (viz. Table 4). The heavy metals are presumed to be dissolved at the ultra trace level, i.e. with total concentrations $\leq 10^{-8}$ M. It should be added that probably the average MW-value for the chelating ligands will be even higher than 100, but this would lead to even higher required amounts in mg l^{-1} of chelating DOM-

Table 3 Chelation of studied heavy metal traces and alkaline earth metals with NTA in sea water.

Metal	Concentration (M)	log K	Required NTA-concentration for 20% chelation degree (M)
Cd	3×10^{-9}	9.0	1×10^{-4}
Pb	2×10^{-8}	10.75	4.3×10^{-6}
Zn	3.2×10^{-8}	10.5	4.0×10^{-7}
Ca	1×10^{-2}	5.76	3.4×10^{-7}
Mg	5.36×10^{-2}	4.76	3.4×10^{-7}

The NTA-values required for 20% chelation of Ca and Mg clearly indicate the strong competition of these salinity components, due to their orders of magnitude higher excess, for the organic ligand.

Table 4 Correlation between the stability constant and required concentrations of organic ligands (MW = 100) for 20% chelation degree of heavy metal traces (10^{-8} - 10^{-9}M) in sea water.

Metal	log K =	7	8	9	10	11	12
		Ligand concentrations in mg l^{-1}					
Cd		1000	100	10	1	0.1	0.01
Pb		2600	260	26	2,6	0.26	0.026
Zn		130	13	1,3	0.13	0.013	0.0013

The values for [L] are to be regarded only as an approximate prediction of the required ligand concentration as function of the stability constant of the respective chelates but they give at any rate an impression of the required order of the ligand concentration. They also reflect in the case of Pb that stronger side effects for a certain metal increase relatively the demanded organic ligand concentration to attain a certain chelation degree[31].

components than given in Table 4. On the other hand, certain polymeric DOM-components, e.g. humates, degraded proteins, etc., will have several chelator sites per molecule.

Based on the at least approximately known stability constants K at sea water ionic strength[15] for a number of potential candidate ligands among the DOM-components and considering the fact that in the open sea usually the total amount of DOM will not exceed 2 mg C l^{-1} [28,32] one concludes that for dissolved Cd, Pb and Zn no significant contribution to their speciation is to be expected for DOM-components belonging to the classes of carbohydrates and of amino acids, except perhaps sulphur containing amino acids and protein degradation products. The situation may be somewhat different in certain shore waters and in estuaries provided there exist locally significantly elevated DOM-levels.

This obvious general prognosis had been more specified with respect to the speciation contribution of dissolved humates for Cd, Pb and Zn in the open sea[27]. On the basis of the stability constants reported for the humates of several transition metals and Ca and Mg by Mantoura et al.[33] and assuming that the DOM-level usually corresponds to < 2 mg C l^{-1} and that its humate content will not exceed about 10%, it had been predicted[27,30], that in the open sea also no noticeable contributions of dissolved humates to the speciation of Cd, Pb and Zn are to occur.

The foregoing examples of the experimental elucidation of general aspects of the trace metal speciation by organic chelates and the resulting far reaching prognostic conclusions feature the significance of such detailed investigations with a carefully selected model ligand as NTA with the outlined voltammetric procedure. This was felt to establish a much safer conclusion basis than the in the past overestimated and rather popular purely theoretical, often rather arbitrary and therefore risky modelling of speciation schemes. Besides these general aspects the speciation studies with NTA have their immediate relevance for the marine and aquatic environment where this compound shows growing abundance as a purely anthropogenic chemical with possibly adversive effects.

Meanwhile the predictions emerging from the NTA-studies have been experimentally confirmed in a number of cases. Studies according to the outlined recording of the dependence of $\Sigma[MeX_j]$ on log [L] with the amino acids glycine[34] for Zn and L-aspartic acid[35] for Cd and Zn by DPP at the DME have shown, that in sea water substantially higher [L] values than present in the sea are required for a noticeable complexation, as was to be expected with respect to the rather moderate stability constants of those complexes. Again Ca and Mg provide due to their competition the most important specific side effects. The required ligand concentrations [L] are not only by orders of magnitude higher than the total amino acid content to be expected for DOM[28,36] but exceed even the total DOM level by about two orders of magnitude or more (viz. Table 2).

Therefore, they can be safely ruled out as contributors to the speciation of the studied metals. As the other amino acids have comparable stability constants[15] the same conclusion is drawn for them. An exception, which suggest future experimental testing, are sulphur containing amino acids, particularly cysteine.

Recent investigations[37] with dissolved humic acids of marine origin in sea water have established that for Pb and Cd up to concentrations of 0.8 mg l^{-1} humate the chelation degree remains below 10% yielding for Zn, however, at 1 mg l^{-1} humate already a chelation degree of 20%. Such high humic acid concentrations will hardly exist in the open sea (see e.g. Table 2) and therefore here contributions to the speciation of the studied trace metals are unlikely.

These results are in a general sense also supported by the findings from electrophoretic studies[38].

Kinetics and Mechanism of the Chelation of Heavy Metal Traces

The outlined voltammetric procedure of recording the $\Sigma[MeX_j]$-log[L] relationship can be also applied for kinetic studies[30,39,40]. Then the time function of the attainment of the equilibrium value of $\Sigma[MeX_j]$ is recorded by DPASV or ASV after the addition of the ligand [L] (viz. Fig. 7). To establish 2nd order kinetics the initial values of $[Me]_{tot}$ and [L] have to be equal. The evaluation of the resulting time function according to the well established principles of reaction kinetics yields then the chelate formation rate constant k_f'. It is again a conditional entity referring to the respective medium. Again several trace metals can be studied simultaneously as their responses corresponding to their respective $\Sigma[MeX_j]$-values appear at different electrode potentials. Measurements have been performed with the model ligand EDTA in sea water and model solutions of its major salinity components. It is obvious, that in the previous section preconditions concerning the avoidance of interferences due to noticeable dissociation of formed MeL during the total time required for the voltammetric measurement of each point in Fig. 7 have to be met. This can be readily achieved for the k_f' range of interest. The example in Fig. 7 and the data in Table 5 reflect also for k_f' the pronounced specific competition of the alkaline earth ions. Actually Ca is of predominant significance, as the coincidence of the time function for the 0.01 M Ca model solution and artificial sea water in Fig. 7 shows.

The k_f'-values and further evidence given elsewhere[30,39,40] revealed as the most important result the mechanism of trace metal

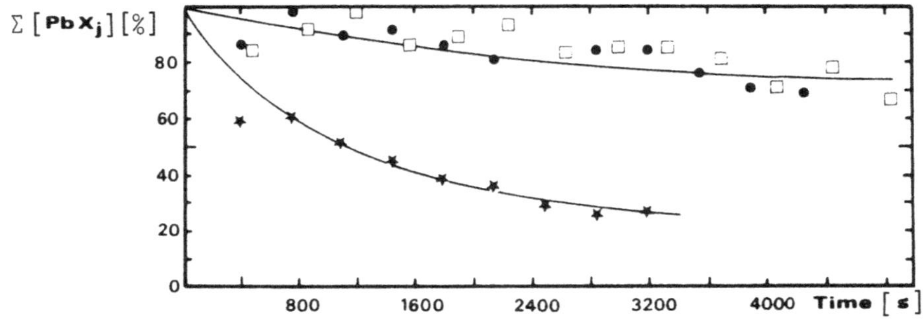

Fig. 7 Time dependence of unchelated $\Sigma[PbX_j]$ after addition of chelating ligand EDTA to measure kinetics of chelate formation. Initial concentrations: 2×10^{-8} M Pb(II) and Na_3HEDTA (1:1). The points correspond to peak heights of reversible Pb-responses measured with DPASV (viz. Ref. 40) in different media: ★ 0.0536 M $MgCl_2$; ● 0.01 M $CaCl_2$; □ artificial sea water.

Table 5 Conditional rate constants k_f' of chelate formation with EDTA in model media and sea water.

Medium	k_f' (l mole^{-1} s^{-1}) Cd	Pb	Zn	Remark
0.55 M NaCl	unmeasurably fast			recombination mechanism
0.0536 M $MgCl_2$	2.8×10^3	5.0×10^4	3.3×10^3	
0.01 M $CaCl_2$	4.2×10^2	2.5×10^3	1.9×10^2	
Artificial sea water	3.3×10^2	4.0×10^3	3.2×10^2	
North Sea water	3.9×10^2	3.9×10^3	-	
Ligurian Sea water	-	2.0×10^3	2.7×10^2	

chelation in the sea. It is not a simple recombination mechanism between hydrated Me^{2+} and ligand L^- but a ligand exchange mechanism as show subsequently.

$$MgL \rightleftharpoons CaL + MeX_j \xrightarrow{k_f'} MeL + Ca^{2+} + jX^- \qquad (6)$$

with $X = Cl^-, OH^-, CO_3^{2-}$ etc.

The reasons are the particular concentration conditions operative for the involved reactants. Ca and Mg are in substantial excess to the chelating components L of DOM and the latter are again in considerable excess to the overall concentrations of the heavy metal traces present predominantly as labile MeX_j. The consequence is that the alkaline earth ions tend to chelate preferentially with the available chelating ligands L. An equilibrium predominantly shifted to the stronger CaL-chelates tends to adjust. If, however, a labile heavy metal species MeX_j approaches CaL ligand exchange has to occur with the respective rate constant k_f', because the stability of the trace metal chelate MeL is still higher than that of CaL. In accordance with this ligand exchange mechanism k_f' is unmeasureably fast in 0.55 M NaCl, as there Ca is absent and then the simple recombination mechanism with k_f-values in the order of 10^9 l mol^{-1}s^{-1} or more prevails. Evidence in the literature[41] shows that in media where the Ca-concentration and that of the ligand [L] become of comparable order the recombination mechanism takes progressively over. With respect to the key role of Ca again attention is drawn on the coincidence of the time functions in 0.01 M Ca and sea water in Fig. 7.

COMPLEXATION CAPACITY

The complexation capacity is an important empirical diagnostic parameter reflecting the global potentiality of sea water from the respective sampling area and depth to speciate in nonlabile species a certain amount of the respective heavy metal added as titrant. The complexation capacity corresponds to the sum of the concentrations of all relevant materials. As the concentrations of those substances forming inert heavy metal species (respective DOM components, organic and inorganic colloids[4] and virtually continuously suspended ultrafine grain particulates[45] with a diameter below 0.45 μm) are usually rather low in the sea, voltammetry, particularly in the DPASV-mode, is a convenient and powerful approach to determine complexation capacities. Most investigators[4,42] have hitherto used standard solutions of Cu, some in addition also Pb, Cd and Zn[42,43,46]. The sea water samples have to be measured without any pretreatment, except filtration (0.45 μm), to avoid alterations in the speciation potentialities.

Fig. 8 shows the example of a typical titration graph obtained by DPASV. The points of the titration plot correspond to the height of the reversible peak of the added heavy metal. Measurements have been made after sufficient equilibration time (about 30

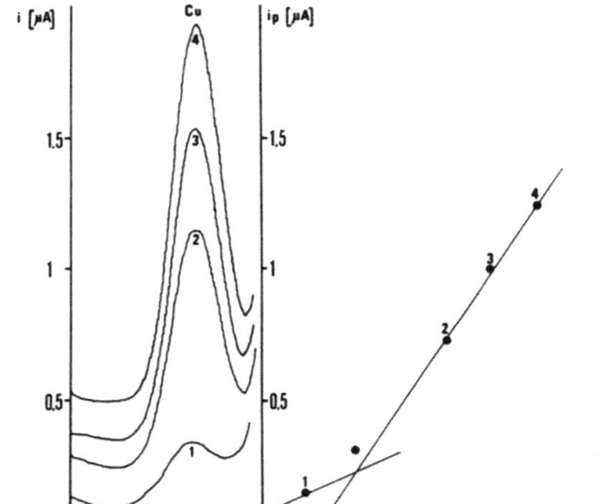

Fig. 8 Determination of complexation capacity by voltammetric titration (DPASV at HMDE). Left: reversible voltammetric responses corresponding to metal present in labile species. Right: graph of titration with 1 mg l^{-1} Cu(II), sample volume 11.6 ml. Deep sea water sample above sea bottom (5500 m depth) at subtropical convergence in southern Pacific 41°55'S, 144°58'W). Complexation capacity 3.6 μg l^{-1} Cu(II).

min) had elapsed. In the initial branch with the lower slope a considerable fraction of the added heavy metal is consumed by the complexing material in the sample. The higher the stability of the formed nonlabile species the lower the slope of the first branch of the titration plot. An apparent global overall stability constant can be even evaluated as an empirical parameter characteristic for the studied sample[46]. The reversible voltammetric responses in this first branch correspond to those fractions of the added heavy metal concentrations which remain in the state of labile complex species MeX_j capable to undergo a reversible electrode process. Once the second branch with the steeper slope has been reached all material in the water sample able to bind the titrant heavy metal in inert species has been consumed. Therefore, all further amounts of added heavy metal remain now totally in the state of labile complexes MeX_j and lead consequently to a steeper increase of the height of the reversible voltammetric peak. The intersection point of the two branches corresponds to the complexation capacity. In the vicinity of this endpoint of the titration one observes in practice usually a rather gradual change of the slope. To ensure sufficient accuracy in the determination of the intersection point and thus of the complexation capacity the first and second branch of the titration plot have to be extrapolated. This implies that a sufficient number of points has been measured in the second branch to obtain the correct slope. An alternative approach to control the correctness of the slope of the second branch of the titration plot consists in the following procedure. A further sample of the investigated water is acidified to pH 2 and subjected to UV-irradiation. By this treatment the metal is leached from the ultrafine suspended particles and the colloids and inert complexes forming DOM-components are decomposed. A linear relationship of the voltammetric peak height on the added titrant heavy metal with a slope corresponding to the second steeper branch of the complexation capacity titration plot is now to be obtained over the whole range of the titrant heavy metal concentration.

As the tendency to form nonlabile species depends also strongly on the nature of the heavy metal, the slope of the initial branch and the heavy metal concentration corresponding to the respective complexation capacity will be different for various heavy metals. Generally the tendency for forming nonlabile species follows the sequence: Cu > Pb > Cd.

Thus, we found recently[43] for deep sea water samples 1 to 2 m above the sea bottom on a pattern of stations from Tahiti to the East Pacific Rise and from there along the subtropical convergence ($\sim 42^{o}$S) to New Zealand values in the following ranges for the complexation capacity: Cu 0.7-5.0 $\mu g\ l^{-1}$; Pb 0.0-1.0 $\mu g\ l^{-1}$; Cd nil. Compared with the ligand concentration requirements of Table 4 these complexation capacities are very small. It is obvious that in sea water and hard inland waters the complexation sites of the complexing material are originally occupied by the in substantial excess present alkaline earth ions Ca(II) and Mg(II). These complexed alkaline earth ions are progressively substituted by the added titrant heavy metal according to the ligand exchange mechanism discussed in the previous section.

Reliable complexation capacities, which can be accurately determined in a convenient manner by voltammetry, usually to be applied in the DPASV-mode, might become in the future a rather valuable diagnostic parameter, particularly if correlations with DOM-levels, biological productivity or colloid content can be established.

GENERAL CONCLUSIONS AND FUTURE TASKS

Accurate and reliable analysis of the total dissolved concentration of heavy and transition metal traces in sea water remains a demanding and difficult task requiring expertise, skill and great care. Voltammetry has substantially contributed to the satisfactory solution of this task and has during the last years opened literally new dimensions to marine trace metal analysis for such important metals as Cu, Pb, Cd, Zn, Hg, Ni and Co[5-13]. Methodological developments for several other trace metals of interest are in progress. Of particular significance is that the convenience of voltammetry enables the analysis of large sample numbers within reasonable time periods. This large scale analysis is a prerequisite to explore the distribution of the concentration levels of these trace metals at various depths down to the sea bottom in the oceans[44].

With respect to the growing significance of speciation it is suggested that future analytical campaigns are performed more speciation minded including on board of the research vessel, where possible 24 hrs. after sampling, complexation capacity determinations preferentially with several different heavy metals. This not too elaborate analysis scheme would yield valuable and essential global informations on geographical variation in speciation potentiality and remain with respect to the involved effort much more feasible than other[4] even more elaborate analysis schemes. Nevertheless, it has to be emphasized in this context, that these exploratory studies on global speciation aspects and that on complexation capacity cannot substitute the detailed physicochemical investigations on defined individual species discussed in the previous sections if substantial further progress in the speciation field is to be achieved.

The treatment of the topic was hitherto mainly focussed on the situation in the open sea, an area for which to a significant extent generalisations of findings seem possible. The speciation situation will become obviously much more divergent, complicated and specific in coastal waters, lagoons or even estuaries. Here the discrepancies in the total concentrations of the considered trace metals and the DOM-levels may differ and fluctuate locally much more than in the oceans and also biological productivity is frequently much larger. Moreover, frequently substantially higher levels of suspended particulate matter including virtually continuously suspended ultrafine grain particulates[45] are involved, interactions with sediments become important and pollution will locally cause significant enhancements of the trace metal concentrations and the input of anthropogenic organics will add to DOM and the colloids.

In contrast to oceanic ecosystems in coastal waters and estuaries suspended matter and sediment surfaces are significantly involved in the species distribution of trace metals. Particle and sediment surfaces coated with films of organics, which are not efficient ligands in the dissolved phase, either due to the moderate stability of their heavy metal complexes and/or with respect to their limited concentration in solution can act, due to their enormous interfacial concentration enhancement by their fixation at the interface, as quite efficient metal scavengers. Moreover, interstitial waters can be really a very special sea water medium.

From our viewpoint future research efforts directed to the speciation situation in the oceans should put priority on the following items. Under fundamental investigations on defined individual complexes the labile complex distribution of Cu, Zn and Hg has to be elucidated. For Hg and also Cu the gold electrode would be the suitable working electrode[10]. For Cu and Hg representative ligands from the various types of DOM compounds have to be tested. For both trace metals much more significant contributions by organics to their speciation are to be expected on the basis of their respective stability constants tabulated in the literature for aqueous media. These stability constants are usually several to many orders of magnitude larger[15] than for analogous complexes with organics of Cd, Pb and Zn.

Determination of complexation capacities should become common practice for research cruises in order to assemble and complete the mapping of the oceans with respect to the distribution of their trace metal levels and their global speciation. For correlations to biological productivity, DOM-levels and colloid levels the determined data are to be tested.

The experiences from recent years and the results hitherto achieved have established that voltammetry provides a most powerful, versatile and convenient methodological approach for this research on the speciation of the under ecotoxicological aspects most significant heavy and transition metals in the sea.

REFERENCES

1. Venugopal, B. and T.D. Luckey, 1978: Metal toxicity in mammals, Vol 2. Plenum Press, London-New York.
2. Friberg, L., G.F. Nordberg and B. Vouk, 1979: Handbook on the toxicology of metals. Elseviere/North Holland Biomedical Press, Amsterdam-New York-Oxford.
3. Nguyen, V.D., P. Valenta and H.W. Nürnberg, 1979: The determination of toxic trace metals in rain water and snow by differential pulse stripping voltammetry. Sci. Tot. Environm., 12, 151-167.
4. Florence, T.M. and G.E. Batley, 1980: Chemical speciation in natural waters. Crit. Rev. Anal. Chem. 9, 219-296.
5. Nürnberg, H.W. and P. Valenta, 1975: Polarography and voltammetry in marine chemistry. In: "The nature of sea water", E.D. Goldberg, ed. Dahlem-Konferenzen, Berlin. pp. 87-136.
6. Nürnberg, H.W., 1978: Potentialities and applications of advanced polarographic and voltammetry methods in aquatic and marine trace metal chemistry. Acta Univ. Upsaliensis Symp. Univ. Upsaliensis Annum Quingentesimum Celebrantis, Almquist and Wiksell International, Stockholm, 12, 270-307.
7. Nürnberg, H.W., 1979: Potentialities of the voltammetric approach in trace metal chemistry of sea water. Proc. Cours Internat. Post Universitaires, Gent 1977, Ministère de l'Education Nationale et de la Culture Francaise, Bruxelles. pp. 1-36.
8. Nürnberg, H.W., 1979: Polarography and voltammetry in studies of toxic trace metals in man and his environment. Sci. Tot. Environm., 12, 35-60.
9. Mart, L., H.W. Nürnberg and P. Valenta, 1980: Voltammetric ultra trace analysis with a multicell system designed for clean bench working. Fresenius Z. Anal. Chem., 300, 350-362.
10. Sipos, L., J. Golimowski, P. Valenta and H.W. Nürnberg, 1979: New voltammetric procedure for the simultaneous determination of copper and mercury in environmental samples. Fresenius Z. Anal. Chem., 298, 1-8.
11. Mart, L., H.W. Nürnberg and D. Dyrssen: Low level determination of trace metals in Arctic sea water and snow by differential pulse anodic stripping voltammetry. This volume.
12. Pilhar, B., P. Valenta and H.W. Nürnberg, 1981: A new high performance analytical procedure for the voltammetric determination of nickel in routine analysis of waters,

biological materials and food. Fresenius Z. Anal. Chem., 307, 337-346.
13. Nürnberg, H.W., P. Valenta, L. Mart, B. Raspor and L. Sipos, 1976: The polarographic approach to the determination and speciation of toxic metals in the marine environment. Fresenius Z. Anal. Chem., 282, 357-367.
14. Nürnberg, H.W., 1980: Features of voltammetric investigations on trace metal speciation in sea water and inland waters. Thalassia Jugosl., 16, 95-110.
15. Sillén, L.G. and A.E. Martell, 1971: Stability constants of metal complexes. Spec. Publ. 17, Chem. Soc., London.
16. Simões Goncalves, M.L.S., P. Valenta and H.W. Nürnberg, in press: Voltammetric and potentiometric investigations on the chelation of Cd(II) by glycine in sea water. J. Electroanal. Chem.
17. Eigen, M. 1963: Ionen- und Ladungsübertragungsreaktionen in Lösung. Ber. Bunsenges. Phys. Chem., 67, 753.
18. Crow, D.R., 1964: Polarography of metal complexes. Academic Press, London.
19. Heyrovský, J. and J. Kuta, 1968: Principles of polarography. Academic Press, New York.
20. de Ford, D.D. and D.N. Hume, 1951: The determination of consecutive formation constants of complex ions from polarographic data. J. Am. Chem. Soc., 73, 5321-5322.
21. Bubić, S. and M. Branica, 1973: Voltammetric characterization of ionic state of cadmium present in sea water. Thalassia Jugosl., 9, 47-53.
22. Sipos, L., P. Valenta, H.W. Nürnberg and M. Branica, 1980: Voltammetric determination of the stability constants of the predominant labile lead complexes in sea water. In: "Lead in the marine environment", M. Branica and Z. Konrad, eds. Pergamon Press, Oxford. pp. 61-76.
23. Sipos, L., B. Raspor, H.W. Nürnberg and R.M. Pytkowicz, 1980: Interaction of metal complexes with coulombic ion pairs in aqueous media of high salinity. Mar. Chem., 9, 37-47.
24. Pytkowicz, R.M. and I.E. Hawley, 1974: Bicarbonate and carbonate ion pairs and a model of sea water at 25 C. Limnol. Oceanogr., 19, 223-234.
25. Mart, L., H.W. Nürnberg, P. Valenta and M. Stoeppler, 1978: Determination of levels of toxic trace metals dissolved in sea water and inland waters by differential pulse anodic stripping voltammetry. Thalassia Jugosl., 14, 171-188.
26. Raspor, B., P. Valenta, H.W. Nürnberg and M. Branica, 1978: The chelation of cadmium with NTA in sea water as a model for the typical behaviour of trace heavy metal chelates in natural waters. Sci. Tot. Environm., 9, 87-109.

27. Raspor, B., H.W. Nürnberg, P. Valenta and M. Branica, 1980: The chelation of Pb by organic ligands in sea water. In: "Lead in the environment", M. Branica and Z. Konrad, eds. Pergamon Press, Oxford., 181-195.
28. Duursma, E.K., 1965: In: "Chemical Oceanography", J.P. Riley and G. Skirrow, eds. Vol. 1, Academic Press, New York.
29. Davison, W., 1978: Defining the electroanalytically measured species in a natural water sample. J. Electroanal. Chem., 87, 395-404.
30. Raspor. B., H.W. Nürnberg and P. Valenta, 1981: Voltammetric studies on the stability of the Zn(II)-chelates with NTA and EDTA and the kinetics of their formation in Lake Ontario water. Limnol. Oceanogr., 26, 54-66.
31. Nürnberg, H.W. and B. Raspor, 1981: Applications of voltammetry in studies of the speciation of heavy metals by organic chelators in sea water. Environm. Technol. Letters, 2, 457-483.
32. Ogura, N., 1972: Rate and extent of decomposition of dissolved organic matter in surface sea water. Mar. Biology, 13, 89-93.
33. Mantoura, R.F.C., A. Dickson and J.P. Riley, 1978: The complexation of metals with humic materials in natural waters. Estuar. Coastal Mar. Sci., 6, 387-408.
34. Simões Goncalves, M.L.S. and P. Valenta, in press: Voltammetric and potentiometric investigations on the chelation of Zn(II) by glycine in sea water. J. Electroanal. Chem., 132, 357-375.
35. Valenta, P. and M. Sugawara, in press: Voltammetric studies on the speciation of trace metals by amino acids in sea water. Rapp. Comm. Int. Mer Médit. (Monaco).
36. Brockmann, U., V. Ittekkot, K.D. Hammer and K. Eberlein: Generation of chelating organic substances by marine phytoplankton. This volume.
37. Raspor, B., H.W. Nürnberg, P. Valenta and M. Branica: Unpublished work.
38. Musani, Lj., P. Valenta, H.W. Nürnberg, Z. Konrad and M. Branica, 1980: On the chelation of toxic trace metals by humic acid of marine origin. Estuar, Coast. Mar. Sci., 11, 639-649.
39. Raspor, B., P. Valenta, H.W. Nürnberg and M. Branica, 1977: Polarographic studies on the kinetics and mechanism of Cd(II)-chelate formation with EDTA in sea water. Thalassia Jugosl., 13, 79-91.
40. Raspor, B., H.W. Nürnberg, P. Valenta and M. Branica, 1980: Kinetics and mechanism of trace metal chelation in sea water. J. Electroanal. Chem., 115, 293-308.
41. Kuempel, J.R. and W. Schaap, 1968: Cyclic voltammetric study on the rate of ligand exchange between cadmium ion and CaEDTA. Inorg. Chem., 7, 2435-2442.

42. Duinker, J.C. and C.J.M. Kramer, 1977: An experimental study on the speciation of dissolved zinc, cadmium, lead and copper in the river Rhine and North Sea water by differential pulse anodic stripping voltammetry. Mar. Chem., 5, 207-228.
43. Klahre, P, J. Golimowski, H.W. Nürnberg: Unpublished work.
44. Mart, L., H. Rützel, P. Klahre, L. Sipos, U. Platzek, P. Valenta and H.W. Nürnberg: Comparative stuides on the distribution of trace metals in the oceans and in coastal waters. Sci. Tot. Environ., submitted.
45. Duinker, J.C.: Dissolved and particulate metals in coastal and off-shore waters. This volume.
46. Plavsic, M., D. Krznarić and M. Branica, in press: Determination of the apparent copper complexing capacity of sea water by anodic stripping voltammetry. Mar. Chem.

STUDIES OF CADMIUM, COPPER AND ZINC INTERACTIONS WITH MARINE FULVIC AND HUMIC MATERIALS IN SEAWATER USING ANODIC STRIPPING VOLTAMMETRY

Stephen R. Piotrowicz, George R. Harvey, M. Springer-Young, Reinier A. Courant and Deborah A. Boran

Ocean Chemistry and Biology Laboratory
Atlantic Oceanographic and Meteorological Laboratories
National Oceanic and Atmospheric Administration
4301 Rickenbacker Causeway
Miami, Florida 33149 U.S.A.

ABSTRACT

Humic and fulvic acids isolated from seawater were found to interact with Cd, Cu and Zn in different ways at natural levels of these elements and natural pH in seawater. Fulvic acids exhibit strong interaction with Zn while Cd and Cu have little or no interaction on the time scale of the diurnal cycles of plankton or bacteria. The Zn-fulvic acid interactions in surface waters probably occur as part of a steady-state cycle of less than 40 hrs duration controlled by photooxidation and bacterial processes. The interaction of Cd, Cu and Zn with humic acids is much more complex.

It appears that the natural association of metals and dissolved humic and fulvic acids is so dynamic that once a seawater sample is taken into a closed container for analysis, natural productive and destructive equilibria slow and finally cease. Thus, our perception of how metal-organic interactions occur in the ocean depends upon how quickly samples can be analyzed because true oceanic conditions cannot be duplicated. The use of synthetic complexers to study trace metal chemistry in seawater is discouraged.

INTRODUCTION

We are conducting a program to understand the interactions between trace metals and naturally occurring organic matter and the effect of those interactions on the biota of the sea. Working with the assumption that any such relationships will vary with the degree of biological activity in a region, the extent to which anthropogenic impacts may be occurring, the extent of terrestrial influence, and with season, a variety of sampling sites were selected in the Gulf of Mexico to be sampled over all seasons. The locations of some of these stations are given in Figure 1, together with the sampling periods to date. The Gulf Loop Intrusion station is always sampled, with the aid of satellite imagery, within the Loop Current itself. This station is considered oligotrophic and free from local terrestrial and most anthropogenic inputs of trace constituents to the water. The Mississippi outflow is the opposite extreme from the Gulf Loop Intrusion in that it is productive and subject to large terrestrial and anthropogenic inputs. The remaining three stations, off Cape San Blas, Florida, on the Yucatan Shelf, and in the Bay of Campeche, are intermediate between these two extremes. In order to understand the interaction of trace metals and naturally occurring organics, sampling for these materials takes place as synoptically as possible and at similar depths.

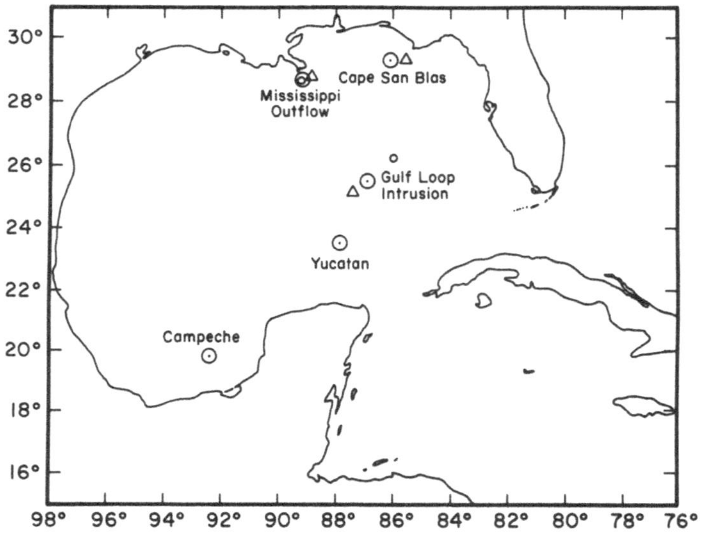

Fig. 1 Location of sampling stations. O - December 1979; ⊙ - April-May 1980; Δ - January-February 1981.

Dissolved humic and fulvic acids were chosen as the most likely class of organic matter in seawater to strongly interact with metals for two reasons. First, they are known to contain many different kinds of carboxyl and hydroxyl groups[1,2] which are known metal complexers. Second, the dissolved humus fraction is the second or third largest fraction of organic matter in seawater comprising up to 50% of the total.

EXPERIMENTAL METHODS

Trace Metals

Recent evidence indicates that much of the existing data on trace metals in seawater may reflect significant levels of contamination in the collection and analysis of open-ocean samples[3-5]. Contamination control, above all, is a prerequisite for this work. Samples were collected, processed, and analyzed using equipment designed and used only for trace metal research. They were collected using an acid-cleaned, Teflon®-lined, Go-Flo® bottle suspended on Kevlar® line and tripped with a Teflon® messenger. The Kevlar® line is spooled on a small, self-contained, electrohydraulic hydrographic winch, without a level-wind. Line out is monitored using a solvent-rinsed, dilute-acid-cleaned half-meter wheel. Dirty or worn cable is discarded frequently. After a sample bottle arrives on deck, its ends are bagged in plastic and it is taken as rapidly as possible into a portable laboratory designed for trace metal analysis. This facility is a 2.4 m x 6.1 m portable fiberglass shell (Grasis Corp., Kansas City, MO) with an interior designed and constructed by a clean room fabricator (Environmental Air Control, Inc., Hagerstown, MD). One-half of the facility is a Class 1000 clean room with 2.9 m of Class 100 vertical, laminar-flow clean bench space. The remaining half of the facility is a changing room containing a sample transfer area. Samples are transferred from the water sampling bottle to Teflon® sample containers inside a floor to ceiling clean cabinet installed in this anteroom. This clean cabinet is a vertical laminar-flow facility with Plexiglas® side and front panels, a polypropylene drain trough at its base, and a PVC water bottle rack mounted on the wall enclosed by the Plexiglas® panels. This design has resulted in an average residence time of water in a sampling bottle of approximately five minutes for shallow casts with a maximum of approximately seven minutes for casts collected at 50 m depth. Deeper casts increase this contact time approximately one minute for every 25 m of cable out. All casts are single bottle to further minimize the time between sampling and transfer of sample to Teflon® storage vessels.

Anodic Stripping Voltammetry (ASV) was used to study the speciation of Cd, Cu and Zn in seawater at natural pH. This technique has been extensively used to study metal-organic interactions[6-10]. The development of the rotating glassy carbon electrode using a pre-plated or co-plated mercury film gave the method the sensitivity and resolution required for use in seawater[11-15]. The instrumentation is adaptable to use at sea and does not generally require any chemical pretreatment of samples prior to analysis. This permits rapid analysis in <u>in situ</u> pH, thus minimizing any changes in speciation due to storage.

The entire ASV operation takes place within the Class 100 clean bench area of the clean room. Measurements are made using a Tacussel EDI rotating electrode interfaced with a Princeton Applied Research Model 384 polarograph. The electrode tip consists of a 6 mm glassy carbon disc (Tokai Mfg. Corp., c/o IMC Corp., NY), sealed with Teflon®, that had been polished with diamond polishing compound (Buehler, Inc., Evanston, IL). The reference electrode is a Ag/AgCl type inserted into an acid-cleaned Vycor® tip Teflon® bridge tube containing clean seawater. The purge gas is high-purity Ar which is passed through a high-temperature catalytic scrubber to remove oxygen and then is rehumidified using an in-line natural seawater bubbler prior to use. The electronic interface between the Model 384 and the electrode permits the polarograph to automatically control all steps of the analysis including purging. Approximately 15 g of sample are placed in an acid-cleaned Teflon® polarography cell (Savillex Corp., Minnetonka, MN) that has been copiously rinsed with the sample to be analyzed. Actual sample size is quantified using a time-averaging electronic balance. The polarography cell sits in a specially designed Teflon® head containing the electrodes, the purge gas inlet, and a small hole to allow additions using micropipettors.

Samples are analyzed as quickly as possible after collection at natural pH for free and other easily reducible species using a pre-plated mercury-film technique. Because of differing sensitivities and natural levels of free or ASV-labile metal, Cd and Cu are analyzed using a ten-minute plating time, a -1.0 V plating potential, and scanning in 6.67 mv s^{-1} increments. Zinc determinations are made on a fresh aliquot of sample in case some contamination may have been introduced in the analysis of Cd and Cu and/or to eliminate any possible effects due to Cu-Zn intermetallic complex formation. Zinc is analyzed by plating at -1.25 V for five minutes. The remaining operating conditions are the same as those for Cd and Cu. Lead, although not quantified in this work, serves as a useful indicator of contamination during the entire sampling and analytical process. Significant increases in the Pb signal indicate potential contamination. The detection limits of this system are approximately 0.02 nmol Zn kg^{-1}, 0.02 nmol Cd kg^{-1}, and 0.3 nmol Cu kg^{-1}.

Organic

Seawater from the desired depth was brought up into 240 l stainless steel drums under nitrogen by an all stainless steel gas lift system[16]. The seawater was acidified and passed through an Amberlite XAD-2 column to remove the dissolved humic and fulvic materials. Elution of the Amberlite XAD-2 columns, fractionation and purification were performed exactly as described by Stuermer and Harvey[1]. The extraction efficiency for these materials is greater than 90%[1,17]. Humic and fulvic acids are always operationally defined. In our case they are the dissolved substances present in seawater which adsorb to a hydrophobic resin, are soluble in aqueous alkali and insoluble in organic solvents. Humic substances extracted from soils and sediments have identical operational behavior. The other two major classes of DOC, carbohydrates and proteins, cannot be isolated by this procedure.

Proton magnetic resonance spectra were run in D_2O-NaOD solution (20 mg ml^{-1}) on a Varian EM-360A spectrometer equipped with a Varian V-2048 signal averager. Sixty-four scans were accumulated for each spectrum. Using identical instrumental conditions the spectrometer was calibrated with n-tetradecane and glucose to determine the moles H cm^{-2} of peak area.

Elemental analyses were performed by Galbraith Laboratories (Knoxville, TN).

Dissolved organic carbon analyses were performed by A. Mendez, University of Miami using the UV-photooxidation method.

RESULTS AND DISCUSSION

Humic and Fulvic Materials

The concentrations of marine fulvic and humic acids (MFA and MHA) found at various times at the Gulf stations (Fig. 1) are shown in Table 1. Since MFA's and MHA's are about 50% C, their combined weights account for 4-50% of the DOC. With few exceptions, there is generally much less MHA than MFA.

The proton magnetic resonance spectra (PMR) of the MFA and MHA isolated from 1,500 l samples collected at a Gulf Loop and a near shore station are shown in Figure 2. The two MHA spectra reveal about 5% of their protons as aromatic appearing at 6.5-7.5 ppm in the spectrum. The MFA's have no aromatic protons. Otherwise, all four spectra reveal very similar proton environments in the MFA's and MHA's from the two contrasting locations. The spectra differ only in the relative amount of saturated aliphatic protons (0.8-2.5 ppm) and protons on hydroxylated carbon ($\sim$ 3.5

Table 1 Summary of Organic Concentrations in Gulf of Mexico

Station	Day/ Night	Season (Fall/ Spring)	Depth (m)	DOC (mg kg^{-1})	Humic (μg kg^{-1})	Fulvic (μg kg^{-1})
Mississippi	D	F	3	1.32	114	550
Outflow	N	F	3	1.50	126	1270
	N	Sp	4	1.10	176	904
Gulf Loop	D	F	20	--	26	724
Intrusion	N	F	20	0.96	3	190
	D	Sp	20	0.50	14	234
	D	Sp	55	0.54	9	58
Yucatan	N	Sp	10	0.63	29	223
Campeche	N	Sp	5	0.98	105	754
Cape San	D	Sp	4	3.19	336	577
Blas	D	Sp	55	2.20	10	139

ppm). The large peak centered at 4.8 ppm is due to traces of HDO in the D_2O solvent and exchangeable OH protons in the sample. Although the PMR spectra do show differences in the relative amounts of proton types, our results to date indicate that this class of organic matter is similar in structure throughout the Gulf of Mexico and like the fulvic acid isolated from the North Atlantic Ocean by Stuermer and Payne[18].

Trace Metals

Figures 3 and 4 are the standard addition curves at natural pH for Zn and Cu for two samples collected at different stations on different cruises. There are two striking points in these plots. The first is that it required additions of metal equivalent to >1.5 nmol kg^{-1} to give detectable stripping current signals at natural pH. The second is that the plots for natural pH appear to be straight lines with a positive x-intercept. The plots indicate that no ASV-labile metal is initially present in the samples and that interaction is occurring such that the free metal added is not initially ASV-labile. Amperometric titrations of coastal seawater, using ASV, generally yield curves through which two lines of different slopes can be drawn with the inflection point interpreted as the metal complexing ability[19]. The slope of the initial portion of the curve is attributable to some type of coordination (complexation and/or adsorption) which causes some of the added metal to be non-ASV-labile. At all of the stations in Figure 1, except the Mississippi, the ASV titration curves illustrated by Figures 3 and 4 were the rule for Zn and Cu

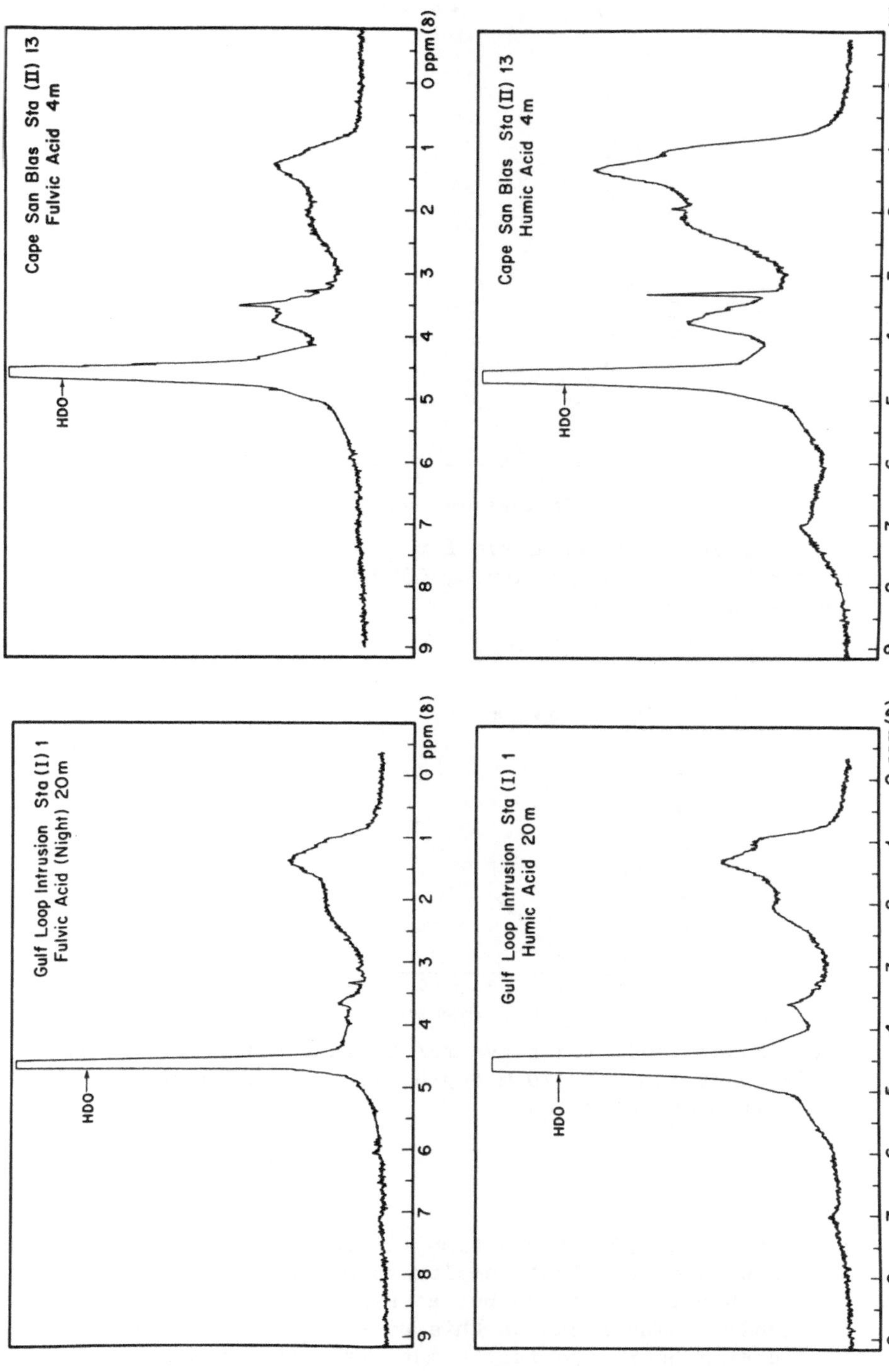

Fig. 2 Proton magnetic resonance spectra of dissolved marine humic and fulvic acids from the Gulf of Mexico.

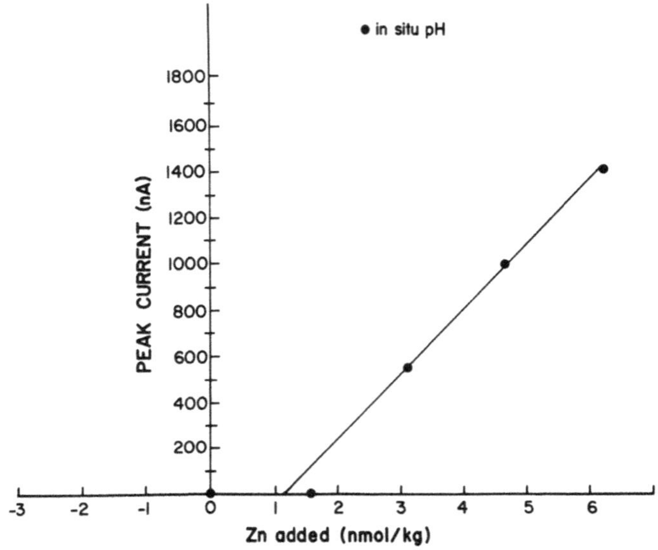

Fig. 3 ASV standard addition curve for Zn in a sample collected in March 1980 at 130 m depth, 0600 hours local time, in the Bay of Campeche.

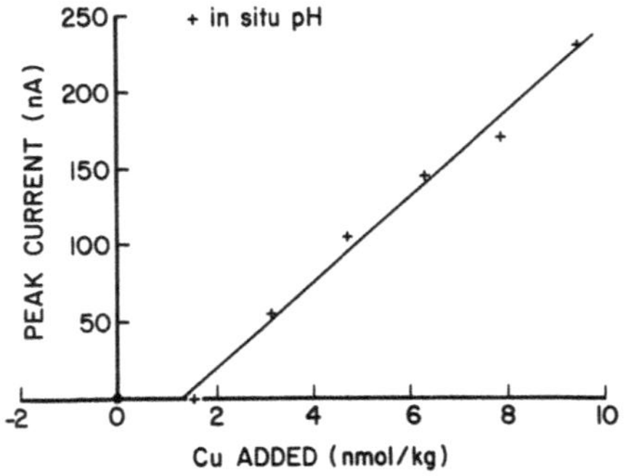

Fig. 4 ASV standard addition curve for Cu in a sample collected in February 1981 at 170 m depth, 1100 hours local time, in the Gulf Loop Intrusion.

and occurred more than 50% of the time for Cd. The interaction of particles with strong artifical complexers has been shown to generate ASV curves similar to these but at higher concentrations of metal and ligand[20]. The focus of this work was to see whether naturally occurring humic and fulvic materials isolated from seawater, which comprise up to 50% of the dissolved organic carbon in seawater (Table 1), could give rise to plots like Figures 3 and 4 in the laboratory and in freshly collected field samples at natural levels of organic matter.

Figure 5 shows the ASV standard addition curves for Zn in a sample of Gulf Stream water (collected ∿ 50 km offshore from Miami) and analyzed at natural pH twenty hours after collection. The sample was then acidified and stored at pH 2. A subsample of the acidified Gulf Stream water was placed in a small Teflon bottle and a concentrated marine fulvic acid (MFA) solution added, using a micropipettor, such that the final concentration of MFA was ∿ 1 mg kg^{-1}. A 15 g aliquot of the Gulf Stream water plus MFA solution was immediately placed in a polarography cell, made basic to pH ∿ 8.3 by micropipetting a NaOH solution and analyzed by the standard addition technique. For comparison, stock Gulf Stream water only (no MFA added) was analyzed at pH ∿ 8.3 by the method of standard additions. The results for Zn are presented in Figure 5. The curve for the Gulf Stream water plus MFA in Figure 5 indicates, when compared with the stock Gulf Stream water, that MFA can coordinate Zn. Similar curves for Gulf Stream water plus MFA but with different inflection points were generated at pH's of 4.8, 7.8, 8.0, and 8.6. The lower the pH, the less positive the x-intercept with a minimum of ∿ 0.8 nmol Zn kg^{-1} at pH 4.8. Performing the experiment by adding an MFA that had not been acidified to stock Gulf Stream water resulted in a curve in which no differences were observed when compared to the stock Gulf Stream water.

Figures 6 and 7 show the results of experiments conducted similarly to that for Zn with MFA but with Cd and Cu. The Gulf

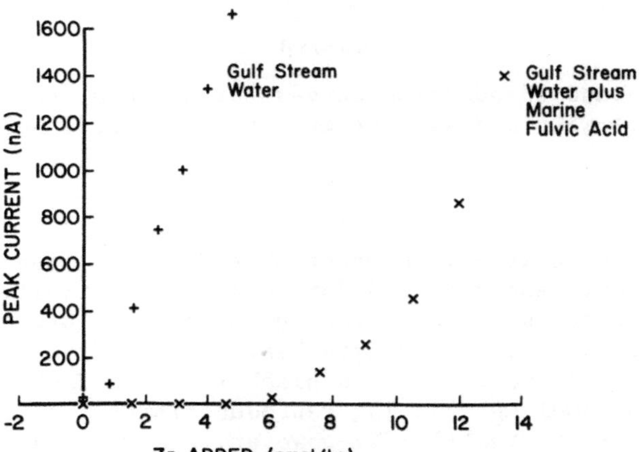

Fig. 5 ASV standard addition curves for Zn in a Gulf Stream water sample with and without marine fulvic acid added.

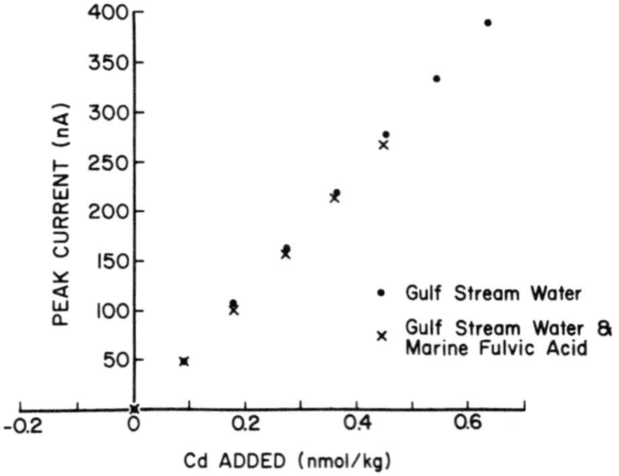

Fig. 6 ASV standard addition curves for Cd in a Gulf Stream water sample with and without marine fulvic acid added.

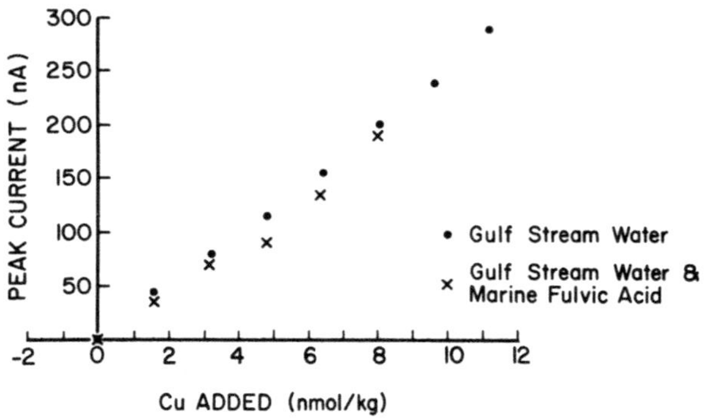

Fig. 7 ASV standard addition curves for Cu in a Gulf Stream water sample with and without marine fulvic acid added.

Stream water and Gulf Stream water plus MFA were made basic in the polarography cell and analyzed for both metals simultaneously. The dilution factors employed for the already dilute NaOH stock solution used to make the analyte basic resulted in an undetectable blank for Cd and Cu (the Zn blank from the NaOH was measureable at ~ 0.02 nmol kg^{-1}). The reproducibility for Cd and Zn in our system is approximately 7%; however, the reproducibility for Cu ranges from $\sim 20\%$ at levels < 1 nmol kg^{-1} to $\sim 7\%$ at 6 nmol kg^{-1} and greater. Within these limits there appears to be no difference in the standard addition curves for Cd and Cu in Gulf Stream water with and without added marine fulvic acid. We do not have sufficient data at this time to conclude that differences in the Cu signal at 5.0 and 6.3 nmol Cu added kg^{-1} are statistically significant.

Marine humic acids (MHA) are, by definition, insoluble in acid solution; therefore, repeating the above experiments could not be conducted using MHA. The MHA's, therefore, were prepared in seawater at natural pH. Freshly prepared solutions of MHA appeared to have no affect on the standard addition plots for all these metals. If the MHA/Gulf Stream water solution was allowed to sit for one to four days, deviations indicating coordination appeared for Cu but not for Cd or Zn. Since aging of the MHA appeared important (see later discussion), aging of the Gulf Stream water might also be affecting the standard addition curves. Laboratory experiments might not, therefore, be representative of what is occurring in nature. It was felt that these experiments had to be conducted with water as freshly captured as possible, but first we had to understand how natural samples varied with time. Samples for this work were processed slightly differently than those for routine analysis (Figures 3 and 4). The plots in Figures 3 and 4 and the laboratory MFA work involved standard additions to a single aliquot. If sufficient sample is available our system generates plots with sharper inflection points and linear sections with higher correlation coefficients (more points on the lines) if separate aliquots are used with a single standard addition of metal to the level desired. Generally, the two methods yield similar results, especially if the sample analyzed using a single aliquot has been freshly collected. A number of time series experiments was performed, therefore, using separate 125 ml subsamples of a single sample, each of which was placed in a clean Teflon® bottle and the standard added to the bottle immediately prior to the analysis at time zero. The samples were kept in the clean laboratory under constant fluorescent light at 20°C.

Figure 8 presents a time series for analysis of Zn in a sample collected at a depth of 5 m at 1100 local time off Cape San Blas, Florida. It appears that there is an increase in the degree to which added metal is coordinated up to 15 hours after innoculation. The trend reverses itself sometime between 15 and 18 hours after innoculation. Figure 9 is for a sample collected from 70 m at 2300 local time at the same station. Both samples have ASV-labile Zn initially present in them which disappears with time and, in fact, the samples gain some capacity to coordinate metal which is then lost. All six samples collected at this station gave similar results. Day samples were collected at 5, 30 and 70 m while night samples were collected at 10, 30 and 70 m (Table 2). The 5 and 30 m samples collected during the day had ASV-labile metal present while the 10 and 30 m samples at night had no ASV-

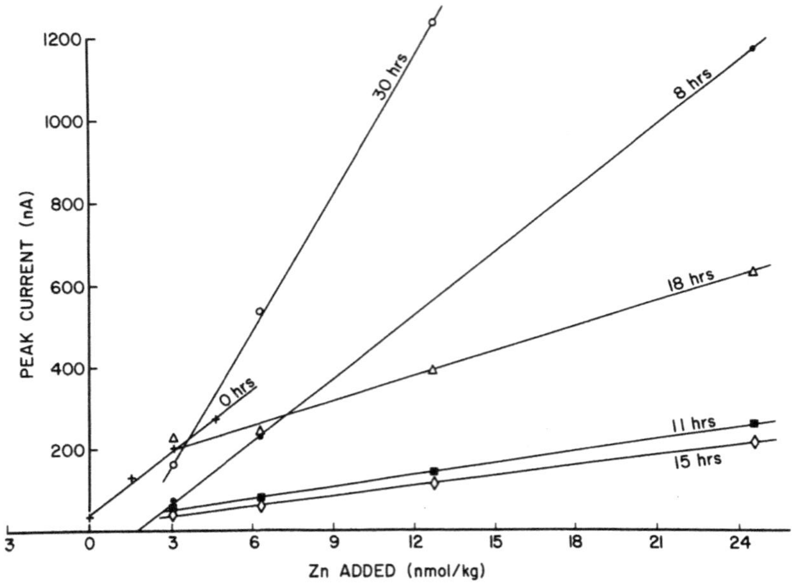

Fig. 8 ASV standard addition curves for Zn in separate aliquots of water in a sample collected in February 1981 off Cape San Blas, Florida at 5 m depth, 1100 hours local time, and analyzed at the times after addition as indicated.

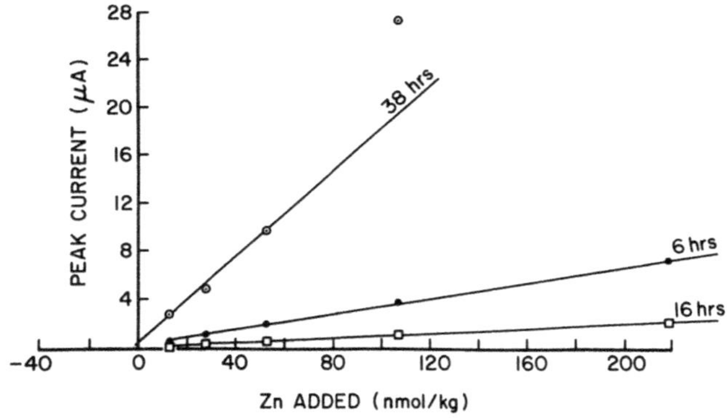

Fig. 9 ASV standard addition curves for Zn in separate aliquots of water in a sample collected in February 1981 off Cape San Blas, Florida at 70 m depth, 2300 hours local time, and analyzed at the times after addition as indicated.

labile Zn and some residual binding capacity. The day 70 m sample had no free Zn present and some binding capacity while the night sample (Figure 9) had free Zn present.

Table 2 ASV-labile Cd, Cu and Zn in Cape San Blas Samples

Local Time	Depth (m)	Concentration Found ($nmol\ kg^{-1}$) Natural pH Cd	Cu	Zn	Residual Binding Capacity ($nmol\ kg^{-1}$) Cd	Cu	Zn
1100	5	0.12	1.9	0.84	--	--	--
1100	30	0.12	0.3	0.15	--	--	--
1100	70	0.05	0.8	nsd	--	--	0.56
2300	10	nsd	1.0	nsd	*	--	0.89
2300	30	0.03	nsd	nsd	--	*	0.31
2300	70	0.04	1.0	0.47	--	--	--

nsd = no signal detected
* = standard addition curves pass through origin

The slope of the 38-hour plot in Figure 9 is approximately the average response (sensitivity) of our system for free Zn. The ∿ 90 nmol Zn added kg^{-1} point probably represents contamination of the sample as the ∿220 nmol Zn added kg^{-1} (not shown) at 38 hours falls on the line formed by the first three points shown. The 30-hour plot in Figure 8 has a somewhat lower slope but is at the low end of the day to day variability in the response of the system. The peak current intercept divided by the free response is the amount of ASV labile metal initially present in the sample[19]. If all the metal is free, then the absolute value of the negative x-intercept is the amount initially present. In all of the determinations made from 0 to ∿30 hours the free response was never attained. This would be indicated by an inflection-point and change in slope of these plots. The changing slopes and the data in Table 2 indicate that changes are occurring with time that affect the concentration of ASV labile metal in these samples. These changes could be due to: (1) Formation of coordinating ligand by biological processes (excretion or decay) in the captured sample; (2) formation of coordinating ligand from simple ligands released by organisms; (3) reaction kinetics of metal with ligand; (4) degradation of metal/ligand complexes (by $h\nu$, bacteria, OH^-, O_2); (5) differing reaction kinetics involving several different ligands; or (6) any combination of the above.

We cannot, a priori, rule out any one of these possibilities. The ASV standard addition data for natural samples and MFA in the laboratory indicate that at least some ligands have fast reaction kinetics with Zn at low levels. It appears that MFA can extensively coordinate Zn. While we have not isolated other classes of organics, we are working with a major portion of the organic material in seawater which, if the reaction is as rapid as indicated, would indicate that the concentration of ASV-labile Zn is governed by the concentration of ligand and the degradation of the metal-ligand complex.

Figure 10 is a summary of the first six hours of time series results for Zn for the three samples collected at 1100 local time at Cape San Blas. There is a very rapid decrease in ASV-labile Zn with time and an apparent increase in binding capacity with depth. In the 70 m sample no signal was detected after three hours for an addition of ~ 13 nmol kg^{-1}. As in Figures 8 and 9, the signal returns to a level above zero hours at 30 hours. This station and the Mississippi station yield the highest concentrations of MFA and MHA of all of the stations sampled. The apparent high binding capacities of this water, therefore, may be primarily a function of concentration of these substances.

The situation is much more complex in conducting these experiments with MHA for Cd and Cu. The time dependence for coordination of Cd and Cu in natural samples must first be evaluated. If the rate of production of coordinating materials is important for these two, the chemical form of the MHA added to seawater is very critical. The slow rates of dissolution of the purified MFA and

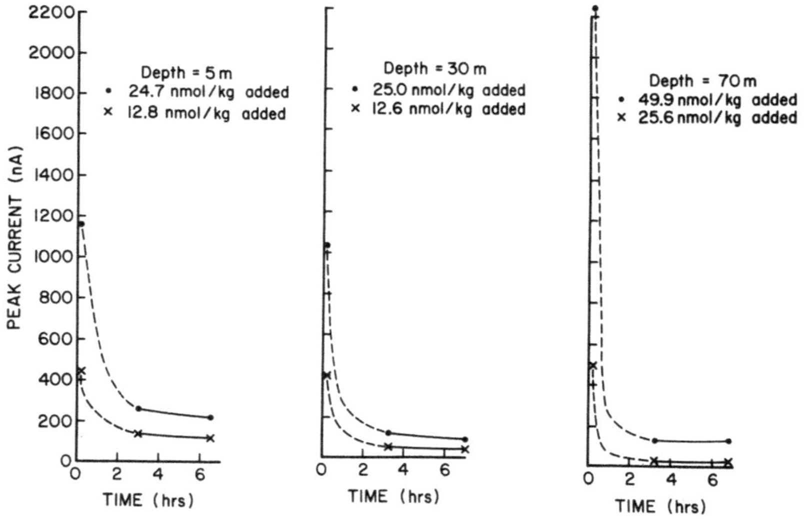

Fig. 10 Portions of ASV standard addition curves for Zn in separate aliquots from three samples collected in February 1981 off Cape San Blas, Florida at 1100 hours local time, and analyzed at the times after addition indicated.

MHA and the lack of any binding capacity for several hours after dissolution which we observed earlier probably has two causes. The final freeze-drying process from acid solution may cause inter- and intra-molecular esterification of the carboxyl groups. The PMR spectra show that there are abundant hydroxyl groups to react with the acid groups. Thus, the esterified material has to hydrolyze before it is active. Second, the PMR spectra and other data indicate that MFA and MHA are composed of hydroxylated long-chain fatty acid moieties. Such molecules would require time to unfold and unwind in water from their isolated dry state to become resolvated in seawater in the most energetically favorable configuration. Another problem with time series studies for Cd and Cu is that the analysis (plating and stripping steps) takes considerably longer for these elements than for Zn, making extensive studies difficult to accomplish when investigating short reaction times. We are presently working on increasing our sensitivity and changing our humic acid purification procedure to avoid dehydration/esterification.

It appears that metal/organic interactions in seawater are very dynamic processes. Field and laboratory studies with marine fulvic and humic materials can only give insights into steps in these cycles. We believe that a discussion of these cycles can be accomplished by the steps in Figure 11. In Figure 11 we assume that the ligands responsible for metal/organic interactions have a biological source and the actual ligands responsible for coordination are condensation and/or reaction products of simpler molecules actually released by organisms. In Figure 11, marine fulvic

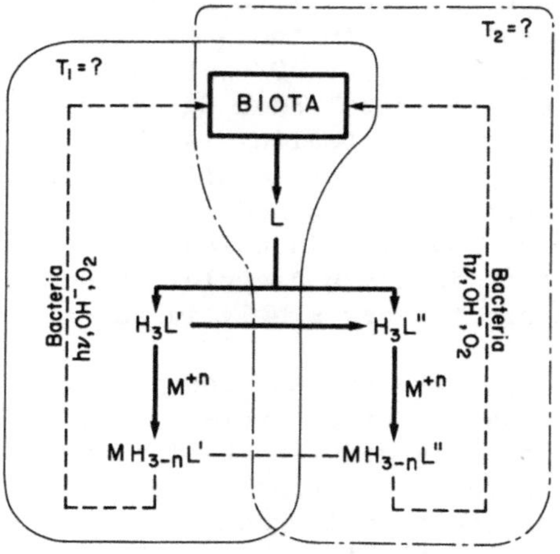

Fig. 11 The steady-state cycle of biological organisms, dissolved organic matter, and trace metals in the marine environment.

materials are represented as H_3L' while humic materials are represented by H_3L''. This notation implies that the basic structure of all of the ligands involved is similar and that the functionality is acidic. Figure 11 implies nothing about the structures of L' and L". In fact, H_3L' may be of higher molecular weight than H_3L''. We cannot say whether H_3L'' is derived directly from L or H_3L' as given in the Figure; however, we feel the preferred route is $L \rightarrow H_3L' \rightarrow H_3L''$.

This work suggests that the loop involving marine fulvic acid (H_3L') might have a time constant, T_1, on the order of one to two days in these waters for Zn. It also suggests that the step $M^{+n} + H_3L' \rightarrow MH_{3-n}L' + nH^+$ is rapid (on the order of minutes). The rate of degradation of $MH_{3-n}L'$ is unknown but probably related to light either through biological activity or photolysis. Cadmium is apparently not involved in the H_3L' cycle to an extent we could measure. The data on Cu in Figure 4 and some of our other field data (unpublished) suggests some copper participation in a cycle similar to that of Zn but which cycle is uncertain.

The cycle involving marine humic acid H_3L'' is delineated as T_2. We cannot shed a great deal of light on this cycle with the results presented here. We have performed some analyses of samples stored at natural pH for up to two weeks in the light and four months in the dark. Samples having no ASV-labile Zn at the time of collection always had ASV-labile Zn present within hours of collection, even samples stored in the dark. Cadmium signals are quite stable with time. Copper signals vary with time. Samples having no ASV-labile Cu at the time of collection from Cape San Blas continued to show no ASV-labile Cu when stored in the dark for four months. Gulf Stream water samples stored in the light and dark always result in those samples exposed to light having higher concentrations of ASV-labile Cu than those stored in the dark. The extent to which added Cu can be coordinated always decreases with storage. Our feeling is that T_2 is quite a bit longer than T_1 for any metal, especially Cu, and may be on the order of weeks in surface waters and longer in deep waters. Zinc does not appear to be involved in a long time constant cycle and, thus, may not interact in the MHA cycle. Cadmium, if involved with any cycle, interacts very slowly in an extremely long cycle.

In nature the cycle(s) represented in Figure 11 are in a steady state following a diurnal cycle. Once a sample is captured, the production of L is going to be interrupted at some time after capture. We expect this time to vary with a variety of factors such as rate of biological production, nutrient content of the water, concentration of biota, etc. The rate of breakdown of H_3L', H_3L'', $MH_{3-n}L'$ and $MH_{3-n}L''$ could be dependent on at least the parameters indicated. A hypothesis for the Cape San Blas data based on Figure 11 would be that Figures 8-10 are the result of the production of L exceeding degradation rates until about 12 hours after collection and degradation of the Zn species is complete after about two days. The presence of ASV-labile Zn during the day in the 5 and 30 m samples and not at night in Table 2 would suggest that $h\nu$ is the primary degradation mechanism in the euphotic zone. Since Cd and Cu do not vary like Zn, the rate of ligand formation is either not important or that different degradation mechanisms are operating on Zn as compared to Cd and Cu. The data also could be explained by Cd and Cu not being involved in a cycle involving the same ligand that Zn interacts with. It is clear that the system is not simple, but by having an understanding of the general MFA/MHA/metal cycle and some of the factors influencing it allows us to better understand what is occurring in the marine environment and to design experiments to understand it.

A solid arrow is given in Figure 11 for the transition $H_3L' \rightarrow H_3L''$. We are unable at this time to evaluate a rate constant for this transition. We are at present conducting extensive research in the areas represented by the solid arrows. Whether a connection exists between $MH_{3-n}L'$ and $MH_{3-n}L''$ is only speculation and is presented as such. This work suggests many areas of research that must be accomplished prior to understanding the interactions between trace metals and naturally occurring organic materials and the role biology plays in affecting the speciation of trace metals and is affected by trace metals. Finally, this work indicates that the use synthetic complexers to elucidate marine trace metal chemistry may be of minimal relevance to sea water.

ACKNOWLEDGEMENTS

We wish to thank the officers and crew of the NOAA ship RESEARCHER for their assistance throughout the collection of these samples. We wish to thank Dr. Peter B. Ortner for many helpful discussions in the course of this work. This program "The Role of Organics in the Marine Environment" (ROME), was sponsored by the Office of Marine Pollution Assessment/Long-Range Effects Research Program of the U.S. National Oceanic and Atmospheric Administration.

REFERENCES

1. Stuermer, D.H. and G.R. Harvey, 1977: The isolation of humic substances and alcohol-soluble organic matter from seawater. Deep-Sea Res., 24, 303-309.
2. Huizenga, D.L. and D.R. Kester, 1979: Protonation equilibria of marine dissolved organic matter. Limnol. Oceanogr., 24, 145-157.

3. Boyle, E.A. and J.M. Edmond, 1975: Copper in surface waters south of New Zealand. Nature, 253, 107-109.
4. Bruland, K.W., G.A. Knauer and J.H. Martin, 1978: Zinc in northeast Pacific waters. Nature, 271, 741-743.
5. Patterson, C.C., 1974: Lead in seawater. Science, 183, 553-554.
6. Bubic, S., L. Sipos and M. Branica, 1973: Comparison of different electroanalytical techniques for the determination of heavy metals in sea water. Thalass. Jugo., 9, 55-63.
7. Chau, Y.K., R. Gächter and K. Lum-Shue-Chan, 1974: Determination of the apparent complexing capacity of lake waters. J. Fish. Res. Board Canada, 31, 1515-1519.
8. Nürnberg, H.W., P. Valenta, L. Mart, B. Raspor and L. Sipos, 1976: Applications of polarography and voltammetry to marine and aquatic chemistry II. The polarographic approach to the determination and speciation of toxic trace metals in the marine environment. Z. Anal. Chem., 282, 357-367.
9. O'Shea, T.A. and K.H. Mancy, 1976: Characterizations of trace metal-organic interactions by anodic stripping voltammetry. Anal. Chem., 48, 1603-1607.
10. Florence, T.M. and G.E. Batley, 1977: Determination of the chemical forms of trace metals in natural waters with special reference to copper, lead, cadmium and zinc. Talanta, 24, 151-158.
11. Florence, T.M., 1970: Anodic stripping voltammetry with a glassy carbon electrode mercury plated in situ. J. Electroanal. Chem., 27, 273-281.
12. Batley, G.E. and T.M. Florence, 1974: An evaluation and comparison of some techniques of anodic stripping voltammetry. J. Electroanal. chem., 55, 23-43.
13. Lund, W. and M. Salberg, 1975: Anodic stripping voltammetry with the Florence mercury film electrode. Determinations of copper, lead and cadmium in sea water. Anal. Chim. Acta., 76, 131-141.
14. Lund, W. and D. Onshus, 1976: The determination of copper, lead and cadmium in sea water by differential pulse anodic stripping voltammetry. Anal. Chim. Acta., 86, 109-122.
15. Valenta, P., L. Mart and H. Rützel, 1977: New potentialities in ultra trace analysis with differential pulse anodic stripping voltammetry. J. Electroanal. Chem., 82, 327-343.
16. Tokar, J.M., G.R. Harvey and L.A. Chesal, 1981: A gas lift system for large volume water sampling. Deep-Sea Res., 28, 1395-1399.
17. Mantoura, R.F.C. and J.P. Riley, 1975: The analytical concentration of humic substances from natural waters. Anal. Chim. Acta., 76, 97-106.

18. Stuermer, D.H. and J.R. Payne, 1976: Investigation of seawater and terrestrial humic substances with carbon-13 and proton nuclear magnetic resonance. Geochim. Cosmochim. Acta., 40, 1109-1114.
19. Duinker, J.C. and J.M. Kramer, 1977: An experimental study on the speciation of dissolved zinc, cadmium, lead and copper in river Rhine and North Sea water, by differential pulse anodic stripping voltammetry. Mar. Chem. 5, 207-228.
20. Plavsic, M., S. Kozak, D. Krznaric, H. Bilinski and M. Branica, 1980: The influence of organics on adsorption of Copper (II) on α -Al_2O_3 in seawater. Model studies with EDTA. Mar. Chem., 9, 175-182.

CHEMICAL PERIODICITY AND THE SPECIATION AND CYCLING OF THE ELEMENTS

M. Whitfield and D.R. Turner

Marine Biological Association of the United Kingdom
The Laboratory, Citadel Hill
Plymouth PL1 2PB, England

ABSTRACT

In the periodic table the elements are arranged in order of increasing atomic weight in a sequence which reflects the progressive occupation of the available ground state electron energy levels. The periodic chemical behaviour revealed when the elements are arranged in this way is a reflection of both the nature of the electrons available for bonding and the extent to which their behaviour is influenced by the nuclear charge. Numerous attempts have been made to derive parameters which take due account of these effects and provide a quantitative summary of the chemical affinity of the various elements.

Recent studies have shown that partition coefficients describing the distribution of the elements between sea water and crustal rock can be related to a number of parameters which reflect the chemical periodicity of the elements. Furthermore the partition coefficients themselves are related in a simple way to the mean oceanic residence times of the elements so that these parameters, too, exhibit periodic behaviour. Since the partition coefficient/residence time correlation also applies to biogenic particles these relationships might have wide implications.

It has also been shown that the chemical speciation of the elements in natural waters and their biological function can be summarized using a grid whose ordinates are the ionisation potential (Z^2/r) and a 'softness' parameter which can be related to the electronegativities of the elements. Marked periodic correlations can also be noted here.

By taking due consideration of these effects it is possible to bring some semblance of order into studies of the cycling and speciation of the elements.

INTRODUCTION

It is a fundamental tenet of inorganic chemistry that the electronic structures of the elements are a periodic function of their atomic number. Chemical periodicity provides a rationalisation of the inorganic chemistry of the elements and all undergraduate chemistry courses belabour the links between electronic structure and chemical behaviour so that is should be an integral part of our chemical thinking. However, relatively few papers or reviews have sought to relate the information obtained from environmental studies to the organisation of the elements in the periodic table. It appears that we tend to lose sight of the fundamental regularities in the chemistry of the elements once we become involved in detailed studies of the behaviour of particular elements in natural systems.

It is not our intention here to provide a refresher course in undergraduate inorganic chemistry since a number of detailed texts are available which place particular emphasis on the periodicity of the chemical behaviour of the elements[1-5]. We intend instead to show how useful rationalisations concerning the behaviour of the elements in the marine environment can be derived from the consideration of a few parameters which reflect the periodicity of the electronic structures of the elements.

SELECTION OF PARAMETERS TO REPRESENT CHEMICAL PERIODICITY

During its passage through the oceans an element will become involved in the formation of a variety of chemical bonds. Within the aqueous phase it will experience hydration, complexation and possibly redox reactions. It may be adsorbed on to particle surfaces via interactions with surface hydroxyl groups or by specific complexation reactions and will eventually find its way into mineral lattices either as a major building block or as a substituent in the host lattice. To help us unravel the details of such interactions we need to know something of the propensity of the elements to form chemical bonds and of the extent to which the valence electrons are likely to be redistributed during their formation. First we will consider ionic bonding - the situation in which no redistribution of valence electrons takes place during bond formation.

The Ionic Model

The ionic model in its simplest form ignores altogether the electronic structure of atoms and considers that chemical species can be represented as hard charged spheres (ions) and that solvents such as water can be represented as structureless media characterised only by their dielectric constant ε. According to the simple ionic model, chemical bonding can therefore be described by elementary electrostatics. Each ion is characterised by its charge $Z_i e$ and its radius r_i. While $Z_i e$ is known precisely the ionic radius r_i is necessarily an approximation since the electron density of any atom extends indefinitely into space. Ionic radii are assigned empirically so that the sum of anionic and cationic radii reproduce with reasonable accuracy the interionic distances observed in ionic crystals. The ionic model is successful in predicting crystal lattice energies and in rationalising ion hydration energies and complex formation constants, although in some cases the success owes more to the self-compensating features of the model than to its correctness (for a full discussion see ref. 6, Ch. 5). The areas of particular importance to marine chemistry are the hydration energies of cations and complex formation constants.

The Born model of hydration[7] assumes water to be a featureless medium of dielectric constant ε. The energy required to hydrate one mole of gaseous ions of charge $Z_i e$ is then given by

$$\Delta G = N_o \int_{Z_i e}^{o} v_{vacuum}\, dq + N_o \int_{o}^{Z_i e} v_{water}\, dq \qquad (1)$$

where q is the charge and N_o is Avogardro's number. With the potentials $v_{vacuum} = q/r_i$ and $v_{water} = q/\varepsilon r_i$

$$\Delta G = N_o \int_{Z_i e}^{o} \frac{q dq}{r_i} + N_o \int_{o}^{Z_i e} \frac{q dq}{\varepsilon r_i} \qquad (2)$$

$$\Delta G = \frac{-N_o (Z_i e)^2}{2 r_i} (1 - 1/\varepsilon) = -164 \frac{Z_i^2}{r_i} \text{ kcal mol}^{-1} \qquad (3)$$

Differentiating with respect to T to obtain ΔS[6,7], we obtain at 25 °C

$$\Delta H = \Delta G + T\Delta S = -167\, Z_i^2/r_i \text{ kcal mol}^{-1} \qquad (4)$$

It is found that this equation agrees well with experiment if instead of the ionic radius r_i one uses $r_{eff} = r_i + 0.85 Å$.

Although explanations have been advanced for the use of the value 0.85Å it is best taken as an empirical correction[6]. We will call Z_i^2/r_i (or Z_i^2/r_{eff}) the electrostatic energy of the ion.

Complex formation in aqueous solution can be written as:

$$M(H_2O)_n^{Z_+} + X^{Z_-} \rightleftharpoons MX\,(H_2O)_m^{(Z_+ - Z_-)} + (n-m)\,H_2O$$

The terms contributing to the overall free energy change (and hence log K) for this reaction can be derived from a thermodynamic cycle (Fig. 1). The contributing terms are the hydration energies of X, M and MX and the energy of the MX interaction in vacuo. For many ligands the values of log K are linearly related to the Z_i^2/r_{eff} values of the cations[6,8,9] suggesting that the various $\Delta G_{hydration}$ terms are dominant. However Hefter[8] has noted that in the case of the fluoride complexes an equally good linear correlation can be obtained between log K and $Z_+/(r_+ + r_-)$, a term which is derived from an electrostatic treatment[7] of the $\Delta G_{complexation}$[(9)] term in Fig. 1. The two hard charged spheres experience a mutual attraction $(Z_+e).(Z_-e)/r^2$ in a vacuum so that the energy gained by bringing them to a separation $(r_+ + r_-)$ is

$$\Delta G = \int_{\infty}^{(r_+ + r_-)} \frac{Z_+Z_-e^2}{r^2}\,dr = -\frac{Z_+Z_-e^2}{r_+ + r_-} \qquad (5)$$

The detailed interpretation of these correlations is therefore complicated by the similarity of the electrostatic parameters describing different processes (Fig. 2, Table 1). We will simply note that two parameters are of value in summarising electrostatic interactions: (i) the ionic electrostatic energy Z_i^2/r_i (or Z_i^2/r_{eff}) is proportional to the hydration energy of a cation; (ii) the ionic potential Z_i/r_i is proportional to the work required to bring a unit charge to the periphery of the ion (cf eq. 5).

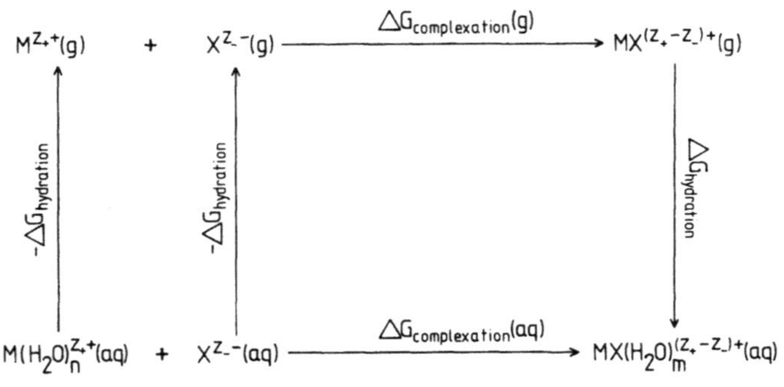

Fig. 1 Born-Haber free energy cycle for aqueous complex formation.

Table 1. Parameters used in the correlations discussed in the text.

Element	r_c/A	Z*	χ_{AR}	Z_i	r_i/A	χ_P	R_5	ΔH/kcalmol^{-1}
Ag	1.339	3.35	1.41	1	1.15	1.93	0.64	-110
Al	1.300	3.15	1.41	3	0.53	1.61	0.00	-1104
As	1.220	5.95	2.18	3	0.58	2.18	0.10	
				5	0.50		0.10	
Au	1.336	3.35	1.42	1	1.37	2.54	0.92	-148
				3	0.85			
B	0.900	2.25	1.74	3	0.23	2.04	0.00	
Ba	1.981	2.50	0.97	2	1.36	0.89	0.00	-304
Be	1.250	1.60	1.11	2	0.35	1.57	0.00	-587
Bi	1.520	5.95	1.67	3	1.02	2.02	0.12	-154
				5	0.74		0.12	
Br	1.140	7.25	2.75	-1	1.96	2.96	0.04	-80
				7	0.39		0.04	
C	0.770	2.90	2.50	4	0.16	2.55	0.00	
Ca	1.736	2.50	1.04	2	1.00	1.00	0.00	-370
Cd	1.413	4.00	1.46	2	0.95	1.69	0.37	-425
Ce	1.647	2.00	1.01	3	1.01	1.12	0.00	-819
Cl	0.990	5.75	2.85	-1	1.81	3.16	0.00	-91
				7	0.27		0.00	
Co	1.260	3.55	1.55	2	0.75	1.88	0.30	-485
Cr	1.176	2.60	1.42	3	0.62	1.60		-1087
				6	0.52		0.00	
Cs	2.350	1.85	0.86	1	1.70	0.79	0.00	-62
Cu	1.173	3.35	1.62	1	0.96	1.90	0.55	-133
				2	0.73	2.00	0.49	-495
Dy	1.589	2.50	1.10	3	0.91	1.22	0.00	
Er	1.567	2.50	1.11	3	0.89	1.24	0.00	
Eu	1.850	2.50	1.01	3	0.95	1.18	0.00	
F	0.710	4.85	4.20	-1	1.33	3.98	0.00	-123
				7	0.08		0.00	
Fe	1.250	3.40	1.53	2	0.78	1.83	0.26	-456
				3	0.65	1.96		-1054
Ga	1.200	4.65	1.90	3	0.62	1.81	0.23	-1106
Gd	1.614	2.65	1.11	3	0.94	1.20	0.00	
Ge	1.220	5.30	2.02	4	0.54	2.01	0.16	
Hf	1.442	2.80	1.23	4	0.71		0.00	
Hg	1.440	4.00	1.44	2	1.02	2.00	0.49	-429
Ho	1.580	2.50	1.10	3	0.90	1.23	0.00	
I	1.330	7.25	2.22	-1	2.20	2.66	0.10	-67
				5	0.95		0.10	
				7	0.50		0.10	
In	1.497	4.65	1.49	3	0.80	1.78	0.24	-977
Ir	1.265	3.55	1.54	4	0.63			
K	1.960	1.85	0.92	1	1.38	0.82	0.00	-73
La	1.690	2.65	1.08	3	1.05	1.10	0.00	-775
Li	1.340	0.95	0.93	1	0.74	0.98	0.00	-119
Lu	1.557	2.65	1.14	3	0.86	1.27	0.00	
Mg	1.450	2.50	1.17	2	0.72	1.31	0.00	-452
Mn	1.390	3.25	1.35	2	0.83	1.55	0.23	-433
				4	0.54			
				7	0.46	2.30	0.00	

Table 1. (continued)

Element	r_c/A	Z*	χ_{AR}	Z_i	r_i/A	χ_P	R_5	ΔH/kcalmol^{-1}
Mo	1.296	2.60	1.30	6	0.60	2.35	0.00	
N	0.750	3.55	3.01	5	0.13		0.00	
Na	1.540	1.85	1.02	1	1.02	0.93	0.00	-94
Nb	1.342	2.45	1.23	5	0.64		0.00	
Nd	1.642	2.50	1.08	3	0.98	1.14	0.00	
Ni	1.210	3.70	1.65	2	0.69	1.91	0.35	-495
O	0.730	4.20	3.57	-2	1.40	3.44	0.00	
Os	1.260	3.40	1.51	4	0.63			
P	1.100	4.45	2.06	5	0.35		0.00	
Pb	1.538	5.30	1.55	2	1.18	1.87	0.20	-347
				4	0.78	2.33	0.20	
Pd	1.283	2.70	1.33	4	0.62			
Pr	1.648	2.50	1.07	3	1.00	1.13	0.00	
Pt	1.295	3.20	1.43	4	0.63		0.88	
Rb	2.160	1.85	0.89	1	1.49	0.82	0.00	-69
Re	1.283	3.25	1.45	7	0.57		0.00	
Rh	1.252	3.05	1.44	4	0.62			
Ru	1.246	2.90	1.41	4	0.62			
S	1.020	5.10	2.50	6	0.30		0.00	
Sb	1.430	5.95	1.79	3	0.76	2.05	0.10	
				5	0.61	2.10	0.10	
Sc	1.439	2.65	1.20	3	0.75	1.36	0.00	-929
Se	1.170	6.60	2.47	4	0.50		0.07	
				6	0.42		0.07	
Si	1.180	3.80	1.72	4	0.40	1.90	0.00	
Sm	1.660	2.50	1.07	3	0.96	1.17	0.00	
Sn	1.400	5.30	1.71	4	0.69	1.96	0.15	-1803
Sr	1.914	2.50	0.99	2	1.13	0.95	0.00	-338
Ta	1.343	2.95	1.33	5	0.64		0.00	
Tb	1.592	2.50	1.10	3	0.92	1.21	0.00	
Te	1.350	6.60	2.04	6	0.56		0.05	
Th	1.652	2.80	1.11	4	1.00		0.00	
Ti	1.324	2.80	1.32	4	0.61		0.00	
Tl	1.549	4.65	1.44	1	1.50	1.62	0.30	-74
				3	0.89	2.04	0.30	-966
Tm	1.562	2.50	1.11	3	0.88	1.25	0.00	
U	1.420	2.65	1.22	4	0.97	1.30		
				6	0.73		0.00	
V	1.224	2.95	1.45	5	0.54	1.80	0.00	
W	1.304	3.10	1.40	6	0.60	2.10	0.00	
Y	1.616	2.65	1.11	3	0.90	1.22	0.00	-854
Yb	1.699	2.50	1.05	3	0.87	1.26	0.00	
Zn	1.200	4.00	1.74	2	0.75	1.65	0.34	-480
Zr	1.454	2.80	1.22	4	0.72		0.00	

Sources: Covalent radii from refs. 10,13,21, ionic radii from refs. 22 & 23; Pauling electronegativities from refs. 10,14&15. Screening loss parameter from ref. 24; enthalpies of hydration from refs. 6&25.

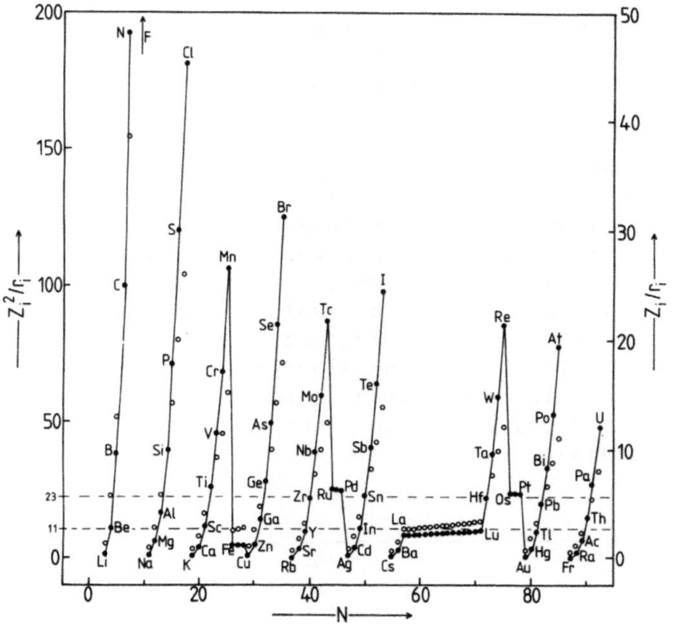

Fig. 2 Plots of electrostatic work functions (Z_i^2/r_i, ● and Z_i/r_i, O) as a function of atomic number (data given in Table 1). All elements are in their group oxidation states except for iron, cobalt and nickel (+ 2) and the platinum metals (+ 4). Data are taken from Table 1 where values for other oxidation states are also given. The horizontal markings indicate regions discussed in the text.

Covalent Interactions

To consider covalent interactions we have to abandon the simple concepts of the ionic model and take account of the electronic structure of the atoms involved. The degree of covalent bonding between two atoms is determined to a significant extent by the relative energies of their valence orbitals. If the valence orbitals are of very different energies little covalent interaction takes place and bonding can be adequately described by the ionic model described above. As the energies of the valence orbitals become closer the importance of ionic bonding declines and that of covalent bonding increases until in the case of homonuclear diatomic molecules the energies of the valence orbitals are exactly equal and the bonding is entirely covalent. Clearly a parameter which describes the energy of the valence orbitals of an atom would be of great value in summarising the significance of covalent interactions. Such a parameter is provided by the electronegativity X which was first proposed by Pauling[10] as a means of rationalising a large body of thermochemical data, and defined as "the power of an atom in a molecule to attract electrons to itself".

The significance of X can be most readily appreciated by considering the electrostatic derivation proposed by Allred and Rochow[11]. To estimate the electronegativity of an element we must take into account its electronic structure and consider the electrostatic force acting on electrons at a distance r_C from the nucleus, where r_C is the covalent radius of the atom. Covalent radii depend on the detailed structure of the atoms and show a clear periodicity (Fig. 3). The electrostatic force exerted by the nucleus on a valence electron at the distance r_C is given by $E_C e$ where e is the charge on the electron and E_C is the electric field at the distance r_C.

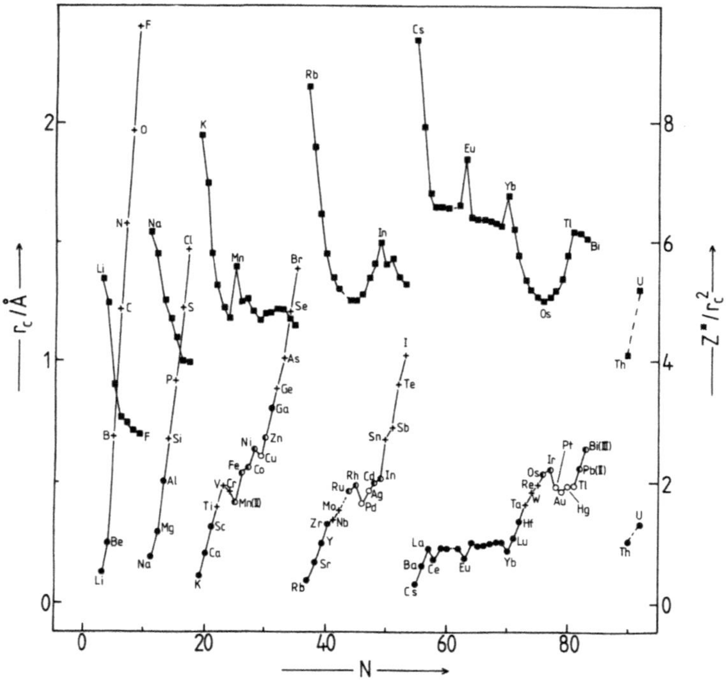

Fig. 3 Plots of covalent radius (r_C, squares) and Allred-Rochow electronegativity (Z^*/r_C^2, circles and crosses) as a function of atomic number (data are given in Table 1). Full circles indicate (a)-type cations, half shaded circles intermediate cations and open circles (b)-type cations. Crosses indicate anionic elements.

At a distance r_c from a charge q, E_c is given by

$$E_c = q/r_c^2$$

In the present case q is less than the nuclear charge $Z_N e$ due to shielding of the valence electron by other electrons. We therefore set

$$E_c = Z^*e/r_c^2 \qquad (6)$$

where Z* is the effective nuclear charge. Z* is estimated using Slater's rules (Table 2) which allow empirically for the effects of shielding. The quantity Z^*/r_c^2 therefore represents the electrostatic force exerted on a valence electron at the covalent radius, and also shows a clear periodic variation (Fig. 3). This electrostatic measure of "the power of an atom in a molecule to attract electrons to itself", can be related directly to Paulings thermochemical electronegativity scale[10] (Fig. 4). A variety of alternative electronegativity scales have been introduced based on work functions, force constants and bond lengths, the compactness of the elements relative to inert gas structures, ionisation potentials and electron affinities. All these different scales correlate quite closely with one another[1,13,14] underlining the fact that they represent some fundamental elemental parameter related to the energy of the valence orbitals. This of course implies that strictly each oxidation state of an element and indeed each individual valence orbital should have its own electronegativity value, and some scales have addressed this problem[14,15]. For our present purposes we will assign a single electronegativity value to each element.

The most useful application of electronegativity in the present context is in the prediction of the polarity of chemical

Table 2 Slater's rules for the calculation of the effective nuclear charge for valence electrons. [a]

Electron group	S_i for electrons in			
	All higher groups	Same group	Groups with n' = n-1	Groups with n'<n-1
1s	0	0.30	-	-
(ns, np)	0	0.35	0.85	1.00
(nd), (nf)	0	0.35	1.00	1.00

The effective nuclear charge, $Z^* = N - \sum_i S_i$ where N is the atomic number, n is the principal quantum number.

[a] Adapted from refs. 4 & 12

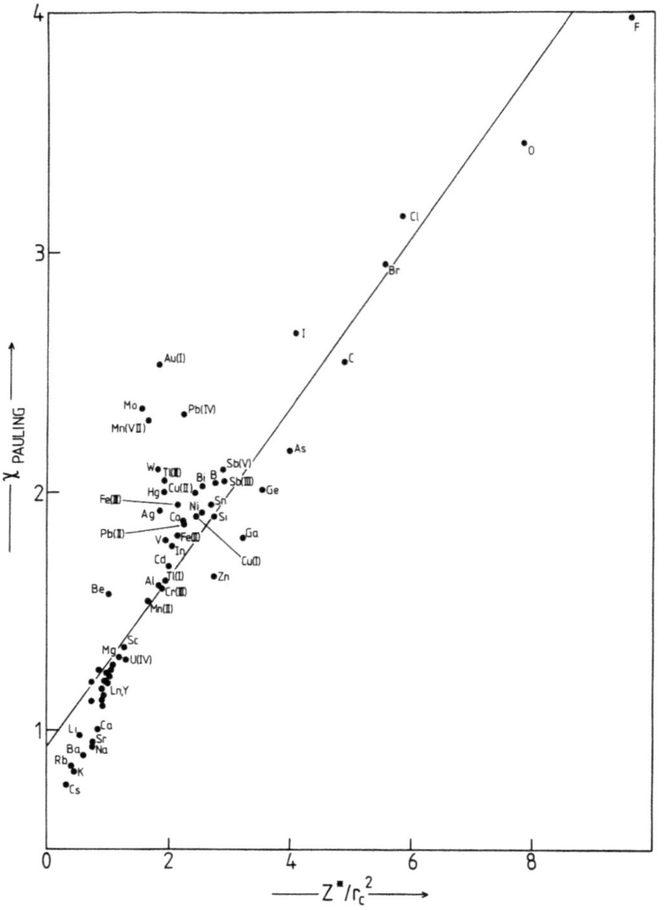

Fig. 4 Comparison between the electrostatic function $Z^*/r_c^{\ 2}$ and Pauling's thermochemical electronegativities. Data are taken from Table 1.

bonds. Consider a heteronuclear diatomic molecule, AB. If A and B have very different electronegativities then the electric field produced at the covalent radius will be very different for each atom (or put another way the energies of the valence orbitals will be very different), so that the valence electrons will reside almost exclusively on one atom and we are back to the simple ionic model. As the electronegativities χ_A and χ_B become closer the valence electrons will be shared more evenly between A and B and the importance of covalent interactions will increase.

The size of the difference $|\chi_A - \chi_B|$ thus gives an estimate of the relative importance of ionic and covalent interactions[16], while the ionic model can be used to estimate the absolute magnitude of the ionic interaction. The problem remains of estimating the intensity of covalent interaction. Even in the simple case of homonuclear diatomic molecules (e.g. N_2, O_2) the parameters described so far are unable to provide an estimate of the strength of the covalent bond produced. A combination of an ionic parameter (Z_i^2/r_i, Z_i/r_i) and the electronegativity χ should therefore be useful in modelling interactions where the ionic content of the bonding is significant, but of little value in modelling strong covalent interactions. In subsequent sections we will consider the use of these parameters to provide correlations which indicate underlying regularities in the behaviour of the elements in the oceans. In considering these correlations it is as well to keep in mind the following principle where environmental parameters are being correlated; usefulness and validity often go hand-in-hand, but they do not always do so.

CHEMICAL SPECIATION AND TOXICITY

The problem of rationalising the chemical speciation of the elements in sea water has been addressed by Ahrland[17] and by Stumm and Brauner[18], and Langmuir[19] has considered the application of simple electrostatic theories to the estimation of stability constants. Here we would like to draw these lines of development together to show how the chemical periodicity of the elements can provide an overview of the chemical speciation of the elements and of the relationships between chemical speciation and biological utilisation.

Hydrolysis

Since the predominant ligand in sea water is H_2O itself any consideration of chemical speciation should begin with hydrolysis[20]. The oxygen in the atmosphere ensures that all but the most electronegative elements (F, Cl, Br, Table 1) are present in positive oxidation states in sea water. Consequently the pattern of hydrolysis can be most easily discerned by the application of Born's electrostatic theory to the interaction of the elemental cations with water molecules. If the extent of hydrolysis at pH 8.2 is expressed in terms of the overall side reaction coefficient[26] ($\overline{\alpha} = [M]_T/[M^{n+}]$) then three distinct bands can be delineated in the Z_i^2/r_i versus N plot[9] (Fig. 2). In the region where $Z_i^2 r_i >$ 23 the elements are fully hydrolysed and do not exist as the free cation even in acidic solutions ($\log \overline{\alpha} \gg 25$). These elements are largely restricted to Groups III to VII of the periodic table (Fig. 2) and are mainly present in their group oxidation states. The nature of the hydrolysis products produced will depend on Z_i^2/r_i and also on the coordination number of the central cation.

These influences can be summarised by a few simple rules [9,10]. In addition to protolytic equilibria, weaker electrostatic interactions with the major seawater cations will also influence the speciation of the fully hydrolysed elements. These interactions have only been investigated in detail for the most abundant oxyanions (HCO_3^-, CO_3^{2-}, $B(OH)_4^-$, SO_4^{2-}, HPO_4^{2-}).

Elements for which $Z_i^2/r_i < 11$ (Fig. 2) are very weakly hydrolysed with $\log \alpha < 1$. The speciation of these elements depends on the extent to which they interact with other ligands in sea water. Within the relatively narrow band $11 < Z_i^2/r_i < 23$ there is a rapid rise in $\bar{\alpha}$ of nearly twenty five orders of magnitude. The elements within this region (Fig. 2) will tend to be strongly hydrolysed in seawater but their speciation will depend not only on pH but also on the stability of the complexes they form with other ligands in solution. Three elements (Hg (II), Bi (III) and Tl (III)) also fall within this category although they have $Z_i^2/r_i < 11$. For these elements the simple electrostatic theory is inadequate and a significant degree of covalent bonding must be invoked.

The term Z_i^2/r_i, derived from the Born theory, therefore provides a simple means for delineating those areas of the periodic table where hydrolysis might be expected to dominate the speciation of the elements in seawater and in other natural waters [9,27]. To provide a rationalisation of the speciation of the elements in the remaining sections of the periodic table we must consider the interactions of these cations with the other ligands in sea water.

Complexation

The nature of the ion-ligand interactions involved may be related to the concept of (a)-type and (b)-type cations developed by Ahrland [17,28,29]. According to this empirical scheme, (a)-type cations are those which bond mainly via electrostatic interactions and which consequently form their strongest complexes with electron donors from the first row of the periodic table, their ligand preferences are in the sequence

$$N >> P > As > Sb \; : \; O >> S > Se > Te \; : \; F >> Cl > Br > I$$

with $F > O > N$. (b)-type cations, which bond mainly via covalent interactions, form their strongest complexes with donors from the second and subsequent rows of the periodic table. Their ligand preferences are in the sequence

$$N << P > As > Sb \; : \; O << S \simeq Se \simeq Te \; : \; F << Cl < Br < I$$

with S>N>O>F. These sequences will also apply when the donor groups are incorporated in organic molecules but structural considerations may cause deviations from the expected order.

Since there is some difficulty in defining a suitable elemental parameter to describe covalent interaction we will introduce an empirical parameter:

$$\Delta\beta = \log\beta^{\theta}_{MF} - \log\beta^{\theta}_{MCl}$$

where β^{θ}_{MF} and β^{θ}_{MCl} are the infinite dilution thermodynamic stability constants for the formation of the mono-fluoro and mono-chloro complexes respectively[9]. For (a)-type cations $\Delta\beta>2$ and for (b)-type cations $\Delta\beta<-2$. The intermediate cations may be further subdivided into borderline-(a) cations ($0<\Delta\beta<2$) and borderline-(b) cations ($0>\Delta\beta>-2$). These four categories form coherent groupings on the periodic table (Fig. 3).

In sea water we are only considering a handful of inorganic ligands (OH^-, CO_3^{2-}, SO_4^{2-}, Cl^-, F^-). Where the cation-ligand interactions are essentially electrostatic in nature we would expect a linear correlation between log β_{MX} and Z_i^2/r_i (see discussion of the ionic model above). Good correlations are observed for 1:1 complexes with F^- and SO_4^{2-} irrespective of the cation type[9]. These ligands may be considered as 'hard' ligands[30] in that they are binding essentially via electrostatic forces to all cations. If a similar plot is prepared for mono-chloro complexes then a linear correlation is again observed for the (a)-type and borderline-(a) cations. The slope of the plot is smaller than for fluoride because of the larger size, and hence weaker electrostatic interactions, of the chloride ion. The stabilities of the mono-chloro complexes of the (b)-type and borderline-(b) cations are, however, much greater than would be expected from electrostatic considerations and the correlation breaks down. The chloride ion is acting as a 'soft' ligand towards these cations and a significant degree of covalency is introduced into the bond. For the weak acid anions (OH^-, CO_3^{2-}) a similar plot reveals a significant correlation for the (a)-type and borderline-(a) cations with slopes comparable to those observed for the 'hard' ligands. Although the information is sparser, it is apparent that the stability constants for the (b)-type and borderline-(b) cations are greater than would be expected from the electrostatic model and again the correlation breaks down. These ligands are therefore 'intermediate' in behaviour in that they act as 'hard' ligands to (a)-type cations and 'soft' ligands to (b)-type cations. Although attempts have been made to provide a quantitative index of ligand softness[28,30-32] no satisfactory scale has been produced and the qualitative terms must suffice.

Speciation

The influence of (a) and (b)-type cations and 'hard' and 'soft' ligands on the complexation of cations in sea water may be summarised by the complexation field (CF) diagram (Fig. 5) where the covalent index ($\Delta\beta$) is plotted against the electrostatic index (Z_i^2/r_i). Elements falling within the same zones on this diagram will exhibit similar solution chemistry and the stability constants of complexes formed with the three classes of ligands show clearly defined trends. Within the CF diagram four general groupings of cations may be discerned. The first (Area I, Fig. 5) consists of the (a)-type cations with a low polarising power (electrostatic energy Z_i^2/r_i, cf Fig. 2) which are so weakly complexed that their speciation is dominated by the free cation. The speciation of (a)-type cations with a high polarising power (Area IIA, Fig. 5) is dominated by hydrolysis and this intense interaction with the water molecules is experienced at progressively lower values of Z_i^2/r_i as the degree of covalency in the M-O bond increased with increasing (b)-character (Area IIB, Fig. 5). For the (b)-type and borderline-(b) cations with a low polarising power the mass action effect and the increasing covalency of the M-Cl bond ensures that chloro-complexes are dominant (Area III,

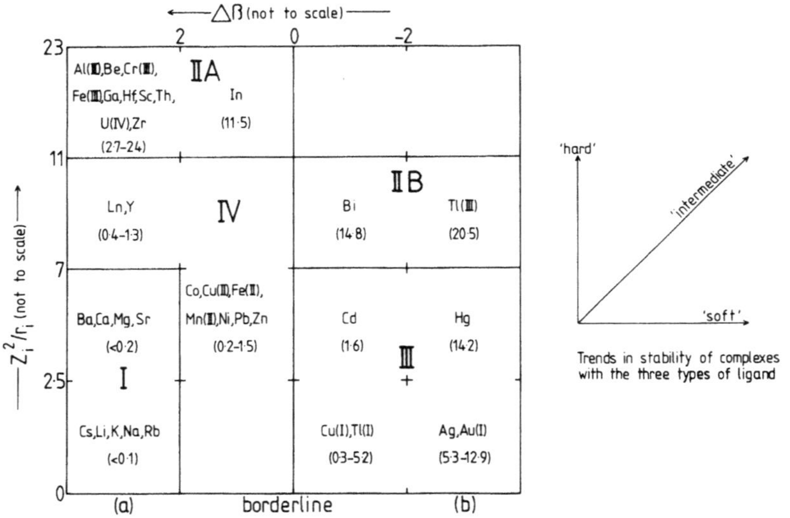

Fig. 5 Complexation field diagram for the cationic elements. The cations are divided into four groups on the horizontal and vertical axes. The figures given below each group of elements show the range of log $\bar{\alpha}_M$ values calculated for a model sea water at pH 8.2. I: very weakly complexed cations, II: hydrolysis-dominated cations, III: chloro-dominated cations, IV: intermediate cations.

Fig. 5). The remaining cations (Area IV, Fig. 5) are mainly (a)-type and borderline-(a) type cations with intermediate polarising power. These elements are able to form stable complexes with a wide variety of ligands and so exhibit complicated speciation patterns.

Although the CF diagram has been prepared from information on the complexes formed with the major sea water anions, similar trends are likely to be observed for other ligands which can be grouped into the same three categories. For example organic compounds that complex mainly via phenolic or carboxylic groups (e.g. humic acids) are likely to behave as intermediate ligands [9,33] (Fig. 5) whereas sulphide ions or organic complexes binding via sulphur sites are likely to behave as soft ligands. This concept has been treated in some detail by Nieboer and Richardson [34] and they provide a more comprehensive summary of the ligand preference of the various cations (Table 3).

(a)-type cations will bind preferentially to oxygen sites in organic molecules such as those provided by carboxylate, carbonyl

Table 3 Classification of ligands encountered in biological systems [34].

I	II	III
F^-, O^{2-}, OH^-, H_2O	NO_2^-, Cl^-, Br^-	I^-, CN^-
CO_3^{2-}, SO_4^{2-}, $ROSO_3^-$, NO_3^-	SO_3^{2-}, NH_3, RNH_2	CO, S^{2-}, RS^-, R_2S, R_3As
HPO_4^{2-}, –O–P(=O)(–O$^-$)–O etc.	R_2NH, R_3N, =N–, –CO–N=R	
ROH, RC(=O)O$^-$, –C(=O)–, ROR	O_2^-, O_2^{2-}	

R represents an organic radical (e.g. methyl or phenyl). The stability of complexes with (b)-type metals increases in the sequence I<II<III. The stability of complexes with the (a)-type metals follows the opposite trend. In general the complexes of (b)-type cations with a particular ligand are stronger than those with (a)-type cations with a similar ionic potential.

and phosphate ester groups. Typical examples are provided by the selective binding of calcium to anionic sites in membrane phospholipids and the specific uptake of monovalent and divalent (a)-type cations by cyclic polyethers (e.g. valinomycin). (a)-type cations will also bond, although less strongly, to nitrogen sites. The weakness of such bonds is typified by the ease with which magnesium can be displaced from the chlorophyll molecule by hydrogen ions and by borderline ions such as Cu^{2+}.

(b)-type cations bind preferentially to sulphur sites and, to a lesser degree, nitrogen sites. A clear example of typical (b)-type behaviour is provided by the preference exhibited by Hg^{2+} ion for disulphide and sulphydryl groups in proteins[35]. As might be expected, the borderline cations are able to form stable complexes with a wide range of ligands and preference for a given donor group will be dictated in part by the degree of (b)-character, in part by the electrostatic term and in part by the steric environment of the reaction site itself.

Cation Toxicity

The different binding preferences of the various cation types will influence their biological functions and these effects can be summarised with reference to the complexation field diagram (Fig. 5). The macro-nutrient elements are all found in Area I which encloses (a)-type cations with a weak polarising power. The fact that these cations are also the dominant sea water components is probably no coincidence. The toxic elements are largely confined to areas II and III and the micronutrient elements to the intermediate grouping (Area IV, Fig. 5). Furthermore it should be noted that the toxicities of the elements for a wide range of organisms[34,36] tend to increase in the sequence.

Area I<Area IV<Area IIA<Area IIB<Area III

or more approximately[34]

(a)-type < borderline < (b)-type

Ochiai[37] suggests that interference by extraneous elements in the functioning of biomolecules can give rise to toxic effects by (i) modifying their active conformation, (ii) blocking essential functional groups and (iii) displacing essential metal ions. Cations from Areas III and IIB bind preferentially with sulphur and, to a lesser degree, nitrogen sites that are important in the functioning and conformation of proteins and enzymes. The bonds in which they participate are stronger than those formed by the other groups of cations with a similar Z_i^2/r_i value, even where these bonds have a significant electrostatic character (i.e. with donors from category I, Table 3). These cations are therefore able to participate in all three of Ochiai's toxicity mechanisms and thus will displace the micronutrient elements (Area IV, Fig. 5, with the exception of the lanthanides) from the active sites in which they are involved. It is likely that for cations from Area IIA (with the exception of indium) toxicity results from the more intense electrostatic interactions that they can exert on the active sites than the cations they displace (e.g. toxic effect of Be^{2+}). The toxic effects exerted by high concentrations of the intermediate cations from Area IV (Fig. 5) probably arise via the second mechanism as (a)-type cations (Area I) are displaced from active centres although for some borderline elements (e.g. Cu) sulphur bonding might also be significant.

Alternative Measures of (b) Character

The general patterns that emerge from the use of the CF diagram are important both chemically and biologically. However, when we come to look at the detailed differences between individual elements, problems arise because of the rather arbitrary nature of the covalent index used. We have used $\Delta\beta$ which was based on the relative stabilities of the fluoro- and chloro-complexes for the following reasons [9]; (i) the differences between the first and second row donors are the most pronounced, (ii) F^- and Cl^- are both important ligands in natural waters, and (iii) a fairly complete set of data is available for fluoro- and chloro-complexes.

While these might be persuasive reasons for using $\Delta\beta$ to clarify speciation patterns in natural waters they can hardly be considered as adequate grounds for using this parameter to study the relative toxicity of the elements. Nieboer and Richardson [34] used the parameter $\chi^2 r_i$ on the basis of an earlier study on the trends observed in the stability of metal complexes [38]. This parameter is loosely correlated with $\Delta\beta$ and produces a very similar subdivision between (a)-type, borderline and (b)-type cations (see reference 9, Fig. 6). Nieboer and Richardson[34] indicate that they 'have been able to demonstrate empirically .... that the index $\chi^2 r_i$ is an estimate of the quotient obtained by dividing the valence orbital energy by the ionic energy'. This, they suggest, provides an estimate of the importance of covalent interactions relative to ionic interactions in the complexation of a metal ion. A number of other approaches based on ionisation potentials or redox potentials have been reviewed [16,28,38,39]. Although they are all in general agreement about the existence of the three metal categories they differ in detail over the allocation of a number of borderline elements. This is essentially a restatement of the problem noted earlier; that electronegativity and related functions relate to the balance between ionic and covalent interac-

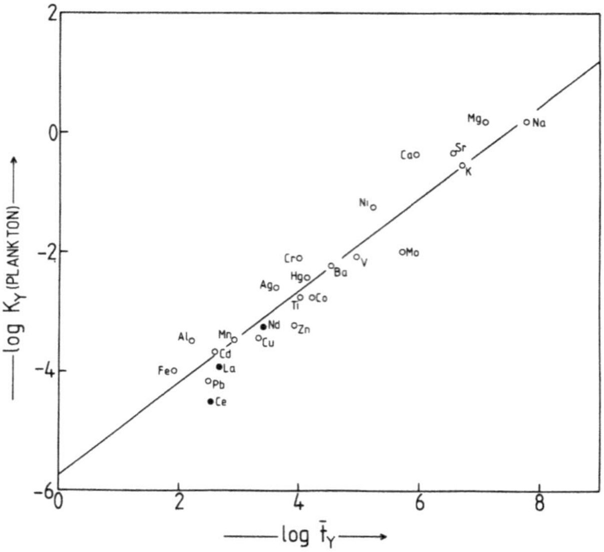

Fig. 6 A plot of the logarithm of the partition function ($\mu g\ kg^{-1}$ seawater/$\mu g\ kg^{-1}$ cell wet weight) for plankton against the logarithm of the mean oceanic residence time. Data for zooplankton[62] (open circles) and total plankton[63] (closed circles).

tions but give little information concerning the intensity of the covalent interactions particularly where they are dominant. These parameters should therefore be used only in a semiquantitative sense to divide the elements into different categories (Fig. 5); it is unwise to place undue emphasis on the exact sequence of elements within each category.

PARTITIONING OF THE ELEMENTS AT THE PARTICLE-WATER INTERFACE

We have already seen in the discussion of the complexation field diagram that there is a clear relationship between the factors governing chemical speciation and those governing the toxicity of the elements. We also noted, in passing, that there was an apparent relationship between the toxicity of the elements and their concentration in seawater. This coincidence has been discussed by Egami[40] and can be dramatically illustrated with reference to the periodic table[41]. Since life evolved in the sea it is reasonable to assume that the evolving metabolic systems would utilise those elements most readily to hand, provided that they could be persuaded to perform the necessary biological functions. To some extent therefore the relationship between biological function and seawater concentration is an unavoidable by-product of the development of life in the oceans since the organisms would need to ensure that they would not suffer a deficiency of essential materials under normal conditions [37]. This raises the question of how the composition of sea water is itself controlled.

Residence Times and Partition Coefficients

The simplest dynamic model of the oceans, on a timescale appropriate to the discussion of the geochemical control of sea water composition, is that of a homogeneous, continuously-stirred tank reactor, or ocean-bucket [42]. Material is added to the reservoir of dissolved solids in the bucket by river flow and, to a lesser extent, by hydrothermal inputs. The elements are removed by incorporation into particles (usually of biological origin) in the water column or by hydrothermal reactions at the active spreading centres along the mid-ocean ridges. If the system is at a steady state with respect to the flux of material then the rate of input should match the rate of removal. In such a system the mean oceanic residence time ($\bar{t}_Y$ = total amount of Y in the oceans/ rate of input or removal) represents the average time that a particular element spends in the ocean. If we assume that the rate of removal of an element Y is directly proportional to the total of Y in the oceans (Y_T) then it is easy to show that

$$\frac{dY}{dt} = - \bar{t}_Y^{\,-1} \cdot Y_T \tag{7}$$

$\bar{t}_Y$ therefore provides a direct measure of the ease with which the elements are removed to the sediment by incorporation into the solid phase either in the water column or at the sea bed. A useful parameter for describing the affinity of a particular element for the solid phase is the partition coefficient (K_Y) which may be described as

$$K_Y = \frac{\text{mean concentration of Y in sea water}}{\text{mean concentration of Y in solid phase}} = \frac{[Y]_{SW}}{[Y]_{\text{solid phase}}} \tag{8}$$

It can be shown [43,44] that for the solid phases responsible for controlling the composition of sea water we should find that at steady state

$$\log K_Y = a_1 + b_1 \log \bar{t}_Y \tag{9}$$

The validity of this relationship has been confirmed for water/crustal rock partition coefficients [45] and for water/sediment partition coefficients [46] (see Table 4). It is likely however that at the present time the main route for the removal of elements from the water column is by inclusion into suspended particulates

Table 4 Fitting parameters for the equation $Y = a_o X + b_o$

Y	X	Elements	n	a_o	b_o	$\lvert r \rvert$	Ref.
Log K_Y (crustal rock)	log $\bar{t}_Y$	a	55	1.55 ± 0.10	-10.05 ± 0.48	0.90	45
Log K_Y (deep sea clay)	log $\bar{t}_Y$	a	56	1.55	-9.83	0.73	46
Log K_Y (faecal pellets)	log $\bar{t}_Y$	a	15	1.20 ± 0.25	-9.2	0.91	43
Log K_Y (phytoplankton)	log $\bar{t}_Y$	a	11	1.19	-8.9	0.95	48
Log K_Y (zooplankton)	log $\bar{t}_Y$	a	21	0.78 ± 0.06	-5.75 ± 0.27	0.95	62
Log K_Y (crustal rock)	ΔH_Y	cationic + halides	31	0.042 ± 0.008	-0.38 ± 0.55	0.69	b
Log K_Y (crustal rock)	$\lvert \Delta Z_i / r_i \rvert$	cationic + halides	47	1.13 ± 0.16	-6.49 ± 0.50	0.72	b
Log K_Y (crustal rock)	Q_{YO}	Na, K, Rb, Cs	4	-3.54 ± 0.84	25.09 ± 6.14	0.95	45
				-4.30 ± 0.76	29.48 ± 5.36	0.97	b
		Li, Mg, Ca, Sr, Ba	5	-1.80 ± 0.55	10.37 ± 3.41	0.89	45
				-1.55 ± 1.27	9.31 ± 8.35	0.57	b
		B, Br, Cl, S	4	-1.55 ± 0.20	3.69 ± 0.25	0.98	45
				-1.01 ± 0.23	3.59 ± 0.42	0.95	b
		main sequence	56	-1.02 ± 0.08	-0.12 ± 0.29	0.88	45
				-0.84 ± 0.10	0.10 ± 0.49	0.74	b
Log $\bar{t}_Y$	ΔH_Y	cationic + halides	26	0.024 ± 0.007	6.12 ± 0.43	0.59	b
Log $\bar{t}_Y$	$\lvert \Delta Z_i / r_i \rvert$	cationic + halides	35	0.75 ± 0.13	2.32 ± 0.46	0.70	b
Log $\bar{t}_Y$	Q_{YO}	Na, K, Rb, Cs	4	-2.10 ± 0.36	22.02 ± 2.62	0.97	45
		Li, Mg, Ca, Sr, Ba	5	-1.68 ± 0.49	16.34 ± 3.06	0.89	45
		B, Br, Cl	3	-0.75 ± 0.14	8.59 ± 0.19	0.98	45
		main sequence	45	-0.40 ± 0.06	5.48 ± 0.24	0.70	45

a All elements for which data are available.
b This paper. Allred-Rochow electronegativities used to calculate Q_{YO}.

which are largely of biological origin. Cherry et al.[43] have provided a neat summary of the implications of an ocean whose composition is controlled by the flux of faecal pellets from the ocean surface into the sediments. They redefine $\bar{t}_Y$ so that

$$\bar{t}_Y = \frac{\text{(Total amount of Y in the oceans)}}{\text{(Flux of Y in faecal pellets into sediment)}}$$

$$= [Y]_{SW}.V/[Y]_{FP}.J_{FP}.A \qquad (10)$$

where J_{FP} is the flux of faecal pellets into the sediment (kg m^{-2} y^{-1}), A is the area of sediment (m^2) and [Y] is the concentration of Y in the faecal material. Since $K_Y(FP) = [Y]_{SW}/[Y]_{FP}$ (i.e. the reciprocal of the 'enrichment factor' normally used to describe biological fractionation) we have, taking logarithms

$$\log K_Y(FP) = \log (A.\ J_{FP}/V) + \log \bar{t}_Y \qquad (11)$$

which is directly analogous to equation (9). Sure enough a linear correlation is observed between log K (FP) and log $\bar{t}_Y$ (Table 4) and a flux of 0.60 x 10^{-3} kg m^{-2} y^{-1} for wet faecal material gives an intercept (9.8) which is tolerably close to the fitted value (7.1 to 9.2 depending on the source data); good linear correlations can also be observed, however, between log $\bar{t}_Y$ and the partition coefficients of the elements between sea water and marine phytoplankton and zooplankton (Figure 6, Table 4), seaweeds[47] and even human blood, rain water and limnetic weeds[48]. The causal relationship between the composition of biogenic particles and the composition of sea water is therefore far from proven.

In an ocean at steady state K_Y and $\bar{t}_Y$ are therefore both expressions of the ease with which the various elements can be incorporated into the solid phase and removed from solution. Consequently it is interesting to try and relate these terms to parameters that might be used to express the intensity of bonding between the elements and the solid phase. According to one hypothesis the uptake of the majority of the elements by particulate matter can be described in terms of a complexation reaction with surface hydroxyl groups since most minerals formed in the oceans have oxygen-dominated lattices or are coated with organic matter which will form complexes via carboxylic and phenolic groups[24,33,49-51]. The problem of defining the processes that control the composition of sea water therefore reduces to one of providing a rationalisation of the intensity of interaction of the various elements with a surface group of the form $\rightarrow \underline{R\,\delta^+}\,O\,\delta^-$.

Electrostatic Models

In the simplest case we are considering a process similar to that illustrated in Figure 1 where the ligand, X, is now replaced by a complexing or bonding site on the solid surface. In such a cycle the hydration energies are often much larger than the complexation energies themselves so that, as a first approximation if we are considering reactions with a common surface site (X) the solid/solution partition coefficient should be related to the free energy of hydration of the corresponding cation (see discussion of equation (4)). Since the entropy of hydration makes a relatively small contribution to the free energy change in most instances, Chang et al.[24] use an equation of the form

$$\log \bar{t}_Y = a_2 + b_2 \log ([Y]_{SW}/[Y]_{RW}) + c_2 \Delta H_Y \qquad (12)$$

where $[Y]_{RW}$ is the global mean concentration of Y in river water and ΔH_Y is the conventional single-ion relative partial molar enthalpy of hydration of the cation Y. From the definition of $\bar{t}_Y$ it can be shown that equation (12) reduces to

$$\log \bar{t}_Y = a_3 + c_3 \Delta H_Y \qquad (13)$$

A similar relationship can be deduced for log K_Y. Plots of these parameters produce broad, but statistically significant, correlations (Table 4) with a rather better fit being observed for log K_Y vs ΔH_Y (Fig. 7A). In general elements with the largest negative enthalpies of hydration tend to favour the solid phase. However the corrrelation is broadened by Group IA and IIA cations and Group VII anions which form separate plots across the main trend. For these groups, the elements that can most easily lose their hydration sheaths are most readily incorporated into the solid phase.

The success of the Born theory in rationalising the heats of hydration of the cationic elements[6] suggests that log K and log $\bar{t}_Y$ should also be correlated with simple electrostatic parameters such as Z_i^2/r_i and Z_i/r_i. Such parameters have indeed been widely used in the geochemical literature to describe the intensity of interaction between a substituted cation and the host lattice (e.g. (Z_i/r_i)[52,53], (Z_i/r_i^2)[54], (Z_i^2/r_i)[55]). Chang et al.[24] have considered the relationship between log $\bar{t}_Y$ and $|\Delta Z_i/r_i|$ (= $|(Z_i/r_i) - 5.5|$, where 5.5 represents the mean value of Z_i/r_i for Al, Fe and Ti). Both log $\bar{t}_Y$ and log K_Y(CR) give broad but statistically significant correlations with $|\Delta Z_i/r_i|$ (see Table 4, Fig. 7B) for the cationic elements. Similar plots for log K_Y have also been reported for crustal rock[56] and for a variety of organisms[57] (bacteria, fungi, pond plants and land animals). These papers also considered cations with very high ionic potentials (e.g. P^{5+}, S^{6+}, Fig. 2) that are fully hydrolysed in solution to produce oxyanions

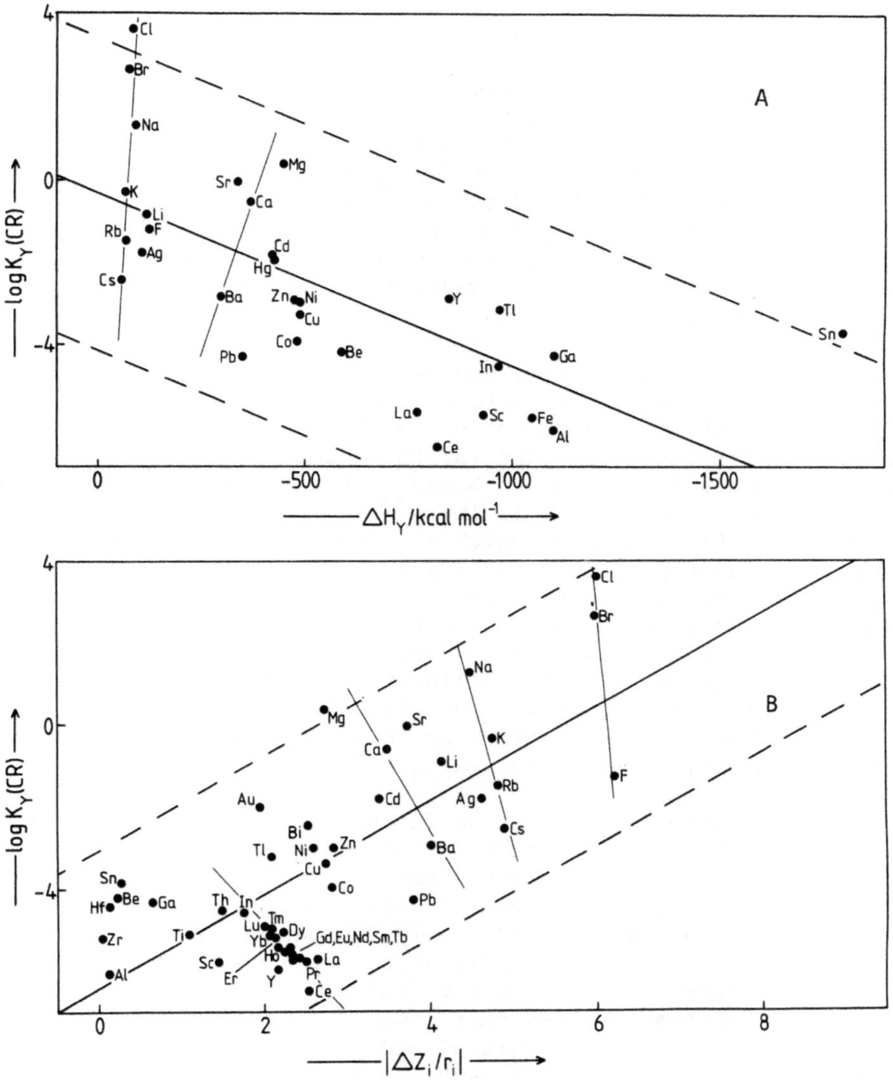

Fig. 7 The correlation of the sea water/crustal rock partition coefficient for the cationic elements against (A) Enthalpy of hydration and (B) relative ionic potential. Data are taken from Table 1. The solid line gives the linear regression fit and the dashed lines encompass a spread of $\pm$ 2Sy,x. The significance of the feint lines plotted across the trend is discussed in the text.

which fall on a separate correlation. When the two correlations are combined a characteristic U-shaped curve is obtained[56,57] with a minimum in log K_Y corresponding to the 'crustal' elements such as Fe, Al and Ti.

Electronegativity Correlations

As an alternative to the simple electrostatic model we can consider the use of electronegativity (χ) to express the relative importance of covalent and ionic contributions to the energy of formation of the Y-O bond. According to Pauling[10] the electrostatic contribution is related to the term Q_{YO} (= $(\chi_Y - \chi_O)^2$). Significant correlations are observed between log K and Q for the sea water/crustal rock[45], river water/crustal rock[45], sea water/deep sea clay[46] and river water/river suspended matter[46] partition coefficients (Table 4).

The correlations observed show more coherent behaviour than the simple electrostatic plots (e.g. ref. 46, Figures 1 and 5). In addition to the main sequence of 56 elements there are separate correlations for the alkali and for the alkaline earth metal cations which are less strongly bound to the oxygen sites than their electronegativities would imply[55] and for the excess volatile elements (Cl, Br, S, B) which are probably enriched relative to the main sequence by selective distillation into the ocean by volcanic activity[58]. Plots of log K_Y vs Q_{YO} for biogenic particles are inconclusive because the data rarely cover more than twenty elements and the scatter is too great to observe any significant correlations. A set of linear relationships between log $\bar{t}_Y$ and Q_{YO} has been used to provide a rationalisation of the mean oceanic residence time concept[59] and to give a useful explanation of the periodicity of the observed $\bar{t}_Y$ values[60,61].

The correlations described so far have used Pauling electronegativities taken from a variety of sources[45]. There are however some differences in the literature between the electronegativity values ascribed to the different oxidation states of the elements. If we restrict the correlations solely to thermochemical (or Pauling) electronegativities applicable to the stable oxidation states of the elements in sea water (Table 1) we obtain plots which, although they have somewhat fewer data points, are still statistically significant (Table 5) and still show the same characteristic groupings of the elements. If we look at the relationship between the Pauling electronegativities[10] and the electrostatically derived values of Allred and Rochow[11] (Fig. 4) we find that a number of elements fall significantly above the linear regression fit. Consequently for the main sequence of elements (sixty in all) the fit for log K_Y relative to Allred-Rochow electronegativities is not as good as that relative to Pauling electronegativities although it does remain statistically highly significant (Tables 4 & 5). It should be emphasised that log K_Y and log $\bar{t}_Y$ also give significant correlations with the same general patterns when plotted against χ_Y or χ_Y^2 on both electronegativity scales (Table 5) so that some caution should be exercised when ascribing mechanistic interpretations to the Q_{YO} plots.

Table 5 Summary of electronegativity-based regressions[a] for log $\overline{t_Y}$ and crustal rock partition coefficients.

Y-axis	X-axis	Pauling electronegativity							Allred-Rochow electronegativity						
		IA/IIA, n=9			main sequence				IA/IIA, n=9			main sequence			
		|r|	F	Sy,x	n	|r|	F	Sy,x	|r|	F	Sy,x	n	|r|	F	Sy,x
log $\overline{t_y}$	Q_{YO}	0.25	0.5	1.13	37	0.71	34.8	0.97	0.32	0.8	1.1	48	0.65	34	1.01
	χ^2	0.26	0.5	1.12	37	0.71	36	0.96	0.33	0.8	1.11	48	0.56	21	1.11
	χ	0.25	0.5	1.13	37	0.74	42	0.92	0.32	0.8	1.11	48	0.62	29	1.05
log K (CR)	Q_{YO}	0.49	2.2	1.25	45	0.85	111	1.26	0.59	3.7	1.16	60	0.80	105	1.40
	χ^2	0.46	1.9	1.27	45	0.76	59	1.55	0.57	3.4	1.18	60	0.63	39	1.81
	χ	0.48	2.1	1.26	45	0.83	94	1.34	0.58	3.6	1.16	60	0.73	66	1.60
log K (RW/CR)[b]	Q_{YO}	0.63	4.8	0.49	37	0.86	97	0.75	0.73	7.8	0.44	48	0.82	96	0.82
	χ^2	0.55	3.1	0.53	37	0.69	31	1.05	0.68	6.1	0.47	48	0.60	26	1.16
	χ	0.61	4.2	0.51	37	0.78	54	0.91	0.71	7.3	0.45	48	0.71	48	1.01

[a] For the purpose of comparing the goodness of fit for different electronegativity scales and different electronegativity parameters the group IA and IIA elements have been treated as one group and the excess volatile elements have been incorporated into the main sequence. Electronegativity values are listed in Table 1 and K_Y values calculated from the data given in reference 45. r = correlation coefficient, F = F- ratio, Sy,x = std. error of estimate of y on x.

[b] River water/crustal rock partition coefficient.

Composite Models

It is reasonable to suppose that just as in the discussion of chemical speciation we really require a combination of an electrostatic parameter (Z_i^2/r_i, Z_i/r_i or Z_i/r_i^2) and a parameter representing the balance between covalent and ionic interactions (e.g. χ_Y or Q_{YO}) to provide an adequate rationalisation of the chemical behaviour of the elements. This approach has been developed in a series of papers by Chang and his co-workers[24,49] where the concepts of electronegativity (χ) and ionic potential (Z_i/r_i) have been combined with a nuclear screening loss correction ($R_§$, Table 1) to provide a series of useful relationships that enable heats of hydration, hydroxide solubilities and the mean oceanic residence times each to be represented by a single correlation[24]. They rationalise their selection of parameters by suggesting that their effective ionic potential (Z_i/r_i corrected by $R_§$) represents the electrostatic potential experienced by a valence electron at the surface of the ion while their electronegativity term (χ) incorporates the influence of electrostatic potential energy, exchange energy and correlation energy[49].

For the mean oceanic residence time they use an equation of the form,

$$\log \bar{t}_Y = a_4 + b_4 \log [Y]_{SW} + c_4 \left|\Delta(Z_i/r_i)_Y\right| + d_4 (\chi_Y - \chi_{Me}) + e_4 R_§ \qquad (14)$$

where χ_{Me} is the mean of the electronegativities of aluminium, iron and titanium and is given the value 1.55. They state that the values of $(Z_i/r_i)_Y$ and χ_Y used must be appropriate to the oxidation state and chemical form of the element (for example for Si, As and W the radius of the appropriate anion should be used) but give no sources for the appropriate values of r_i and χ_Y.

The authors give values for the parameters a_4 - e_4 in equation (14) (see Table 6) but they given no indication of the procedures used to fit the data nor of the goodness of fit. We have therefore refitted equation (14) to the data given in Table 1 for the cationic elements and the halides using a stepwise multiple linear regression routine. The full equation provides a good description of the mean oceanic residence times of the elements (Fig. 8) but this is mainly due to the inclusion of the term log $[Y]_{SW}$ which is strongly covariant with log $\bar{t}_Y$. In the fit to equation (14) only a_4 and b_4 are significant at the 99% level (Table 6). Coefficients c_4 and e_4 are significant at the 95% level but d_4 (representing the electronegativity contribution) is not significant. If the term in log $[Y]_{SW}$ is omitted the resulting correlation is much broader (see Table 6) and d_4 is still not significant. The contradiction between this finding and the

Table 6 Stepwise multiple linear regression fit to Chang's multiparametric equation (equation 14).

Ref.	a_4	b_4	c_4	d_4	e_4	|r|	SE
24	1.5	0.2	1.1	-1.0	1.0	-	-
Recalculated	2.32	0.0	0.74	0.0	0.0	0.70	1.18
using Pauling	±0.46		±0.13				
electronega-	2.28	0.0	0.76	-0.074	0.0	0.70	1.20
tivities [a]	±0.49		±0.15	±0.31			
	2.61	0.0	0.72	0.13	-2.61	0.74	1.15
	±0.50		±0.14	±0.32	±1.30		
	4.14	0.34	0.22	-0.05	-2.00		
	±0.34	±0.04	±0.10	±0.18	±0.74	0.93	0.65
Recalculated	2.38	0.0	0.73	0.0	0.0	0.72	1.14
using Allred-	±0.39		±0.12				
Rochow electro-	2.40	0.0	0.73	0.27	0.0	0.72	1.15
negativities [a]	±0.44		±0.13	±0.32			
	2.64	0.0	0.71	0.15	-2.55	0.76	1.10
	±0.44		±0.12	±0.31	±1.19		
	3.80	0.32	0.30	-0.04	-1.79	0.92	0.66
	±0.30	±0.04	±0.09	±0.19	±0.72		

a Using data from Table 1.

success of the electronegativity correlations discussed earlier is more apparent than real since there is a significant covariance between Z_i/r_i and χ as we noted earlier (cf Figures 2 and 3). Further developments of this approach must await a more detailed model of the chemical cycle involved (Fig. 1) that will hopefully provide some guidance as to the form and relative importance of the various terms to be used in such multiparametric equations.

CONCLUSIONS

Although each element has its own idiosyncracies many of the trends in chemical behaviour can be rationalised by reference to the periodic table. The study of chemistry is the study of chemical bonding and any summary of the chemical periodicity of the elements must revolve around parameters that are related to the intensity of electrostatic and of covalent bond formation. A number of parameters (e.g. Z_i/r_i, Z_i^2/r_i) can be derived which reflect periodic changes in the intensity of electrostatic bonding

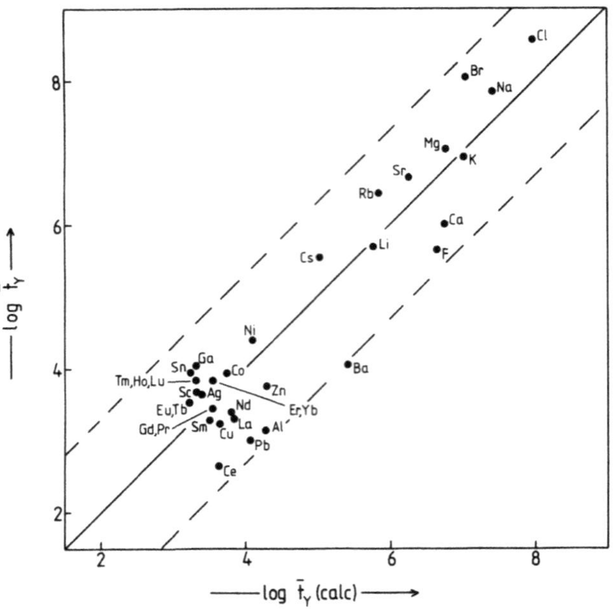

Fig. 8 Comparison of the observed mean-oceanic residence times[45] with values calculated from the full multiparametric equation of Chang et al. (equation 14). The parameters of the fit are given in Table 6. The solid line gives the multiple linear regression fit and the dashed lines encompass a spread of $\pm$ 2SE.

between the elements. When it is no longer adequate to treat the elements as rigid charged spheres some allowance must be made for the degree of covalency or electron sharing exhibited by chemical bonds. The electronegativity term can be used to assess the degree of covalency likely to be encountered in a particular bond but at present there is no satisfactory index of the <u>intensity</u> of covalent bonding that may be experienced. Bearing this limitation in mind we can use the chemical periodicity of the elements, as reflected in terms such as the ionic potential Z_i/r_i, ionic electrostatic energy (Z_i^2/r_i) and the electronegativity, to provide some rationalisation of the behaviour and occurrence of the elements in the marine environment. Despite the incomplete and frequently unreliable data to hand, a general pattern of behaviour is discernible in the chemical speciation of the elements that is also reflected in their biological functions. Furthermore the partitioning of the elements between the solid phase and sea water, and consequently the residence times of the elements in the oceans, can be related in a systematic, if somewhat scattered fashion to simple periodic functions. It is possible that similar trends may be exhibited by biogenic particles but it is not yet possible to isolate any significant trends from the statistical noise.

As our knowledge of the speciation and distribution of the elements improves, the rough correlations and summaries presented here are likely to be refined and made more specific. To ensure that new data are used to their greatest effect it will be necessary to look more closely at the various combinations of parameters that have been used and to attempt to define an integrated set of parameters that can provide a common explanation for the observed trends in the behaviour of the elements in the marine environment.

REFERENCES

1. Sanderson, R.T., 1967: "Inorganic Chemistry". Reinhold Publishing Corporation, New York.
2. Sanderson, R.T., 1960: "Chemical Periodicity". Reinhold Publishing Corporation, New York.
3. Rich, R., 1965: "Periodic Correlations". W.A. Benjamin, Inc., New York.
4. Purcell, K.F. and J.C. Kotz, 1977: "Inorganic Chemistry". W.B. Saunders, Philadelphia.
5. Puddephatt, R.J., 1972: "The Periodic Table of the Elements". Oxford University Press.
6. Phillips, C.S.G. and R.J.P. Williams, 1965: "Inorganic Chemistry". Vol. 1, Clarendon Press, Oxford.
7. Bockris, J. O'M. and A.K.N. Reddy, 1970: "Modern Electrochemistry". Vol. 1, pp. 49-59.
8. Hefter, G., 1974: Simple electrostatic correlations of fluoride complexes in aqueous solution. Coord. Chem. Rev., 12, 221-239.
9. Turner, D.R., M. Whitfield and A.G. Dickson, 1981: The equilibrium speciation of dissolved components in freshwater and seawater at 25°C and 1 atmosphere pressure. Geochim. Cosmochim. Acta., 45, 855-881.
10. Pauling, L., 1960: "The Nature of the Chemical Bond". University Press, Cornell.
11. Allred, A.L. and E. Rochow, 1958: A scale of electronegativity based on electrostatic force. J. Inorg. Nucl. Chem., 5, 264-268.
12. Slater, J.C., 1930: Atomic shielding constants. Phys. Rev., 36, 57-64.
13. Little, E.J. and M.M. Jones, 1960: A complete table of electronegativities. J. Chem. Educ., 37, 231-233.
14. Gordy, W. and W.J.O. Thomas, 1956: Electronegativities of the elements. J. Chem. Phys., 24, 439-444.
15. Allred, A.L., 1961: Electronegativity values from thermochemical data. J. Inorg. Nucl. Chem., 17, 215-221.
16. Williams, A.F., 1979: "A Theoretical Approach to Inorganic Chemistry". Springer-Verlag, Berlin.

17. Ahrland, S., 1975: Metal complexes in sea water. In: "The Nature of Seawater", E.D. Goldberg, ed. Dahlem Konferenzen, Berlin.
18. Stumm, W. and P.A. Brauner, 1975: Chemical speciation. In: "Chemical Oceanography", J.P. Riley and G. Skirrow, eds. Vol. 1, Academic Press, London.
19. Langmuir, D., 1979: Techniques for estimating thermodynamic properties for some aqueous complexes of geochemical origin. In: "Chemical Modelling in Aqueous Systems", E.A. Jenne, ed. ACS Symposium Series 93, Am. Chem. Soc., Washington, D.C.
20. Baes, C.F. and R.E. Mesmer, 1976: "The Hydrolysis of Cations". Wiley, Interscience, New York.
21. Huheey, J.E., 1978: "Inorganic Chemistry". Harper and Row, New York.
22. Shannon, R.D. and C.T. Prewitt, 1969: Effective ionic radii in oxides and fluorides. Acta. Cyrstallogr., B25, 925-946.
23. Shannon, R.D. and C.T. Prewitt, 1969: Revised values of effective ionic radii. Acta. Crystallogr., B26, 1046-1048.
24. Chang, C-P., L-S Liu and N. Chen, 1979: A $\emptyset(Z/1,\chi)$ rule for chemical processes in oceans and its application. VII The transport of elements in oceans and the screening loss parameter. Oceanolog. Limnolog. Sinica, 10, 214-229, (in Chinese).
25. Basolo, F. and R.G. Pearson, 1969: "Mechanisms of Inorganic Reactions". J. Wiley, New York.
26. Ringbom, A., 1963: "Complexation in Analytical Chemistry". Interscience, New York.
27. Whitfield, M., D.R. Turner and A.G. Dickson, 1981: Speciation of dissolved constituents in estuaries. In: Proc. SCOR Workshop "River Inputs in Ocean Systems". UNESCO, Paris, 132-148.
28. Ahrland, S., 1968: Thermodynamics of complex formation between hard and soft acceptors and donors. Struct. and Bonding (Berlin), 5, 118-145.
29. Ahrland, S., J. Chatt and N.R. Davies, 1958: The relative affinities of ligand atoms for acceptor molelcules and ions. Quart. Rev. Chem. Soc., 12, 265-276.
30. Pearson, R.G.: Hard and soft acids and bases. Surv. Progr. Chem., 5, 1-52.
31. Klopman, G., 1968: Chemical reactivity and the concept of charge- and frontier-controlled reaction. J. Am. Chem. Soc., 90, 223-234.
32. Williams, R.J.P. and J.D. Hale, 1966: The classification of acceptors and donors in inorganic reactions. Struct. and Bonding (Berlin), 1, 249-281.
33. Balistrieri, L., P.G. Brewer and J.W. Murray, 1981: Scavenging residence times of trace metals and surface

chemistry of sinking particles in the deep ocean. Deep-Sea Res., 28A, 101-121.
34. Nieboer, E. and D.H.S. Richardson, 1980: The replacement of the nondescript term 'heavy metals' by a biologically and chemically significant classification of metal ions. Environ. Poll., (Ser. B), 1, 3-26.
35. Vallee, B.L. and D.D. Ulmer, 1972: Biochemical effects of mercury, cadmium and lead. Ann. Rev. Biochem., 41, 91-128.
36. Bowen, H.J.M., 1979: "Environmental Chemistry of the Elements". Academic Press, London.
37. Ochiai, E., 1977: "Bioinorganic Chemistry: an Introduction". Allyn and Bacon, Boston.
38. Nieboer, E. and W.A.E. McBryde, 1973: Free energy relationships in coordination chemistry. III A comprehensive index to complex stability. Can. J. Chem., 51, 2512-2524.
39. Jørgensen, C.K., 1975: Continuum effects indicated by hard and soft anti-bases (Lewis acids) and bases. Topics in Current Chemistry , 56, 1-66.
40. Egami, F.: Minor elements and evolution. J. Mol. Evol., 4, 113-120.
41. Whitfield, M., 1981: World ocean - mechanism or machination? Interdisc. Sci. Rev., 6, 12-35.
42. Whitfield, M., 1976: The evolution of the oceans and the atmosphere. In: "The Environmental Physiology of Animals", J. Bligh, J.L. Cloudsley-Thompson and A.G. McDonald, eds. Blackwells, Oxford.
43. Cherry, R.D., J.J. Higgo and S.W. Fowler, 1978: Zooplankton faecal pellets and element residence times in the ocean. Nature (London), 274, 246-248.
44. Whitfield, M. and D.R. Turner, 1979: Water-rock partition coefficients and the composition of seawater and river water. Nature (London), 278, 132-137.
45. Turner, D.R., A.G. Dickson and M. Whitfield, 1980: Water-rock partition coefficients and the composition of natural waters - a reassessment. Marine Chem., 9, 211-218.
46. Martin, J.M. and M. Whitfield, this volume: The significance of river inputs of chemical elements to the ocean system.
47. Yamamoto, T., 1972: Chemical studies on sea weeds, 27. The relations between concentration factors in seaweeds and residence time of some elements in seawater. Rec. Oceanogr. Works Japan, 11, 65-72.
48. Yamamoto, T., Y. Otsuka, M. Okazaki and K. Okamoto, 1980: A method of data analysis on the distribution of chemical elements in the biosphere. In: "Analytical Techniques in Environmental Chemistry", J. Albaiges, ed. 401-408, Pergamon Press, Oxford.

49. Chang, C-P. and L-S. Liu, 1978: A Ø(Z/1,X) rule of inorganic ion-exchange reactions in seawater and its applications. Oceanic Selections, 1, 52-78, (in Chinese).
50. Schindler, P.W., 1975: Removal of trace metals from oceans: a zero order model. Thalassia Jugoslavica, 11, 101-111.
51. Stumm, W., H. Hohl and F. Dalang, 1976: Interaction of metal ions with hydrous oxide surfaces. Croatica Chimica Acta., 48, 491-500.
52. Goldschmidt, V.M., 1937: The principles of distribution of chemical elements in minerals and rocks. J. Chem. Soc., (London), 1937, 655-673.
53. Goldschmidt, V.M., 1954: "Geochemistry". Clarendon Press, Oxford.
54. Ahrens, L.H., 1953: The use of ionisation potentials. Part 2. Anion affinity and geochemistry. Geochim. Cosmochim. Acta., 3, 1-22.
55. Turner, D.R. and M. Whitfield, 1979: Control of seawater composition. Nature (London), 281, 468-469.
56. Ishibashi, M., 1969: Distribution and regular relationships of quantities of chemical elements in sea water, including so-called nutrient elements. In: "Morning Review Lectures, 2nd International Oceanographic Congress. Moscow, 1966". UNESCO, Paris.
57. Banin, A. and J. Navrot, 1975: Origin of life: clues from relations between chemical compositions of living organisms and natural environments. Science, N.Y., 189, 550-551.
58. Rubey, W.W., 1951: Geologic history of seawater. An attempt to state the problem. Geol. Soc. Am. Bull., 62, 1111-1148.
59. Whitfield, M., 1979: The mean oceanic residence time (MORT) concept - a rationalisation. Marine Chem., 8, 101-123.
60. Brewer, P.G., 1975: Minor elements in seawater. In: "Chemical Oceanography", J.P. Riley and G. Skirrow, eds. Vol. 1, Academic Press, London.
61. Holland, H.D., 1978: "The chemistry of the atmosphere and oceans". J. Wiley, New York.
62. Martin, J.H. and G.A. Knauer, 1972: A comparison of inshore vs offshore levels of 21 trace and major elements in marine phytoplankton. In: "Baseline studies of pollutants in the marine environment", E.D. Goldberg, ed. Brookhaven National Laboratory.
63. Elderfield, H., C.J. Hawkesworth, M.J. Greaves and S.E. Calvert, 1981: Rare earth element geochemistry of oceanic ferromanganese nodules and associated sediments. Geochim. Cosmochim. Acta., 45, 513-528.

POTENTIOMETRIC AND CONFORMATIONAL STUDIES OF THE ACID-BASE PROPERTIES OF FULVIC ACID FROM NATURAL WATERS

M.S. Varney*+#, R.F.C. Mantoura+, M. Whitfield#, D.R. Turner#, and J.P. Riley*

*Department of Oceanography, University of Liverpool Liverpool L69 3BX, U.K.
+NERC Institute for Marine Environmental Research Prospect Place, The Hoe, Plymouth, PL1 3DH, U.K.
#Marine Biological Association, U.K., The Laboratory Citadel Hill, Plymouth, PL1 2PB, U.K.

ABSTRACT

Gram quantities of reference fulvic acid (FA) have been extracted from natural waters for a systematic physico-chemical investigation of proton and metal binding properties. Electrophoretic and gel filtration chromatographic studies of FA indicate that during the course of an acid-base titration, marked conformational changes occurred with the molecular radius increasing from 0.65 nm (6.5Å) at pH 1.15 to 1.32 nm (13.2 Å) at pH 9.26. These conformational charges arise from intra-molecular electrostatic replusive forces associated with the build up of charge (Z) on the flexible FA polyanion. This in turn has a marked effect on the acid association constant (k). For this reason, the broad polydisperse titration curves which are characteristic of FA could not be adequately explained by simple acid-base models which assume single values of k. However, by incorporating both electrostatic and conformational terms into an expanded Tanford model, we obtained a good fit to the titration data. The value of the intrinsic association constant k_i (when Z = 0) derived from this model was rather acidic (log k_i = 2.3-2.7) suggesting that the FA is strongly electron withdrawing. However, the combined effects of electrostatic and conformational changes is to spread the apparent constants over a wide range with the greatest number of sites having apparent log k = 4-5.

INTRODUCTION

Recent simulations of trace metal speciation in natural waters have succeeded in including metal-organic interactions[1,2] but have not taken into account the competitive effects of proton binding. The acid base properties of dissolved humic compounds which constitute a major proportion of the organic material in most natural waters are important not only in influencing speciation of trace metals but also in effecting conformational changes, rates of binding reactions, surface activity, adsorption and solubilization reactions[3]. Even the pH of some organically-rich fresh waters are apparently controlled by the acid-base properties of humic compounds[29]. Although many papers have been published on the chemical properties of the humic components of natural water[3-6] the operational nature of their extraction, purification, etc., has meant that results from various reports cannot necessarily be used to complement one another. The unavailability of large quantities of reference fulvic material has further exacerbated the situation. The use of conventional analytical techniques together with the non-ideal, polydisperse and macromolecular nature of these ligands, have also contributed to the difficulties of obtaining a rigorous physico-chemical treatment of these complex binding reactions.

Gram quantities of fulvic acid have been isolated from river waters to enable an extensive investigation to be made into acid-base, metal-binding and electrochemical properties in order to better understand the physico-chemical basis for the measurement of metal complexing capacities in natural waters. In addition, conformational and structural studies have been used to supplement these experiments to assist in the selection of the most suitable binding model for the experimental data.

MATERIAL AND METHODS

Fulvic acid (FA) was extracted from 1400 l of River Tamar water by hydrophobic adsorption[7] onto column of Amberlite XAD-7. The eluted FA was desalted and rendered trace metal-free and finally converted to the protonated form by repeated passage of the FA solutions through columns of Chelex-100 (in the hydrogen and ammonium ion forms). The purified FA solutions were then lyophilized and stored dry over P_2O_5 *in vacuo*. Final yield of FA was 2.46 g.

The elemental composition of the FA (C, 41.4%; H, 5.1%; N, 2.8%; O, 49.5%; Ash, 1.18%) was similar to those reported by others[8,9] for freshwater FA. The trace metal content of the FA was determined by wet ashing with Aristar HNO_3 and analysis by atomic absorption spectrophotometry (Cu 6 nmol g^{-1}; Zn 4 nmol g^{-1};

Fe 8 nmol g^{-1}; Al 55 nmol g^{-1}, and Pb and Cd below detection limit of 0.5 and 0.1 nmol g^{-1}, respectively). These levels of residual trace metals are sufficiently low to be unlikely to interfere with the potentiometric and polarographic studies of metal binding reactions. Although the ultra violet-visible spectra were monotonic and generally featureless, their extinction coefficients were highly pH dependent. The infra red spectra, obtained using the KBr disc method showed the typical FA absorption bands at 1720 cm^{-1} and 1610 cm^{-1} which are characteristic of carboxyl and aromatic groups[10]. Preliminary mass spectrometric studies using Curie point pyrolysis were kindly carried out by Dr. M. Kütter, Edgenossische Technische Hochschule, Zurich. An abundance of acetic acid and monocyclic aromatic residues, characteristic of soil derived FA[11] were the prominent fragmentation feature of the mass spectra. The absence of methoxyphenols and polysaccharide derived furan rings is indicative of advanced stages in the humification processes.

CONFORMATIONAL STUDIES

Gel filtration chromatography was performed on 1.6 x 40 cm columns of Sephadex G-10, G-25 and G-75 (Pharmacia products, Uppsala, Sweden) eluted with buffers at pH 1.15, 2.15, 4.00, 6.89, 7.44 and 9.26 and at constant ionic strength of 0.1. Nominal molecular weights were determined by the procedure described by Determann . Electrophoretic separations of FA were carried out on an LKB apparatus using 5 x 15 cm cellulose paper strips at 200 V, 10 mA for 30 minutes. The resultant separations at various pH values (1.0, 2.0, 3.3, 4.0) were examined visually under UV light to assess electrophoretic mobility.

POTENTIOMETRY

All solutions were prepared in CO_2-free distilled water using Aristar grade reagents. Because of possible complexing of K^+ by fulvate anions[30], 0.1 M sodium perchlorate was used as the inert background electrolyte in the potentiometric titrations. Solutions of FA were prepared in CO_2-free 0.1 M $NaClO_4$ and allowed to equilibrate for a fixed period prior to titration. Reproducible results could be obtained only when the preparation of FA solutions was carefully replicated. All glassware was washed in chromic acid and rinsed with distilled water.

EQUIPMENT

The potentiometric titrations were performed using the automatic titration apparatus ("Automat") shown schematically in Fig. 1.

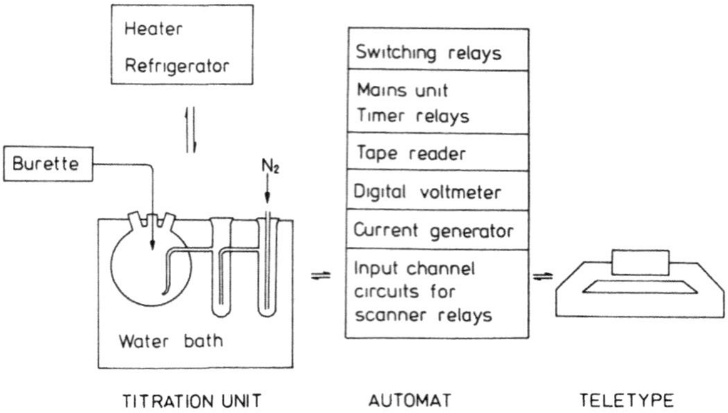

Fig. 1 Schematic diagram for the automated potentiometric titration system.

The essential features of this assembly are the Data Precision mV meter, the electrode scanner capable of switching inputs for up to six high input impedance electrodes and six low impedance electrodes, the paper tape control for operation of the scanner and other relay operations (burette additions, stirrer, on/off and so on), water bath temperature control ($\pm$ 0.02°C), and the N_2 gas supply. The Automat outputs results onto a hardcopy printer and if required, to punched paper tape for transfer to a PDP 11/34 computer. The titrant could be delivered in preset or variable increments from a 10 ml Metrohm autoburette. The titration cell (200 ml) was fitted with a N_2 inlet and outlet burette, two glass electrodes (Corning Triple purpose electrodes) and two double junction calomel reference electrodes (Orion 90-02-00). The design of the Automat is such that cross readings between all electrode pairs could be compared in order to monitor electrode drift or failure. In order to avoid the ingress of CO_2 as well as the characteristic foaming of these surface active FA solutions, a N_2 blanket was maintained throughout the titration. A PTFE coated magnetic stirring bar was used for stirring.

PROCEDURE

Prior to and following each FA titration, the electrode assembly was calibrated in 0.1 M $NaClO_4$ by titrating, in the same cell, standard solutions of 0.1000 M HCl prepared from 'Calibrated Volumetric Standard' (BDH) stock with sodium hydroxide titrant. By utilizing the titration data obtained in the region pH 3-4 the constants for the various electrode combinations were calculated where each electrode potential, E, is expressed as

$$E = E^{o} + E_{j} + RTF^{-1} \ln [H^{+}] \quad (1)$$

Where E^{o} is the standard electrode potential, E_j is the potential due to the liquid junction, and $[H^+]$ is the hydrogen ion concentration expressed on a constant ionic medium scale. However, the value of E_j is dependent on the hydrogen ion concentration[12] and thus equation (1) can be rewritten as:

$$E = E^{o'} + E_{j}^{'} + RTF^{-1} \ln[H^{+}]$$

where $E^{o'} = E^{o}$ + constant part of E_j

and $E_{j}^{'} = j_{H}[H^{+}] + j_{OH}[OH^{-}]$

The values of the proportionality terms, j_H and j_{OH}, depend only on the ionic medium and the particular design of reference electrode used. The value of j_H, can be obtained from a plot of $(E - RTF^{-1} \ln [H^+])$ against $[H^+]$, and then the constant value of

$$E^{o'} = E - RTF^{-1} \ln [H^{+}] - j_{H} [H^{+}]$$

can be determined for any set of measurements. The calibration titrations, which took ∿1 hour to perform could also be used to estimate the amount of protolytic impurities in the background medium. The titration procedure for FA consisted of preparation of 100 mg FA dissolved in 100 ml of 0.1 M $NaClO_4$ equilibrated at $24.68 \pm 0.02°C$ for 30 min. prior to the addition of titrant. This was found to be particularly important for reproducible titrations. Both forward and reverse titrations of FA (comprising 70-80 points) were performed away from direct sunlight, and took ∿3 hours. Following the addition of titrant, the electrodes were allowed to equilibrate for 2.5 minutes after which the volume and electrode potentials were automatically recorded. No coagulation of FA was observed throughout the titrations.

RESULTS AND DISCUSSION

Gel filtration chromatography

The conformational properties of FA were investigated by gel filtration chromatography (GFC) on columns of Sephadex G-10, G-25 and G-75 elution being carried out with constant ionic strength buffers (pH 1.15, 2.15, 4.00, 6.89, 7.44 and 9.26). The molecular weight polydispersity of the FA was determined by integrating the absorbance (λ = 254 nm) chromatograms in terms of nominal molecular weights. The results are listed in Table 1 with examples of the G-25 GFC at selected pH values shown in Figure 2. However,

Table 1 Gel filtration chromatography of fulvic acid at various pH's; ionic strength = 0.1.

Sephadex molecular weight ranges M	Equivalent Stokes' radii[1] R(Å)	Mean $\bar{R}$ (Å)	Percentage distribution (D) pH 1.15	pH 2.15	pH 4.00	pH 6.89	pH 7.44	pH 9.26
<700	<6.50	5.8[2]	79	70	21	18	0	0
700-5,000	6.50-12.52	9.5	21	30	72	68	82	49
5,000-10,000	12.52-15.77	14.2	0	0	7	14	16	36
10,000-70,000	15.77-30.14	23.0	0	0	0	0	2	15
	Weight average[3] radii (R_w) in Å		6.58	6.91	9.06	9.49	10.52	13.21

[1]Equation (2)
[2]Since molecular weights < 400 have rarely been reported,[5,18], it is reasonable to assume that M < 700 corresponds to M ∿ 500, therefore $\bar{R}$ = 5.8 Å, 0.58 nm.
[3]From $R_w = \frac{1}{100} \sum_i \left[D_i\ (\%) \bar{R}_i \right]$

since the mechanism of GFC is based on the steric exclusion principle (i.e. molecular size) rather than on molecular weights[13,14] it was necessary to convert the reported molecular weight ranges and exclusion limits of the gels which are based on spherical proteins, into their equivalent Stokes' radii by the equation:

$$\frac{4}{3} \pi R^3 = \frac{M}{N} (\bar{\upsilon}_2 + \delta_1 \upsilon_1^0) \qquad (2)$$

where R, M, $\bar{\upsilon}_2$ and δ_1 are the Stokes' radius, molecular weights, partial specific volume and degree of solvation of the protein macromolecule respectively, N is Avogadro's number and υ_1^0 is the partial specific volume of the solvent (H_2O, $\upsilon_1^0 = 1.0$). Using an averaged value of $\bar{\upsilon}_2 = 0.730$ for a range of proteins[14], $\upsilon_1^0 = 1.0$ and $\delta_1 = 0.2$[15], equation (2) was then used to estimate the equivalent values of R for the various gels and the resultant weight average Stokes' radii (R_w) of the polydisperse FA at various pH's (see Table 1). It is clear from both Table 1 and the plot of R_w vs pH in Figure 3, that there are marked configurational changes resulting from changes in pH. R_w increases approximately linearly

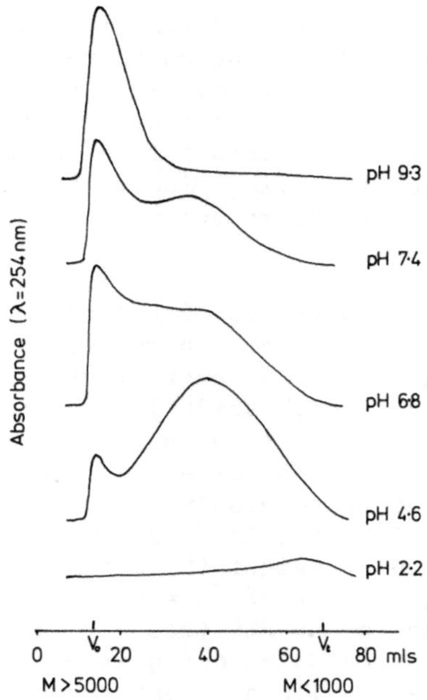

Fig. 2 Gel filtration chromatograms of FA on Sephadex G-25 at various pH (I = 0.10). Protein calibrants of M > 5,000 elute at V_o, whereas those of M < 1,000 emerge at V_t.

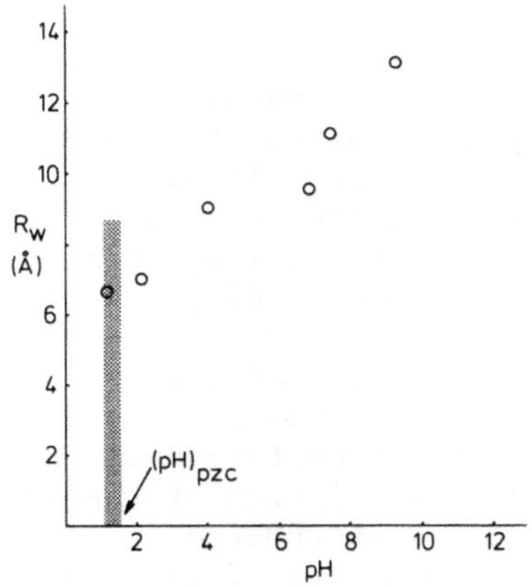

Fig. 3 The weight average molecular radius (R_w) of FA at various pHs. The pH of the hatched zone corresponds to the point of zero charge (PZC) derived from paper electrophoresis.

in the pH ranges 1.15 - 4.00 and 6.89 - 9.26, with relatively little expansion of the FA molecule in the range 4.00 - 6.89.

Increasing the pH from 1.15 to 9.26 results in a doubling of the FA radius (0.658 nm - 1.321 nm) which corresponds to an eight-fold increase in molecular volume of FA. Thus at pH 8 more than 70% of the volume of the solvated FA is occupied by water and couterions. With increased pH, both acid dissociation and molecular charge on the FA polyanion increase, giving rise to intra-molecular electrostatic repulsion and hence to an expansion in the molecular structure.

The electrostatic properties of the FA in acidic conditions were investigated by paper electrophoresis. Although the migrating front of FA was diffuse, the overall electrophoretic mobility toward the anode decrease with acidity, with mobility ceasing altogether at pH 1.0 - 1.2. At this pH, the FA molecule is probably completely protonated and electrostatically discharged (point of zero charge, PZC). The corresponding compact molecular structure (R_w = 6.58 Å; 0.658 nm) at PZC is thus free from electrostatic effects making a molecular weight (M) estimate from GFC at this pH more accurate. Using a value of $\bar{\upsilon}_2$ = 0.67 typical for humic compounds[16], and R_w = .658 nm at PZC (see Fig. 3), the weight average molecular weight can be calculated (equation 2) to be 790. This compares quite favourably with a recent vapour pressure osmometric measurement for riverine FA of M = 614[17] and soil FA of M = 952[18]. It follows from the elemental composition of FA and the molecular weight of 790 that the molecular formula of FA approximates to $C_{28}H_{42}O_{26}N_{1.6}$.

ELECTROSTATICS

If the FA macromolecule is viewed as a flexible polymer-like coil, then, it may, by means of rotation about its single bonds, take on any configuration compatible with its fixed bond lengths and angles, and any relevant steric considerations. When such a molecule is electrically charged, i.e. when it becomes a flexible polyanion, repulsion between charges gives rise to a relatively high electrostatic free energy (ΔG_e) for compact configurations and relatively low ΔG_e for expanded ones. In order to minimize ΔG_e the charged FA polyanion must expand. Since the dissociation of FA is accompanied by a stoichiometric build up of charges which tend to inhibit further dissociation of the acidic sites, then ΔG_e and those factors affecting it (charge, radius) must profoundly influence the acid-base chemistry of FA (see also ref. 19, 20).

Tanford[15] has reviewed the various classes of charged macromolecules and the calculation of ΔG_e; the most suitable model for FA is the expanding flexible coil model of Hermans and Overbeek[15]. The assumptions of this model are that water and small mobile ions can freely penetrate the internal volume of the coil, that there is a uniform distribution of charge throughout this region, and that the Debye-Hückel assumptions regarding the distribution of mobile univalent ions apply. The Hermans-Overbeek model is described by:

$$\Delta G_e = \frac{3NZ^2\varepsilon^2}{2DR_w}\left[\frac{1}{\kappa^2R_w^2} - \frac{3}{2\kappa^5R_w^5}\left(\kappa^2R_w^2 - 1 + (1+\kappa R_w)^2\exp(-2\kappa R_w)\right)\right] \quad (3)$$

where Z and R_w are the net charge and effective radius of the flexible coil, ε is the protonic charge, D the dielectric constant of the medium and κ is the Debye parameter defined by:

$$\kappa = \left(\frac{8\pi N\varepsilon^2}{1000\ DKT}\right)^{1/2} I^{1/2} \quad (4)$$

where I and T are the ionic strength and absolute temperature of the medium, and K the Boltzmann constant. The results of these model calculations using conditions similar to the gel filtration and potentiometric experiments are plotted in Fig. 4 in terms of ΔG_e as a function of R_w and Z. It is clear that the effect of a two fold expansion of a flexible polyelectrolyte is to reduce ΔG_e by 60–70%.

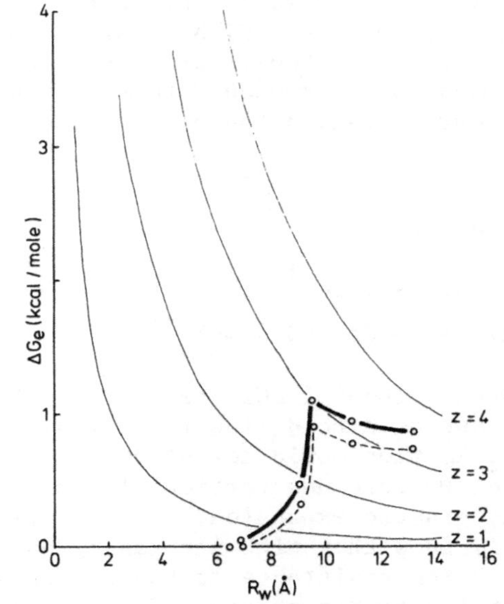

Fig. 4 Thin lines (———) show relationship between electrostatic free energy (ΔG_e) and the radius (R_w) of flexible coil polyanions based on the Hermans-Overbeek model (equation 3) at various charges (Z). The corresponding values of ΔG_e and R_w for FA are shown as (- - - -) with ΔG_e values referred to the PZC shown as bold line.

It can be shown that the bracketted terms in equation 3 remain very nearly constant (0.38 - 0.36), over the range of R_w (6-12 Å; 0.1-1.2 nm) encountered in this s dy, and thus the Herman-Overbeek equation reduces to the form:

$$\Delta G_e = \frac{3\pi N Z^2 \varepsilon^2}{2DR_W} \times [0.37] = 3.49 \; \frac{N Z^2 \varepsilon^2}{2DR_W} \tag{5}$$

In order to calculate the corresponding values of ΔG_e for FA the average molecular charge $\bar{Z}$ must be known at the pH for which R_w has been determined. As will be shown later, the formation function δn_{OH}, which is the quantity of titrated protons on FA (meqv. g^{-1}), may also be equated to the resultant average charge $\bar{Z}$.

$$\bar{Z} = \delta n_{OH} \times M \times 10^{-3} \tag{6}$$

thus by deriving δn_{OH} from the formation function curves for FA (see Fig. 7) ΔG_e can be calculated using equation (3). The results are listed in Table 2 and also superimposed on the Hermans-Overbeek model curves in Figure 4. However, it should be noted that since δn_{OH} is defined as equal to zero at the starting pH of the titrations (pH 2.2) this does not necessarily correspond to the fully protonated state of the FA, since, as indicated earlier, the PZC is at pH 1.1 - 1.2. However, by extrapolating the ΔG_e vs R_w FA curve in Fig. 4 from R_w = 6.91 Å to 6.58 Å which corresponds to the PZC, and at which ΔG_e should formally be equal to zero, then the charge Z_o and the corrected formation function $\delta n_{OH}{}^o$ at the starting pH of our titration can be estimated (eq. 3 and 6) to be:

$$\delta n_{OH}{}^o = 0.45 \text{ meqv g}^{-1}$$

$$Z^o = 0.35 \text{ charges mole}^{-1}$$

$$\Delta G_e{}^o = 26 \text{ cals mole}^{-1} \text{ or } 109 \text{ J mole}^{-1}$$

The recalculated values of $\Delta G_e'$ referred to PZC are included in Table 2 and redrawn as a bold line in Fig. 4. Vertical transitions in the ΔG_e vs R_w plot would correspond to the charging of a rigid macromolecule, whereas non vertical linear transitions represent perfectly elastic expansion. The ΔG_e vs R_w curves for FA in Fig. 4 clearly show that with increasing pH, there is a progressively increasing resistance to the electrostatic expansion of the FA polyanion (isosteric phase). This gives rise to rapid build up of electrostatic free energy. A critical point is reached (ΔG_e = 4600 J mole^{-1}, R_w = 0.95 nm, pH 6.9) when major configurational changes (e.g. uncoiling) are triggered which subsequently allow relatively unhindered expansion of the FA macromolecule with further accumulation of charge. This is more

Table 2 Electrostatic free energy of FA, based on Hermans-Overbeek model.

pH	R_w (Å)	δn_{OH} (meqv g^{-1})	Z	ΔG_e (cals $mole^{-1}$)	$Z + Z_o$	$\Delta G'_e$ [1] (cals $mole^{-1}$)	δn_{OH}^{c} [2] (meqv g^{-1})
1.15	6.58	-	-	-	0.00	0.00	0.00
2.15	6.91	0.00	0.00	0.00	0.35	26	0.45
4.00	9.06	1.91	1.50	314	1.85	477	2.36
6.89	9.49	3.32	2.62	888	2.97	1139	3.77
7.44	11.13	3.52	2.78	758	3.13	961	3.97
9.26	13.21	4.00	3.16	717	3.51	884	4.45

(1) recalculated ΔG_e with reference to PZC
(2) δn_{OH} corrected to PZC by $\delta n_{OH}^{c} = \delta n_{OH} + \delta n^{o}_{OH}$

clearly seen in Fig. 5 in which the R_w values from GFC are plotted against δn_{OH}^{c}, showing a relatively small expansion in the region δn_{OH}^{c} 2.3 – 3.8. It should be added, that the estimations of ΔG_e using other macromolecular models from Tanford[15] result in trends similar to those shown in Fig. 4.

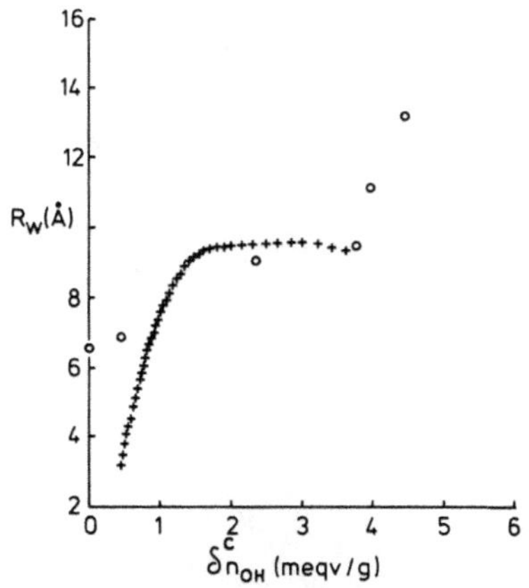

Fig. 5 Increase in molecular radius R_w with formation function during the course of FA titration. Data points (O) originate from the GFC experiments whereas (X) correspond to calculated values of R_w derived from linearizing the Tanford plot (Fig. 9) using equation 16.

POTENTIOMETRY

Potentiometric analysis of ionic equilibria provides information on the binding of ionic species to the organic matter by measuring the changing concentration of one of the reacting species as a function of the total composition of the solution. However, as the number of species formed increase a unique identification becomes more difficult. Fig. 6 shows the titration curve of FA which is typical of several other titrations carried out in this study. The broad, unresolved titration curve in Fig. 6 is indicative of the chemically disperse nature of acid dissociating sites on the FA. It is not possible to unequivocally specify equivalence points, total acidity and binding constants that are necessary for a classical description of the titration data. The formation function, δn_{OH}[21, 22], provides an alternative and unambiguous measure of the progress of the titration in terms of the number of acidic sites titrated.

For the general reaction $FA\text{-}H \rightleftharpoons FA^- + H^+$ the charge balance at any point in the titration is given by :

$$Na^+ + H^+ = OH^- + A^- \tag{7}$$

where Na^+ is equivalent to the mole of alkali titrant added (OH_a), A^- (= At - HA) is equivalent to the quantity of proton dissociation (n_{OH}) formally equal to "mmole OH^- bound"[23] and assuming that the background electrolyte does not react with any of the above species. The formation function (δn_{OH}) in the forward direction is then given by:

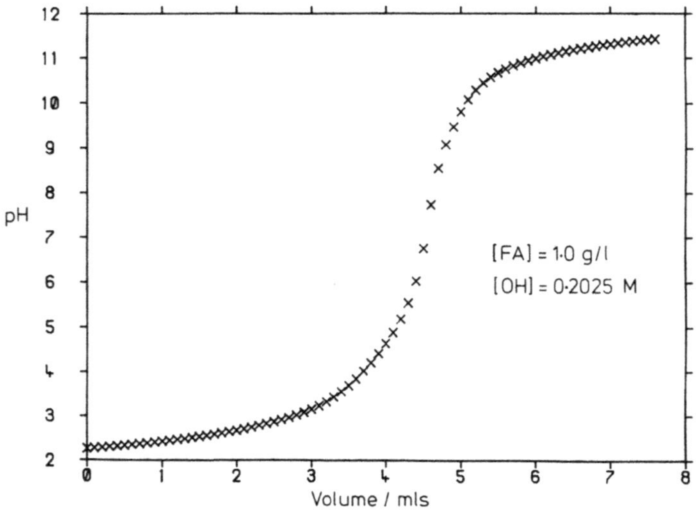

Fig. 6 pH vs volume data for the forward titration of FA in 0.1 M $NaClO_4$ at 25°C.

$$\delta n_{OH} = \frac{(OH_a - OH_a^o) - (OH - OH^o) - (H^o - H)}{FA} \quad (8)$$

where OH and H are the mmoles of OH^- and H^+ in solution, FA is the weight of FA in grams, and the superscript ^{o} refers to the beginning of the titration or any other reference point. A similar six parameter expression can be derived for the reverse formation function, δn_H, and for perfectly reversible titrations:

$$\delta n_{OH} = \delta n_H \quad (9)$$

The six parameter equation is sufficiently general to quantitatively describe any portion of the titration curve and thus allow comparative analysis of any number of titrations.

Both the forward (δn_{OH}) and reverse (δn_H) formation functions for the titration of FA are shown in Fig. 7. Marked hysteresis occurs in the mid pH regions between forward and reverse titrations indicating that the titrations are not completely reversible and that $\delta n_{OH} \neq \delta n_H$. (see also: Sposito and Holtzclaw[22]). This possibly arises from conformational changes which consecutively occlude and expose the functional sites to the bulk solution. Fig. 3 suggests that the extent of steric hindrance is apparently related to the type of groups being titrated so that the sequence in which groups are titrated will affect the accessibility of sites and hence the apparent acidity of the FA and will give rise to the hysteresis. Between pH 2.2 and pH 11.4 there are approximately 6.5 meqv g^{-1} FA of sites (5.14 sites per mole) which is remarkably similar to that observed for FA extracted from Satilla River water[22] and from soils[23]. The inflection point at pH 7.2

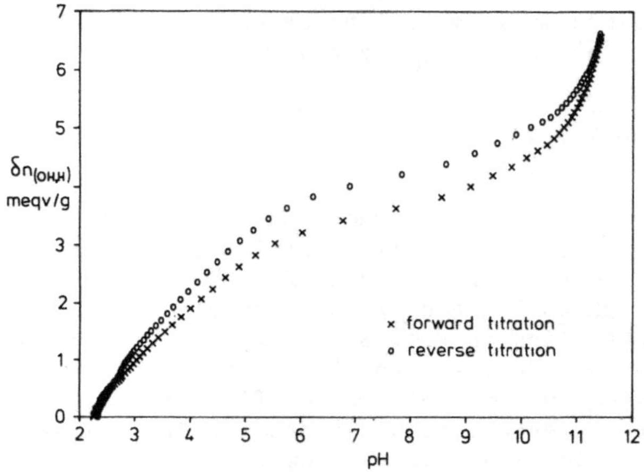

Fig. 7 Forward (δn_{OH}) and reverse (δn_H) formation function with respect to ph for the titration of FA.

separates two pH regions each containing appreciable numbers of titratable sites and suggests the presence of two classes of incompletely titrate sites. These may well correspond to the now classically differentiated carboxylic and phenolic sites[8,19,22] common to humic compounds. Our infrared and pryolysis mass spectra confirm that the FA used in this study is probably similar in nature (see methods). However, our data cannot be used to further identify the complete structural nature of these sites (e.g. salicylate, polyphenolate, etc.). The formation function curves of ideal diprotic acids whose pK's are separated by typically two pK units are usually steep, sigmoid and, naturally, characterised by two inflection points. Thus, the broad nature of δn_{OH} over the pH range 2.2 - 7.2 indicates a polydisperse distribution of carboxyl site acidities, some of which are quite acidic.

Since the electrophoretic studies have shown the PZC to be at pH 1.0 - 1.2 then clearly the starting point of the titration does not correspond to the fully protonated state, but nevertheless its proximity to the PZC would make the carboxyl region (pH 2.2 to pH 7.2) of the titration curve more amenable to binding analysis. Since the alkaline portion of the δn_{OH} curve is clearly incomplete, analysis will henceforth be confined to the acidic portion of the forward titration curve.

The most elementary treatment of acid base data is the Henderson-Hasselbalch equation for the dissociation of simple acids[11].

$$pH = pK_a + \log \frac{\alpha}{1 - \alpha}$$

where K_a and α are the dissociation constant and degree of dissociation respectively. It can be shown that an equivalent relationship expressed in terms of the association constant (k) and the formation function (δn_{OH}) is given by

$$pH = \log k + \log \frac{\delta n_{OH}}{\Delta \delta n_{OH}} \tag{10}$$

where $\Delta\delta n_{OH} = \Delta\delta n_{OH}^{T} - \delta n_{OH}$ and where δn_{OH}^{T} is the total number of titrated carboxyl groups and is operationally defined as equal to δn_{OH} at the inflection pH of 7.2 in Fig. 7 ($\delta n_{OH}^{T} = 3.64$ meqv g^{-1}). Thus a plot of pH vs log ($\delta n_{OH}/\Delta\delta n_{OH}$) should be linear with unit slope (see dashed line in Fig. 8). Instead, both corrected and uncorrected FA data deviate from linearity at low pH (Fig. 8) but with δn_{OH}^{c} showing considerably better fit than δn_{OH}. Since δn_{OH}^{c} corrects for the presence of untitrated acidic groups, the corresponding value of log k decreases from 3.80 to 3.58. It is important to realize that the derived value of k is not a unique constant for all carboxyl sites but is an overall constant $\bar{k}$. The

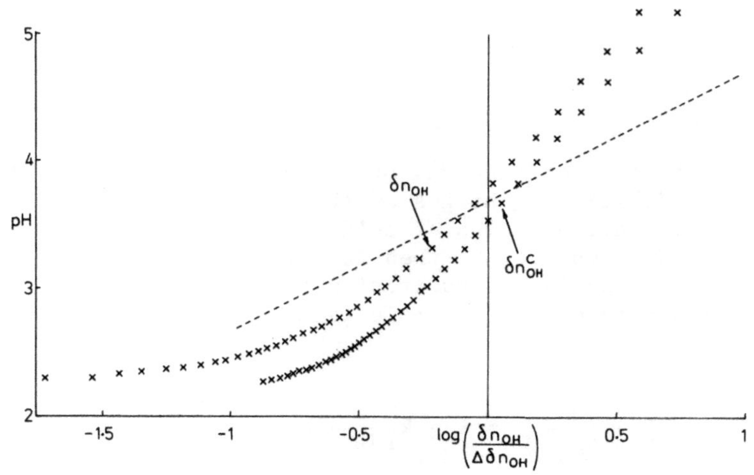

Fig. 8 Henderson-Hasselbalch plots of ideal mono protic acid (- - -) and FA data (X). Both corrected (δn_{OH}^{c}) and uncorrected (δn_{OH}) data are included.

linear portions of Fig. 8 appear to conform to a 'randomly linked polymer model' as described by the Katchalsky-Spitnik[24] equation:

$$pH = p\bar{k} + s \log \frac{\delta n_{OH}}{\Delta\delta n_{OH}} \quad ; \; s > 1.0 \tag{11}$$

The slopes of the FA plots in Fig. 8 have a value of s = 2.0 and 2.2 for the corrected and uncorrected data instead of unity, indicating a polyprotic nature to the FA. Similar values of s have been reported for marine organic material[25] and soil humic acids[20] but contrary to the conclusions of Huizenga and Kester[25] the value of s is not an empirical fitting parameter but is fundamentally related to the number and size of 'monomer' units and the degree of molecular stretch[26].

It is commonly observed that deviations from linearity in the Henderson-Hasselbalch plots arise from site-site interactions e.g. electrostatic repulsions. Tanford[15] has shown that the equation

$$pH = \log\left(\frac{\alpha}{1-\alpha}\right) + \log k_{INT} - 0.868\, WZ_i\, \bar{Z} \tag{12}$$

successfully accounts for the effect of electrostatic interaction on the acid-base chemistry of rigid protein-like molecules. In eq. (12) k_{INT} is the intrinsic proton association constant for the conditions when the charge Z is zero, Z_i is the charge on the proton (+1) and W is related to the electrostatic free energy ΔG_e

by the equation:

$$W = \frac{\Delta G_e}{RT\bar{Z}^2} \tag{13}$$

where R is the gas constant and other symbols have been defined previously. If all the charges on the FA are assumed to arise solely from the dissociation of acid groups and if condensation of counter ions does not occur, then it follows that $\bar{Z} = M \times \delta n_{OH} \times 10^{-3}$ (i.e. equation 6). Substituting this into equation 12 to eliminate $\bar{Z}$ and re-expressing α in terms of δn_{OH}, equation 12 is then given by

$$\mathrm{pH} - \log\left(\frac{\delta n_{OH}}{\Delta\delta n_{OH}}\right) = \log k_{INT} - 8.68 \times 10^{-4}\, WM\, \delta n_{OH} \tag{14}$$

The value of any electrostatic interaction parameter W can then be obtained from the slope of a plot of pH - log $(\delta n_{OH}/\Delta\delta n_{OH})$ vs M δn_{OH}. The corresponding plot for FA, shown in Fig. 9 is curvilinear but the effect of including the correction term δn^{o}_{OH} clearly improves the linear fit for the data (coefficient of regression r^2 increases from 0.51 to 0.94 for a least squares regression of all data points).

The linear regions of the Tanford plots (Fig. 9) occur in the pH range 3.2 - 7.2 indicating that over this portion of the titration, the FA polyanion must be structurally rigid. Indeed, this agrees with the presence and general location of the isosteric phase identified from the gel filtration experiments (Fig. 3). Extrapolation of the linear portions of the Tanford plots in Fig. 9, results in intercepts of log k_{INT} = 2.26 and 2.70 for the

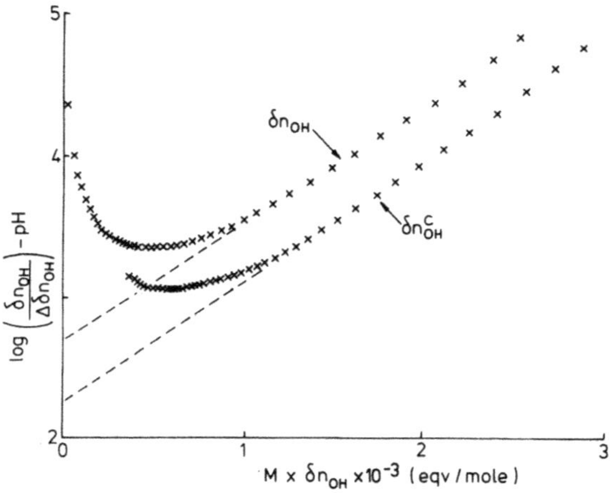

Fig. 9 Tanford plots of the FA data.

corrected and uncorrected data respectively. The log k_{INT} values correspond to the condition when $\bar{Z} = 0$ (i.e., free from electrostatic contributions) and partly explains why they are lower (more acidic) than the log k values obtained from the Katchalsky-Spitnik treatment (see also Table 3). Nevertheless, these constants are quite acidic relative to monocarboxylic compounds (e.g. acetic acid log k = 4.73, benzoic acid 4.21) possibly suggesting an electron withdrawing or inductive role to the FA backbone.

In order to check for consistency between the results of the GFC experiments and potentiometry, the values of the electrostatic interaction parameter W derived from these two approaches should be compared. Combining the Hermans-Overbeek expression for ΔG_e (equation 5) with equation 13 it follows that

$$W = \frac{1.11 \pi \boldsymbol{N} \varepsilon^2}{2 D R T} \quad \frac{1}{R_w} = \frac{L}{R_w} \tag{15}$$

which when applied to the isosteric region of the GFC where R_w = 9.5 Å or 0.95 nm (see Fig. 3 and 4), yields a value of W = 1.48. This compares quite favourably with the value W = 1.01 derived from the slope of the Tanford plot (Fig. 9) particularly in view of the approximations necessitated in the derivation of R_w and δn^o_{OH}.

Although the curvature evident at low δn_{OH} may have arisen from uncertainty in the precise value of δn^o_{OH}, it is more likely that some of the assumptions underlying the Tanford model do not

Table 3 List of acid association constants for FA resulting from various models.

Model	Notation & Log association constant		Number of sites per mole of FA
Katchalski-Spitnik	log $\bar{k}$	3.83 [1]	2.98
		3.58 [2]	3.43
Tanford	log k_{int}	2.70 [1]	2.98
		2.26 [2]	3.43
Scatchard	log k_1	5.43	0.76
	log k_2	3.23	2.03
	log k_o	3.34	2.79
Distribution plot	log k_n	2.29	3.43

1 and 2; derived from uncorrected and corrected (with respect to PZC) FA data, respectively.

hold. In particular, it is clear from the GFC experiments (see Fig. 3) that it is not valid to assume constant R_w at low values of δn_{OH}. However, by substituting equation 15 into equation 14, it is possible to expand the Tanford model to take into account conformational charges in FA. This yields:

$$pH - \log \frac{\delta n_{OH}}{\Delta\delta n_{OH}} = \log k_{INT} - 8.68 \times 10^{-4} L \frac{M\delta n_{OH}}{R_w} \qquad (16)$$

The constant L may determined (as in W) either from equation 15 or from the slope of Fig. 9 with $R_w = 9.5$ Å. Equation 16 was then applied to the FA data to estimate the values of R_w necessary to linearize the Tanford plot and compared with the experimentally derived values of R_w. The results which are shown in Fig. 5 not only reproduce the isosteric region but also predict a decrease in R_w at low δn_{OH} which is in qualitative agreement with the GFC data. A better fit could undoubtedly have been obtained if more precise estimates of R_w were available, particularly those pertaining to separations on Sephadex G-10.

An alternative approach to the analysis of complex binding reactions is to use the Scatchard formulation[27] expressed here in terms of δn_{OH}.

$$\bar{\nu} = \sum_{i}^{m} \frac{n_i k_i [H^+]}{1 + k_i [H^+]} = M\,(\Delta\delta n_{OH})\,10^{-3} \qquad (17)$$

where $\bar{\nu}$ is the number of proton equivalents bound per mole of FA and n_i is the number of sites of class i (i.e. 1, 2,....m) characterized by a site specific association constant k_i. The results of the Scatchard analysis of the FA data (Fig. 10) have been resolved into two linear components based on the limiting slope technique[28]. The overall association constant $\log k_0 = 3.34$ was derived from n_1, n_2, k_1, and k_2 (see Fig. 10, Table 3) and is in good agreement with the log k value of 3.58 obtained from the Katchalsky-Spitnik treatment.

Finally, the distribution of binding sites with respect to their apparent log k values can conveniently be obtained from a plot of $\Delta(\delta n_{OH})/\Delta(\log k)$ <u>vs</u> log k as shown in Fig. 11 where

$$\frac{\Delta(\delta n_{OH})}{\Delta(\log k)} = \frac{\Delta(\delta n_{OH})}{\Delta(pH)} \times \frac{\Delta(pH)}{\Delta(\log k)} \qquad (18)$$

and where differences (Δ) are calculated between successive points, utilizing equation (10) for estimating log k. The broad distribution curve clearly displays the polydisperse nature of these acidic groups.

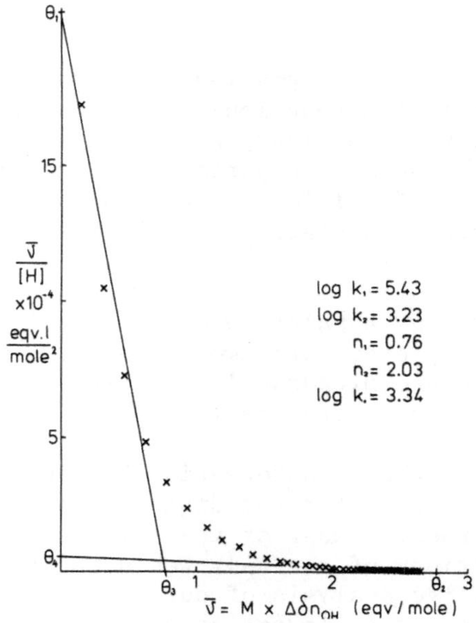

Fig. 10 Scatchard analysis of proton binding by FA; $\theta_1 - \theta_4$ are the limiting slope intercepts required for the derivation of n_1, n_2, k_1, k_2, binding parameters using Hunston's[28] approach.

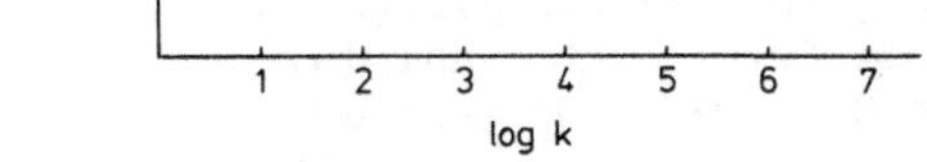

Fig. 11 Distribution of sites as a function of log k. Area below curve is equal to δn^T_{OH}.

Although the greatest number of sites appears to be centred at log k of 4.8, they are outweighed by the greater acidity of the less abundant sites with the result of producing a lower number average value of log k_n = 2.29.

CONCLUSION

The combined use of conformational and potentiometric studies of the complex equilibrium reactions of FA has provided new insights into the solution chemistry and structure of these ubiquitous metal-binding ligands. In particular, the FA macromolecule was found to possess a flexible structure allowing considerable expansion during acid dissociation and the resultant accumulation of charge.

Electrostatics play a major role in controlling the availability of sites for proton exchange, mainly via sterically restricted occlusion and withdrawal of these sites in concert with molecular expansion and contraction.

By incorporating these conformational changes into a conventional electrostatic model for the dissociation of acidic sites, it was possible to resolve most of the polydispersity evident in the acid base titrations of the FA. We intend to apply these models to a comparative analysis of our other titrations as well as those published by other investigators.

ACKNOWLEDGEMENTS

The authors are grateful for advice from Dr. A. Dickson, and to Dr. C. Fay for assistance in computing. M.S.V. was supported by a NERC Special Topic grant GST/02/06 throughout this study.

REFERENCES

1. Mantoura, R.F.C., A. Dickson and J.P. Riley, 1978: The complexation of metals with humic materials in natural waters. Est. Coast. Mar. Sci., 6, 387-408.
2. Sunda, W.G. and P.J. Hanson, 1979: Chemical Speciation of Copper in River Water; Effects of total Cu, pH, Carbonate and dissolved Organic Matter. In: "Chemical Modelling in Aqueous Systems", E.A. Jenne, ed. ACS Symposium Series 93. pp. 147-180.
3. Mantoura, R.F.C., 1981: Organo-metallic interactions in natural waters. Ch. 7 in: "Marine Organic Chemistry: evolution, composition, interactions and chemistry of organic matters in seawater", E.K. Dursma and R. Dawson, eds. Elsevier, Amsterdam.
4. Rashid, M.A. and L.H. King, 1971: Chemical characteristics of fractionated humic acids associated with marine sediments. Chem. Geol., 7, 37-43.
5. Reuter, J.H. and E.M. Perdue, 1977: Importance of heavy metal-organic matter interactions in natural waters. Geochim. Cosmochim. Acta., 41, 326-334.

6. Jenne, E.A., ed., 1979: "Chemical modelling in aqueous systems". ACS Symposium Series No. 93.
7. Mantoura, R.F.C. and J.P. Riley, 1975: Analytical concentration of humic substances from natural waters. Anal. Chim. Acta., 76, 97-106.
8. Schnitzer, M. and S.U. Khan, 1972: Humic substances in the marine environment. Marcel Dekker, H.Y.
9. Weber, J.H. and S.A. Wilson, 1975: The isolation and characterisation of fulvic acid and humic acid from river water. Water Res., 9, 1079-1084.
10. Stevenson, F.J. and J.H.A. Butler, 1969: Chemistry of humic acids and related pigments. Ch. 22 in: "Organic Geochemistry: Methods and results", G. Eglington and M.T.J. Murphy, eds. Springer, N.Y.
11. Bracewell, J.M., G.W. Robertson and B.L. Williams, 1980: Pyrolysis - mass spectrometry studies of humification in a peat and a peaty podzol. J. Anal. App. Pyrolysis, 2, 53-62.
12. Biederman, G. and L.G. Sillen, 1953: The hydrolysis of metal ions. IV Liguid junction potentials and constancy of activity factors in NaClO - HClO ionic medium. Ark. Kemi., 5, 425-440.
13. Determann, H., 1968: Gel chromatography - a laboratory handbook. Springer Verlag, N.Y., 195 p.
14. Andrews, P., 1970: Estimation of molecular size and molecular weights of biological compounds by gel filtration. In: "Methods of Biochemical Analysis", D. Glick, ed. 18, 1-53.
15. Tanford, C., 1961: Physical chemistry of macromolecules. Wiley, N.Y., 710 p.
16. Cameron, R.S., R.S. Swift, B.K. Thornton and A.M. Posner, 1972: Calibration of gel permeation chromatography for use with humic acid. J. Soil Sci., 23, 342-349.
17. Reuter, J.H. and E.M. Perdue, 1981: Calculation of molecular weights of humic substances from colligative data: application to aqueous humus and its molecular size fractions. Geochim. Cosmochim. Acta., in press.
18. Hansen, E.H. and M. Schnitzer, 1969: Molecular Weight measurements of polycarboxylic acids in water by vapour pressure osmometry. Anal. Chim. Acta., 46, 247-254.
19. Wilson, D.E. and P. Kinney, 1977: Effects of polymeric charge variations on the proton-metal ion equilibria of humic materials. Limnol. Oceanogr., 22, 281-289.
20. Huizenga, D.L. and D.R. Kester, 1979: Protonation equilibria of marine dissolved organic matter. Limnol. Oceanogr., 24, 145-150.
21. Rossotti, H., 1978: The Study of Ionic Equilibria: an introduction. Longman, London.
22. Perdue, E.M., J.H. Reuter and M. Ghosal, 1980: The operational nature of acidic functional group analyses and

its impact in mathematical descriptions of acid-base equilibria in humic substances. Geochim. Cosmochim. Acta., 44, 1841-1851.

23. Sposito, G. and K.M. Holtzclaw, 1977: Titration studies on the polynuclear, polyacidic nature of fulvic acid extracted from sewage sludge-soil mixtures. Soil Sci. Soc. Am. J., 41, 330-336.

24. Katchalsky, A. and P. Spitnik, 1947: Potentiometric titrations of polymethacrylic acid. J. Polymer. Sci., 2, 432-447.

25. Pommer, A.M. and I. A. Breger, 1960: Potentiometric titration and equivalent weight of humic acid. Geochim. Cosmochim. Acta. , 20, 30-44.

26. Katchalsky, A. and J. Gillis, 1949: Theory of the potentiometric titration of polymeric acids. Rec. Trav. Chim., 68, 879-897.

27. Fletcher, J.E., 1977: A generalized approach to equilibrium models. J. Phys. Chem., 81, 2374-2387.

28. Hunston, D.L., 1975: Two techniques for evaluating small molecule-macromolecule binding in complex systems. Anal. Biochem., 63, 99-109.

29. Beck, K.C., J.H. Reuter and E.M. Perdue, 1974: Organic and inorganic geochemistry of some coastal plain rivers of the south eastern United States. Geochim. Cosmichim. Acta., 38, 341-364.

30. Gamble, D.S., 1973: Na^+ and K^+ binding by fulvic acid. Can. J. Chem., 51, (19), 3217-3222.

COPPER SPECIATION IN MARINE WATERS

Richard W. Zuehlke and Dana R. Kester

Graduate School of Oceanography
University of Rhode Island
Kingston, RI 02881 U.S.A.

ABSTRACT

The properties of a naturally occurring copper-organic complex have been examined. Two binding sites for copper were found and their protonation and copper-binding properties characterized. The parameters thus obtained have been used to develop a pH-dependent speciation model for dissolved copper in seawater. The resulting predictions of free and labile copper concentrations are examined and discussed with respect to copper's toxicity, adsorption, and voltammetric analysis.

INTRODUCTION

Numerous experimental investigations have shown that copper in seawater is extensively associated with marine dissolved organic matter, MDOM. Many investigators have assumed that copper-binding MDOM is of the humic acid type (i.e., polymeric, some aromaticity, with active carboxylic, phenolic, and amine functionality)[1]. Laboratory studies of copper binding by soil humic materials in acidic media have been summarized by Buffle et al.[2] but are of marginal value in interpreting the seawater observations. Several years ago, Mantoura and coworkers[3,4] developed a gel filtration scheme which allowed them to measure an overall conditional stability constant at pH 8.0:

$$\beta^{c}_{CuL} = [CuL]/[Cu^{2+}]\,(T(L) - [CuL]) \qquad (1)$$

where CuL refers to the copper-MDOM complex, and T(L) is the total concentration of ligand sites (including free, protonated, and complexed). Humic materials from various sources were used and introduced into a copper-containing seawater sample. The value of log β^{c}_{CuL} was found to be 9.30 ± 0.59. Stability constants for calcium and magnesium humates were also determined.

Van den Berg and Kramer[5] measured the stability constant for freshwater fulvic acids at pH 7.6 and ionic strength 0.01 using an adsorption technique for determining $[Cu^{2+}]$. Their method is strongly dependent on inorganic copper speciation which appears to have been neglected, and their fulvic acid sources seem not amenable to comparison with those referred to in this study. An important aspect of their work, however, was the apparent ability to resolve the copper binding into two components, mono- and bidentate.

Buffle et al.[2] also studied freshwater humic acid-copper interactions, but recognized the importance of inorganic copper speciation and directly determined $[Cu^{2+}]$ using a specific ion electrode. Again, since their humic acids are of freshwater origin, their results are not directly comparable to those of this work. The significance of their work to this study lies in their detection of two types of copper binding sites, also identified as mono- and bidentate in character.

Protonation equilibria are important in regulating copper binding. An overall reaction can be symbolized by the following equilibrium statement:

$$Cu^{2} + HL \rightleftharpoons CuL^{+} + H^{+} \cdot \tag{2}$$

which may in turn be expressed as the sum of two reactions:

$$HL \rightleftharpoons H^{+} + L^{-} \tag{3}$$

$$Cu^{2+} + L^{-} \rightleftharpoons CuL^{+} \tag{4}$$

The conditional constants referred to above relate most closely (but not precisely) to Reaction (4). In order to obtain a complete picture of copper-MDOM binding, the equilibria in Reactions (3) and (4) must be addressed separately and explicitly.

Huizenga and Kester[6] reviewed the pH titration studies of marine dissolved organic matter and noted the separate effects of intrinsic functional group acid strength and polyelectrolyte charging, and their roles in defining an apparent dissociation constant. In their experimental work on a variety of MDOM samples, they found a well defined group of sites (denoted by them as Type 1) with a pK^{*}_{HL} of 3.6. It is important to note that the functional form they required to define the equilibrium constant applicable to Reaction (3) is:

$$K^*_{HL} = [H^+]\cdot([L^-]/[HL])^{2.0} \qquad (5)$$

This unusual form results from the polyelectrolytic character of the MDOM.

These investigators also found evidence for a Type 2 site whose pK* is greater than that for Type 1. Further, the concentration of Type 1 sites was found to be 1.1 x 10^{-6} mol (mg C)$^{-1}$. Since MDOM is not a well characterized single compound, the concentration and nature of these sites frequently vary with the MDOM source. Each site type is therefore a broad classification and may, in fact, include a range of similar sites.

Perdue[7] performed calorimetric acid-base titrations of soil-derived humic acids and was able to clearly differentiate between carboxylic acid groups (pK^*_{carb} could not be determined quantitatively) and phenolic groups, for which pK^*_{ph} was found to be 10.5 $\pm$ 0.3.

These acid-base studies did not relate group functionality to copper binding, but Gamble et al.[8] have described the relation in some theoretical detail.

Recent work by some of our colleagues[9] combines the copper binding and protonation components into one experimental study based on <u>naturally occurring</u> copper-organic complexes. Since the results are generally in accord with the previously listed results, we here derive quantitative parameters from them and use those parameters to broaden a dissolved copper speciation model to include an organic component. The implications of the speciation predictions are then discussed relative to recent analytical studies on copper.

SPECIATION MODEL

In order to avoid the problems associated with the use of thermodynamic stability constants and the requirement that activity coefficients be estimated for all species as arises in many speciation models[10], we have written a major ion speciation program, SEASPE, which utilizes stoichiometric stability constants obtained at seawater ionic strength, 25°C, and one atm pressure. The program further involves a new feature, chloride ion pairing among the major cations, as recently suggested by Johnson and Pytkowicz[11,12,13]. Recent evidence[14] makes this approach even more appealing, so we have implemented it in this study, after correcting those component stability constants which were determined in sodium chloride media for chloride ion pairing. As noted by Johnson and Pytkowicz[12], chloride ion pairing reduces the ionic strength of 35 ‰ seawater from the commonly used value of 0.69 to 0.53.

SEASPE computes the free ion and ion-pair concentrations by a process of iteration in which the equilibrium requirements for all species are satisfied. The free ligand concentrations are then used in CUSPE, a program which computes copper speciation. The parameters used in CUSPE are shown in Table 1. A critical assessment of these parameters may be found in Kester et al.[15].

In their work, Mills et al.[9] filtered samples of Narragansett Bay water and separated each into a number of subsamples. The subsamples were adjusted to pH's between 1.5 and 8.0 and allowed to equilibrate for one hour. The copper-organic complexes were isolated using a procedure developed by Mills and Quinn[16] which employs bonded C_{18} - reverse-phase liquid chromatography on a C_{18} SEP-PAK (TM Waters Associates) cartridge. Elution was with a 50:50 water:methanol mixture. The results are reproduced in Figure 1. Complete details may be found in Mills et al.[9] and in Mills[17].

RESULTS

For pH < 5, we assume that virtually all the active MDOM sites are protonated, and therefore we approximate Equation (5) as:

$$K^*_{HL} = [H^+] \cdot [L^-]^2 / T(L)^2$$

or

$$[L^-] = (K^*_{HL}/[H^+])^{0.5}\, T(L) \qquad (6)$$

Table 1. Stability constants for copper speciation (I = 0.5277)

Ligand	Stability Constant	Source
$-OH^-$	9.2×10^{-9}	Paulson and Kester[29]
$-(OH^-)_2$	2.3×10^{-17}	Paulson and Kester[29]
$-CO_3^{2-}$	2.29×10^{6}	Zuehlke and Kester[30]
$-HCO_3^-$	584	Zuehlke and Kester[30]
$-SO_4^{2-}$	2.7	Matheson[31]
$-Cl^-$	0.5	Byrne and Kester (TBP)
$-(Cl^-)_2$	0.1	Byrne and Kester (TBP)
-MDOM (Site 1)	2.5×10^{4}	This work
-MDOM (Site 2)	1.0×10^{10}	This work

TBP: To be published.

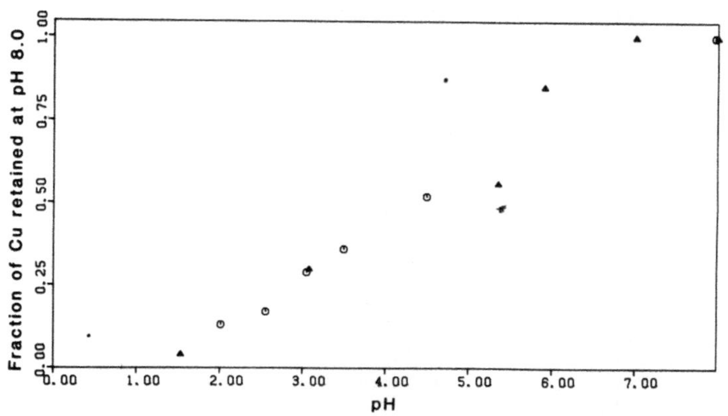

Fig. 1 Summary of data of Mills et al.[9] (Data Set I = ⊙ ; T(Cu) = 2.46 μg kg $^{-1}$, DOC = 6.20 mg C/l, Cu retained by SEP-PAK at pH 8 = 0.84 μg kg^{-1}: Data Set II = △ ; T(Cu) = 1.76 μg kg^{-1}, DOC = 1.18 mg C/l, Cu retained by SEP-PAK at pH 8 = 0.27 μg kg^{-1}, organic carbon retained by SEP-PAK at pH 1.53 = 0.30 mg C kg^{-1}).

We next define the stoichiometric stability constant for the copper-MDOM interaction based on Reaction (4):

$$\beta^*_{CuL} = [CuL^+]/[Cu^{2+}][L^-] \tag{7}$$

At low pH, the only forms of copper which need be considered in a speciation scheme are Cu^{2+}, $CuCl^+$, $CuCl_2{}^o$, and CuL. We may therefore write:

$$\begin{aligned} T(Cu) &= [Cu^{2+}] + [CuCl^+] + [CuCl_2{}^o] + [CuL] \\ &= [Cu^{2+}]\ (1 + \beta^*_{CuCl}\ [Cl^-] + \beta^*_{CuCl_2{}^o}\ [Cl^-]^2 \\ &\quad + \beta^*_{CuL}\ [L^-]) \\ &= [Cu^{2+}]\ (Q + \beta^*_{CuL}\ [L^-]) \end{aligned} \tag{8}$$

where Q is defined as:

$$Q = 1 + \beta^*_{CuCl^+}\ [Cl^-] + \beta^*_{CuCl_2{}^o}\ [Cl^-]^2 \tag{9}$$

and can be computed from known stability constants (Table 1), the composition of seawater, and a major ion speciation model.

Combining (6), (8), and (9) and rearranging:

$$\frac{T(Cu)}{Q\ [Cu^{2+}]} = 1 + \left\{ \frac{\beta^*_{CuL}\ K^*_{HL}{}^{0.5}\ T(L)}{Q} \right\} \cdot \frac{1}{[H^+]^{0.5}} = \frac{1}{1 - F} \tag{10}$$

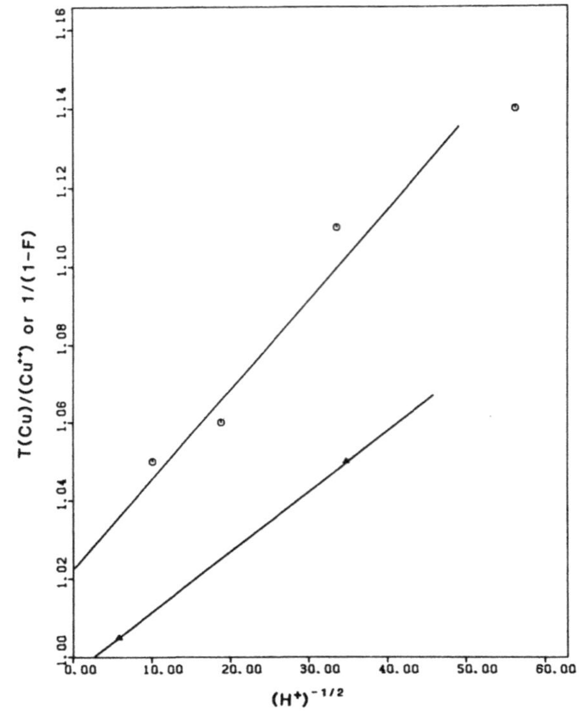

Fig. 2 Test fitting to Equation 10 of low pH data sets of Mills et al. ⊙ = Data Set I; △ = Data Set II.

where F is the fraction of organically bound Cu (i.e., that retained in the SEP-PAK ™ experiment of Mills et al.[9]. Equation (10) is only valid at low pH.

The test of adherence of the data to the model is through a plot of 1/(1-F) against $1/[H^+]^{0.5}$. The plot should be linear with an intercept of unity and a slope of $\beta^*_{CuL} K^{*\,0.5}_{HL}$ T(L)/Q. Such a plot of each of the two lowest pH data sets is shown in Figure 2, and it shows a linear relationship with the required intercept.

The lower plot has a slope of 0.0016; total organic carbon retained on the SEP-PAK ™ at low pH is 0.30 mg C kg^{-1}. This approximates conditions under which Huizenga and Kester's[6] work was done, so we use their Type 1 site distribution to compute T(L):

$$T(L) = 0.30 \text{ mg C/kg} \times 1.1 \times 10^{-5} \text{ mol (mg C)}^{-1} = 3.3 \times 10^{-6} \text{ mol kg}^{-1} \quad (11)$$

Using Huizenga and Kester's[6] K^*_{HL} (2.5×10^{-4}), T(L), the computed Q factor (1.26), and the slope, we find that $\beta^*_{CuL} = 3.8 \times 10^4$.

For the upper plot of Figure 2, repetition of the above computation yields $\beta^*_{CuL} = 1.1 \times 10^4$. Averaging these two values:

$$\beta^*_{CuL} \text{ (average)} = 2.5 \times 10^4 \qquad (12)$$

This stability constant can only be taken as a crude estimate, for there is undoubtedly some competition for the same sites by Ca^{2+} and Mg^{2+}. In the absence of major cation data in the work of Mills et al.[9], however, we must regard this as an apparent constant which includes the effect of other MDOM-binding ions.

Our next goal therefore is to examine the higher pH portion of the data set. This is clearly not a simple extension of the low pH analysis, for the distinct inflection point around pH 5 implies a different set of binding sites, perhaps Type 2 in the Huizenga and Kester[6] notation. We symbolize the associated ligands as M.

An analytical fit of the Type 2, high pH data to a simple model is an intractable mathematical problem. Our approach therefore is to modify SEASPE and CUSPE by introducing the following new parameters and assumptions:

1. Type 1 sites bind only copper, so β^*_{CuL}, T(L), and K^*_{HL} appear only in CUSPE.
2. Type 2 sites can bind Ca and Mg as well as Cu. Thus β^*_{CaM} and β^*_{MgM} (as determined by Mantoura et al.[4]) and T(M) and K^*_{HM} (adjustable parameters) appear in SEASPE to set [M], and β^*_{CuM} (also an adjustable parameter) appears in CUSPE.
3. β^*_{CuM} and K^*_{HM} have the same functional forms as their CuL and HL counterparts.

Table 2 Copper-MDOM parameters determined in the model

Type 1 Sites	Type 2 Sites
$T(L) = 3.3 \times 10^{-6}$ mol kg^{-1}	$T(M) = 8 \times 10^{-7}$ mol kg^{-1} (I), 2.5×10^{-6} mol kg^{-1} (II)†
$\beta^*_{CuL} = 2.5 \times 10^4$	$\beta^*_{CuM} = 1.0 \times 10^{10}$
$K^*_{HL} = 2.5 \times 10^{-4}$	$K^*_{HM} = 3.0 \times 10^{-10}$

†I and II refer to Mills et al.[9] data sets.

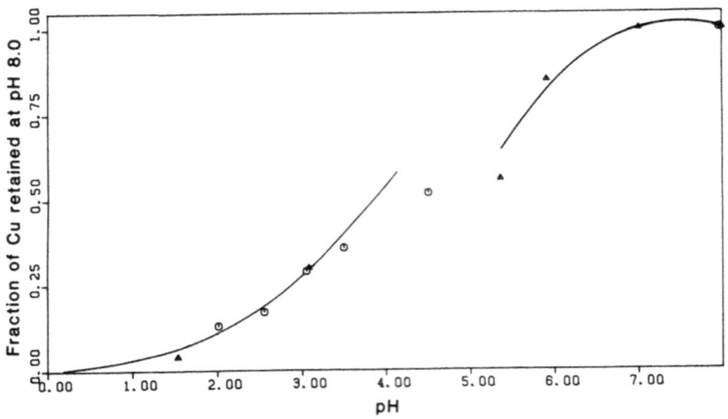

Fig. 3 Results of use of derived stability constants in copper speciation model to fit data of Mills et al.[9]. The high pH curve is derived from Data Set II parameters (△). Low pH parameters are derived from Data Set I (⊙). When plotted in fractional form, both data sets behave similarly.

The model is then run with the two data sets of Mills et al.[9] and the adjustable parameters β^*_{CuM}, K^*_{HM}, and T(M) are independently varied until the best fit to the high pH data is achieved. The best fit parameters are shown in Table 2, and a graphical fit is displayed in Figure 3. In Figure 3, the low pH curve represents the fitting procedure described earlier; the curve above pH 5 results from the modeling fit. Since SEASPE is not currently functional below pH 6, it is impossible to fit the entire data set with the model.

DISCUSSION

The β^* values determined above are not directly comparable to the overall stability constants determined by other investigators. The best comparable Type 2 conditional constant which we can derive is:

$$\begin{aligned}\beta^c_{CuM} &= [CuM]/[Cu^{2+}]\ T(M)\\ &= [CuM]/[Cu^{2+}]\ [M]\cdot[M]/T(M)\\ &= [CuM]/[Cu^{2+}]\ [M]\cdot[M]/[HM] \text{ at pH} < pK^*_{HM}\\ &= {}^*\beta_{CuM}\,(K^*_{HM}/[H+])^{0.5} \qquad (13)\end{aligned}$$

Inserting the Type 2 parameters from Table 2 into Equation (13), we determine at pH 8.0 that:

$$\beta^{c}_{CuM} = 5.2 \times 10^{9}$$

and

$$\log \beta^{c}_{CuM} = 9.72.$$

This result is statistically indistinguishable from that found by Mantoura et al.[4] of 9.30 $\pm$ 0.59.

Since the study by Mills et al.[9] contains the only available pH dependent data for Cu-MDOM interactions, it is difficult to compare the derived K^{*}_{HM} with any other work. Suffice to say, it is close to the value found by Perdue[7] for phenolic groups in soil humic acids.

Our Type 1 parameters are also difficult to compare with those from other studies, both because of the paucity of data on MDOM, and the relative insensitivity of our speciation model to Type 1 interactions.

Use of the Huizenga and Kester[6] K^{*}_{HL} value to derive β^{*}_{CuL} gives a value of 2.5 x 10^{4}, close to the values found by Buffle et al.[2] for soil humic acid binding of copper. This value is certainly not unusual in light of the analysis of Gamble et al.[8] who point out that this figure is a weighted average over many sites and is therefore a function of humic acid structure. Although useful in equilibrium models, these derived stability constants are not true thermodynamic values because of the variability of humic acid structures. The K^{*}_{HL} value is typical of carboxylic acid sites[7] which are commonly assumed to be active in binding copper[8].

A brief comment can also be made concerning the nature of the Type 2 binding sites. The K^{*}_{HM} value suggests that the associated protonation occurs on phenolic or amine type sites. The ratio of T(M) to T(L) in this study, however, is 0.25; the nitrogen to carboxyl ratio in various humic acids[18,19] varies from 2 to 0.64. On the other hand, Schnitzer[18] gives a phenolic-to-carboxylic ratio of 0.75 for a variety of humic acids. Our derived value is smaller than all those listed above. Given the variety and differences in the methods of sample isolation among these investigators, this is not surprising, but it does not allow us to assign a specific functionality to the two sites.

The T(M) ratio for the two data sets is in the same ratio as the T(Cu) values; it does not follow the ratio of the DOC's. This implies that a large fraction of the MDOM in Data Set I is inactive in binding copper, an observation which may be coupled to the fact that the high DOC value in this data set is associated with an active phytoplankton bloom.

The nature of the copper binding sites is much more obscure. As Gamble et al.[8] point out, the copper binding sites are probably all multidentate in character; in our case, they would be formed by variously available stereochemical positions of both carboxylic and phenolic ligands. The two broad classes of copper stability constants identified by us might result from two uniquely available ligand environments. The ESR data presented in the Mills et al.[9] paper suggest that copper is bound in oxygen-rich sites, but that nitrogen chelation cannot be ruled out.

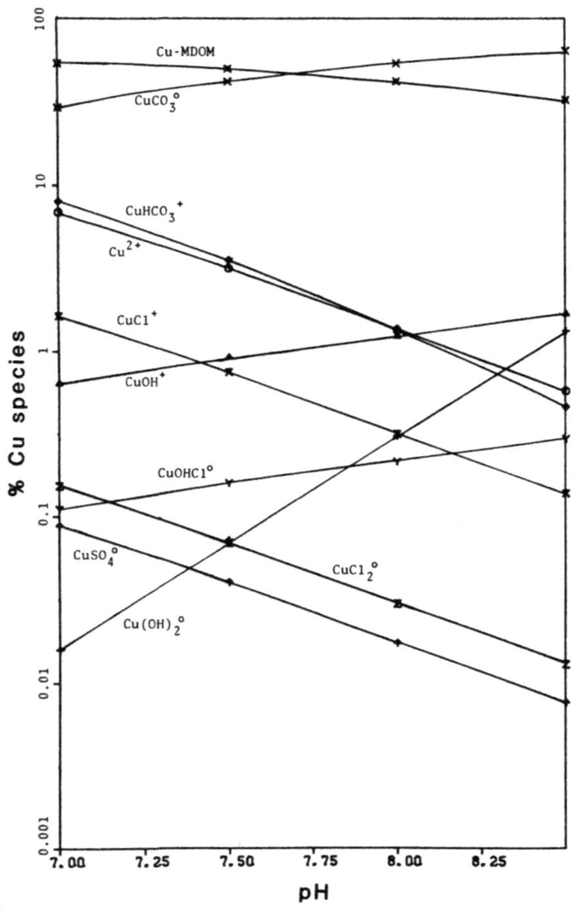

Fig. 4 Complete speciation plot for copper in 35 $^o/oo$ seawater. $T(Cu) = 9.4$ nmol kg^{-1}, $T(MDOM) = 1.0 \times 10^{-6}$ mol kg^{-1}.

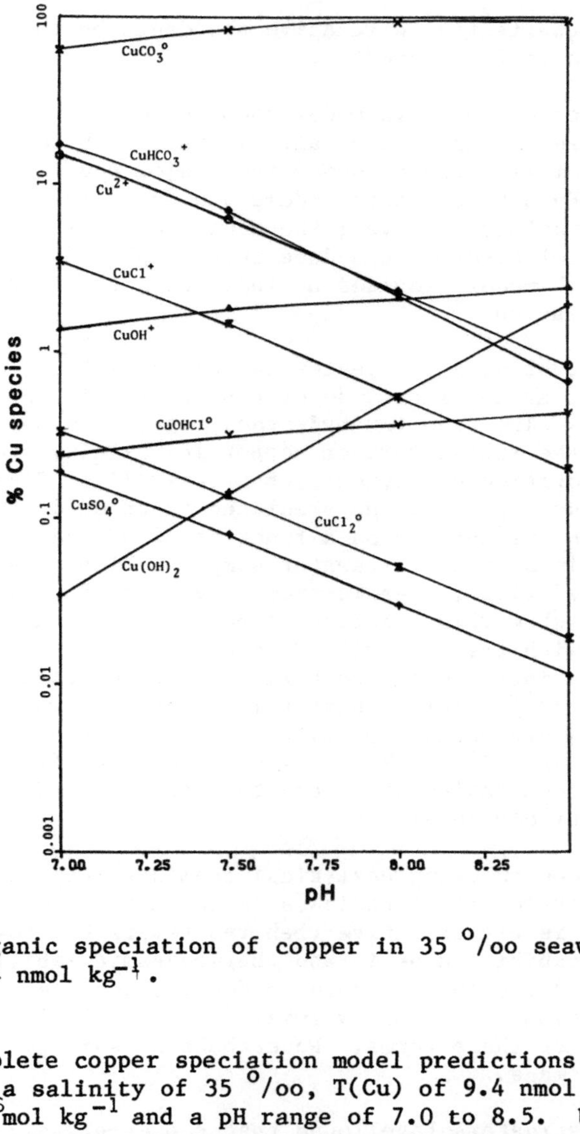

Fig. 5 Inorganic speciation of copper in 35 ^{o}/oo seawater. T(Cu) = 9.4 nmol kg^{-1}.

The complete copper speciation model predictions are shown in Figure 4 for a salinity of 35 ^{o}/oo, T(Cu) of 9.4 nmol kg^{-1}, T(M) of 1.0×10^{-6}mol kg^{-1} and a pH range of 7.0 to 8.5. Under the conditions given, Type 1 sites bind a negligible fraction of the total copper; the Cu-MDOM contribution shown is thus almost entirely from Type 2 sites. For comparison, the inorganic speciation alone is shown in Figure 5.

The dominance and effects of $CuCO_3^o$ and Cu-MDOM (i.e., copper-humate, copper-organic complexes) are clearly shown. The carbonate ion competes so strongly for Cu^{2+} that it is able to largely overcome the effects of MDOM at normal seawater pH's. In fact, these two species are so dominant that the remainder of the speciation is unaffected in relative abundance by the addition of MDOM to an organic-free seawater.

The predictions of this model compare well in a qualitative sense with those of Mantoura et al.[4] in that both considerations find that organically bound copper is a major equilibrium species in seawater. Even though both models use the same total organic ligand concentrations, however, the Mantoura et al.[4] model gives a lower fraction of Cu-MDOM than does ours at pH 8.0. Further, the Mantoura et al.[4] model provides no indication of the pH dependence of copper speciation.

The effect of organic species on the concentration of free Cu^{2+} is as important as the magnitude of copper-organic species concentrations. The free (or hydrated) cupric ion has been implicated as the active form of copper in toxicity studies[20], as well as in adsorption on various substrates[21,22]. Figure 6, therefore, shows a plot of the predicted fraction of free Cu^{2+} (labeled "C" on the curves) as a function of pH and organic ligand concentration in a typical seawater sample. Bioassays of $[Cu^{2+}]$ in UV-irradiated (i.e., organic-free) seawater (Sunda, personal communication) show this fraction to be about 2.5% at pH 8, which is consistent with the T(MDOM) = 0 value shown in Figure 6. As organic ligands increase in ocean waters, copper toxicity at fixed T(Cu) declines[23,24]. This behavior is clearly shown in Figure 6 in which $[Cu^{2+}]$ decreases from 2.3% to 0.2% between T(MDOM) = 0 and T(MDOM) = 4×10^{-6} mol kg^{-1}. The toxicity at fixed T(Cu) increases with decreasing pH[25], and this trend in $[Cu^{2+}]$ also is shown quantitatively in Figure 6.

A major concern among analytical chemists measuring copper in seawater by voltammetric techniques is the extent to which a given copper species is electroactive (behaves reversibly toward the electrode and exhibits kinetic and thermodynamic lability). It is difficult to address this concern in detail, because, as is pointed out elsewhere in this volume, there is little agreement on the definition of these terms. Nevertheless, some general observations can be made.

Some investigators have found that copper-organic complexes are slow to dissociate over a period of many minutes when acidified, heated, and ultrasonified. Likewise several authors[26,27] find that the lack of electroactivity of copper-organic species makes it possible to use voltammetric methods as one component of an operational speciation scheme. On the other hand, Mills et al.[9] find that 70% of their organically bound copper undergoes rapid exchange. For this consideration, we will assume that all the copper-organic species in our model are "non-labile".

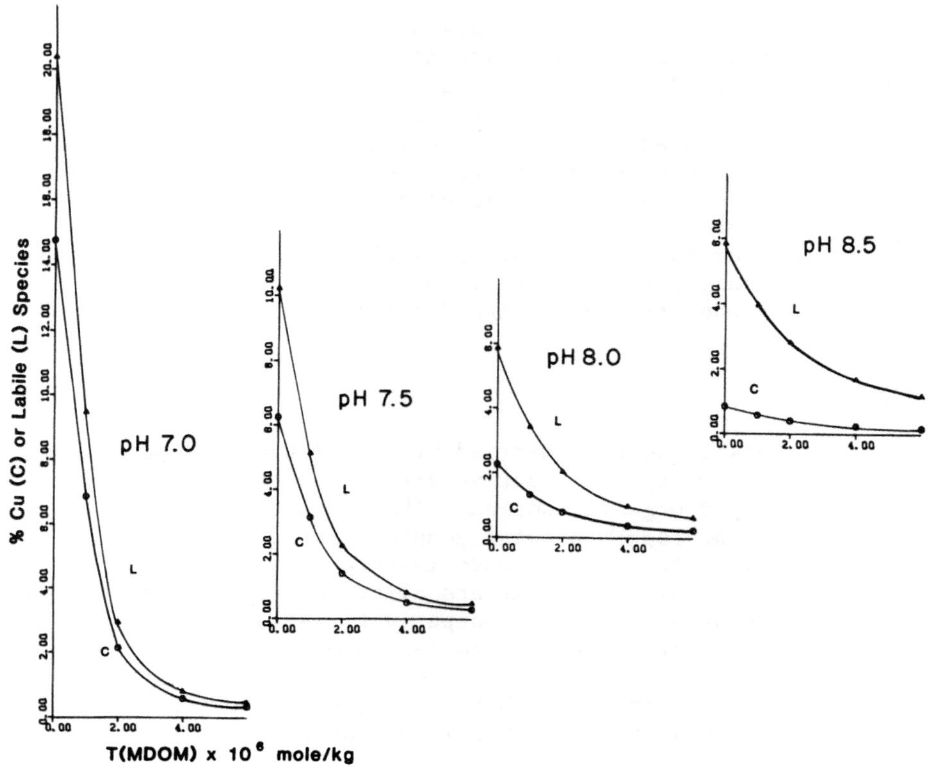

Fig. 6 Predicted fractions of free Cu^{2+} (C) and labile copper (L) as functions of pH and total organic ligand concentration in 35 $^{o}/oo$ seawater. T(Cu) = 9.4 nmol kg^{-1}.

We have related our model calculations to electrochemical measurements of copper by grouping the copper species into two categories: electrochemically labile and non-labile. The copper-carbonate (and possibly copper-bicarbonate) complexes have been regarded as probably behaving irreversibily at an anodic stripping voltammetery (ASV) electrode[28]. In our own laboratory we have found that these species may dissociate slowly, over a period of hours, when solutions containing them are acidified. Thus, we will include both the carbonate and bicarbonate complexes as "non-labile" copper species. For this analysis we therefore assume that Cu^{2+} and the chloride, hydroxide, and sulfate species are labile (Figure 6).

An interesting pattern is observed as one examines the variation of the labile fraction with pH and total organic ligand concentration. If a sample of seawater containing 9.4 nmol kg^{-1} total copper were measured at its normal pH of 8, and if it contained 4.0×10^{-6} mol kg^{-1} of dissolved organic ligands (about 0.7 mg DOM kg^{-1}), only about one percent of the total copper would be expected to be seen in an ASV analysis. Associated with this hypothetical sample is an apparent complexing capacity (the concentration of free Type 2 ligands) of 18 nmol kg^{-1}.

The effect on lability of decreasing pH is shown in Figure 6. As pH decreases, the fraction of copper which is labile increases dramatically when the total ligand concentration is low.

SUMMARY

In this work we have derived protonation and copper binding constants for two types of active sites in a naturally occurring dissolved copper-organic complex. The protonation constant for one site and the copper binding constant for another site are consistent with observations made in other systems; our determination of the remaining parameters allows us to develop the most complete speciation model for copper yet available. The model's predictions of copper-organic complex formation, free copper available for toxic effects, and the extent of non-labile copper formation agree favorably with bioassay, gel filtration, and amperometric analyses for copper in seawater.

ACKNOWLEDGEMENTS

This work benefitted greatly from discussions with many of our colleagues, especially G.L. Mills, A.K. Hanson, D.L. Huizenga, and J.G. Quinn. We owe much thanks to them. Our efforts were supported by National Science Foundation Grant No. OCE 78-26172.

REFERENCES

1. Harvey, G.R., D.A. Boran, L.A. Chesal and J.M. Tokar, 1981: The structure of marine fulvic and humic acids. Mar. Chem., submitted.
2. Buffle, J., P. Deladoey, F.L. Greter and W. Haerdi, 1980: Study of the complex formation of copper (II) by humic and fulvic substances. Anal. Chim. Acta, 116, 255-274.
3. Mantoura, R.F.C. and J.P. Riley, 1975: The use of gel filtration in the study of metal binding by humic acids and related compounds. Anal. Chim. Acta, 78, 193-200.

4. Mantoura, R.F.C., A. Dickson and J.P. Riley, 1978: The complexation of metals with humic materials in natural waters. Estuar. Coast. Mar. Sci., 6, 387-408.
5. van den Berg, C.M.G. and J.R. Kramer, 1979: Determination of complexing capacities of ligands in natural waters and conditional stability constants of the copper complexes by means of manganese dioxide. Anal. Chim. Acta, 106, 113-120.
6. Huizenga, D.L. and D.R. Kester, 1979: Protonation equilibria of marine dissolved organic matter. Limnol. Oceanogr., 24, 145-150.
7. Perdue, E.M., 1978: Solution thermochemistry of humic substances --- I. Acid-base equilibria of humic acid. Geochim. Cosmochim. Acta, 42: 1351-1358.
8. Gamble, D.S., A.W. Underdown and C.H. Langford, 1980: Copper (II) titration of fulvic acid ligand sites with theoretical potentiometric, and spectrophotometric analysis. Anal. Chem., 52, 1901-1908.
9. Mills, G.L., A.K. Hanson, Jr., J.G. Quinn, W.R. Lamela and N.D. Chasteen, 1981: Chemical studies of copper-organic complexes isolated from estuarine waters using C_{18} reverse-phase liquid chromatography. Mar. Chem., submitted.
10. Nordstrom, D.K. and 18 others, 1979: Comparison of computerized chemical models for equilibrium calculations in aqueous systems, in: "Chemical Modeling in Aqueous Systems", E.A. Jenne, ed., American Chemical Society Symposium Series 93, Washington, D.C.
11. Johnson, K.S. and R.M. Pytkowicz, 1978: Ion association of Cl^- with H^+, Na^+, K^+, Ca^{2+}, and Mg^{2+} in aqueous solutions at 25°C. Am. J. Sci., 278, 1428-1447.
12. Johnson, K.S. and R.M. Pytkowicz, 1979: Ion association of chloride and sulphate with sodium, potassium, magnesium and calcium in seawater at 25°C. Mar. Chem., 8, 82-93.
13. Johnson, K.S. and R.M. Pytkowicz, 1981: The activity of NaCl in seawater of 10-40 °/oo salinity and 5-25°C at 1 atmosphere. Mar. Chem., 10, 85-91.
14. Johnson, K.S., 1981: The calculation of ion pair diffusion coefficients: a comment. Mar. Chem., 10, 195-208.
15. Kester, D.R., R.W. Zuehlke, A.J. Paulson and R.H. Byrne, Jr., 1981: Inorganic speciation of copper in seawater. To be submitted.
16. Mills, G.L. and J.G. Quinn, 1981: Isolation of dissolved organic complexes from estuarine waters using reverse-phase liquid chromatography. Mar. Chem., 10, 93-102.
17. Mills, G.L., 1981: An investigation of dissolved copper-organic complexes in estuarine and coastal waters using reverse-phase liquid chromotography. Ph.D. thesis, University of Rhode Island.

18. Schnitzer, J., 1976: The chemistry of humic substances, in: "Environmental Biogeochemistry", Vol. 1, J.S. Nsriagu, ed., Ann Arbor Science Publishers.
19. Gagosian, R.B. and D.H. Stuermer, 1977: The cycling of biogenic compounds and their diagenetically transformed products in seawater. Mar. Chem., 5, 605-632.
20. Huntsman, S.A. and W.G. Sunda, 1980: The role of trace metals in regulating phytoplankton growth with emphasis on Fe, Mn, and Cu, in: "The Physiological Ecology of Phytoplankton", I. Morris, ed., Blackwell Scientific Publishers, Oxford.
21. O'Connor, T.P. and D.R. Kester, 1975: Adsorption of copper and cobalt from fresh and marine systems. Geochim. Cosmochim. Acta, 39, 1531-1543.
22. Murray, J.W., 1975: The interaction of metal ions at the manganese dioxide-solution interface. Geochim. Cosmochim. Acta, 39, 505-519.
23. Sunda, W.G. and P.A. Gillespie, 1979: The response of a marine bacterium to cupric ion and its use to estimate cupric ion activity in seawater. J. Mar. Res., 37 (4), 761-777.
24. Jackson, G.A. and J.J. Morgan, 1978: Trace metal-chelator interactions and phytoplankton growth in seawater media: Theoretical analysis and comparison with reported observations. Limnol. Oceanog., 23, 268-282.
25. Sunda, W.G. and R.L. Guillard, 1976: The relationship between cupric ion activity and the toxicity of copper to phytoplankton. J. Mar. Res., 34 (4), 511-529.
26. Batley, G.E. and T.M. Florence, 1976: Determination of the chemical forms of dissolved cadmium, lead, and copper in seawater. Mar. Chem., 4, 347-363.
27. Sugai, S.F. and M.L. Healy, 1978: Voltammetric studies of the organic association of copper and lead in two Canadian inlets. Mar. Chem., 6, 291-308.
28. Shuman, M.S. and L.C. Michael, 1978: Reversibility of copper in dilute aqueous carbonate and its significance to anodic stripping voltammetry of copper in natural waters. Anal. Chem., 50, 2104-2108.
29. Paulson, A.J. and D.R. Kester, 1980: Copper(II) ion hydrolysis in aqueous solution. J. Sol. Chem., 9, 269-277.
30. Zuehlke, R.W. and D.R. Kester, 1981: Ultraviolet spectroscopic determination of the stability constants for copper carbonate and bicarbonate complexes up to the ionic strength of seawater. Mar. Chem., submitted.
31. Matheson, R.A., 1965: A spectrophotometric study of the association of Cu^{2+} and SO_4^{2-} ions in aqueous solutions of constant ionic strength. J. Phys. Chem., 69, 1537-1545.

PLANKTON COMPOSITIONS AND TRACE ELEMENT FLUXES FROM THE SURFACE OCEAN

Robert W. Collier[1] and John M. Edmond

Department of Earth and Planetary Sciences
Massachusetts Institute of Technology
Cambridge, MA 02139 U.S.A.

[1]School of Oceanography
Oregon State University
Corvallis, OR 97331 U.S.A.

ABSTRACT

Plankton samples have been carefully collected from a variety of marine environments under the rigorous conditions necessary to prevent contamination for major and trace-chemical analysis. Immediately after collection, the samples were subjected to a series of physical and chemical leaching - decomposition experiments designed to identify the major and trace element composition of particulate carrier phases. Elements examined through some or all of these experiments include: C, N, P, Ca, Si, Fe, Mn, Ni, Cu, Cd, Al, Ba and Zn. Emphasis was placed on the identification of trace element/major element ratios in the biogenic materials.

The majority of the trace elements in the samples were directly associated with the non-skeletal organic phases of the plankton. This included a very labile, rapidly-recycled fraction, as well as more refractory, metal-specific binding sites. Calcium carbonate and opal were not significant carriers for any of the trace elements studied. A refractory phase containing Al and Fe in terrigenous ratio was present in all samples - even from the more remote marine locations. The aluminosilicates contributed insignificant concentrations to the other trace elements studied.

A variety of processes affecting the geochemical cycles of specific trace elements were identified. As much as 50% of the particulate Cd, Ni, Mn and P are rapidly released from the plankton and recycled within the surface ocean. During this process, the metal/P ratio in the residual particles must decrease by 10-30% for Cd and increase by a factor of 2-4 for Ni and Cu in order to balance their deep ocean enrichments. Although Mn is taken up and regenerated by plankton, the magnitude of this process is small with respect to other non-biogenic Mn fluxes and has little influence on its dissolved distribution. The Ba content of all known surface carriers is insufficient to account for the deep enrichment of dissolved Ba. A secondary concentration process results in the formation of significant particulate Ba within the upper thermocline.

INTRODUCTION

The marine geochemistry of trace elements has been a primary focus of oceanographic research for the past decade. With the development of improved sampling and analytical techniques it has been demonstrated that many of the trace elements show large concentration variations in both vertical profile and areal distribution. Because of this heterogeneity, these elements are extremely sensitive tracers of physical, geochemical, and biological processes in the oceans. The vertical and horizontal segregation of the trace elements often parallels that of the major nutrient elements involved in biological processes. These gradients are driven by the production, transport, and remineralization of particulate organic matter. Most of the recent research on the geochemistry of trace elements has only indirectly studied these processes through the examination of their resultant "signatures" in the water column and sedimentary record.

The trace elements Cd and Zn have dissolved distributions which closely follow those of phosphate and silicate, respectively[1,2]. These metals and nutrients are depleted in oligotrophic surface waters to concentrations which are less than 1% of their deep Pacific values. Cd is regenerated rapidly such that it correlates linearly with phosphate in the water column. Zn is regenerated more slowly and has a deep distribution which is similar to that driven by the production and dissolution of opal and carbonate. Another group of trace elements, represented by Cu, Ni and Ba, also show deep enrichment - but to a smaller degree[2,3,4] Their surface water concentrations are never depleted below 10% of their deep concentrations and their correlations with the nutrients show significant complexity - especially near the surface and sediment interfaces. Other trace element distributions, including those of Mn and Al, show little relationship to the nutrient cycles[5,6] - which is consistent with their individual chemistries. In order to understand better the variety of processes controlling the distributions of trace elements, the specific biogenic vectors must be independently determined.

Nearly all published chemical analyses of plankton and other marine particulate matter have been determined on bulk samples. Most include only the major or minor element composition - rarely both. Very few of the investigations demonstrated that the elements were quantitatively recovered and not contaminated during sampling or analysis. Experience gained in the collection of water samples for dissolved trace elements suggests that it is very likely that many of the reported plankton analyses are seriously contaminated. This fact, combined with other sampling and analytical problems discussed in this work, gives one very little confidence in using currently reported plankton data in trace element geochemical models. Notable exceptions include analyses by Martin and Knauer[7] who made serious attempts to address the problems of contamination in their samples. Martin et al.[8] have also published one of the only sets of high-quality plankton analyses which include major components (P, Si, Ca) along with the trace analyses on the bulk samples.

There is a large body of data and a relatively detailed understanding of the processes controlling the flux of major biologically cycled elements. The intention here is to understand and quantify the trace element cycles by the extension of these major element cycles. This will be accomplished through the careful examination of the ratios and chemical relationships between the trace elements and major elements representing biogenic carrier phases.

Numerous carrier phases and types of associations are possible between trace elements and marine particulate matter. These include: terrigenous material scavenged by biogenic particles; specific biochemical functions associated with metabolic processes; inclusion within structural-skeletal materials such as calcite, opal, or celestite; and scavenging processes at active surfaces such as hydrous metal-oxide precipitates or organics. Here we examine the significance of these different "carriers" in open-ocean surface plankton samples. The correlations between plankton compositions and the carrier ratios reflected in the water column and sediments will be linked to the known processes and fluxes determining the major element cycles.

There is a serious need for quantitative estimates of the role of organisms in determining the distributions and fluxes of trace elements in the oceans and sediments. The experiments reported here were directed at this problem and were not designed to examine specific trace element functions in the physiology or ecology of marine plankton. The complex biochemical and nutritional relationships between organisms and trace elements still need to be studied under simplified and controlled laboratory conditions; the geochemical problem is best approached through actual measurements in the field.

We present a comprehensive set of chemical analyses on a variety of plankton samples. These include the major element compositions as well as the concentrations of a group of trace elements. The specific trace elements were selected either because their dissolved distributions have been determined or because of some anticipated relationship to particulate carriers. Immediately after collection, the samples were split for total concentration determinations and were subjected to a series of chemical leaching experiments designed to separate carrier phases and associated trace elements. Two demands were imposed on the design of all experiments: the minimization of every possibility of trace element contamination and the prevention of excessive dilution of the trace element signals in the leaching solutions. To satisfy the goal of relating the major and minor element cycles, within these necessary experimental constraints, numerous trade-offs were made between increasing the experimental complexity and decreasing the handling and splitting of the samples.

EXPERIMENTAL METHODS

Over the past five years, there has developed an increasing awareness of the problems of sample contamination during collection, handling, and analysis. Many of these problems have been exhaustively detailed in recent publications of high quality trace element analyses, and we will not retrace those developments here[8,14] Specific details which are unique or particularly important to this research will be covered, but it should be noted that every step in the preparation, collection, storage, and analysis of these samples has been executed with "continuous contamination consciousness".

Seawater and plankton samples were collected from four open-ocean sites: a north-south transect between Capetown, South Africa and the Antarctic (STN E - 48°S, STN J1 - 60°S, STN M - 67°S); an eastern equatorial Pacific site, east of the Galapagos Islands (86°W); another equatorial Pacific site at 140°W (MANOP C); and a tropical north Pacific site at 11°N, 140°W (MANOP S).

Collection of uncontaminated particulate samples at sea is one of the most demanding sampling tasks. No method has been devised that will provide both significant freedom from contamination and large quantities of material. Practical compromises between these goals have to be made.

Two basic towing methods were used. Most samples (all Galapagos and most MANOP) were collected by manual vertical daytime tows from an inflated rubber raft which was located at least several hundred meters upstream (upwind) of the main research vessel. The methods were essentially those of Martin et al.[8].

The hand-towing method, although far superior to normal ship tows in preventing contamination, was severely limited by wind. sea-state, and the low concentration of plankton found in the surface waters of oligotrophic environments. Therefore, some plankton samples were collected from the main research vessels. The same basic towing rig used in the rubber raft was set up on a long boom which extended away from the ship over waters which were uncontaminated by the ship in its direction of travel. In cases where a comparison between raft and ship tows could be made, the elemental composition of materials collected from the same site gave no indications of contamination due to the ship.

All towing equipment was constructed of non-contaminating materials and the care taken in its handling was equivalent to that demanded by other trace element procedures. The plankton nets were conical, 3:1 in length to width ratio with a 0.5 meter mouth opening, and were made of 44 μm Nitex nylon. The mouth-ring was epoxy-coated and completely sewn into the front edge of the net. The dacron net-harness was spliced around the mouth-ring and all other fittings were nylon or polypropylene. The nets and polypropylene towline were carefully cleaned before going to sea and were further cleaned before and after each use by towing without the cod end installed.

All preparations and sample processing at sea were carried out within a HEPA-filtered air, laminar flow bench. The cod end was cleaned before each use and loaded into a polyethylene container. Immediately before deployment, the cup was tied into the cleaned net. At the end of the tow, the cod end was returned to the container and suspended in surface seawater. To reduce the rates of bacterial activity and zooplankton excretion, the samples were kept cold until processed on the ship.

For each tow, a set of splits was immediately collected to represent the total untreated plankton. The subsamples of the original plankton-seawater suspension were collected by filtration and centrifugation and were not washed. All filtrates and supernates were saved for analysis. The rest of the splits were resuspended and centrifuged from leaching solutions designed to selectively solubilize the particulate sample. An outline of the procedure is presented in Figure 1 and Table 1. The specific reagents and solutions used in these leaches are listed in Table 2.

The solubilization of all particulate matter samples involved oxidative dissolution with hot nitric acid in teflon bombs followed by dissolution of opal and silicate phases with hydrofluoric acid. To avoid volatilization losses, all processing steps were done at temperatures below 120°C and, when possible, within a closed system. The samples were not taken to dryness in order to avoid the formation of secondary precipitates.

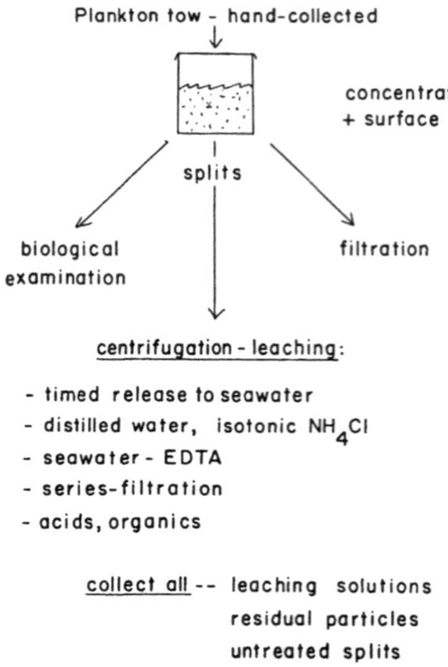

Fig. 1 Shipboard sample processing outline.

Table 1 Specific samples and experiments

Antarctic:	3 tows w/ship - bulk determinations; isolation of pure opaline material; distilled water, peroxide, 0.1N NCl washes.
Galapagos:	3 tows w/Zodiac - bulk determinations; distilled water, isotonic ammonium chloride, ethanol washes; acid leaching series - distilled water, .001N HCL, 0.1N HCl, HNO_3; timed release to seawater suspension.
MANOP Site C:	2 tows w/Zodiac - bulk determinations; distilled water, isotonic ammonium chloride, hot ethanol, and acid series (as above); series filtration of seawater suspension - 1.0 um, 0.4 μm; 0.1 μm, 0.05 μm, ultrafiltration.
MANOP Site S:	3 tows w/Zodiac - bulk determinations; 2 tows w/ship EDTA, APDC, and solvent extractions; timed release to seawater suspension - 72 hours in dark at 4 deg. (C) with and without antibiotics; 72 hours on deck (in light) at surface water temp.

Concentrations determined in some or all of the subsamples:
organic C, N, P, Si, Na, K, Ca, Mg, Fe, Al, Zn, Ni, Cu, Mn, Cd, Ba

Table 2 Solutions used in leaching and analysis.

Solution or reagent	Notes
Surface Seawater	Hand collected at the time and site of the plankton tow. this water is analyzed for nutrients and trace elements and these values represent the blanks for further processing.
Distilled water	Distilled at MIT and transported to sea in polyethylene containers. Processing involves boiling distillation, deionization, followed by another quartz-glass distillation. Blanks are usually below detection and never significant to these analyses.
Ammonium chloride	Synthesized from 6N HCl (vycor distilled) which was bubbled with clean NH_3 to a pH of 5.5. The resulting solution was desiccated and the collected crystals dried at 105 degrees. Solutions were made 0.56N with distilled water (which is iso-osmotic with seawater - 34.8^o/oo) and pH adjusted to 8 with NH_3.
Ethanol, chloroform	Reagent grade solvents, redistilled 2X in vycor glass and stored in teflon.
HCl (6N)	Distilled 2X in vycor and checked for blanks.
HNO (16N)	Reagent grade acid distilled 3X in vycor and checked for blanks.
HF	Baker Ultrex.
H O (30%)	Baker Ultrex, blanks checked.
HClO (70%)	GFS Co., 2X distilled from vycor glass.
APDC	Reagent grade prepared to 2% w/w in H_2O. Solution purified by repeated extraction with chloroform.
EDTA	Aldrich Chemical, Gold Label, blanks checked.
Ascorbic acid	Grand Isl. Biol. Co., blanks checked.

Analysis of major elements components followed methods outlined by Bishop et al.[15]. The trace elements in the solubilized plankton and leaching solutions were analyzed by atomic absorption spectrometry using electrothermal atomization (Perkin Elmer 5000 with HGA 500). The wide variety of sample matrices encountered in this work required that all analyses be made by methods using known additions of standards to the actual sample matrix[16]. Details of all analytical techniques are given by Collier[17].

RESULTS

The total bulk concentration of elements in the original plankton material must always be calculated from determinations on several fractions. In the simplest case, this involves the sum of concentrations in an untreated subsample of plankton and in the seawater that has been separated from the plankton. For other samples, it involves summing the analyses of a series of leaching solutions and particulate residues. The concentrations determined in various subsamples from a single sample are averaged to calculate an estimated total composition of the tow. The calculated totals for all of the analyzed plankton tows are summarized in Table 3.

One of the first results observed in the leaching experiments was the high concentration of many of the elements in the seawater which was separated from the plankton suspensions. These elements were initially associated with the particles when the plankton were collected and were released to the seawater during the two hours it took to bring the samples from the tow site into the ship's laboratory for processing. The time history of this release from samples stored in the dark at 4°C is shown in Figure 2. The data are expressed as the percentage of the estimated total particulate element which had been released to the seawater by the time of separation. The most dramatic release is for P where over 40% of the total was released within a few hours and over 70% within 24 hours. In parallel to this release, the trace elements Cd, Mn, and Ni were released to a somewhat smaller, but significant extent. There was always an initial pulse of Cu released, but its concentration then decreased by some secondary process. The release of Zn, Ba, Fe, and Al was always low with respect to their total concentrations.

Suspension of the samples in distilled water accelerated the release of P, Ni, Cd, and Mn up to the level of their maximum release into seawater after 72 hours (Table 4). The implications of this rapid remineralization process must be considered during the sampling and handling of organic particulate matter by towing, filtration, and trapping. The rapid release of elements requires careful containment of the sample and complete mass balancing from the time of collection if any systematics of their chemistry is to be understood. Variations in sampling techniques will surely result in large variations in the collected concentrations of P, Ni, Cd, Mn, and perhaps other trace components. Washing the samples with distilled water causes an even more extensive release. This rapid solubilization, coupled with sample contamination, may account for much of the variability in concentrations seen over the long history of reported plankton elemental analyses.

Table 3. Bulk plankton compositions, all stations. Estimates based on averages of subsamples of each tow. Values listed immediately below concentration values are the standard deviation of the mean when several subsamples were used. If no error is listed then the concentration is based on a single subsample only. Concentration units: Ca, P, Si, C, N - nmol g^{-1} dry plantkon; Fe, Zn, Al - μmol g^{-1} dry plankton; Cu, Ni, Cd, Mn, Ba - nmol g^{-1} dry plankton.

		Antarctic		Galapagos		MANOP C		MANOP S	
	Stn M	Stn J1	Stn E	Tow 1	Tow 3	Tow 1	Tow 2	Tow 3	Tow 6
Ca	0.011	0.23	0.3	0.84	1.39	1.34	1.5	3.66	1.59
				(0.03)	(0.04)	(0.27)	(0.24)	(0.26)	
P	0.024		0.22	0.42	0.35	0.26	0.3	0.44	0.36
				(0.04)	(0.02)	(0.03)	(0.01)	(0.01)	
Si	12.3			1.28	0.68	0.56	0.37	0.15	0.1
				(0.01)				(0.01)	
C				33.4	36.8				
				(1.8)					
N				6.2	6.25				
				(0.2)					
Cu	47	300	410	206	223	125	166	291	151
				(33)	(10)	(4)	(14)	(17)	
Ni	16.3	20	63	232	215	255	220	392	433
				(5)	(21)	(25)	(12)	(16)	
Cd	1.7	20	98	482	311	159	143	316	214
				(48)	(27)	(7)	(7)	(11)	
Fe	0.17	2.53	2.73	6.8	2.93	1.31	1.23	7.4	1.96
				(0.17)		(0.02)	(0.12)	(0.7)	
Mn				165	146	96	95	209	131
				(17)	(1)	(7)	(4)	(11)	
Zn	0.32	6.1	3.75	1.36	2.26	1.02	0.64	1.35	1.26
				(0.11)		(0.07)	(0.02)	(0.02)	
Al	0.39			9	7.7	0.72	0.94	12.5	3.48
						(0.09)	(0.1)	(0.3)	
Ba	205				490		516		

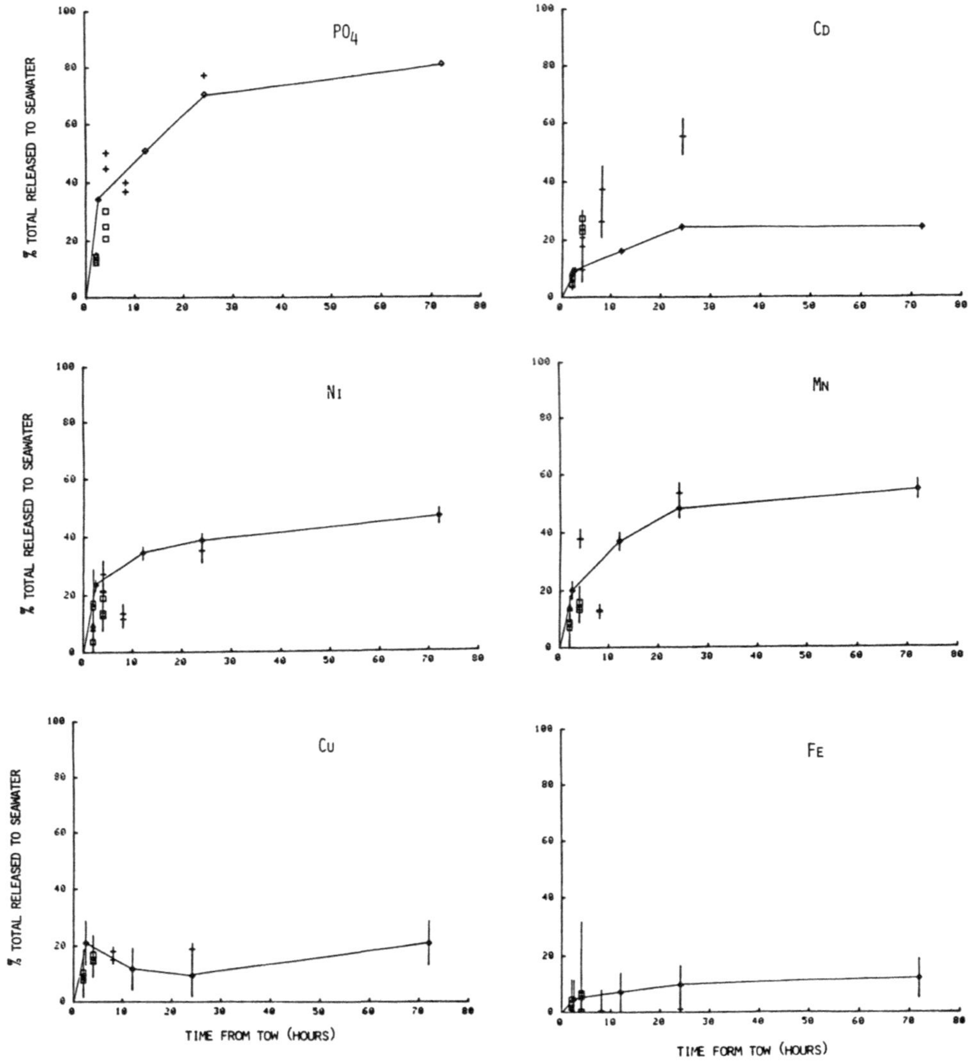

Fig. 2 Seawater release percentages vs time. Vertical lines on points represent 95% confidence limits on the percentage. Points connected by lines are from subsamples of a single tow.

Numerous other leaching experiments were performed, the results of which can only be summarized in this report (Figure 1). The rapid release process appears to be due to a combination of active excretion, cell lysis, and bacterial decomposition of the organic matter. No direct identifiable relationship exists

Table 4 Seawater and distilled water suspensions. Percentage of each element released to each solution. Time listed in last row indicates the time that each solution was exposed to the sample.

	Galapagos Tow 3 (sw)*	(dw)	(dw)	MANOP C Tow 1 (sw)	(dw)	Tow 2 (sw)	(dw)	(dw)
Ca	1.2	3.8	4.6	0	1.2	0	0	6.7
P	47	27	29	28	36	13	33	69
Cu	14	8.9	9.7	15	16	17	38	57
Ni	23	12	13	15	32	17	38	57
Cd	19	10	10	25	17	6.7	26	34
Fe	4.6	4.0	4.3	6.3	7.1	3.7	6.8	10
Mn	38	21	22	15	30	8.7	30	54
Zn	5.8	3.3	4.1				21	22
Si	.47	.38	.41					
Al	.19	.14	.22			5.4	2.0	2.2
Ba	3.4		4.9					
time	4hrs.	1hr.		4hr.	1/2hr.	2hr.	1/2hr.	8hr.

* "(sw)" indicates a seawater suspension supernate.
"(dw)" indicates a distilled water supernate.

between these labile trace metals and P in the release products. The solubilized metals have major fractions which are small enough to pass through 0.4 μm filters and are not strongly bound to non-polar, extractable organic matter or photosynthetic pigments. The stability of any surface organic-metal binding is such that significant fractions of Cd, Mn, and Ni as well as smaller fractions of the Cu and Fe are exchangeable with 0.56N ammonium chloride (isotonic with the seawater suspension) and are directly chelateable by EDTA and APDC. This rapidly solubilized fraction should be returned to the dissolved pool of these elements soon after the particulate material begins to decompose.

Significant fractions of the Cu, Ni, Cd, Ba, Mn, and Zn are bound to organic sites and are not rapidly exchanged with seawater or distilled water. Empirically, this pool of elements is rapidly extracted from the organic matrix at pH 1 (0.1N HCl) along with the release of Ca (Figure 3). However, direct examination of the calcium carbonate (hand-picked foraminifera) in these samples reveals that the concentration of these trace elements in that phase is not significant with respect to the total HCl-soluble fraction (Table 5, and ref. 18). Based on a low level of exchange in EDTA leaches, hydrous metal-oxides did not appear to be major carriers in these samples. It is likely that this acid-exchange-

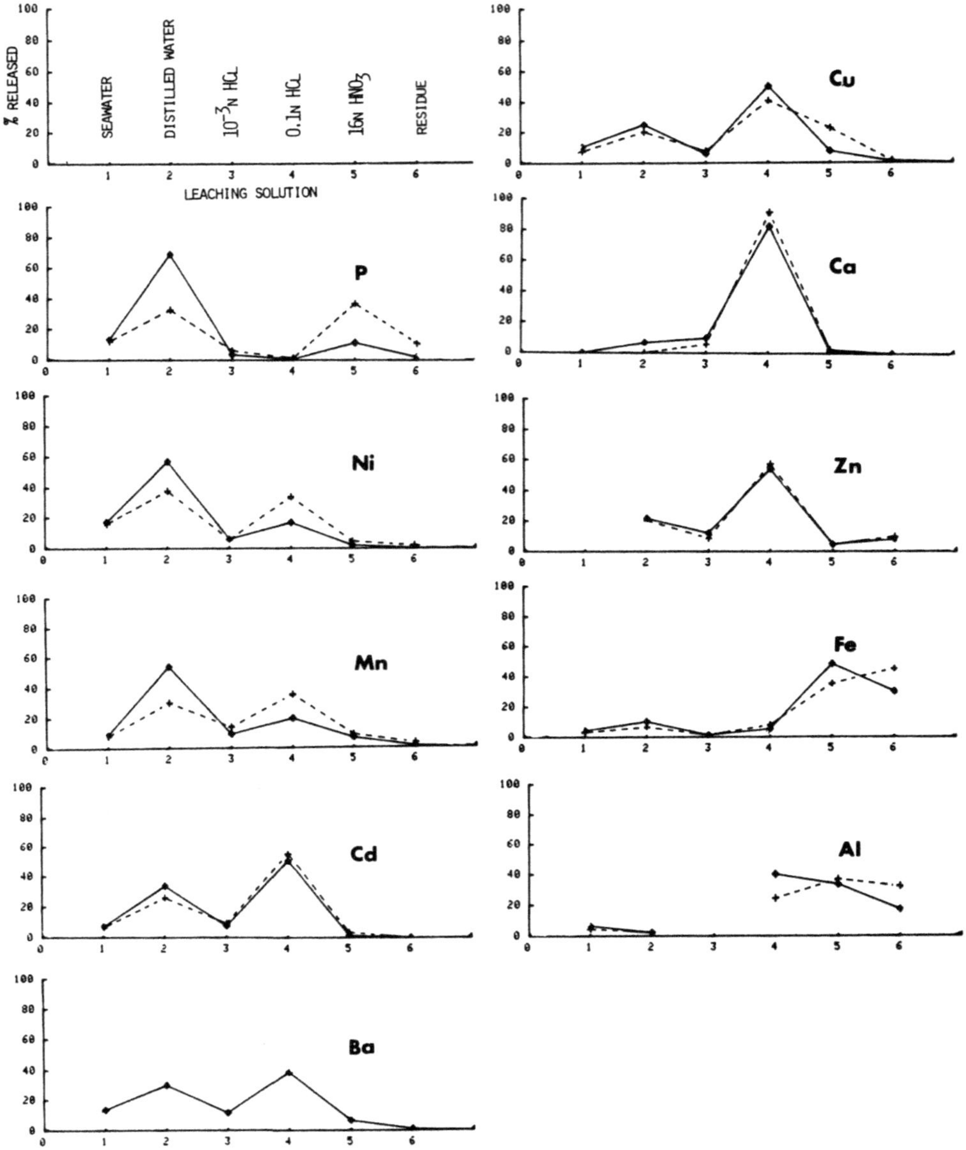

Fig. 3 Acid leaching series, MANOP C, tow 2. The points plotted and connected with lines on the graph represent the percentage of the total element released to each leaching solution. Leaching solutions are listed in upper-lefthand corner. Dotted line is a short-term exposure to the solution (∿ 30 minutes), solid line is an 8-hour exposure to each solution.

able fraction will undergo a much slower rate of remineralization, relative to the labile trace elements and P, as the particles settle into the deep ocean and sediments.

There is an even more refractory pool of trace elements associated with the plankton which contain high concentrations of Si, Al, Fe, and variable amounts of Zn. Examination of one of the major biogenic phases in this fraction, opal, shows that this is not the carrier for these elements (Table 6). The high concentrations of Al and Fe are consistent with the presence of lithogenous material. It is not likely that these residual components are solubilized in the water column, and their transport to the sediments must be, essentially, a "single-pass" conservative process.

DISCUSSION

A goal of this research was to quantify the biogenic fluxes of trace elements by using surface plankton compositions and leaching behavior. The flux estimates made from a "carrier model" will be compared with values derived from a two-box vertical model describing the dissolved distribution of trace elements.

The carrier model makes use of independent predictions of the flux of major organically-cycled elements. Minor elements are coupled to these biogenic fluxes through their ratios to major elements which represent the biogenic carrier phase. The major elements that can be used include organic C, N, P, Ca (carbonate), and Si (opal). Since P was measured in all samples and leaches along with the trace elements, it will be used to represent the primary flux of organic matter. Independent estimates of the P

Table 5 Metal/Ca ratios in bulk samples and forams.

Samples: X169 MANOP C, Tow 2, 0.1N HCl, 8 hour exposure; Foram sample "A" - hand picked from MANOP samples.

Ratios to Ca ($x10^{-6}$)	Cu	Ni	Cd	Fe	Mn	Zn	Al	Ba
X169 bulk	63	32	60	51	15	270	270	100
Forams	.31	.4	.025	(0)	.4	.5	3.5	3

Water column "regeneration" ratios (x 10^{-6}):
Ni/Ca - 130 Zn/Ca - 100 Ba/Ca - 1000

Table 6 Elemental ratios in opal samples.

Antarctic diatoms, Stn. M:	(Ratios to Si - x 10^{-5}, to Al - x 1) Fe/Si	Zn/Si	Al/Si	Ba/Si	Fe/Al
bulk (untreated)	1.3	2.6	3.3	1.6	.38
peroxide-water washed	0.6	1.9	2.7	1.1	.23
HCl washed	(1.9)	.28	2.5	0.2	(.73)

Water column "regeneration" ratios (x 10^{-5}): Zn/Si - 5 Ba/Si - 70
Antarctic diatom filter sample A11. Element ratios (x 10^{-5}):
Cu/Si - 0.38 Ni/Si - 0.13
Radiolaria hand-picked from Pacific Eocene sediment samples
(x 10^{-5}): Ba/Si - 0.5

flux are derived from inventories of total surface-ocean carbon fixation rates (550-650 μmol C cm^{-2} yr^{-1})[19,20]. Empirical relationships derived from calculations of nitrogen fluxes in the surface ocean[21] are then used to estimate the fraction of the total production which is not recycled within the surface and actually constitutes a net flux to the deep ocean. This "new production"[21] is approximately 100 μmol C cm^{-2} yr^{-1}. It is further assumed that this organic carbon flux is accompanied by a phosphorus flux at a "Redfield" C:P ratio of 106:1[22]. The resulting P flux is approximately 1 μmol P cm^{-2} yr^{-1}, which agrees well with the box model prediction (Table 7) and with other inventories[23].

Trace elements which are rapidly recycled with the organic matter (Cd, Ni, Mn) must have a net particulate flux out of the surface ocean which is lower than their total rate of production in biogenic particles. Elements which are fixed into biogenic particles and not recycled (Cu, Zn, Ba, Fe, Al) must be transported out of the surface at a rate equivalent to their total production. This concept is directly analogous to that developed for carbon and nitrogen[21].

The fractions of elements which are rapidly regenerated in the seawater release experiments (Figure 2) are taken to represent the fraction that would be rapidly recycled in the surface ocean. A relative enrichment factor, β, for the residual metal:P ratio in the particles is estimated from the steady-state releases. The residual metal:P ratio is applied to the "new production" estimate for the net P flux to calculate the biogenic carrier flux out of the surface ocean for each trace element. A diagram of the carrier model is presented in Figure 4. Although the parameters of the model are poorly constrained, it represents one of the few methods of estimating trace element fluxes from compositional data for surface particulate matter. The plankton collected for these experiments were made up of a mixture of phytoplankton and microzooplankton and may not directly represent the composition of the primary producers. However, the composition of this bulk surface particulate matter _is_ representative of some intermediate stage of the formation - recycling process. Since the empirical release coefficient, β , is also measured on this material, its application in the carrier model is still valid.

Another estimate of surface fluxes can be derived by the application of a vertical two-box ocean model to the distribution of the trace elements[24]. The particulate flux out of the surface layer, F, is calculated to balance the sum of the other input and output fluxes:

$$F = V_m(C_d - C_s) + V_r C_r + A;$$

where V_m (3.5 m yr^{-1}) is the exchange rate between the surface and deep boxes of water with concentrations C_s and C_d, respectively; $V_r C_r$ represents the river input[25,26]; A is the atmospheric deposition flux[27,28]. The last two terms are poorly known (at best) and, in some cases, introduce significant uncertainty in the box

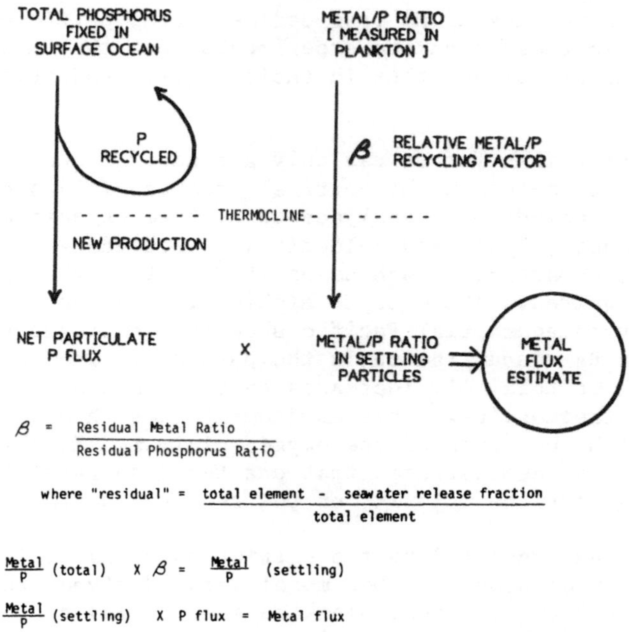

Fig. 4 Outline of carrier model.

model flux estimate (e.g. Mn). In other cases, where there is a significant enrichment of the deep waters ($C_d >> C_s$) then the cyclic term or "mixing flux" dominates all of the inputs to the surface ocean and a relatively good estimate of the required particulate flux is derived (e.g. Ba or P).

The results of the carrier model and box model are summarized in Table 7 along with the parameters chosen as input for each element. The two models agree, to within a factor of 2, on fluxes for Cd, Ni, and Cu. This suggests that the biogenic material sampled and considered with the carrier model can represent the major particulate flux for these elements which maintains their vertical gradient.

Other flux estimates can be made and compared to the results of these models. Direct measurements of fluxes have been made through the use of sediment traps[29,30]. In general, the results of these experiments are consistent with those of the carrier model, where the data were collected from similar locations. Further corroboration of the results can be seen by their agreement with estimates derived from vertical advection-diffusion models, suspended matter compositions, and sediment accumulation rates[17].

The box model flux for Mn is completely dependent on the values chosen for the primary atmospheric and river fluxes and is difficult to compare with the carrier flux. Martin and Knauer[31] showed a rapid release of Mn from sediment trap particles, similar to that seen in these plankton experiments, but they measured a much higher total surface flux in their highly productive coastal environment.

The carrier flux for Ba can only account for one third of that required to maintain the vertical gradient in dissolved Ba. This estimate includes all analyses of surface organic matter, calcium carbonate, opal, and celestite. A secondary process forming Ba-rich particles such as barite[32] within the upper thermocline is required. The GEOSECS Niskin-filtered particulate data[33] from the eastern equatorial Pacific show this enrichment process clearly. The Ba concentration in the particles, when normalized to total mass or total Al, increases by more than a factor of 20 in the upper thermocline. This enrichment goes through a maximum coincident with the depth of the oxygen minimum. In this case the carrier model has demonstrated that unaltered surface plankton cannot account for the majority of the required surface flux.

The Zn flux predicted by the carrier model is at least an order of magnitude above the box model value derived from the dissolved distributions[2]. These plankton samples may not, then, be representative of the Zn composition of the open-ocean particulate matter whose flux determines the distribution. Under the assump-

Table 7 Carrier Model, Box Model Results.

Element	Metal/P (measured) x 10^{-3}	Cyclic enrichment factor (β)	Metal/P (settling) x 10^{-3}	C_s* (nmol kg^{-1})	C_d** (nmol kg^{-1})	Flux Estimates (nmol cm^{-2} yr^{-1}) via organic carrier (P flux estimate)	Flux Estimates (nmol cm^{-2} yr^{-1}) box model estimate
P				50	2500		870
Cd	0.4 -0.7	1.6	0.6-1.1	0.0	0.9	0.5-1.0	0.34
Ni	0.6 -1.0	2.4	1.4-2.6	2	9	1.2-2.3	2.7
Cu	0.5 -0.6	3.4	1.7-2.0	1.2	4	1.5-1.7	1.2-1.4
Mn	0.35-0.43	2	0.7-0.9	0.8-3.0	0.5-1.2	0.6-0.8	2.1-2.3
Ba	1.4 -1.8	6	8-11	35	135	7-10	37
Zn	2 -5	6	12-30	0.8	7	10-25	2.4
Fe	5 -15	6	30-90	?		25-80	--

*C_s = average surface water concentration used in box model.
**C_d = average deep water concentration used in box model - weighted between deep Atlantic and Pacific according to their relative volumes [ref. 18].

tion that the dissolved distributions are correct, this may also indicate contamination of the plankton samples with Zn. The Fe and Al concentrations in the plankton are relatively variable and their refractory leaching behavior supports the importance of terrigenous carriers for these elements. At the levels present, using a possible range of crustal metal:Al ratios, this carrier could not be significant for any of the other trace elements studied. At very low concentrations of particulate Al, a significant fraction of the total Al was solubilized at pH 1. This fraction is probably not terrigenous and may represent biogenic Al which has been removed from the surface water.

CONCLUSIONS

This research presents a set of major and minor element analyses on open-ocean plankton samples which have been used to predict geochemically consistent fluxes for trace elements. It is shown that significant fractions of some elements are very weakly associated with the plankton and that their loss during sampling is very likely. Unless extreme care is taken to contain all fractions and avoid washing the sample, errors will result in estimating the role of plankton in the surface cycles of these elements.

The utility of the carrier model is demonstrated by comparison to box model results, but the need for direct flux measurements is clear. Sediment trap experiments which are designed for trace element collection are currently underway in several geochemical research programs and some of their results are discussed in this volume. The release of labile elements still occurs within the sediment traps[31], and this must be considered in their design and handling and in data interpretation. Sediment traps respond directly to the variations in surface productivity above them[34] and if their results are to be applied on a global scale then carrier models will be needed to normalize the measured trace element fluxes.

ACKNOWLEDGEMENTS

Support for R. Collier was provided by an NSF Graduate Traineeship at the M.I.T. - W.H.O.I. Joint Program and by NSF Grant DES 75-03826 and ONR Grant N00014-80-C-0Z73. Samples were also collected during the NSF-IDOE Galapagos Hydrothermal and MANOP programs.

REFERENCES

1. Boyle, E.A., F. Sclater and J.M. Edmond, 1976: On the marine geochemistry of Cd. Nature, 263, 42-44.
2. Bruland, K.W., 1980: Oceanographic distribution of Cd, Zn, Ni, and Cu in the North Pacific. Earth Planet. Sci. Lett., 47, 177-198.
3. Chan, L.H., J.M. Edmond, R.F. Stallard, W.S. Broecker, Y.C. Chung, R.F. Weiss and T.L. Ku, 1976: Radium and barium at GEOSECS stations in the Atlantic and Pacific. Earth Planet. Sci. Lett., 32, 258-267.
4. Boyle, E.A., F. Sclater and J.M. Edmond, 1977: The distribution of dissolved Cu in the Pacific. Earth Planet. Sci. Lett., 37, 38-54.
5. Klinkhammer, G.P. and M.L. Bender, 1980: The distribution of Mn in the Pacific Ocean. Earth Planet. Sci. Lett., 46, 361-384.
6. Hydes, D.J., 1979: Aluminum in seawater: Control by inorganic processes. Science, 205, 1260-1262.
7. Martin, J.H. and G.A. Knauer, 1973: The elemental composition of plankton. Geochim. Cosmochim. Acta., 37, 1639-1653.
8. Martin, J.H., K.W. Bruland and W.W. Broenkow, 1976: Cadmium transport in the California current. In: "Marine Pollutant Transfer", H.L. Windom and R.A. Duce, eds. Lexington Books, pp. 159-184.
9. Boyle, E.A., 1976: "The marine geochemistry of trace metals". Doctoral dissertation, WHOI-MIT Joint Program in Oceanography.
10. Bender, M.L., G.P. Klinkhammer and D.W. Spencer, 1977: Manganese in seawater and the marine manganese balance. Deep-Sea Res., 24, 799-812.
11. Schaule, B. and C. Patterson, 1978: The occurence of lead in the northeast Pacific and the effects of anthropogenic inputs. In: "Proceedings of an International Experts Discussion of Lead: Occurence, Fate, and Pollution in the Marine Environment, Roving Yugoslavia, October 1977", M. Branica, ed. Pergamon Press.
12. Bruland, K.W., G.A. Knauer and J.H. Martin, 1978: Cadmium in the northwest Pacific waters. Limnol. Ocean., 23, 618-625.
13. Measures, C.I., R.E. McDuff and J.M. Edmond, 1980: Selenium redox chemistry at GEOSECS I re-occupation. Earth Planet. Sci. Lett., 49, 102-108.
14. Andreae, M.O., 1979: Arsenic speciation in seawater and interstitial waters: The influence of biological-chemical interactions on the chemistry of a trace element. Limnol. Ocean., 24, 440-452.
15. Bishop, J.K., R.W. Collier, D.R. Ketten and J.M. Edmond, 1980: The chemistry, biology, and vertical flux of

particulate matter from the upper 1500 m of the Eastern Equatorial Pacific. Deep-Sea Res., 27, 615-640.
16. O'Haver, T.C., 1976: Analytical considerations. In: "Trace Analysis: Spectroscopic Methods for Elements", J.D. Winefordner, ed. John Wiley and Sons, pp. 15-62.
17. Collier, R., 1981: "Trace element geochemistry of marine biogenic particulate matter", Ph.D. thesis, WHOI-MIT Joint Program in Oceanography, WHOI-81-10.
18. Boyle, E.A., 1981: Cadmium, zinc, copper, and barium in foraminifera tests. Earth Planet. Sci. Lett., 53, 11-35.
19. Koblentz-Mishke, O.J., V.V. Volkovinsky and J.G. Kabanova, 1970: Plankton primary production of the world ocean. In: "Scientific Exploration of the South Pacific", W.S. Wooster, ed. National Academy of Science, Washington, pp. 183-193.
20. Platt, T. and D.V. Subba Rao, 1973: Fisheries Research Board of Canada, Technical Report No. 370.
21. Eppley, R.W. and B.J. Peterson, 1979: Particulate organic matter flux and planktonic new production in the deep ocean. Nature, 282, 677-680.
22. Redfield, A.C., 1934: On the proportions of organic derivatives in seawater and their relation to the composition of plankton. James Johnstone Memorial Volume, Liverpool, pp. 117-192.
23. Froelich, P.N., 1979: "Marine phosphorus geochemistry", Doctoral dissertation, U. of Rhode Island.
24. Broecker, W.S., 1971: A kinetic model for the chemical composition of seawater. Quarternary Research, 1, 188-207.
25. Edmond, J.M., E. Boyle and R. Stallard: unpublished data.
26. Martin, J.M. and M. Meybeck, 1979: Elemental mass-balance of material carried by major world rivers. Mar. Chem., 7, 173-206.
27. Wallace, G.T., G.L. Hoffman and R.A. Duce, 1977: Influence of organic matter and atmospheric deposition on the particulate trace metal concentration of northwest Atlantic surface seawater. Mar. Chem., 5, 143-170.
28. Hodge, V., S.R. Johnson and E.D. Goldberg, 1978: Influence of atmospherically transported aerosols on surface ocean water composition. Geochem. J., 12, 7-20.
29. Cobler, R. and J. Dymond, 1980: Sediment trap experiment on the Galapagos Spreading Center, Equatorial Pacific. Science, 209, 801-803.
30. Brewer, P.G., Y. Nozaki, D.W. Spencer and A.P. Fleer, 1981: Sediment trap experiments in the Deep North Atlantic: isotopic and elemental fluxes. J. Mar. Res., 38, 703-728.

31. Martin, J.H. and G.A. Knauer, 1980: Manganese cycling in the northeast Pacific waters. Earth Planet. Sci. Lett., 51, 266-274.
32. Dehairs, F., R. Chesselet and J. Jedwab, 1980: Discrete suspended particles of barite and the barium cycle in the open ocean. Earth Planet. Sci. Lett., 49, 528-550.
33. Brewer, P.G. and D.W. Spencer, (in preparation): GEOSECS Pacific - Neutron activation data report for suspended particulate matter.
34. Dueser, W.G. and E.H. Ross, 1980: Seasonal change in the flux of organic carbon to the deep Sargasso Sea. Nature, 28, 364-365.

METALS IN SEAWATER AS RECORDED BY MUSSELS

Edward D. Goldberg* and John H. Martin†

*Scripps Institution of Oceanography
University of California at San Diego
La Jolla, California, 92093, U.S.A.

†Moss Landing Marine Laboratory
Moss Landing, California, 95039, U.S.A.

ABSTRACT

Metal concentrations in the soft tissues of bivalves from U.S. coastal waters reveal seawater distributional patterns which repeat themselves year after year. Such a situation may arise from long biological half-lives of the metal or uniform concentrations in the waters over the time period studied or a combination of the two. Not only are anthropogenic inputs of the metals revealed by bivalve metal concentrations but also the occurence of upwelling which brings into surface waters high metal concentrations. Both U-234/U-238 and Am-241/Pu-239+240 ratios have characteristic values for the dissolved and particulate states in seawaters. The values in bivalves can indicate which states are important in biological uptake.

INTRODUCTION

The concentrations of metals in bivalves such as mussels and oysters provide information about their levels and availability in seawaters. This concept has been extensively developed through recent programs concerned with monitoring marine waters for metals introduced as a consequence of industrial, agriculture and social activities. The difficulties in the direct analyses of the extremely low levels of many metals in coastal waters led some investigators to the use of sentinel organisms which were markedly enriched in these substances. As a consequence, assays could be carried out with relative ease and modest expense. Although the concentrations within a given species of organisms are not normally related to an exact seawater concentration, they are of great use in comparing relative concentrations in coastal regions, such as the Southern California and New York Bights and adjacent waters.

The sentinel organisms through their body parts can provide relative metal concentrations averaged over different time periods. A sense of these times can be obtained from the biological half-lives of metals in mussels (Table 1). For example, a single reservoir model[1] has been developed for Po-210 and Pb-210 in the soft tissues of mussels and biological half-lives for these two nuclides have been computed to be of the order of two years. Three pools have been found in a Mediterranean mussel for vanadium with turn-over rates of 1, 7 and 103 days[2]. The depuration curves were similar for both the shells and soft tissues. Clearly, the relative seawater concentrations of metals induced from levels in organisms will reflect periods of time longer than those measured directly in water samples recovered from a single cast.

The interpretation of metal results from sentinel organisms can be complexed by a variety of factors. For example, concentrations can be a function of bivalve size or age, season of sampling, sex of bivalve, its vertical position on the shoreline and the salinity and temperature of the ambient environment[3]. Many of these parameters may be interrelated. In estuarine zones the salinity variations may correspond to the differences in river flow at different times of the year. Thus, in any comparative study of seawater concentrations using mussels, recognition must be given to these parameters. The sentinel organisms should be sampled at the same time of the year, be of the same size, and be taken from similar locations.

Table 1. Biological Half-lives of Metals in Mussels

Substance	Model	Half-life	Reference
Pb-210 Po-210	Single reservoir	≃ 2 years	1
V	Three pools	1, 7 and 103 days	2
Pu-239+240	Two pools	≃ 2 years for slowly exchanging pool	3

Herein we shall review some recent results relating bivalve metal concentrations to those in seawater and to the form of the metal in the environment as well as to the sources and modes of transport of the metal from the continents to the oceans.

THE MUSSEL WATCH

Perhaps the most extensive set of data relating metal concentrations in sentinel organisms to those presumed in the environment have come from Mussel Watch Programs[3]. Both geographical and temporal trends in seawater concentrations are sought through soft-tissue analyses. In the U.S. Program[4] bivalves (oysters and mussels) were collected at over 100 stations along the East, West and Gulf coasts of the United States. Animals of uniform size were sought, where possible, approximately 5 to 8 cm long, although oysters were slightly larger. Rock environments were favored over man-made structures such as pilings or metal buoys where paints, creosote or other materials were applied. Immediately after collection, the animals were placed in a freezer and maintained in a frozen state during air-shipment to the participating laboratories. Organisms for metal or radionuclide analyses were placed in plastic bags. A single collector carried out the operation during each year of the program.

Two laboratories carried out the metal analyses, the Scripps Institution of Oceanography and the Moss Landing Marine Laboratories, both in California. The differences in metal concentrations (Pb, Cd, Ag, Zn, Cu, and Ni) from the atomic absorption assays carried out on duplicate sets of mussels collected at the same time were usually of the order of 10% or less. In the 1978 collection of the U.S. Mussel Watch, the laboratories analyzed alternately taken samples and there was a concord in the data. Finally, a composite sample of around twenty-five to thirty individuals was taken for the soft tissue or shell analyses.

The general picture that emerges for most of the metals is a distributional pattern that repeats itself year after year. An example is shown in Figure 1 of the results from years 1 and 3 for zinc in oysters and mussels collected along the coasts of the U.S. Such a situation can result from long biological half-lives of metals in the organisms (half-lives of the order of a year or more) or from uniformly, uniform levels of the metals in the seawater or a combination of the two.

We present here several examples of the records of environmental levels in mussels from the U.S. program. Detailed data will be presented elsewhere.

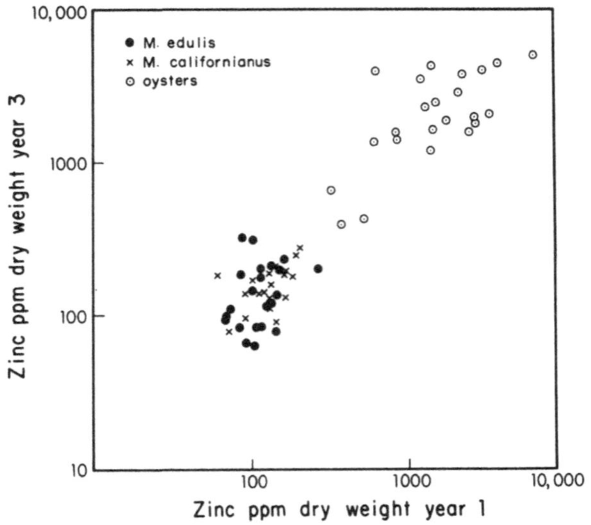

Fig. 1. Zinc contents in bivalves sampled during 1976 and 1978 in U.S. Mussel Watch Program.

Lead

The high lead concentrations in seawaters adjacent to urban areas results from the combustion of lead alkyls as anti-knock agents in gasolines[5]. Both direct atmospheric input as well as sewage, storm and river runoff contribute to the anthropogenic lead burden of surface waters. It has been estimated that the annual Pb inputs to the southern California coastal waters are 310 metric tons from the atmosphere, 200 tons from sewage, 190 tons from storm runoff and 40 tons from natural sources[6]. As lead alkyls are phased out from usages in gasolines, it is expected that lead concentrations in the waters, and in the bivalves, will decrease with time.

Mussel analyses lead to the identity of "hot spots", where the lead concentrations in the mussels, and presumably in their environmental waters, are raised over adjacent areas as a consequence of fluxes from highly populated industrial areas. The regional variations may be seen in the data from the U.S. west coast from its northern boundary in the state of Washington to its more southerly parts in California (Figure 2). Low lead concentrations were found in mussels taken in the central California stations, San Simeon to San Francisco, with the exception of those samples taken from the Farallon Islands where the levels were high (3.3 to 9.3 ppm, dry weight). In contrast, mussels from southern California stations had high lead concentrations, greater than 2.5

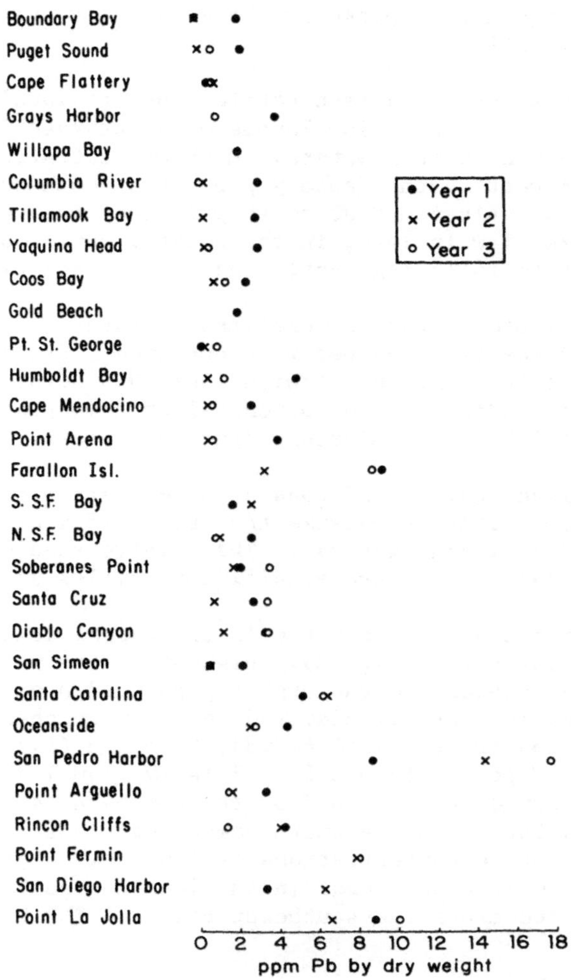

Fig. 2. Lead contents in mussels taken from U.S. West Coast stations going from Boundary Bay, Washington, south to Point La Jolla, California.

ppm dry weight, with the exception of Point Arguello and Rincon Cliffs. Levels were especially high at Point La Jolla (6.5 to 10.0 ppm), Point Fermin (7.9 to 8.0 ppm), Santa Catalina (5.2 to 6.4) and San Pedro Harbor (8.8 to 17.7). These high mussel concentrations are attributed to high influxes of lead in the Los Angeles/San Diego/Santa Barbara region primarily from automotive exhausts. The lead is transported principally through the atmosphere and accommodated in the seawaters following wet and dry fallout. The subsequential uptake by the mussels is evidenced by their unusual concentrations.

A similar program had been carried out previously along the California coast[7]. Again, low levels were observed in mussels from central California coastal waters. High concentrations were found in mussels adjacent to the densely populated southern California coasts. The low values may arise in part from the entry of upwelled waters, low in lead, in the coastal region off central California. (See following section on cadmium and plutonium.)

The lead isotopic ratios were also measured in the soft and hard parts and identical values were obtained. But of greater significance is their agreement with 1961-1969 aerosol lead isotopic ratio, indicating the potential of atmospheric transport of the element following emissions from motor cars.

Clearly lead analyses of coastal waters are highly desireable. The few data that exist emphasize the great difficulties in collecting and analyzing waters[5]. The bivalve studies are important in indicating where significant samples should be taken.

A similar situation was noted in the mussel lead concentrations on the east coast (Fig. 3). Highest values were found in animals living adjacent to the highly populated areas. For the three year sampling period, elevated levels were observed at Cape Newagen (4.4-9.5), Portland (4.6-5.3), Cape Ann (8.7-15.6), Boston (5.9-14.2) and Cape Cod (3.6-6.5). Relative low lead concentrations were observed in mussels from the northernmost (Blue Hill Falls, Sears Island) and the southernmost stations (Atlantic City to Assateague) where concentrations of less than 2 ppm were recorded. There were no trends in the lead concentrations of oysters collected along the southeast coast of the United States.

Silver

In comparison with lead, the sources of silver to the coastal waters of the U.S. have not been clearly identified. The photographic industry is perhaps the largest consumer of silver and its discards probably enter the oceans via sewage. Inputs from the plating industry to sewage are a secondary source. With the trends toward recovery of increasingly valuable waste silver, the sewer fluxes of anthropogenic silver will probably decrease in the future.

There appears to be no significant atmospheric input of silver into the marine environment. Thus, elevated levels of lead in mussels living adjacent to urban areas may or may not be accompanied by complementary increased concentrations of silver. Such

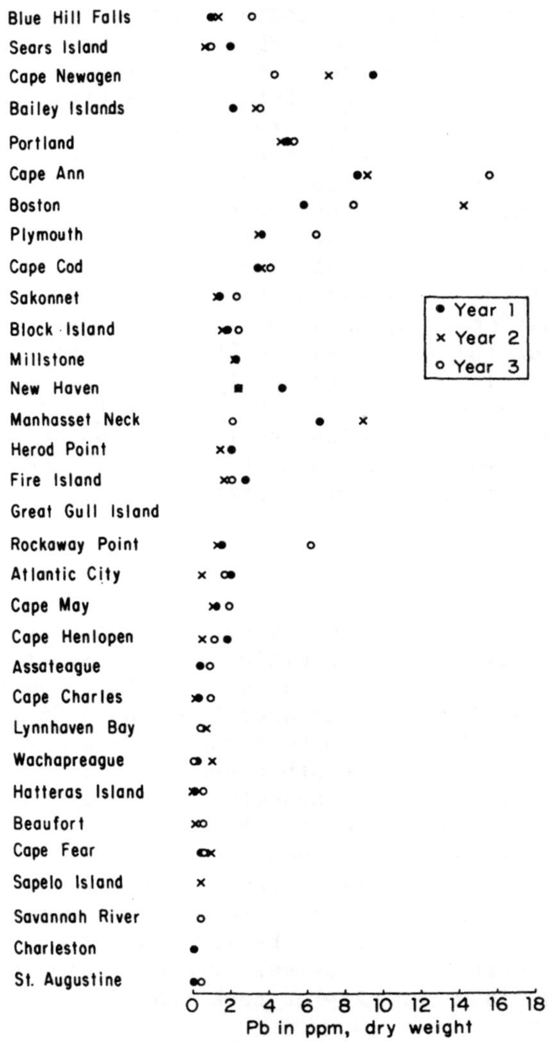

Fig. 3. Lead contents in mussels taken from U.S. East Coast stations going from Blue Hills Falls, Maine, south to St. Augustine, Florida.

appears to be the case in two west coast stations, the Farallon Islands and the San Pedro Harbor. Both have high lead concentrations in mussels and low silver values for the three year sampling period. Neither of these stations receive sewerage. The San Pedro Harbor area does receive outflow from the Los Angeles River, which undoubtedly carried anthropogenic lead, washed into it from the storm run-off.

In contrast, mussels from stations having exposure to sewerage show elevated amounts of both Pb and Ag. West coast samples include Point La Jolla, San Diego Harbor, Point Fermin and Santa Catalina Island. Of these four stations, the first three are located in the vicinity of major urban sewer outfalls and the elevated levels of silver and lead are not unexpected. However, the Santa Catalina station is located well offshore (40 km across the sea) and the high values there are puzzling. In comparison with their west coast counterparts, the east coast mussels had consistently low Ag levels. This may relate to lower levels of silver in the waters, to different bio-accumulating abilities of the mussels of different species, or to the differences in the silver contents of the food consumed by the mussels. We suspect the primary reason is the species difference between the east coast *M. edulis* and the west coast *M. californianus* since in San Francisco Bay where both species are taken from the same environment, there is a lower silver concentration in *M. edulis* (Stephenson, M.D. personal communication).

Plutonium

In addition to lead, plutonium serves as an example of the usefulness of the isotopic composition for the identification of sources. The Pu-238/Pu-239+240 ratio resulting from the entry of nuclear weapons debris ranges between 0.03 and 0.08 in the byssal threads of mussels taken from waters where there are no localized nuclear point sources. On the other hand, the Pu-238/Pu-239+240 ratio in the byssal threads of mussels taken near the site of nuclear reactors in San Onofre, California had values of 0.21 and 0.16. The Pu-238 is used as a fuel and probably leaked into the marine environment in the cooling water discharges. Thus, the plutonium burden of these coastal waters has been increased by a factor of 3 or 4 over that of the background fallout the plutonium on this basis. Monitoring of byssal threads, in which the plutonium is enriched, is a far simpler task than the monitoring of the waters themselves.

Cadmium and Plutonium

Metal concentrations in mussels can achieve unusual values from natural processes. Such appears to be the case with cadmium and plutonium. Mussel cadmium levels are generally higher on the west coast than the east. The high values on the west coast are most probably the result of upwelling. This can be illustrated with the mussels from Diablo Canyon and Soberanes Point, California located near Point Sur, at the midsection of the U.S. west coast an area in which upwelling occurs during much of the year. The upwelling process brings cadmium rich waters to the surface.

Table 2. Cadmium and Plutonium in Mussels from the U.S. West Coast

Location	Year 1		Year 2		Year 3	
	Cd	Pu	Cd	Pu	Cd	Pu
Pt. La Jolla, California	1.7	1.0	0.8	0.7	1.8	
San Pedro Harbor, California	2.3	0.3	1.3	0.4	2.0	
Diablo Canyon, California	7.7	3.7	5.9	5.7	9.2	2.1
Soberanes Point, California	9.4	5.5	20.2	11.2 13.9	9.7	3.9
Columbia River, Washington	1.4	1.0	1.0	1.0	3.8	
Puget Sound, Washington	2.6	0.7	0.8	0.5	2.7	0.3

Normally, cadmium in surface waters exhibits a depletion as a consequence of its transfer to depths by fast sinking biogenous particles[8,9]. The higher concentrations in mussels from these two sites can be compared with those to the north (Columbia River and Puget Sound) and those to the south (Pt. La Jolla and San Pedro Harbor) (Table 2). The latter values are substantially lower. Similarly, the transuranic element plutonium, introduced to the marine environment in fallout from nuclear weapons testing, shows surface depletion and mid-depth enrichment in the Pacific[10]. There is a strong co-variance between Pu and Cd in mussels from the Pacific coast.

There is an apparent permutation upon this picture. Mussels living in the high intertidal zone accumulate more Cd than mussels living in the low intertidal[11]. The mussels collected during the second year had substantially higher values at Soberanes Point than those during the first and third years. During the second year collection, sea conditions were very rough at Soberanes Point and the mussels were collected from the high intertidal rather than from the low intertidal range, the case during years one and three. The reason for this cadmium relationship with site in the tidal zone is yet to be elucidated.

The Ag, Pb, Cd and Pu data illustrate the importance of bivalve monitoring programs. First of all, they provide evidence of metal pollution along parts of the conterminous U.S. We can define clean environments, without actual measurements, within the water column. The U.S. Mussel Watch suggests a lead baseline of 1.0 ppm, a west coast Ag baseline (<u>Mytilus californianus</u>) of 0.1 ppm and an east coast Ag baseline (<u>M. edulis</u>) of 0.05 ppm for organisms inhabiting a clean environment.

Secondly, without expensive and time-consuming water analyses, natural phenomena influencing metal concentrations in seawater can be identified. Clearly, there is a crucial importance for a confirmation of such hypotheses through actual water studies. Without systematic surveys, elevated Cd and Pu values might have been interpreted as the result of a localized anthropogenic input rather than a natural physical phenomenon such as upwelling.

THE BIO-AVAILABILITY OF METAL STATES IN SEAWATER

The state of a metal in seawater (solid, liquid, gaseous or colloidal) may govern its involvement with members of the biosphere. The uptake or uranium appears to involve primarily the particulate state[12]. In coastal waters off California, the dissolved and particulate uranium can be distinguished on the basis of the U-234/U-238 ratio. Dissolved uranium species enter the oceans with ratios greater than one. The average in seawater is 1.15 and results from a preferential dissolution of U-234 into weathering fluids. On the other hand, the residual particulate phases have a U-234/U-238 ratio close to unity.

Mussels taken off the California coast have average values for the ratio of 1.13 which indicates that about 11% of their uranium was taken up as solid phases (Table 3). The mussels have about ten times more uranium in their byssal threads than in their soft parts and this uranium seems to have arisen primarily from the dissolved state in seawater. On the other hand, tunicates have values of the U-234/U-238 ratio markedly lower than those of the mussels. These filter feeders have in their digestive systems low pH value[13]. This may result in a leaching of uranium from ingested solid phases which entered their bodies. In the analyses of the guts and its contents, the ratio was always lower than, but near those of the bodies. One sample of scallops had a ratio of 0.96, indistinguishable from that of the particulate phases.

The Am-241/Pu-239+240 ratio has a characteristic value in the particulate phases of California coastal waters higher (Table 4) than the values previously reported for unfiltered waters[14]. The ratios varied between 0.5 and 3.2 for particulate phases filtered from the seawaters taken off the southern California coast as well as from the Ross Sea and Northeast Pacific Ocean. Unfiltered waters had values below 0.2. Apparently there is a fractionation of plutonium, away from the americium, once these transuranics are introduced into the oceans primarily from weapons testing. The Am-241/Pu-239+240 ratio in mussels from the U.S. Pacific coast have values clustering around 1.0-2.0. We infer that the particulate phases of seawater containing these transuranics influenced the animal's body burden.

Table 3. Uranium Concentrations and Isotopic Compositions in Marine Organisms from the California Coast. Concentrations are on a dry weight basis.

Location	U-238 d.p.m. per kg	U-234/U-238 Activity ratio
Mussels (*Mytilus* sp.)		
Pt. Fermin	130±1	1.12±0.01
Pt. Arguello	118±1	1.13±0.01
Bodega head	121±1	1.13±0.01
Oceanside	224±1	1.09±0.01
San Simon	138±1	1.12±0.01
Diablo Canyon	200±1	1.15±0.01
Santa Cruz	177±1	1.11±0.01
Soberanes Pt.	241±1	1.15±0.01
N. San Francisco	251±1	1.15±0.01
S. San Francisco	172±1	1.10±0.01
Farallon Is.	246±1	1.15±0.01
San Diego	125±1	1.13±0.01
Byssal threads		
Bodega Head	9,100±40	1.12±0.01
Cape Flattery	3,860±30	1.15±0.02
Farallon Islands	4,640±20	1.13±0.01

Table 4. Americium, Plutonium and Uranium Nuclides in Seawater

Activity ratio	Dissolved	Particulate
U-234/U-238	1.15	1.00
Am-241/Pu-239+240	<0.2	0.5 - 3.2

CONCLUSIONS

Bivalves are useful as monitors for pollutant metals introduced into the oceans. Their metal concentrations and isotopic compositions can provide measures of environmental levels, can identify anthropogenic sources, and can indicate whether the particulate or dissolved state is important in biological uptake.

ACKNOWLEDGEMENT

This research was carried out under a contract with the Environmental Protection Agency (EPA R-804215010, National Marine Pollution Monitoring Program) administered by the Environmental Research Laboratory, Naragansett, Rhode Island.

REFERENCES

1. Griffin, J.J., M. Koide, V. Hodge and E.D. Goldberg, 1980: Estimating the ages of mussels by chemical and physical methods. In "Isotope Marine Chemistry", edited by E.D. Goldberg, M. Horibe and M. Saruhashi. Uchida Rokakuho Publishing Co. Ltd., Tokyo, Japan, 193-209.
2. Miramand, P., J.C. Guary and S.W. Fowler, 1980: Vanadium transfer in the mussel Mytilus galloprovincialis. Marine Biology 56, 281-293.
3. NAS, 1980: The International Mussel Watch. U.S. National Academy of Sciences, xvi + 248 pp.
4. Goldberg, E.D., V.T., Bowen, J.W. Farrington, G. Harvey, J.H. Martin, P.L. Parker, R.W. Risebrough, W. Robertson, E. Schneider and E. Gamble, 1978: The Mussel Watch, Environ. Conserv. 5, 101-125.
5. Patterson, D., D. Settle, B. Schaule and M. Burnett, 1976: Transport of pollutant lead to the oceans and within ocean ecosystems. In: "Marine Pollutant Transfer", edited by H.L. Windom and R.H. Duce, Lexington Books, 23-38.
6. Elias, R., Y. Hirao and C. Patterson, 1975: Impact of the present levels of aerosol Pb concentrations on both natural ecosystems and humans. International conference on heavy metals in the environment. Toronto, October 27-31, 1975, 257-272.
7. Chow, T.J., H.G. Snyder and C.B. Snyder, 1978: Mussels (Mutilus sp.) as an indicator of lead pollution. Science of the Total Environment 6, 55-63.
8. Bruland, K.W., G.A. Knauer and J.H. Martin, 1978: Cadmium in northeast Pacific waters. Limnol. Oceanogr. 23, 618-625.
9. Knauer, G.A. and J.H. Martin, 1981: Phosphorous-cadmium cycling in northeast Pacific waters. J. Mar. Res., 39, 65-76.
10. Noshkin, V.E., K.M. Wong, T.A. Jokela, R.J. Eagle and J.L. Brunk, 1978: Radionuclides in the marine environment near the Farallon Islands. UCRL-52381. Lawrence Livermore Laboratory, Livermore, California.
11. Stephenson, M.D., M. Martin, S.E. Lange, A.R. Flegal, and J.H. Martin, 1979: California Mussel Watch 1977-1978, v. II. Trace metal concentrations in the California

mussel, Mytilus californianus. State Water Resources Control Board Water Quality Monitoring Report No. 79-22, Sacramento, 110 pp.

12. Hodge, V.F., M. Koide and E.D. Goldberg, 1979: Particulate uranium, plutonium and polonium in the biogeochemistries of the coastal zone. Nature, 277, 206-209.

13. Younge, C.M., 1935: On some aspects of digestion in ciliary feeding animals. J. Mar. Biol. Assoc. U.K., 20, 341-346.

14. Koide, M., P.W. Williams and E.D. Goldberg, 1981: Am-241/Pu239+240 ratios in the marine environment. Marine Environmental Research, 5, 241-246.

TRACE ELEMENTS AND PRIMARY PRODUCTION: PROBLEMS, EFFECTS AND SOLUTIONS

George A. Knauer and John H. Martin

Moss Landing Marine Laboratories
Moss Landing, CA
U.S.A.

ABSTRACT

The measurement of primary production in the ocean is basic to biological oceanographic processes. Factors which control primary production are varied, but include micronutrients (i.e., N and P) in sufficient supply, as well as many nanonutrients such as Cu, Mn and Zn. High levels of many of these trace constituents are toxic. Recent studies have shown that ambient concentrations of a number of nanonutrients in sea water are lower by an order of magnitude than previously believed. These findings present the potential for serious problems in terms of primary production measurements, since techniques used to measure production often employ various chemical reagents and sampling procedures that can contribute significantly to overall trace metal levels. We have been looking into various aspects of this problem, and will present our results concerning metal levels associated with accepted techniques used in primary production and many physiological studies that employ unialgal populations. We also present our results concerning the effects of Cu on primary production and total adenylate levels. Total adenylates have been used as a measure of the health of microbial populations using the concept of energy charge. Specifically, open-ocean phytoplankton were innoculated with Cu in the range of 0.012-5 μg Cu l^{-1}. Estimates of primary production derived using metal-free collection and processing techniques as well as ATP, ADP, AMP and energy charge were determined. Copper additions as low as 0.25 μg l^{-1} (approximately 2 times ambient) reduced ^{14}C uptake by 20%. Total adenylates were reduced 30% at a Cu concentration of 1 μg l^{-1}. However, the energy charge remained essentially unchanged over the range of Cu additions. The implications of these results are discussed.

INTRODUCTION

In recent years, there has been a great deal of research involving the effects of trace elements on marine organisms[1]. However, many of the results are conflicting and few general conclusions can be drawn. In this respect, the phytoplankton have been intensely studied due to their importance as primary producers, fixing almost all of the carbon in the world's oceans. As in terrestrial plant systems, the phytoplankton require various nutrients which are essential for growth. These essential compounds and elements are generally divided into the micronutrients (e.g., nitrate, phosphate, silicate) and nanonutrients such as vitamins and trace elements (e.g., Cu, Co, Mn, Mo, Zn). It is the latter category that concerns us here. Many of the nanonutrient trace metals can become toxic at high conceentrations[2,3,4,5], while at low concentrations, they can limit growth[6,7,8,9]. However, in spite of a considerable body of literature concerning phytoplankton-metal relationships, upper (toxicity) and lower (growth enhancement) boundaries remain vague.

In our opinion, one of the major reasons for such confusion derives strictly from problems associated with metal (and perhaps organic) contamination. In the past, metal contamination in relation to phytoplankton research was not seriously considered. Indeed, many of the culture media involved in phytoplankton rate experiments (e.g., primary production, nutrient uptake and effect studies) specifically included "biologically active" trace metals in order to provide a complete medium. The metal concentrations used in such experiments were derived from values taken from the currently available chemical oceanography literature. However, during the past five years, chemical oceanographers measuring trace elements in sea water have adopted clean sampling and analytical techniques[10,11]. Implementation of non-contaminating procedures has resulted in studies showing that several trace elements (e.g., Cd, Cu, Ni, Mn, Pb, Zn) have oceanographically consistent distributions and that concentrations are one to three orders of magnitude lower than the previously believed levels of 1-10 μg l^{-1} [12,13,14,15,16,17,18].

These findings have important implications for marine scientists studying phytoplankton processes. They must deal with the same potential problems of metal contamination that so long plagued chemical oceanographers. The use of contaminating methods in plankton sample collection and manipulation can be expected to negatively affect various phytoplankton rate measurements. From the new knowledge concerning trace element concentrations in sea water, it is apparent that the whole area of phytoplankton-trace element interrelationships must be re-examined using non-contaminating techniques. We have been looking into various aspects of this problem. This paper presents our results concerning metal levels associated with accepted techniques used in primary production (i.e., Strickland and Parsons[19]) as well as many physiological studies employing laboratory cultures. We also present our initial results concerning the effects of copper on primary production and total adenylate levels using open-ocean phytoplankton populations. Although the use of primary production as the dependent variable in bioassays is commonplace, the use of adenylate measurement is relatively new. Total adenylates and the corresponding energy charge have been used as a measure of the health of microbial populations and may prove to be an effective bioassay tool in future pollution studies[20,21].

GENERAL METHODS AND MATERIALS

The preparation of all labware and reagents used in these experiments was performed in either a portable (for shipboard use) or landbased HEPA filtered-air clean laboratory. All sample containers were thoroughly cleaned according to the following general procedure: Rinse with deionized water (DIW). Fill with warm Micro^R (a laboratory detergent) and soak for five days. Thoroughly rinse with DIW, fill with DIW and soak overnight. Empty, rinse with Milli-Q^R grade water, fill with 0.25 N redistilled, quartz-distilled HNO_3 ($RQHNO_3$) and soak for three days. Acids greater than 0.25 N are not recommended for plastics such as polycarbonate, since wall surfaces may be damaged. Empty and thoroughly rinse in Milli-Q water. Containers can then be filled with Milli-Q water until used. A more detailed description of our procedures for clean collection and processing of phytoplankton samples can be found in Fitzwater, Knauer and Martin[22]. All salts tested for metal contamination were obtained from unopened bottles of either Fisher or Baker reagent grade salts. These were diluted with Milli-Q water to the appropriate concentration, and analyzed on a Perkin-Elmer HGA 500 graphite furnace interfaced with a Perkin-Elmer 603 atomic absorption spectrophotometer (AAS). Metal concentrations were determined in the salt solutions using the method of standard additions.

Phytoplankton samples used in the bioassay experiments were collected from north-eastern equatorial Pacific surface waters (25 m) using 30-liter teflon-coated PVC Go-Flo^R bottles (General Oceanics) suspended on synthetic hydro-line (Phillystran^R, Philadelphia Resins). The bottles were closed using solid teflon messengers. Copper additions used for both the ^{14}C and adenylate bioassays were prepared from 1000 mg l^{-1} stock Harleco^R atomic absorption standards diluted to appropriate concentrations with 0.05 N $RQHNO_3$. Copper innoculations never exceeded 100 μl. Polycarbonate bottles (250 ml) were used for the ^{14}C bioassays and were incubated for six hours in sea water-cooled, deck-top incubators using neutral density screens to simulate the 25 m light level. Productivity samples were innoculated with 25 μCi 250 ul injections of [^{14}C] bicarbonate diluted in 0.3 g Na_2CO_3 l^{-1} using Eppendorf^R micropipettes fitted with acid-rinsed plastic tips. Following a six-hour incubation, the ^{14}C samples were filtered through 25 mm HA Millipore filters washed with 100 ml of filtered sea water, placed in 10 ml of Aquasol^R scintillation cocktail and counted on a Beckman LSC-100 liquid scintillation counter. Possible quench was corrected using external standardization. Five hundred ml samples incubated concurrently with productivity samples were used for the total adenylate bioassays. Following incubation, the samples were filtered onto 24 mm GF/C glass fiber filters. The filters were immediately placed in 4 ml of 1 M H_3PO_4 and extracted for thirty minutes followed by adjustment of the pH to 7.8-7.9 using NaOH. Adenylate levels (i.e., ATP, ADP, AMP) were determined using the enzymatic conversion methods described in Karl and Holm-Hansen[23]. Adenylate assays (ATP) were performed using an SAI Technology model 2000 ATP photometer. Energy charge values were calculated using the formula presented in Atkinson[20].

RESULTS AND DISCUSSION

Determination of Selected Metal Concentrations in Reagents Used in the ^{14}C Primary Productivity Technique and in Typical Culture or Enrichment Media

Twenty years ago, J.D. Strickland[24] stated in part that "... the use of non-metallic sampling equipment is essential when studying the metabolism of plankton..." Ten years ago, Steeman Nielsen and Wium-Anderson[25] stated in part that "we cannot be too careful in avoiding the influence of any heavy metal...It is difficult to elucidate how much harm Cu containing ampoules in the past have done in productivity work..." Unfortunately, these warnings continue to be ignored and without doubt, significant amounts of contamination occur not only during routine sample collections, but also as a consequence of the preparation and innoculation of [^{14}C] bicarbonate working solutions[22]. For example, one of the most widely used techniques for estimating primary production in the marine environment is described in Strickland and Parsons[19]. Briefly, the method calls for diluting a concentrated [^{14}C] labeled bicarbonate stock (e.g., 0.5 m Ci ml^{-1}diluted to 25 μCi · 2 ml^{-1}) with a solution containing reagent grade NaCl, Na_2CO_3 and NaOH. Two ml of the diluted ^{14}C solution are then sealed in soft glass ampoules until used. The amount of metal in 2 ml of the diluted ^{14}C stock divided by the total sample volume (ml) gives a final effective concentration of metal contamination (e.g., in this case, 2 ml of the ampulated ^{14}C solution are injected into a 250 ml phytoplankton sample).

The degree of metal contamination was determined for the ^{14}C salts by analyzing replicates prepared from two new bottles of reagent grade Fisher and Baker salts. The concentrated ^{14}C solution that we analyzed was obtained from Amersham/Searle without activity. The results are given in Table 1. It is obvious that of the three reagents used to dilute the concentrated ^{14}C stocks, NaCl contains large quantities of all metals analyzed with the exception of Pb, which was below our detection limit. In terms of the NaCl, it is also interesting to note that Mn levels between Fisher (= 560 pg • ml^{-1}) and Baker (= 38000 pg • ml^{-1}) were considerably different. Such differences could potentially lead to variations in terms of enrichment if, for example, Mn was limiting growth.

In terms of the concentrated ^{14}C solution, there are large quantities of Cu[25], Mn, Zn, Pb and particularly high levels of Ni and Fe present (Table 1). The addition of metals from these components to a 250-ml productivity flask using the Strickland and Parsons' method will lead to significant increases in the "final effective concentration", particularly if productivity measurements are conducted in open-ocean waters, where metal levels are low. For example, Cu levels can be increased by 60 or 50 pg • ml^{-1} (= parts per trillion (ppt)) depending on which brand of reagent is used. As mentioned in the introduction, these levels would not be considered significant before current sea water metal levels were discerned (i.e., compare "open ocean surface" with "pre-1975", Table 1). However, when the above Cu example is compared with current open-ocean dissolved sea water Cu values, it is apparent that additions using Strickland and Parsons' methods can double ambient Cu levels. Concentrations of Mn (Baker), Zn, Ni and Pb can also be increased over ambient levels. In contrast, "final effective concentrations" using the Moss Landing method (Table 1; described in a later section) contribute metals below open-ocean levels. If the ^{14}C stock solution is stored in glass ampoules[19], further contamination may occur. This is illustrated in Table 2. We rinsed the walls of five randomly-selected ampoules with 0.25 N $RQHNO_3$ in order to determine the potential for metal contamination from this source. The wall rinses were analyzed by direct injection into a Perkin-Elmer HGA 2100 graphite furnace coupled to a Perkin-Elmer 603 AAS. "Final effective concentrations" were elevated for all four metals with Fe contributing significant quantities relative to open-ocean values. Note that metal variability between ampoules was also pronounced, especially for Fe and Zn, which can certainly be expected to reduce precision, particularly if these elements are limiting.

Phytoplankton culture media have micronutrient (e.g., N, P, Si) as well as nanonutrient (e.g., trace metals) additions, in order to assure adequate growth rates. One popular recipe is the f-1 medium described in Guillard and Ryther[26]. We have analyzed

Table 1 Trace metal concentrations in reagents used in the ^{14}C primary productivity procedure outlined in Strickland and Parsons (1972). Final effective concentrations** in 250 ml productivity bottles using standard ^{14}C procedures are also compared with those using Moss Landing modified procedures. Pre-1975 dissolved sea water metal levels are contrasted with currently accepted values (modified from Fitzwater, Knauer and Martin, 1982).

		Mn	Zn	Cu	Ni	Pb	Fe
		pg ml^{-1}					
NaCl	F*	560	1400	5100	117000	< 300	5100
(5% w/V)	B*	38000	1400	4100	100000	< 300	4600
Na_2CO_3	F	< 100	< 100	< 200	500	< 300	< 100
(0.3 g liter^{-1})	B	< 100	< 100	< 200	400	< 300	600
NaOH	F	< 100	< 100	< 200	400	< 300	< 100
(0.2 g liter^{-1})	B	< 100	< 100	< 200	500	< 300	< 100
^{14}C Stock solution		15000	61000	86000	230000	25000	260000
Final Effective Concentration**							
Strickland and Parsons	F	10	25	60	970	5	90
	B	300	25	50	840	5	90
Moss Landing		1	2	3	16	1	12
Recent Open-Ocean Surface Values		45 †	1-15 ††	40 ††	160 ††	5 §	15 §§
Pre-1975 Values §§§		200	4900	500	1600	41	3000

* F and B indicate Fisher and Baker reagent grade salts.
** See text for explanation of "final effective concentrations".
† Moss Landing, Chelex 100 ion exchange.
†† From Bruland (1980), Pacific open ocean.
§ R. Flegal and C. Patterson, pers. comm.
§§ Moss Landing, organic extraction, after Bruland.
§§§ From Brewer, 1975.

the micro- and nano-nutrients used in this medium plus the organic chelator ethylenediamine tetraacetic acid (EDTA) and the buffer Tris (hydroxymethyl) aminomethane and hydrochloride (Trisma) for their metal contributions. The results are presented in Table 3.

Table 2 Concentrations of Cu, Fe, Mn and Zn washed from standard soft glass ampoules with 2 ml of 0.25 N quartz-distilled, redistilled HNO_3. Values in parentheses are final effective concentrations if 2 ml are injected into 250 ml productivity bottles (from Fitzwater, Knauer and Martin[22]).

Element	Ampoule 1	2	3	4	5
	ng liter^{-1}				
Cu	700	1800	600	600	700
	(6)	(14)	(5)	(5)	(6)
Fe	66000	22000	40000	7200	9900
	(528)	(176)	(320)	(58)	(79)
Mn	800	1700	1000	500	500
	(6)	(14)	(8)	(4)	(4)
Zn	4800	1100	5200	300	500
	(38)	(9)	(42)	(2)	(4)

Except for the Trisma, most of the contaminant metals associated with the individual components are relatively low. However, when total additions of contaminants from all components are considered, significant amounts of Zn, Cu, Ni and Pb can be introduced into the sample medium relative to open-ocean metal levels. In addition, specific nanonutrient enrichments from the recipe are extremely high (e.g., Mn - 100000 ng liter^{-1}). Although the problems of associated metal contaminants can be alleviated by the use of EDTA (i.e., chelation eliminates or reduces metal toxicity to marine phytoplankton) and by diluting enrichment additions, it is our opinion that the use of any chemical not normally found in sea water should not be used unless absolutely necessary. The point is that any chemical reagent used in culture work should be considered potentially contaminated unless analyzed and found otherwise.

Plastic Versus Glass Experimental Containers

Both field and laboratory measurements involving phytoplankton usually require the use of some sort of vessel in which to maintain the sample population. The type of vessel used may affect rate measurements in either a positive or negative sense by the removal or addition of inorganic (or organic) materials due to adsorption or desorption from container walls. Glass has been shown to contain relatively high levels of metal impurities, as well as providing sites for adsorption (i.e., many trace elements

Table 3 Concentrations of metals associated with some of the components used to prepare Guillard's f-1 medium (Guillard and Ryther[26]). Values represent final effective concentration of metals in the sea water medium.

Components*		Mn	Zn	Cu	Ni	Pb	Cd
		ng l^{-1}					
Micro-Nutrients	mg l^{-1}						
$NaNO_3$	150	1.8	3.9	5.0	380	4.2	<0.2**
NaH_2PO_4 H_2O	10	2.3	0.47	14.0	80	44.0	<0.2
$NaSiO_3 \cdot 9$ H_2O	45	5.5	4.6	1.6	38	1.1	<0.2
Chelators and Buffers	mg l^{-1}						
EDTA	10	0.48	0.92	1.0	51	5.0	<0.2
Trisma	500	13	72	8.9	190	130	<1.3
Nano-Nutrients	µg l^{-1}						
$CuSO_4 \cdot 5$ H_2O	19.6	0.004	0.13	NA†	0.015	0.02	<0.0002
$ZnSO_4 \cdot 7$ H_2O	44	0.001	NA	0.002	0.037	0.02	0.01
$CoCl_2 \cdot 6$ H_2O	20	3.5	0.0003	0.002	5.3	0.01	5.6
$MnCl_2 \cdot 4$ H_2O	360	NA	0.009	0.03	0.51	0.9	<0.0009
$Na_2MoO_4 \cdot 2$ H_2O	12.6	0.0003	0.001	0.001	0.016	0.002	<0.0002
Total Additions of Metals to Media							
Metal Contamination from Component Stocks		27	82	31	750	190	5.6
Nano-Nutrient Enrichment		100000	10120	5100	--	--	--

* All reagents used here were Baker analyzed reagents except for the Trisma, which was purchased from Sigma Chemical Company.

**Less than values indicate instrument sensitivity.

† Not analyzed.

such as Fe may be rapidly adsorbed from sea water onto glass surfaces[27]). On the other hand, plastics such as polyethylene are relatively pure[28]. It would appear then, that where primary productivity measurements are concerned, transparent plastic incubation vessels such as polycarbonate would be preferred over glass. This can be demonstrated when adsorption/desorption characteristics of polycarbonate versus glass are compared using Cu as the sorbed species. In this experiment, both glass and polycarbonate containers were thoroughly cleaned according to the procedures described above (also see Fitzwater, Knauer and Martin[22]) and filled with pH adjusted Milli-Q water (pH = 8). The pH adjusted solutions were allowed to incubate for four hours, following which desorption effects were evaluated by direct injection of the solutions into a Perkin-Elmer HGA-500 graphite furnace coupled to a Perkin-Elmer 603 AAS. Following the analyses for desorbed Cu, the solutions were spiked with ionic Cu and allowed to stand for eighteen hours. The containers were then emptied, thoroughly rinsed with pH adjusted Milli-Q water, and allowed to dry. The walls were then rinsed with 5 ml of 0.25 N RQHNO_3 and analyzed for Cu as above. The results are presented in Table 4. Even after rigorous cleaning, the glass still continued to leach significant and variable (variable = less precision) amounts of Cu (i.e., $\bar{X}$ = 210 ng • 250 ml^{-1}, which is 20X higher than open-ocean levels; Table 4). In contrast, no detectable Cu appeared in the solutions contained in the polycarbonate vessels (i.e., < 50 ng • 250 ml^{-1}). In terms of adsorption, glass also tends to remove considerably more Cu from solution than does polycarbonate (Table 4). If other

Table 4 Desorption/adsorption of Cu from the walls of 250 ml polycarbonate and glass incubation flasks cleaned according to text (from Fitzwater, Knauer and Martin[22]).

ng Cu Desorbed			ng Cu Adsorbed	
Polycarbonate	Glass	ng Cu Added	Polycarbonate	Glass
< 50*	320	125	1.5	4
< 50	90		1.7	61
< 50	160	500	2.6	90
< 50	< 50**		2.4	3
< 50	260	1250	5.4	210
< 50	180		4.3	150
< 50	200			
< 50	230			
$\bar{X}$ < 50	210			

* Less than values indicate instrument sensitivity.
**Not included in calculation of $\bar{X}$.

trace metals adsorb in a similar fashion, it is easy to see how this could decrease growth rates or under-estimate toxicity effects. We realize that in actual measurements, adsorption/desorption reactions are occurring simultaneously and therefore, this experiment represents a new result. In any case, both processes occur in sea water, although their actual magnitude is unknown.

Comparison of Primary Production Using Standard Techniques Versus Clean Techniques

It is apparent that the various reagents used in the preparation of ^{14}C stock solutions plus ampulation can increase metal levels considerably. On the other hand, the use of glass containers to maintain cultures may cause problems in either a positive or negative sense, due to adsorption/desorption processes. The effects of contaminated reagents and labware on primary productivity can be illustrated by experiments conducted by us on open-ocean populations and by Carpenter and Lively[29] using coastal populations (Table 5).

The use of "clean" ^{14}C with glass productivity bottles cleaned using old procedures (e.g., Strickland and Parsons[19]) show that productivity results derived from glass bottle incubations are lowered by 25-35% relative to values obtained using "clean" ^{14}C and acid-cleaned polycarbonate productivity bottles. The effect of improperly prepared ^{14}C stored in glass ampoules is even more pronounced (i.e., 60-70% lower, Table 5). In addition,

Table 5 Primary productivity comparisons using "clean" ^{14}C versus standard ^{14}C methods in glass and polycarbonate 250 ml containers (from Fitzwater, Knauer and Martin[22]).

	dpm*			dpm**		
	Clean ^{14}C Polycarb.	Clean ^{14}C Glass	Std. ^{14}C Polycarb.	Clean ^{14}C Polycarb.	Clean ^{14}C Glass	Std. ^{14}C Polycarb.
	4230	2750	1140	17700	14700	7240
	4150	2740	2100	18800	10400	6140
	4260	2820	975	16100	18900	5170
	4780	2710	1050	16800	13400	7530
$\bar{X}$	4350	2750	1320	17400	13300	6520
C.V.	7	2	40	7	15	17

* Open ocean, North Pacific Gyre, from Knauer, unpubl. data.
**Coastal samples; modified from Carpenter and Lively, 1980.

coefficients of variation are also higher when standard ^{14}C procedures are used. (Again, differential contamination implies less precision as well as less accuracy.)

Reduction of Metal Contamination Associated with the ^{14}C Primary Productivity Technique

The method we use to reduce metal contamination associated with primary production measurements is relatively simple. First, we recommend that all reagents except Na_2CO_3 (0.3 g l^{-1}) be omitted. The use of this reagent alone is sufficient to maintain a pH of approximately 9.7, and we have stored such stocks for up to six months without loss of ^{14}C. Second, in order to reduce contamination derived from concentrated manufactured stocks, we suggest that "high activity stocks (e.g., 10 m Ci• ml^{-1}) be diluted with the sodium carbonate solution and stored as a single stock in acid-washed teflon bottles. The use of glass ampoules should be avoided. Finally, innoculation of the ^{14}C solution should be performed with Eppendorf^{R} (or equivalent) pipettes fitted with acid-washed plastic tips. Injected volumes should be kept as low as possible relative to total sample volume to avoid any possibility of sample alteration. We never exceed 250 µl/250 ml samples.

The adoption of such methods will reduce metal contamination by as much as an order of magnitude in many cases (Table 1).

The Effects of Cu on Primary Production, Total Adenylates and Energy Charge

The effects of Cu on primary production [^{14}C], total adenylates (A_{tot} = ATP + ADP + AMP) and energy charge (EC_A) are shown in Figure 1. Open-ocean phytoplankton have been treated with ionic copper additions ranging from 0 (controls) to 5000 ng l^{-1} using clean techniques. The first process to be effected is primary production, where supression (relative to control) occurs somewhere between 100-250 ng l^{-1}. (It should be noted that actual ionic Cu concentrations or "activities" are probably lower, due to *in vivo* complexation.) In contrast, total adenylates are not affected by Cu additions until somewhere between 250-1000 ng l^{-1}. However, it is interesting that the $EC_{(A)}$ shows no significant change relative to the control, even at added Cu concentrations as high as 5000 ng • l^{-1}. $EC_{(A)}$ is considered a linear measurement of the amount of energy momentarily stored in the adenine nucleotide pool and ranges from 0.0 (all AMP) to 1.0 (all ATP). Past work, primarily with bacterial cultures, suggests that values above approximately 0.7 represent a reasonably healthy population, while values at the lower end of the range are indicative of

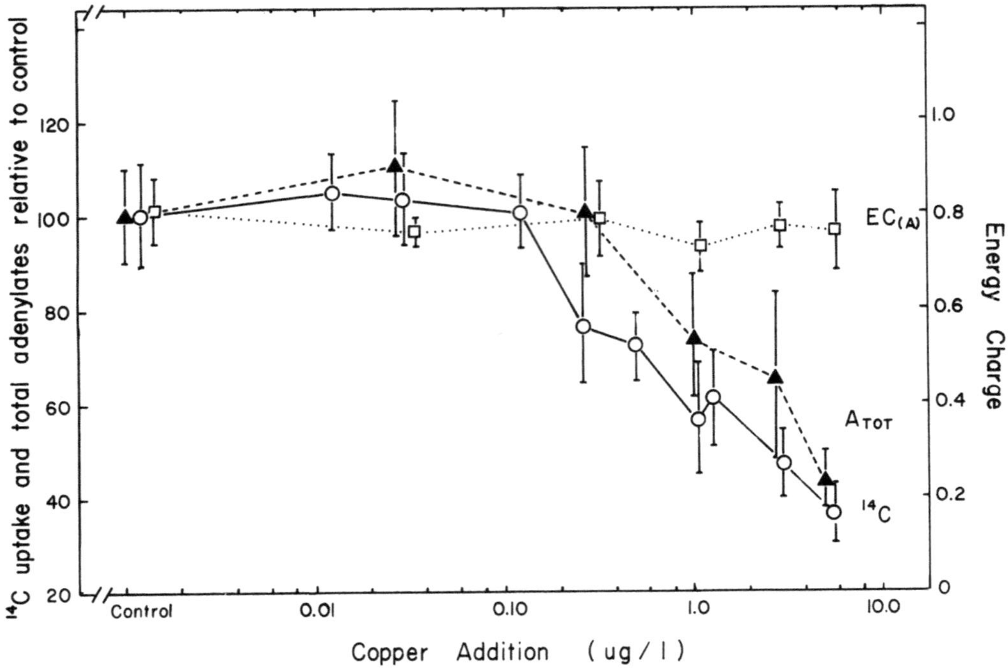

Fig. 1 The effects of Cu additions on ^{14}C uptake (O) and total adenylates (▲) as relative percent of controls. Symbols indicate means $\pm$ S.D. (n = 20 for ^{14}C uptake; n = 12 for A_{tot}). Energy charge (□) was calculated from ratios of the three adenylate nucleotides, ATP, ADP, AMP.

stress. Thus, although there is a considerable drop in ^{14}C uptake as well as a considerable loss in A_{tot}, $EC_{(A)}$ shows no corresponding change, indicating that at least over the time constraints of this experiment (six hours), the population still appears to be "coping". The implications of this finding are interesting. Since the apparent acute affect of Cu on ^{14}C and A_{tot} does not apparently extend to $EC_{(A)}$, the obvious question is: What will occur over the long-term? The problem to solve now becomes one of some sort of combination between acute and chronic effects in terms of describing total impact. In other words, how long can a population maintain $EC_{(A)}$ at constant Cu levels and can a population recover if such levels are decreased? Experiments are currently underway in our laboratory to answer these questions.

It is also instructive to speculate on the reason why many previous metal studies did not detect effects until well into the $\mu g \cdot l^{-1}$ range. The reason may lie in the fact that the controls and treatments are already severely contaminated (e.g., with Cu), due to contaminating collection and experimental techniques. If this were the case, controls may already be inhibited, as illustrated by the dashed line shown in Figure 2. Here, lower Cu additions would obviously not be resolvable and the erroneous conclusion would be made that inhibition does not occur until levels exceeding 2500 ng $\cdot$ l^{-1} are reached. Potential errors of this sort would result in a serious under-estimation of the effects of Cu or any other pollutant.

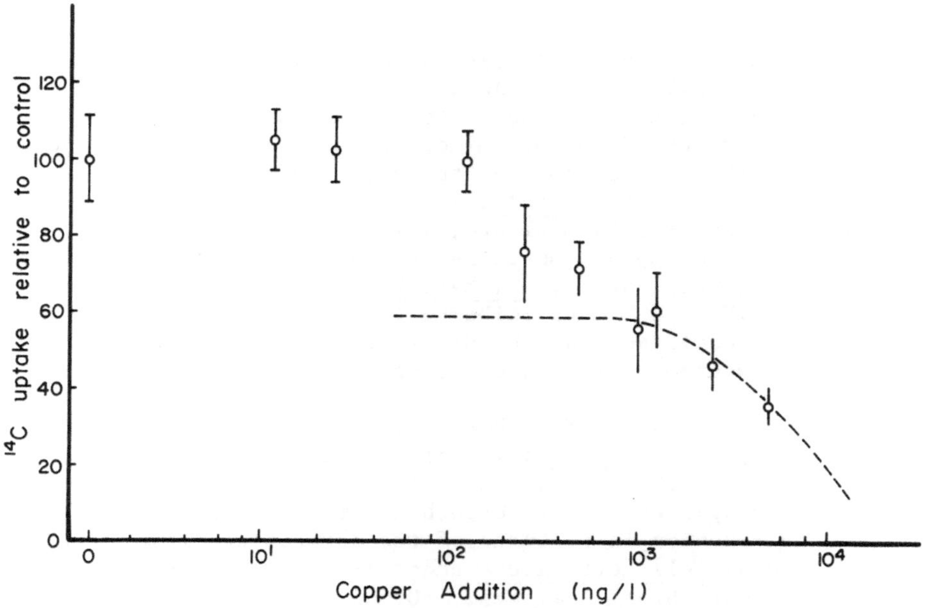

Fig. 2 The effects of Cu additions on ^{14}C uptake relative to controls. Open circles indicate the mean ($\pm$ 1 S.D.; n = 20) of a series of six-hour bioassays done over a five-day period. Dashed line represents the possible results if controls were contaminated with up to 1000 ng Cu l^{-1} (from Fitzwater, Knauer and Martin[22]).

CONCLUSIONS

As we have suggested in this paper, the three most serious sources of metal contamination are dirty sampling procedures, reagents of doubtful purity and the use of glass ampoules and culture vessels. The use of clean water sampling procedures, polycarbonate incubation flasks, ommission of the reagents NaCl and NaOH in preparing the ^{14}C stocks, storage of ^{14}C solutions as a single stock in teflon bottles, together with small-volume injections of ^{14}C using clean pipettes will not only result in more accurate and precise primary production measurements, but will also enable the detection of pollutant effects at realistically low levels, and facilitate the culturing of sensitive oceanic species.

REFERENCES

1. Davies, A.G., 1978: Pollution studies with marine plankton. II. Heavy Metals. Adv. Mar. Biol., 15, 381-508.
2. Jensen, A., B. Rystead and S. Melsom, 1974: heavy metal tolerance of marine phytoplankton. I. The tolerance of three algal species to Zn in coastal sea water. J. Exp. Mar. Biol. Ecol., 15, 145-157.
3. Sunda, W. and R.R.L. Guillard, 1976: The relationship between cupric ion activity and the toxicity of copper to phytoplankton. J. Mar. Res., 24, 511-529.
4. Thomas, W.H. and D.L.R. Seibert, 1977: Effects of copper on the dominance and the diversity of algae: Controlled ecosystem pollution experiment. Bull. Mar. Sci., 27, 23-33.
5. Anderson, D.M. and F.M.M. Morel, 1978: Copper sensitivity of Gonyaulax tamarensis. Limnol. Oceanogr., 23, 283-295.
6. Davies, A.G., 1970: Iron chelation and the growth of marine phytoplankton. I. Growth kinetics and chlorophyll production in cultures of the euryhaline flagellata Dunalliela tertiolecta under iron-limiting conditions. J. Mar. Biol. Ass. U.K., 50, 65-86.
7. Thomas, W.H., D.L.R. Seibert and A.N. Dodson, 1974: Phytoplankton enrichment experiments and bioassays in natural coastal sea water and in sewage outfall receiving waters off southern California. Estuarine Coast. Mar. Sci., 2, 191-206.
8. Anderson, M.A., F.M.M. Morel and R.R.L. Guillard, 1978: Growth limitation of a coastal diatom by low zinc ion activity. Nature, London, 276, 70-71.
9. Bowen, H.J.M., 1979: Environmental chemistry of the elements. Academic Press, N.Y.
10. Patterson, C.C. and D.M. Settle, 1976: The reduction of orders of magnitude errors in lead analyses of biolo-

gical materials and natural waters by evaluating and controlling the extent and sources of industrial lead contamination introduced during sample collection and analysis, p. 321. In: P.D. LeFleuer (ed.). "Accuracy in Trace Analysis: Sampling, Sample Handling, Analysis". U.S. Nat. Bur. Stand. Special Pub. 322.
11. Bruland, K.W., R.P. Franks, G.A. Knauer and J.H. Martin, 1979: Sampling and analytical methods for the determination of copper, cadmium, zinc and nickel at the nanogram per liter level in seawater. Anal. Chim. Acta., 105, 233-245.
12. Boyle, E.A. and J.M. Edmond, 1975: Copper in surface waters south of New Zealand. Nature, 253, 107-109.
13. Bender, M.L. and C.L. GAgner, 1976: Dissolved copper, nickel and cadmium in the Sargasso Sea. J. Mar. Res., 34, 327-339.
14. Moore, R.M. and J.D. Burton, 1976: Concentrations of dissolved copper in the eastern Atlantic Ocean 23°N to 47°N. Nature, 264, 242-243.
15. Schlater, F.R., E.A. Boyle and J.M. Edmond, 1976: On the marine geochemistry of nickel. Earth Planet. Sci. Lett., 31, 119-128.
16. Bruland, K.W., 1980: Oceanographic distributions of cadmium, zinc, nickel and copper in the north Pacific. Earth Planet. Sci. Lett., 47, 176-198.
17. Landing, W.M. and K.W. Bruland, 1980: The distribution of manganese in the eastern north Pacific. Earth Planet. Sci. Lett., 49, 45-56.
18. Martin, J.H. and G.A. Knauer, 1980: Manganese cycling in northeast Pacific waters. Earth Planet. Sci. Lett., 51, 266-274.
19. Strickland, J.D.H. and T.R. Parsons, 1972: A practical handbook of seawater analysis. 2nd Ed. Fish. Res. Bd. Canada, Bull. 167.
20. Atkinson, D.E., 1977: Cellular energy metabolism and its regulation. Academic Press, N.Y.
21. Karl, D.M. and O. Holm-Hansen, 1978: Methodology and measurement of adenylate energy charge ratios in environmental samples. Mar. Biol., 48, 185-197.
22. Fitzwater, S.E., G.A. Knauer and J.H. Martin, 1982: Metal contamination and primary production: Field and laboratory methods of control. Limnol. Oceanogr. 27(3), 544-551.
23. Karl, D.M., 1980: Cellular nucleotide measurements and applications in microbial ecology. Microbiol. Rev., 44(4), 739-796.
24. Strickland, J.D.H., 1960: Measuring the production of marine phytoplankton. Fish. Res. Bd. Canada, Bull. 122.
25. Steeman-Nielsen, E. and S. Wium-Anderson, 1970: Copper ions as poison in the sea and in fresh water. Mar. Biol., 6, 93-97.

26. Guillard, R.L. and J.H. Ryther, 1962: Studies of marine planktonic diatoms. I. Cyclotella nana hustedt and Detonula confervacea (cleve) gran. Can. J. Micro., 8, 229-239.
27. Paasche, E., 1977: Growth of three plankton diatom species in Oslo fjord water in the absence of artificial chelators. J. Exp. Mar. Biol. Ecol., 29, 91-106.
28. Robertson, D.W., 1968: The absorption of trace elements in sea water on various container surfaces. Anal. Chim. Acta., 42, 533-536.
29. Carpenter, E.J. and J.S. Lively, 1980: Review of estimates of algal growth using ^{14}C tracer technique, pp. 161-178. In: P.G. Falkowski (ed.). "Primary Productivity in the Sea." Plenum Press, N.Y.

TRACE METALS AND PLANKTON IN THE OCEANS: FACTS AND SPECULATIONS

F.M.M. Morel and N.M.L. Morel-Laurens

Ralph M. Parsons Laboratory
Department of Civil Engineering, Bldg. 48-425
Massachusetts Institute of Technology
Cambridge, Massachusetts 02139, U.S.A.

INTRODUCTION

Understanding the mechanisms of biological and chemical interactions among trace metals and planktonic organisms is the key to elucidating the role of trace metals in the ecology of the oceans and the role of organisms in the geochemistry of metals.

The availability of trace metals as essential micronutrients to phytoplankton and their toxicity to all planktonic organisms are dependent upon their chemical speciation in the water. For example, the availability of zinc and iron and the toxicity of zinc, copper and cadmium are controlled by their free ionic activities in aquatic systems[1,2,3,4,5,6].

In turn, organisms play an important, probably a dominant, role in controlling trace metal chemistry in the ocean. They may increase trace metal solubilities by releasing complexing agents in the medium or, on the contrary, they may enhance the incorporation of metals into particles and thus foster metal sedimentation in aquatic systems.

This paper is an attempt at organizing some of the known facts regarding trace metal-microorganisms interactions into a coherent picture. To do so we have focused on a few topics that were particularly tractable and selected experimental results that can be interpreted quantitatively. In this way we hope to obtain a sound factual basis from which to launch into reasonable speculations. Our goal is to present an extremely simple conceptual model with which to interpret experimental data and organize future research. To be useful, such a model need not account for all possible variability in biological systems or turn out to be strictly correct. By providing a conceptual framework it should help understand observations that deviate from the model predictions as well as those that conform to them.

The principal focus of this paper is the interaction of metals (chiefly cationic metals such as Cu, Cd, Fe, Zn) and phytoplankton, a subject recently reviewed thoroughly by Huntsman and Sunda [6]. We hope to show that the biological effects of the metals are controlled by a simple reaction with a cellular ligand on the cell surface and that such reaction accounts for (and predicts) the critical role of medium chemistry as well as the synergistic and antagonistic effects among metals. In terms of the role of the organisms in controlling metal chemistry, it is argued that, except for copper and iron, complexing agents should have a relatively small effect on metal complexation in solution, but that organic compounds adsorbed on the surface of inorganic particles may control the adsorption of metals. Depending on the concentration of dissolved organic carbon, copper may be largely present as organic complexes (either with planktonic exudates or humic material), while some fraction of the iron is probably bound to planktonically produced siderophores.

Perhaps more important to metal speciation in surface waters are the possible redox - often photochemical - reactions involving trace metals and organic compounds. Insufficient attention has been paid so far to such redox processes - driven partly by autochtonous organic material and hence by biological activity - that may dominate the chemistry of metals such as iron and manganese in surface waters. While our progress in understanding the interactions between trace metals and organisms is largely due to our efforts to elucidate the equilibrium speciation of metals in the water, we must now focus our attention on the non-equilibrium processes and determine the biological and geochemical role of thermodynamically unstable and possibly ephemeral metallic species.

EFFECTS OF METALS ON ORGANISMS

Role of Metal Speciation

A dominant fact to emerge from the recent work on the relationship between trace metals and planktonic organisms is the paramount importance of the chemical speciation of the metals in the organisms' external milieu. According to the modern paradigm, and disregarding a few known exceptions, the chemical parameters controlling metal-organism interactions are the activities of the free metal ions. This is most easily demonstrated by assay experiments spanning a range of metals and complexing agent concentrations as illustrated in Fig. 1. For example, in Fig. 1b, different growth rates are obtained for the same total zinc concentrations in the presence of different concentrations of chelators (EDTA), while similar responses result from identical metal ion activities obtained with various combinations of total zinc and EDTA concentrations. Although the bulk of the available data involves phytoplankton, this result appears generally applicable to most aquatic organisms, having been demonstrated in some bacteria[7,8], zooplankton[9], shrimp[10] and even to some degree in fish[11]. In phytoplankton the same dependence on free ionic activities is observed whether metals act as limiting micronutrients (e.g., Zn, Fig. 1b), or as toxicants (e.g., Cu, Fig. 1a). Only the complexing characteristics, not the precise nature of the complexing agents, appear to be important: artificial chelating agents[1,2,3,4], natural humic material[12], even inorganic ligands[10,11] can be used to vary the free metal ion activities and modulate the biological effects.

Methodological Issues

Before examining the specifics of the biological effects of metals, let us first focus on some methodological problems. If the free ionic activities of metals are indeed the controlling chemical parameters, it is obviously necessary to know these free ionic activities in order to interpret experiments on metal-organism interactions. Direct measurements of free metal ion activities in seawater systems are presently impossible, or at least, unreliable. Potentiometric techniques using metal ion sensitive electrodes are limited in sensitivity (typically total metal concentrations have to be in excess of 10^{-8} - 10^{-7}M) and are often inapplicable in seawater due to the presence of interfering major ions[15]. There is a great deal of disagreement regarding the usefulness of the various voltammetric techniques for measuring metal speciation in natural waters[14,15]. Even if one were to overlook the difficulties in voltammetric data interpretation for systems as complex as (unacidified) seawater, it remains that the measurement of total labile metal is far from synonymous with a measurement of free metal ion activity.

Most studies of trace metal effects on aquatic organisms have thus relied on thermodynamic calculations of free metal ion activities in the experimental medium. The experimental requirements for such calculations to be valid are quite stringent and failure to meet these requirements makes much of the existing bioassay data uninterpretable quantitatively. (For example, all data reported as toxic concentrations of total metals are basically useless unless the solution composition is precisely known and activities can be calculated with some confidence). Above all, it is necessary to use a chemically defined medium with relatively

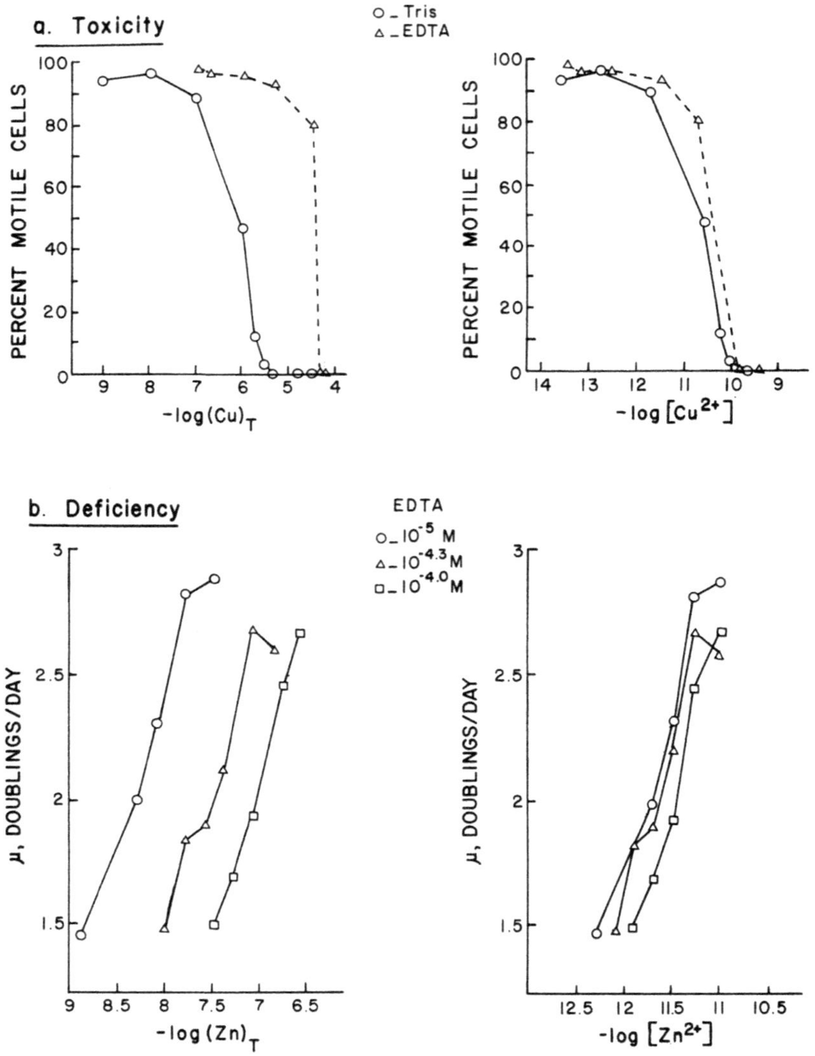

Fig. 1 Experiments with different chelators and a wide range of trace metal concentrations demonstrate that trace metal toxicity and deficiency are determined by metal ion activity and not total metal concentration. a. Motility data of Gonyaulax tamarensis as a function of total copper $(Cu)_T$ and cupric ion activity $[Cu^{2+}]$ for 2 chelators: Tris and EDTA; after D.M. Anderson and F.M.M. Morel Limnol. Oceanogr. 23, 283 (1978). b. Growth rate of Thalassiosira weissflogii as a function of total zinc $(Zn)_T$ and zinc ion activity $[Zn^{2+}]$ for 3 concentrations of EDTA; after M.A. Anderson, F.M.M. Morel and R.R.L. Guillard, Nature, 276, 70 (1978).

low levels of metal impurities (artificial media may contain up to micromolar concentrations of various trace metal contaminants), low concentrations of unknown and potentially complexing organic compounds, and negligible adsorption of the metals of interest - hence the use of special containers and the strict requirement to avoid formation of precipitates. In addition, the metal ion activities should be suitably buffered so as not to be significantly affected by the uptake of metals by the organisms. Finally, the experimental design should be such that chemical equilibrium is actually a good approximation for the metal speciation in the experimental medium: either the medium should be pre-equilibrated or the kinetics of the reactions that control metal ion activities must be fast compared to the measured physiological response[2].

The most common and practical solution to meet all these requirements is to use rather high concentrations of metals and complexing ligands[1,16]. It is an often poorly appreciated paradox that to mimic the effects of metals at nanomolar and lower concentrations in oceanic waters with very low organic content it is in fact necessary to use culture media containing micromolar concentrations of metals and artificial organic ligands. This is a central consequence of the role of free ionic activities in controlling the biological effects of trace metals. Relatively high concentrations of metals and chelating agents in experimental cultures provide chemically the type of buffering capacity for trace metals afforded in nature by the large solution volume relative to the biota concentration. Note, however that the kinetics of metal-organism interactions may be profoundly influenced by the buffering process and that one should be particularly careful in interpreting laboratory data and extrapolating to field conditions. For example, the kinetics of iron uptake by phytoplankton can be conveniently studied in the presence of chelating agents, but the results may bear little resemblance to the same process *in situ*[4].

Growth Responses

With these methodological caveats in mind, let us look at some of the available data and focus on the physiological effects of metals on phytoplankton. Most of the data are phenomenological, relating, for example, the growth rate of a batch culture to the activity of a toxic metal. Such data do not lead directly to a mechanistic interpretation, yet details of the phenomonology may provide us with some important clues. Consider, for example, the great variability in the time course of growth inhibition by toxic metals in batch cultures (Fig. 2). In some cases extended lag phases are observed, with ultimate resumption of normal exponential growth rates (Fig. 2a); in many cases, exponential growth rates are affected, sometimes immediately (Fig. 2b), sometimes

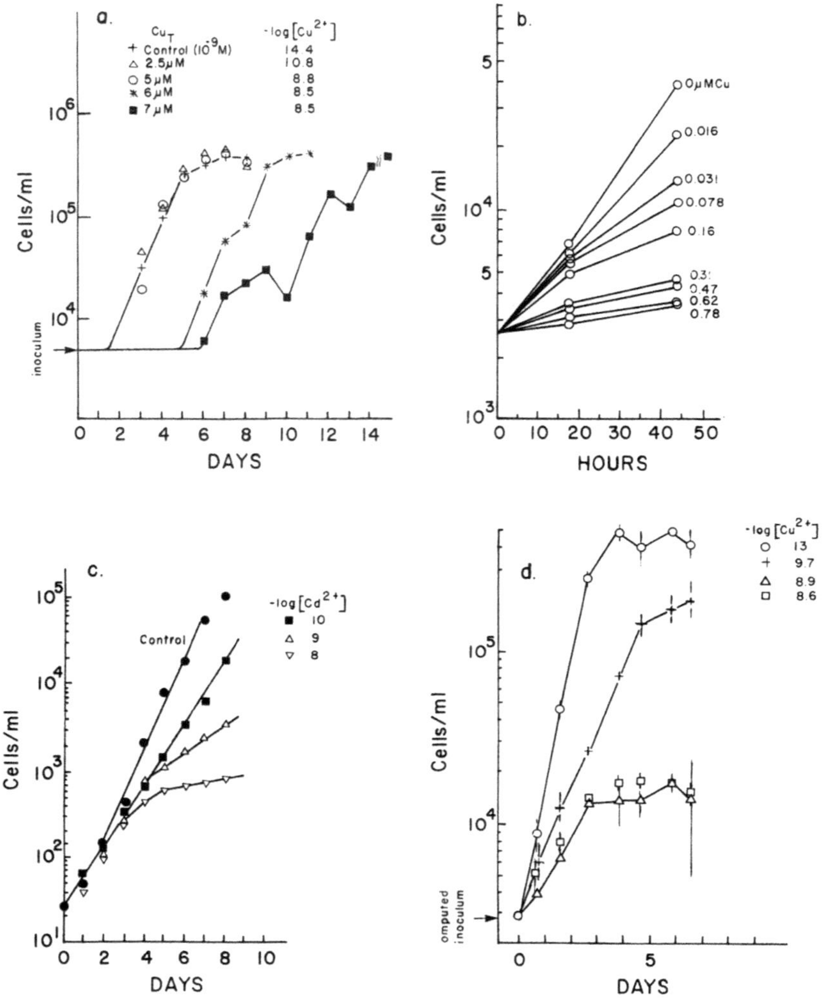

Fig. 2 Example of various growth responses to metal ion activities. a. Extension of lag phase. Organism: Skeletonema costatum; from N.M.L. Morel, J.G. Rueter, Jr. and F.M.M. Morel, J. Phycol., 11:43 (1978). b. Instantaneous reduction of exponential growth rate. Organism: Thalassiosira pseudonana (clone 13-1); after E.W. Davey, M.J. Morgan and S.E. Erickson, Limnol. Oceanogr., 18:993 (1973). c. Reduction of exponential growth rate after a delay. Organism: Thalassiosira weissflogii; after P.L. Foster and F.M.M. Morel, Limnol. Oceanogr., 27:745 (1982) d. Reduction of exponential growth rate and of plateau level. Organism: Thalassiosira pseudonana (clone 3H); J.G. Rueter, S.W. Chisholm and F.M.M. Morel, J. Phycol., 17:270-278 (1981).

after a lag phase, often after a delay during which metal containing cultures grow at the same exponential rates as the controls (Fig. 2c); in a few cases maximum cell densities (phase) are decreased (Fig. 2d). The different characteristics of these various growth responses suggest that different mechanisms of toxicity may be at play.

Let us focus on the delay that is commonly observed in the growth rate inhibition due to metals such as copper or cadmium (Fig. 2c) [1,17]. Such delay may be caused either by a slow accumulation of the toxic metal and effectively represent an equilibration time between the cell and the medium, or by a delay in the toxic response itself after rapid action of the metal on a sensitive cellular site. In the second case a tempting hypothesis is that the effect of the metal is to inhibit the uptake (or the assimilation) of some critical nutrient. The delay in growth response to the toxic metal would then be explained by utilization of internal storage of the nutrient.

Inhibition of Nutrient Uptake

Not very much is known of the effect of toxic metals on nutrient uptake. The best documented case is the inhibition of silicic acid uptake by copper in diatoms [18,19,20]. Nitrate or phosphate uptake have been reported not to be affected by copper [18] while there is some evidence of ammonium uptake inhibition by mercury [21]. Consider, then, the effect of copper on silicic acid uptake and growth in *Thalassiosira pseudonana*. At low silicic acid concentrations, the Si uptake inhibition is rapid and is, in fact, measurable before any decrease in growth rate [18]. However, the reduced Si uptake rate never accounts quantitatively for the ultimate slow growth of the diatom. On the contrary, as shown in Fig. 3, in the presence of copper, the slower growing cells accumulate more Si than do the control - i.e., growth rate is relatively more inhibited by Cu than Si uptake rate. The feature in the growth curve that may in fact be explained by the Cu-Si interaction is not the delay in growth inhibition that is sometimes observed, but the decrease in maximum cell yield; the cellular Si requirement of diatoms may be higher at higher cupric ion activities resulting in a lower maximum cell concentration in Si limited batch cultures (Fig. 3).

Although the delay in growth rate inhibition by copper in *T. pseudonana* is not explained by an effect of Cu on macronutrient uptake (i.e., NO_3^-, PO_4^{3-}, $Si(OH)_4$), it is possible that it results from inhibition of uptake or assimilation of essential trace elements. We have preliminary data showing that, in some cases at least, the toxicity of metals such as cadmium or copper is affected by the iron nutrition of the algae, and that the delay in

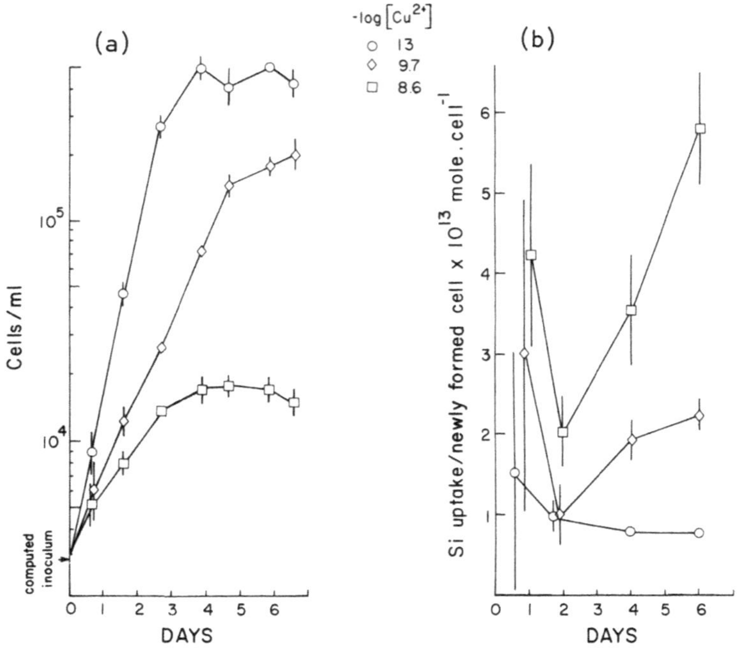

Fig. 3 Effect of copper on growth and silicic acid uptake by Thalassiosira pseudonana. a. Growth curves with reduced exponential rates and reduced plateau levels. b. Silicic acid uptake per newly formed cell. At stationary phase cell quota is 3 times larger in cells most inhibited by copper; after J.R. Reuter, Jr., S.W. Chisholm and F.M.M. Morel, J. Phycol., 17:270-278 (1981).

growth inhibition is very much reduced in iron stressed cells. According to recent evidence[4], the uptake of Fe by marine algae involves a membrane site with an affinity constant of ca 10^{20} for Fe^{3+}. Since all known strong iron complexing agents (including siderophores and transferrins) also complex other metals, such as Cu or Cd with reasonably high affinities (log K=7-14), it stands to reason that these metals could compete with Fe for the uptake site and interfere with Fe nutrition either at the level of uptake or further in the accumulation and assimilation process. Since the ratio $\{Cu^{2+}\}/\{Fe^{3+}\}$ ranges from 10^6 to 10^{11} in culture media (a similar range is obtained for $\{Cd^{2+}\}/\{Fe^{3+}\}$ when cadmium is added to the system) inhibition of Fe nutrition due to toxic metals is certainly a possibility. Although the necessary direct evidence is lacking, one may speculate that such inhibition may well be one of the major mechanisms by which toxic metals decrease the growth rate of phytoplankton, and the observed delays in growth rate inhibition may result from utilization of stored iron.

It has proved difficult to clearly demonstrate that the duration of lag phases induced by toxic metals in batch cultures of algae was indeed dependent on a medium conditioning mechanism as has been hypothesized for a long time[22,23]. Nonetheless, in at least one instance, it now appears rather clear cut that the duration of the lag phase induced by copper in various strains of Chlorella is determined by the release of copper complexing agents in the medium[24]. As shown in Fig. 4, such release of chelating agents gradually decreases the cupric ion activity in the cultures which resume exponential growth when some critical value is reached ($\{Cu^{2+}\} = 10^{-7.5}$ and $10^{-6.5}$ M for the Cu sensitive and Cu resistant strains respectively). Time course of the cellular copper concentration (obtained with the same strains under differ-

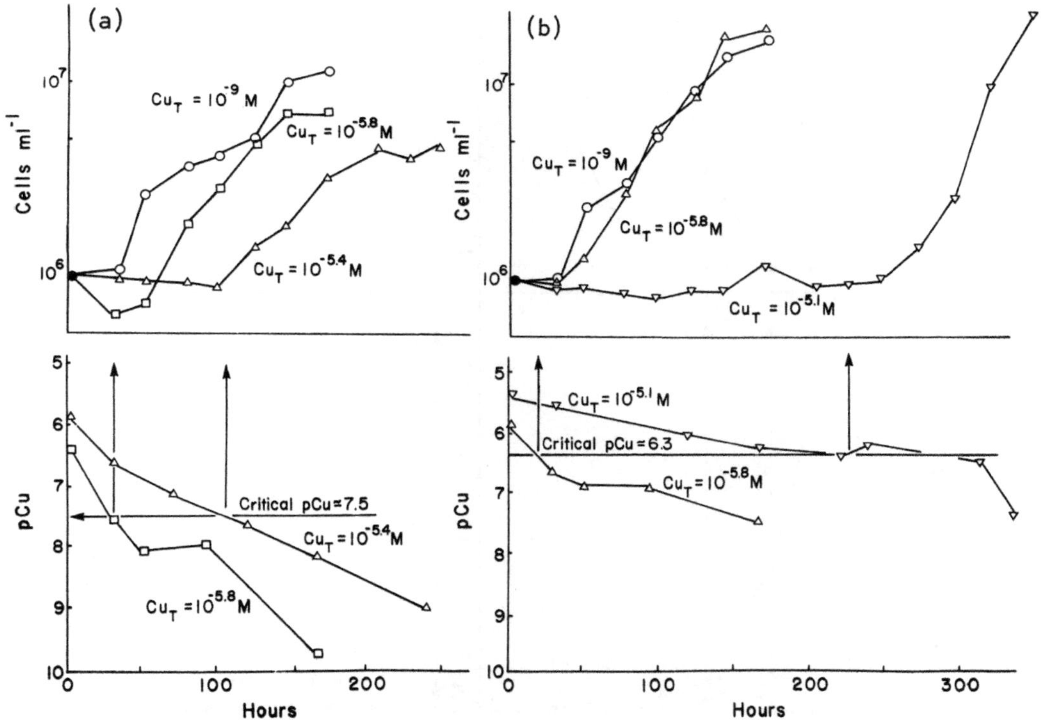

Fig. 4 Evolution of the cupric ion activity during the growth of copper sensitive (a) and copper resistant (b) strains of Chlorella. When the cupric ion activity (pCu = $-\log [Cu^{2+}]$) reaches a critical value (ca. $10^{-7.5}$ M in (a) and ca. $10^{-6.3}$ M in (b)), the cells resume exponential growth. The decrease in the cupric ion activity with time is accounted for by the release of an extracellular complexing agent; after P.L. Foster and F.M.M. Morel, in preparation.

ent conditions, by a different investigator[25] Fig. 5), shows that copper is gradually lost from the cells during the lag phase after

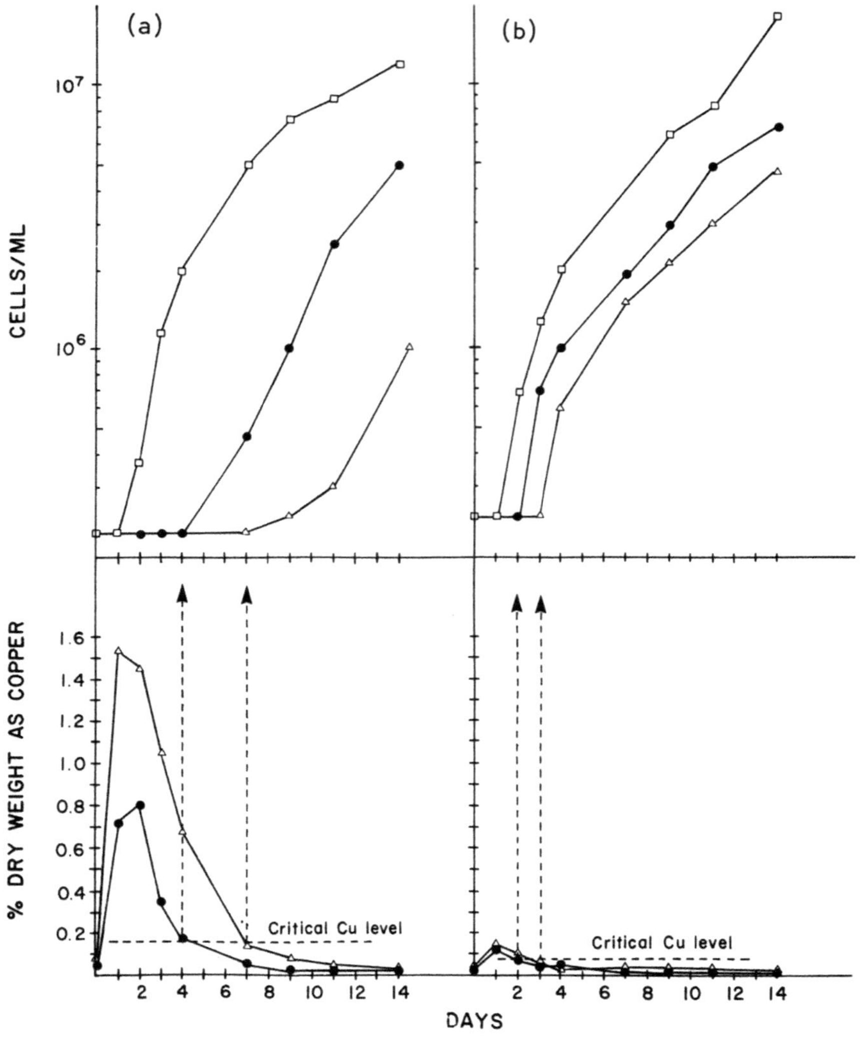

Fig. 5 Evolution of the cellular copper content (as % of dry weight) during the growth of copper sensitive (a) and copper resistant (b) strains of <u>Chlorella</u>. After an initial uptake, copper is released from the cells. When a critical level is reached the cultures start growing again; after M. Butler, A.E.J. Hasken and M.M. Young, <u>Plant Cell and Environ.</u>, 3:119 (1980).

an initial rapid uptake. Hence the accumulation of extracellular complexing agents in the batch cultures induces a release of copper from the cells; when the critical cupric ion in the medium is reached the cellular concentration of copper is low enough to allow cell division. The complexing agents do not possess a particularly high affinity for copper nor do they seem to be exuded in response to copper toxicity[24]. In other words, the release of this particular type of conditioning agent seems to have little adaptive or teleological meaning, unlike the release of siderophores for iron uptake by cyanophytes. Nonetheless, the presence of such extracellular complexing agents in oceanic water may alter the chemistry, and hence the biological effects of some metals, copper in particular (see second section).

Rapidity and Reversibility of Effects

In the few cases where the measurements have permitted a better time resolution than that afforded by the ultimate decrease in growth rate, the effects of toxic metals on phytoplankton have generally been found to be rapid. This is true, for example, of the inhibition of Si uptake rate in T. pseudonana[18] or the loss of motility in G. tamarensis[2], both of which occur within minutes (<1 hour) after exposure to copper. The effects are also usually reversible, resulting ultimately in resumption of normal growth rates (or regain of motility). This is examplified in Fig. 6 which shows how the addition of EDTA to decrease the activity of cadmium (and simultaneous addition of other trace metals to maintain their activities constant) reverses Cd toxicity as demonstrated by the growth response of T. weissflogii[5,17]. Although there are certainly some exceptions to this rule - and there is a question of semantics regarding what the effect of a metal actually is - it is probably reasonable to regard the effects of toxic metals on microorganisms as being fast and reversible, in general.

A General Model

The general mechanistic implications of the dependence of the biological effects of a metal on its free ionic activity seems rather clear: the effects are probably controlled by a reaction between the metal and some cellular component. Considering, for simplicity, a 1:1 stoichiometry between a divalent metal, Me, and a divalent cellular ligand, X, we can write the reaction:

$$Me^{2+} + X^{2-} = MeX \quad ; \quad K_x \qquad (1)$$

The cellular ligand may be inside the cell or on its surface, and the equilibrium state of the reaction or its forward kinetics may be, in principle, controlled by the free metal ion activity. Since the external ionic activity of the metal is the controlling parameter, it may be hypothesized that the reactive cellular ligand is at the surface, although location of the reactive site is not critical to our chemical understanding. We can also assume that the reaction is in fact at equilibrium (pseudo-equilibrium) since the biological effects appear unaffected by the presence of complexing ligands with notably slow dissociation kinetics. For example, in the presence of EDTA, the cupric ion activity could not be maintained at its equilibrium value in the medium if reaction (1) were to proceed to the right in the course of a copper toxicity experiment. The steady state cupric ion activity which we hypothesize to control the forward kinetics of (1) and the toxic response of the organism would then be dependent on the slow dissociation kinetics of the chelate, and, contrary to observation, different chelators would yield different responses. (Note that the equilibrium between the metal chelate in solution and the cellular ligand need not depend on the dissociation of the metal chelate and be limited by its slow kinetics.)

Both the rapidity and the reversibility of metal toxicity to phytoplankton bolsters our general image of an equilibrium reaction with cellular sites. In the example of Fig. 6, the added EDTA effectively outcompetes the cellular ligand and reaction (1) has been forced backward, decreasing the extent of cadmium complexation with the cellular ligand, and hence, the intensity of the biological effect. The natural complexing agents released by Chlorella achieved a similar result in the examples of Fig. 4 and 5.

In terms of the actual mechanism of toxicity, the reactive site X may be the sensitive site itself or it may only be a transport site for the metal, allowing it to reach a sensitive intracellular site. In the first case - which pertains at least when the uptake rate of a metallic nutrient is the observed physiological response - the dependence of the biological effect on the free metal ion activity is a direct result of the equilibrium mass law for reaction (1). For the second case to be possible, one has to hypothesize that the inactivation of the intracellular site that determines the measured physiological response (e.g., blockage of cell division) is controlled by the rate of metal accumulation, which is in turn controlled by the pseudo-equilibrium of reaction (1).

In either case, both synergistic and antagonistic effects among metals are expected on the basis of this general model. If two metals are present at activities that result in appreciable formation of both their complexes with the cellular ligand, then one should observe a synergism if both metals are toxic and an antagonism if only one metal is toxic. An antagonistic effect

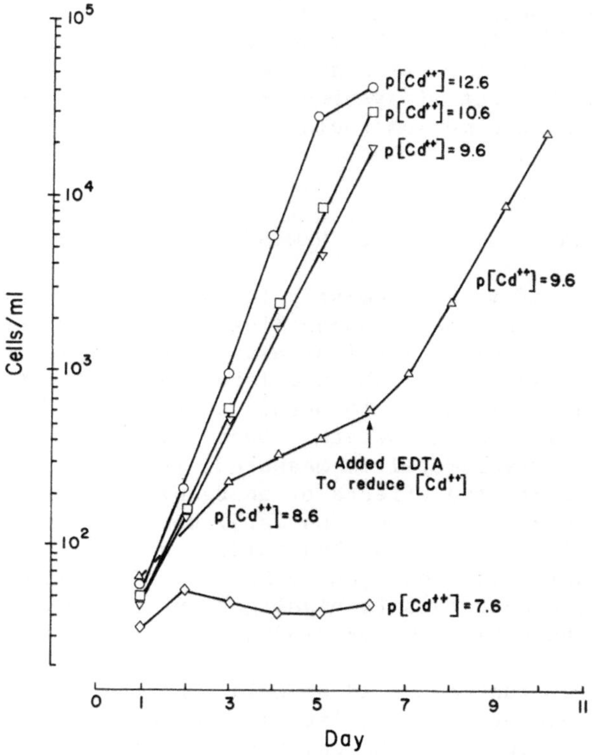

Fig. 6. Growth of Thalassiosira fluviatilis exposed to Cd^{2+} ions. Addition of EDTA reverses apparent Cd toxicity. $p[Cd^{2+}] = -\log [Cd^{2+}]$. After P.L. Foster and F.M.M. Morel, Limnol. Oceanogr., 27:745 (1982).

between Zn and Cu for Si uptake has actually been observed in T. pseudonana[19]. As expected on the basis of the simplest possible model of competition for the same ligand site, the Si uptake rate appears modulated by the ratio of metal ion activities $\{Zn^{2+}\}/\{Cu^{2+}\}$. The extent of the inhibition is also dependent on the silicic acid concentration. If one wishes to extrapolate laboratory data to oceanic systems, one should then consider not only individual metal ion activities, but also their relative values and, in some cases, ambient concentrations of limiting nutrients.

Although the basis for the argument has been somewhat refined by demonstrating the importance of the free metal ion activities in the medium, the final conceptual model is strictly similar to the earliest attempts at correlating the toxicity of metals to their chemical properties. Metal toxicity to organisms has been correlated to the electronegativities of the metals[26,27], the stabilities of their complexes[28,29], and the insolubilities of their sulfides[30]. Hence, the reactivities of metals with various ligands have always been seen as the basis of their biological effects. What has not always been realized is the degree to which the chemistry in the medium could control the (re)activities of individual metals.

EFFECTS OF ORGANISMS ON METAL CHEMISTRY

Comparison of metal chemistry in beakers with that in the ocean is clearly the most tenuous link in the chain of assumptions necessary to extrapolate metal bioassay laboratory data to natural conditions. If we are to take the laboratory data at face value, a *sine qua non* condition is to obtain an estimation of free metal ion activities in oceanic waters. Once a reliable analytical concentration of a given metal is measured, it is a relatively easy matter to calculate the effects of inorganic speciation. Although there may be some disagreement on metal complexation constants for the principal ligands such as hydroxide, carbonate, chloride or borate (see Table 1.1), our confidence in this fundamental thermodynamic information serves to highlight our embarrassing ignorance regarding the possible complexation by organic ligands.

Recognizing that educated speculation may not be less informative than voltametric data, let us examine what is known of oceanic organic ligands. The major part of the organic content of marine waters is known to be autochtonous and we can take as our starting point the existing studies on the production of complexing agents in laboratory cultures - including data on freshwater organisms for lack of a better alternative[31,32,33,34,35]. Two major facts emerge from these: all organisms release relatively weak complexing agents in their medium and some organisms, namely cyanophytes, bacteria, and fungi, produce strong iron complexing agents - iron siderophores - under iron stress. Notably lacking from such data is any evidence for the production of very strong metal complexing agents that are not iron siderophores.

Weak Complexing Agents

Consider the possible effect of the weak complexing material. Copper complexation has been used to characterize the coordination properties of planktonic exudates and - leaving iron aside for the moment - it is certainly the most reactive of the metals that concern us here[36,37,38]. Most of the available data indicates that an upper limit for the affinity of the extracellular ligands for copper is about ten times their affinity for the hydrogen ion (log $^*K=1$) leading to an effective constant in seawater of about 10^9:

Table 1 Speciation of Some Trace Metals in Seawater.

The results presented in the following tables are derived from calculations on a model seawater system containing the following components.

o Major Species	Log (total concentration, M)
bromide	-3.08
borate	-3.31
calcium	-1.98
carbonate	-2.62
chloride	-0.25
fluoride	-4.15
magnesium	-1.26
sodium	-0.32
strontium	-4.20
potassium	-1.99
sulfate	-1.54
o Minor Species	**Log (total concentration, M)**
copper	-8.5
cadmium	-9.0
zinc	-8.2
iron	-8.0

o pH 8.2, ionic strength 0.5

Unless indicated, all constants are from: Sillen, L.G. and Martell, A.E. (1964), Stability constants of metal ion complexes, 2nd. ed., Chem. Soc. (London), Spec. Publ. 17. Smith, R.M. and Martell, A.E. (1975), Critical stability constants, Plenum Press, New York.

$$Cu^{2+} + Y = CuY \ ; \quad K_{Cu} = 10^9$$

A total ligand concentration of the order of a few namomoles - i.e., slightly in excess of the total copper - would then be sufficient to complex the metal in oceanic waters. Such concentration is about what one would expect on the basis of the relative concentrations of organisms in nature and in culture media where the ligand concentration can reach a few micromoles. Recognizing then that we have considered an upper limit for K_{Cu} and that such ligands typically exhibit a lower affinity for other metals than

Table 1.1 Equilibrium Inorganic Speciation

Copper	Log Conc.	%	Cadmium	Log Conc.	%
TOT Cu	-8.50[1]		TOT Cd	- 9.00[3]	
$CuCO_3^o$	-8.60	79	$CdCl_2^o$	- 9.29	51
$Cu(CO_3)_2^{2-}$	-9.61	8	$CdCl^+$	- 9.42	38
Cu^{2+}	-9.83	5	$CdCl_3^-$	-10.14	7
$CuOH^+$	-9.85	4	Cd^{2+}	-10.55	3
Zinc	**Log Conc.**	**%**	**Iron**	**Log Conc.**	**%**
TOT Zn	-8.20[4]		TOT Fe	- 8.00[6]	
Zn^{2+}	-8.48	52	$Fe(OH)_3(s)$ [7]	- 8.02	96
$ZnCO_3^o$ [5]	-8.65	35	$Fe(OH)_2^+$ [8]	- 9.70	2
$ZnSO_4^o$	-9.35	7	$Fe(OH)_4^-$	- 9.79	2
$ZnCl^+$	-9.86	3	Fe^{3+}	-20.66	--

[1] Boyle, E.A. et al. (1977). Earth planet, Sci. Lett., 37:38-54. Bruland, K.W. (1980). Earth planet, Sci. Lett., 47:176-98.
[2] Copper borate complexes not considered.
[3] Boyle, E.A. et al. (1976). Nature, 263:42-44. Bruland, K.W. (1980). Earth planet, Sci. Lett., 47:176-98.
[4] Bruland, K.W. (1980). Earth planet, Sci. Lett., 47:176-98.
[5] Constant estimated by Zirino and Yamamoto (1972). Limnol. Oceanogr., 17:661-671.
[6] Chester, R. and Stoner, J.H. (1974). Marine Chem., 2:17-32.
[7] K_{sp} $Fe(OH)_3(s) = 10^{-38.8}$ used (Smith and Martell, 1976).
[8] $Fe(OH)_3^o$ not considered. See discussion by Byrne and Kester (1976). Marine Chem., 4:255-74. Zafiriou and True (1980). Marine Chem., 8:281-88.

for copper, one is led to the tentative conclusion that most metals would not normally be complexed by weak organic ligands in oceanic surface waters. Copper is likely to be the most notable exception to this rule with a range of organic complexation from 0 to 100% of the total metal depending on the particular conditions. In all cases the depression in the free cupric ion activity due to organic complexation should be rather modest.

This argument is made somewhat more quantitative by considering the effect of adding citrate or ethylenediamine as model organic ligands to an equilibrium model of seawater: as seen in Table 1.2, copper becomes appreciably complexed to the organics when their concentrations exceed a few namomoles. Cd, Zn, or Fe

Table 1.2 Complexation by Weak Organic Ligands (log $*K_{Cu}$ #1)

Total Citrate*	CuCit* (Citrate)	% Cu as CuCit	Cu^{2+}
-9.5	-10.77	1	- 9.84
-9.0	-10.28	2	- 9.84
-8.5	- 9.79	5	- 9.86
-8.3	- 9.60	8	- 9.87
-8.0	- 9.33	15	- 9.90
-7.5	- 8.95	35	-10.02
-7.0	- 8.69	66	-10.28
-6.5	- 8.57	85	-10.66

Total En	CuEn*	$CuEn_2$* (Ethylenediamine)	%Cu as CuEn + $CuEn_2$	Cu^{2+}
-9.5	-10.44	-12.54	1	- 9.84
-9.0	- 9.95	-11.55	4	- 9.85
-8.5	- 9.49	-10.59	11	- 9.89
-8.3	- 9.32	-10.22	17	- 9.92
-8.0	- 9.10	- 9.70	31	-10.00
-7.5	- 8.92	- 9.02	68	-10.33
-7.0	- 9.10	- 8.67	93	-11.02
-6.5	- 9.49	- 8.55	99	-11.94

*All concentrations expressed as LOG_{10} ; (Cu^{2+}) expressed as free concentration (subtract .6 to obtain activity). TOT Cu = $10^{-8.5}$ M complexation of Cd, Zn and Fe by citrate and ethylenediamine insignificant.

are not calculated to be significantly complexed for ligand concentrations up to 500nM. On the basis of what is known of the stability of the metal-humic complexes[39,40,41,42], this result, which is applicable to the material continuously released, and presumably continuously degraded, by the planktonic biota, is probably applicable to the refractory fraction of the dissolved organic matter as well.

Iron Siderophores

Although the data are still quite sparse, there is little doubt that iron siderophores must be produced by some oceanic organisms[43,44]. While they exhibit an extremely large affinity for Fe^{3+} ($K_{Fe} \simeq 10^{30}$), siderophores are also strong complexing agents for transition metals (e.g., $K_{Cu} > 10^{10}$) and they could, in principle, play a role in complexing metals other than iron in oceanic waters. Calculations with the trihydroxamate siderophore nocardamin do not show this effect. As the nocardamin concentration is increased in the model seawater, the iron becomes bound (note that a large excess of nocardamin is necessary if one considers a rather stable form of ferric hydroxide) but magnesium effectively outcompetes the other trace metals (Table 1.3). Even if one were to consider a siderophore with a relatively lower affinity for magnesium and calcium, the organic ligand would have to be present in excess of the ambient iron concentration before complexation of other metals became appreciable. Although this condition is often encountered in iron starved laboratory cultures and may be found in lakes afflicted with cyanophyte blooms, it seems an unlikely situation for open ocean waters. In fact, since iron siderophore complexes are known to be unavailable to eukaryotic algae[4,45], the very presence of such taxa as diatoms, coccolithopores or dinoflagellates as important primary producers in oceanic systems implies that iron is not totally sequestered in siderophore complexes. _A fortiori_, other metals cannot be significantly bound to the siderophores produced by marine organisms.

The final image that is projected from these speculations on complexing ligands in the open ocean is schematized in Table 1.4: most metals are typically controlled by inorganic complexation, copper being partly bound to weak organic ligands and some fraction of the iron being tied up in a siderophore complex.

Incorporation into Particles

The production of extracellular complexing agents is a mechanism by which organisms may enhance the solubility of metals. From the point of view of the geochemist, the principal question is just the opposite and concerns the role of organisms in incorporating metals into particles and sedimenting them out of surface waters. In effect, one is interested in explaining the known correlation between vertical profiles of trace metals and nutrients in the ocean, and providing a mechanistic interpretation of the biological processes at play. Although this is an area for which laboratory studies of metal phytoplankton interactions have only limited relevance, it may be useful just to mention some ideas that have received heretofore little attention among geochemists.

Possible direct effects of organisms on trace metals cycling due to cellular uptake, adsorption on hard and soft parts, or incorporation in fecal pellets have been discussed by various authors[46,47,48,49] as has the adsorption of trace metals on inorganic particles whose sedimentation may be controlled by organ-

Table 1.3 Complexation by Strong Organic Ligand - Nocardamin

Total	K_{sp}[1] $Fe(OH)_3(s) = 10^{-38.8}$				K_{sp}[2] $Fe(OH)_3(s) = 10^{-37.5}$			
Nocardamin	$Fe(OH)_3(s)$	% ppt.	FeNOC	% FeNOC	$Fe(OH)_3$	% ppt.	FeNOC	% FeNoc
-9.5	-8.02	96	-10.25	1	-8.60	25	-9.59	3
-9.0	-8.02	96	- 9.75	2	-8.71	20	-9.09	8
-8.5	-8.04	91	- 9.25	6	-9.73	2	-8.59	26
-8.3	-8.06	87	- 9.05	9	--	0	-8.41	39
-8.0	-8.10	79	- 8.75	18	--	0	-8.18	66
-7.5	-8.40	40	- 8.25	56	--	0	-8.03	93
-7.0	--	0	- 8.01	98	--	0	-8.01	98
-6.5	--	0	- 8.00	100	--	0	-8.00	100

Notes: 1. Only iron is significantly complexed by Nocardamin. Other metals cannot outcompete magnesium for the ligand ($K_{(MgNoc)} = 10^{15.90}$).

2. The proportion of total iron bound by Nocardamin is very dependent on the solubility of $Fe(OH)_3(s)$ (see above).

TOT Fe = $10^{-8.0}$M

[1]Smith and Martell (1976).

[2]Byrne, R.H. and Kester, D.R. (1976). Marine Chem., 4:255-74.

Table 1.4 Effect of Strong and Weak Ligands

Major Species	Log Conc.	% of Total Metal
FeNoc	- 8.41	39
$Fe(OH)_2{}^+$	- 8.48	33
$Fe(OH)_4{}^-$	- 8.56	28
Fe^{3+}	-19.23	--
CuCit	- 8.70	63
$CuCO_3{}^o$	- 9.04	29
Cu^{2+}	-11.02	--
$CdCl_x$	- 9.01	97
Cd^{2+}	-10.55	3
Zn^{2+}	- 8.48	52
$ZnCO_3$	- 8.65	35
$ZnSO_4$	- 9.35	7
$ZnCl^+$	- 9.86	3

TOT Fe = $10^{-8.0}$M; TOT Cu = $10^{-8.5}$M; TOT Cd = $10^{-9.0}$M; TOT Zn = $10^{-8.2}$M; TOT Nocardamin = $10^{-8.3}$M; TOT Citrate = $10^{-7.0}$M. (K_{sp} $Fe(OH)_3(s)$ = $10^{-37.5}$ used)

isms[50,51,52,53]. What is not always realized is how physiological interactions at the cellular level could regulate such processes. For example, the incorporation of copper by diatoms in culture appears to be largely dependent upon, and inversely correlated with, the silicic acid concentration in the water[18]. This has been interpreted as an antagonistic interaction between copper and silicate at the level of the Si transport site. Such a process, and similar ones involving other metals and nutrients, could modulate the coupling between the geochemical cycles of trace metals and major nutrients in oceanic waters.

The role of the planktonic biota in controlling the sedimentation of metals may also be quite indirect. For example, it is becoming increasingly clear that suspended inorganic solids in natural waters behave as if they were in fact organic; their physical and chemical surface properties - including adsorption of trace metals and coagulation - are determined by the organic matter adsorbed on their surfaces[54,55,56]. This interfacial role of organic compounds may be more important than simple complexation in solution for mediating metal organism interactions and controlling geochemical cycles of metals in aquatic systems.

Redox Cycles

In laboratory cultures, it appears that an important role of artificial chelating agents as well as extracellular metabolites may be to serve as electron donors and that they may not just complex but also reduce some metals. For example, Fe(II) has been shown to be produced in oxic culture medium and to enhance the availability of iron to phytoplankton[4,57]. Such reduction is partly dependent upon light in accordance with the known photochemistry of iron chelates[58].

Since natural humic compounds have been shown to exert the same reducing effect on iron[59,60,61], one would expect a similar disequilibrium redox process to take place in surface waters. The process should be particularly intense in the upper layers which receive high energy photons and near the oxygen minimum where the reducing tendency should be maximum (and the oxidation tendency minimum). Such a mechanism would result in a dynamic iron cycle in surface waters: in the absence of strong chelating agents, the reduced iron should reoxidize and precipitate as a hydrous ferric oxide. This would directly affect the geochemical cycle of iron by maintaining it in a colloidal suspension and avoiding its settling out of the euphotic zone; it would indirectly affect the geochemical cycle, and the physiological effect of other metals by continuously providing a fresh and finely dispersed oxide surface with high adsorption affinity and capacity. The cycles of other metals - particularly manganese - could also be controlled by a similar redox process; either directly as primary electron acceptors, or indirectly as electron acceptors for the reoxidation of iron, according to the overall reaction:

$$Fe^{2+} + \frac{1}{2} Mn\ O_2\ (s) + 2\ H_2O \rightarrow \frac{1}{2} Mn^{2+} + H^+ + Fe(OH)_3\ (s)$$

Since iron availability may be a critical aspect of oceanic plant life, and the (photo)reduction of iron clearly enhances iron availability to eukaryotic algae[4], from a biological point of view the major role of extracellular ligands may be to serve as electron donors in redox processes rather than electron sharers in metal complexation.

METALS AND ORGANISMS IN THE OCEANS

Making a courageous leap from laboratory beakers into the open seas, let us consider the possible effects of trace metals on the oceanic biota. There seemingly are only few trace metals that are sufficiently active biologically to affect living organisms at their low natural concentrations in the oceans. Comparing the range of calculated metal ion activities in oceanic waters (Table

1.4) with those found to affect organisms in culture[5], one is led to the conclusion that three cationic trace metals may be playing a direct biological role in oceanic ecosystems: copper may be marginally toxic to some organisms and zinc and iron may be somewhat limiting plant growth. Although reliable laboratory data are not available for manganese, there is evidence that it may affect plant growth in the sea[62] and Mn should be added to the list of "ecologically interesting" metals.

Following our reasoning that metal activity ratios rather than absolute values may be the relevant biological parameters, we see that the calculated ratio $\{Cu^{2+}\}/\{Zn^{2+}\} \simeq 10^{-2.5}$ is within the range found to affect silicic acid uptake in _T. pseudonana_. However, differences in sensitivites among species should be considered, and oceanic species may actually be _less_ sensitive than coastal species such as _T. pseudonana_ which presumably thrives in an environment more affluent with complexing agents. Note also that the $\{Cu^{2+}\}/\{Fe^{3+}\}$ ratio is in excess of 10^8, high enough to suggest a possible antagonism between the two metals for iron uptake.

Direct bioassays with surface ocean water would seem the best way to verify these implications regarding the possible biological effects of trace metals. However, such assays present enormous difficulties as they require the combination of sterile and metal clean techniques. In addition, the very necessity to contain the water in a small volume creates the possibility of dramatic and unknown changes in metal speciation (e.g., due to adsorption on the surface of the container) in a poorly buffered system such as seawater. A few recent experiments have been reported that demonstrate important differences in ^{14}C uptake between incubations with and without metal clean techniques[63,64]. Certainly such results support the general contention that trace metals are important to oceanic ecosystems; however, they are difficult to interpret with a causative mechanism since the metal speciation in the incubation system is unresolved. For example, if toxic metals that naturally depressed photosynthesis in the oceans were adsorbed on the walls of an ultra-clean incubator, the measured carbon uptake rate could become artificially high. _Vice versa_, adsorption of limiting micronutrients would lead to an underestimation of the productivity. Such incubation data would then be no closer to measuring the real rate of primary productivity than a typical ^{14}C incubation in the "aura" of a rusty wire. Clean techniques are insufficient to resolve experimentally the question of trace metal organism interactions in the open ocean. We must resolve the chemical speciation problem as well.

While the possible role of trace metals in controlling oceanic productivity has been historically the prime motivation for studying metal phytoplankton interactions, over the past few years, the major emphasis has been instead on the possible role of trace metals in controlling species assemblages. This is a logical inference of the laboratory observation that different species exhibit wide differences in sensitivities to toxic metals. To some degree, this hypothesis is confirmed by the principal results of the CEPEX experiments [65,66]. In large plastic enclosures in Saanich Inlet (Victoria, B.C., Canada), the major response of natural populations to copper additions was a shift of dominant species from centric diatoms and dinoflagellates to pennate diatoms and microflagellates over a period of a few days, following a rapid decrease in productivity. However, the time scale of such experiments is still quite short compared to that of oceanic processes and the observed shift in dominant species may have no more relevance to natural systems than the short term inhibition of photosynthesis by toxic metals. The problem here is that phytoplankton exhibit large intraspecific differences in metal sensitivities, as illustrated in Fig. 7 for two isolates of Biddulphia aurita. Clone B-1 isolated from the Delaware Bay is some two and a half orders of magnitude more sensitive to the cupric ion activity than clone STX-88 isolated from St. Croix. Since metal resistant strains are typically found in metal polluted areas the implication is that species actually adapt to metal stress - presumably if the change in metal chemistry is appropriately slow. If we are to understand the ecological role of metals in aquatic systems, we must then understand the adaptation mechanism itself.

The little that we know regarding adaptation of phytoplankton to high activities of toxic metals implies that an exclusion mechanism is at play: by and large, an alga that is more metal toler-

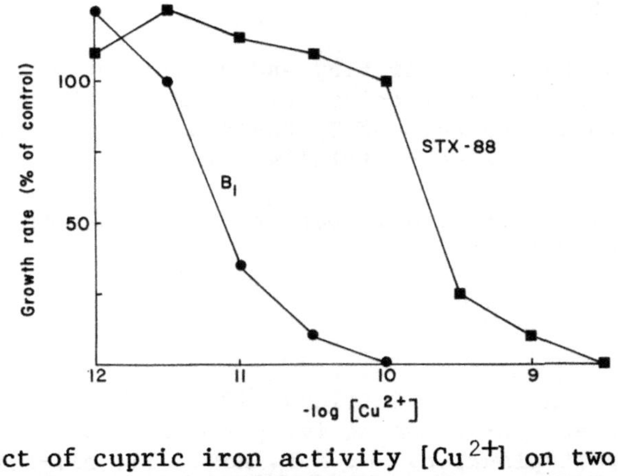

Fig. 7 Effect of cupric iron activity $[Cu^{2+}]$ on two clones of Biddulphia aurita; after D.J. Hughes, MIT Technical Note #25.

ant is one that takes up less of the toxic metals either in comparison with an alga of the same/or different species[67,68,69]. [As shown in Figs. 4 and 5, the affinity for copper of the copper tolerant strain of Chlorella - K_x in reaction 1 - is actually lower than that of the copper sensitive strain.] We must then understand the nature of the role of the reaction between metals and cellular ligands before we can understand how the affinity of the organism for metals can be modified and what the possible physiological and ecological consequences of such modification might be.

The fields of Chemical and Biological Oceanography are both in a fair state of transition. The chemists have now repudiated all their past measurements of trace metals in the sea and begun a slow and contentious acquisition of "oceanographically consistent" data sets. The biologists are now also beginning to doubt the validity of their past measurements of productivity and critical nutrients and are wondering whether they may have underestimated the possible role of the "picoplankton." Under these conditions, discussing the relation between the geochemistry of trace metals in the oceans and the ecology of the planktonic biota is certainly foolhardy. However, we have to start focussing on the critical problems of the oceans, not just those amenable to easy solutions. Even if the techniques are difficult and the problems seemingly unwieldy, the chemists can no longer ignore biological processes and the biologists cannot overlook the importance of trace elements. Central to both their preoccupations is the issue of chemical speciation. For too long, like the fabled drunk under the street light, we have searched where we could see, perhaps not where the answers lie.

ACKNOWLEDGEMENTS

We wish to thank S.J. Tiffany and particularly S.W. Chisholm who provided us with constructive criticisms of the various drafts of the manuscript. The help of T. David Waite in performing the model calculations is also gratefully acknowledged.

This work was supported in part by NSF grant OCE-7919549 and by NOAA grant NA79AA-D-00077.

REFERENCES

1. Sunda, W.G. and R.R.L. Guillard, 1976: The relationship between cupric ion activity and the toxicity of copper to phytoplankton. J. Mar. Res., 34, 511-529.
2. Anderson, D.M. and F.M.M. Morel, 1978: Copper sensitivity of Gonyaulax tamarensis. Limnol. Oceanogr., 23, 283-295.

3. Anderson, M.A., F.M.M. Morel and R.R.L. Guillard, 1978: Growth limitation of a coastal diatom by low zinc ion activity. Nature, 276, 70-71.
4. Anderson, M.A. and F.M.M. Morel, in press: The influence of aqueous iron chemistry on the uptake of iron by the coastal diatom Thalassiosira weissflogii. Limnol. Oceanogr.
5. Hughes, D.J., 1981: An interspecific comparison of trace metal toxicity to marine phytoplankton. Tech. Note No. 25, Dept. Civil Eng., Mass. Inst. Technol., Cambridge, MA.
6. Huntsman, S.A. and W.G. Sunda, 1980: The role of trace metals in regulating phytoplankton growth in natural waters. In: "The physiological ecology of phytoplankton", I. Morris, ed. Studies in Ecology, Vol. 7, Blackwell Scientific, Boston.
7. Gillespie, P.A. and R.F. Vaccaro, 1978: A bacterial bioassay for measuring the copper-chelation capacity of seawater. Limnol. Oceanogr., 23, 543-548.
8. Sunda, W.G. and P.A. Gillespie, 1979: The response of a marine bacterium to cupric ion and its use to estimate cupric ion activity in seawater. J. Mar. Res., 37, 761-777.
9. Andrew, R.W., K.E. Biesinger and G.E. Glass, 1977: Effects of inorganic complexing on the toxicity of copper to Daphnia magna. Water Res., 11, 309-315.
10. Sunda, W.G., D.W. Engel and R.M. Thuotte, 1978: Effect of chemical speciation on toxicity of cadmium to grass shrimp, Palaemonete pugio: Importance of free cadmium ion. Environ. Sci. Technol.12, 409-413.
11. Chakoumakos, C., R.C. Russo and R.V. Thurston, 1979: Toxicity of copper to cutthroat trout (Salmo darki) under different conditions of alkalinity, pH, and hardness. Environ. Sci. Technol., 13, 213-219.
12. Sunda, W.G. and J.H. Lewis, 1978: Effect of complexation by natural organic ligands on the toxicity of copper to a unicellular alga, Monochrysis lutheri. Limnol. Oceanogr., 23, 870-876.
13. Westall, J.C., F.M.M. Morel and D.N. Hume, 1979: Chloride interference in cupric ion selective electrode measurements. Analyt. Chem., 51, 1792-1798.
14. Brezonik, P.L., P.A. Brauner and W. Stumm, 1976: Trace metal analysis by anodic stripping voltammetry: Effect of sorption by natural and model organic compounds. Water Res., 10, 605-612.
15. Stolzberg, R.J., 1977: Potential inaccuracy in trace metal speciation measurements by differential pulse polarography. Anal. Chim. Acta., 92, 193-196.
16. Morel, F.M.M., J.G. Rueter, Jr., D.M. Anderson and R.R.L. Guillard, 1979: Aquil: A chemically defined phyto-

plankton culture medium for trace metal studies. J. Phycol., 15, 135-141.

17. Foster, P.L. and F.M.M. Morel, 1982: Reversal of cadmium toxicity in the diatom Thalassiosira weissflogii. Limnol. Oceanogr., 27, 745.
18. Rueter, J.G., Jr., S.W. Chisholm and F.M.M. Morel, 1981: The effect of copper toxicity on silicic acid uptake and growth in Thalassiosira pseudonana (Bacillariophyceae). J. Phycol., 17, 270-278.
19. Rueter, J.G., Jr. and F.M.M. Morel, 1981: The interaction between zinc deficiency and copper toxicity as it affects the silicic acid uptake mechanisms in Thalassiosira pseudonana. Limnol. Oceanogr., 26, 67-73.
20. Goering, J.J., D. Boisseau and A. Hattori, 1977: Effects of copper on silicic acid uptake by a marine phytoplankton population: Controlled ecosystem pollution experiment. Bull. Mar. Sci., 27, 58-65.
21. Cloutheir-Mantha, L. and P.J. Harrison, 1980: Effects of sublethal concentrations of mercuric chloride on ammonium-limited Skeletonema costatum. Mar. Biol., 56, 219-231.
22. Fogg, C.E. and D.F. Westlake, 1955: The importance of extracellular products of algae in freshwater. Proc. Int. Assoc. Theor. Appl. Limnol., 12, 219.
23. Barber, R.T., 1973: Organic ligands and phytoplankton growth in nutrient-rich seawater. In: "Trace metals and metal-organic interactions in natural waters", P.C. Singer, ed. Ann Arbor Science, Ann Arbor, MI, pp. 321-338.
24. Foster, P.L. and F.M.M. Morel, in preparation.
25. Butler, M., A.E.J. Haskew and M.M. Young, 1980: Copper tolerance in the green algae, Chlorella vulgaris. Plant, Cell and Environ., 3, 119-126.
26. Danielli, J.F. and J.T. Davies, 1951: Reactions at interfaces in relation to biological problems. Adv. Enzymol., 11, 35-89.
27. Somers, E., 1960: Fungi toxicity of metal ions. Nature, 187, 427-428.
28. Bowen, H.J.M., 1966: "Trace elements in biochemistry", Academic Press, Inc., London.
29. Shaw, W.H.R., 1961: Cation toxicity and the stability of transition-metal complexes. Nature, 192, 754-755.
30. Shaw, W.H.R., 1954: Toxicity of cations toward living systems. Science, 120, 361-363.
31. Swallow, K.C., J.C. Westall, D. M. McKnight, N.M.L. Morel and F.M.M. Morel, 1978: Potentiometric determination of copper complexation by phytoplankton exudates. Limnol. Oceanogr., 23, 538-542.
32. McKnight, D.M. and F.M.M. Morel, 1979: Release of weak and strong copper complexing agents by algae. Limnol. Oceanogr., 24, 823-837.

33. McKnight, D.M. and F.M.M. Morel, 1980: Copper complexation by siderophores from filamentous blue-green algae. Limnol. Oceanogr., 25, 62-71.
34. Ragan, M.A., O. Smidsrod and B. Larsen, 1969: Chelation of divalent metal ions by brown algal polyphenols. Marine Chem., 7, 265-271.
35. Van den Berg, C.M.G., P.T.S. Wong and Y.K. Chau, 1979: Measurement of complexing material excreted from algae and their ability to ameliorate copper toxicity. J. Fish Res. Board Can., 36, 901. Interfaces, Ann Arbor Science, Ann Arbor, MI (1981).
36. Irving, H. and R.J.P. Williams, 1953: The stability of transition-metal complexes. J. Chem. Soc., 3192.
37. Pearson, R.G., 1963: Hard and soft acids and bases. J. Amer. Chem. Soc., 85, 3533-3539.
38. Stumm, W. and J.J. Morgan, 1981: "Aquatic Chemistry", John Wiley and Sons, Inc., New York.
39. Van den Berg, C.M.G. and J.R. Kramer, 1979: Determination of complexing capacities and conditional stability constants for copper in natural waters using MnO_2. Anal. Chim. Acta., 106, 113-120.
40. Shum, M.S. and G.P. Woodward, 1977: Stability constants of copper-organic chelates in aquatic samples. Environ. Sci. Technol., 11, 809-813.
41. Mantoura, R.F.C., A. Dickson and J.P. Riley, 1978: The complexation of metals with humic materials in natural waters. Estuar. and Coast. Mar. Sci., 6, 387-408.
42. Van den Berg, C.M.G. and J.R. Kramer, 1979: Conditional stability constants for copper ions with ligands in natural waters. In: "Chemical modeling in Aqueous Systems", E.A. Jenne, ed. American Chemical Society, Washington.
43. Estep, M., J.E. Armstrong and C. van Ballen, 1975: Evidence for the occurrence of specific iron(III)-binding compounds in near-shore marine ecosystems. Appl. Microbiol., 30, 186-188.
44. Gonye, E.R. and E.J. Carpenter, 1974: Production of iron-binding compounds by marine microorganism. Limnol. Oceanogr., 19, 840-841.
45. Murphy, T.P., D.R.S. Lean and C. Nalewajko, 1976: Blue-green algae: Their excretion of iron-selective chelators enables them to dominate other algae. Science, 192, 900-902.
46. Martin, J.H., 1970: The possible• transport of trace metals via molted copepod exoskeletons. Limnol. Oceanogr., 15, 756-761.
47. Wright, D.A., 1978: Heavy metal accumulation by aquatic invertebrates. Adv. Appl. Biol., 3, 331.
48. Renfro, W.C., S.W. Fowler, M. Heyraud and J. La Rosa, 1975: Relative importance of food and water in long term

zinc-65 accumulation by marine biota. J. Fish. Res. Bd. Can., 32, 1339.

49. Kavanaugh, M.C. and J.O. Leckie, 1980: "Particulates (sic) in water", Advances in Chemistry Series, American Chemical Society, Washington.

50. Boyle, E.A., 1979: Copper in natural waters. In: "Copper in the environment", Part I., J.O. Nriagu, ed. John Wiley and Sons, Inc., New York, pp. 77-86.

51. Schindler, P.W., 1974: Removal of trace metals from the oceans: A zero order model. Thalassia Yugosla., 11, 01.

52. O'Connor, T.P. and D.R. Kester, 1975: Adsorption of copper and cobalt from fresh and marine systems. Geochem. Cosmochim. Acta., 39, 1531-1543.

53. Anderson, M.A. and A.J. Rubin, 1981: "Adsorption of inorganics at the solid/liquid interfaces". Ann Arbor Science, Ann Arbor, MI.

54. Tipping, E., in press: The surface chemistry of hydrous iron oxides in Esthwaite water (U.K.). Environ. Sci. Technol.

55. Davis, J.A. and J.O Leckie, 1978: Effects of adsorbed complexing ligands on trace metal uptake by hydrous oxides. Environ. Sci. Technol., 12, 1309-1315.

56. Davis, J.A., in press: Adsorption of natural organic matter from freshwater environments by aluminum oxide. In: "Contaminants and Sediments", R.A. Baker, ed. Vol. 2, Ann Arbor Science, Ann Arbor, MI.

57. Anderson, M.A. and F.M.M. Morel, 1980: Uptake of Fe(II) by a diatom in oxic culture medium. Mar. Biol. Letters, 1, 263-268.

58. Hill-Cottingham, D.G., 1955: Photosensitivity of iron chelates. Nature, London, 175, 347-348.

59. Theis, T.L. and P.C. Singer, 1973: The stabilization of ferrous iron by organic compounds in natural waters. In: "Trace metals and metal organic interactions", P.C. Singer, ed. Ann Arbor Science, Ann Arbor, MI, pp. 303-320.

60. Theis, T.L. and P.C. Singer, 1974: Complexation of iron(II) by organic matter and its effect on iron(II) oxygenation. Env. Sci. Technol., 8, 569-573.

61. Miles, C.J. and P.L. Brezonick, in press: Oxygen consumption in humic-colored waters by a photochemical ferric-ferrous catalytic cycles. Envir. Sci. Technol.

62. Sunda, W.G., R.T. Barber and S.A. Huntsman, in preparation: Phytoplankton growth in nutrient rich seawater: Importance of copper-manganese cellular interactions.

63. Carpenter, E.J. and J.S. Lively, 1980: Review of estimates of algal growth using ^{14}C tracer technique. In: "Primary productivity in the sea", P.G. Falkowski, ed. Plenum Press, N.Y., pp. 161-178.

64. Fitzwater, S.E., G.A. Knauer and J.H. Martin, in preparation: Metal contamination and primary production: Field and laboratory methods of control.
65. Thomas, W.H., O. Holm-Hansen, D.L.R. Seibert, F. Azam, R. Hodson and M. Takahashi, 1977: Effects of copper on phytoplankton standing crop and productivity: Controlled ecosystem pollution experiment. Bull. Mar. Sci., 27, 34-43.
66. Thomas, W.H. and D.L.R. Seibert, 1977: Effects of copper on the dominance and the diversity of algae: Controlled ecosystem pollution experiment. Bull. Mar. Sci., 27, 23-33.
67. Foster, P.L., 1977: Copper exclusion as a mechanism of heavy metal tolerance in a green alga. Nature, 269, 322-323.
68. Nakajima, A., T. Horikoshi and T. Sakaguchi, 1979: Uptake of copper ion by green microalgae. Agric. Biol. Chem., 43, 1455-1460.
69. Hall, A., A.H. Fielding and M. Butler, 1979: Mechanisms of copper tolerance in the marine fouling alga Ectocarpus siliculosus: Evidence for an exclusion mechanism. Mar. Biol., 54, 195-199.

SENSITIVITY OF NATURAL BACTERIAL COMMUNITIES TO ADDITIONS OF COPPER AND TO CUPRIC ION ACTIVITY: A BIOASSY OF COPPER COMPLEXATION IN SEAWATER

William G. Sunda and Randolph L. Ferguson

National Marine Fisheries Service, NOAA
Southeast Fisheries Center
Beaufort Laboratory
Beaufort, North Carolina 28516 U.S.A.

ABSTRACT

Cupric ion bioassays with natural bacterial communities were conducted in seawater samples at five stations in the Gulf of Mexico, ranging from low-productivity oceanic to high-productivity coastal. In these bioassays, we measured the incorporation of ^{3}H-labeled amino acids by marine microbial communities following serial additions of ionic copper ($CuSO_4$) and cupric ion buffers (combinations of $CuSO_4$ and nitrilotriacetic acid, a synthetic chelator). The bioassays were designed to yield relations between (1) amino acid incorporation and total copper concentration, (2) incorporation and cupric ion activity, and (3) cupric ion activity and total copper concentration.

Microbial uptake of ^{3}H-labeled amino acids in low-productivity seawater was reduced by 2 nmol l^{-1} additions of $CuSO_4$, an addition at or below ambient copper concentrations. Added copper was appreciably less toxic in high-productivity coastal seawater samples. Differences in the effect of added copper among samples could be attributed both to differences in the relationships between inhibition of labeled amino acid incorporation and cupric ion activity and to differences in relationships between cupric ion activity and total copper concentration. The bioassays indicated higher complexation of added copper in high-productivity coastal samples. Complexation data was modeled using chemical equilibrium theory to estimate conditional stability constants and total concentrations of strong copper complexing ligands and to estimate copper speciation at ambient copper concentrations. Conditional stability constants ranged from 10^9 to $\geq 10^{11}$ and ligand concentrations ranged from 5 to 150 nmol l^{-1}. Cupric ion activity at ambient levels of copper are estimated to be in the range $\leq 10^{-12}$ to 10^{-11} mol l^{-1}.

INTRODUCTION

The chemical speciation of trace metals in seawater strongly influences their biogeochemical behavior. It has long been recognized that the complexation of metals by both synthetic and natural ligands can detoxify or render metals biologically unavailable [1,2]. Recent experimental evidence has shown that the basis for this detoxification is a dependence of biological availability on free metal ion activity[3,4,5]. Such free ion dependence appears to be fairly general and has been shown to apply for a number of different metals (e.g., Cu, Cd, Zn, Mn), complexing ligands (EDTA, NTA, Tris, natural organic ligands, carbonate ions), and organisms (phytoplankton, bacteria, crustaceans, fish).

Measurement of trace metal chemical speciation in seawater has been extremely difficult experimentally because of low total metal concentrations (often in the nannomolar or subnannomolar range) and lack of sufficient sensitivity or selectivity of existing techniques. Although much progress has been made recently in expanding our knowledge of total trace metal concentrations in the ocean[6,7], the speciation of these metals, particularly with respect to free metal ion activities, is still open to question. Uncertainty in trace metal ion activities is especially problematic for metals, such as copper, that have a high tendency to undergo complex formation.

In this paper we discuss experiments that use cupric ion toxicity to marine bacterial communities to bioassay cupric ion activity and copper complexation in seawater. The bioassay consists of measurements of microbial uptake of tritium-labeled amino acids[8], following parallel serial additions of a copper salt ($CuSO_4$) and cupric ion buffers (combinations of $CuSO_4$ and nitrilotriacetic acid (NTA), a well characterized synthetic chelator). The bioassay yields the relation between (1) inhibition of bacterial incorporation of amino acids and total copper concentration, (2) incorporation of amino acids and cupric ion activity, and (3) curpic ion activity and total copper concentration. The last of these represents a measure of the extent of copper complexation in seawater.

MATERIALS AND METHODS

Seawater was collected at five stations in the Gulf of Mexico during April 1980 (Fig. 1; Table 1). The seawater sampled ranged

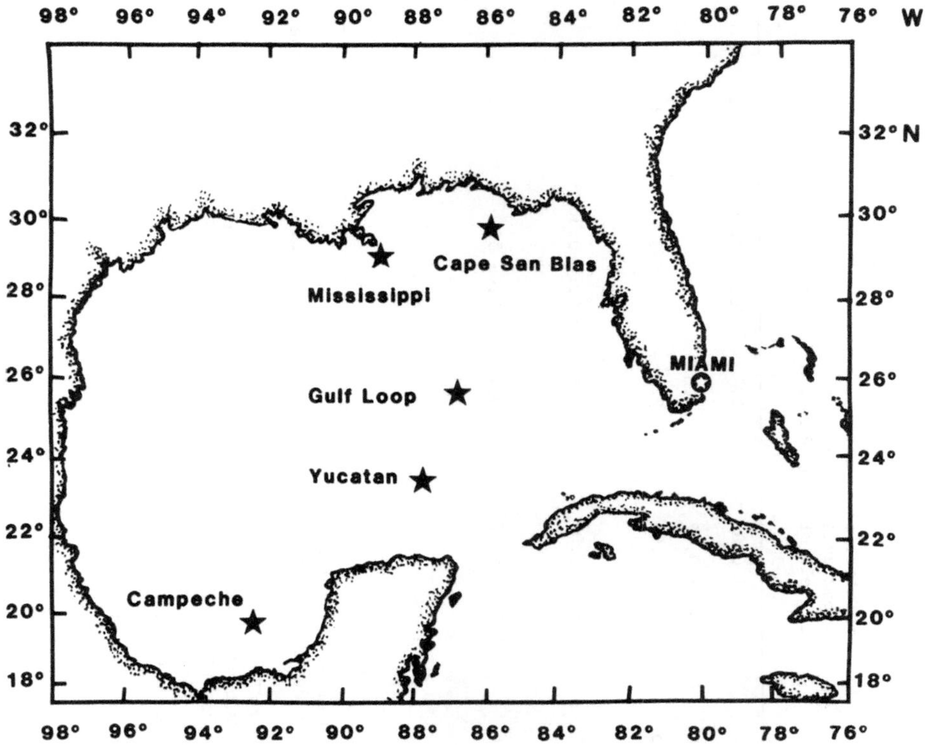

Fig. 1 Location of sampling stations.

from high-productivity, low-salinity coastal water from the Mississippi River discharge plume and from off Cape San Blas, Florida, to low-productivity, high-salinity oceanic water from within the loop of the Gulf Loop Current. Seawater from two other stations, one off the Yucatan Peninsula and the other in the Bay of Campeche, had high salinity and low to intermediate productivity.

Trace metal clean procedures were used to collect and process samples. We collected seawater with a 30 l Teflon-lined sampler (General Oceanics GO-Flo*) suspended on a plastic hydrowire (Dupont Kevlar) and closed with a Teflon messenger. We also used a concrete weight on the hydrowire, a stainless steel pulley and a portable winch with a hand-held Teflon level wind. On deck the ends of the sampling bottle were covered with polyethylene bags. Water was transferred from the sampling bottle to a 5 l Teflon bottle in a laminar flow hood and was stirred continuously with a

*Reference to trade names does not imply product endorsement by the National Marine Fisheries Service.

Table 1. Descriptive data for stations, experimental samples, and experimental conditions

Station	Depth (m)	Salinity (°/oo)	Experimental temperature (°C)	pH	Exposure to Cu & CuNTA (h)	Concentration of bacteria (10^6 cell ml^{-1})	Amino acid turnover rate* (% label h^{-1})	Primary productivity** (μg C l^{-1} h^{-1})	Ambient total Cu*** (nmol l^{-1})
Gulf Loop 25° 25'N 86° 59'W	50	36.1	24 ± 1	8.20	5.2	0.32 ± 0.02	1.7	0.04-0.08	∿2.7
	18	36.1	24 ± 1	8.18	5.0	0.30 ± 0.02	7.3		2.4
Yucatan 23° 34'N 87° 49'W	10	36.4	23 ± 1	8.21	5.5	0.54 ± 0.04	3.2	0.05-0.14	∿3.6
Campeche Bay 19° 46'N 92° 25'W	5	35.4	22 ± 1	8.20	5.6	0.57 ± 0.03	9.9	0.18-0.36	4.8
Mississippi 28° 51'N 89° 16'W	4	28.9	22 ± 1	8.09	4.6	1.08 ± 0.10	69.4	3.0- 32	8.8
Cape San Blas 29° 51'N 86° 08'W	4	26.8	23 ± 1	8.16	4.4	1.69 ± 0.51	52.1	0.5 - 1.8	4.4

* Determined for unamended controls tested at 6.2 to 8.0 h after collection of water.

** Determined from daily measurements over a 2-4 d period at depth of maximum productivity using a modified ^{14}C technique with trace metal clean experiment procedures.

*** All values are from data of Piotrowicz[22] as determined by ASV measurements at pH 2 using a standard addition technique. Samples were acidified immediately after collection and stored in clean Teflon bottles for several months before analyses. Mississippi and Cape San Blas samples, which had high concentrations of particulate matter, were centrifuged and decanted before analysis. Campeche, Gulf Loop (18 m), Mississippi, and Cape San Blas total copper concentrations were measured directly on subsamples of seawater used in our experiments. The Gulf Loop (50m) total copper concentration listed in the table represents a mean value for measurements made at 18 m (2.4 and 3.3 nmol l^{-1}) and 200 m (2.2 and 3.1 nmol l^{-1}). Copper concentration for our 10 m Yucatan sample is assumed to equal that for a similar sample collected at 7 m.

Teflon magnetic stir bar to prevent particle settling. Fifteen ml portions of the seawater sample were dispensed with a Teflon-lined dispenser (Brinkman Dispensette) into 30 ml Teflon experimental bottles. This water, which served as a final rinse, was then poured out and the bottles were refilled immediately with 25 ml of the seawater sample. The dispensing of rinse and experimental seawater and the pipetting of copper solutions and labeled amino acids were conducted either in the anteroom of a clean van or in a laminar flow hood in the ship's oceanographic laboratory. The caps of the Teflon experimental bottles were removed for only 2-3 seconds during rinse, filling, and pipeting, further minimizing the possibility of sample contamination from particles in the air. After conventional cleaning, all Teflon bottles were soaked in 1.2 N HCl for at least one week and then rinsed three times with redistilled water from a quartz subboiling still. The 30 ml experimental bottles were dried in a laminar flow hood with their caps slightly ajar and used once. The same 30 l Go-Flow bottle, magnetic stir bar, dispensette, and 5 l Teflon sample bottle were used for each experiment. The sampling bottle was rinsed between stations with 1.2 N HCl. The dispensette, stir bar, and 5 l bottle were thoroughly rinsed with 1.2 N HCl, deionized water (Milli-Q), and then sample water before each use.

The 25 ml subsamples of seawater were arranged into three series. One series received additions of $CuSO_4$ and the second and third received cupric ion buffers consisting of mixtures of copper and NTA. The first series of buffers gave a final NTA concentration of 1 μmol l^{-1} while the second gave a concentration of 5 μmol l^{-1}. Two different NTA concentrations were used to insure that the bacterial response to cupric ion activity was independent of the concentration of NTA. All treatments were run in triplicate.

Individual stock solutions were prepared for each treatment such that a 100 μl addition to a 25 ml sample yielded the appropriate concentration of $CuSO_4$ and NTA. These solutions were prepared with redistilled water from a quartz subboiling still. The water was adjusted to pH 4.2 with HCl to minimize adsorptive loss of copper to the walls of the linear polyethylene stock solution bottles. Stock solutions were prepared just before the cruise and were checked by anodic stripping voltametry measurements during the cruise against an independent set of pH 2 standard copper solutions.

A mixture of tritium-labeled amino acids (New England Nuclear NET-250) was added to the seawater subsamples following a 2-4.5 h exposure to copper and to cupric ion buffers. Amino acids were added at a concentration of 1 or 2 nmol l^{-1}, below values reported for open ocean seawater ($\geq$ 10 nmol l^{-1})[9]. Therefore, the added amino acids should have been at or near true tracer levels. Bacteria in the seawater were allowed to take up the labeled amino acids for either 0.5 or 1.0 h. Uptake was terminated by the addition 0.2 ml of 37% formaldehyde. Bacteria were collected on 47 mm diameter 0.2 μm pore size Nuclepore filters and the amount of ^3H-amino acids that they incorporated was measured using standard liquid scintillation counting. All measurements are reported as radioactive disintegrations per minute (dpm) and are corrected for blanks in which the bacteria were killed with formaldehyde before addition of label. Blanks were low, only 0.4 to 5% of the sample values.

Temperature and pH were measured in subsamples carried through the entire experimental procedure, but without addition of label or formaldehyde. Temperature was measured periodically during the experiments. The pH was measured just before the addition of label with a glass electrode (Beckman 39301) coupled to a single junction Ag/AgCl reference electrode (Orion 90-91 with Orion 90-00-01 filling solution).

Bioassay of cupric ion activity and copper complexation

Previous investigations with a bacterial isolate have shown that the toxicity of copper is determined by the free cupric ion activity and that complexes of copper with EDTA, NTA, and natural organic ligands (humic and fulvic acids) are not directly toxic[10,11]. These observations along with similar subsequent ones[12] form the basis for the present experiments, which use the sensitivity to cupric ions of natural bacterial communities to bioassay cupric ion activity in seawater. The theoretical considerations and calculations used in the cupric ion bioassays are described in detail in two other manuscripts[11,12], but also will be described briefly here.

We measured cupric ion inhibition of microbial incorporation of amino acids in the presence of three series of additions: (1) ionic copper (as $CuSO_4$), (2) copper plus 1 μmol l^{-1} NTA, and (3) copper plus 5 μmol l^{-1} NTA. The addition of the copper-NTA cupric ion buffers in parallel with serial additions of ionic copper allowed us to compute the relationship between inhibition of amino acid incorporation and cupric ion activity. These computations are as follows:

Higher concentrations of copper must be added to elicit the same inhibition in the presence of added NTA than in its absence because most of the copper added with NTA exists as a $CuNTA^-$ complex, which is not directly toxic[11]. Equal levels of inhibition should occur at equal levels of cupric ion activity, and at constant activity the concentration of copper species in seawater other than $CuNTA^-$, such as complexes with inorganic and natural organic ligands, should be constant. Therefore, for each combination of added copper plus NTA, the concentration of $CuNTA^-$ complex should be equal to the total concentration of copper minus the concentration of copper that causes the same level of inhibition in the absence of NTA.

The addition of NTA alone had no apparent effect on microbial amino acid incorporation in all but the Yucatan sample. Therefore, in computing CuNTA concentrations for all but this sample, equal cupric ion inhibition at different NTA concentrations was based on equal absolute levels of ^{3}H-amino acid incorporation. For the Yucatan sample, the addition of 1 and 5 μmol l^{-1} NTA caused 9 and 14% inhibition respectively, relative to the no addition control, and for this sample we normalized all amino acid incorporation values relative to appropriate control values.

The computed concentration of $CuNTA^-$ associated with each added copper-NTA buffer was used to calculate cupric ion activities using equilibria between copper and NTA in seawater[11,13]. Based on these calculations, the activity of cupric ion, A_{Cu}, is related to the concentration of $CuNTA^-$ chelate, $[CuNTA^-]$, and the total NTA concentration, $[NTA_{TOT}]$, by the expression:

$$[CuNTA^-] = \frac{[NTA_{TOT}]\ A_{Cu}K_{CuNTA}}{A_{Cu}K_{CuNTA} + A_{Mg}K_{MgNTA} + A_{Ca}K_{CaNTA}}$$

where K_{CuNTA}, K_{CaNTA}, and K_{MgNTA} are stability constants selected from Sillen and Martell[14]. Their respective values are $10^{12.96}$, $10^{6.41}$, and $10^{5.41}$ at 20°C. The terms A_{Ca} and A_{Mg} are the activities of calcium and magnesium ions as computed from total concentrations of these metals in seawater and from total ion activity coefficients[16]. Total concentrations were estimated from values compiled by Goldberg[15] adjusted for differences in seawater salinity.

Relationships between inhibition of microbial amino acid incorporation and computed cupric ion activity and between inhibition and the total concentration of copper in the absence of NTA (i.e., ambient copper plus added $CuSO_4$) were combined to yield bioassayed relationships between cupric ion activity and total copper concentration. These latter relationships provide a measure of the extent to which copper is complexed in seawater. They were analyzed using chemical equilibrium theory (Scatchard plots) to estimate concentrations and conditional stability constants for copper complexing ligands[17,18]. This analysis assumed that reactions between copper and ligands are 1:1. The conditional stability constant, K_c, is valid only at a given pH (in this case 8.1 to 8.2) and is defined by the equation:

$$K_c = \frac{[CuL]}{[Cu^{2+}][L^*]}$$

where [CuL] is the concentration of copper-ligand complex, $[Cu^{2+}]$ is the concentration of free cupric ion, and [L*] is an operationally defined free ligand concentration equal to the total ligand concentration minus [CuL]. K_c is defined in terms of cupric ion concentration rather than cupric ion activity; an activity coefficient of 0.25 was used to convert cupric ion activities to free cupric ion concentrations.

RESULTS

Microbial amino acid incorporation

Incorporation of tritium-labeled amino acids by bacteria was much more rapid than we had anticipated from previously published data, particularly in oligotrophic seawater[19]. Agreement among replicates was also very good. We attribute both high incorporation rates and good replication to the use of trace metal clean techniques that minimized or eliminated random sample contamination. At the two lowest productivity stations (Gulf Loop and Yucatan, Table 1), the bacteria incorporated 1.7 to 7.3% of the added ^{3}H-labeled amino acids during a 1 h incubation, and from these values we estimate turnover times of 14 to 60 h. Incorporation was more rapid at the two highest productivity stations (Cape San Blas and Mississippi). At these stations 26 and 35% of the label was incorporated during a 0.5 h incubation, yielding turnover time estimates of 1.4 and 1.9 h. Therefore, at the two highest productivity stations the amino acid turnover time was shorter than the time frame of the experiments (4.5 ± 0.1 h).

Response to added copper and copper-NTA buffers

Microbial incorporation of amino acids was inhibited by very low additions of copper, particularly in low-productivity waters (Fig. 2). For samples from the Campeche and Yucatan stations, 15% inhibition was observed with a 5.5 h exposure to copper at the lowest level of addition, 2 nmol l^{-1} (0.12 ppb). Inhibition was even more pronounced at the lowest productivity station (Gulf Loop), where a 5.2 h exposure to copper at a 2 nmol l^{-1} addition caused 27% inhibition in the sample collected from 18 m and 53% in the sample from 50 m. Copper additions caused appreciably less inhibition in low-salinity 4 m samples from the two high-productivity coastal northern Gulf of Mexico stations (Fig. 2). The copper inhibition curves at these two stations were similar with about 20% inhibition at 30 nmol l^{-1} $CuSO_4$ and about 75% inhibition at 200 nmol l^{-1}.

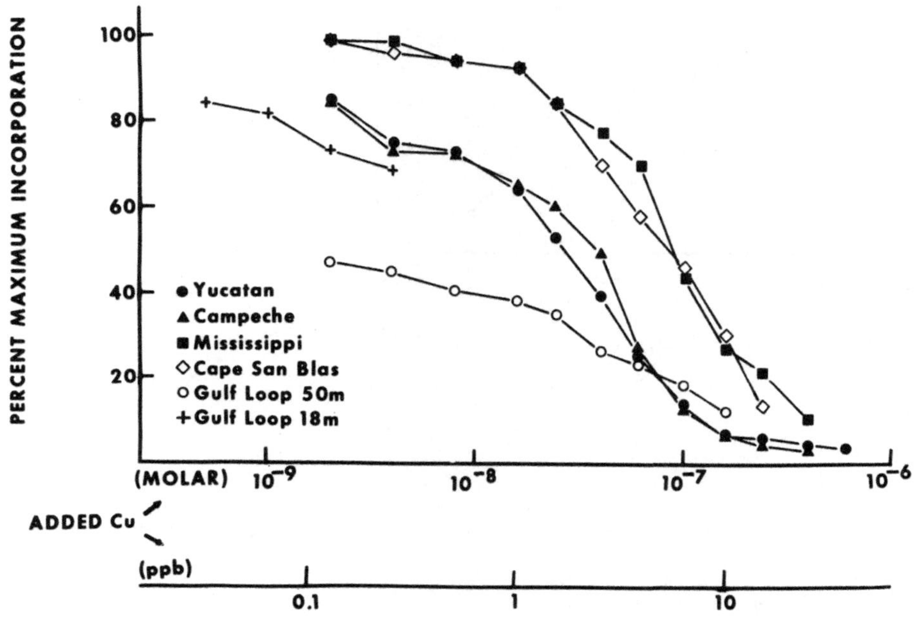

Fig. 2 Effect of copper additions on the incorporation of labeled amino acids by bacterial communities in different waters of the Gulf of Mexico. Incorporation values are given as percent of that in controls receiving no added copper.

An intriguing observation was that at the three intermediate to low productivity stations, the lowest level of added copper that caused a significant effect (2 nmol l^{-1} and in one case 0.5 nmol l^{-1}, see Fig. 2) was lower than the measured ambient total concentration of copper (Table 1). For the 18 m Gulf Loop sample 16% inhibition was recorded at an addition (0.5 nmol l^{-1}) that was only 21% of the ambient concentration (2.4 nmol l^{-1}).

The addition of NTA decreased the toxicity of added copper to natural bacterial communities (Fig. 3), similar to results found previously with a bacterial isolate [11]. Relationships between inhibition and total copper concentration differed systematically with 0, 1 and 5 μM NTA. In all but the Campeche sample, however, relationships at different concentrations of NTA approached one another at the lowest additions of copper.

Relationships between inhibition of microbial incorporation of amino acids and cupric ion activity

Relationships between amino acid incorporation and the computed cupric ion activity were independent of the concentration of

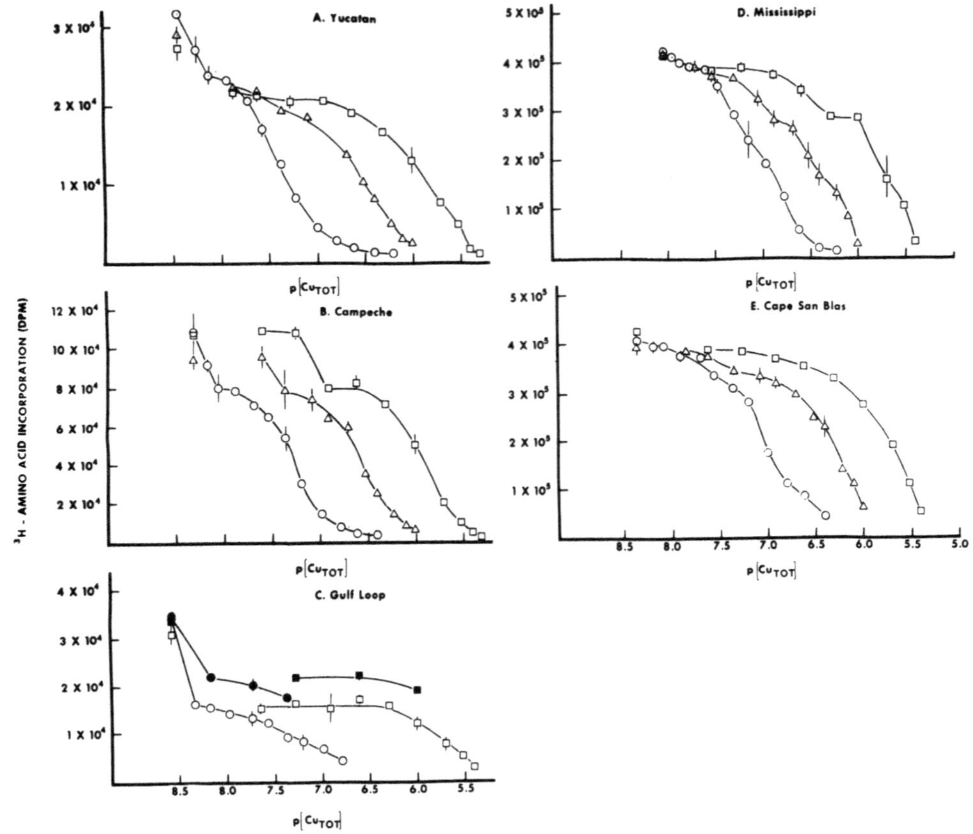

Fig. 3 Effect of additions of copper and combinations of copper and NTA (i.e., cupric ion buffers) on the incorporation of ^{3}H-labeled amino acids by bacterial communities in the Gulf of Mexico. Y-axis values are in disintegrations per minute (DPM) for radioactively labeled amino acids taken up by the bacteria. p[Cu$_{TOT}$] is the negative log of the total copper concentration in molar units. Total copper concentration equals the added plus initial ambient concentration. Symbols are defined as follows: ○ ● No NTA added; Δ 1.0 μmol l^{-1} NTA: □ ■ 5.0 μmol l^{-1} NTA. Open and closed symbols for the Gulf Loop seawater indicate 5.2 and 3.2 h exposure to copper. All other experimental times are listed in Table 1.

NTA used to buffer cupric ion activity (Fig. 4). These relationships showed differences among samples at cupric ion activities below $10^{-10.0}$ mol l^{-1}, but converged at activities above this level (Fig. 5). Fifty percent inhibition occured at cupric ion activities ranging from $10^{-10.0}$ to $10^{-9.4}$ mol l^{-1} and inhibition

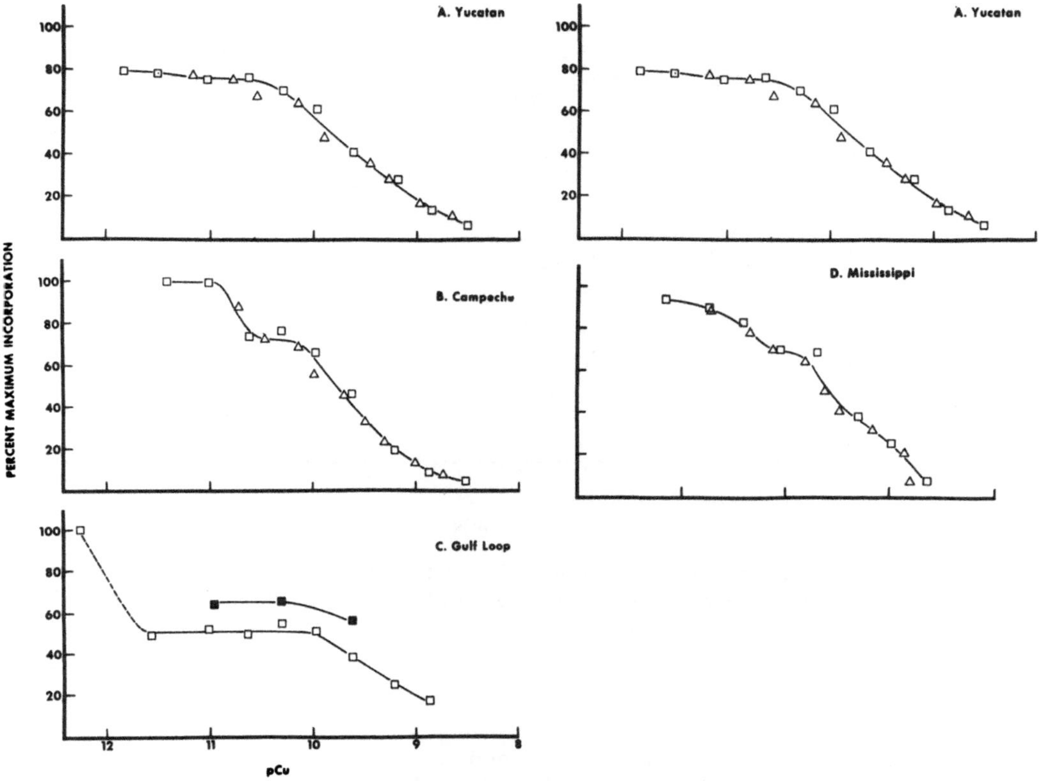

Fig. 4 Relationship between the incorporation of labeled amino acids and the negative log of the cupric ion activity (pCu). The relationships were derived from data plotted in Fig. 3. Symbols are defined as follows: Δ Values determined from addition of cupric ion buffers giving an NTA concentration of 1.0 μmol l^{-1} in seawater; □ ■ Same as above, but for 5.0 μmol l^{-1} NTA. Open and closed symbols are as defined in legend of Fig. 3.

approached 100% at similar values (ca. $10^{-8.5}$ mol l^{-1}). All relationships between inhibition and curpic ion activity showed some degree of bimodal behavior, with this bimodal character being most pronounced for samples from the two lowest productivity stations, Yucatan and Gulf Loop. For the Gulf Loop sample, the lowest cupric ion activity tested ($10^{-11.6}$ mol l^{-1}) caused 50% inhibition of amino acid incorporation and inhibition remained at 50% until the activity exceeded 10^{-10} mol l^{-1}. Similarly, for the Yucatan sample the lowest activity tested ($10^{-11.8}$ mol l^{-1}) caused 20%

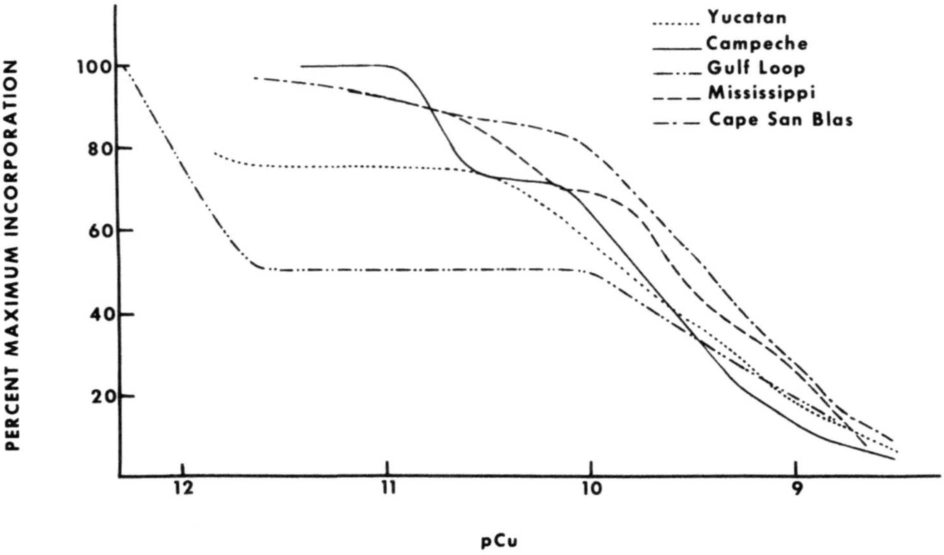

Fig. 5 Comparison of the sensitivities of bacterial communities to pCu (-log cupric ion activity) in different waters of the Gulf of Mexico.

inhibition, and inhibition remained at ca. 25% in the cupric ion activity range of $10^{-11.5}$ to $10^{-10.6}$ mol l^{-1}. This strong bimodal behavior at the Gulf Loop and Yucatan stations was associated with a greater inhibitory effect of cupric ion activity at these two stations at activities below 10^{-10} mol l^{-1}.

The exact nature of the two step toxicity curves is not known, but similar bimodal curves have been observed previously for the toxicity of copper to individual phytoplankton clones[3,20]. With the phytoplankton clones the most likely explanation is the existence of different cellular sites with different sensitivities to cupric ion poisoning. But, in the present experiments, bimodal toxicity curves may result from differential sensitivity among different species or groups of species within the microbial community.

Bioassayed relationships between curpic ion activity and total copper concentration

Relationships between cupric ion activity and total copper concentration indicated complexation of copper in excess of that predicted solely from complex formation with inorganic ligands (e.g. CO_3^{2-}, OH^-, Cl^-) for all sea water samples (Fig. 6). There were, however, differences among samples with positive trends between copper complexation and the level of primary productivity, the concentrations of bacteria and amino acid turnover rates (Table 1). Copper appeared to be least complexed in the 50 m sample at the lowest productivity station, Gulf Loop. The highest apparent level of complexation was observed for the highest primary productivity sample collected from the Mississippi discharge plume.

Scatchard analysis of bioassay data indicated both similarities and differences among seawater samples with regard to number of complexing ligands and their concentrations and conditional stability constants (Table 2). For the 50 m Gulf Loop sample our analysis was consistent with complexation of copper only by inorganic ions in the copper concentration range 10^{-8} to 10^{-7} mol l^{-1}. Below 10^{-8} mol l^{-1} there was a sharp decrease in the ratio of cupric ion activity to total copper consistent with complexation to a ligand present at ca. 5 nmol l^{-1} with a conditional stability constant $\geq$ ca. 10^{11}. There was a somewhat higher level of complexation in the 10 m Yucatan sample in which there appeared to be at least two complexing ligands: one was similar in concentration and stability constant to that in the Gulf Loop sample and the second was present at a concentration of ca. 15 nmol l^{-1} and had a conditional stability constant of $10^{9.8}$. Data from the Campeche Station could be modeled using a single ligand (L_{TOT} = 40 nmol l^{-1}, $K_C = 10^{9.3}$), but that for the two highest complexing stations, Cape San Blas and Mississippi, again required two ligand models. We should point out that a two ligand model does not necessarily imply that there are only two ligands present. Instead there may be mixtures of many ligands that cannot be resolved by the present technique. In such a case, the ligand concentrations and stability constants would represent mean values for groups of ligands having similar stability constants.

Estimates of ambient copper speciation and cupric ion activities

Bioassayed relationships between cupric ion activity and total copper concentration, along with the speciation models fit to these relationships, were extrapolated to ambient copper concentration to estimate ambient cupric ion activities and copper speciation (Table 3). These estimates predict ambient cupric ion activities of $\leq 10^{-12}$ mol l^{-1} for all samples except the 5 m Campeche Bay sample which had an estimated activity of 10^{-11} mol l^{-1}. Major soluble inorganic species of copper (i.e., Cu^{2+}, $CuCO_3$, $CuOH^+$, etc.) were estimated to represent only 0.6 to 18% of the ambient copper. The Mississippi and Cape San Blas samples appeared to have the highest levels of complexation and the Campeche Bay sample the least. Ambient speciation of copper for samples from

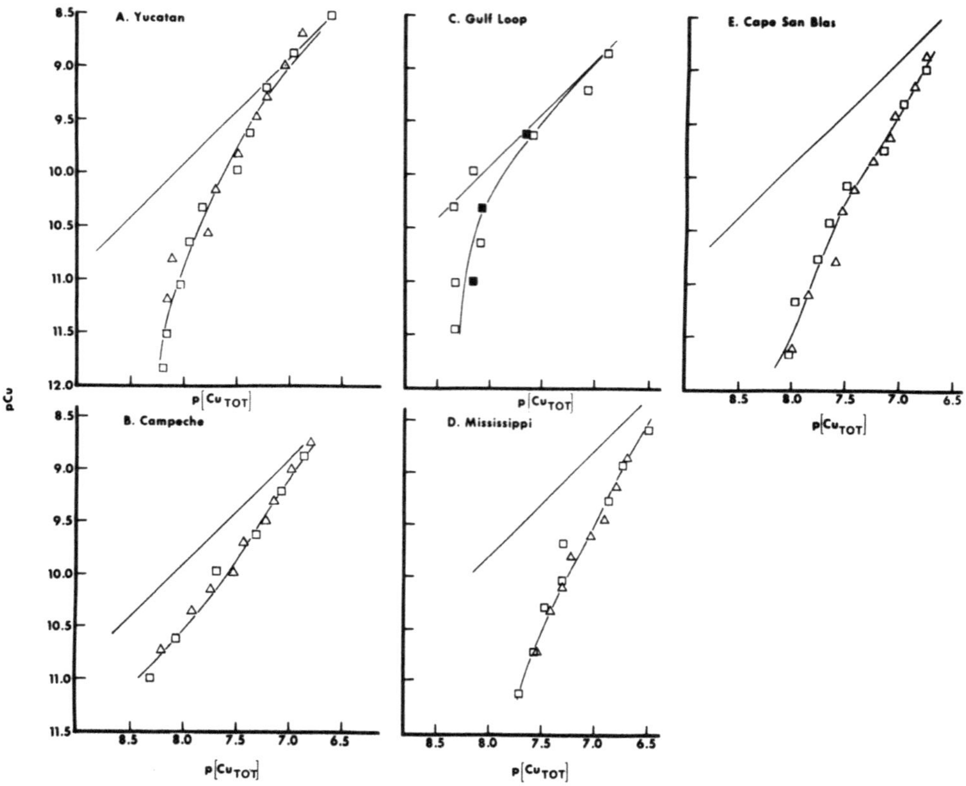

Fig. 6 Relationships between the negative log of cupric ion activity (pCu) and the negative log of the total copper concentration ($p[Cu_{TOT}]$) as determined from bioassay data. Solid lines of unity slope are the estimated relationships for copper complexation by inorganic ions (CO_3^{2-}, OH^-, Cl^-, etc.). Deviation of data points below this line indicates the extent of copper chelation or copper binding to colloidal or particulate matter. Solid lines fit to the data are modeled curves computed on the basis of ligand concentrations and conditional stability constants given in Table 2. △ Data for 1.0 μmol l^{-1} NTA; □ ■ Data for 5.0 μ mol l^{-1} NTA.

two low-productivity stations, Gulf Loop and Yucatan, was particularly difficult to estimate because of steepness and uncertainty in curves for cupric ion activity vs. total copper concentration at copper concentrations approaching ambient (Fig. 6).

Table 2 Total ligand concentrations and conditional stability constants estimated from Scatchard analysis of bioassay data for cupric ion activity vs. total copper concentration.

Station	Ligand concentration (nmol l^{-1}) L_1	L_2	Log K* K_1	K_2
Gulf Loop	-	5	-	>11
Yucatan	15	5	9.8	≥12
Campeche	40	-	9.3	-
Cape San Blas	80	13	9.0	11.2
Mississippi	130	20	8.9	11.1

* $$K_c = \frac{[CuL]}{[Cu^{+2}]\ [L^*]}$$

where [L*] is the total ligand concentration minus [CuL].

Table 3: Estimated ambient cupric ion activity and copper speciation.

Station (depth)	Ambient total Cu (nmol l^{-1})	Estimated pCu (-log cupric ion activity)	Estimated Cu Fractionation (%) Cu^{2+}*	Dissolved inorganic species**	Organic and particulate bound
Gulf Loop (50 m)	∿ 2.7	≥12***			≥96
Yucatan (10 m)	∿ 3.6	≥12***			≥98
Campeche (5 m)	4.8	11.0	0.8	18	82
Mississippi (4 m)	8.8	11.9	0.06	0.9	99.1
Cape San Blas (4 m)	4.4	12.1	0.07	1.3	98.7

* Values based on the estimated ambient cupric ion activity and an activity coefficient for Cu^{2+} of 0.25.

** Values based on estimated cupric ion activity and the ratio of cupric ion activity to the concentration of soluble inorganic species. This ratio is $10^{-1.8}$ at pH 8.1 and $10^{-1.9}$ at pH 8.2 as based on previous calculations and experimental data[11,12] and the cupric ion bioassay data in this paper.

***Estimates of cupric ion activity in these samples are order of magnitude values.

DISCUSSION

Natural communities of marine bacteria in low productivity seawater are exceedingly sensitive to additions of copper, and significant effects occur with additions at or below ambient copper concentrations. For the 18 m Gulf Loop sample, 14% inhibition at an addition of one fifth the ambient copper concentration suggest that some portion of the added ionic copper (as $CuSO_4$) may not equilibrate rapidly and consequently may be bound to a lesser extent than the copper initially present in the seawater. Such observations would appear to be consistent with those of Batley and Florence[21] who found that a significant portion of the copper in their seawater samples was present in chemical forms that did not readily adsorb to Chelex resin (although equivalent amounts of added ionic copper did adsorb to the Chelex) and that did not readily exchange with additions of ionic ^{64}Cu. If we are indeed observing slow reaction kinetics, then our bioassay results may underestimate equilibrium levels of copper complexation. We note, however, that when we varied the time of exposure within a single experiment (Gulf Loop), the bioassay results at 3 h were consistent with those at 5 h, showing no evidence of slow kinetics (Fig. 6). Similar results have been obtained with a Gulf Loop sample collected from a depth of 30 m during a subsequent cruise in February 1981; i.e., the same bioassayed relationship between cupric ion activity and total copper concentration was observed with experimental times of either 2 or 5 h.

The toxicity of copper at additions at or below ambient concentrations also suggests the possibility that ambient levels of copper themselves may exert some inhibitory effect. The fact that the addition of NTA to seawater does not stimulate amino acid incorporation would appear, at first glance, to argue against this hypothesis. The hypothesis, however, must be left open because the addition of the lowest concentrations of copper in combination with NTA were for the Yucatan, Gulf Loop, and Mississippi samples, not much less toxic that the addition of copper alone. Such observations are consistent with the presence of ligands in seawater that can out-compete the added NTA. In such a case, it would be these naturally occurring ligands rather than the added NTA that would primarily complex copper, thereby controlling cupric ion activity and hence copper toxicity.

Cupric ion activity caused the greatest relative inhibition of microbial incorporation of ^{3}H-labeled amino acids in samples from the two lowest productivity stations, Gulf Loop and Yucatan, and the least inhibition in samples from the two highest productivity coastal stations, Mississippi and Cape San Blas. These observations appear to indicate a higher sensitivity to cupric ion poisoning for the microbial communities in lower productivity seawater. The bimodal shape of the inhibition vs. pCu curves, particularly for Gulf Loop and Yucatan samples, further suggests that only a portion of the microbial community exhibits increased copper sensitivity. There are, however, some difficulties in attempting to compare the copper sensitivities of bacteria in different samples. For example, the experimental times were not exactly the same in all cases (Table 1), and even if they were, uncertainty would exist in comparisons among microbial communities with different amino acid assimilation rates. Slightly shorter experimental times were used at the two highest productivity stations because of the extremely rapid incorporation of amino acids at these stations. This rapid accumulation may have resulted in either high or low bias in our estimates of microbial sensitivity to copper in these samples. Incorporation of a significant fraction of the ^{3}H-amino acids in the Mississippi and Cape San Blas samples (35 and 25% incorporated in the controls during the 0.5 h incubation) would tend to decrease the apparent toxicity of copper due to a higher depletion of dissolved labeled amino acids in controls relative to copper treated subsamples. On the other hand, rapid accumulation and resulting depletion of free amino acid pools during the 4 h period before the addition of label may have caused an increase in the apparent toxicity of copper. If depletion occurred, subsamples in which microbial amino acid incorporation was reduced by copper would, after the addition of label, contain a lower ratio of labeled amino acids to total free amino acids. This decreased ratio could decrease the amount of label taken up in these treatments relative to control values[19] and therefore would tend to increase the apparent toxicity of copper. That depletion occurred in the Mississippi sample is suggested by the decrease in measured free amino acid turnover time in the no addition control (1.4 h) at 4.6 h compared to that (5.3 h) measured initially. For the Cape San Blas sample, however, the turnover time of the control was the same at 4.4 h as it was at the start of the experiment. The above potential biases should not have affected bioassayed relationships between cupric ion activity and total copper concentration because any bias within a single experiment should have occurred to the same extent for treatments with the same cupric ion activity.

Our observations indicating the presence of ligands in seawater that strongly complex copper show similarities with anodic stripping voltametry (ASV) measurements made on either the same samples or on similar samples collected at the same stations[22]. Piotrowicz[22] found much less ASV-labile copper at ambient pH ($\sim$ 8.2) than at pH 2, similar to previous observations in seawater [23,24]. There was no detectable ASV signal (indicating $\leq$ 0.3 nmol l^{-1} labile copper) for an unfiltered 7 m sample from the Yucatan station and for 18 and 300 m samples from the Gulf Loop station, although pH 2 measurements indicated the presence of 3.6 nmol l^{-1} copper for the Yucatan sample and 2.4 and 3.0 nmol l^{-1} for the two Gulf Loop samples. These observations are consistent with ours in that both indicate strong complexation of copper by ligands other than inorganic ions**. ASV complexation capacities based on the difference in the X-intercepts of curves for stripping current vs. added copper concentration at pH 8 and pH 2 were consistent with the presence of strong complexing ligands at a concentration of 4 nmol l^{-1} for both the Yucatan and Gulf Loop samples. Our estimated concentrations for very strong complexing ligands (log $K_C \geq 11$) for 10 and 50 m samples from these two stations, about 5 nmol l^{-1}, are consistent with the ASV values.

The 5 m Campeche sample provides another interesting comparison between ASV data of Piotrowicz[22] and our bioassay data. Measurements at pH 8 indicated 0.6 nmol l^{-1} ASV-labile copper in this sample, while total copper in this sample, measured at pH 2, was 4.8 nmol l^{-1}. Thus, 88% of the copper was in a form that was not detected by ASV at pH 8. A similar series of measurements on a subsequent 5 m sample gave a value of 83% non-labile copper. Our bioassay value for excess complexation (i.e., complexation above and beyond that with inorganic ions) is 82%, in agreement with the ASV values.

Not all comparisons between ASV data and complexation bioassays were favorable. The bioassay indicated the highest level of copper complexation in the two high-productivity coastal samples (Mississippi and Cape San Blas). By contrast, significant ASV-labile copper (3 and 1.3 nmol l^{-1}) was measured at pH 8 in these samples which would suggest a low level of copper complexation. The reasons for the discrepancies between the two techniques are not clear. A possible explanation may be found in the fact that copper complexes with sufficiently fast dissociation kinetics are ASV-labile; that is, they are reduced at the Hg-film electrode. This, for example, is the case with inorganic complexes and organic complexes with stability constants less than about 10^{10} [25]. Where there is an appreciable presence of labile complexes, ASV can significantly underestimate the total level of copper complexation.

What is the nature of ligands that strongly complex copper? The high stability constants ($K_C \geq 10^9$) suggest organic chelation either at dissolved or surface sites. Also, since the samples were not filtered, we cannot exclude uptake of copper by organisms. There is a trend between total binding capacity and dissol-

**A decrease in pH from 8 to 2 would decrease copper complexation due to hydrogen ion competition; an increase in the concentration of ASV-labile copper would occur with the dissociation of non-labile complexes. Since inorganic complexes are completely ASV-labile, pH-induced shifts in inorganic complexation would not be detected by ASV.

ved organic carbon, particulate organic carbon, and primary productivity which further supports an organic nature for the binding agents (Tables 1 and 4). The highest binding capacities were found in the Mississippi and Cape San Blas samples whose low salinities, 28.9 and 26.8 °/oo, suggest that these samples could contain at least some binding agents of estuarine or terrestrial origin.

CONCLUSION

Copper inhibited the incorporation of amino acids by marine bacteria in low productivity waters at additions of 2 nmol l^{-1}, equal to or below ambient copper concentrations. Similar additions of copper to high-productivity, low-salinity coastal seawater had much less of an inhibitory effect. These differences in the response of bacteria could be attributed in large measure to two factors: (1) copper added to the low salinity coastal seawater was more highly complexed, resulting in lower cupric ion activities and therefore lower toxicity, and (2) amino acid incorporation was inhibited at lower cupric ion activities in the low-productivity, high-salinity seawater samples. Although this second observation suggests a greater sensitivity to cupric ion activity for bacterial communities in oligotrophic seawater, other factors such as possible bias due to differences in amino acid turnover rates between high-productivity and low-productivity seawater, and to short amino acid turnover times in high-productivity seawater, relative to experimental time scales, make this hypothesis tentative at present.

Table 4. Camparison of dissolved organic carbon (DOC) and particulate organic carbon (POC) with values for copper binding capacity.

Station	DOC* (mg l^{-1})	POC* (mg l^{-1})	Total Cu binding capacity (nmol l^{-1})**
Gulf Loop	0.69	0.045	5
Yucatan	1.11 ± 0.03	0.024 ± 0.06	20
Campeche	1.15 ± 0.04	0.034 ± 0.001	40
Mississippi	1.56 ± 0.03	> 0.2	150
Cape San Blas	2.36 ± 0.01	> 0.2	80

* DOC and POC were determined by peroxydisulfate oxidation.
**Binding capacity equals the concentration of L_1 plus L_2 (Table 2)

ACKNOWLEDGEMENTS

This research was supported by a contract from NOAA's Office of Marine Pollution Assessment's (OMPA) Long Range Effects Research Program and represents field application of a bioassay technique developed under contract from OMPA's Ocean Dumping Complementary Research Program. The research was conducted on the NOAA ship Researcher April 9 - May 5, 1980, in cooperation with scientists from the Ocean Chemistry and Biological Laboratory of NOAA's Atlantic Oceanographic and Meterological Laboratories, (AOML), Miami, Fla. We thank Drs. D.A. Atwood and G. Harvey, Chief Scientists, Dr. S. Piotrowitz and Mr. T. Puig for sampling protocol and equipment, clean van use and anodic stripping voltammetry analyses of trace metals and Dr. P. Ortner for primary productivity data, all from AOML. We also thank Dr. P. Hanson for the organic carbon analyses and acknowledge the assistance of Mrs. M.B. Murdoch and Mrs. J.M. Lewis, from our laboratory and Dr. A.V. Palumbo of Massachusetts Institute of Technology, Cambridge, Mass. Contribution No. 81-42B of the National Marine Fisheries Service, Southeast Fisheries Center, NOAA.

REFERENCES

1. Spencer, C.P., 1957: Utilization of trace elements by marine unicellular algae. J. Gen. Microbiol., 16, 282-285.
2. Fogg, G.E. and P.F. Westlake, 1955: The importance of extracellular products of algae in fresh water. Verh. Int. Ver. Limnol., 12, 219-232.
3. Sunda, W.G. and R.R.L. Guillard, 1976: The relationships between cupric ion activity and the toxicity of copper to phytoplankton. J. Mar. Res., 34, 511-529.
4. Anderson, D.M. and F.M.M. Morel, 1978: Copper sensitivity of Gonyaulax tamarensis. Limnol. Oceanogr., 23, 283-295.
5. Zevenhuizen, L.P.T.M., J. Dolfing, E.W. Eshuis and I.J. Scholten-Koerselman, 1979: Inhibitory effects of copper on bacteria related to the free ion concentration. Microb. Ecol., 5, 139-146.
6. Boyle, E. and J.M. Edmond, 1975: Copper in surface waters south of New Zealand. Nature, 253, 107-109.
7. Bruland, K.W., 1980: Oceanographic distributions of cadmium, zinc, nickel, and copper in the North Pacific. Earth Planetary Sci. Lett., 47, 176-198.
8. Azam, F. and H. Holm-Hansen, 1973: Use of tritiated substrates in the study of heterotrophy in seawater. Mar. Biol., 23, 191-196.
9. Lee, C. and J.L. Bada, 1977: Dissolved amino acids in the equatorial Pacific, the Sargasso Sea, and Biscayne Bay. Limnol. Oceanogr., 22, 205-210.

10. Gillespie, P.A. and R.F. Vaccaro, 1978: A bacterial bioassay for measuring the copper-chelation capacity of seawater. Limnol. Oceanogr., 23, 543-548.
11. Sunda, W.G. and P.A. Gillespie, 1979: The responses of a marine bacterium to cupric ion and its use to estimate cupric ion activity. J. Mar. Res., 37, 761-777.
12. Sunda, W.G., A.V. Palumbo and R.L. Ferguson, unpublished manuscript: Bioassay of copper complexation in seawater.
13. Sunda, W.G., 1975: Relationship between cupric ion activity and the toxicity of copper to phytoplankton. Ph.D. thesis, Massachusetts Institute of Technology, Cambridge. 167 p.
14. Sillen, L.G. and A.E. Martell, 1964: Stability constants of metal-ion complexes. Chem. Soc. Publ., 17, 754 p.
15. Goldberg, E.W., 1965: Minor elements in seawater. In: "Chemical Oceanography", v. 1, J.P. Riley and G. Skirrow, eds. Academic Press, New York, p.163-193.
16. Whitfield, M., 1973: A chemical model for the major electrolyte component of seawater based on the Bronsted-Guggenheim hypothesis. Mar. Chem., 1, 251-266.
17. Mantoura, R.F.C. and J.P. Riley, 1975: The use of gel filtration in the study of metal binding by humic acids and related compounds. Anal. Chim. Acta., 78, 193-200.
18. Sunda, W.G. and P.J. Hanson, 1979: Chemical speciation of copper in river water: Effect of total copper, pH, carbonate and dissolved organic matter. In: "Chemical Modeling in Aqueous Systems". E.A.Jenne, ed., Chem. Soc. Symp. Ser. 93, p. 147-180.
19. Ferguson, R.L., and W.G. Sunda, unpublished manuscript: Rapid utilization of ^{3}H-labeled amino acids in seawater: importance of trace metal clean technique and low substrate additions.
20. Gavis, J., R. Guillard and B. Woodward, 1981: Cupric ion activity and the growth of phytoplankton clones isolated from different marine environments. J. Mar. Res., 39, 315-333.
21. Batley, G.E. and T.M. Florence, 1976: Determination of chemical forms of dissolved cadmium, lead and copper in seawater. Mar. Chem., 4, 347-363.
22. Piotrowicz, S.R., unpublished data.
23. Duinker, J.C. and C.J.M. Kramer, 1977: An experimental study on the speciation of dissolved zinc, cadmium, lead and copper in River Rhine and North Sea water, by differential pulsed anodic stripping voltammetry. Mar. Chem., 5, 207-228.
24. Sugai, S.F. and M.L. Healy, 1978: Voltammetric studies of the organic association of copper and lead in two Canadian inlets. Mar. Chem., 6, 291-308.
25. Chau, U.K., R. Gachter and K. Lum-Shue-Chan, 1974: Determination of the apparent complexing capacity of lake waters. J. Fish. Res. Board. Can., 31, 1515-1519.

Participants in the NATO Advanced Research Institute on Trace Metals in Sea Water held in "Ettore Majorana" Centre for Scientific Culture at Erice on the island of Sicily, Italy, March 30- April 3, 1981.

CONFERENCE PARTICIPANTS

Adler, D.
Lamont-Doherty Geological Observatory of Columbia University, Palisades, New York, 10964, U.S.A.

Ambe, M.
Sagami Chemical Research Center, Nishiohnuma 4-4-1, Sagamihara 229, Japan.

Amdurer, M.
Lamont-Doherty Geological Observatory of Columbia University, Palisades, New York, 10964, U.S.A.

Andreae, M.O.
Department of Oceanography, Florida State University, Tallahasee, Florida, 52306, U.S.A.

Bacon, M.
Department of Chemistry, Woods Hole Oceanographic Institution, Woods Hole, Massachusetts, 02543, U.S.A.

Bertine, K.K.
Department of Geological Sciences, San Diego State University, San Diego, California 92182, U.S.A.

Bewers, J.M.
Chemical Oceanography Division, Bedford Institute of Oceanography, Dartmouth, Nova Scotia, B2Y 4A2, Canada.

Boulègue, J.
Laboratoire de Geochimie des Eaux, Université de Paris 7, 2 Place Jussieu, 75221 Paris Cedex 05, France.

Bourg, A.C.M.
Department of Oceanography I.G.B.A., University of Bordeaux I, 351 Cours de la Liberation, F-33405, Talence, France.

Boyle, E.
Department of Earth and Planetary Sciences, Massachusetts Institute of Technology, Cambridge, Massachusetts, 02139, U.S.A.

Brinckman, F.E.
National Bureau of Standards Chemical and Biodegradation Group, Division 561, Room 329, Washington, D.C. 20234, U.S.A.

Brockman, H.
Institute for Organic and Biochemistry, University of Hamburg, Martin Luther King Platz 6, 2000 Hamburg 13, F.R. Germany.

Bruland, K.W.
University of California 34-115, Center for Coastal Marine Studies, Division of Natural Sciences, Santa Cruz, California 95064, U.S.A.

Burton, J.D.
Department of Oceanography, The University, Southampton, SO9 5NH, U.K.

Chatt, A.
Trace Analysis Research Center, Department of Chemistry, Dalhousie University, Halifax, Nova Scotia, B3H 4J1, Canada.

Chen, C.G.
National Bureau of Oceanography, Oregon State University, Corvallis, Oregon, 97330, U.S.A.

Cossa, D.
Fisheries and Oceans, Quebec Region, C.P. 15500, Quebec, Que. G1K 7Y7, Canada.

Crecelius, E.A.
Battelle Pacific Northwest Division, Marine Research Laboratory, Washington Harbour Road, Sequim, Washington, 98383, U.S.A.

Danielsson, L-G.
Department of Analytical and Marine Chemistry, Chalmers University of Gothenburg, S-412 96 Goteborg, Sweden.

Duinker, J.C.
Netherlands Institute for Sea Research, P.O. Box 59, Tesel, Netherlands.

Dyrssen, D.
Chalmers University of Technology and University of Gothenburg, Department of Analytical and Marine Chemistry, S-412, 96, Goteborg, Sweden.

Elderfield, H.
Department of Earth Sciences, The University, Leeds, LS2 9JT, U.K.

Emerson, S.
University of Washington, Department of Oceanography, Seattle, Washington, 98195, U.S.A.

Feeley, R.A.
Pacific Marine Environmental Laboratory, NOAA Building Number 32, 7600 Sand Point Way, N.E., Seattle, Washington, 98115, U.S.A.

Fitzgerald, W.F.
Marine Sciences Department, University of Connecticut, Avery Point, Groton, Connecticut 06340, U.S.A.

Fukai, R.
Scientific Center of Monaco, 16 Boulevard de Suisse, Monaco.

Goldberg, E.D.
Scripps Institution of Oceanography, University of California at San Diego, La Jolla, California, 92093, U.S.A.

Green, E.J.
Naval Ocean Research and Development, Code 420, Bay St. Louis, Mississippi, 39520, U.S.A.

Heggie, D.
Graduate School of Oceanography, University of Rhode Island, Kingston, R.I. 02881, U.S.A.

Honjo, S.
Woods Hole Oceanographic Institution, Woods Hole, Massachusetts 02543, U.S.A.

Huang, Z-Q.
Sea water Chemistry, Third Institute of Oceanography, Xiamen University, Xiamen, People's Republic of China.

Huested, S.S.
Department of Earth and Sciences, Massachusetts Institute of Technology, Cambridge, Massachusetts, 02139, U.S.A.

Imber, B.
Department of Oceanography, The University, Southampton, SO9 5NH, U.K.

Kijviriya, V.
Faculty of Sciences
Ramkamhaeng University
Banggapi, Bangkok 24,
Thailand.
Jagner, D.
Department of Analytical
and Marine Chemistry,
Chalmers University of
Technology and University
of Gothenburg, S-412 96
Goteborg, Sweden.
Klinkhammer, G.P.
Graduate School of
Oceanography, University
of Rhode Island, Narragansett
Bay Campus, Kingston, RI,
02881, U.S.A.
Knauer, G.A.
Moss Landing Marine
Laboratories, P.O. Box 223,
Moss Landing, California,
95039, U.S.A.
Kremling, K.
Institut für Meereskunde an
der Universität Kiel, Abt.
Meereschemie, Dusternbrooker
Weg 20, 2300 Kiel,
F.R. Germany.
Latouche C.
Institut de Geologie du
Bassin d'Aquataine,
Université de Bordeaux 1-351,
Cours de la Liberation,
33405 Talence, Cedex,
France.
Lyons, Wm. B.
Department of Earth Sciences,
University of New Hampshire,
Durham, New Hampshire 03824,
U.S.A.
Macchi, C.
Istituto di Ricerce sulle
Acque, Consiglio Nazionale
delle Ricorohe, Via Reno 1,
00198, Rome, Italy.
Macdonald, R.W.
Ocean Chemistry Division,
Institute of Ocean Sciences,
P.O. Box 6000, Sidney,
B.C. V8L 4B2, Canada.
Magnusson, B.
Department of Analytical
and Marine Chemistry,
Chalmers University of
Technology and University
of Gothenburg, S-412, 96
Goteborg, Sweden.
Mantoura, F.C.
Institute for Marine
Environmental Research,
Plymouth, U.K.
Mart, L.
Institut für Chemie der
Kernforschungsanlage Jülich
GmbH, Institut für Angewandte
Physikalische Chemie, D-5170
Jülich 1, Postfach 1913,
F.R. Germany.
Martin, J.M.
Laboratoire de Geology,
Ecole Normale Superieure,
75230 Paris, Cedex 05, France
Measures, C.I.
Massachusetts Institute of
Technology, Cambridge,
Massachusetts, 02139, U.S.A.
Millero, F.
Rosentiel School of Marine
and Atmospheric Science,
University of Miami, Miami,
Florida 33149, U.S.A.
Moore, R.M.
Department of Oceanography,
Dalhousie University,
Halifax, Nova Scotia, B3H
4J1, Canada.
Morel, F.M.
Massachusetts Institute of
Technology, Civil Engineer-
ing Department, 48-423,
Cambridge, Massachusetts,
02139, U.S.A.

Murray, J.
Department of Oceanography, University of Washington, Seattle, Washington, 98195, U.S.A.

Nürnberg, H.W.
Direktor am Institut für Chemie der Kernforschungsanlage Jülich GmbH, Institut für Angewandte Physikalische Chemie, D-5170 Jülich 1, Postfach 1913, F.R. Germany.

Olafsson, J.
Hafrannsokn Marine Research Institut, Skulagata 4, Reykjavik, Iceland.

Piotrowicz, S.R.
U.S. Department of Commerce, NOAA, Atlantic Oceanographic and Meteorological Laboratories, Ocean Chemistry and Biology Laboratory, 15 Rickenbacker Causeway, Miami, Florida 33149, U.S.A.

Piuze, J.
Fisheries and Oceans, Quebec Region, C.P. 15500, Quebec, Quebec G1K 7Y7, Canada.

Santiago, V.
Department of Chemical Engineering, University of Exeter, North Park Road, Exeter, U.K.

Schaule, B.
California Institute of Technology, Geology 170-25, Pasedena, California, 91125, U.S.A.

Schwartz, U.
Geologisch-Paläontologisches Institut, Universität Hamburg Bundesstrasse 55, D-2000. Hamburg, F.R. Germany.

Scoullos, M.J.
University of Athens, Department of Chemistry, Inorganic Chemistry Laboratory, Navarinoy 13A, Athens, Greece.

Spivack, A.J.
Department of Earth and Planetary Sciences, Massachusetts Institute of Technology, Cambridge, Massachusetts, 02139, U.S.A.

Stebbing, A.R.D.
Natural Environment Research Council, Institute for Marine Environmental Research, Prospect Place, Plymouth, PL1 3AS, U.K.

Stukas, V.J.
Seakem Oceanography Limited, 9817 West Saanich Road, Sidney, B.C. V8L 3S1, Canada.

Sunda, W.G.
Southeast Fisheries Center Beaufort Laboratory, Beaufort, North Carolina 28516, U.S.A.

Topping, G.
Department of Agriculture and Fisheries for Scotland Marine Laboratory, P.O. Box 101, Victoria Road, Aberdeen, AB9 8DB, Scotland.

Westerlund, F.G.
Department of Analytical and Marine Chemistry, Chalmers University of Technology and University of Gothenburg, S-412 96 Goteborg, Sweden.

Whitfield, M.
Marine Biological Association of the U.K., The Laboratory, Citadel Hill, Plymouth, U.K. PL1 2PB.

Wong, C.S.
Ocean Chemistry Division, Institute of Ocean Sciences, P.O. Box 6000, Sidney, B.C., V8L 4B2, Canada.

Zuehlke, R.W.
University of Rhode Island, Graduate School of Oceanography, Narragansett Bay Campus, Kingston, Rhode Island, 02881, U.S.A.

INDEX

INDEX OF AUTHORS

GPSR Compliance

The European Union's (EU) General Product Safety Regulation (GPSR) is a set of rules that requires consumer products to be safe and our obligations to ensure this.

If you have any concerns about our products, you can contact us on ProductSafety@springernature.com

In case Publisher is established outside the EU, the EU authorized representative is:

Springer Nature Customer Service Center GmbH
Europaplatz 3
69115 Heidelberg, Germany

Batch number: 10615115

Printed by Printforce, the Netherlands